T0360798

Fundamentals of Theoretical Plasma Physics

Mathematical Description of Plasma Waves

Fundamentals of Theoretical Plasma Physics

Mathematical Description of Plasma Waves

Hee J Lee
Professor of Physics, Emeritus
Hanyang University, South Korea

NEW JERSEY • LONDON • SINGAPORE • BEIJING • SHANGHAI • HONG KONG • TAIPEI • CHENNAI • TOKYO

Published by

World Scientific Publishing Co. Pte. Ltd.
5 Toh Tuck Link, Singapore 596224
USA office: 27 Warren Street, Suite 401-402, Hackensack, NJ 07601
UK office: 57 Shelton Street, Covent Garden, London WC2H 9HE

British Library Cataloguing-in-Publication Data
A catalogue record for this book is available from the British Library.

FUNDAMENTALS OF THEORETICAL PLASMA PHYSICS
Mathematical Description of Plasma Waves

ISBN 978-981-3276-75-8

For any available supplementary material, please visit
https://www.worldscientific.com/worldscibooks/10.1142/11168#t=suppl

Typeset by Stallion Press
Email: enquiries@stallionpress.com

Printed in Singapore

Introduction

In this Introduction, previews of some chapters are addressed to guide or motivate the readers before embarking on the details of the relevant subjects. Some topics are touched upon rather heavily.

This book presents a coherent review of the plasma state (the fourth state of matter) in regard to its various characteristics, the modes of its description, and its interaction with electromagnetic waves. The mathematical steps leading to the desired or established conclusions are kindly exposited with mindful details, often with alternative routes. I believe that knowing two different methods that solve the same problem widens and deepens the understanding of the problem under consideration. In the theoretician's view, physical reasoning is a shortcut for the solution but its validity and its conformity with the governing equations should be checked against painstaking mathematical steps to make sure of the concomitance between the insight and the analysis. Some author's claim that a certain problem can be answered by means of sheer physical reasoning is often his or her physical interpretation of the result that has been acquired by clandestine effort of analytic investigation. This is the reason why we should risk the boredom of scrutinizing the algebraic steps in the relevant development. In my opinion, theoretical plasma physics is a discipline which requires the ability to *read and grasp* the mathematical equations without going through the detailed steps, but ironically this ability is acquired only at the expense of diligent algebraic work and necessary patience. This book aims to help the students to cultivate this ability and make their road easy, the road of trekking the rugged analytical terrain.

Putting aside the application side, the nature of plasma physics is theoretical; plasma physics is modeled by the Liouville equation on its foremost foundation, apart from the collaboration with the Maxwell equations. Depending upon various time and parameter scales, the Liouville equation is *evolved* into the Lenard–Balescu equation, or the Landau equation, or the

Vlasov equation, or two-fluid equations, or magnetohydrodynamic equations. Therefore, the plasma physics is inherently theoretical, and its mathematical aspects cannot be downgraded. For this reason, this book touches upon some salient features of the plasma kinetic theory (Chapter 12). My reason why the kinetic theory is introduced in the rather later stage of the book is pedagogical; the simplest version of the kinetic equation is the Vlasov equation, which can be handled by various thoroughly-known analytical methods, and the methods are rather easily accessible. A wide variety of mathematical methods that are learnt from analyzing the Vlasov equation can be applied without much difficulty to the problem of higher degree of complexity. For example, the quasilinear theory for the Vlasov equation can be immediately applied to the Liouville equation (Chapter 11).

In this book, familiarity with complex variables, Fourier transformation, and Dirac delta function is required. However, important mathematical background materials that warrant independent introduction are not simply relegated to references but have been explained to the extent that the exposition becomes self-sufficient, so that the readers can follow the relevant physical development without hindrance. For example, I spent some pages on the mathematical theory of characteristics which is useful in dealing with hyperbolic partial differential equations. The Liouville equation, the Vlasov equation, and the Euler equation all belong to the category of hyperbolic partial differential equations. Thus the familiarity with the mathematical theory of the characteristics is immensely useful in theoretical plasma physics. I am greatly indebted to the book by Hildebrand [Hildebrand, F. B. (1976), *Advanced Calculus for Applications*, Prentice-Hall.] in the analysis of the Vlasov equation in Chapter 4. The usefulness of the characteristic method can be demonstrated in the study of the Liouville equation. The Liouville equation and Klimontovich equation are equivalent to the dynamical equation that comprises Maxwell's equations plus Newton's law of motion. It is well known that the Liouville equation describes the manner of how the ensemble density (the N-particle distribution function) of the particles composing the system in Γ-space ($6N$-dimensional space with N being the number of particles of the system) behaves. The Liouville equation states that the ensemble of the system points in the Γ-space flows as if the ensemble is an incompressible fluid, representing the conservation of the probability density:

$$\frac{\partial F^{(N)}}{\partial t} + \sum_{i=1}^{N} \left(\frac{\partial H}{\partial \mathbf{p}_i} \cdot \frac{\partial F^{(N)}}{\partial \mathbf{r}_i} - \frac{\partial F^{(N)}}{\partial \mathbf{p}_i} \cdot \frac{\partial H}{\partial \mathbf{r}_i} \right) = 0 \tag{1}$$

with

$$F^{(N)} = F^{(N)}(\mathbf{r}_1, \mathbf{r}_2, \ldots, \mathbf{r}_N, \mathbf{p}_1, \mathbf{p}_2, \ldots, \mathbf{p}_N, t) \tag{2}$$

being the N-particle distribution function; H, the Hamiltonian; $\mathbf{r}$, position vector; and $\mathbf{p}$, momentum vector. The solution of (1), which is a hyperbolic partial differential equation, satisfies the following characteristic equations [see the characteristic equation written in the vicinity of (12.12), Chapter 12]

$$\frac{d\mathbf{r}_i}{dt} = \frac{\partial H}{\partial \mathbf{p}_i}, \quad \frac{d\mathbf{p}_i}{dt} = -\frac{\partial H}{\partial \mathbf{r}_i} \quad (i = 1, 2, \ldots, N). \tag{3}$$

(3) is Hamilton's equations of motion for each of the N particles. (1) is solved by an arbitrary function of the constants of motion that is obtained by integrating (3). Thus it is clear that $F^{(N)}$ is constant along the trajectory of the system point in the Γ-space. In fact, the trajectory represented by (3) is the curve in Γ-space that minimizes the action integral I taken along that trajectory:

$$\delta I = \delta \int_{t_0}^{t_1} L dt = 0 \tag{4}$$

with L being the Lagrangian, the sum of the kinetic energy of all the particles minus the potential energy. The variational principle expressed by (4) yields the Hamilton's equation of motion in (3). We summarize: the ensemble in Γ-space moves about in a fluid fashion, in conformity with the variational principle (4). [About variational mechanics, see W. Yourgrau and S. Mandelstam (1968) *Variational principles in dynamics and quantum theory* Dover; R. Weinstock (1952) *Calculus of variations* McGraw-Hill] The foregoing synopsis is introduced in this preface to highlight the power of the characteristic method which in this example laborlessly connects two physical ideas: conservation of probability and the principle of the least action.

If we go into the realm of applications of plasma physics, soon our endeavor becomes interdisciplinary. Over the last 70 years (1950–) plasma physics has expanded its application to such areas as controlled thermonuclear fusion, astrophysics, space physics (solar, magnetospheric, ionospheric physics), surface physics such as semiconductor and metal processing; all of them are closely inter-woven with human interests, immediate or remote, since our environment surrounding the Earth as well as the deep cosmos consists of plasma state, and the fourth state of matter gradually becomes coexistent with the ordinary states as the human utilities require more ingenuity. This book is not directly concerned with those applications but

I hope that solid theoretical understanding of plasma will be eventually beneficial to the students who intend to embark on these applications.

This book is partly based on the plasma physics course that I taught for many years to senior undergraduate students and graduate students at Hanyang University. In addition to the usual materials that constitute the basics of plasma physics, as can be seen in the table of contents, this book includes the materials which are the outcome of my research conducted during my seclusion at rural areas in Korea after retirement from Hanyang. The pertinent topics are briefly introduced in this Introduction to guide or motivate the readers for non-casual reading of the materials of the relevant later chapters. I advise coming back to these introductory explanations, if necessary, at the beginning or end of each chapter.

• Landau damping is interpreted to be an outcome of the *causality* (Chapter 4). Here the causality means the general physical notion that the response of a medium (effect) cannot be precedent to the cause. Specifically, the electric susceptibility of a medium, $\chi(t)$, is constrained by the condition $\chi(t) = 0$ for $t < 0$. In Chapter 4, Landau damping as well as the Kramers–Kronig relations are derived by applying this algebraic constraint to the susceptibility. We show that the Kramers–Kronig relations, if applied to the electrostatic plasma wave, result in Landau damping. We might say that the causal interpretation of the Landau damping is more fundamental than the mechanical interpretation that it is a consequence of a wave-particle interaction. As the famous Landau paper [Landau, L. (1946) *On the vibrations of the electronic plasma,* J. Physics (USSR) **10** p. 25] shows, Landau damping is obtained by expressing the solution of the Vlasov–Poisson equation in the complex plane of the dielectric permittivity, with the aid of the mathematical abstraction of analytic continuation. This abstract mathematical approach somewhat obscures the implication of the causal constraint which defines the direction of time. The causal consequence is clearly and directly revealed when one employs the usual Fourier transform approach if the causal algebraic constraint is enforced.

• The electric fluctuation and the Cerenkov emission, being phenomena that depend on discreteness in plasma, require a higher order description than Vlasov equation which is the lowest order crude approximation of the exact kinetic equation. A test particle approach which is known as *dressed particle* is adopted to calculate these quantities (Chapter 11). The Vlasov description of a plasma is basically akin to the fluid description which is appropriate for a continuum. The physical origin of the Cerenkov emission, which is the inverse process of the Landau damping, can be traced to the causality. This is anticipated since the fundamental reason for the Landau damping is the causality. The Cerenkov emission due to the thermal motions of plasma particles is the source of thermal noise (spontaneous emission), and its universality may be traced to causality. The fluctuations are the cause of the thermal emissions and wave scattering. The incoherent scattering of electromagnetic wave is due to the plasma density fluctuations.

• The fluctuation-dissipation theorem is derived classically with the aid of the characteristic method (Chapter 11). In this derivation the Liouville operator is not used; the linearized Liouville equation is *integrated along the unperturbed orbit.* The fluctuation-dissipation theorem connects the fluctuations of plasma density or electric field with energy dissipation for a system in equilibrium. The energy dissipation of a wave propagating in a medium is determined by the dielectric permittivity. The latter can be found by means of the inversion of the fluctuation-dissipation theorem. Therefore, the theorem offers an alternative means of describing the electromagnetic plasma properties, independently of the consideration of the kinetic equation. In regard to the mathematical aspect, the fluctuation-dissipation theorem is obtained by solving the linearized Liouville equation, and the method learned from the analysis of the Vlasov equation can be directly applied.

• The surface waves are treated in depth by means of the fluid and kinetic equations (Chapters 13 and 14). An infinite plasma (no boundary) supports various plasma normal modes whose characteristics depend upon the plasma parameters and the electromagnetic polarizations and refractive indexes. The presence of the boundary in a plasma changes the wave characteristics and excites new modes that have no corresponding modes in the infinite plasma. Mathematically speaking, the boundary imposes constraints, electromagnetic and fluid or kinetic constraints to be satisfied on the boundary. Thus inevitably, the analysis for a bounded plasma has extra complication as compared with that for an infinite plasma. In a cold (warm) fluid plasma, the surface charges are formed (not formed) on the boundary. The existence of the surface charges on the interface yields a jump of the electric displacement across the interface, which is the electromagnetic boundary condition that must be satisfied in the surface wave problem. The absence of surface charges on the boundary of a warm plasma gives the boundary condition that the normal component of the fluid velocity on the interface is zero, which corresponds to the *specular reflection condition* in the kinetic plasma. In a Vlasov plasma, the specular reflection boundary condition is commonly used for the distribution function (another boundary condition is the diffusive boundary condition). In semi-bounded plasmas, the linear dispersion relations of the surface waves are well-known and thoroughly documented. In this book, we also investigate the slab geometry (the plasma is confined between two parallel infinite planes) to obtain the kinetic surface wave dispersion relations. The simultaneous specular reflection boundary conditions on the two parallel planes are successfully enforced to obtain the slab dispersion relations in closed forms. In a slab, the plasma wave has a two-mode structure, symmetric and anti-symmetric modes. Other interesting surface-wave-related subjects are: Landau damping of surface waves, plasma echoes in a bounded plasma, which are addressed in Chapter 14. Although not attempted in this book, the variational formulation of the surface wave problem appears to be a useful alternative because the boundary conditions can be elegantly dealt with in terms of the Lagrange undetermined multipliers.

• Parametric instabilities. The existence of various eigen modes of plasma wave is a manifestation of the collective plasma properties exhibiting medium-like continuum. For the description of such a plasma medium, a self-consistent theoretical model has been developed on the basis of the Maxwell and Vlasov equations or the Maxwell and fluid equations. The electromagnetic properties of the plasma can be investigated under the governance of these equations. Parametric instabilities are the important nonlinear interactions due to the quadratic terms in the governing plasma equations — $\mathbf{v} \cdot \nabla \mathbf{v}$ term in the fluid equation or $\mathbf{E} \cdot \frac{\partial f}{\partial \mathbf{v}}$ in the Vlasov equation. They represent the nature of plasma as a highly nonlinear dielectric medium. In particular, *nonlinear Landau damping* in unmagnetized plasma or *nonlinear cyclotron damping* in a magnetized plasma is one of the fundamental mechanisms for energy transfer from wave to particles due to wave-particle interaction, which is believed to be the cause of enhanced diffusion and anomalous resistivity of plasma. A simple picture of nonlinear Landau damping derives from the resonance condition

$$\omega - \omega' = (\mathbf{k} - \mathbf{k}') \cdot \mathbf{v}_i \tag{5}$$

where ω is the frequency of electron wave (Langmuir wave), ω' is the frequency of the scattered electron wave of slightly lower frequency, scattered by ion density perturbation, and $\mathbf{v}_i$ is the ion thermal velocity. (5) is a resonance condition which indicates that the beat wave of frequency $(\omega - \omega')$ satisfies the Landau wave-particle resonance condition on ions. This ion nonlinear Landau damping is responsible for the parametric decay instability.

Due to this resonance, the beat wave (ion acoustic wave) gets heavily damped, discharging its wave energy to the scattered side-band wave (eventually to the ions). Because of the heavy damping, the beat wave loses its identity as a wave, and is often called a quasi mode. The ion nonlinear Landau damping transfers power from one Langmuir wave to another via an ion acoustic beat wave. Specifically, the nonlinear coupling responsible for the energy transfer is the nonlinear current of high frequency $\mathbf{J} = -en_i\mathbf{v}_e$ (subscript e denotes electron, the high frequency) and the *ponderomotive force* at the beat frequency which is derived from the nonlinear quadratic term $\mathbf{v}_e \cdot \nabla\mathbf{v}_e$ in the fluid equation or $\mathbf{E} \cdot \frac{\partial f}{\partial \mathbf{v}}$ in the Vlasov equation (Chapter 9).

• Under the same namesake of *nonlinear Landau damping*, a repetitive damped oscillation of wave in a collisionless plasma can occur due to repetitive Landau dampings. This phenomenon was predicted by O'Neil (1965) in terms of the elliptic function solution of the nonlinear orbit [Reference: O'Neil, T. (1965) *Collisionless damping of nonlinear plasma oscillations*, Phys. Fluids **8**, p. 2255]. Without resorting to the elliptic function, we deal with the equation of motion for an electron in a sinusoidal electric field in the wave frame as follows:

$$\frac{d^2x}{dt^2} = -\frac{e}{m}E(k,\omega)\sin[kx(t)] = \frac{dv}{dt} \tag{6}$$

$$\text{or} \quad \frac{d^2v}{dt^2} = -akv\cos[kx(t)] \quad \left(a = \frac{eE}{m}\right). \tag{7}$$

Multiplying by dv/dt and using $\cos\ kx = \pm\sqrt{1 - \frac{1}{a^2}(\frac{dv}{dt})^2}$, (7) becomes

$$\frac{d}{dt}\left(\frac{dv}{dt}\right)^2 = \pm ka\sqrt{1 - \frac{1}{a^2}\left(\frac{dv}{dt}\right)^2}\ \frac{dv^2}{dt}. \tag{8}$$

Equation (8) can be integrated to yield

$$\sqrt{1 - \frac{1}{a^2}\left(\frac{dv}{dt}\right)^2} = \pm\frac{k}{2a}v^2 + c_1 \tag{9}$$

where c_1 is an integration constant. After squaring, this equation can be arranged in the form

$$\left(\frac{dv}{dt}\right)^2 + V(v) = a^2 \tag{10}$$

where

$$V(v) = \left(\frac{kv^2}{2} + c_1 a\right)^2 \tag{11}$$

is the potential when we view (10) as the energy integral of a particle motion with correspondence $v \to x$, the particle position. By inspecting the shape of the potential $V(v)$, we can figure out qualitatively the behavior of $v(t)$. $V(v)$ is W-shaped, symmetric with respect to the axis $v = 0$. $v(t)$ draws anharmonic motion between two turning points or even a solitary pulse behavior, depending upon the value of "energy" a^2. While the particles execute a round trip between two turning points, the Landau wave-particle resonance condition can be met, and the electric field Landau damps again over a somewhat lowered energy level, and the repetitive dampings bring about damped oscillations of the electric field, i.e., *nonlinear Landau damping.* O'Neil (1965) performed detailed nonlinear analysis of the Vlasov–Poisson plasma impressed by the sinusoidal electric field in the form of (6) in terms of the elliptic function, and derived the nonlinear Landau damping.

• An *adiabatic invariant* remains to be an approximate constant of motion of a nearly periodic system when a parameter of the system changes slowly (Chapter 3). The word "adiabatic" appears to be coined from quantum mechanical situation where the energy states distribution is unchanged even though the Hamiltonian is changed by external means without adding heat to the system — say, by changing the volume. When we have a particle moving in a potential well (trapped particle), the *adiabatic change* of the motion is brought about by infinitesimal dislocation of turning point. If the particle encounters a grazing turning point during its motion, the subsequent motion is subject to nonadiabatic abrupt change. The change in parameter should be slow as compared to the periods of motion and should not be related to the periods if it is an adiabatic change of motion. A well-known invariant for a particle moving in a magnetic field is its magnetic moment $\mu = W_\perp/B$ where $W_\perp$ is the kinetic energy of the particle perpendicular to B direction. With B changing slowly, the motion of the particle can be viewed as small perturbations from the constant field, which can be solved exactly. Here the usable small parameter is

$$\epsilon = \frac{T}{B}\frac{\partial B}{\partial t} \ll 1, \quad T: \text{ period}, \quad \text{cf. (3.61)} \tag{12}$$

in terms of which we can work on the perturbation series for the orbit. (12) states that the fractional change of B over one period is $\ll 1$. We are interested in knowing the accuracy of μ's invariance. Being a constant of motion in the lowest order, i.e. in ϵ-order, is not enough. Since we are dealing with an asymptotic series, the true qualification for an invariant requires that the higher order deviation from invariance goes to zero rapidly as ϵ goes to zero. Let us designate $\Delta\mu$ the deviation of μ from the invariance. We need $\Delta\mu \ll \epsilon^N$, for all N, with N being the order. When $\epsilon = 10^{-3}$ and $N = 1$, $\Delta\mu = \epsilon^N$ gives the estimation that appreciable change of μ is given rise to after 10^3 oscillations. If the experiment time is longer than 10^3 periods, the adiabatic invariant for $N = 1$ cannot be considered to be an invariant. We need a higher order invariant in the cosmic or laboratory scheme where the particles bounce back and forth for a large number of periods. It is desirable to prove the existence of higher order invariants in order to explain or support the validity of the scheme utilizing magnetic invariants, such as magnetic mirror confinement or Fermi acceleration. Oscillatory motion between two turning points can be depicted in a phase plane (p, q) by a closed loop, where p and q are the generalized momentum and coordinates, respectively. Depending upon the values of the total energy, $W = mv^2/2 + V(q, \lambda)$, with V the potential energy, the trajectories represent particles with different energies. The distribution function f of particles can be expressed in terms of W and θ, where θ is the angular coordinate along the loop: $f = f(W, \theta)$. f is time-independent only if it does not depend on θ (f is constant along particle trajectories). When $V(q, \lambda(t))$ is slowly changing through a parameter λ, the orbit will not be closed completely but spiralling with tiny pitches, showing a nearly periodic trajectory. According to Liouville equation, the points in the phase space behave as an incompressible fluid: the area enclosed by the almost closed loop is conserved. That is, the phase plane area enclosed by the orbit corresponding to any one of the orbit is an approximate constant of motion:

$$\oint pdq = J. \tag{13}$$

J in the above is known as the action integral and is an adiabatic invariant. Adiabatic invariance is a fundamental property of nearly periodic Hamiltonian system. In the literature, its general discussion has been done by using perturbed trapped particle motion [Chandrasekhar, S. *Plasma Physics*; Notes compiled by S. K. Trehan, Phoenix Books, University of Chicago Press (1960)] or perturbed Hamiltonian [Landau, L. D. and Lifshitz, E. M. (1960) *Mechanics*, Pergamon Press Oxford].

• Magnetohydrodynamics deals with the coupling between electromagnetic fields and motion of electrically conducting fluid in *large scale*. We find its applications in various cosmic and geomagnetic conditions. In thermonuclear research, magnetohydrodynamic equations are the governing equations of the plasma at hand. We must emphasize the role of the magnetic field, in the magnetohydrodynamic domain, which gives rise to the Lorentz force ($\frac{e}{c}\mathbf{v}\times\mathbf{B}$ or $\mathbf{J}\times\mathbf{B}$) for the charged particles. This force changes the motion of the electrically conducting medium hydrodynamically, and we have an interplay between two disciplines: hydrodynamics and electromagnetism. One fast way to derive the magnetohydrodynamic governing equation is to include the electromagnetic stress tensor, in addition to the fluid stress tensor, to the hydrodynamic kinematics. Magnetohydrodynamics governs the realm of large scale and slow process whose time scale is much greater than the ion cyclotron period. In Chapter 2, we show that the magnetohydrodynamic linear waves can be derived from the two-fluid plasma equations, which are more intimate to the Boltzmann equation, by employing the approximation $\omega \ll \omega_{ci}$, the ion cyclotron frequency.

• Intimacy between the Vlasov and Euler equations. The Euler equation, the momentum equation of the cold fluid plasma is a close kin of the Vlasov equation, closer than casual conjecture supposes. We know well that the fluid equation can be derived from the Boltzmann equation by taking the velocity moment; but it is easily missed that the Euler equation is contained in the Vlasov equation as a special class of the solution of the latter. This can be seen by considering the characteristics of both equations. We have the Vlasov equation for electrons written in the form

$$\frac{\partial}{\partial t}g(x,v,t) + v\frac{\partial g}{\partial x} - \frac{e}{m}E(x,t)\frac{\partial g}{\partial v} = 0. \tag{14}$$

The characteristic equations of (14) read

$$\frac{dt}{1} = \frac{dx}{v} = \frac{dv}{-eE(x,t)/m} = \frac{dg}{0} \tag{15}$$

which state that the distribution function $g(x,v,t)$ is constant along the characteristics or orbit represented by

$$\frac{dx}{dt} = v, \tag{16}$$

$$\frac{dv}{dt} = -\frac{e}{m}E(x,t), \tag{17}$$

$$v\,\frac{dv}{dx} = -\frac{e}{m}E(x,t). \tag{18}$$

The solutions of (14) can be constructed with the constants of motion obtained by solving the characteristic equations (16) to (18) [in fact, (18) is not independent]. Thus, let the implicit relation $g(x, v(x,t), t) = \text{const.} = c$ be an integral of (14). Then, we have per partial differentiation

$$\frac{\partial g}{\partial t} + \frac{\partial g}{\partial v}\frac{\partial v}{\partial t} = 0, \quad \frac{\partial g}{\partial x} + \frac{\partial g}{\partial v}\frac{\partial v}{\partial x} = 0. \tag{19}$$

If the above relations are substituted into (14), we obtain

$$\frac{\partial v}{\partial t} + v\frac{\partial v}{\partial x} = -\frac{e}{m}E(x,t) \tag{20}$$

which is the Euler equation of motion. The characteristic equations of the Euler equation (20) read

$$\frac{dt}{1} = \frac{dx}{v} = \frac{dv}{-eE(x,t)/m}. \tag{21}$$

Therefore, (17), the particle orbit expressed in Lagrangian derivative, is the characteristic of the Vlasov equation as well as the Euler equation.

Conversely, we can assume that the solution of (20) is obtained in implicit form, $f(x, t, v(x,t)) = \text{const.}$ in the sense that the function f determines v as a function of x and t. Then by partial differentiation, we have similar equations as (19) for f. Solving for $\partial v/\partial x$ and $\partial v/\partial t$, and substituting the results into (20) give

$$\frac{\partial}{\partial t}f(x,v,t) + v\frac{\partial f}{\partial x} - \frac{e}{m}E(x,t)\frac{\partial f}{\partial v} = 0$$

which is the Vlasov equation. Therefore, we may suppose that the solutions of the Euler equation form a subset of the solutions of the Vlasov equation. Landau damping can be derived from the cold fluid equation by regarding a plasma as a collection of beams distributed with different beam velocities (Chapter 4).

• Landau's collision integral (1.90) is a type of *Fokker–Planck equation* which can be derived from the *Smoluchowski law.* The latter applies to *Markovian processes* and can be expressed in the form

$$f(\mathbf{v}, t + \Delta t) = \int d\Delta\mathbf{v} f(\mathbf{v} - \Delta\mathbf{v}, t) P(\mathbf{v} - \Delta\mathbf{v}, \Delta\mathbf{v}) \tag{22}$$

where $f(\mathbf{v}, t)$ is the distribution function at time t and $P(\mathbf{v} - \Delta\mathbf{v}, \Delta\mathbf{v})$ is the transition probability that the velocity changes from $(\mathbf{v} - \Delta\mathbf{v}) \to \mathbf{v}$

by a small amount $\Delta\mathbf{v}$. The time interval Δt can be assumed long enough for a large number of collisions to take place, but short enough to warrant only small $\Delta\mathbf{v}$. The Markovian change of f in the above is appropriate to describe the collisions in plasma where numerous small angle Coulomb scattering is prevalent. Generally, in a Markovian process, the rate of change of a time-dependent quantity $\partial a(t)/\partial t$ depends on the instantaneous value of the quantity $a(t)$, but not on its previous history. The change is dictated by the transition probability of the event as in the random walk. Let us expand $f(\mathbf{v}, t + \Delta t)$, $P(\mathbf{v} - \Delta\mathbf{v}, \Delta\mathbf{v})$ in Taylor series and define the mean values

$$\langle \Delta\mathbf{v} \rangle = \int \Delta\mathbf{v} P(\mathbf{v}, \Delta\mathbf{v}) d\Delta\mathbf{v}, \tag{23}$$

$$\langle \Delta v_i \Delta v_j \rangle = \int \Delta v_i \Delta v_j P(\mathbf{v}, \Delta\mathbf{v}) d\Delta\mathbf{v}. \tag{24}$$

Then we can obtain from (22)

$$\frac{\partial f}{\partial t} = -\frac{\partial}{\partial v_i} \frac{\langle \Delta v_i \rangle f}{\Delta t} + \frac{1}{2} \frac{\partial^2}{\partial v_i \partial v_j} \frac{\langle \Delta v_i \Delta v_j \rangle f}{\Delta t} + \cdots \tag{25}$$

which is known as the Fokker–Planck equation. Earlier Chandrasekhar applied (22) to guiding center distribution to obtain its time rate of change in plasma (§51.1 *Plasma Physics* S. Chandrasekhar). In the derivation of the Landau kinetic equation (§1.7), which belongs to the Fokker–Planck class, we shall not pursue the analysis based on the Smoluchowski law, but directly engage Taylor expansion of the Boltzmann collision integral. The result is the same.

• Fourier transform, Laplace transform, causality, and the analytic domain of the susceptibility function $\chi(\omega)$. Fourier–Laplace transform is very useful for solving initial-value problems in plasma physics. In particular, a wave analysis of plasma equations requires Fourier transform of a physical function $F(x, t)$ to $F(k, \omega)$ in order to derive dispersion relations. Unfortunately, the phasor in Fourier transforms is not of one unified form but depends on the individual author's choice. This book adopts the following convention for the pair of Fourier transforms of a function $g(x)$ or $f(t)$:

$$g(k) = \int_{-\infty}^{\infty} dx g(x) e^{-ikx}, \tag{26}$$

$$g(x) = \int_{-\infty}^{\infty} \frac{dk}{2\pi} g(k)e^{ikx}, \tag{27}$$

$$f(\omega) = \int_{-\infty}^{\infty} dt f(t)e^{i\omega t}, \tag{28}$$

$$f(t) = \int_{-\infty}^{\infty} \frac{d\omega}{2\pi} f(\omega)e^{-i\omega t}, \tag{29}$$

$$F(k,\omega) = \int_{-\infty}^{\infty} dx \int_{-\infty}^{\infty} dt F(x,t)e^{-ikx+i\omega t}, \tag{30}$$

$$F(x,t) = \int_{-\infty}^{\infty} \frac{dk}{2\pi} \int_{-\infty}^{\infty} \frac{d\omega}{2\pi} F(k,\omega)e^{ikx-i\omega t}. \tag{31}$$

A function and its Fourier transform will be written with the same symbol; the functional dependence is expressed by k or ω in the parenthesis. Therefore, $f(t)$ and $f(\omega)$ have different dimensions.

Let us look into the condition for $f(t)$ to have its Fourier transform in (28). In order for this integral to exist, the integrand should not blow up at the lower and upper limits. At $t = -\infty$ we can assume $f(-\infty) = 0$; we assume that nothing was given rise to at the remote past. This agrees with the notion of causality. At $t = +\infty$, the integrand can be tamed by endowing ω an infinitesimal *positive* imaginary frequency: $\omega = \omega_r + i\omega_i$ with ω_i *positive* infinitesimal. In this way, $t = +\infty$ and $t = -\infty$ can be distinguished in the Fourier transform. $f(-\infty) = 0$ and $\omega = \omega_r + i\epsilon$ are the conditions for $f(t)$ to have its Fourier transform. Fourier transformed functions can be used with great facility in analyzing plasma waves because differential operators become algebraic quantities. As to the spatial coordinate x, we need not or cannot distinguish $x = \pm\infty$. We assume the boundary conditions $g(x = \pm\infty) = 0$. If one uses the phasor of the form $f(x,t) \sim e^{-ikx+i\omega t}$, a small negative imaginary frequency should be assumed. Laplace transform is useful when one transforms a function that is nonzero only for $t \geq 0$, as in an initial-value problem. Naturally, Laplace transform is defined like "half-range Fourier transform":

$$LT[f(t)] \equiv F(\omega) = \int_0^{\infty} e^{i\omega t} f(t)dt. \tag{32}$$

The inversion formula is derived in Chapter 5, and is written as

$$f(t) = \frac{1}{2\pi} \int_{-\infty+ic}^{\infty+ic} e^{-i\omega t} F(\omega)d\omega \tag{33}$$

where c must be large enough so that the straight line path in the complex ω-plane runs *above* all the singularities of $F(\omega)$ in order that $f(t)$ be zero for $t < 0$. This is easy to see because the inversion contour in (33) must be closed by surrounding the upper-half ω-plane for $t < 0$. Then, $f(t) = 0$ since $F(\omega)$ is analytic in the interior of that closed contour.

• The susceptibility function $\chi(t)$ represents the response of a plasma, which is a dielectric medium, when it is acted on by a certain electric field. The causality requires $\chi(t) = 0$ for $t \leq 0$. Therefore its Fourier transform is written

$$\chi(\omega) = \int_0^\infty e^{i\omega t}\chi(t)dt. \tag{34}$$

The inverse transform is

$$\chi(t) = \int_{-\infty}^{\infty} \frac{d\omega}{2\pi}\,\chi(\omega)e^{-i\omega t}. \tag{35}$$

This integral should yield that $\chi(t) = 0$ for $t \leq 0$. As in (33), we use the infinite semi-circle in the upper-half ω-plane for $t \leq 0$. Therefore, $\chi(\omega)$ *should be analytic in the upper-half ω-plane.* Using (35) in (34) yields, after carrying out integration,

$$\chi(\omega) = \frac{1}{2\pi i}\oint \frac{\chi(\omega')}{\omega' - \omega}d\omega' \quad (\text{Im}\ \omega > 0). \tag{36}$$

This is the Cauchy integral formula valid in the indicated analytic region (Im ω means the imaginary part of ω), and $\chi(\omega)$ is represented by the Cauchy formula. We are ready to obtain the Kramers–Kronig relations (§ 4.6) from (36) by separating it into real and imaginary parts.

• **The Van Kampen mode** of plasma wave drew attention because it was discovered by employing a different mathematical approach to solve the Vlasov equation — not by using Fourier transform. In the Fourier transform approach, if applied to *homogeneous* linear equations, we end up with a dispersion relation $\omega = \omega(k)$, which is the solvability condition for the algebraic set of equations. In Van Kampen's approach, the inherent singularity in the governing equation is surfaced in the form of a "source term" and the subsequent solution need not have a dispersion relation. Instead, the Van Kampen "eigen functions" are complete so that they can be used for expansion of certain functions. Specifically, we should pay attention to the singularity in the form $1/(\omega - kv)$ that appears in the Vlasov or fluid equation (Eq. (4.49) or Eq. (40)). Historically, the singularity at $\omega = kv$ has been taken care of in three different ways: 1) Vlasov simply disregarded it by taking only the principal value. 2) Landau elaborated it by analytic continuation in the Fourier domain. 3) Van Kampen paid a full account of it algebraically by introducing Dirac δ-function. Van Kampen's normal mode solutions of the Vlasov equation, which are simple harmonic waves in progressive wave form, have been proven to be capable of recovering Landau damping. Van Kampen included the δ-function term as an essential complementary solution to the Vlasov equation. In order to apply the Van Kampen eigen functions to the Vlasov initial value problem, a singular integral equation must be solved. Van Kampen's eigen functions are analogous to the eigen functions in the Sturm–Liouville theory of a differential equation.

• **Convective instability, absolute instability, evanescent wave, amplifying wave**. Fourier–Laplace transform is useful in the investigation of the asymptotic behavior of the waves excited by some initial perturbation. Let us consider an initial value problem posed by Vlasov and Poisson equations (§5.3.2):

$$\frac{\partial}{\partial t}g(x,v,t)+v\frac{\partial g}{\partial x}-\frac{e}{m}E(x,t)\frac{dg_0(v)}{dv}=0, \tag{37}$$

$$\frac{\partial E}{\partial x}=-4\pi e\int_{-\infty}^{\infty}g(x,v,t)dv, \tag{38}$$

with the initial condition $g(x,v,0)=\delta(x)I(v)$. Fourier–Laplace transform of the above two equations with the given initial condition yields (see Eq. (5.71))

$$E(x,t)=\frac{en_0}{\pi}\int_{-\infty+i\sigma}^{\infty+i\sigma}d\omega e^{-i\omega t}\int_{-\infty}^{\infty}\frac{dk}{k^2}\,e^{ikx}\,\frac{\int_{-\infty}^{\infty}\frac{I(v)}{v-\omega/k}dv}{\varepsilon(\omega,k)} \tag{39}$$

where

$$\varepsilon(\omega,k)=1-\frac{\omega_p^2}{k^2}\int_{-\infty}^{\infty}\frac{\frac{dg_0}{dv}}{v-w/k}dv \tag{40}$$

is the dielectric function.

If we simply look into the normal modes of plasma, we should solve $\varepsilon(\omega,k)=0$ to derive the dispersion relations. (This is seen from (39) by putting $I(v)=0$ in the numerator.) Solutions of the algebraic equation $\varepsilon(\omega,k)=0$ connect complex k to complex ω, generally. We talk about instability or damping for real k; amplifying or evanescent wave for real ω. However, convective and absolute instabilities can be distinguished only after examination of the asymptotic behavior of the response function $E(x,t)$.

A few criteria have been discovered to distinguish these two instabilities (Chapter 8).

(39) contains all the information in regard to the asymptotic behavior of the waves excited by the initial perturbation. It is convenient to introduce the Green function in terms of which the response in (39) can be investigated:

$$G(x,t) = \int_{-\infty+i\sigma}^{\infty+i\sigma} \frac{d\omega}{2\pi} e^{-i\omega t} \int_{-\infty}^{\infty} \frac{dk}{2\pi}\, e^{ikx}\, \frac{1}{D(\omega,k)} \quad (D(\omega,k) = k^2\varepsilon(\omega,k)). \tag{41}$$

The integration paths in the above indicate that k is real and ω is complex, but we can or have to deform, if necessary, the contours as far as the contours remain inside analytic regions. Thus we are free from any such restrictions for k and ω. We have two-fold integral in (41). We shall consider two model dispersion relations $D(k,\omega) = 0$ (Eqs. (8.19) and (8.20)):

$$D(k,\omega) = \left(k - \frac{\omega}{c}\right)\left(k - \frac{\omega}{c}\right) + k_0^2 = 0, \tag{42}$$

$$D(k,\omega) = \left(k - \frac{\omega}{c}\right)\left(k + \frac{\omega}{c}\right) - k_0^2 = 0. \tag{43}$$

Assume $x > 0$, then the k-contour should encircle the upper-half k-plane. The k-integration can be done by calculating the residues at the poles:

$$G(x,t) = \int_{-\infty+i\sigma}^{\infty+i\sigma} \frac{d\omega}{2\pi} e^{-i\omega t}\, i \sum_{k_s} \frac{e^{ik_s x}}{[\partial D/\partial k]_{k_s}} \tag{44}$$

where the k-poles (k_s) are obtained from $D(\omega,k) = 0$. Let us first consider the wave in (42),

$$k_s = \frac{\omega}{c} \pm ik_0. \tag{45}$$

We have $\partial D/\partial k = 2(k - \omega/c) = \pm 2ik_0$. The imaginary part of $\omega = \omega_r + i\omega_i$ can be large, so both signs should be included. Using these values in (44) gives

$$G(x,t) = \int_{-\infty+i\sigma}^{\infty+i\sigma} \frac{d\omega}{2\pi} e^{-i\omega t}\, e^{i\frac{\omega}{c}x} \frac{1}{2k_0}\left(e^{-k_0 x} - e^{k_0 x}\right) \tag{46}$$

which yields

$$G(x,t) = \frac{1}{2k_0}\left(e^{-k_0 x} - e^{k_0 x}\right)\delta\left(t - \frac{x}{c}\right). \tag{47}$$

This gives an amplifying wave along the trajectory $x = ct$.

Next, we consider the wave in (43), from which we have

$$\frac{\partial D}{\partial k} = 2k = \pm 2\sqrt{\frac{\omega^2}{c^2} + k_0^2} = \pm\frac{2}{c}\sqrt{\omega - \omega_s}\sqrt{\omega + \omega_s} \tag{48}$$

with

$$\omega_s = ick_0. \tag{49}$$

Clearly, the k-pole, $k_s = 0$, in (44) obliterates the x-dependence. We have ω-poles at $\omega = \pm\omega_s$. We shall consider the pole (branch point) at $\omega = +\omega_s$. We put $\partial D/\partial k = \frac{2}{c}\sqrt{2\omega_s}\sqrt{\omega - \omega_s}$ in (44) to write

$$G(x,t) = \int_{-\infty+i\sigma}^{\infty+i\sigma} \frac{d\omega}{2\pi} e^{-i\omega t}\, \frac{ic}{2\sqrt{2\omega_s}}\, \frac{1}{\sqrt{\omega - \omega_s}} \tag{50}$$

or

$$G(x,t) = \frac{ic\, e^{-i\omega_s t}}{2\sqrt{2\omega_c}} \int_{-\infty+i\sigma}^{\infty+i\sigma} d\omega\, \frac{e^{-i(\omega-\omega_s)t}}{\sqrt{\omega - \omega_s}}. \tag{51}$$

The $\int d\omega$ integral in the above has the value $i^{-\frac{1}{2}}\, 2\, \sqrt{\pi}/\sqrt{t}$. This can be proven by a contour integration along the contour in the cut ω-plane as shown in Fig. 8.10(b). Thus we finally obtain

$$G(x,t) = \frac{\sqrt{i}}{2}\, \frac{c}{\sqrt{2\pi}}\, \frac{e^{-i\omega_s t}}{\sqrt{\omega_s t}} \sim \frac{e^{ck_0 t}}{\sqrt{t}} \tag{52}$$

which indicates an absolute instability, independent of x. Note that

$$\int \frac{dk}{2\pi}\, \frac{e^{ikx}}{D(\omega,k)} = \int \frac{dk}{2\pi}\, \frac{e^{ikx}}{(k - \frac{1}{c}\sqrt{\omega^2 + c^2k_0^2})(k + \frac{1}{c}\sqrt{\omega^2 + c^2k_0^2})}$$

in which we have two poles, but only one of them is in the upper k-plane. Picking up the pole at $k = \frac{1}{c}\sqrt{\omega^2 + c^2k_0^2}$, the $\int dk$ integral becomes

$$\int \frac{dk}{2\pi}\, \frac{e^{ikx}}{D(\omega,k)} = \frac{ic}{2}\, \frac{exp\left(\frac{ix}{c}\sqrt{\omega^2 + c^2k_0^2}\,\right)}{\sqrt{\omega^2 + c^2k_0^2}} = \frac{ic}{2}\frac{1}{\sqrt{\omega - \omega_s}\sqrt{\omega + \omega_s}}$$

The exponential factor becomes unity in the vicinity of $\omega = ick_0$.

Alternatively, let us expand $D(\omega,k)$ in (44) as a Taylor series in the vicinity of $\omega = \omega_s$ and $k = k_s$, which are the values satisfying $\partial D/\partial k = 0$ and $D(\omega_s, k_s) = 0$: we find $k_s = 0$ and $\omega_s = ick_0$,

$$\begin{aligned} D(\omega,k) = D(\omega_s,k_s) &+ \left.\frac{\partial D}{\partial k}\right|_s (k - k_s) + \left.\frac{\partial D}{\partial \omega}\right|_s (\omega - \omega_s) \\ &+ \frac{1}{2}(k - k_s)^2 \left.\frac{\partial^2 D}{\partial k^2}\right|_s + \cdots = 0. \end{aligned} \tag{53}$$

The first two terms are zero. $\frac{\partial^2 D}{\partial k^2}|_s = 2$ per (43). Thus (53) reduces to $(\omega - \omega_s) \frac{\partial D}{\partial \omega}\big|_s + k^2 = 0$. We have $\frac{\partial D}{\partial \omega}\big|_s = -2\omega_s/c^2$. Then, (53) gives

$$k^2 - \frac{2}{c^2}\omega_s(\omega - \omega_s) = 0 \tag{54}$$

which again yields

$$\frac{\partial D}{\partial k} = 2k = \frac{2}{c}\sqrt{2\omega_s}\sqrt{\omega - \omega_s}. \tag{55}$$

The critical point of observation is that, as the ω's on the Laplace contour move onto a deformed ω-contour, the k-poles emigrate correspondingly on the complex k-plane. When the emigrating two k-poles merge, convective or absolute instabilities take place, depending upon the direction of the merging poles. The pole movement is discussed in Chapter 8, following Briggs [Briggs, R. J. (1964) *Electron-stream interaction with plasmas*, MIT Press, Cambridge]. Also general discussion of (44) will be continued to Chapter 8.

• **Kinematics of wave.** The asymptotic state of a wave packet (multi-chromatic wave) can be studied by two approaches: 1) by asymptotically ($t \to \infty$) calculating the Fourier integral, 2) by assuming a slowly varying *eikonal* (phase). §7.1 and §7.2 undertake the mathematical steps accompanying the first approach. Here, we consider the second approach for an electric field, varying slowly in space and time, which we represent in the form

$$\mathbf{E}(\mathbf{r}, \mathbf{t}) = \mathbf{e_k}\tilde{E}e^{-i\omega_\mathbf{k}t + i\mathbf{k}\cdot\mathbf{r}} + c.c. \tag{56}$$

where $\omega_\mathbf{k}$ is the eigen frequency satisfying the dispersion relation $\Delta(\omega_\mathbf{k}, \mathbf{k}) = 0$ and $e_\mathbf{k}$ is the unit vector indicating the wave polarization. $e_\mathbf{k}$ can be obtained from the Fourier-space wave equation written in terms of ω and $\mathbf{k}$ by putting $\omega = \omega_\mathbf{k}$. (The solvability condition for the Fourier-space wave equation is the dispersion relation, $\triangle(\omega, \mathbf{k}) = 0$.) The vector $\mathbf{e_k}$ may have multiplying factor, the constant phase shift $\exp(i\varphi_0)$, thus $\mathbf{e_k}$ is complex. More generally than (56), we can introduce eikonal $\varphi(\mathbf{r}, t)$, and write

$$\mathbf{E}(\mathbf{r}, \mathbf{t}) = \mathbf{e_k}\tilde{E}e^{-i\varphi(\mathbf{r},\, t)} + c.c. \tag{57}$$

Let us expand the phase φ about a local point $(\mathbf{r}_0,\ t_0)$,

$$\varphi(\mathbf{r},\ t) = \varphi(\mathbf{r}_0, t_0) + (\mathbf{r} - \mathbf{r}_0) \cdot \nabla\varphi\big|_0 + (t - t_0)\frac{\partial \varphi}{\partial t}\bigg|_0 \tag{58}$$

where the subscript 0 means $(\mathbf{r}_0,\ t_0)$. (56) and (58) should be equivalent, and we identify, as follows [for convenience, we put the reference point

$(\mathbf{r}_0, t_0) = (0,0)]$

$$\omega = \frac{\partial \varphi}{\partial t}, \quad \mathbf{k} = -\nabla \varphi. \tag{59}$$

Therefore, we have

$$\frac{\partial \mathbf{k}}{\partial t} + \nabla \omega = 0. \tag{60}$$

(59) and (60) are the fundamental equations relating the eikonal with local ω and $\mathbf{k}$.

In a homogeneous medium, $\omega = \omega(\mathbf{k})$, and (60) becomes

$$\frac{\partial \mathbf{k}}{\partial t} + \mathbf{v}_g \cdot \nabla \mathbf{k} = 0 \tag{61}$$

where $\mathbf{v}_g = \partial\omega/\partial\mathbf{k}$ is the group velocity. Thus $\mathbf{k}$ is constant along the characteristics $d\mathbf{r}/dt = \mathbf{v}_g$.

In an inhomogeneous medium, $\omega = \omega(\mathbf{k}, \mathbf{r})$, i.e. ω has explicit dependence on $\mathbf{r}$. In this case, (60) becomes

$$\frac{\partial \mathbf{k}}{\partial t} + \mathbf{v}_g \cdot \nabla \mathbf{k} = -\frac{\partial \omega}{\partial \mathbf{r}} = -\nabla \omega \tag{62}$$

which shows that $\mathbf{k}$ is not conserved along the curve $d\mathbf{r}/dt = \mathbf{v}_g$. Alternatively, we can write

$$\frac{\partial \omega}{\partial t} = \frac{\partial \omega}{\partial \mathbf{k}} \cdot \frac{\partial \mathbf{k}}{\partial t}. \tag{63}$$

Then, use of (63) in the scalar product of (60) by $\partial\omega/\partial\mathbf{k}$ gives

$$\frac{\partial \omega}{\partial t} + \mathbf{v}_g \cdot \nabla \omega = 0. \tag{64}$$

This equation holds in inhomogeneous medium as well as in homogeneous medium. ω is conserved along the characteristics $d\mathbf{r}/dt = \mathbf{v}_g$.

Another important quantity associated with the propagation of wave packet is the energy. In plasma, the energy is the sum of the electromagnetic field energy plus the particle energy; they are all contained in the Maxwell equations. The particle energy manifests itself through the current, which also produces the electromagnetic field. So the distinction between the two is not compulsory. The usual mathematical technique in calculating the wave packet energy is two-time and two-space scale analysis: the wave packet has a slowly changing amplitude while its phase is rapidly varying, space and time-wise. The two-scale perturbation calculation is also used for deriving the *ponderomotive force* which is a potential force resulting from mixing of two waves of slightly different frequencies. See §7.4 and §6.3. For the topic of wave packet energy, useful references are:

Kadomtsev, B. B. *Plasma Turbulence*, Academic Press, 1965, p. 18.
Kadomtsev, B. B. *Cooperative effects in plasmas* in *Reviews of Plasma Physics* Vol. **22**. Edited by Shafranov, V. D. KluerAcademic/Consultant Bureau, 2001, p. 202.
Bekefi, G. *Radiation Processes on plasmas*, Wiley, 1966.

Conservation of wave action (wave energy/frequency) can be derived from the argument of adiabatic invariance. It can be expressed in the form of a continuity equation akin to the Boltzmann equation.

Reference: Camac, M., Kantrowitz, A. R., Litvak, M. M., Patrick, R. M., and Petschek, H. E. *Shock waves in collision-free plasmas*, Nuc. Fusion Supplement **Pt. 2** (1962) p. 423–444.

• **Solution of the Korteweg–de Vries (KdV) equation by inverse scattering method.** See §7.7. The solution $u(x,t)$ of the KdV equation

$$\frac{\partial u}{\partial t} - 6u\frac{\partial u}{\partial x} + \frac{\partial^3 u}{\partial x^3} = 0 \tag{65}$$

with the initial boundary condition $u(x,0)$ was obtained by employing the quantum mechanical inverse scattering method [Gardner, C. S., Greene, J. M., Kruskal, M. D., Miura, R. M. *Method for solving the Korteweg-de Vries equation*, *Phys. Rev. Lett.*, **19** (1967) p. 1095]. The solution is related to a function $K(x,z,t)$ by the differentiation

$$u(x,t) = -2\frac{d}{dx}K(x,x,t) \tag{66}$$

where K satisfies the Gel'fand–Levitan integral equation

$$K(x,z,t) + B(x+z,t) + \int_x^\infty K(x,y,t)B(y+z,t)dy = 0 \tag{67}$$

with the kernel B given by

$$B(\xi,t) = \sum_{n=1}^{N} C_n^2(t)e^{-\kappa_n(t)\xi} + \frac{1}{2\pi}\int_x^\infty R(k,t)e^{ik\xi}dk \tag{68}$$

where the scattering coefficients $\kappa_n(t)$, $C_n(t)$, and $R(k,t)$ are determined in terms of the corresponding initial values by equations

$$\kappa_n(t) = \kappa_n(0), \tag{69}$$

$$C_n(t) = C_n(0)e^{4\kappa_n^3 t}, \tag{70}$$

$$T(k,t) = T(k,0), \tag{71}$$

$$R(k,t) = R(k,0)e^{8ik^3 t}. \tag{72}$$

The initial scattering coefficients, $\kappa_n(0)$, $C_n(0)$, and $R(k,0)$ are obtained in terms of the initial potential (the initial KdV function) by solving the Schrödinger equation for continuous spectrum

$$\left[\frac{d^2}{dx^2} + k^2 - u(x,0)\right]\psi = 0 \tag{73}$$

with the boundary conditions

$$x \to -\infty : \ \psi = T(0)e^{-ikx}$$

$$x \to +\infty : \ \psi = e^{-ikx} + R(0)e^{+ikx}$$

and for discrete spectrum

$$\left[\frac{d^2}{dx^2} - \kappa_n^2 - u(x,0)\right]\psi_n = 0 \tag{74}$$

with the boundary condition $\psi_n = 0$ at $|x| \to \infty$, thus $\psi_n(x,t) = C_n(t)e^{-\kappa_n x}$.

• The topics on the **surface waves** in Chapter 14 are heavily drawn from my research papers published over the last 10 years. I am happy to have the opportunity to reorganize them in a single chapter, thus providing a more collected view of the subject. Particularly, the surface wave dispersion relation in slab plasma would be useful because the closed form of the dispersion relation is rather unknown. The kinetic surface wave dispersion relations for *magnetized slab plasmas* are obtained with details of the algebraic steps, and are shown to agree with the fluid equation results. Landau (1946) [Reference: Landau, L. *On the vibrations of the electronic plasma*, J. Physics (USSR) **10** (1946) p. 25] investigated how an external electric field penetrates in a plasma when the plasma occupying a half-space is placed into an external periodical electric field. We address a similar problem *when a slab plasma is inserted into an external electric field.* This problem which is of practical interest can be solved in the framework of bounded plasma formulated in Chapter 14.

• **Dusty plasma**. *"for dust you are and to dust you will return"* [Genesis 3:19]. A dusty plasma is the fourth state of matter mixed with small grains which interact electromagnetically with electrons and ions by charging process. At the same time, a dusty plasma is subject to the gravitational force due to the grain's massiveness. Dusty plasmas represent the usual state of matter in cosmos as well as in laboratory plasmas. In a dusty plasma with sufficiently large size of grains, the gravitational and electric forces are equally important — a self gravitational plasma. Dust grains are usually negatively charged (10^3–10^4 electronic charge) and can modify the dispersion of the plasma normal modes in the frequency range $\omega \geq \omega_{pi}$. Grains can participate in the collective behavior of the plasma, thus producing a new mode of wave. A very low frequency electrostatic mode called a *dust acoustic wave* can propagate in a dusty plasma of thermalized electron and ion background. Chapter 15 investigates in detail the charging process of grains by capturing the electrons and ions in the plasma. As the consequence, the electron and ion continuity equations have heavy damping terms due to the depletion (sink) of the charges tending to attach to the grains. Also, the continuity equations of electrons and ions are coupled.

References: Whipple, E. C., Northrop, T. G., and Mendis, D. A. *J. Geophys. Res* [Oceans] **90** (1985) p. 7405.
Shukla, P. K. in *The physics of dusty plasma* Ed. Shukla, P. K., Mendis, D. A., and Chow, V. W. World Scientific, Singapore (1996).

• **A useful observation and its verification**. We have the phase-mixing integral (Chapter 5)

$$\lim_{u \to \pm\infty} \int_{-\infty}^{\infty} f(x)\, e^{iux} dx = 0, \tag{75}$$

if $f(x)$ is *integrable* (not blowing up) over $(-\infty, \infty)$. Intuitively, we agree with (75), because the exponential function e^{iux} becomes rapidly oscillating with tiny interval of repetition and the integral becomes vanishingly small as $u \to \infty$ by successive cancellations. This can be proved formally as follows.

$$\begin{aligned} \int_{-\infty}^{\infty} f(x)\, e^{iux} dx &= \frac{1}{2} \int_{-\infty}^{\infty} f(x) \left[e^{iux} - e^{iu(x-\pi/u)} \right] dx \\ &= \frac{1}{2} \int_{-\infty}^{\infty} \left[f(x) - f\left(x + \frac{\pi}{u}\right) \right] e^{iux} dx \\ &\leq \frac{1}{2} \int_{-\infty}^{\infty} \left| f\left(x + \frac{\pi}{u}\right) - f(x) \right| dx. \end{aligned} \tag{76}$$

Clearly, the last integral vanishes as $u \to \infty$ since the integrand vanishes. In the above steps, we made a change of variable $x - \pi/u \to x'$.

• **Wiener–Hopf method**. This book deals with this useful subject in a pedagogical way in Appendix E. To motivate the mathematics underlying this method, a preliminary is presented in this space. It begins with the half-range Fourier transforms which we designate by $\pm$ subscripts,

$$f_+(k) = \int_0^\infty e^{-ikx} f(x)dx, \tag{77}$$

$$f_-(k) = \int_{-\infty}^0 e^{-ikx} f(x)dx, \tag{78}$$

$$f(k) = f_+(k) + f_-(k) = \int_{-\infty}^\infty e^{-ikx} f(x)dx. \tag{79}$$

Let $f(x) \sim e^{-\alpha x}$ for $x > 0$ and sufficiently large, and $f(x) \sim e^{-\beta|x|}$ for $x < 0$ and $|x|$ sufficiently large, with $\alpha,\ \beta$ real constants. Then we can show that

$$f(x) = \int_{-\infty+i\alpha_1}^{\infty+i\alpha_1} e^{ikx} f_+(k)\ \frac{dk}{2\pi} + \int_{-\infty-i\beta_1}^{\infty-i\beta_1} e^{ikx} f_-(k)\ \frac{dk}{2\pi} \tag{80}$$

where $\alpha_1,\ \beta_1$ are arbitrary but $\alpha_1 < \alpha$ and $\beta_1 < \beta$.

Proof: Let us take a look at (77). If one substitutes the asymptotic expression $f(x) \sim e^{-\alpha x}$, it follows

$$f_+(k) \sim \int_0^\infty e^{-ik_r x}\ e^{(k_i - \alpha)x} dx \quad (k = k_r + ik_i).$$

In order for this integral to converge at $x = +\infty$, we need

$$\text{Im}\ k < \alpha.$$

For the integral in (78), it is convenient to change $x \to -x$ to write

$$f_-(k) = \int_0^\infty e^{ikx} f(-x)dx.$$

If one substitutes the asymptotic expression $f(-x) \sim e^{-\beta x}$ into (78), it follows

$$f_-(k) \sim \int_0^\infty e^{ik_r x}\ e^{-(k_i+\beta)x} dx.$$

In order for this integral to converge at $x = +\infty$, we need

$$\text{Im } k > -\beta.$$

Therefore the analytic region for $f(k)$ is the strip $-\beta < \text{Im } k < \alpha$, and the inversion integral can be written as (80). Q. E. D.

The Wiener–Hopf technique is useful for certain integral equation and half-range boundary-value problems (data given on only $0 < x < \infty$) in a semi-infinite medium ($y > 0$).

Another useful relation is

$$f_+(k) - f_-(k) = \frac{i}{\pi} P \int_{-\infty}^{\infty} dk' \, \frac{f(k')}{k - k'} \tag{81}$$

where P denotes the principal value. The above equation can be proven by Cauchy's integral formula

$$f(k) = \frac{1}{2\pi i} \oint \frac{f(k')}{k' - k} dk' \tag{82}$$

where $f(k)$ should be analytic in the interior of the closed contour. $f_+(k)$ ($f_-(k)$) is analytic in the lower (upper) plane. Therefore for $f_+(k)$ ($f_-(k)$) the contour should be the real k'-axis plus the infinite semi-circle encircling the lower (upper) plane. When k is real, the integral is analytically continued to the real k': k should move toward the real k'-axis from below (above) for $f_+(k)$ ($f_-(k)$). The integral along the real k'-axis is therefore a sum of the principal value plus the δ- function part:

$$f_+(k) = \frac{1}{2\pi i} \left[P \int_{-\infty}^{\infty} \frac{f(k')}{k' - k} dk' + i\pi f(k) \right], \tag{83}$$

$$f_-(k) = -\frac{1}{2\pi i} \left[P \int_{-\infty}^{\infty} \frac{f(k')}{k' - k} dk' - i\pi f(k) \right], \tag{84}$$

which yields (81) upon subtraction. (81) is useful to solve Van Kampen's integral equation, Chapter 5. Decomposition of a Fourier integral ($f(k)$) into two half-range integrals ($f_\pm(k)$) is an essential ingredient in Wiener–Hopf method. It is useful in solving Van Kampen's integral equation. $f_\pm(k)$ have a common analytic region but their asymptotic analytic regions are different.

- **Integral equation of Voltera type**. We encounter a Voltera-type integral equation when we consider the causality in electrodynamics (see Chapter 5). In the Voltera integral equation, one of the integration limits is the variable, and it is generally dealt with by half-range Fourier transform.

Here we solve an easy exemplary problem which can be solved by Fourier transform but requires algebraic fundamental caution that can easily escape our attention. Solve

$$\psi(t) = C + A\int_t^\infty e^{\alpha(t-t')}\psi(t')dt' \tag{85}$$

where C, A, α are constants. Fourier transforming (85) gives

$$\psi(\omega) = 2\pi C\delta(\omega) + A\int_{-\infty}^\infty dt e^{i\omega t}\int_t^\infty e^{\alpha(t-t')}\psi(t')dt'. \tag{86}$$

The integral following A is

$$\begin{aligned} &A\int_{-\infty}^\infty dt e^{(i\omega+\alpha)t}\int_t^\infty e^{-\alpha t'}\psi(t')dt' \\ &= \frac{A}{i\omega+\alpha}\int_{-\infty}^\infty dt \frac{d}{dt}e^{(i\omega+\alpha)t}\int_t^\infty e^{-\alpha t'}\psi(t')dt'. \end{aligned} \tag{87}$$

This can be integrated by parts; the integrated term vanishes provided $\alpha >$ Im ω. Assuming this, (86) becomes

$$\psi(\omega)\left[1 - \frac{A}{i\omega+\alpha}\right] = 2\pi C\delta(\omega) \tag{88}$$

which is solved for ψ in the form

$$\psi(\omega) = \frac{2\pi C\delta(\omega)}{1-\frac{A}{i\omega+\alpha}} + C'\delta\left(1 - \frac{A}{i\omega+\alpha}\right) \tag{89}$$

where C' is constant. The C'-term should be included to obtain the complete solution. Recall $x\delta(x) = 0$ identically. It remains to invert (89). One obtains

$$\psi(t) = \frac{C\alpha}{\alpha - A} + C_1 e^{t(\alpha-A)} \quad (C_1:\ \text{const.}). \tag{90}$$

The undetermined constant C_1 can be determined in terms of the initial value $\psi(0)$. If $C = \psi(0)$, the final solution is obtained

$$\psi(t) = \frac{C\alpha}{\alpha - A} - \frac{CA}{\alpha - A}\, e^{(\alpha-A)t}. \tag{91}$$

This problem can be solved by Laplace transform by dividing the integral into half-range integrals [see Morse, P. M. and Feshbach, H., *Methods of theoretical physics* McGraw-Hill, New York, 1953 Part 1 §8.5]. We shall consider a similar Voltera-type integral equation for the causal relations of the electric field in Chapter 5.

- As another example which can be solved by Laplace transform, we consider the integral equation [Morse and Feshbach p. 973]

$$\psi(t) = \psi_0 + v_0 t + k^2 \int_0^t (t' - t)\psi(t')dt' \quad (t > 0). \tag{92}$$

Direct differentiation shows that (92) is equivalent to the differential equation $d^2\psi/dt^2 + k^2\psi = 0$ with boundary condition $\psi(0) = \psi_0$, $\frac{\partial \psi}{\partial t}|_0 = v_0$. (92) can be solved by Laplace transform or the half-range Fourier transform. Performing $\times \int_0^\infty \exp(i\omega t)dt$ yields

$$\psi(\omega) = \frac{i\psi_0}{\omega} - \frac{v_0}{\omega^2} + \frac{k^2}{\omega^2}\psi(\omega) \tag{93}$$

where

$$\psi(\omega) = \int_0^\infty \psi(t)\; e^{i\omega t} dt \quad (\text{Im } \omega > 0). \tag{94}$$

In deriving (93), we integrated by parts after setting $\exp(i\omega t) = \frac{1}{i\omega}\frac{d}{dt}\exp(i\omega t)$. Solving (93) for $\psi(\omega)$ gives

$$\psi(\omega) = \int_0^\infty \psi(t')\; e^{i\omega t'} dt' = \frac{i\omega\psi_0 - v_0}{\omega^2 - k^2} + C\delta(\omega^2 - k^2) \tag{95}$$

where C is a constant. (95) can be inverted by performing on both sides $\times \int_{-\infty}^\infty \exp(-i\omega t)d\omega$. Since $t > 0$, the $\int d\omega$-integral on the LHS yields $\delta(t-t')$, giving

$$\psi(t) = \int_{-\infty}^\infty \frac{d\omega}{2\pi}\; e^{-i\omega t}\left[\frac{i\omega\psi_0 - v_0}{\omega^2 - k^2} + C\delta(\omega^2 - k^2)\right]. \tag{96}$$

The above line integral path runs above the real ω-axis (Im $\omega > 0$) and can be extended by surrounding the lower infinite semi-circle (since $t > 0$). Calculating the residues at the pole $\omega = \pm k$, one obtains

$$\psi(t) = \psi_0 \cos kt + \frac{v_0}{k} \sin kt + \frac{C}{k} \cos kt. \tag{97}$$

The last term represents the homogeneous solution.

- **A useful integral**. Let us consider the integral

$$I_0 \equiv P \int_{-\infty}^\infty \frac{F(\omega)\; e^{-i\omega t}}{\omega - \omega_0} d\omega \tag{98}$$

where $F(\omega)$ is a function which renders the Fourier transform integral possible, and the symbol P stands for the principal value of the integral. The integral is singular since the singularity $\omega = \omega_0$ is on the integration path. The principal value defines the integral by chopping out the infinitesimal segment about $\omega = \omega_0$. Or

$$I_0 = \lim_{\epsilon \to 0} \left[\int_{-\infty}^{\omega_0 - \epsilon} \frac{F(\omega)\, e^{-i\omega t}}{\omega - \omega_0}\, d\omega + \int_{\omega_0 + \epsilon}^{\infty} \frac{F(\omega)\, e^{-i\omega t}}{\omega - \omega_0}\, d\omega \right].$$

It is convenient to write I_0 in the form

$$I_0 = \int_{-\infty}^{\infty} \frac{F(\omega) - F(\omega_0)}{\omega - \omega_0}\, e^{-i\omega t}\, d\omega + I$$

with

$$I = F(\omega_0)\; P \int_{-\infty}^{\infty} \frac{e^{-i\omega t}}{\omega - \omega_0} d\omega.$$

The first integral in I_0 is *not* singular (so we omitted the symbol P), and thus we focus discussion on the integral I. The integral I can be associated with a contour integration by completing a closed contour: the chopped out section can be connected by either convexed-up infinitesimal semi-circle or concave-down infinitesimal semi-circle around $\omega = \omega_0$ (see Fig. 1). When $t < 0$ ($t > 0$) the contour should encircle the upper (lower) ω-plane, so that the contribution from the infinite circle vanishes. According to residue theorem (also note that $F(\omega)$ is analytic over the entire plane), we have for $t < 0$

$$\textit{path } a: \; I - \pi i e^{-i\omega_0 t} F(\omega_0) = 0;$$
$$\textit{path } b: \; I + \pi i e^{-i\omega_0 t} F(\omega_0) = 2\pi i e^{-i\omega_0 t} F(\omega_0).$$

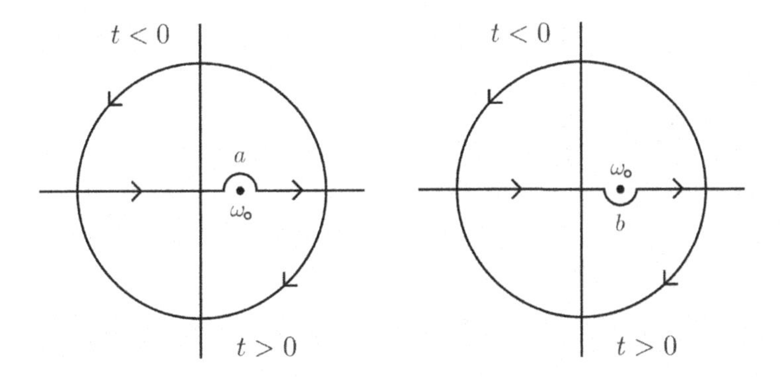

Fig. 1. Contours in complex ω-plane.

Hence, either path yields the value for $t < 0$

$$F(\omega_0)P\int_{-\infty}^{\infty}\frac{e^{-i\omega t}}{\omega-\omega_0}\,d\omega = \pi i e^{-i\omega_0 t}F(\omega_0) \quad (t < 0). \tag{99}$$

When $t > 0$, we use the lower contour.

$$\textit{path a}: \; I - \pi i e^{-i\omega_0 t}F(\omega_0) = -2\pi i e^{-i\omega_0 t}F(\omega_0);$$

$$\textit{path b}: \; I + \pi i e^{-i\omega_0 t}F(\omega_0) = 0.$$

Hence, either path yields the value for $t > 0$

$$F(\omega_0)P\int_{-\infty}^{\infty}\frac{e^{-i\omega t}}{\omega-\omega_0}\,d\omega = -\pi i e^{-i\omega_0 t}F(\omega_0) \quad (t > 0). \tag{100}$$

In particular, setting $F(\omega) = 1$ gives formula (A.41) in Appendix A.

About the Author

Hee J Lee was born in Korea 1941. After graduating from Seoul National University (1963), he received a Ph.D. degree from the University of California at San Diego in plasma physics 1975. Earlier he attended Cornell University with a major in space plasma physics (M.S.). He worked at NASA Goddard Institute for Space Studies, New York, for two years, in the project of general circulation of atmosphere. He was a visiting research professor at Clarkson University and University of Michigan (Ann Arbor). He has been a Professor of physics at Hanyang University (1980-2006) and a Professor Emeritus (2006-). He is interested in basic theoretical plasma physics, and has published more than 100 refereed papers in international journals. Throughout his research career, he has been grateful to Professor Chul-Soo Kim at Seoul National University, Professors W. Ian Axford and Jules A. Fejer at UCSD, and Professor L. Stenflo at Umea University, Sweden, and Professor Richard V. Lovelace at Cornell University.

Contents

Chapter 1

Boltzmann equation

1.1. Heuristic derivation of Boltzmann equation

The Boltzmann equation is of fundamental importance in the kinetic theory of gases and plasmas. It expresses a mathematical description of the distribution function $f(\mathbf{r}, \mathbf{v}, t)$ of the particles at $(\mathbf{r}, \mathbf{v})$ in the phase space and at time t. It takes the form of (1.14) below, and gives the rate of change of f along the characteristics, the particle trajectory. The Boltzmann equation becomes the Vlasov equation when we regard the change of f due to particle collisions is zero in (1.14). The *Vlasov fluid* behaves like an incompressible fluid in the 6-dimensional phase space, just as the Liouville ensemble behaves likewise in the $6N$-dimensional Γ-space. This Chapter may be read with a quick consulting of Chapter 12. Boltzmann invented the collision term in Section (1.2) below, which is his great accomplishment. The following assumptions are made as a preliminary to the development of Boltzmann's equation:

(a) We assume that the state of the gas is described by a one-body distribution function $f(\mathbf{r}, \mathbf{v}, t)$.
(b) The density of particles is low enough that only two-body interactions need to be considered, i.e., the range of inter-particle forces (r_0) is much smaller than the mean inter-particle distance.
(c) The duration of an encounter between two particles (interaction time) is much smaller than the duration of free motion of the particles, i.e., $r_0/v_{av} \equiv t_{int} \ll \lambda/v_{av}$ where λ is the mean free path and v_{av} is the mean speed of the particles.
(d) The particles are assumed to be point centers of spherically symmetric fields, so that the one-body distribution function depends only on the position $\mathbf{r}$, velocity $\mathbf{v}$ of the particles and time t. In case of exceptional models for the particles, other variables, such as the angular velocity, may be introduced.

Let $f(\mathbf{r}, \mathbf{v}, t)\ d^3rd^3v$ be the expected number of particles to be found in a volume element d^3rd^3v of phase space about $\mathbf{r}$ and $\mathbf{v}$ at time t. The volume element $\Delta\mu \equiv d^3rd^3v$ must be large enough to contain a sufficient number of particles in order that probability concepts can be applied at all. Thus

$$f(\mathbf{r}, \mathbf{v}, t) = \frac{1}{\Delta\mu} \int_{\Delta\mu} F(\mathbf{r}, \mathbf{v}, t) d^3r_1 \ldots d^3r_N d^3v_1 \ldots d^3v_N \tag{1.1}$$

$$F(\mathbf{r}, \mathbf{v}, t) = \sum_{i=1}^{N} \delta(\mathbf{r} - \mathbf{r}_i(t))\delta(\mathbf{v} - \mathbf{v}_i(t)) \tag{1.2}$$

where i is a particle index. Moreover, the changes in f will be observed over a time Δt which is much larger than the interaction time $t_{int} = r_o/v_{av}$.

We are concerned with developing an equation which determines the temporal evolution of f given its value at some initial time t_0 for all $\mathbf{r}$ and $\mathbf{v}$. By definition the total number of particles

$$N = \int f(\mathbf{r}, \mathbf{v}, t) d^3rd^3v \tag{1.3}$$

where the integration is carried over the volume V in configuration space occupied by the particles and over the accessible region of velocity space. Furthermore, we define a density n

$$n(\mathbf{r}, t) = \frac{1}{V} \int_{all\ v} f(\mathbf{r}, \mathbf{v}, t) d^3v \tag{1.4}$$

with the configuration volume V vanishingly small but large enough to contain many enough particles.

If we assume that those particles contained in d^3rd^3v do not interact with each other, then at a time $t + dt$ ($dt \gg t_{int}$) we expect these particles to be in the volume element $d^3r'd^3v'$ about $\mathbf{r}'$ and $\mathbf{v}'$ where

$$\mathbf{r}' = \mathbf{r} + \mathbf{v}dt + O(dt)^2 \tag{1.5}$$

$$\mathbf{v}' = \mathbf{v} + \mathbf{a}dt + O(dt)^2 \tag{1.6}$$

where $\mathbf{a}$ is the acceleration suffered by the particles due to the fields that may be applied by external means or those generated by the collective action of of all the particles excluding those whose trajectories (internal) are under examination. Thus

$$\mathbf{a} = \mathbf{a}_e + \mathbf{a}_i \tag{1.7}$$

where the subscripts e and i correspond to the cause of the acceleration. The new volume element $d^3r'd^3v'$ is related to the old volume element d^3rd^3v by the relation

$$d^3r'd^3v' = J\left(\frac{\mathbf{r}',\mathbf{v}'}{\mathbf{r},\mathbf{v}}\right)d^3rd^3v \tag{1.8}$$

where J is the Jacobian determinant of the transformation written out in full as:

$$J = \begin{vmatrix} \frac{\partial r_1'}{\partial r_1} & \frac{\partial r_2'}{\partial r_1} & \frac{\partial r_3'}{\partial r_1} & 0 & 0 & 0 \\ \frac{\partial r_1'}{\partial r_2} & \frac{\partial r_2'}{\partial r_2} & \frac{\partial r_3'}{\partial r_2} & 0 & 0 & 0 \\ \frac{\partial r_1'}{\partial r_3} & \frac{\partial r_2'}{\partial r_3} & \frac{\partial r_3'}{\partial r_3} & 0 & 0 & 0 \\ 0 & 0 & 0 & \frac{\partial v_1'}{\partial v_1} & \frac{\partial v_2'}{\partial v_1} & \frac{\partial v_3'}{\partial v_1} \\ 0 & 0 & 0 & \frac{\partial v_1'}{\partial v_2} & \frac{\partial v_2'}{\partial v_2} & \frac{\partial v_3'}{\partial v_2} \\ 0 & 0 & 0 & \frac{\partial v_1'}{\partial v_3} & \frac{\partial v_2'}{\partial v_3} & \frac{\partial v_3'}{\partial v_3} \end{vmatrix}$$

This expression for the Jacobian assumes that the after-collision position of a molecule is uncorrelated with the before-collision velocity of the particle, which is the assumption of *molecular chaos.* Owing to this assumption, the chain of hierarchy of distribution functions can be terminated at the one-particle distribution function, and the Boltzmann equation becomes self-closed with the collision term that is here derived and is expressed by the Boltzmann distribution function itself. Making use of (1.5) and (1.6),

$$J = 1 + \frac{\partial a_i}{\partial v_i}dt + O(dt)^2 \tag{1.9}$$

where we note that the determinant J is equal to the product of the two 3 by 3 sub-determinants located at the upper-left and lower-right corners. The upper-left determinant is unity. The lower-right determinant is diagonal with elements

$$1 + \frac{\partial a_1}{\partial v_1}dt,$$

and the like. Here the repeated indexes are summed over.

$$d^3r'd^3v' = \left[1 + \frac{\partial a_i}{\partial v_i}dt + O(dt)^2\right]d^3rd^3v \tag{1.10}$$

and further we have

$$f(\mathbf{r}', \mathbf{v}', t + dt) d^3 r' d^3 v' = f(\mathbf{r}, \mathbf{v}, t) d^3 r d^3 v + \left(\frac{\delta f}{\delta t} \right)_c d^3 r d^3 v dt \tag{1.11}$$

which means that the same number of particles are in the new volume element as in the old element except for those gained or lost by interaction among the particles themselves denoted by the last term of (1.11). Expanding the left hand side (LHS) of (1.11) in a Taylor series about $(\mathbf{r}, \mathbf{v}, t)$, we have

$$LHS(1.11) = \left[f(\mathbf{r}, \mathbf{v}, t) + \left(\frac{\partial f}{\partial t} + \frac{\partial f}{\partial r_i} \frac{dr_i}{dt} + \frac{\partial f}{\partial v_i} \frac{dv_i}{dt} \right) dt \right]$$
$$\times\, d^3 r d^3 v dt \left(1 + \frac{\partial a_i}{\partial v_i} dt \right) + O(dt)^2. \tag{1.12}$$

Thus

$$\frac{\partial f}{\partial t} + \frac{\partial f}{\partial r_i} \frac{dr_i}{dt} + \frac{\partial f}{\partial v_i} \frac{dv_i}{dt} + f \frac{\partial a_i}{\partial v_i} dt = \left(\frac{\delta f}{\delta t} \right)_c. \tag{1.13}$$

As we restrict ourselves to forces that are either independent of particle velocity or the Lorentz force $e\, \mathbf{v} \times \mathbf{B}$, if they depend on velocity, $\frac{\partial a_i}{\partial v_i} = 0$ always, and we finally have

$$\frac{\partial f}{\partial t} + v_i \frac{\partial f}{\partial r_i} + a_i \frac{\partial f}{\partial v_i} = \left(\frac{\delta f}{\delta t} \right)_c. \tag{1.14}$$

1.2. Collision term $(\frac{\delta f}{\delta t})_c$

Consider a two-body elastic collision. Denote the particle velocities before and after collision as $(\mathbf{v}, \mathbf{v}_1)$ and $(\mathbf{v}', \mathbf{v}'_1)$, respectively, where the primed are after-collision velocities, which can be expressed in terms of the pre-collision velocities if the force law governing the interaction is known. The collision event is more conveniently described in the center-of-mass frame of the two particles, in which the collision is equivalent to a deflection of a fictitious particle of reduced mass (moving with the relative velocity) from the scattering center. Figure 1.1 shows the geometry of the scattering process. The incident particle approaches the scattering center with an *impact parameter* b and is deflected by a *scattering angle* θ. The orbit of the particle involves two angles (φ, θ) where φ is the azimuthal angle and the polar angle θ corresponds to the scattering angle. The flux of particles passing through an elemental area $b\, db\, d\varphi$ is

$$b\, db\, d\varphi |\mathbf{v} - \mathbf{v}_1|\, f(\mathbf{r}, \mathbf{v}_1, t) d^3 v_1 \quad (sec^{-1}). \tag{1.15}$$

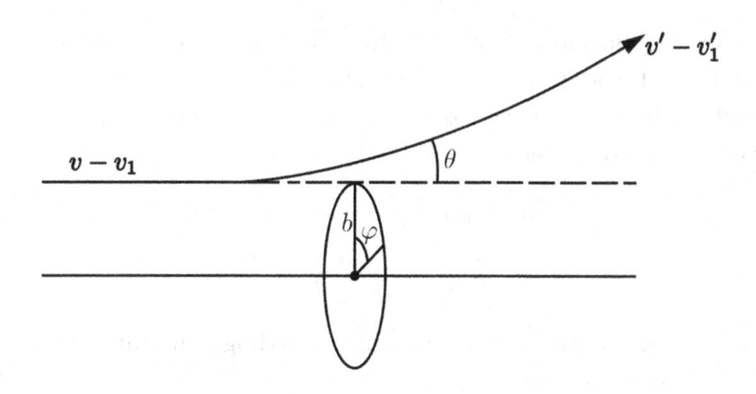

Fig. 1.1. Collision in the center-of-mass frame.

Then the number of collisions (per unit time) of the particles in the above flux with the particles of velocity $\mathbf{v}$ is (1.15) $\times f(\mathbf{v}, \mathbf{r}, t)d^3v$:

$$b\ db\ d\varphi|\mathbf{v} - \mathbf{v}_1|\ f(\mathbf{r}, \mathbf{v}_1, t)d^3v_1\ f(\mathbf{v}, \mathbf{r}, t)d^3v.$$

Then integrating the above equation over all $\mathbf{v}_1$ and dividing by d^3v gives the total rate of change of $f(\mathbf{v}, \mathbf{r}, t)$ due to collisions which scatter $\mathbf{v}$-particles out of the range $(\mathbf{v}, \mathbf{v} + d\mathbf{v})$:

$$\left(\frac{\delta f}{\delta t}\right)_{out} = \int d^3v_1 \int_0^\infty db\ b \int_0^{2\pi} d\varphi|\mathbf{v} - \mathbf{v}_1|\ f(\mathbf{v}, \mathbf{r}, t)f(\mathbf{r}, \mathbf{v}_1, t). \quad (1.16)$$

Note that $|\mathbf{v} - \mathbf{v}_1| \equiv v_R$, the relative speed, is a function of the scattering angle θ, which in turn is determined by b and φ if the force law is known. It is customary to introduce the differential scattering cross section

$$d\sigma(v_R, \theta) = b\ db\ d\varphi; \quad \theta = \theta(v_R, b).$$

For more about the scattering angle θ, §1.7 and Fig. 1.3 should be consulted. If we go into the center of mass frame, θ is the angle between the pre and post relative velocities, and is a function of b and the initial relative velocity. Then (1.16) is written as

$$\left(\frac{\delta f}{\delta t}\right)_{out} = \int d^3v_1 \int d\Omega\ \frac{d\sigma}{d\Omega}|\mathbf{v} - \mathbf{v}_1|\ f(\mathbf{v}, \mathbf{r}, t)f(\mathbf{r}, \mathbf{v}_1, t),$$

$$d\Omega = sin\theta\ d\theta\ d\varphi. \quad (1.17)$$

The rate of scattering *into* the range $(\mathbf{v}, \mathbf{v} + d\mathbf{v})$ can be calculated in the same manner, giving

$$\left(\frac{\partial f}{\partial t}\right)_{in} = \int d^3v'_1 \int \left(d\Omega\frac{d\sigma}{d\Omega}\right)' |\mathbf{v}' - \mathbf{v}'_1| f(\mathbf{r}, \mathbf{v}', t)f(\mathbf{r}, \mathbf{v}'_1, t). \quad (1.18)$$

In the above equation, $\mathbf{v}', \mathbf{v}_1'$ are functions of $\mathbf{v}, \mathbf{v}_1$ because these are the velocities of the interacting particles before and after the collision, respectively. Since we have assumed an elastic collision the following conservation relations hold,

$$m\mathbf{v}' + m\mathbf{v}_1' = m\mathbf{v} + m\mathbf{v}_1, \tag{1.19}$$

$$\frac{m}{2}\mathbf{v}'^2 + \frac{m}{2}\mathbf{v}_1'^2 = \frac{m}{2}\mathbf{v}^2 + \frac{m}{2}\mathbf{v}_1^2. \tag{1.20}$$

The collision may be thought of as a linear orthogonal transformation from $(\mathbf{v}, \mathbf{v}_1)$ to $(\mathbf{v}', \mathbf{v}_1')$, and we have

$$|\mathbf{v} - \mathbf{v}_1| = |\mathbf{v}' - \mathbf{v}_1'|.$$

Furthermore, for this transformation, the Jacobian is unity and therefore

$$d^3v_1' d^3v' = d^3v_1 d^3v.$$

Because of the symmetry of the interaction, we also have

$$\left(\frac{d\sigma}{d\Omega}d\Omega\right)' = \left(\frac{d\sigma}{d\Omega}d\Omega\right).$$

Collecting the above equations for the collision integral, the Boltzmann equation becomes

$$\frac{\partial f}{\partial t} + v_i \frac{\partial f}{\partial v_i} + a_i \frac{\partial f}{\partial v_i} = \left(\frac{\delta f}{\delta t}\right)_c$$

where

$$\left(\frac{\delta f}{\delta t}\right)_c = \int d^3v_1 \int d\Omega \frac{d\sigma}{d\Omega} |\mathbf{v} - \mathbf{v}_1| \Big[f(\mathbf{r}, \mathbf{v}', t) f(\mathbf{r}, \mathbf{v}_1', t) - f(\mathbf{r}, \mathbf{v}, t) f(\mathbf{r}, \mathbf{v}_1, t) \Big] \tag{1.21}$$

The actual evaluation of this integral for small angle scattering will be performed in §1.7. If the fluid contains several kinds of particles, then each kind of particle has its own distribution function and we readily generalize Boltzmann's equation to

$$\frac{\partial f_\alpha}{\partial t} + v_i \frac{\partial f_\alpha}{\partial r_i} + a_i \frac{\partial f_\alpha}{\partial v_i} = \sum_\beta C_{\alpha\beta}(f_\alpha f_\beta) \tag{1.22}$$

where α and β refer to the particle species, $C_{\alpha\beta}(f_\alpha f_\beta)$ denotes the collision operator and the summation includes α.

1.3. Chapman–Enskog solution of Boltzmann equation

In this section, we shall discuss the solution of the Boltzmann equation for a simple gas (consisting of one kind of spherical molecules having only the translational energy) which is not subject to any external forces. The Chapman and Enskog treatment of the Boltzmann equation is known as the method of moments.

Summation invariant. Let Q be any particle property such that the sum of the Q's of two particles involved in collision remains unchanged in the collision, i.e., a *summation invariant.* Important examples of Q are 1, $m\mathbf{v}$ (momentum), and $mv^2/2$ (kinetic energy) (see Eqs. (1.19) and (1.20)).

$$Q(\mathbf{v}) + Q(\mathbf{v}_1) = Q(\mathbf{v}') + Q(\mathbf{v}_1') \tag{1.23}$$

Then ΔQ defined below vanishes. This is the property of a summation invariant which is found useful in later discussion.

$$\begin{aligned} \Delta Q &= \int Q(\mathbf{v}) \left(\frac{\delta f}{\delta t}\right)_c d^3v \\ &= \int Q(\mathbf{v}) \left(f'f_1' - ff_1\right) g \frac{d\sigma}{d\Omega} d\Omega \, d^3v_1 \, d^3v = 0. \end{aligned} \tag{1.24}$$

In the above equation, we abbreviated

$$\mathbf{g} = \mathbf{v} - \mathbf{v}_1, \tag{1.25}$$

$$f = f(\mathbf{v}, \mathbf{r}, t), \quad f_1 = f(\mathbf{v}_1, \mathbf{r}, t), \quad f' = f(\mathbf{v}', \mathbf{r}, t), \quad f_1' = f(\mathbf{v}_1', \mathbf{r}, t).$$

The proof of $\Delta Q = 0$ goes as follows. Owing to the symmetry of the integrand, we have

$$\Delta Q = \int Q(\mathbf{v}_1) \left(f'f_1' - ff_1\right) g \frac{d\sigma}{d\Omega} d\Omega d^3v_1 d^3v. \tag{1.26}$$

Adding Eqs. (1,24) and (1.26) gives

$$\Delta Q = \frac{1}{2} \int \Big(Q(\mathbf{v}) + Q(\mathbf{v}_1)\Big)(f'f_1' - ff_1) \, g \frac{d\sigma}{d\Omega} d\Omega d^3v_1 d^3v. \tag{1.27}$$

Due to the symmetry of the integrand again, (1.27) can be written as

$$\Delta Q = \frac{1}{2} \int \Big(Q(\mathbf{v}') + Q(\mathbf{v}_1') - Q(\mathbf{v}) - Q(\mathbf{v}_1)\Big) ff_1 \, g \frac{d\sigma}{d\Omega} d\Omega d^3v_1 d^3v. \tag{1.28}$$

Hence, $\Delta Q = 0$.

Exercise. In writing (1.28), we used specifically

$$\int \Phi(\mathbf{v}_a) f_a(\mathbf{v}'_a) f_b(\mathbf{v}'_b) g \frac{d\sigma}{d\Omega} d\Omega d^3 v_a d^3 v_b$$
$$= \int \Phi(\mathbf{v}'_a) f_a(\mathbf{v}_a) f_b(\mathbf{v}_b) g \frac{d\sigma}{d\Omega} d\Omega d^3 v_a d^3 v_b \tag{1.29}$$

where Φ is a summation invariant. Prove this by using (1.19) and (1.20).

Let us consider a gas in equilibrium (steady state in time and uniform in space). Then LHS of (1.14) vanishes and the distribution function should satisfy

$$\left(\frac{\delta f}{\delta t}\right)_c = \int d^3 v_1 d\sigma g \left(f' f'_1 - f f_1\right) = 0 \tag{1.30}$$

which is satisfied if and only if $f' f'_1 = f f_1$. Or

$$log f(\mathbf{v}') + log f(\mathbf{v}'_1) = log f(\mathbf{v}) + log f(\mathbf{v}_1).$$

Hence the quantity $log f(\mathbf{v})$ is a summation invariant and it should be expressible as a linear combination of the three summation invariants, 1, $m\mathbf{v}$, and $mv^2/2$:

$$log f(\mathbf{v}) = m a_1 + \mathbf{a}_2 \cdot m\mathbf{v} - a_3 \frac{m}{2} v^2$$

where a_1, $\mathbf{a}_2$, a_3 are arbitrary constants. By completing the squares, one obtains

$$log f(\mathbf{v}) = m\left(a_1 + \frac{a_2^2}{2a_3}\right) - \frac{m}{2} a_3 \left(\mathbf{v} - \frac{\mathbf{a}_2}{a_3}\right)^2.$$

By a simple transformation, $f(\mathbf{v})$ can be put into the form

$$f(\mathbf{v}) = N\left(\frac{m}{2\pi T}\right)^{\frac{3}{2}} exp - \frac{m(\mathbf{v} - \mathbf{V})^2}{2T} \tag{1.31}$$

where N, $\mathbf{V}$, T are related with a_1, $\mathbf{a}_2$, a_3 and defined by

$$N = \int f d^3 v, \quad \mathbf{V} - \frac{1}{N} \int \mathbf{v} f d^3 v, \quad \frac{3}{2} T = \frac{1}{N} \int \frac{m}{2} v^2 f d^3 v - \frac{m}{2} V^2. \tag{1.32}$$

Equation (1.31) is a moving Maxwell–Boltzmann distribution function. If the gas has no directional preference ($\mathbf{V} = 0$) and is distributed isotropically,

$$f(\mathbf{v}) = N\left(\frac{m}{2\pi T}\right)^{\frac{3}{2}} exp - \frac{m\mathbf{v}^2}{2T}. \tag{1.33}$$

[In the above and following, temperature T should be read as kT where k is the Boltzmann constant]. Thus the particles in equilibrium state are Maxwell–Boltzmann distributed. Particles in non-equilibrium state approach the Maxwell–Boltzmann distribution in the course of time to establish equilibrium state (*relaxation*).

Let us consider a special case where an electrostatic force derivable from a potential,

$$\mathbf{F}(\mathbf{r}) = -e\nabla\phi(\mathbf{r})$$

acts on a plasma consisting of immobile ions and warm electrons. We also assume that a steady state is maintained ($\frac{\partial f}{\partial t} = 0$). In this case the distribution

$$f(\mathbf{r}, \mathbf{v}, t) = N_0 \, e^{-e\phi(\mathbf{r})/T} \left(\frac{m}{2\pi T}\right)^{\frac{3}{2}} e^{-\frac{m\mathbf{v}^2}{2T}} \tag{1.34}$$

solves the Boltzmann equation

$$v_i \, \frac{\partial f}{\partial r_i} + \frac{F_i}{m} \frac{\partial f}{\partial v_i} = \left(\frac{\delta f}{\delta t}\right)_c . \tag{1.35}$$

It is readily seen that LHS of (1.35) vanishes per (1.34), and thus $(\frac{\delta f}{\delta t})_c = 0$ since the *Boltzmann factor* $e^{-e\phi(\mathbf{r})/T}$ can be taken out of the velocity integral in (1.30). Now the particle number density of the plasma which maintains an equilibrium under the action of an electrostatic force is calculated as

$$N(\mathbf{r}) = \int (1.34) \; d^3v = N_0 e^{-e\phi(\mathbf{r})/T}. \tag{1.36}$$

The inhomogeneous distribution (1.36) is the Boltzmann distribution when the particles are in an electrostatic force field.

Boltzmann's H-theorem. Boltzmann introduced into his collision integral the famous quantity called H:

$$H(t) = \int d^3v f(\mathbf{v}, t) \; log f(\mathbf{v}, t).$$

This is the kinetic theory version of the entropy, which is in fact the negative entropy, $-S(t)$. We have

$$\frac{dH}{dt} = \int d^3v \; \frac{\partial f}{\partial t}\Big(1 + log \; f(\mathbf{v}, t)\Big).$$

For a spatially homogeneous gas, using (1.21) in the above equation gives

$$\frac{dH}{dt} = \int d^3v_1 \int d^3v_2 \int bdb \int d\varphi |\mathbf{v_1} - \mathbf{v_2}| \left(f_2' f_1' - f_2 f_1\right) (1 + log f_1).$$

The above integral is unchanged upon the interchange of the subscripts 1 and 2, which is the property of the collision integral that we used to derive (1.30). Also the integral is symmetric with respect to the interchange of the prime and no prime. Also note that $\int bdb \int d\varphi$ is invariant in the reverse event. Hence, we can deduce

$$\frac{dH}{dt} = \frac{1}{4}\int d^3v_1 \int d^3v_2 \int bdb \int d\varphi |\mathbf{v_1} - \mathbf{v_2}| \left(f_2'f_1' - f_2f_1\right) \times \left[log\left(f_1f_2\right) - log(f_i'f_2')\right]$$

where we note that $(f_2'f_1' - f_2f_1)\left[log\left(f_1f_2\right) - log(f_i'f_2')\right] \leq 0$ always. Therefore, we conclude that $\frac{dH}{dt} \leq 0$ (the equality holds for equilibrium state) always, which is the Boltzmann's H-theorem.

1.4. Simple relaxation model of the collision term

In a weakly ionized plasma whose major population is neutrals, the following model equation called the BGK model (after Bhatnagar, Gross, and Krook, in Phys. Rev. **94** 1954 p. 511) is often made use of for the Boltzmann collision term:

$$\left(\frac{\delta f}{\delta t}\right)_c = -\nu\left[f(\mathbf{r}, \mathbf{v}, t) - f_0(v)\right]. \tag{1.37}$$

Here $f_0(v)$ is a suitable Maxwell–Boltzmann distribution, and $1/\nu \equiv \tau$ has the dimension of time and is termed as the relaxation time. A *Maxwellian* $f_0(v)$ makes the collision integral vanish, and the negative sign signifies that the collisions reduce the deviation of the distribution function from the equilibrium Maxwell–Boltzmann distribution. Assuming that the external force is absent from the uniform plasma under consideration, $f(\mathbf{v}, t)$ is governed by

$$\frac{\partial}{\partial t} f(\mathbf{v}, t) = -\nu[f(\mathbf{r}, \mathbf{v}, t) - f_0(v)].$$

Integrating this equation gives

$$f(\mathbf{v}, t) - f_0(v) = \Big(f(\mathbf{r}, \mathbf{v}, t) - f_0(v)\Big)\, e^{-\nu t}.$$

As time t increases, $f(\mathbf{v}, t)$ approaches to the Maxwell–Boltzmann distribution $f_0(v)$ as the model equation (1.37) intends to.

1.5. Moment equations

Reference: Chapman, S. and T. G. Cowling, *The mathematical theory of non-uniform gases* 2nd Ed. Cambridge University Press (1952).

Instead of solving the Boltzmann equation as an initial-value problem, Chapman and Enskog reduced the description of $f(\mathbf{r}, \mathbf{v}, t)$ in the six-dimensional phase space to the description in terms of the following moments: $n(\mathbf{r}, t), \mathbf{V}(\mathbf{r}, t), T(\mathbf{r}, t)$ in the ordinary coordinate space. In this reduction, there involves a time-wise *coarse graining* of the Boltzmann equation by distinguishing two time scales. The relaxation time t_0 is the time for $f(\mathbf{r}, \mathbf{v}, t)$ to approach the local Maxwellian $f^{(0)}$ in (1.51). That time scale is assumed to be much shorter than the time scale for the moments, i.e., $n(\mathbf{r}, t), \mathbf{V}(\mathbf{r}, t), T(\mathbf{r}, t)$ to change appreciably. Therefore the fluid equations describe slow process by suppressing the fast time involved in Boltzmann's f. In terms of the three moments, the Boltzmann equation is transformed to a series of inhomogeneous integral equations, which we shall not step into for detailed discussion. But we will summarize the salient features of the Chapman and Enskog solution.

Let us first define the following macroscopic variables which are the first three moments of the distribution function:

$$n(\mathbf{r}, t) = \int f(\mathbf{r}, t, \mathbf{v}) d^3 v \quad (\textit{density}), \tag{1.38}$$

$$\mathbf{V}(\mathbf{r}, t) = \frac{1}{n} \int \mathbf{v} \; f(\mathbf{r}, t, \mathbf{v}) d^3 v \quad (\textit{fluid velocity}), \tag{1.39}$$

$$T(\mathbf{r}, t) = \frac{m}{3n} \int \Big(\mathbf{v} - \mathbf{V}(\mathbf{r}, t)\Big)^2 f(\mathbf{r}, t, \mathbf{v}) d^3 v \quad (\textit{temperature}). \tag{1.40}$$

In order to obtain the transport equations for these quantities of a simple gas (a gas consisting of one species), we multiply (1.21) by 1, $m\mathbf{v}$, $mv^2/2$, respectively, and integrate over velocity. The integrals of RHS of (1.21) vanish since these quantities are summation invariants. Carrying out the integrals for LHS, we obtain

$$\frac{\partial n}{\partial t} + \nabla \cdot (n\mathbf{V}) = 0 \quad (\textit{continuity equation}) \tag{1.41}$$

$$\frac{\partial}{\partial t}(mnV_i) + \frac{\partial}{\partial x_j}(mn\langle v_i v_j \rangle) - nF_i = 0 \quad (F_i = ma_i)$$
$$(\textit{momentum equation}) \tag{1.42}$$

$$\frac{\partial}{\partial t}\left(\frac{mn}{2}\langle v^2 \rangle\right) + \frac{\partial}{\partial x_j}\left(\frac{mn}{2}\langle v^2 v_j \rangle\right) - nF_i V_i = 0$$
$$(\textit{energy equation}) \tag{1.43}$$

where the bracket $\langle \cdots \rangle$ denotes the average value:

$$\langle \cdots \rangle = \frac{1}{n} \int (\cdots) f d^3 v.$$

It is convenient to transform (1.42) and (1.43) as follows. Let $\mathbf{w}$ be the random velocity (the '*peculiar velocity*' in Chapman and Cowling's book) defined by

$$\mathbf{w} = \mathbf{v} - \mathbf{V}(\mathbf{r}, t).$$

Then evidently, $\langle \mathbf{w} \rangle = 0$ and $\langle v_i v_j \rangle = V_i V_j + \langle w_i w_j \rangle$. Using (1.41) in (1.42) yields

$$mn \frac{dV_i}{dt} = -\frac{\partial P_{ij}}{\partial x_j} + nF_i \tag{1.44}$$

where

$$\frac{d}{dt} = \frac{\partial}{\partial t} + V_j \frac{\partial}{\partial x_j}$$

$$P_{ij} = mn \langle w_i w_j \rangle = m \int w_i w_j f d^3 v. \tag{1.45}$$

The pressure tensor P_{ij} is usually split into two parts:

$$P_{ij} = p\delta_{ij} + \pi_{ij} \tag{1.46}$$

where p is the scalar pressure, and π_{ij} is the *viscous stress tensor* with the following definitions

$$p = \frac{nm}{3} \langle w^2 \rangle = nT(\mathbf{r}, t) = \frac{1}{3} P_{ii}, \tag{1.47}$$

$$\pi_{ij} = nm \left(\langle w_i w_j \rangle - \frac{\langle w^2 \rangle}{3} \delta_{ij} \right) \quad (\mathit{trace},\ \pi_{ii} = 0). \tag{1.48}$$

If the velocity distribution function is isotropic, then $\langle w_x^2 \rangle = \langle w_y^2 \rangle = \langle w_z^2 \rangle = \frac{1}{3} \langle w^2 \rangle$, $\langle w_x w_y \rangle = \langle w_x w_z \rangle = \langle w_y w_z \rangle = 0$, so $P_{ij} = p\ \delta_{ij}$. Also note that π_{ij} is traceless ($\pi_{ii} = 0$). The stress tensor π_{ij} represents the part of P_{ij} that arises as a result of the deviation of the distribution function from spherical symmetry. Carrying out similar transformations for (1.43), we can reduce it to the form

$$\frac{d}{dt} \left(\frac{3}{2} nT \right) + \frac{3}{2} nT \nabla \cdot \mathbf{V} + \nabla \cdot \mathbf{q} + P_{ij} \frac{\partial V_i}{\partial x_j} = 0, \tag{1.49}$$

$$\mathbf{q} = nm \left\langle \frac{w^2}{2} \mathbf{w} \right\rangle \quad (\mathit{heat\ flux\ vector}). \tag{1.50}$$

In obtaining (1.49), we used (1.41) and (1.44).

So far we have obtained the first three moment equations, (1.41), (1.44), and (1.49). Our purpose is to obtain hydrodynamic equations to govern the hydrodynamic variables $n, \mathbf{V}, T$. The three moment equations contain π_{ij} and $\mathbf{q}$, which must be expressed in terms of $n, \mathbf{V}, T$. Since the quantities π_{ij} and $\mathbf{q}$ involve the integrations of the distribution function f, an approximate solution for f must be found. If f is obtained approximately in terms of $(n, \mathbf{V}, T)$, π_{ij} and $\mathbf{q}$ can be expressed in terms of $(n, \mathbf{V}, T)$, and a closed set of equations for $(n, \mathbf{V}, T)$ is obtained. The Chapman-Enskog solution for the Boltzmann equation is a successive approximation for f. At each stage of approximation f is found in terms of the three hydrodynamic variables, thereby the three moment equations form a closed set of equations for the three hydrodynamic variables. The solutions for the three moment equations then in turn serve to determine the next approximation for f, and so on. Here we shall not delve into this successive steps, but will write down the results of primary importance. The lowest order solution of the distribution function is the local Maxwellian

$$f^{(0)}(\mathbf{v}, \mathbf{r}, t) = n(\mathbf{r}, t)\left(\frac{m}{2\pi T(\mathbf{r}, t)}\right)^{\frac{3}{2}} exp\ -\frac{m(\mathbf{v} - \mathbf{V}(\mathbf{r}, t))^2}{2T(\mathbf{r}, t)}. \tag{1.51}$$

The moment equations for the three hydrodynamic variables are as follows.

$$\frac{\partial n}{\partial t} + \nabla \cdot (n\mathbf{V}) = 0, \tag{1.52}$$

$$mn\left(\frac{\partial \mathbf{V}}{\partial t} + (\mathbf{V} \cdot \nabla)\mathbf{V}\right) = -\nabla p + n\mathbf{F}, \tag{1.53}$$

$$p = nT, \tag{1.54}$$

$$\left(\frac{\partial}{\partial t} + \mathbf{V} \cdot \nabla\right) T = -\frac{2p}{3n}\ \nabla \cdot \mathbf{V} \quad or \quad \frac{d}{dt}(nT^{-3/2}) = 0. \tag{1.55}$$

Note that the energy equation (1.55) implies the adiabatic law of a perfect gas, $pn^{-\gamma} = const.$ with $\gamma = 5/3$. Recall that the specific entropy $s = \frac{3k}{2m} ln(pn^{-\gamma})$, and (1.55) gives $ds/dt = 0$. A quick way of getting (1.55) is to put $\mathbf{q} = 0$ and $P_{ij} = p\delta_{ij}$ in (1.49).

The next higher order solutions are as follows.

$$\frac{dV_i}{dt} = \frac{F_i}{m} - \frac{1}{mn}\frac{\partial P_{ij}}{\partial x_j} \tag{1.56}$$

where

$$P_{ij} = p\delta_{ij} + \pi_{ij} = p\delta_{ij} - \mu\left(\frac{\partial V_i}{\partial x_j} + \frac{\partial V_j}{\partial x_i} - \frac{2}{3}\delta_{ij}\nabla \cdot \mathbf{V}\right) \tag{1.57}$$

and μ is the *coefficient of shear viscosity.* Equation (1.56) together with (1.57) is the *Navier–Stokes equation* for a viscous fluid. The quantity in the parenthesis in (1.57) is called the *rate of strain tensor.* The viscous stress tensor π_{ij} which represents the internal friction due to shear gives rise to the irreversible viscous transfer of momentum between the particles in the fluid, i.e. the exchange of momentum between the particles due to collisions. When such momentum transfer is absent from a fluid, the fluid is called *ideal* (or perfect).

Using (1.57) in (1.56) yields

$$\frac{dV_i}{dt} = \frac{F_i}{m} - \frac{1}{mn}\left[\frac{\partial p}{\partial x_i} - \mu\nabla^2 V_i - \frac{\mu}{3}\frac{\partial}{\partial x_i}\nabla\cdot\mathbf{V}\right]. \tag{1.58}$$

The energy equation takes the form

$$\frac{3}{2}n\frac{dT}{dt} = -\lambda\nabla^2 T - p\nabla\cdot\mathbf{V} - \frac{1}{2}\pi_{ij}\left(\frac{\partial v_i}{\partial x_j} + \frac{\partial v_j}{\partial x_i}\right) \tag{1.59}$$

where λ is *the coefficient of thermal conduction*: $\mathbf{q} = -\lambda\nabla T$.

1.6. Ideal two-fluid equations

The lowest order moment equations (1.52)–(1.55) are obtained when the collisional interactions between the constituent particles of a plasma are ignored. They should be compatible with the collisionless Boltzmann equation or Vlasov equation. In this context, we often use *the ideal two-fluid equations* which hold independently for each species, electron fluid and ion fluid, in the following form.

$$\frac{\partial n_\alpha}{\partial t} + \nabla\cdot(n_\alpha\mathbf{V}_\alpha) = 0 \tag{1.60}$$

$$m_\alpha n_\alpha\left(\frac{\partial\mathbf{V}_\alpha}{\partial t} + \mathbf{V}_\alpha\cdot\nabla\mathbf{V}_\alpha\right) = -\nabla p_\alpha + n_\alpha\mathbf{F} \tag{1.61}$$

$$p_\alpha = n_\alpha T_\alpha \tag{1.62}$$

$$\left(\frac{\partial}{\partial t} + \mathbf{V}_\alpha\cdot\nabla\right)(n_\alpha T_\alpha^{-3/2}) = 0 \tag{1.63}$$

where the subscript α takes e(i) for electron (ion) fluid.

Phenomenological derivation of π_{ij}. The pressure tensor P_{ij} is a surface force: $P_{ij}ds_j$ (no sum on j) is the force directed along i-direction, acted upon the surface whose normal is j-direction. If $i = j$, it is normal stress (p, pressure); if $i \neq j$, it represents the shear stress (here π_{ij}). For a fluid at rest or in uniform motion, the three normal stresses have the same value, i.e. the fluid is isotropic:

$$P_{11} = P_{22} = P_{33} = p. \tag{1.64}$$

Thus the stress tensor (pressure tensor) can be written

$$P_{ij} = p\delta_{ij} + \pi_{ij}.$$

The shear stress π_{ij} should be traceless, because, by definition, it makes zero contribution to the mean normal stresses. Furthermore, the conservation of angular momentum (see Eq. (2.49)) demands

$$\pi_{ij} = \pi_{ji}. \tag{1.65}$$

We also assume that in general flow the shear stress components are proportional to the corresponding time rates of angular deformation. Consulting Fig. 1.2, we have the relations

$$dx_1 \frac{d\phi_1}{dt} = dV_2, \quad \left(\pi_{21} \sim \frac{d\phi_1}{dt}\right)$$

$$dx_2 \frac{d\phi_2}{dt} = dV_1. \quad \left(\pi_{12} \sim \frac{d\phi_2}{dt}\right)$$

So we can write

$$\pi_{12} = \frac{1}{2}(\pi_{12} + \pi_{21}) = \mu\left(\frac{d\phi_1}{dt} + \frac{d\phi_2}{dt}\right) = \mu\left(\frac{dV_2}{dx_1} + \frac{dV_1}{dx_2}\right).$$

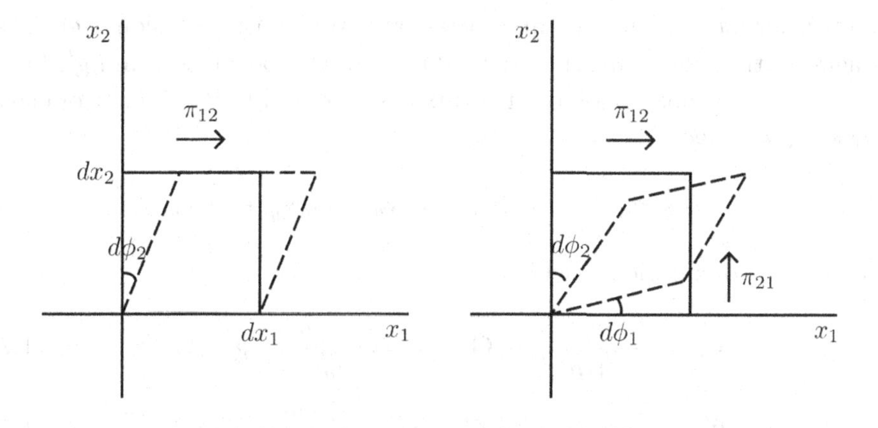

Fig. 1.2. Deformation under shear force π_{ij}.

Generalizing this to all directions, we have

$$\pi_{ij} = \mu\left(\frac{dV_i}{dx_j} + \frac{dV_j}{dx_i}\right).$$

The complete form of π_{ij} which is traceless can be easily written down as

$$\pi_{ij} = \mu\left(\frac{dV_i}{dx_j} + \frac{dV_j}{dx_i} - \frac{2}{3}\nabla\cdot\mathbf{V}\delta_{ij}\right). \tag{1.66}$$

1.7. Fokker–Planck equation, Landau collision integral

In a fully ionized plasma consisting of electrons and ions, the meaningful collisional event occurs by accumulation of small angle scatterings, and the collision term can be appropriately approximated by Fokker–Planck equation, which will be derived from the Boltzmann collision integral. The collision term $C_{\alpha\beta}(f_\alpha f_\beta)$ in (1.22) can be considerably simplified for small angle scatterings.

Kinematics of elastic binary collision. We rewrite

$$C_{ab}(f_a f_b) = \int d^3v_b \int d\Omega \frac{d\sigma(g,\theta)}{d\Omega}\,|\mathbf{v}_a - \mathbf{v}_b| \times \left[f_a(\mathbf{v}_a')f_b(\mathbf{v}_b') - f_a(\mathbf{v}_a)f_b(\mathbf{v}_b)\right]. \tag{1.67}$$

Introduce the following notations for the relative velocities,

$$\mathbf{g} = \mathbf{v}_a - \mathbf{v}_b, \quad \mathbf{g}' = \mathbf{v}_a' - \mathbf{v}_b' \tag{1.68}$$

where unprimed (primed) quantities are pre-(post-)collision velocities. Remember that the scattering angle θ is the angle between $\mathbf{g}$ and $\mathbf{g}'$. Let $\mathbf{G}$ be the center-of-mass velocity of two particles a and b. By the law of elastic collisions, we have

$$(m_a + m_b)\mathbf{G} = m_a\mathbf{v}_a + m_b\mathbf{v}_b = m_a\mathbf{v}_a' + m_b\mathbf{v}_b'. \tag{1.69}$$

The above two equations yield

$$\mathbf{v}_a = \frac{m_b}{m_a + m_b}\mathbf{g} + \mathbf{G}, \quad \mathbf{v}_b = \frac{-m_a}{m_a + m_b}\mathbf{g} + \mathbf{G}, \tag{1.70}$$

$$\mathbf{v}_a' = \frac{m_b}{m_a + m_b}\mathbf{g}' + \mathbf{G}, \quad \mathbf{v}_b' = \frac{-m_a}{m_a + m_b}\mathbf{g}' + \mathbf{G}. \tag{1.71}$$

Hence the pre-collision velocities are determined by $\mathbf{G}$ and $\mathbf{g}$; the post-collision velocities by $\mathbf{G}$ and $\mathbf{g}'$. It is readily shown that

$$\frac{1}{2}(m_a v_a^2 + m_b v_b^2) = \frac{1}{2}[(m_a + m_b)^2 G^2 + \mu g^2]$$

$$\frac{1}{2}(m_a v_a'^2 + m_b v_b'^2) = \frac{1}{2}[(m_a + m_b)^2 G^2 + \mu g'^2]$$

where $\mu = \frac{m_a m_b}{m_a + m_b}$ is the reduced mass. Since the kinetic energy is conserved, the above relations give $g = g'$. The relative velocity $\mathbf{g}$ is deflected through the scattering angle θ by collision. The scattering angle θ depends in general on the magnitude of the initial relative velocity and the impact parameter b: $\theta = \theta(g, b)$. Introducing a unit vector $\hat{\mathbf{k}}$ defined by

$$\hat{\mathbf{k}} = \frac{\mathbf{g}' - \mathbf{g}}{|\mathbf{g}' - \mathbf{g}|} \quad (\textit{see } \text{Fig. 1.3})$$

Equations (1.70) and (1.71) can be written in the form

$$\mathbf{v}_a' = \mathbf{v}_a - \frac{2m_b}{m_a + m_b}(\mathbf{g} \cdot \hat{\mathbf{k}})\hat{\mathbf{k}}, \quad \mathbf{v}_b' = \mathbf{v}_b + \frac{2m_a}{m_a + m_b}(\mathbf{g} \cdot \hat{\mathbf{k}})\hat{\mathbf{k}}, \tag{1.72}$$

$$\mathbf{v}_a = \mathbf{v}_a' - \frac{2m_b}{m_a + m_b}(\mathbf{g}' \cdot \hat{\mathbf{k}})\hat{\mathbf{k}}, \quad \mathbf{v}_b = \mathbf{v}_b' + \frac{2m_a}{m_a + m_b}(\mathbf{g}' \cdot \hat{\mathbf{k}})\hat{\mathbf{k}}. \tag{1.73}$$

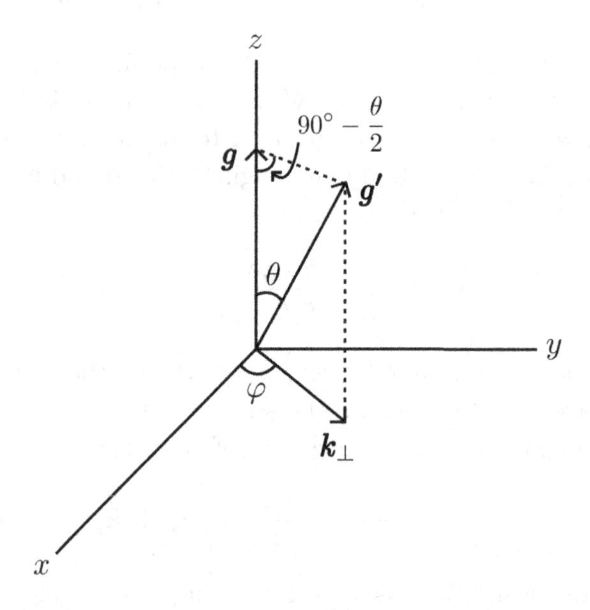

Fig. 1.3. The unit vector $\hat{\mathbf{k}} = (\mathbf{g}' - \mathbf{g})/|\mathbf{g}' - \mathbf{g}|$.

Thus when $\mathbf{v}_a$, $\mathbf{v}_b$, $\hat{\mathbf{k}}$ are given, the post-collision velocities are determined and $\hat{\mathbf{k}}$ is determined by the impact parameter b and the azimuthal angle φ. Equations (1.72) and (1.73) can be used to derive a theorem concerning the Jacobian determinant needed in transformation of the velocity integral from the pre-collision variables to post-collision variables. The Jacobian determinant is defined by

$$J = \frac{\partial(\mathbf{v}_a', \mathbf{v}_b')}{\partial(\mathbf{v}_a, \mathbf{v}_b)} = \begin{vmatrix} \frac{\partial v_{ax}'}{\partial v_{ax}} & \frac{\partial v_{ay}'}{\partial v_{ax}} & \frac{\partial v_{az}'}{\partial v_{ax}} & \frac{\partial v_{bx}'}{\partial v_{ax}} & \frac{\partial v_{by}'}{\partial v_{ax}} & \frac{\partial v_{bz}'}{\partial v_{ax}} \\ \frac{\partial v_{ax}'}{\partial v_{ay}} & \frac{\partial v_{ay}'}{\partial v_{ay}} & \frac{\partial v_{az}'}{\partial v_{ay}} & \frac{\partial v_{bx}'}{\partial v_{ay}} & \frac{\partial v_{by}'}{\partial v_{ay}} & \frac{\partial v_{bz}'}{\partial v_{ay}} \\ \frac{\partial v_{ax}'}{\partial v_{az}} & \frac{\partial v_{ay}'}{\partial v_{az}} & \frac{\partial v_{az}'}{\partial v_{az}} & \frac{\partial v_{bx}'}{\partial v_{az}} & \frac{\partial v_{by}'}{\partial v_{az}} & \frac{\partial v_{bz}'}{\partial v_{az}} \\ \frac{\partial v_{ax}'}{\partial v_{bx}} & \frac{\partial v_{ay}'}{\partial v_{bx}} & \frac{\partial v_{az}'}{\partial v_{bx}} & \frac{\partial v_{bx}'}{\partial v_{bx}} & \frac{\partial v_{by}'}{\partial v_{bx}} & \frac{\partial v_{bz}'}{\partial v_{bx}} \\ \frac{\partial v_{ax}'}{\partial v_{by}} & \frac{\partial v_{ay}'}{\partial v_{by}} & \frac{\partial v_{az}'}{\partial v_{by}} & \frac{\partial v_{bx}'}{\partial v_{by}} & \frac{\partial v_{by}'}{\partial v_{by}} & \frac{\partial v_{bz}'}{\partial v_{by}} \\ \frac{\partial v_{ax}'}{\partial v_{bz}} & \frac{\partial v_{ay}'}{\partial v_{bz}} & \frac{\partial v_{az}'}{\partial v_{bz}} & \frac{\partial v_{bx}'}{\partial v_{bz}} & \frac{\partial v_{by}'}{\partial v_{bz}} & \frac{\partial v_{bz}'}{\partial v_{bz}} \end{vmatrix}$$

The partial differentiations involved in J are performed regarding $\hat{\mathbf{k}}$ as a constant. Since (1.72) which gives $\mathbf{v}_a', \mathbf{v}_b'$ in terms of $\mathbf{v}_a, \mathbf{v}_b$ is linear, J depends only on $\hat{\mathbf{k}}$, m_a, m_b. But $\mathbf{v}_a, \mathbf{v}_b$ are given in terms of $\mathbf{v}_a', \mathbf{v}_b'$ by (1.73) which can be obtained from (1.72) by interchanging primed and unprimed letters. Thus the Jacobian J' defined by

$$J' = \frac{\partial(\mathbf{v}_a, \mathbf{v}_b)}{\partial(\mathbf{v}_a', \mathbf{v}_b')}$$

is equal to the same function of $(\hat{\mathbf{k}}, m_a, m_b)$ as J, and therefore $J = J'$. Since $JJ' = 1$, we have $J = \pm 1$, where (-1) is irrelevant.

For small angle scattering, the change of velocity

$$\mathbf{v}_a' - \mathbf{v}_a \equiv \mathbf{u} = \frac{-2m_b}{m_a + m_b}(\mathbf{g} \cdot \hat{\mathbf{k}})\hat{\mathbf{k}} \tag{1.74}$$

will be small for most collisions, so that we may expand (1.67) in powers of $\mathbf{u}$ and keep only the first and second powers. This will be carried out. We multiply both sides of (1.67) by an arbitrary function $\Phi(\mathbf{v}_a)$ and integrate over $\mathbf{v}_a$. In the final result, Φ can be eliminated owing to the fact that it is an entirely arbitrary function. We have

$$\int \Phi(\mathbf{v}_a)\, C_{ab}(\mathbf{v}_a) d^3 v_a$$
$$= \int \Phi(\mathbf{v}_a) \Big[f_a(\mathbf{v}'_a) f_b(\mathbf{v}'_b) - f_a(\mathbf{v}_a) f_b(\mathbf{v}_b) \Big] g\; b\; db\; d\varphi d^3 v_b d^3 v_a. \quad (1.75)$$

In the first integral in the above equation, we change the integration variables from $(\mathbf{v}_a, \mathbf{v}_b) \to (\mathbf{v}'_a, \mathbf{v}'_b)$. In doing this, $\mathbf{v}_a$ in Φ should be expressed as a function of $\mathbf{v}'_a$ and $\mathbf{v}'_b$:

$$\int \Phi(\mathbf{v}_a) f_a(\mathbf{v}'_a) f_b(\mathbf{v}'_b) g\; b\; db\; d\varphi d^3 v_b d^3 v_a$$
$$= \int \Phi(\mathbf{v}'_a - \mathbf{u}') f_a(\mathbf{v}'_a) f_b(\mathbf{v}'_b) g\; b\; db\; d\varphi d^3 v'_b d^3 v'_a$$

where $\mathbf{u}' = \frac{2m_b}{m_a+m_b}(\mathbf{g}' \cdot \hat{\mathbf{k}})\hat{\mathbf{k}}$ (see (1.73)) and we used $d^3 v_a d^3 v_b = d^3 v'_a d^3 v'_b$. In the RHS integral, we rename the integration variables by dropping the primes (then $\mathbf{g}'$ goes to $\mathbf{g}$) to obtain

$$\int \Phi(\mathbf{v}_a) f_a(\mathbf{v}'_a) f_b(\mathbf{v}'_b) g\; b\; db\; d\varphi d^3 v_b d^3 v_a$$
$$= \int \Phi(\mathbf{v}'_a) f_a(\mathbf{v}_a) f_b(\mathbf{v}_b) g\; b\; db\; d\varphi d^3 v_b d^3 v_a. \quad (1.76)$$

Using (1.76) in (1.75) gives

$$\int \Phi(\mathbf{v}_a)\, C_{ab}(\mathbf{v}_a) d^3 v_a$$
$$= \int \Big[\Phi(\mathbf{v}'_a) - \Phi(\mathbf{v}_a) \Big] f_a(\mathbf{v}_a) f_b(\mathbf{v}_b) g\; b\; db\; d\varphi d^3 v_b d^3 v_a. \quad (1.77)$$

Expanding $\Phi(\mathbf{v}'_a)$ in the powers of $\mathbf{u}$ (see (1.69)), RHS of (1.77) becomes

$$\int \left([[u_j]] \frac{\partial \Phi(\mathbf{v}_a)}{\partial v_{aj}} + \frac{1}{2} [[u_i u_j]] \frac{\partial^2 \Phi(\mathbf{v}_a)}{\partial v_{ai} \partial v_{aj}} \right) f_a(\mathbf{v}_a) f_b(\mathbf{v}_b)\; g\; d^3 v_a d^3 v_b \quad (1.78)$$

where

$$[[\cdots]] = \int (\cdots) b\; db\; d\varphi.$$

By integration by parts, (1.78) can be cast into the form

$$\int \Phi(\mathbf{v}_a) \left[-\frac{\partial}{\partial v_{ai}}([[u_i]] f_a(\mathbf{v}_a) f_b(\mathbf{v}_b) g) \right.$$
$$\left. + \frac{1}{2} \frac{\partial^2}{\partial v_{ai} \partial v_{aj}}([[u_i u_j]] f_a(\mathbf{v}_a) f_b(\mathbf{v}_b) g) \right] d^3 v_a d^3 v_b.$$

This is equal to LHS of (1.77) for arbitrary choice of the function Φ, and thus we conclude

$$C_{ab}(f_a f_b) = -\frac{\partial}{\partial v_{ai}} \left[f_a(\mathbf{v}_a) \int [[u_i]] f_b(\mathbf{v}_b) g d^3 v_b \right]$$
$$+ \frac{1}{2} \frac{\partial^2}{\partial v_{ai} \partial v_{aj}} \left[f_a(\mathbf{v}_a) \int [[u_i u_j]] f_b(\mathbf{v}_b) g d^3 v_b \right]. \tag{1.79}$$

This is the *Fokker–Planck collision term*. Now the Boltzmann equation with the Fokker–Planck collision term, or the Fokker–Planck equation, is not an integro-differential equation but a differential equation.

To evaluate $[[u_i]]$ and $[[u_i u_j]]$, we introduce the spherical coordinates as shown in Fig. 1.3, in which the $\hat{\mathbf{z}}$ direction is taken to be parallel to $\mathbf{g}$, i.e. $\hat{\mathbf{z}} = \mathbf{g}/|\mathbf{g}|$. Then $\hat{\mathbf{k}}$ in (1.74) is resolved as (note that $\hat{\mathbf{k}}_\parallel = -sin\frac{\theta}{2}$ and $\hat{\mathbf{k}}_\perp = cos\frac{\theta}{2}$, where $\parallel$ and $\perp$ are referred to the direction of $\hat{\mathbf{g}}$)

$$\hat{\mathbf{k}} = \hat{\mathbf{x}} cos\frac{\theta}{2} cos\varphi + \hat{\mathbf{y}} cos\frac{\theta}{2} sin\varphi - \hat{\mathbf{z}} sin\frac{\theta}{2}.$$

Therefore, we have

$$\mathbf{u} = \frac{2m_b}{m_a + m_b} g \; sin\frac{\theta}{2} \left(\hat{\mathbf{x}} cos\frac{\theta}{2} cos\varphi + \hat{\mathbf{y}} cos\frac{\theta}{2} sin\varphi - \hat{\mathbf{z}} sin\frac{\theta}{2} \right). \tag{1.80}$$

Performing the φ-integration gives

$$[[u_i]] = -4\pi \frac{m_b}{m_a + m_b} g_i \int sin^2\frac{\theta}{2} \; b \; db. \tag{1.81}$$

The tensor $[[u_i u_j]]$ will be diagonal because terms like $u_x u_y$ vanish upon performing the φ-integration. We have

$$[[\mathbf{uu}]] = 2\pi \left(\frac{2m_b}{m_a + m_b} \right)^2$$
$$\times \int \left[\hat{\mathbf{z}}\hat{\mathbf{z}} \; g^2 \; sin^4\frac{\theta}{2} + (\hat{\mathbf{x}}\hat{\mathbf{x}} + \hat{\mathbf{y}}\hat{\mathbf{y}}) \frac{g^2}{2} sin^2\frac{\theta}{2} cos^2\frac{\theta}{2} \right] bdb. \tag{1.82}$$

To change the dyad above into tensor notation, note that

$$\hat{\mathbf{x}}\hat{\mathbf{x}} + \hat{\mathbf{y}}\hat{\mathbf{y}} + \hat{\mathbf{z}}\hat{\mathbf{z}} = \delta_{ij}, \quad \hat{\mathbf{x}}\hat{\mathbf{x}} + \hat{\mathbf{y}}\hat{\mathbf{y}} = \delta_{ij} - \frac{g_i g_j}{g^2}.$$

Thus we obtain

$$[[u_i u_j]] = 4\pi g^2 \left(\frac{m_b}{m_a + m_b}\right)^2$$
$$\times \int b\, db \left[2\frac{g_i g_j}{g^2} sin^4\frac{\theta}{2} + \left(\delta_{ij} - \frac{g_i g_j}{g^2}\right) sin^2\frac{\theta}{2} cos^2\frac{\theta}{2}\right]. \tag{1.83}$$

In the above equation, we neglect the terms of $sin^4\frac{\theta}{2}$ as compared to the term of $sin^2\frac{\theta}{2}$, which is permissible for small angle scattering. Then we finally obtain

$$[[u_i u_j]] = 4\pi g^2 \left(\frac{m_b}{m_a + m_b}\right)^2 (\delta_{ij} - \frac{g_i g_j}{g^2}) \int sin^2\frac{\theta}{2}\, b\, db. \tag{1.84}$$

Using the Rutherford scattering formula for Coulomb collisions,

$$tan\frac{\theta}{2} = \frac{e_a e_b}{b\, \mu\, g^2}, \quad \left(e:\ charge, \quad \mu = \frac{m_a m_b}{m_a + m_b}\right)$$

the integral in (1.81) is obtained

$$\int_0^\infty sin^2\frac{\theta}{2}\, b\, db = \frac{e_a e_b}{\mu g^2}\, J \tag{1.85}$$

$$J = \int_0^\infty \frac{\lambda d\lambda}{1 + \lambda^2} \left(\lambda = \frac{b\, \mu g^2}{e_a e_b}\right). \tag{1.86}$$

Using the above equations in (1.81) and (1.83) gives

$$[[u_i]] = -4\pi e_a^2 e_b^2 J\, \frac{g_i}{m_a g^4} \left(\frac{1}{m_b} + \frac{1}{m_a}\right) \tag{1.87}$$

$$[[u_i u_j]] = \frac{4\pi e_a^2 e_b^2}{g^4 m_a^2}\, J\, (g^2 \delta_{ij} - g_i g_j). \tag{1.88}$$

Now the collision term (1.79) becomes, upon using the above equations,

$$c_{ab}(f_a f_b) = 4\pi e_a^2 e_b^2 \frac{J}{m_a} \left(\frac{1}{m_a} + \frac{1}{m_b}\right) \frac{\partial}{\partial v_{ai}} \left[f_a \int \frac{g_i}{g^3} f_b\, d^3 v_b\right]$$
$$+ 2\pi \frac{e_a^2 e_b^2}{m_a^2} J \frac{\partial^2}{\partial v_{ai} \partial v_{aj}} \left[f_a \int \frac{g^2 \delta_{ij} - g_i g_j}{g^3} f_b d^3 v_b\right]. \tag{1.89}$$

Use the following identity

$$\frac{g_i}{g^3} = \frac{1}{2}\frac{\partial}{\partial v_{bj}} \left(\frac{g^2 \delta_{ij} - g_i g_j}{g^3}\right) = -\frac{1}{2}\frac{\partial}{\partial v_{aj}} \left(\frac{g^2 \delta_{ij} - g_i g_j}{g^3}\right)$$

in (1.89) and integrate by parts the result to finally obtain the **Landau collision integral**

$$c_{ab}(f_a f_b) = 2\pi \frac{e_a^2 e_b^2}{m_a} J \frac{\partial}{\partial v_{ai}} \int d^3 v_b \, U_{ij} \left[\frac{f_b}{m_a} \frac{\partial f_a}{\partial v_{aj}} - \frac{f_a}{m_b} \frac{\partial f_b}{\partial v_{bj}} \right] \tag{1.90}$$

where

$$U_{ij} = \frac{g^2 \delta_{ij} - g_i g_j}{g^3}. \tag{1.91}$$

The integral J diverges as $\lambda \to \infty$ (λ should run from 0 to ∞ as b does so). This is a consequence of the long range of the Coulomb potential. However, the actual interaction does not extend far beyond a Debye length since the Coulomb potential is screened off. It is therefore reasonable to cut off the integral at b_{max}, equal to the Debye length, corresponding to

$$\lambda_{max} = \frac{\mu g^2}{e_a e_b} \sqrt{\frac{T}{4\pi n_0 e^2}}.$$

With this cut-off, J in (1.86) yields $J = log\sqrt{1 + \lambda_{max}^2}$. As will become clear shortly, 1 in the square root can be neglected as compared to λ_{max}^2. Since the logarithmic term changes slowly, we take the mean value of g^2 by replacing it with $3T/\mu$, We then obtain

$$J \simeq log\Lambda, \quad \Lambda = 12\pi n_0 \frac{e^2}{e_a e_b} \left[\frac{T}{4\pi n_0 e^2} \right]^{3/2}. \tag{1.92}$$

Apart from a numerical factor of order 10, Λ is the number of particles in a Debye sphere, and hence much greater than 1.

Landau [Reference: Landau, L. D. *The transport equation in the case of Coulomb interactions* in Collected papers of L. D. Landau, Ed. D. Ter Haar Pergamon Oxford (1965) p. 163] did not make use of the exact formula for the Rutherford scattering, but used the *impulse approximation.* One replaces the hyperbolic paths of the particles by rectilinear motion with constant velocity, and computes the force which they exert on each other during their rectilinear motion. The time integral of this force is then taken as the total momentum transfer:

$$sin\theta = \frac{1}{g} \int F_\perp dt = \frac{1}{g^2} \int F_\perp dx = \frac{e_a e_b}{\mu g^2} \int_{-\infty}^{\infty} \frac{b \, dx}{(x^2 + b^2)^{3/2}} = \frac{2 e_a e_b}{b \mu g^2} \simeq \theta.$$

Using the above relation, the integral in (1.85) becomes

$$\int_0^\infty sin^2\frac{\theta}{2}\; b\; db \simeq \left(\frac{e_a e_b}{\mu g^2}\right)^2 \int \frac{db}{b} = \left(\frac{e_a e_b}{\mu g^2}\right)^2 log\left(\frac{b_{max}}{b_{min}}\right)$$

where we have taken a second cut-off: $b_{min} = \frac{e_a e_b}{3T}$, which corresponds to a deflection through an angle of $\theta = \pi/2$ (in Rutherford scattering formula, g^2 is replaced by $3T/\mu$). Taking b_{max} again to be a Debye length, one arrives at exactly the same Landau equation in the impulse approximation.

Fokker–Planck equation. One can easily verify the following relations by direct calculation.

$$\frac{\partial}{\partial v_i}\frac{1}{g} = -\frac{g_i}{g^3}$$

$$\frac{g^2\delta_{ij} - g_i g_j}{g^3} = \frac{\partial}{\partial v_i}\left(\frac{g_j}{g}\right) = \frac{\partial^2 g}{\partial v_i \partial v_j}.$$

Then (1.89) can be put into the *Fokker–Planck form* of the collision integral:

$$c_{ab}(f_a f_b) = A\,\frac{\partial}{\partial v_{ai}}\left[-f_a\,\frac{\partial}{\partial v_{ai}}\int \frac{1}{g}\, f_b\, d^3v_b\right] + B\,\frac{\partial^2}{\partial v_{ai}\partial v_{aj}}\left[f_a\,\frac{\partial^2}{\partial v_{ai}\partial v_{aj}}\int g\, f_b\, d^3v_b\right]$$

where A and B are constants. The integrals in the above equation, $\int(\frac{1}{g}\ or\ g) f_b\ d^3v_b$, are called *Rosenbluth potential.* The Landau collision integral in (1.90) is a variation of the above form obtained after integration by parts. The above expression applies to the event when two particles a and b collide. Let us change the notaions: $\mathbf{v}_a \to \mathbf{v},\ \ \mathbf{v}_b \to \mathbf{v}'$. Then $\mathbf{g} = \mathbf{v} - \mathbf{v}'$, and the above equation can be written in the form

$$\left.\frac{\delta f(\mathbf{v},t)}{\delta t}\right|_c = -A\,\frac{\partial}{\partial v_i}\left[f(\mathbf{v})\,\frac{\partial}{\partial v_i}\int \frac{f(\mathbf{v}')}{|\mathbf{v}-\mathbf{v}'|}\, d^3v'\right] + B\,\frac{\partial^2}{\partial v_i \partial v_j}\left[f(\mathbf{v})\,\frac{\partial^2}{\partial v_i \partial v_j}\int |\mathbf{v}-\mathbf{v}'|\, f(\mathbf{v}')\, d^3v'\right].$$

This equation gives the change of the distribution function due to the collisions between the composing particles. The effect of prime and no prime particle collisions are summed over the distribution. The first term is termed the *dynamic friction* and the second the *diffusion term.* This equation, although derived from the Boltzmann collision integral here, is valid for *Markovian process* of weak scattering and can be derived independently. For this, see References: Krall, N. A. and Trivelpiece, A. W., *Principles of Plasma Physics*, San Francisco Press, 1986; Sturrock, P. A. *Plasma Physics* Cambridge University Press (1994). Also the Fokker–Planck equation above can be derived from Lenard–Balescu equation (Chapter 12) by taking an appropriate approximation [see Reference: Schram, P. P. J. M., *Kinetic Theory of Gases and Plasmas*, Kluer Academic, Dordrecht (1991)].

1.8. Braginskii's two-fluid equations

Reference: S. I. Braginskii, *Transport process in a plasma* in Reviews of plasma physics Vol. 1: Consultant's Bureau, New York (1965).

Braginskii derived two-fluid equations for a plasma consisting of electrons and ions from the set of Boltzmann equations with the Landau collision terms (1.90):

$$\frac{\partial f_e}{\partial t} + \mathbf{v} \cdot \frac{\partial f_e}{\partial \mathbf{r}} - \frac{e}{m_e}\left(\mathbf{E} + \frac{1}{c}\mathbf{v} \times \mathbf{B}\right) \cdot \frac{\partial f_e}{\partial \mathbf{v}} = C_{ee} + C_{ei}, \tag{1.93}$$

$$\frac{\partial f_i}{\partial t} + \mathbf{v} \cdot \frac{\partial f_i}{\partial \mathbf{r}} + \frac{e}{m_i}\left(\mathbf{E} + \frac{1}{c}\mathbf{v} \times \mathbf{B}\right) \cdot \frac{\partial f_i}{\partial \mathbf{v}} = C_{ii} + C_{ie}. \tag{1.94}$$

(1.93) and (1.94) are two coupled equations through the cross-collision terms C_{ei} and C_{ie}. Since the Landau collision terms are appropriate for plasma of frequent Coulomb collisions, Brakinskii's two-fluid equation is widely used for collisional plasma investigation.

In contrast with a simple gas, there are no summation invariants. But we have the following relations:

$$\int C_{ab} d^3v = 0 \quad (a = e,\ i;\quad b = e,\ i) \tag{1.95}$$

which implies that the total number of particles of each species is not changed due to collisions between like particles as well as between unlike particles. We have for like particles

$$\int m_a \mathbf{v} C_{aa} d^3v = 0, \tag{1.96}$$

$$\int m_a v^2 C_{aa} d^3v = 0. \tag{1.97}$$

The above two equations denote, respectively, the conservation of momentum and energy in collisions between like particles. For unlike particles, we have

$$\int (m_i C_{ie} + m_e C_{ei}) \mathbf{v} d^3 v = 0, \tag{1.98}$$

$$\int (m_i C_{ie} + m_e C_{ei}) v^2 d^3 v = 0. \tag{1.99}$$

The above two equations express the conservation of momentum and energy for the total pair of collisions involving unlike particles.

In order to derive the moment equations, multiply both sides of (1.93) with 1, $m_e \mathbf{v}$, $\frac{m_e}{2} v^2$, respectively, and integrate over the velocity. Doing the same operation with the ion equation (1.94), we obtain the three moment equations:

$$\frac{\partial n_a}{\partial t} + \nabla \cdot (n_a \mathbf{V}_a) = 0 \tag{1.100}$$

$$m_a n_a \left(\frac{\partial}{\partial t} + \mathbf{V}_a \cdot \nabla \right) \mathbf{V}_a = -\nabla p_a - \nabla \cdot \pi_a^{ij} + e_a n_a \left[\mathbf{E} + \frac{\mathbf{V}_a}{c} \times \mathbf{B} \right] + \mathbf{R}_a \tag{1.101}$$

$$\frac{3}{2} \frac{\partial}{\partial t} (n_a T_a) + \frac{3}{2} \nabla \cdot (n_a T_a \mathbf{V}_a) + n_a T_a \nabla \cdot \mathbf{V}_a + \pi_a^{ij} \frac{\partial V_{ai}}{\partial x_j} + \nabla \cdot \mathbf{q}_a = Q_a \tag{1.102}$$

where $a = e, i$, and the hydrodynamic variables $n_a, \mathbf{V}_a, T_a$ are defined for each species of particles similarly to (1.38)–(1.40):

$$n_a(\mathbf{r}, t) = \int f_a(\mathbf{r}, \mathbf{v}, t) d^3 v, \tag{1.103}$$

$$\mathbf{V}_a(\mathbf{r}, t) = \frac{1}{n_a} \int \mathbf{v} f_a d^3 v, \tag{1.104}$$

$$P_a^{ij} = m_a \int (v_i - V_{ai})(v_j - V_{aj}) f_a d^3 v = p_a \delta_{ij} + \pi_a^{ij} \quad (\textit{pressure tensor}) \tag{1.105}$$

$$T_a = \frac{m_a}{3 n_a} \int (\mathbf{v} - \mathbf{V}_a)^2 f_a d^3 v. \tag{1.106}$$

Exercise. Show that (1.106) can be arranged into the form of (1.59) with $Q = 0$.

The appearance of the quantities $\mathbf{R}_a$ and Q_a in (1.101) and (1.102) is due to the cross-collisional terms in (1.93) and (1.94); the corresponding terms in a simple gas do not exist. The quantity $\mathbf{R}_a$ represents the mean change of the momentum of particles of species 'a' due to collisions with particles of other species:

$$\mathbf{R}_e = \int C_{ei} m_e (\mathbf{v} - \mathbf{V}_e) d^3 v, \tag{1.107}$$

$$\mathbf{R}_i = \int C_{ie} m_i (\mathbf{v} - \mathbf{V}_i) d^3 v. \tag{1.108}$$

Due to the relation (1.98) we have

$$\mathbf{R}_e = -\mathbf{R}_i. \tag{1.109}$$

The quantity Q_a in (1.102) is the heat generated in the particles of species 'a' as a consequence of collisions with particles of other species.

$$Q_e = \int C_{ei} \frac{m_e}{2} (\mathbf{v} - \mathbf{V}_e)^2 d^3 v, \tag{1.110}$$

$$Q_i = \int C_{ie} \frac{m_i}{2} (\mathbf{v} - \mathbf{V}_i)^2 d^3 v. \tag{1.111}$$

Adding the above two equations give

$$Q_e + Q_i = -\mathbf{R}_e \cdot (\mathbf{V}_e - \mathbf{V}_i) \tag{1.112}$$

where we used (1.97)–(1.99). The foregoing formulas can be obtained without actually solving the coupled equations (1.93) and (1.94). Braginskii decoupled these equations by making use of the smallness of the mass ratio of electron and ion. Then the problem is reduced to that of solving the kinetic equation of single species, for which the successive approximation method of Chapman and Enskog can be applied. The key step to separate the electron and ion kinetic equations is to expand the relative velocity $\mathbf{g}$ in powers of ion velocity. We shall not go into the details of the Braginskii solution, but will summarize the important results.

Example. Expand $U_{ij} = \left(g^2 \delta_{ij} - g_i g_j\right) / g^3$ in powers of ion velocity.
This is a problem of Taylor expanding a three variable function in the form

$$\begin{aligned}\psi(\mathbf{r} + \mathbf{a}) &= \sum_{n=0}^{\infty} \frac{1}{n!} (\mathbf{a} \cdot \nabla)^n \psi(\mathbf{r}) \\ &= \psi(\mathbf{r}) + a_i \frac{\partial}{\partial x_i} \psi(\mathbf{r}) + \frac{1}{2} a_i a_j \frac{\partial}{\partial x_i} \frac{\partial}{\partial x_j} \psi(\mathbf{r}) + \cdots .\end{aligned}$$

Note $g_\alpha = v_\alpha - v'_\alpha$ where α is the Cartesian index, no prime denotes electron velocity, prime denotes ion velocity. In the above formula, put $a_\alpha = -v'_\alpha$. Then we obtain

$$U_{\alpha\beta} = V_{\alpha\beta} - \frac{\partial V_{\alpha\beta}}{\partial v_\gamma} v'_\gamma + \frac{1}{2} \frac{\partial^2 V_{\alpha\beta}}{\partial v_\gamma \partial v_\kappa} v'_\gamma v'_\kappa + \cdots$$

$$where\ \ V_{\alpha\beta} = \frac{1}{v^3}\left(v^2 \delta_{\alpha\beta} - v_\alpha v_\beta\right).$$

In the lowest order, the electron and ion distributions are the local Maxwellians

$$f_a(\mathbf{v}, \mathbf{r}, t) = n_a(\mathbf{r}, t)\left(\frac{m_a}{2\pi T_a(\mathbf{r}, t)}\right)^{\frac{3}{2}} exp - \frac{m_a(\mathbf{v} - \mathbf{V}_a(\mathbf{r}, t))^2}{2T_a(\mathbf{r}, t)} \qquad (a = e, i) \tag{1.113}$$

which yield the following two-fluid equations:

$$\frac{\partial n_a}{\partial t} + \nabla \cdot (n_a \mathbf{V}_a) = 0 \tag{1.114}$$

$$m_a n_a \left(\frac{\partial}{\partial t} + \mathbf{V}_a \cdot \nabla\right) \mathbf{V}_a = -\nabla p_a + e_a n_a \left[\mathbf{E} + \frac{\mathbf{V}_a}{c} \times \mathbf{B}\right] + \mathbf{R}_a^{(0)} \tag{1.115}$$

$$\frac{3}{2} n_a \left(\frac{\partial}{\partial t} + \mathbf{V}_a \cdot \nabla\right) T_a + n_a T_a \nabla \cdot \mathbf{V}_a = 0. \tag{1.116}$$

In the momentum equation (1.115),

$$\mathbf{R}_e^{(0)} = -\frac{m_e n_e}{\tau_e}\mathbf{u} = -\mathbf{R}_i^{(0)} : frictional\ force, \tag{1.117}$$

$$\mathbf{u} = \mathbf{V}_e - \mathbf{V}_i, \quad \tau_e = \frac{3\sqrt{m_e} T_e^{3/2}}{4\sqrt{2\pi}\ n_i\ J\ Z^2 e^4} \ : collision\ time,$$

where Z is ionic charge number and J is given by (1.86). The energy equation (1.116) is the adiabaticity: the entropy change, $ds_a/dt = 0$.

As the higher order solutions, the momentum and energy equations take the forms of (1.101) and (1.102), respectively. But the frictional force is reduced by a factor of 0.51 and, in addition, a thermal force $\mathbf{R}_T$, which arises owing to temperature gradients, appears. So we write in (1.101)

$$\mathbf{R}_e = \mathbf{R}_u + \mathbf{R}_T \tag{1.118}$$

$$\mathbf{R}_u = -\frac{m_e n_e}{\tau_e}(0.51\mathbf{u}_{\parallel} + \mathbf{u}_{\perp}) \tag{1.119}$$

$$\mathbf{R}_T = -0.71 n_e \nabla_{\parallel} T_e - \frac{3}{2}\frac{n_e}{\omega_{ce}\tau_e}\frac{\mathbf{B}_0}{B_0} \times \nabla T_e. \tag{1.120}$$

The symbols $\parallel$ and $\perp$ are referred to the direction of the ambient magnetic field ($\mathbf{B}_0$). If the plasma is unmagnetized, the $\perp$-components are discarded. The thermal forces are given rise to, due to the temperature gradients. The heat Q_a is a plasma-proper term which represents the heat generated by collisions between electrons and ions:

$$Q_i = \frac{3m_e}{m_i}\frac{n_e}{\tau_e}(T_e - T_i) \equiv Q_\Delta. \tag{1.121}$$

The heat acquired by the electrons is obtained by using (1.121) in (1.112). Another point worthy of attention in Braginskii's solution is that a heat flux ($\mathbf{q}_u^e$) is given rise to by the relative motion between electrons and ions, in addition to the usual heat flux($\mathbf{q}_T^e$) generated by temperature gradients: $\mathbf{q}_e = \mathbf{q}_u^e + \mathbf{q}_T^e$.

$$\mathbf{q}_u^e = 0.71 n_e T_e \mathbf{u}_{\parallel} + \frac{3}{2}\frac{n_e T_e}{\omega_{ce}\tau_e}\frac{\mathbf{B}_0}{B_0} \times \mathbf{u}. \tag{1.122}$$

The ion heat flux $\mathbf{q}_i$ takes the familiar form $\propto \nabla T_i$. Viscosity π_{ij}, heat flux $\mathbf{q}$, heat Q, and frictional and thermal force $\mathbf{R}$ all contribute to the production of entropy in the plasma. Define the entropy per electron as

$$s_e = \frac{3}{2} lnT_e - ln\ n_e + const. \tag{1.123}$$

The electron entropy equation is written in the form

$$\frac{\partial}{\partial t}(n_e s_e) + \nabla \cdot \left(s_e n_e \mathbf{V}_e + \frac{\mathbf{q}_e}{T_e}\right) + \frac{Q_\Delta}{T_e} = \vartheta_e \tag{1.124}$$

where ϑ_e is the entropy production per unit volume:

$$T_e \vartheta_e = -\mathbf{q}_e \cdot \nabla lnT_e - \mathbf{R} \cdot \mathbf{u} - \frac{1}{2}\pi_{ij}^e \left(\frac{\partial V_{ei}}{\partial x_j} + \frac{\partial V_{ej}}{\partial x_i} - \frac{2}{3}\delta_{ij} \nabla \cdot \mathbf{V}_e\right). \tag{1.125}$$

The electron entropy production ϑ_e is shown to be a positive definite quantity.

Exercise. Show that (1.124) is equivalent to (1.102). Derive the corresponding ion entropy equation.

The entropy equation for the entire plasma can be written in the form (see Braginskii 1965)

$$\frac{\partial}{\partial t}(s_e n_e + s_i n_i) + \nabla \cdot \left(s_e n_e \mathbf{V}_e + s_i n_i \mathbf{V}_i + \frac{\mathbf{q}_e}{T_e} + \frac{\mathbf{q}_i}{T_i} \right)$$
$$= \Theta : positive\ definite. \tag{1.126}$$

Example. In the Boltzmann equation of electron distribution function, we use the BGK type collision integral: $(\frac{\delta f_e}{\delta t})_c = -\nu(f_e - f_i)$. This model implies that the electron and ion distributions become the same after infinite time. Derive the momentum equation for the fluid velocity $\mathbf{V}_e(\mathbf{r}, t)$. Use (1.93) for the left hand side. This example demonstrates the necessary algebra. Multiply (1.93) by $\mathbf{v}$ and integrate over d^3v.

$$m_e \int d^3v \mathbf{v} \frac{\partial f_e}{\partial t} = m_e \frac{\partial}{\partial t} \int \mathbf{v} f_e d^3v = \frac{\partial}{\partial t}(m_e n_e \mathbf{V}_e(\mathbf{r}, t))$$

$$m_e \int d^3v \mathbf{v}\ \mathbf{v} \cdot \frac{\partial f_e}{\partial \mathbf{r}} = m_e \int d^3v v_i v_j \frac{\partial f_e}{\partial x_j} = m_e \frac{\partial}{\partial x_j} \int d^3v\ v_i v_j f_e$$
$$= m_e \frac{\partial}{\partial x_j}(n_e \langle v_i v_j \rangle) \tag{1.127}$$

where $\mathbf{v} = v_i$ (in the following 'i' is the free index indicating the vector component). We define the random velocity

$$w_i = v_i - V_i(\mathbf{r}, t). \tag{1.128}$$

Then we have

$$= \frac{\partial}{\partial x_j}(m_e n_e V_i V_j + m_e n_e \langle w_i w_j \rangle) = \frac{\partial}{\partial x_j}(m_e n_e V_i V_j + P_{ij})$$

$$-e \int d^3v \mathbf{v} \left(\mathbf{E} + \frac{1}{c} \mathbf{v} \times \mathbf{B} \right) \cdot \frac{\partial f_e}{\partial \mathbf{v}} d^3v$$
$$= -e \int d^3v v_i \left(E_j + \frac{1}{c} e_{jlm} v_l B_m \right) \frac{\partial f_e}{\partial v_j}.$$

Integrating by parts and using $\partial v_i / \partial v_j = \delta_{ij}$ give

$$= e \int d^3v \left(E_j \delta_{ij} + \frac{B_m}{c} e_{jlm}(v_i \delta_{lj} + v_l \delta_{ij}) \right) \quad (e_{jlm}\delta_{lj} = 0)$$
$$= e n_e \left(\mathbf{E} + \frac{1}{c} \mathbf{V}_e \times \mathbf{B} \right). \tag{1.129}$$

The collision term gives

$$-\nu m_e n_e(\mathbf{V}_e - \mathbf{V}_i).$$

Collecting the above results, we write

$$\frac{\partial}{\partial t}(m_e n_e \mathbf{V}_e) + \frac{\partial}{\partial x_j}(m_e n_e V_i V_j + P_{ij}) + e n_e\left(\mathbf{E} + \frac{\mathbf{V}_e}{c}\times\mathbf{B}\right)$$
$$= -\nu m_e n_e(\mathbf{V}_e - \mathbf{V}_i).$$

Using the continuity equation in the above yields [Krall and Trivelpiece 1986 p. 89]

$$m_e n_e\left(\frac{\partial}{\partial t} + \mathbf{V}_e\cdot\nabla\right)\mathbf{V}_e + \nabla\cdot P_{ij} + e n_e\left(\mathbf{E} + \frac{\mathbf{V}_e}{c}\times\mathbf{B}\right)$$
$$= -\nu m_e n_e(\mathbf{V}_e - \mathbf{V}_i). \tag{1.130}$$

Note that $\nabla\cdot P_{ij} = \frac{\partial P_{ij}}{\partial x_j} = \frac{\partial P_{ji}}{\partial x_j}$.

Example. Derive the energy equation by taking the moments of (1.93) and (1,94) with $\frac{1}{2}m_\alpha v_\alpha^2$. The first two terms give

$$\frac{\partial}{\partial t}\left(\frac{1}{2}m_\alpha n_\alpha\langle v^2\rangle_\alpha\right) + \frac{\partial}{\partial x_j}\left(\frac{1}{2}m_\alpha n_\alpha\langle v_j v^2\rangle_\alpha\right).$$

The electromagnetic force term yields

$$\frac{e_\alpha}{2}\int d^3v v^2\left(E_j + \frac{1}{c}e_{jlm}v_l B_m\right)\frac{\partial f_\alpha}{\partial v_j}$$
$$= -\frac{e_\alpha}{2}\int d^3v\left[2v\frac{\partial v}{\partial v_j}\left(E_j + \frac{1}{c}e_{jlm}v_l B_m\right) + \frac{v^2}{c}e_{jlm}\delta_{jl}B_m\right]$$
$$= -e_\alpha n_\alpha \mathbf{E}\cdot\mathbf{V}_\alpha.$$

Collecting the above results gives for the energy equation in the form

$$\frac{\partial}{\partial t}\left(\frac{1}{2}m_\alpha n_\alpha\langle v^2\rangle_\alpha\right) + \frac{\partial}{\partial x_j}\left(\frac{1}{2}m_\alpha n_\alpha\langle v_j v^2\rangle_\alpha\right) - e_\alpha n_\alpha\mathbf{E}\cdot\mathbf{V}_\alpha = Q_\alpha. \tag{1.131}$$

Exercise. Proceed further with (1.131) to derive (1.102).

References

Braginskii, S. I. Transport process in a plasma, in *Reviews of Plasma Physics* Vol. 1: Consultant's Bureau, New York, 1965.

Braginskii, S. I. Transport phenomena in a completely ionized Two-temperature plasma, *Soviet Phys. JETP*, **6** (1958) p. 358.

Bhatnagar, P. L., Gross, E. P. and Krook, M. *Phys. Rev.* **94** (1954) p. 511

Chapman, S. and Cowling, T. G. *The Mathematical Theory of Nonuniform Gases*, Cambridge University Press, 1952.

Krall, N. A. and Trivelpiece, A. W. *Principles of Plasma Physics*, San Francisco Press, 1986.

Landau, L. D. *The Transport Equation in the Case of Coulomb Interactions*, in Collected papers of L. D. Landau, Ed. D. Ter Haar Pergamon Oxford, 1965, p. 163.

Reif, F. *Fundamentals of Statistical and Thermal Physics*, McGraw-Hill International, 1985, Chapter 13.

Sturrock, P. A. *Plasma Physics*, Cambridge University Press, 1994.

Van Kampen, N. G. and Felderhof, B. U. *Theoretical Methods in Plasma Physics*, John Wiley, 1967, Chapter XV.

Wu, T.-Y. *Kinetic Equations of Gases and Plasmas*, Addison-Wesley, 1966, Chapter 2.

References

Chapter 2

Magnetohydrodynamics

2.1. Equations of plasma as a lumped single fluid

In the low frequency motion of plasma, we can treat it, by suitable combinations of the electron and ion quantities, as a single fluid instead of regarding it as two interacting fluids via collisions and Coulomb force. Under frequent collisions, the lumped volume elements of combined electrons and ions may be considered to form single fluid particles. In this low frequency region of plasma wave, the charge neutrality is often assumed: in a small volume, small but containing many electrons and ions, the net charge is approximately nil ($(n_i - n_e)/n_i \ll 1$). The magnetic field that is imbedded in the plasma or generated by current plays the major role in the dynamics of this *magnetohydrodynamic* plasma. A popular view of magnetohydrodynamics (MHD) is that it is a conducting fluid moving in a magnetic field. If a conductor moves in a magnetic field, electric field is induced as is called a *motional electric field* and electric currents arise. The magnetic field exerts forces on the currents, and the conducting fluid changes its motion under the force. Also the currents modify the magnetic field. The MHD equations describe the dynamical change of the fluid velocity, the magnetic field, and the electric field in the plasma.

We have several different methods to derive the MHD equations. The first method introduced now defines the lumped hydrodynamic variables:

$$\varrho_m = m_i n_i + m_e n_e \quad (mass\ density), \tag{2.1}$$

$$\varrho_c = e(n_i - n_e) \quad (charge\ density), \tag{2.2}$$

$$\mathbf{V} = \frac{m_i n_i \mathbf{V}_i + m_e n_e \mathbf{V}_e}{\varrho_m} \quad (center\ of\ mass\ velocity), \tag{2.3}$$

$$\mathbf{J} = e(n_i \mathbf{V}_i - n_e \mathbf{V}_e) \quad (current). \tag{2.4}$$

Multiply (1.114) by m_a and sum over a to obtain

$$\frac{\partial \varrho_m}{\partial t} + \nabla \cdot (\varrho_m \mathbf{V}) = 0 \quad (continuity\ \ equation). \tag{2.5}$$

Multiply (1.114) by charge 'e' and subtract electron equation from ion equation:

$$\frac{\partial \varrho_c}{\partial t} + \nabla \cdot \mathbf{J} = 0 \quad (charge \;\; conservation). \tag{2.6}$$

Taking the first moment of (1.93) (see (1.127)) gives,

$$\frac{\partial}{\partial t}(m_\alpha n_\alpha \mathbf{V}_\alpha(\mathbf{r},t)) + m_\alpha \frac{\partial}{\partial x_\mu}(n_\alpha \langle v_\mu v_\nu \rangle_\alpha)$$
$$- e_\alpha n_\alpha \left(\mathbf{E} + \frac{\mathbf{V}_\alpha}{c} \times \mathbf{B} \right) = \mathbf{R}_\alpha \tag{2.7}$$

where $\alpha = e, i$ ($e_e = -e$), indicating electrons and ions, μ and ν are Cartesian indexes, ν is the free index indicating the vector components, and $\mathbf{R}_\alpha$ is the momentum exchange rate due to collisions (see (1.115)). Here, we define the random velocity as, in contrast with (1.128),

$$\mathbf{w} = \mathbf{v} - \mathbf{V} \quad or \quad \langle \mathbf{w} \rangle_\alpha = \mathbf{V}_\alpha - \mathbf{V}. \tag{2.8}$$

Then we have

$$\langle v_\mu v_\nu \rangle_\alpha = -V_\mu V_\nu + V_\mu V_{\alpha\nu} + V_\nu V_{\alpha\mu} + \langle w_\mu w_\nu \rangle_\alpha. \tag{2.9}$$

Substituting (2.9) into (2.7) and adding the electron and ion equations, one obtains

$$\frac{\partial}{\partial t}(\varrho_m V_\nu) + \frac{\partial}{\partial x_\mu}(\varrho_m V_\mu V_\nu + P_{\mu\nu}) - \varrho_c E_\nu - \frac{1}{c}(\mathbf{J} \times \mathbf{B})_\nu = 0 \tag{2.10}$$

where $P_{\mu\nu} = P^e_{\mu\nu} + P^i_{\mu\nu}$ is the total pressure tensor. Note that the collision terms add to zero if there are no neutral particles. Using (2.5) in the above equation, it can be written in the form

$$\varrho_m \left(\frac{\partial}{\partial t} + \mathbf{V} \cdot \nabla \right) \mathbf{V} + \nabla \cdot P_{ij} - \varrho_c \mathbf{E} - \frac{1}{c} \mathbf{J} \times \mathbf{B} = 0. \tag{2.11}$$

Exercise. Show that the energy equation (1.131) summed over α takes the form

$$\frac{\partial}{\partial t}\left(\frac{3}{2} p + \frac{1}{2} \varrho_m V^2 \right) + \nabla \cdot \left(\frac{5}{2} p \mathbf{V} + + \frac{1}{2} \varrho_m V^2 \mathbf{V} \right) - \mathbf{J} \cdot \mathbf{E}$$
$$+ \nabla \cdot (\mathbf{q} + \pi_{ij} \cdot \mathbf{V}) = 0 \tag{2.12}$$

where

$$\mathbf{q} = \frac{1}{2}\sum_\alpha m_\alpha n_\alpha \langle w^2 \mathbf{w}\rangle_\alpha$$

and the random velocity is defined by (2.8).

The energy equation (2.12) is too complex to use. We often use the equations of state, in lieu of the energy equation:

$$\frac{d}{dt}(p\varrho_m^{-\gamma}) = 0 \quad (adiabatic), \tag{2.13}$$

$$\frac{d}{dt}\frac{p}{\varrho_m} = 0 \quad (isothermal). \tag{2.14}$$

2.2. Generalized Ohm's law

The lumped variables defined in (2.1)–(2.4) can be used to determine back the two fluid variables. In low frequency phenomena, the charge neutrality holds. Then we have, neglecting m_e/m_i as compared to 1,

$$n_e \simeq n_i = \tfrac{\varrho_m}{m_i} \tag{2.15}$$

$$\mathbf{V}_i \simeq \mathbf{V} + \tfrac{m_e}{m_i}\tfrac{\mathbf{J}}{n_i e}, \quad \mathbf{V}_e \simeq \mathbf{V} - \tfrac{\mathbf{J}}{n_e e}. \tag{2.16}$$

Equation (1.101) can be recast into the form

$$\frac{\partial}{\partial t}(m_a n_a \mathbf{V}_a) + \frac{\partial}{\partial x_\mu}(m_a n_a V_{a\mu} V_{a\nu})$$
$$= -\nabla p_a - \frac{\partial}{\partial x_\mu}\pi_a^{\mu\nu} + e_a n_a \left[\mathbf{E} + \frac{\mathbf{V}_a}{c}\times\mathbf{B}\right] + \mathbf{R}_a. \tag{2.17}$$

Multiply (2.17) by e_a/m_a, use (2.15) and (2.16), and take sum over 'a' to get

$$\frac{m_i m_e}{\varrho_m e^2}\frac{\partial \mathbf{J}}{\partial t} = \mathbf{E} + \frac{\mathbf{V}}{c}\times\mathbf{B} - \frac{1}{en_e c}\mathbf{J}\times\mathbf{B}$$
$$- \frac{1}{em_i n_i}(m_e\nabla p_i - m_i\nabla p_e) - \frac{\mathbf{R}_e}{en_e} \tag{2.18}$$

where we omitted $\pi_{\mu\nu}$ and the quadratic terms, and neglected m_e/m_i as compared to 1. In the last term, we put

$$\frac{\mathbf{R}_e}{en_e} = \eta\mathbf{J}, \quad \eta = \frac{m_e}{e^2 n_e \tau_e} \quad (electrical\ resistivity) \tag{2.19}$$

where we used (1.117) and (2.4). This equation indicates that the momentum exchange through collisions between electrons and ions is proportional to the relative fluid velocity of electrons and ions. Equation (2.18) is referred to as the *generalized Ohm's law.*

Adding the electron and ion terms of (2.17) yields another form of momentum equation in terms of the lumped variables (use (2.15) and (2.16) and $n_e \simeq n_i$):

$$\frac{\partial}{\partial t}(\varrho_m \mathbf{V}) + \frac{\partial}{\partial x_\mu}\left(\varrho_m V_\mu V_\nu + \frac{m_i m_e J_\mu J_\nu}{e^2 \varrho_m}\right)$$
$$= -\nabla p - \frac{\partial}{\partial x_\mu}\pi_{\mu\nu} + \frac{1}{c}\mathbf{J} \times \mathbf{B} + \varrho_c \mathbf{E}. \tag{2.20}$$

The last term in (2.20) can be neglected due to the charge neutrality, $\varrho_c \simeq 0$. Also we have from (2.4)

$$\nabla \cdot \mathbf{J} = 0. \tag{2.21}$$

The above two equations are useful in a current-carrying plasma. The difference between (2.11) and (2.20) should be noted: in (2.11) the random velocity is defined relative to the center of mass velocity while in (2.20) it is defined relative to the average velocity of each species.

2.3. Kinematics of continuum mechanics

In this section, we study the kinematics of continuum mechanics to apply it to MHD flow. Fluid can be defined as a continuum whose motion or state is determined by the properties such as density, fluid velocity, stress tensor, and thermodynamic quantities; temperature, internal energy (ε), entropy, specific heats, energy flux vector, etc. In a magnetohydrodynamic medium, the stress tensor includes not only the intermolecular mechanical stress tensor but also the electromagnetic (Maxwell) stress tensor. Various physical quantities (such as temperature, momentum, or kinetic energy) are associated with each fluid particle; the '*particle*' is meant by macroscopically small volume but containing many fluid molecules. In a sufficiently dense medium, the particles constitute a continuum whose state is determined by the motion of each individual fluid particle. In the *Eulerian* description of fluid, a property of the fluid is represented by a field $\phi(\mathbf{x}, t)$. The time rate of change of $\phi(\mathbf{x}, t)$ is, therefore, due to two changes: explicit temporal change and implicit change resulting from the particle movement. Thus we can write

$$d\phi = \frac{\partial \phi}{\partial t}dt + \nabla\phi \cdot d\mathbf{x}$$

where $d\mathbf{x}$ is the displacement of the particle position that took place during the time interval dt. Recognizing that $d\mathbf{x}/dt = \mathbf{v}$, the particle velocity, we

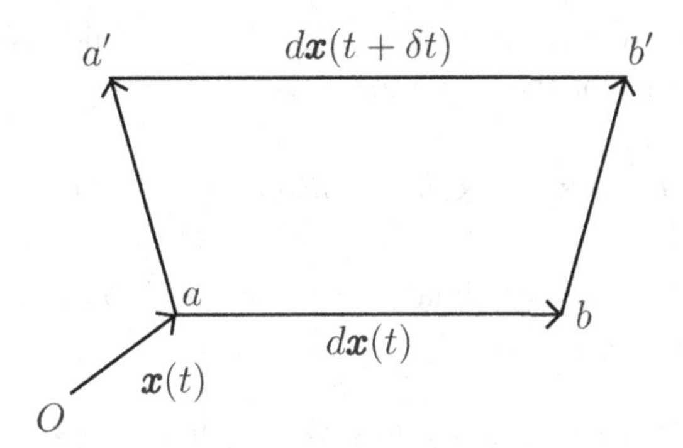

Fig. 2.1. Convective change of material line element.

have

$$\frac{d\phi}{dt} = \left(\frac{\partial}{\partial t} + \mathbf{v} \cdot \nabla\right) \phi \quad \textit{(convective derivative)}. \tag{2.22}$$

Kinematics of fluid motion begins with consideration of the rate of change of any arbitrary *material line element* during the course of fluid motion. The material line element is by definition a line element composed of fluid particles that lie on it. Imagine that a line element $d\mathbf{x}(t)$ has been stretched to $d\mathbf{x}(t + \delta t)$ (see Fig. 2.1) by the corresponding movement of $a \rightarrow a'$ and $b \rightarrow b'$.

We have for the vectors

$$\overrightarrow{aa'} = \mathbf{v}(\mathbf{x}, t)\delta t, \quad \overrightarrow{bb'} = \mathbf{v}(\mathbf{x} + d\mathbf{x}, t)\delta t.$$

Thus we write

$$\mathbf{v}(\mathbf{x}, t)\delta t + d\mathbf{x}(t + \delta t) = d\mathbf{x}(t) + \mathbf{v}(\mathbf{x} + d\mathbf{x}, t)\delta t.$$

Expanding the last term in Taylor series and neglecting the higher order terms, one obtains

$$\frac{d}{dt} d\mathbf{x} = \lim_{\delta t \to 0} \frac{d\mathbf{x}(t + \delta t) - d\mathbf{x}(t)}{\delta t} = (d\mathbf{x} \cdot \nabla)\mathbf{v}. \tag{2.23}$$

The above equation is an important kinematic relation which describes how a material line element changes in the course of fluid motion. Next we consider the total rate of change of a line integral associated with a fluid quantity θ

$$\frac{d}{dt} \int_P^Q \theta(\mathbf{x}, t) d\mathbf{x} \tag{2.24}$$

where P and Q are two arbitrary points in the fluid. By definition the convective derivative of the line integral is

$$\lim_{\delta t\to 0}\frac{1}{\delta t}\left[\int_{P'}^{Q'}\theta(\mathbf{x}+\mathbf{v}(\mathbf{x})\delta t, t+\delta t)d\mathbf{x}(t+\delta t)-\int_{P}^{Q}\theta(\mathbf{x},t)d\mathbf{x}\right]$$

where P' and Q' are infinitesimally displaced position of P and Q, respectively. Expand as

$$\theta(\mathbf{x}+\mathbf{v}(\mathbf{x})\delta t, t+\delta t)=\theta(\mathbf{x},t)+\frac{\partial\theta}{\partial t}\delta t+\nabla\theta\cdot\mathbf{v}\delta t+\cdots$$

$$d\mathbf{x}(t+\delta t)=d\mathbf{x}(t)+\left(\tfrac{d}{dt}d\mathbf{x}\right)\delta t+\cdots.$$

Substituting from (2.23) and neglecting the higher order terms, the line integral (2.24) becomes

$$\frac{d}{dt}\int_P^Q\theta(\mathbf{x},t)d\mathbf{x}=\int_P^Q\frac{d\theta}{dt}\,d\mathbf{x}+\int_P^Q\theta\,(d\mathbf{x}\cdot\nabla)\mathbf{v}. \tag{2.25}$$

Operationally, the total derivative and the integral commute, but it is important to have the line element $d\mathbf{x}$ differentiated. According to the above formula, it goes without proof that the total differentiation can be taken into the integral sign even in a surface or a volume integral. A line integral associated with a vector field convectively changes in a like fashion

$$\frac{d}{dt}\int\mathbf{A}(\mathbf{x},t)\cdot d\mathbf{x}=\int\frac{dA_i}{dt}dx_i+\int A_i(d\mathbf{x}\cdot\nabla)v_i. \tag{2.26}$$

Next we consider a fixed arbitrary surface S surrounding a portion of fluid of volume V. Through the surface S, fluid particles move in and out, and we attempt to calculate the time rate of change of mass occupying the volume V. We assume that no mass is created or destroyed inside the volume V.

Rate of increase of mass in V $=\frac{\partial}{\partial t}\int\varrho d\tau$ (ϱ, *mass density;* $d\tau$, *volume element)* = *rate of net in-flow of mass through S* $=-\int\varrho\mathbf{v}\cdot d\mathbf{S}=-\int\nabla\cdot(\varrho\mathbf{V})d\tau$.

The above equality comes from the divergence theorem, and the negative sign is due to the convention that the surface element takes positive sign when it is directed outwardly. Since the volume is fixed we arrive at

$$\frac{\partial\varrho}{\partial t}+\nabla\cdot(\varrho\mathbf{v})=0\quad or\quad\frac{d\varrho}{dt}+\varrho\nabla\cdot\mathbf{v}=0. \tag{2.27}$$

This relation, representing the principle of mass conservation, is called continuity equation. When there is a source or sink in the fluid medium, that is, there exists any mechanism of production or annihilation of fluid mass, RHS of (2.27) should include the appropriate source or sink term.

Now we specialize (2.27) to an infinitesimal volume element $d\tau$ containing a mass δm. The volume element is envisaged to be enlarged or shrunk as the fluid particles occupying the volume move around. But no particles should cross the boundary of the volume (as the classical law of mechanics testifies). So δm is constant even though the size of the volume element is subject to change during the course of fluid motion. We then have from (2.27)

$$\frac{d}{dt}\left(\frac{1}{\delta\tau}\right) + \frac{1}{\delta\tau}\nabla\cdot\mathbf{v} = 0, \quad or \quad \frac{d}{dt}\delta\tau = \nabla\cdot\mathbf{v}\,\delta\tau. \tag{2.28}$$

This is another kinematic law, along with (2.23), relating how a moving volume element should change as the fluid particles composing the volume element move about. It is easy to show that the continuity equation (2.27) comes out from

$$\begin{aligned}\frac{d}{dt}\int_V \varrho\, d\tau &= \int_V \frac{d\varrho}{dt}d\tau + \int_V \varrho\,\frac{d}{dt}d\tau \\ &= \int_V \left(\frac{d\varrho}{dt} + \varrho\nabla\cdot\mathbf{v}\right) d\tau = 0 \end{aligned} \tag{2.29}$$

where the material volume V now is not fixed but changes along the fluid movement.

Next consider a tiny parallelepiped with sides $\delta x_1, \delta x_2$, and δx_3. Its volume $\delta\tau = \delta x_1\delta S_1 = \delta x_2\delta S_2 = \delta x_3\delta S_3 = \delta x_i\delta S_i$ (no sum on $'i'$) where δS_i is the lateral elemental surface whose normal is directed to $'i'$-direction (see Fig. 2.2). We have

$$\frac{d}{dt}d\tau = \frac{d}{dt}(\delta x_i\;\delta S_i) = \delta S_i\frac{d}{dt}\delta x_i + \delta x_i\frac{d}{dt}\;\delta S_i.$$

Substituting from (2.23) for $(d/dt)\delta x_i$ and making use of (2.28), one can easily obtain

$$\frac{d}{dt}\delta S_i = \delta S_i\;\nabla\cdot\mathbf{v} - \delta S_j\;\frac{\partial v_j}{\partial x_i}. \tag{2.30}$$

This equation relates the way an arbitrary material surface element convectively changes in the course of fluid motion.

2.4. MHD: Dynamics of electrically conducting fluid

MHD is basically classical fluid dynamics of an electrically conducting fluid in the presence of magnetic fields. The magnetic fields can be either imbedded

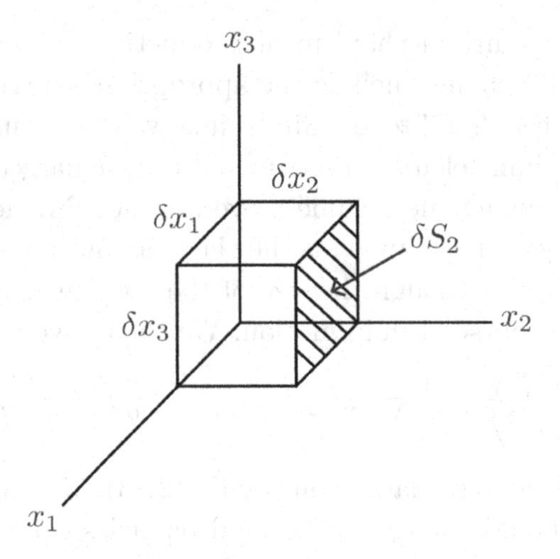

Fig. 2.2. Surface element of parallelepiped.

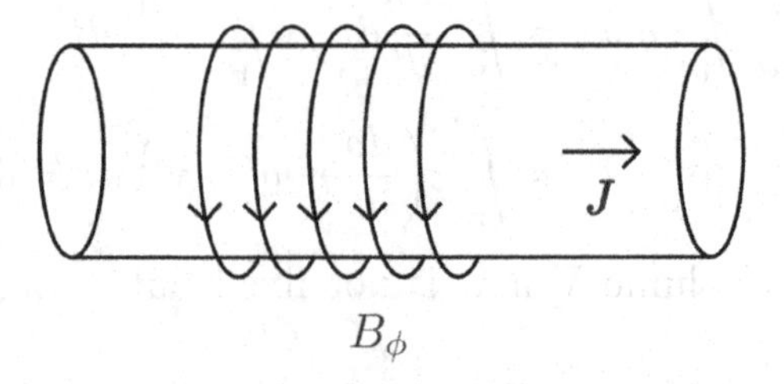

Fig. 2.3. Bennet pinch.

in the space or generated by the bulk motion of the conducting fluid. They are in return capable of generating dynamical effects on the bulk motion of the medium itself. Thus we deal with a complicated coupled system of fluid and magnetic fields. This interplay of conducting fluid and magnetic field can best be illustrated by a device known as z-pinch (Bennet pinch). A cylindrical column of conducting fluid with an axial current density $\mathbf{J}$ is depicted in Fig. 2.3. The axial current in the conducting fluid is generated by an appropriately applied electric field. The current produces azimuthal induction $\mathbf{B} = \hat{\varphi}B$ which in turn gives rise to $\mathbf{J} \times \mathbf{B}$ force. The $\mathbf{J} \times \mathbf{B}$ force acting on the fluid forces the column to be pushed inwardly in all directions, to the effect of confinement.

• **Distinction between plasma and magnetohydrodynamic fluid.** Roughly speaking, plasma is a collection of suitably dense charged particles which interact through electromagnetic forces between them. This definition equally well applies to electrically conducting fluid that is dealt with in MHD. But a clear distinction exists between them: magnetohydrodynamic fluid exhibits only a single fluid characteristics, always preserving the charge neutrality, while plasma is characterized by high frequency collective oscillations known as plasma oscillation which owes to deviation from charge neutrality. What makes a medium containing both positive and negative charged particles behave sometimes as a single charge-neutral fluid and other times as a collective medium exhibiting different inertial effect of the constituent charged particles? It depends on the inter-particle collision frequency. If the collision frequency is sufficiently high, then the electrons and the ions move in such a way that there is no separation of charge, and the medium behaves like a single fluid. If the collision frequency is low, the high frequency electronic motion gives rise to charge separation and the medium exhibits the characteristic collective behavior. However, it may be more appropriate to accept that these two different behaviors coexists in ionized gases. Then it depends on what part of the spectrum of the motion we are looking at. The charge separation effect is important in the time scale of inverse plasma frequency (ω_{pe}^{-1}, $\omega_{pe} = \sqrt{\frac{4\pi Ne^2}{m_e}}$) and in the length scale of $v_{\rm th}/\omega_{pe} = \lambda_e$ ($v_{\rm th}$ is the electron thermal speed and λ_e is electron Debye length). If one is not interested in the fine structure of motion but in the low frequency ($\ll \omega_{pe}$) and large scale ($\gg \lambda_e$) motion, the MHD equations will suffice.

2.5. Magnetohydrodynamic momentum equation

The momentum of a conducting fluid occupying a volume element $d\tau$ is $\varrho_m \mathbf{v} d\tau$. Now we wish to calculate the total time rate of change of momentum of the fluid enclosed by a material surface S:

$$\frac{d}{dt}\int_V \varrho_m v_i \, d\tau$$

with V designating the volume enclosed by the surface S. According to Newton's law this momentum change is brought about by the action of forces exerted on each fluid particle within and on S. The forces are divided into two types, the body force and the surface force. The body force acting on each individual elemental volume $d\tau$ is $\varrho_m F_i \tau$. The surface force acting on the elemental surface dS_j is $(P_{ij} + M_{ij})$ where P_{ij} is the mechanical pressure tensor ($= p\delta_{ij} + \pi_{ij}$) and M_{ij} denotes the Maxwell stress tensor. Thus we write

$$\frac{d}{dt}\int_V \varrho_m v_i d\tau = \int_V \varrho_m F_i d\tau + \int_S (P_{ij} + M_{ij}) dS_j. \tag{2.31}$$

The expression for the Maxwell stress tensor is

$$M_{ij} = \frac{1}{4\pi}\left(E_iE_j - \frac{1}{2}\delta_{ij}E^2 + B_iB_j - \frac{1}{2}\delta_{ij}B^2\right). \tag{2.32}$$

Now we calculate divergence of M_{ij}:

$$\frac{\partial M_{ij}}{\partial x_j} = \frac{1}{4\pi}\left(E_j\frac{\partial E_i}{\partial x_j} - \frac{1}{2}\nabla E^2 + B_j\frac{\partial B_i}{\partial x_j} - \frac{1}{2}\nabla B^2\right) + \varrho_c\, E_i \tag{2.33}$$

where we used

$$\nabla \cdot \mathbf{E} = 4\pi\varrho_c, \tag{2.34}$$

$$\nabla \cdot \mathbf{B} = 0. \tag{2.35}$$

Using the vector identity

$$\frac{1}{2}\nabla A^2 = \mathbf{A}\times\nabla\times\mathbf{A} + (\mathbf{A}\cdot\nabla)\mathbf{A} \tag{2.36}$$

(2.33) becomes

$$\frac{\partial M_{ij}}{\partial x_j} = -\frac{1}{4\pi}(\mathbf{E}\times\nabla\times\mathbf{E} + \mathbf{B}\times\nabla\times\mathbf{B}) + \varrho_c\mathbf{E}. \tag{2.37}$$

We substitute into the above equation the remaining Maxwell equations

$$\nabla\times\mathbf{E} = -\frac{1}{c}\frac{\partial\mathbf{B}}{\partial t} \tag{2.38}$$

$$\nabla\times\mathbf{B} = \frac{1}{c}\frac{\partial\mathbf{E}}{\partial t} + \frac{4\pi}{c}\mathbf{J} \tag{2.39}$$

to obtain

$$\frac{\partial M_{ij}}{\partial x_j} = \frac{1}{4\pi c}\frac{\partial}{\partial t}(\mathbf{E}\times\mathbf{B}) + \frac{1}{c}\,\mathbf{J}\times\mathbf{B} + \varrho_c\mathbf{E}. \tag{2.40}$$

Then (2.31) becomes

$$\varrho_m\frac{d\mathbf{v}}{dt} = \varrho_m\mathbf{F} + \nabla\cdot P_{ij} + \frac{1}{c}\,\mathbf{J}\times\mathbf{B} + \varrho_c\mathbf{E} + \frac{\partial}{\partial t}\frac{\mathbf{E}\times\mathbf{B}}{4\pi c}. \tag{2.41}$$

The last term of (2.41) is the rate of change of the electromagnetic momentum density. It is customary in MHD to neglect the last two terms involving the electric field, which amounts to neglect of $\mathbf{E}$ terms in M_{ij} and the displacement current in (2.39). Thus MHD momentum equation reads

$$\varrho_m\frac{d\mathbf{v}}{dt} = \varrho_m\mathbf{F} + \nabla\cdot P_{ij} + \frac{1}{c}\,\mathbf{J}\times\mathbf{B}. \tag{2.42}$$

The idea behind the neglect of the electric body force $\varrho_c \mathbf{E}$ is that, owing to the high conductivity, small charges cause large currents, so that the magnetic force $\mathbf{J} \times \mathbf{B}$ is much larger than the electric force. Also the Maxwell displacement current in (2.39) can be neglected as compared to the conduction current $\mathbf{J} = \mathbf{E}/\eta$ (resistivity η has a dimension of time, and so conductivity $\sigma = 1/\eta$ has a dimension of frequency) owing to the high σ and the slow temporal change ($\sigma \gg \partial/\partial t$). So Ampere's law used in MHD reads

$$\nabla \times \mathbf{B} = \frac{4\pi}{c} \mathbf{J}. \tag{2.43}$$

By the same token, the rate of change of electromagnetic momentum can be neglected as compared to the $\mathbf{J} \times \mathbf{B}$ force in (2.41).

Some remarks on the pressure tensor P_{ij} of MHD flow are in order. The pressure tensor is the sum of electron and ion terms. Recall that the viscous stress tensor π_{ij} represents the irreversible (entropy-producing) momentum transfer between fluid particles due to inter-particle collisions. Since the ion mass is much greater than the electron mass, the dominant momentum transfer owes to the ion-ion collisions. Thus the ion term of the viscous stress tensor dominates the electron term. When the magnetic field is sufficiently weak (ion-ion collision frequency $\gg$ ion cyclotron frequency), the MHD medium can be regarded as isotropic. Under this condition, the pressure tensor can be written as

$$P_{kl} = (p_e + p_i)\delta_{kl} + \eta_i \left(\frac{\partial v_k}{\partial x_l} + \frac{\partial v_l}{\partial x_k} - \frac{2}{3}\delta_{kl} \nabla \cdot \mathbf{v} \right)$$

where $\eta_i = 1.365 n_i T_i / \nu_i$ is the coefficient of viscosity due to ion-ion collision, with ν_i the ion-ion collision frequency as given in the Reference: I. P. Shkarofsky, T. W. Johnston, and M. P. Bachynsky *The Particle Kinetics of Plasmas*, Addison-Wesley (1966).

$$\nu_i = \frac{4}{3}\sqrt{2\pi} n_i \left(\frac{Z^2 e^2}{T_i} \right)^2 \sqrt{\frac{T_i}{m_i}}\, A \quad (A: \mathit{Coulomb\ logarithm}).$$

When the collision frequency is much less than the ion cyclotron frequency, the double adiabatic theory (see § 2.10) can be used.

2.6. Ohm's law

From (2.18) we write

$$\mathbf{J} = \sigma \left(\mathbf{E} + \frac{\mathbf{v}}{c} \times \mathbf{B} \right) \tag{2.44}$$

neglecting the rest of the terms. It is an empirical law that in a metal the current density $\mathbf{J}$ is linearly proportional to the electric field $\mathbf{E}$. The current and the electric field in a conducting fluid have an analogous relation with an additional complication that a conducting fluid is not stationary but moving. If a conductor is moving in a magnetic field $\mathbf{B}$, then an equivalent electric field $\mathbf{v} \times \mathbf{B}/c$ is induced in the conductor, which is known as the *motional electric field*, hence (2.44). In summary, electric currents in a conducting fluid are generated by a small amount of charge density and by the convective motion of the conducting fluid across the magnetic field. From (2.43) and (2.44), $\mathbf{E}$ can be expressed in terms of $\mathbf{B}$:

$$\mathbf{E} = \frac{c}{4\pi\sigma}\nabla \times \mathbf{B} - \frac{\mathbf{v}}{c} \times \mathbf{B}. \tag{2.45}$$

Using (2.38) in (2.45) gives the equation for $\mathbf{B}$ in a conducting fluid:

$$\frac{\partial \mathbf{B}}{\partial t} = \frac{c^2}{4\pi\sigma}\nabla^2\mathbf{B} + \nabla \times (\mathbf{v} \times \mathbf{B}). \tag{2.46}$$

2.7. Magnetic pressure and magnetic tension

A surface element $d\mathbf{S}$ exerted by the magnetic stress M_{ij} is under the force

$$\mathbf{M}_{ij} \cdot d\mathbf{S} = -\frac{B^2}{8\pi}d\mathbf{S} + \frac{1}{4\pi}\mathbf{B}\,(\mathbf{B} \cdot d\mathbf{S}).$$

The first term $\frac{B^2}{8\pi}$ is the force directed normal to the surface and is naturally called magnetic pressure. The second term which is directed along the magnetic field can be interpreted as a magnetic tension. If converted into body force by divergence theorem, we have

$$\nabla \cdot \mathbf{M}_{ij} = \frac{1}{c}\mathbf{J} \times \mathbf{B} = -\nabla\left(\frac{B^2}{8\pi}\right) + \frac{1}{4\pi}(\mathbf{B} \cdot \nabla)\mathbf{B}.$$

Introducing the unit vector along the field line, $\hat{\mathbf{b}} = \mathbf{B}/B$, the above expression becomes

$$-\nabla\left(\frac{B^2}{8\pi}\right) + \frac{1}{4\pi}B\frac{\partial}{\partial \xi}(\hat{\mathbf{b}}B)$$

where $d\xi$ is infinitesimal length along the field line. Using

$$\frac{\partial \hat{\mathbf{b}}}{\partial \xi} = \frac{\hat{\mathbf{n}}}{R}, \quad (\hat{\mathbf{n}}:\ \textit{unit normal vector}, \quad R:\ \textit{local radius of curvature})$$

we have

$$\frac{1}{c}\mathbf{J} \times \mathbf{B} = -\nabla_\perp\left(\frac{B^2}{8\pi}\right) + \frac{B^2}{4\pi R}\,\hat{\mathbf{n}}:\quad \textit{body force} \tag{2.47}$$

where $\nabla_\perp$ denotes the gradient in the direction perpendicular to the magnetic field. The second term on the RHS of (2.47) operates only when the field line is curved in much the same way tension acts on a curved string (the resultant is always toward the center of curvature). The magnetic pressure force is directed away from the region of high field strength. Bending the field line produces tension, and compressing it produces a restoring force.

• **Conservation of angular momentum.** The principle of conservation of angular momentum leads to the symmetry of the pressure tensor. This will be shown here. Equating the rate of convective change of of angular momentum possessed by the fluid enclosed inside a material surface S to the torque exerted on the same volume of the fluid, we write

$$\frac{d}{dt}\int_V \mathbf{x}\times\mathbf{v}\varrho_m \, d\tau = \int_V \varrho_m \mathbf{x}\times\mathbf{F}d\tau + \int_S \mathbf{x}\times\mathbf{P}_{ij}\cdot d\mathbf{S} \tag{2.48}$$

where the pressure tensor $\mathbf{P}_{ij}$ is understood as the mechanical pressure tensor in ordinary fluid or mechanical plus Maxwell stress tensor in MHD flow. The last term is the torque exerted on the surface of the system through the surface force. Carrying out the convective differentiation and using the continuity equation and $\frac{d\mathbf{x}}{dt}\times\mathbf{v}=0$, the LHS of (2.48) becomes

$$\int_V \varrho_m \mathbf{x}\times\frac{d\mathbf{v}}{dt}d\tau = \int_V \mathbf{x}\times(\varrho_m\mathbf{F}+\nabla\cdot\mathbf{P})d\tau.$$

Thus we are left with

$$\int_V \mathbf{x}\times\nabla\cdot\mathbf{P}_{ij}d\tau = \int_S \mathbf{x}\times\mathbf{P}_{ij}\cdot d\mathbf{S}$$

which gives

$$\begin{aligned}\int_V e_{ijk}x_j\frac{\partial}{\partial x_l}P_{kl}d\tau &= \int_S e_{ijk}x_jP_{kl}dS_l \\ &= \int_V e_{ijk}\frac{\partial}{\partial x_l}(x_jP_{kl})d\tau \\ &= \int_V e_{ijk}\left(P_{kj}+x_j\frac{\partial P_{kl}}{\partial x_l}\right)d\tau.\end{aligned}$$

Thus we have

$$e_{ijk}P_{kj}=0 \quad i.e. \quad P_{kj}=P_{jk}. \tag{2.49}$$

• **Poynting theorem.** The Poynting vector

$$\mathbf{N} = \frac{c}{4\pi}\mathbf{E} \times \mathbf{B}$$

satisfies

$$\frac{\partial}{\partial t}\frac{B^2}{8\pi} + \nabla \cdot \mathbf{N} = -\mathbf{J} \cdot \mathbf{E}. \tag{2.50}$$

Exercise. Using (2.24)–(2.46), prove (2.50).

2.8. Energy equation

The energy equation of fluid is basically the first law of thermodynamics which states that the increase of energy possessed by a certain portion V of the fluid is due to the work done on V by external forces plus the heat absorbed by the portion V through heat conduction, and the heat generated in V by molecular interactions. In addition we shall see that the energy equation implies the second law of thermodynamics: the entropy of the fluid should never decrease. The fluid energy contained in a volume V of MHD flow is the sum of the kinetic energy, internal energy, and the magnetic field energy. Denoting by ε the internal energy per unit mass, we write (here we shall drop the subscript m at the mass density) for a *fixed volume* of V

$$\frac{\partial}{\partial t}\int_V \left(\frac{\varrho}{2}v^2 + \varrho\,\varepsilon + \frac{B^2}{8\pi}\right) d\tau$$

= (1) the out going energy flux (non-electromagnetic) through S + (2) rate of working of the body force **F** in V + (3) rate of working of the surface stresses on S + (4) electromagnetic energy flux (Poynting vector) through S + (5) heat acquired by conduction through S.

$$= -\int_S \varrho\left(\frac{1}{2}v^2 + \varepsilon\right)\mathbf{v} \cdot d\mathbf{S} + \int_V \varrho F_i v_i d\tau + \int_S P_{ij} v_i dS_j$$
$$-\int_s \mathbf{N} \cdot d\mathbf{S} + \int_S \lambda \nabla T \cdot d\mathbf{S}.$$

Converting the surface integrals into volume integrals and using the continuity equation (2.27), and (2.50) yield

$$\varrho\frac{d}{dt}\left(\frac{v^2}{2} + \varepsilon\right) - \frac{\partial}{\partial x_j}(v_i P_{ij}) = \mathbf{J} \cdot \mathbf{E} + \nabla \cdot (\lambda \nabla T) + \mathbf{F} \cdot \mathbf{v}.$$

Using (2.42) in the above equation gives

$$\varrho\frac{d\varepsilon}{dt}-\frac{p}{\varrho}\frac{d\varrho}{dt}=\varrho T\frac{ds}{dt}$$
$$=\pi_{ij}\frac{\partial v_i}{\partial x_j}+\frac{c}{16\pi^2\sigma}(\nabla\times\mathbf{B})^2+\nabla\cdot(\lambda\nabla T) \tag{2.51}$$

where s is the entropy per unit mass. The RHS of (2.51) can be shown to be positive definite.

Exercise. Show that (2.51) can be cast into the form of energy continuity equation

$$\frac{\partial}{\partial t}\left(\frac{1}{2}\varrho v^2+\varrho\,\varepsilon+\frac{B^2}{8\pi}\right)+\nabla\cdot\left[\mathbf{v}\left(\frac{1}{2}\varrho v^2+\varrho\varepsilon\right)+\mathbf{N}\right]$$
$$=\nabla\cdot(\mathbf{P}_{ij}\cdot\mathbf{v}+\lambda\nabla T). \tag{2.52}$$

• **Summary of MHD equations**

$$\frac{\partial\rho}{\partial t}+\nabla\cdot(\rho\mathbf{v})=0:\quad \textit{mass continuity equation} \tag{2.53}$$

$$\rho\frac{d\mathbf{v}}{dt}=\nabla\cdot P_{ij}+\frac{1}{c}\mathbf{J}\times\mathbf{B}+\rho\mathbf{F}:\quad \textit{momentum equation} \tag{2.54}$$

$$\mathbf{J}=\sigma\left(\mathbf{E}+\frac{1}{c}\mathbf{v}\times\mathbf{B}\right):\quad \textit{Ohm's law} \tag{2.55}$$

with $\nabla\times\mathbf{E}=-\frac{1}{c}\frac{\partial\mathbf{B}}{\partial t}$.

$$\frac{\partial\mathbf{B}}{\partial t}=\frac{c^2}{4\pi\sigma}\nabla^2\mathbf{B}+\nabla\times(\mathbf{v}\times\mathbf{B}):\quad \textit{equation of magnetic field.} \tag{2.56}$$

Energy equation is too complicated to be used, and often replaced by entropy conservation:

$$\frac{ds(p,\rho)}{dt}=0:\quad \textit{isentropicity.} \tag{2.57}$$

If the isentropicity is assumed, it would be consistent to discard all the dissipative terms in (2.54)–(2.56), in which case the pressure tensor term in (2.54) is replaced by $\nabla\cdot P_{ij}\to-\nabla p$ and $\sigma\to\infty$. Then the resulting equations constitute the *ideal MHD equations.*

2.9. Kinematics of magnetic field in a conducting fluid

In a conducting MHD fluid of a very high conductivity ($\sigma \to \infty$), (2.46) reduces to

$$\frac{\partial \mathbf{B}}{\partial t} = \nabla \times (\mathbf{v} \times \mathbf{B}) = (\mathbf{B} \cdot \nabla)\mathbf{v} - (\mathbf{v} \cdot \nabla)\mathbf{B} - \mathbf{B}\nabla \cdot \mathbf{v}. \tag{2.58}$$

Substituting from the mass conservation equation (2.27) for the last term above, it can be written as

$$\frac{\partial \mathbf{B}}{\partial t} = (\mathbf{B} \cdot \nabla)\mathbf{v} - (\mathbf{v} \cdot \nabla)\mathbf{B} + \frac{\mathbf{B}}{\rho}\frac{d\rho}{dt} \quad \textit{or}$$

$$\frac{d}{dt}\frac{\mathbf{B}}{\rho} = \frac{\mathbf{B}}{\rho} \cdot \nabla\, \mathbf{v}. \tag{2.59}$$

In view of (2.23) which describes the way of convective change of material line element $d\mathbf{x}$, we see that $\frac{\mathbf{B}}{\rho}$ changes in an identical manner as $d\mathbf{x}$ changes. Therefore, if at $t = 0$ two vectors, $d\mathbf{x}$ and $\mathbf{B}/\rho$, are parallel, they will remain parallel, and their length will remain in the same ratio at later times. In other words, if at $t = 0$ two fluid particles belong to a same field line, then they will always belong to the same field line and the value of B/ρ will be proportional to the distance between the particles. Alfven described this by saying that in a conducting fluid of infinite conductivity the lines of magnetic field are *frozen* in the fluid and move with it (*frozen theorem*). An interesting consequence of the frozen behavior of a conducting fluid is that in a fluid of constant density the magnetic field intensity is increased by the motion of *frozen particles* which lengthen the field lines.

Now consider a portion of a tube of field lines (flux tube) in Fig. 2.4.

While it is carried along by the motion of the fluid, it will remain a portion of the flux tube at later instant. Since the two portions have

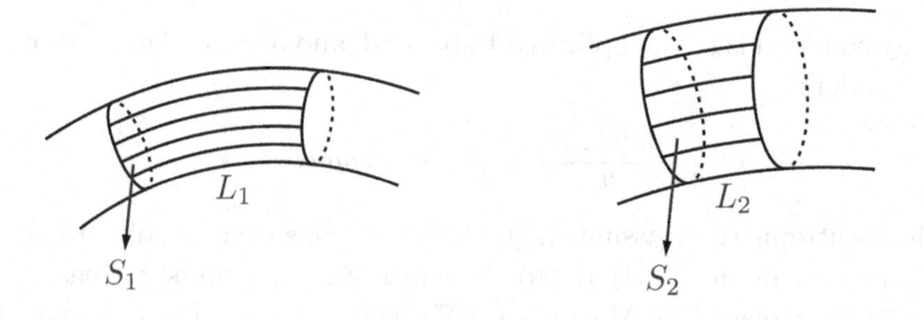

Fig. 2.4. Flux tubes.

boundaries composing of the same fluid particles, the mass contained in it remains the same always. So

$$\rho_1 S_1 L_1 = \rho_2 S_2 L_2.$$

Sine the length L is proportional to B/ρ, we have

$$S_1 B_1 = S_2 B_2. \tag{2.60}$$

Thus the magnetic flux over a material surface remains constant during the fluid motion. This is another way of expressing Alfven's frozen theorem.

Remark. One can prove the flux conservation law (2.60) alternatively. We have

$$\begin{aligned}\frac{d}{dt}\int_S \mathbf{B}\cdot d\mathbf{S} &== \int_S \frac{d\mathbf{B}}{dt}\cdot d\mathbf{S} + \int_S \mathbf{B}\cdot\frac{d\mathbf{S}}{dt}\\ &= \int_S \left[\frac{\partial \mathbf{B}}{\partial t} - \nabla\times(\mathbf{v}\times\mathbf{B})\right]\cdot d\mathbf{S}\end{aligned}$$

where we used (2.30) and S is an arbitrary material surface. Using (2.46) in the above gives

$$\frac{d}{dt}\int_S \mathbf{B}\cdot d\mathbf{S} = \frac{c^2}{4\pi\sigma}\int_S \nabla^2\mathbf{B}\cdot d\mathbf{S}.$$

In the limit of $\sigma \to \infty$, the right hand side vanishes, which means that the number of lines of forces passing through the material surface remains unchanged during the convective motion of fluid. In other words, any closed material contour cuts no field line in the course of motion. Any motion of the fluid perpendicular to the field lines carries them without breaking the field lines or having them slipped out, that is, fluid motion across **B** is prohibited. The motion of a low β plasma is drawn along by magnetic field lines where

$$\beta = \frac{p}{B^2/8\pi} = \frac{\textit{thermal kinetic energy density}}{\textit{magnetic energy density}}.$$

In high β plasma, the magnetic field line is convected by the fluid motion. Putting it another way, in high β (low β) plasma the disturbances are carried by the sound wave (Alfven wave). Since any vector **A** can be resolved along two perpendicular directions as

$$\mathbf{A} = \mathbf{A}_\parallel + \mathbf{A}_\perp = (\hat{\mathbf{e}}\cdot\mathbf{A})\hat{\mathbf{e}} + \hat{\mathbf{e}}\times(\mathbf{A}\times\hat{\mathbf{e}}),$$

the velocity component $\mathbf{v}_\perp$ perpendicular to the magnetic field is written as

$$\mathbf{v}_\perp = \frac{\mathbf{B} \times (\mathbf{v} \times \mathbf{B})}{B^2}.$$

Substituting

$$\mathbf{E} = -\frac{1}{c}\mathbf{v} \times \mathbf{B} \tag{2.61}$$

gives

$$\mathbf{v}_\perp = \frac{c\mathbf{E} \times \mathbf{B}}{B^2} \equiv \mathbf{v}_D \tag{2.62}$$

which is called $\mathbf{E} \times \mathbf{B}$ drift velocity.

Remark. Alfven's frozen theorem has a striking similarity with Kelvin–Helmholtz theorem of vorticity flux conservation in inviscous fluid, which states that the circulation or the vorticity flux

$$K = \oint_C \mathbf{v} \cdot d\mathbf{x} = \int_S \vec{\omega} \cdot d\mathbf{S} = constant$$

where the subscript C indicates *circulation* and $\vec{\omega} = \nabla \times \mathbf{v}$ is vorticity.

When the electrical conductivity is finite, the magnetic flux passing through a material surface is not conserved, but slippage of field lines out of the contour takes place. There is a dimensionless number, analogous to Reynolds number in viscous fluid, which approximately determines whether the behavior of magnetic field line in the fluid is frozen or slipping. The fractional change of magnetic flux (Φ) in a time interval τ is

$$\frac{\frac{d\Phi}{dt}\tau}{\Phi} = O\left(\frac{c^2}{4\pi\sigma VL}\right)$$

with characteristic velocity V and length scale L. The magnetic Reynolds number R_m is defined by

$$R_m = \frac{4\pi\sigma VL}{c^2} \equiv \frac{VL}{\eta}$$

where $\eta = c^2/4\pi\sigma$ is called magnetic viscosity (note the similarity to the kinematic viscosity in Reynolds number of a viscous fluid). When $R_m \gg 1$, the gross behavior of magnetic field lines can be viewed as frozen. In the opposite limit of $R_m \ll 1$, the second term of (2.56) can be neglected, so we have a diffusion equation for the magnetic field

$$\frac{\partial \mathbf{B}}{\partial t} = \eta \nabla^2 \mathbf{B}$$

whose quick solution can be written as

$$\mathbf{B} \sim e^{-\eta t/L^2}$$

which indicates that the magnetic field in a medium of finite conductivity suffers from dissipation with a time scale of L^2/η.

2.10. Fluid description of collisionless plasma in a strong magnetic field

A plasma can be described by MHD fluid equations when the collision time is much shorter than any other characteristic times involved. If inter-particle collisions are frequent, the plasma particles tend to stick together and fluid description as a single fluid is justified. In fact, fluid equations can be derived from the Boltzmann equation assuming that the distribution function is nearly Maxwellian, which in turn requires frequent collisions in order for the fluid to be established as a local Maxwellian. In many important plasmas, the collision time is so long that collisions should be ignored. It would appear that for such collisionless plasma a fluid theory is not appropriate. However, in a plasma with a strong magnetic field such that the ion cyclotron period is much shorter than any other time scale of variation, the required localization can be provided by the magnetic field rather than collisions. The dominant perpendicular motion in a strong mgnetic field is the $\mathbf{E} \times \mathbf{B}$ drift which is the same for both ions and electrons. Now the pressure is anisotropic (the mean thermal kinetic energy per volume will be different for parallel and perpendicular directions), instead of being $P_{ij} = p\delta_{ij}$. We start from the collisionless Boltzmann equation

$$\frac{\partial f}{\partial t} + \mathbf{v} \cdot \frac{\partial f}{\partial \mathbf{r}} + \frac{e}{m}\left(\mathbf{E} + \frac{1}{c}\mathbf{v} \times \mathbf{B}\right) \cdot \frac{\partial f}{\partial \mathbf{v}} = 0$$

for a particle species. The above Boltzmann equation can be expanded in powers of e/m. See Reference: G. F. Chew, M. L. Goldberger, and F. E. Low, *The Boltzmann equation and the one-fluid hydodynamic equations in the absence of particle collisions*, Proc. Royal Soc. London **236A** 112. In the zero order one has

$$\left(\mathbf{E} + \frac{1}{c}\mathbf{v} \times \mathbf{B}\right) \cdot \frac{\partial f_0}{\partial \mathbf{v}} = 0. \tag{2.63}$$

Now introduce the new variable $\mathbf{u}$ such that

$$\mathbf{u} = \mathbf{v} - \mathbf{v}_d = \mathbf{v} - \frac{c\,\mathbf{E} \times \mathbf{B}}{B^2}.$$

Then (2.63) reduces to

$$(\mathbf{u} \times \mathbf{B}) \cdot \frac{\partial f_0}{\partial \mathbf{v}} = 0, \quad or \quad (\mathbf{u}_\perp \times \mathbf{B}) \cdot \frac{\partial f_0}{\partial \mathbf{v}} = 0 \tag{2.64}$$

where we assumed $\mathbf{E}_\parallel = 0$. Taking $\mathbf{B} = \hat{\mathbf{z}}B$, the above equation reads

$$u_y \frac{\partial f_0}{\partial v_x} - u_x \frac{\partial f_0}{\partial v_y} = 0$$

whose characteristic equations (see Section 5.2) are

$$\frac{dv_x}{u_y} = -\frac{dv_y}{u_x} = \frac{df_0}{0}.$$

We have a constant of motion: $u_x^2 + u_y^2 = A$ (note that $dv_x = du_x, dv_y = du_y$) and f_0 is a function of this constant, that is, $f_0 = f_0(A)$. Therefore we have

$$f_0(\mathbf{r}, \mathbf{v}, t) = f_0(u_\perp^2, v_\parallel, \mathbf{r}, t). \tag{2.65}$$

Exercise. Show that the distribution function (2.65) gives for the average fluid velocity (1.39)

$$\mathbf{V}(\mathbf{r}, t) = \mathbf{v}_d.$$

Use the anisotropic Maxwellian of the form $f_0 \sim e^{-\frac{u_\perp^2}{\alpha}} e^{-\frac{v_\parallel}{\beta}}$.

We can show that (2.65) makes the pressure tensor (1.46) anisotropic; the relation $P_{ij} = p\delta_{ij}$ does not hold. If we introduce the unit vector $\hat{\mathbf{b}} = \frac{\mathbf{B}}{B} = (0, 0, 1)$ (taking the third coordinate as the z-direction), the anisotropic pressure tensor which is obtained from (2.65) takes the form

$$P_{ij} = p_\perp \, \delta_{ij} + (p_\parallel - p_\perp) b_i b_j \tag{2.66}$$

$$\mathbf{P_{ij}} = \begin{pmatrix} p_\perp & 0 & 0 \\ 0 & p_\perp & 0 \\ 0 & 0 & p_\parallel \end{pmatrix}.$$

Therefore one obtains

$$\frac{\partial P_{ij}}{\partial x_j} = \frac{\partial p_\perp}{\partial x_i} + \frac{\partial}{\partial x_j}\left[(p_\parallel - p_\perp)\frac{B_i B_j}{B^2}\right]$$

$$= \nabla p_\perp + (\mathbf{B} \cdot \nabla)\left[\frac{(p_\parallel - p_\perp)\mathbf{B}}{B^2}\right].$$

Thus the momentum equation (cf. (2.42)) takes the form

$$\rho \frac{d\mathbf{V}}{dt} = -\nabla\left(p_\perp + \frac{B^2}{8\pi}\right) + (\mathbf{B} \cdot \nabla)\left[\frac{\mathbf{B}}{4\pi} - \frac{(p_\parallel - p_\perp)\mathbf{B}}{B^2}\right]. \tag{2.67}$$

We need the equations for $p_\parallel$ and $p_\perp$. In the original paper by Chew *et al.*, the higher order equations for the distribution function f were solved by equating the two heat fluxes $q_\parallel$ and $q_\perp$ to zero simultaneously (hence the name *double adiabatic theory*). Here we use a little heuristic approach. The pressure p which is the trace of the pressure tensor is the average kinetic energy per unit volume. We use (1.49) and (2.66) with $\mathbf{q} = 0$ and

$$p = nT = p_\parallel + 2p_\perp.$$

Then we have

$$\frac{d}{dt}\left(p_\perp + \frac{p_\parallel}{2}\right) + \left(p_\perp + \frac{p_\parallel}{2}\right)\nabla\cdot\mathbf{V} = -p_\perp\nabla\cdot\mathbf{V} - (p_\parallel - p_\perp)b_i b_j\frac{\partial V_i}{\partial x_j}$$

or

$$\frac{d}{dt}(2p_\perp + p_\parallel) + (4p_\perp + p_\parallel)\nabla\cdot\mathbf{V} + 2(p_\parallel - p_\perp)b_i b_j\frac{\partial V_i}{\partial x_j} = 0.$$

Since $p_\parallel$ and $p_\perp$ should be independent of each other, we can decompose the above equation as

$$\frac{dp_\parallel}{dt} + p_\parallel\nabla\cdot\mathbf{V} + 2p_\parallel b_i b_j\frac{\partial V_i}{\partial x_j} = 0, \tag{2.68}$$

$$\frac{dp_\perp}{dt} + 2p_\perp\nabla\cdot\mathbf{V} - p_\perp b_i b_j\frac{\partial V_i}{\partial x_j} = 0. \tag{2.69}$$

Using

$$\frac{d\mathbf{B}}{dt} = (\mathbf{B}\cdot\nabla)\mathbf{V} - \mathbf{B}\nabla\cdot\mathbf{V} \quad or \quad \frac{d\mathbf{B}}{dt} = (\mathbf{B}\cdot\nabla)\mathbf{V} + \frac{\mathbf{B}}{\rho}\frac{d\rho}{dt}$$

it follows that

$$b_i b_j\frac{\partial V_i}{\partial V_j} = \frac{1}{B^2}\mathbf{B}\cdot(\mathbf{B}\cdot\nabla)\mathbf{V} = \frac{\mathbf{B}}{B^2}\cdot\left(\frac{d\mathbf{B}}{dt} - \frac{\mathbf{B}}{\rho}\frac{d\rho}{dt}\right).$$

Therefore (2.68) and (2.69) become

$$\frac{dp_\parallel}{dt} - \frac{3p_\parallel}{\rho}\frac{d\rho}{dt} + \frac{2p_\parallel}{B}\frac{dB}{dt} = 0,$$

$$\frac{dp_\perp}{dt} - \frac{p_\perp}{\rho}\frac{d\rho}{dt} - \frac{p_\perp}{B}\frac{dB}{dt} = 0,$$

or equivalently,

$$\frac{d}{dt}\left(\frac{p_\parallel B^2}{\rho^3}\right) = 0, \tag{2.70}$$

$$\frac{d}{dt}\left(\frac{p_\perp}{\rho B}\right) = 0. \tag{2.71}$$

Equations (2.70) and (2.71) are known as the double adiabatic equations for a conducting anisotropic fluid in a strong magnetic field.

2.11. Lagrangian description of fluid motion

We have two different methods in describing fluid motion. The *Eulerian* description which has been used in this chapter so far is like the determination of an electromagnetic field in that the field quantities are specified as functions of the spatial coordinate $\mathbf{x}$ and time t. In contrast, *Lagrangian* description chooses to follow the trajectory of each fluid particle which can be expressed as

$$\mathbf{x} = \mathbf{x}(\mathbf{x}_0, t). \tag{2.72}$$

The variable $\mathbf{x}_0$, the Lagrangian coordinate, can be regarded as individualization parameter of each particle, which is usually taken as the initial position of the particle. But other means of individualization can be employed depending upon the problems at hand. Motion of a continuum comprises all the individual motion of fluid particles, so a study of the motion of a volume may be sometimes insufficient. Lagrangian description remedies this shortcoming, but Eulerian description is more convenient and used predominantly in fluid mechanics. In Lagrangian description, the gradient of field quantities are not available immediately, but it has some advantage when one wants to put the mechanical framework into the variational formulation.

The function $\mathbf{x}$ in (2.72) is a single-valued function according to the law of mechanics. So the Jacobian determinant

$$\mathbf{J} = \left|\frac{\partial \mathbf{x_i}}{\partial \mathbf{x_{0j}}}\right| = \begin{vmatrix} \dfrac{\partial x_1}{\partial x_{01}} & \dfrac{\partial x_1}{\partial x_{02}} & \dfrac{\partial x_1}{\partial x_{03}} \\[2ex] \dfrac{\partial x_2}{\partial x_{01}} & \dfrac{\partial x_2}{\partial x_{02}} & \dfrac{\partial x_2}{\partial x_{03}} \\[2ex] \dfrac{\partial x_3}{\partial x_{01}} & \dfrac{\partial x_3}{\partial x_{02}} & \dfrac{\partial x_3}{\partial x_{03}} \end{vmatrix}$$

is not zero, i.e., (2.72) can be solved for $\mathbf{x}_0$ in the form of single-valued continuous function:

$$\mathbf{x}_0 = \mathbf{x}_0(\mathbf{x}, t). \tag{2.73}$$

Mathematically, the communication between the Eulerian coordinates $(\mathbf{x}, t)$ and the Lagrangian coordinates $(\mathbf{x}_0, \tau)$ is effected by the transformations

$$t = t(\mathbf{x}_0, \tau), \quad \mathbf{x} = \mathbf{x}(\mathbf{x}_0, \tau), \tag{2.74}$$

$$\tau = \tau(\mathbf{x}, t), \quad \mathbf{x}_0 = \mathbf{x}_0(\mathbf{x}, t). \tag{2.75}$$

Specifically, the appropriate changes of variables are $t = \tau$,

$$\left.\frac{\partial \mathbf{x}}{\partial \tau}\right|_{\mathbf{x}_0} = \mathbf{v}(\mathbf{x}_0, \tau) \tag{2.76}$$

$$\mathbf{x}(\mathbf{x}_0, \tau = 0) = \mathbf{x}_0. \tag{2.77}$$

The time transformation is identity transformation, but it is more convenient to introduce the Lagrangian time τ since $\partial/\partial\tau$ is not same as $\partial/\partial t$. The spatial transformation is defined by differential equation (2.76) with (2.77) serving as the initial condition. As a matter of fact, the functional relations between the Eulerian and the Lagrangian coordinates as expressed in (2.74) and (2.75) are determined only after solving the law of mechanics.

Integrating (2.76) gives

$$\mathbf{x}(\mathbf{x}_0, \tau) = \mathbf{x}_0 + \int_0^\tau \mathbf{v}(\mathbf{x}_0, \tau')d\tau'. \tag{2.78}$$

By the chain rule of differentiation, we have

$$\frac{\partial}{\partial \tau} = \frac{\partial t}{\partial \tau}\frac{\partial}{\partial t} + \frac{\partial x_i}{\partial \tau}\frac{\partial}{\partial x_i} = \frac{\partial}{\partial t} + \mathbf{v}\cdot\nabla \tag{2.79}$$

$$\frac{\partial}{\partial t} = \frac{\partial \tau}{\partial t}\frac{\partial}{\partial \tau} + \frac{\partial x_{0i}}{\partial t}\frac{\partial}{\partial x_{0i}} = \frac{\partial}{\partial \tau} + \frac{\partial x_{0i}}{\partial t}\frac{\partial}{\partial x_{0i}}. \tag{2.80}$$

The Lagrangian time derivative is the convective derivative in Eulerian description. A simple application of (2.79) is the particle acceleration,

$$\mathbf{a} = \left.\frac{\partial \mathbf{v}}{\partial \tau}\right|_{\mathbf{x}_0} = \frac{\partial \mathbf{v}}{\partial t} + \mathbf{v}\cdot\nabla\mathbf{v}. \tag{2.81}$$

For spatial coordinates, we have

$$\frac{\partial}{\partial x_{0i}} = \frac{\partial t}{\partial x_{0i}}\frac{\partial}{\partial t} + \frac{\partial x_j}{\partial x_{0i}}\frac{\partial}{\partial x_j} = \frac{\partial x_j}{\partial x_{0i}}\frac{\partial}{\partial x_j} \tag{2.82}$$

$$\frac{\partial}{\partial x_i} = \frac{\partial \tau}{\partial x_i}\frac{\partial}{\partial \tau} + \frac{\partial x_{0j}}{\partial x_i}\frac{\partial}{\partial x_{0j}} = \frac{\partial x_{0j}}{\partial x_i}\frac{\partial}{\partial x_{0j}}. \tag{2.83}$$

Let us first consider the case of one-dimensional Eulerian equations

$$\frac{\partial \rho}{\partial t} + \frac{\partial}{\partial x}(\rho v) = 0, \tag{2.84}$$

$$\frac{\partial v}{\partial t} + v\frac{\partial v}{\partial x} = -\frac{1}{\rho}\frac{\partial p}{\partial x} + F. \tag{2.85}$$

Our task is to write the above two equations in terms of Lagrangian variables $(\tau, \mathbf{x}_0)$. Equation (2.84) is written as

$$\frac{\partial \rho}{\partial t} + v\frac{\partial \rho}{\partial x} + \rho\frac{\partial v}{\partial x} = 0. \tag{2.86}$$

Owing to (2.79), the first two terms add to $\partial\rho/\partial\tau$. Now

$$\begin{aligned}\frac{\partial v}{\partial x} &= \frac{\partial v}{\partial x_0}\frac{\partial x_0}{\partial x} = \frac{1}{J}\frac{\partial v}{\partial x_0}\\ &= \frac{1}{J}\frac{\partial}{\partial x_0}\frac{\partial x}{\partial \tau} = \frac{1}{J}\frac{\partial}{\partial \tau}\frac{\partial x}{\partial x_0} = \frac{1}{J}\frac{\partial J}{\partial \tau}\end{aligned}$$

where we used (2.76), $J = \frac{\partial x}{\partial x_0}$, $\frac{1}{J} = \frac{\partial x_0}{\partial x}$ (this is true only in one-dimensional case). Using the above equation in (2.86) gives

$$\frac{\partial \rho}{\partial \tau} + \frac{\rho}{J}\frac{\partial J}{\partial \tau} = 0. \tag{2.87}$$

This can be immediately integrated to yield

$$\rho J = const. = \rho(\mathbf{x}_0, 0) \equiv \rho_0. \tag{2.88}$$

Equation (2.88) is physically obvious. Consider a fluid mass contained in a small volume of d^3x centered about $\mathbf{x}(\mathbf{x}_0, t)$. Let $\rho(\mathbf{x}, t)$ be the fluid density. Since the mass was contained initially in a volume d^3x_0 centered about the position $\mathbf{x}_0$, we should have $\rho(\mathbf{x}, t)d^3x = \rho(\mathbf{x}_0, 0)d^3x_0$. Since the Jacobian is the ratio of the two corresponding volume elements, $J = \frac{d^3x}{d^3x_0}$, we have (2.88). In particular, $J = 1$ for incompressible fluid.

The momentum equation (2.85) can be converted in terms of Lagrangian variables as follows. The left hand side is simply $\partial v/\partial\tau$. Now

$$\frac{1}{\rho}\frac{\partial p}{\partial x} = \frac{1}{\rho}\frac{\partial p}{\partial x_0}\frac{\partial x_0}{\partial x} = \frac{1}{J\rho}\frac{\partial p}{\partial x_0} = \frac{1}{\rho_0}\frac{\partial p}{\partial x_0}.$$

Therefore the momentum equation (2.85) is written in Lagrangian coordinates as

$$\frac{\partial v}{\partial \tau} = -\frac{1}{\rho_0}\frac{\partial p}{\partial x_0} + F. \tag{2.89}$$

- **Three-dimensional treatment.** We have

$$\nabla \cdot \mathbf{v} = \frac{\partial x_{0j}}{\partial x_i} \frac{\partial v_i}{\partial x_{0j}}.$$

Using (2.76) in the above gives

$$\nabla \cdot \mathbf{v} = \frac{\partial x_{0j}}{\partial x_i} \frac{\partial}{\partial x_{0j}} \frac{\partial x_i}{\partial \tau} = \frac{\partial x_{0j}}{\partial x_i} \frac{\partial}{\partial \tau} \frac{\partial x_i}{\partial x_{0j}}. \tag{2.90}$$

The matrices $(\partial x_{0j}/\partial x_i)$ and $(\partial x_i/\partial x_{0j})$ are inverse to each other since

$$\frac{\partial x_{0j}}{\partial x_i} \frac{\partial x_i}{\partial x_{0l}} = \frac{\partial x_{oj}}{\partial x_{0l}} = \delta_{jl}. \tag{2.91}$$

When a matrix $A_{ij}(\lambda)$ is an $n \times n$ matrix whose elements are functions of λ, we have the following identity (See Tensor Analysis by Sokolnikoff, I.S.)

$$\frac{\partial A_{ij}}{\partial \lambda} A_{ji}^{-1} = \frac{1}{\mid A \mid} \frac{\partial \mid A \mid}{\partial \lambda}.$$

Then (2.90) immediately yields

$$\nabla \cdot \mathbf{v} = \frac{1}{J} \frac{\partial J}{\partial \tau}. \tag{2.92}$$

Equation (2.92) is an important relation for the divergence of velocity field which gives, together with (2.78), fundamental connection between the Lagrangian and Eulerian descriptions. The Eulerian form of continuity equation

$$\frac{d\rho}{dt} + \rho \nabla \cdot \mathbf{v} = 0$$

immediately changes into (2.87) upon using (2.92) and $\frac{d}{dt} \to \frac{\partial}{\partial \tau}$.

2.12. Lagrangian operator

In this section, we transform the operator $\nabla_{\mathbf{r}}$ into Lagrangian coordinates. We have

$$\frac{\partial x_{0i}}{\partial x_j} \frac{\partial x_j}{\partial x_{0l}} = \delta_{il}, \quad \textit{or} \quad A\,B = I: \textit{ unit matrix}$$

where

$$\mathbf{A} = \mathbf{a_{ij}} = \begin{pmatrix} \frac{\partial x_0}{\partial x} & \frac{\partial x_0}{\partial y} & \frac{\partial x_0}{\partial z} \\ \frac{\partial y_0}{\partial x} & \frac{\partial y_0}{\partial y} & \frac{\partial y_0}{\partial z} \\ \frac{\partial z_0}{\partial x} & \frac{\partial z_0}{\partial y} & \frac{\partial z_0}{\partial z} \end{pmatrix},$$

$$\mathbf{B} = \mathbf{b_{ij}} = \begin{pmatrix} \frac{\partial x}{\partial x_0} & \frac{\partial x}{\partial y_0} & \frac{\partial x}{\partial z_0} \\ \frac{\partial y}{\partial x_0} & \frac{\partial y}{\partial y_0} & \frac{\partial y}{\partial z_0} \\ \frac{\partial z}{\partial x_0} & \frac{\partial z}{\partial y_0} & \frac{\partial z}{\partial z_0} \end{pmatrix}.$$

Since $A = B^{-1}$, we have

$$a_{11} = \frac{\partial x_0}{\partial x} = \frac{cofactor\ of\ b_{11}}{J} = \frac{1}{J}\left(\frac{\partial y}{\partial y_0}\frac{\partial z}{\partial z_0} - \frac{\partial y}{\partial z_0}\frac{\partial z}{\partial y_0}\right),$$

$$a_{21} = \frac{\partial y_0}{\partial x} = \frac{cofactor\ of\ b_{12}}{J} = -\frac{1}{J}\left(\frac{\partial y}{\partial x_0}\frac{\partial z}{\partial z_0} - \frac{\partial y}{\partial z_0}\frac{\partial z}{\partial x_0}\right),$$

$$a_{31} = \frac{\partial z_0}{\partial x} = \frac{cofactor\ of\ b_{13}}{J} = \frac{1}{J}\left(\frac{\partial y}{\partial x_0}\frac{\partial z}{\partial y_0} - \frac{\partial y}{\partial y_0}\frac{\partial z}{\partial x_0}\right).$$

Therefore we obtain

$$J\,\frac{\partial}{\partial x} = J\left(\frac{\partial x_0}{\partial x}\frac{\partial}{\partial x_0} + \frac{\partial y_0}{\partial x}\frac{\partial}{\partial y_0} + \frac{\partial z_0}{\partial x}\frac{\partial}{\partial z_0}\right) = \nabla_0 y \times \nabla_0 z \cdot \nabla_0 \tag{2.93}$$

$$= \begin{vmatrix} \frac{\partial}{\partial x_0} & \frac{\partial}{\partial y_0} & \frac{\partial}{\partial z_0} \\ \frac{\partial y}{\partial x_0} & \frac{\partial y}{\partial y_0} & \frac{\partial y}{\partial z_0} \\ \frac{\partial z}{\partial x_0} & \frac{\partial z}{\partial y_0} & \frac{\partial z}{\partial z_0} \end{vmatrix},$$

$$J\,\frac{\partial}{\partial y} = J\left(\frac{\partial x_0}{\partial y}\frac{\partial}{\partial x_0} + \frac{\partial y_0}{\partial y}\frac{\partial}{\partial y_0} + \frac{\partial z_0}{\partial y}\frac{\partial}{\partial z_0}\right) = \nabla_0 z \times \nabla_0 x \cdot \nabla_0 \tag{2.94}$$

$$= \begin{vmatrix} \frac{\partial}{\partial x_0} & \frac{\partial}{\partial y_0} & \frac{\partial}{\partial z_0} \\ \frac{\partial z}{\partial x_0} & \frac{\partial z}{\partial y_0} & \frac{\partial z}{\partial z_0} \\ \frac{\partial x}{\partial x_0} & \frac{\partial x}{\partial y_0} & \frac{\partial x}{\partial z_0} \end{vmatrix},$$

$$J\frac{\partial}{\partial z} = J\left(\frac{\partial x_0}{\partial z}\frac{\partial}{\partial x_0} + \frac{\partial y_0}{\partial z}\frac{\partial}{\partial y_0} + \frac{\partial z_0}{\partial z}\frac{\partial}{\partial z_0}\right) = \nabla_0 x \times \nabla_0 y \cdot \nabla_0 \tag{2.95}$$

$$= \begin{vmatrix} \frac{\partial}{\partial x_0} & \frac{\partial}{\partial y_0} & \frac{\partial}{\partial z_0} \\ \frac{\partial x}{\partial x_0} & \frac{\partial x}{\partial y_0} & \frac{\partial x}{\partial z_0} \\ \frac{\partial y}{\partial x_0} & \frac{\partial y}{\partial y_0} & \frac{\partial y}{\partial z_0} \end{vmatrix}.$$

Thus we can write the Lagrangian expression of the Eulerian gradient operator as follows:

$$\nabla_{\mathbf{r}} = \frac{\partial}{\partial \mathbf{r}} = \frac{1}{J}\,(\nabla_0\, y \times \nabla_0\, z \cdot \nabla_0,\ \nabla_0\, z \times \nabla_0\, x \cdot \nabla_0,\ \nabla_0\, x \times \nabla_0\, y \cdot \nabla_0). \tag{2.96}$$

This equation is useful in changing the Eulerian momentum equation into the Lagrangian form.

Exercise. Using (2.93)–(2.95), prove (2.92). Hint. Use (2.76) and formula for differentiating determinant.

Example. Prove (2.23) using Lagrangian variable. We have

$$dx_i = \frac{\partial x_i}{\partial x_{0j}} dx_{0j}$$

$$\frac{d}{dt}dx_i = \frac{d}{dt}\frac{\partial x_i}{\partial x_{0j}}\,dx_{0j} = \frac{\partial}{\partial x_{0j}}\frac{dx_i}{dt}dx_{0j} = dx_{0j}\frac{\partial\, v_i}{\partial x_{0j}}$$

$$dx_i = \frac{\partial x_i}{\partial x_{0j}}\,dx_{0j},\quad \frac{\partial}{\partial x_i} = \frac{\partial x_{0l}}{\partial x_i}\frac{\partial}{\partial x_{0l}}$$

$$dx_i\,\frac{\partial}{\partial x_i} = \frac{\partial x_i}{\partial x_{0j}}\frac{\partial x_{0l}}{\partial x_i}dx_{0j}\,\frac{\partial}{\partial x_{0l}} = dx_{0l}\,\frac{\partial}{\partial x_{0l}},\ \textit{so}$$

$$\frac{d}{dt}dx_i = dx_j\frac{\partial v_i}{\partial x_j},\ \textit{or}\ \frac{d}{dt}d\mathbf{x} = (d\mathbf{x}\cdot\nabla)\mathbf{v}.$$

Remark. In one-dimensional case, the relations between Eulerian and Lagrangian variables read

$$\tau = t,\quad x_0 = x - \int_0^\tau v(x_0,\tau')d\tau' \quad (\textit{cf}.\ (2.78)) \tag{2.97}$$

$$x = x_0 + \int_0^\tau v(x_0, \tau')d\tau'. \tag{2.98}$$

Note that the second relation is an implicit equation for x_0. We have

$$\frac{\partial}{\partial x} = \frac{\partial}{\partial \tau}\frac{\partial \tau}{\partial x} + \frac{\partial x_0}{\partial x}\frac{\partial}{\partial x_0} = \frac{\partial x_0}{\partial x}\frac{\partial}{\partial x_0}$$

$$\frac{\partial x_0}{\partial x} = 1 - \int_0^\tau \frac{\partial v(x_0, \tau')}{\partial x_0}\, d\tau'\, \frac{\partial x_0}{\partial x}$$

$$\frac{\partial x}{\partial x_0} = 1 + \int_0^\tau \frac{\partial v(x_0, \tau')}{\partial x_0}\, d\tau' = \left(\frac{\partial x_0}{\partial x}\right)^{-1}.$$

The last equality is valid in the one-dimensional case. Therefore,

$$\frac{\partial}{\partial x} = \left[1 + \int_0^\tau \frac{\partial v(x_0, \tau')}{\partial x_0} d\tau'\right]^{-1} \frac{\partial}{\partial x_0}, \tag{2.99}$$

$$\frac{\partial}{\partial t} = \frac{\partial x_0}{\partial t}\frac{\partial}{\partial x_0} + \frac{\partial \tau}{\partial t}\frac{\partial}{\partial \tau} = \frac{\partial x_0}{\partial t}\frac{\partial}{\partial x_0} + \frac{\partial}{\partial \tau}.$$

Differentiating (2.98) with respect to t (note $\partial x/\partial t = 0$),

$$0 = \frac{\partial x_0}{\partial t} + \int_0^\tau d\tau' \frac{\partial v}{\partial x_0}\frac{\partial x_0}{\partial t} + v(x_0, \tau)$$

from which one obtains

$$\frac{\partial x_0}{\partial t} = -v(x_0, \tau)\left[1 + \int_0^\tau \frac{\partial v(x_0, \tau')}{\partial x_0} d\tau'\right]^{-1},$$

$$\frac{\partial}{\partial t} = \frac{\partial}{\partial \tau} - v(x_0, \tau)\left[1 + \int_0^\tau \frac{\partial v(x_0, \tau')}{\partial x_0} d\tau'\right]^{-1} \frac{\partial}{\partial x_0}. \tag{2.100}$$

Note that $v(x_0, \tau)$ plays the crucial role in the communication between the two descriptions.

2.13. Waves in ideal MHD flow

We now seek solutions of the *ideal* MHD equations for small amplitude linear waves. The '*ideal*' means that we neglect the dissipation, so we put the electrical conductivity σ to infinity in Ohm's law (see (2.55)). The MHD waves have frequency range $\omega \ll \omega_{ci}$, the ion cyclotron frequency. The governing MHD equations are summarized in (2.53)–(2.57) which we rewrite:

$$\frac{\partial \rho}{\partial t} + \nabla \cdot (\rho \mathbf{v}) = 0 \tag{2.101}$$

$$\rho \frac{d\mathbf{v}}{dt} = -\nabla p + \frac{1}{c}\mathbf{J} \times \mathbf{B} \tag{2.102}$$

$$\mathbf{E} + \frac{1}{c}\mathbf{v} \times \mathbf{B} = 0 \tag{2.103}$$

$$\nabla \times \mathbf{E} = -\frac{1}{c}\frac{\partial \mathbf{B}}{\partial t} \tag{2.104}$$

$$\nabla \times \mathbf{B} = \frac{4\pi}{c}\mathbf{J} \tag{2.105}$$

$$p\rho^{-\gamma} = const. \tag{2.106}$$

where we assumed that the change of state is adiabatic (Eq. (2.106)), and neglected the high frequency displacement current in (2.105) because we are interested in the low frequency domain. We can eliminate $\mathbf{E}$ from (2.103) and (2.104) to write

$$\frac{\partial \mathbf{B}}{\partial t} = \nabla \times (\mathbf{v} \times \mathbf{B}). \tag{2.107}$$

In the equilibrium state, the fluid is at rest. The plasma is assumed to be magnetized by a static magnetic field $\mathbf{B}_0$. Denoting perturbations associated with the wave by subscript 1, the zero order equilibrium quantities by the subscript 0, we write

$$\mathbf{B} = \mathbf{B}_0 + \mathbf{B}_1(\mathbf{r}, t), \quad \rho = \rho_0 + \rho_1(\mathbf{r}, t),$$

$$\mathbf{v} = \mathbf{v}_1(\mathbf{r}, t), \quad p = p_0 + p_1(\mathbf{r}, t).$$

Linearizing the governing equations with respect to the perturbed quantities gives

$$\frac{\partial \rho_1}{\partial t} + \rho_0 \nabla \cdot \mathbf{v}_1 = 0 \tag{2.108}$$

$$\rho_0 \frac{\partial \mathbf{v}_1}{\partial t} + v_s^2 \nabla \rho_1 + \frac{\mathbf{B}_0}{4\pi} \times (\nabla \times \mathbf{B}_1) = 0 \tag{2.109}$$

$$\frac{\partial \mathbf{B}_1}{\partial t} - \nabla \times (\mathbf{v}_1 \times \mathbf{B}_0) = 0 \tag{2.110}$$

where $v_s^2 = \gamma p_0/\rho_0$ is the squared speed of sound. By eliminating ρ_1 and $\mathbf{B}_1$, these equations can be arranged in the form

$$\frac{\partial^2 \mathbf{v}_1}{\partial t^2} - v_s^2 \nabla\nabla \cdot \mathbf{v}_1 + \mathbf{v}_A \times \nabla \times [\nabla \times (\mathbf{v}_1 \times \mathbf{v}_A)] = 0 \tag{2.111}$$

where we introduced the Alfven velocity

$$\mathbf{v}_A = \frac{\mathbf{B}_0}{\sqrt{4\pi\rho_0}}. \tag{2.112}$$

Assuming a plane wave solution for $\mathbf{v}_1$ with wave vector $\mathbf{k}$ and frequency ω, $\mathbf{v}_1(\mathbf{r},t) = \mathbf{v}_1\, e^{i\mathbf{k}\cdot\mathbf{r}-i\omega t}$, (2.111) becomes

$$\begin{aligned}\mathbf{v}_1[-\omega^2 + (\mathbf{k}\cdot\mathbf{v}_A)^2] + \mathbf{k}[(v_s^2+v_A^2)\mathbf{k}\cdot\mathbf{v}_1 - (\mathbf{k}\cdot\mathbf{v}_A)(\mathbf{v}_1\cdot\mathbf{v}_A)]\\ -\mathbf{v}_A(\mathbf{k}\cdot\mathbf{v}_1)(\mathbf{k}\cdot\mathbf{v}_A) = 0.\end{aligned} \tag{2.113}$$

Let the coordinate axis be such that

$$\mathbf{B}_0 = \hat{\mathbf{z}}B_0, \quad \mathbf{k} = (k_x, 0, k_z).$$

Then (2.113) is decomposed into the form

$$v_{1x}(-\omega^2 + k^2v_A^2 + k_x^2v_s^2) + v_{1z}k_xk_zv_s^2 = 0 \tag{2.114}$$

$$v_{1y}(-\omega^2 + k_z^2v_A^2) = 0 \tag{2.115}$$

$$v_{1x}k_xk_zv_s^2 + v_{1z}(-\omega^2 + k_z^2v_s^2) = 0. \tag{2.116}$$

Putting the determinant of the coefficients to zero, we can obtain the normal modes of MHD wave:

$$(\omega^2 - k_z^2v_A^2)\Big[(\omega^2 - k_z^2v_s^2)(-\omega^2 + k^2v_A^2 + k_x^2v_s^2) + v_s^4k_x^2k_z^2\Big] = 0. \tag{2.117}$$

One immediate root is

$$\omega^2 = k_z^2v_A^2 \quad (Alfven\ \ wave). \tag{2.118}$$

The second factor in (2.115) is arranged in the form

$$\omega^4 - \omega^2k^2(v_s^2+v_A^2) + k^2k_z^2v_s^2v_A^2 = 0 \tag{2.119}$$

from which we find

$$\omega_\pm^2 = \frac{k^2}{2}\left[v_s^2 + v_A^2 \pm \sqrt{(v_s^2+v_A^2)^2 - 4v_s^2v_A^2cos^2\theta}\right]$$

$$(magnetosonic\ waves) \tag{2.120}$$

where θ is the angle between the z-axis and $\mathbf{k}$ vector.

• **Conductivity tensor.** It is well-known that MHD equations are suitable for describing the low frequency behavior of a plasma in the region of the frequency much smaller than the ion-cyclotron frequency($\omega \ll \omega_{ci}$). The typical MHD normal modes are Alfven wave and magnetosonic waves as we saw in the last section. In the MHD description of plasma, the magnetic field

$\mathbf{B}$ plays the primary role with the wave electric field $\mathbf{E}$ entirely eliminated as is seen in (2.107). Thus, the MHD field is described by three vectors, $\mathbf{B}$, $\mathbf{J}$, and $\mathbf{v}$. Adopting this view, we lose the sight of the electric field $\mathbf{E}$. The formulation centered about the magnetic field leaves the impression that MHD is somewhat detached from general plasma dynamics, and it appears that MHD is an independent genre isolated from the main body of plasma dynamics. However, the connection between plasma dynamics and MHD is not weak, and it can be demonstrated unequivocally by deriving the MHD normal modes from the two-fluid equations (see next section). In this section, to enlighten the unnegligible role of the electric field, we derive the conductivity tensor of ideal MHD plasma by eliminating $\mathbf{B}$. To begin with, we write the linearized ideal (dissipationless) MHD equations

$$\frac{\partial \rho}{\partial t} + \rho_0 \nabla \cdot \mathbf{v} = 0 \tag{2.121}$$

$$\rho_0 \frac{\partial \mathbf{v}}{\partial t} = -\nabla p + \frac{1}{c}\mathbf{J} \times \mathbf{B}_0 \tag{2.122}$$

$$\mathbf{E} + \frac{1}{c}\mathbf{v} \times \mathbf{B}_0 = 0 \tag{2.123}$$

$$\nabla \times \mathbf{E} = -\frac{1}{c}\frac{\partial \mathbf{B}}{\partial t} \tag{2.124}$$

$$\nabla \times \mathbf{B} = \frac{4\pi}{c}\mathbf{J} \tag{2.125}$$

$$p\rho^{-\gamma} = const; \quad p = \frac{\gamma p_0}{\rho_0}\varrho. \tag{2.126}$$

We eliminate $\mathbf{B}$ in (2.124) and (2.125), and write in terms of Fourier variables:

$$k^2\mathbf{E} - \mathbf{k}(\mathbf{k} \cdot \mathbf{E}) = \frac{4\pi i\omega}{c^2}\mathbf{J}. \tag{2.127}$$

Equations (2.121), (2.122), and (2.126) can be arranged in the form

$$\mathbf{v} = \frac{v_s^2}{\omega^2}\,\mathbf{k}\,\mathbf{k} \cdot \mathbf{v} + i\,\frac{\mathbf{J} \times \mathbf{B}_0}{c\,\omega\rho_0} \quad \left(v_s^2 = \frac{\gamma p_0}{\rho_0}\right), \tag{2.128}$$

which yields in turn

$$\mathbf{v} = i\,\frac{\mathbf{J} \times \mathbf{B}_0}{c\,\omega\rho_0} + \frac{i}{c\,\omega\rho_0}\,\frac{v_s^2}{(\omega^2 - k^2v_s^2)}\,\mathbf{k}\;\mathbf{J} \cdot (\mathbf{B}_0 \times \mathbf{k}). \tag{2.129}$$

We set up the coordinate axis in such a way that $\mathbf{B}_0 = \hat{\mathbf{z}}B_0$, $\mathbf{k} = \hat{\mathbf{x}}k_x + \hat{\mathbf{z}}k_z$. The latter designation of the wave vector $\mathbf{k}$ is not a loss of generality but a useful specification, barring infiltration of unnecessary terms. Using (2.129) in (2.123) gives

$$\mathbf{E} = \frac{B_0^2}{ic^2\rho_0\omega}\left[-\mathbf{J} + \hat{\mathbf{z}}\mathbf{J}\cdot\hat{\mathbf{z}} - \frac{v_s^2}{\omega^2 - k^2v_s^2}\,\mathbf{k}\times\hat{\mathbf{z}}\,\mathbf{J}\cdot(\mathbf{k}\times\hat{\mathbf{z}})\right] \quad or \tag{2.130}$$

$$\mathbf{E} = \frac{4\pi i v_A^2}{c^2\omega}\left[\hat{\mathbf{x}}J_x + \hat{\mathbf{y}}J_y\left(1 + \frac{k_x^2v_s^2}{\omega^2 - k^2v_s^2}\right)\right], \quad v_A^2 = \frac{B_0^2}{4\pi\rho_0}. \tag{2.131}$$

A useful relation is $E_z = 0$ as ascertained from (2.123) or (2.130) by dotting the equation by $\hat{\mathbf{z}}$.

- **Dispersion relation.** Equation (2.127) gives in components:

$$\frac{4\pi i\omega}{c^2}J_x = k_z^2E_x, \tag{2.132}$$

$$\frac{4\pi i\omega}{c^2}J_y = k^2E_y, \tag{2.133}$$

$$\frac{4\pi i\omega}{c^2}J_z = -k_xk_zE_x. \tag{2.134}$$

Substituting (2.132) and (2.133) into (2.131) yields the dispersion relations

$$\omega^2 = v_A^2k_z^2, \tag{2.135}$$

$$\omega^4 - \omega^2k^2(v_s^2 + v_A^2) + k^2k_z^2v_s^2v_A^2 = 0, \tag{2.136}$$

which yields two magnetosonic waves

$$\omega_\pm^2 = \frac{k^2}{2}\left[v_s^2 + v_A^2 \pm \sqrt{(v_s^2+v_A^2)^2 - 4v_s^2v_A^2cos^2\theta}\right] \tag{2.137}$$

where θ is the angle between $\mathbf{B}_0$ and $\mathbf{k}$.

- **Conductivity tensor $\boldsymbol{\sigma_{ij}}$.** Inverting (2.131) for $\mathbf{J}$ yields

$$J_i = \sigma_{ij}\;E_j \quad (i,j = x,y) \tag{2.138}$$

$$where \quad \sigma_{\mathbf{ij}} = \frac{c^2\omega}{4\pi i v_A^2}\begin{pmatrix} 1 & 0 \\ 0 & \dfrac{\omega^2 - k^2v_s^2}{\omega^2 - k_z^2v_s^2}\end{pmatrix}.$$

The above conductivity tensor is valid for the magnetosonic waves (not for the Alfven wave).

2.14. Derivation of MHD waves from two-fluid equations

In this section, we derive the MHD waves from two-fluid equations, taking the low frequency approximation, neglecting ω/ω_{ci} as compared with unity, and neglecting also m_e/m_i, the mass ratio, as compared with unity. Thus we indirectly show that the MHD waves are subset of more wide spectrum of plasma waves predicted by two-fluid equations. We start from linearized two-fluid equations for electron (e) and ion (i) species:

$$\mathbf{v}_i \times \overrightarrow{\omega}_{ci} = -i\omega\mathbf{v}_i - \frac{e}{m_i}\mathbf{E} \quad \left(\overrightarrow{\omega}_{ci} = \frac{e\mathbf{B}_0}{m_i c}, \ \overrightarrow{\omega}_{ce} = -\frac{e\mathbf{B}_0}{m_e c}\right), \tag{2.139}$$

$$\mathbf{v}_e \times \overrightarrow{\omega}_{ce} = -i\omega\mathbf{v}_e + \frac{e}{m_e}\mathbf{E} + i\mathbf{k}\beta^2\frac{n_e}{N} \quad \left(\beta^2 = \frac{\gamma_e T_e}{m_e}\right) \tag{2.140}$$

where we assumed that ions are cold. The essential features of our development are not impaired by the assumption of cold ions. The current is

$$\mathbf{J} = Ne(\mathbf{v}_i - \mathbf{v}_e) \tag{2.141}$$

where N is the equilibrium number density of the plasma. The above equations will be combined to derive the expressions that are valid in MHD region by taking the low frequency approximation ($\frac{\omega}{\omega_{ci}} \ll 1$) and by filtering out the high mobility terms ($\frac{m_e}{m_i} \ll 1$).

Let us multiply (2.140) by m_e/m_i and add to (2.139) to obtain

$$\mathbf{J} \times \overrightarrow{\omega_{ci}} = -ieN\omega\mathbf{w} + ie\mathbf{k}v_s^2 n_e, \tag{2.142}$$

$$v_s^2 = \frac{m_e}{m_i}\beta^2 = \frac{\gamma_e T_e}{m_i},$$

$$\mathbf{w} = \mathbf{v}_i + \frac{m_e}{m_i}\mathbf{v}_e, \tag{2.143}$$

where we used $\frac{m_e}{m_i}\overrightarrow{\omega_{ce}} = -\overrightarrow{\omega_{ci}}$. The first approximation to be enforced for the low frequency slow waves in the MHD region is to take $\frac{m_e}{m_i} \to 0$, and we can set $\mathbf{w} \approx \mathbf{v}_i$. Therefore, (2.144) is written.

$$\mathbf{J} \times \overrightarrow{\omega_{ci}} = -ieN\omega\mathbf{v}_i + ie\mathbf{k}v_s^2 n_e. \tag{2.144}$$

We look at the ion equation, (2.139). Because of $\omega \ll \omega_{ci}$, the inertial term can be neglected as compared with the magnetic field term, and we have

$$\mathbf{E} + \frac{1}{c}\mathbf{v}_i \times \mathbf{B}_0 = 0 \tag{2.145}$$

which is the Ohm's law. An immediate consequence of the above equation is $E_z = 0$. This approximate relation greatly simplifies the analysis.

Solving for $\mathbf{v_i}$, we have

$$\mathbf{v}_i = \frac{e}{m_i}\frac{1}{\omega_{ci}^2}\mathbf{E}\times\overrightarrow{\omega_{ci}}. \tag{2.146}$$

Now we use the charge neutrality $n_e \simeq n_i$, which is not only physically acceptable but also mathematically legitimate, because (2.127) gives $\mathbf{k}\cdot\mathbf{J}=0$, and thus $\mathbf{k}\cdot\mathbf{v}_i = \mathbf{k}\cdot\mathbf{v}_e$. Therefore, we have in place of (2.144),

$$\mathbf{J}\times\overrightarrow{\omega_{ci}} = -iNe\omega\mathbf{v}_i + ie\mathbf{k}v_s^2\frac{N}{\omega}\mathbf{k}\cdot\mathbf{v}_i. \tag{2.147}$$

Equations (2.127), (2.147), and the ion equation (2.146) constitute the complete set for the low frequency two-fluid plasma. If we rewrite (2.147) to reveal $\mathbf{B}_0$, it takes the form

$$\mathbf{v}_i = \frac{i\mathbf{J}\times\mathbf{B}_0}{c\omega N m_i} + \mathbf{k}\frac{v_s^2}{\omega^2}\mathbf{k}\cdot\mathbf{v}_i. \tag{2.148}$$

This equation corresponds to (2.128) for ideal MHD plasma. Solving (2.148) for $\mathbf{v}_i$ yields

$$\mathbf{v}_i = \frac{i}{c\omega N m_i}\left(\mathbf{J}\times\mathbf{B}_0 + \frac{v_s^2}{\omega^2 - k^2 v_s^2}\mathbf{k}\mathbf{J}\cdot\mathbf{B}_0\times\mathbf{k}\right). \tag{2.149}$$

Using Eqs. (2.145) and (2.149) yields $\mathbf{E}$ in terms of $\mathbf{J}$:

$$\mathbf{E} = \frac{iB_0^2}{c^2\omega N m_i}\left[\hat{\mathbf{x}}J_x + \hat{\mathbf{y}}J_y\left(1 + \frac{k_x^2 v_s^2}{\omega^2 - k^2 v_s^2}\right)\right]. \tag{2.150}$$

Equation (2.127) with $E_z = 0$ and (2.150) gives the dispersion relations

$$\omega^2 = v_A^2 k_z^2 \quad \left(v_A^2 = \frac{B_0^2}{4\pi N m_i}\right)$$

$$\omega^4 - \omega^2 k^2(v_s^2 + v_A^2) + k^2 k_z^2 v_s^2 v_A^2 = 0.$$

Note that (2.150) is equivalent to (2.131).

Dispersion relation (2.135) indicates that Alfven wave is a wave which propagates along the direction of the zero order magnetic field $\mathbf{B_0}$: $\mathbf{k}_A = \hat{\mathbf{z}}k_z$ (Subscript A denotes Alfven wave). We have per (2.131) $\mathbf{E} \perp \mathbf{k}_A$. Also $\mathbf{B} = \frac{c}{\omega}\mathbf{k}_A \times \mathbf{E}$ per (2.104). Therefore, Alfven wave is a low frequency electromagnetic wave which propagates along the zero order magnetic field. The magnetosonic waves are of mixed mode; partly electromagnetic, partly electrostatic. *Electromagnetic wave* is often called *transverse electromagnetic wave* to emphasize the fact that the electric and magnetic fields are mainly or entirely perpendicular to the direction of the propagation vector. When the electric field is primarily along the direction of wave propagation, the wave is called *electrostatic* or *longitudinal.*

References

Alfven H. and Fälthammar, C. -G. *Cosmical Electrodynamics*, Clarendon, Oxford, 1963.

Chew, G. F., Goldberger, M. L. and Low F. E. The Boltzmann equation and the one-fluid hydromagnetic equations in the absence of particle collisions, *Proc. Royal Soc. London*, **236A** (1956) p. 112.

Landau L. D. and Lifshitz, E. M. *Electrodynamics of Continuous Media*, Pergamon, London, 1966.

Shkarofsky, I. P., Johnston, T. W. and Bachynski, M. P. *The Particle Kinetics of Plasmas*, Addison-Wesley, London, 1966.

Sokolnikoff, I. S. *Tensor Analysis*, Wiley, Toppan, 1964.

Chapter 3

Single particle motion in electric and magnetic fields

At first sight, a plasma is a collection of charged particles moving in electric and/or magnetic fields. If the density is sufficiently rare, the collective effect can be neglected; in the first approximation, the individual particle motion needs to be considered, neglecting the mutual interaction. In cosmic situation, charged particles gyrate around the imbedded magnetic fields in the space. A good example is the *Van Allen Belt* around the earth. The Van Allen Belt may be considered as a collection of gyrating charged particles along the Earth magnetic fields which also move in circumferential direction without mutual interaction; only being subject to the magnetic force due to the Earth magnetic field. In this Chapter, the electric and magnetic fields are assumed to be prescribed, independently of the charged particles. Thus, the problem is purely mechanical but the necessary analysis requires a good deal of mathematical ingenuity. As a result of the study of particle orbits in a given electric and magnetic field, scholars discovered three adiabatic invariants. A similar mathematical endeavor has been pursued in celestial mechanics. Also familiarity of the characteristics of particle motion in electric and magnetic fields is helpful to gain physical insight for rarefied plasma medium. Some electromagnetic properties of plasma can be derived from superposition of individual particle motion in prescribed electric and magnetic fields. The characteristic equation of the Vlasov equation is particle orbit. The equivalence of the orbit equation and the collisionless Boltzmann equation is referred to as *Jean's theorem* in astronomical literature.

3.1. Radius of curvature vector

Consider a positive charge moving on (x, y)-plane in a space with a static constant magnetic field $\mathbf{B} = \hat{\mathbf{z}}B$ imbedded. We have the equation of motion for single particle,

$$m\frac{d\mathbf{v}}{dt} = \frac{e}{c}\,\mathbf{v}\times\mathbf{B}. \tag{3.1}$$

Since the force acting on the particle in (3.1) is always perpendicular to the velocity, the particle draws a circle. The sense of rotation is clockwise about the magnetic field (negative θ-direction in polar coordinate). The kinetic energy is constant, and the radius of the circle (ρ) is obtained from $\frac{mv^2}{\rho} = \frac{e}{c}\,vB$, giving $\rho = \frac{mcv}{eB}$, or in vector form

$$\overrightarrow{\rho} = \frac{mcv}{eB}\frac{\mathbf{v}\times\mathbf{B}}{|\mathbf{v}\times\mathbf{B}|} = \frac{mc}{eB^2}\,\mathbf{v}\times\mathbf{B}. \tag{3.2}$$

The radius of curvature vector defined above is a vector drawn from the particle position to the center of the circle or *the center of the gyration.* In a varying magnetic field, the center of the gyration, which is called the *guiding center*, drifts. In this case, we define the instantaneous radius curvature vector in analogous way.

• **Kinematics of a particle motion along a space curve.** Let a particle orbit in a plane be represented by [extension to a three-dimensional space curve is straightforward.]

$$(1)\quad \mathbf{r}(t) = \hat{\mathbf{x}}x(t) + \hat{\mathbf{y}}y(t).$$

Let s be the length measured along the curve: $(ds)^2 = (dx)^2 + (dy)^2$. Then the particle velocity can be written as

$$(2)\quad \frac{d\mathbf{r}}{dt} = \left(\frac{dx}{ds}\hat{\mathbf{x}} + \frac{dy}{ds}\hat{\mathbf{y}}\right)\frac{ds}{dt}.$$

Clearly, the expression in parentheses is the unit vector along the tangential direction at the point $(x(t), y(t))$. We denote this tangential unit vector by $\hat{\mathbf{u}}$. Then (2) is written as

$$(3)\quad \frac{d\mathbf{r}}{dt} = \hat{\mathbf{u}}\,\frac{ds}{dt} = v\hat{\mathbf{u}};\quad \hat{\mathbf{u}} = \frac{dx}{ds}\hat{\mathbf{x}} + \frac{dy}{ds}\hat{\mathbf{y}}.$$

Differentiating (3) with respect to t gives the acceleration vector

$$(4)\quad \mathbf{a} = \frac{dv}{dt}\hat{\mathbf{u}} + v^2\frac{d\hat{\mathbf{u}}}{ds};\quad \frac{d\hat{\mathbf{u}}}{ds} = \hat{\mathbf{x}}\frac{d^2x}{ds^2} + \hat{\mathbf{y}}\frac{d^2y}{ds^2}.$$

The vector $\frac{d\hat{\mathbf{u}}}{ds}$ is perpendicular to $\hat{\mathbf{u}}$; normal to the curve. The normal unit vector is denoted by $\hat{\mathbf{n}}$. We write

$$(5)\quad \frac{d\hat{\mathbf{u}}}{ds} = \frac{1}{\rho}\,\hat{\mathbf{n}}$$

with

$$(6)\quad \frac{1}{\rho} = \sqrt{\left(\frac{d^2x}{ds^2}\right)^2 + \left(\frac{d^2y}{ds^2}\right)^2} = \left|\frac{d\hat{\mathbf{u}}}{ds}\right|.$$

ρ is the *radius of curvature*. Then the acceleration in (4) is written as

$$(7)\quad \mathbf{a} = \frac{v^2}{\rho}\,\hat{\mathbf{n}} + \frac{dv}{dt}\,\hat{\mathbf{u}}.$$

In a uniform circular motion, the second term vanishes. We have

$$\frac{dx}{ds} = \frac{dx}{dt}\frac{dt}{ds} = \frac{v_x}{v}, \qquad \frac{d^2x}{ds^2} = etc. = \frac{1}{v^3}\left(va_x - v_x\frac{dv}{dt}\right).$$

We can show

$$(8)\quad \frac{1}{\rho^2} = \frac{1}{v^6}\left[v^2a^2 + v^2\left(\frac{dv}{dt}\right)^2 - 2v\frac{dv}{dt}\mathbf{a}\cdot v\right] = \frac{1}{v^4}(a^2 - a_T^2)$$

where $a_T = dv/dt = \mathbf{v}\cdot\mathbf{a}/v$ is the tangential component of the acceleration. (8) can be put into the form

$$(9)\quad \rho = \frac{v^2}{\sqrt{a^2 - a_T^2}} = \frac{v^3}{\sqrt{v^2a^2 - (\mathbf{v}\cdot\mathbf{a})^2}}.$$

. Now we define vector radius of curvature as follows

$$(10)\quad \overrightarrow{\rho} = \frac{v^2}{R^2}(v^2\mathbf{a} - \mathbf{v}\cdot\mathbf{a}\;\mathbf{v}), \quad R^2 = v^2a^2 - (\mathbf{v}\cdot\mathbf{a})^2.$$

Clearly, the absolute value of $\overrightarrow{\rho}$ agrees with (9) and the direction of $\overrightarrow{\rho}$ is normal to the curve ($\overrightarrow{\rho}\cdot\mathbf{v} = 0$). Check (10) against (3.2).

3.2. $\boldsymbol{E} \times \boldsymbol{B}$ drift

- **E, B** = constant. Consider an electron in constant fields:

$$m_e\frac{d\mathbf{v}}{dt} = -e\left(\mathbf{E} + \frac{1}{c}\,\mathbf{v}\times\mathbf{B}\right) \tag{3.3}$$

where we prescribe the fields as

$$\mathbf{E} = \hat{\mathbf{y}}E_y + \hat{\mathbf{z}}E_z, \quad \mathbf{B} = \hat{\mathbf{z}}B_0.$$

In components, we have

$$\frac{dv_x}{dt} = -\omega_c v_y \quad \left(\omega_c = \frac{eB_0}{m_e c} > 0,\ cyclotron\ frequency\right)$$

$$\frac{dv_y}{dt} = \omega_c v_x - \frac{e}{m_e}E_y, \quad \frac{dv_z}{dt} = -\frac{e}{m_e}E_z, \tag{3.4}$$

$$\frac{d^2v_x}{dt^2} = \omega_c^2\left(-v_x + c\frac{E_y}{B_0}\right), \quad \frac{d^2v_y}{dt^2} = -\omega_c^2 v_y \tag{3.5}$$

which are solved by

$$v_x = v_\perp cos(\omega_c t + \beta) + v_d, \quad v_y = v_\perp\ sin(\omega_c t + \beta), \tag{3.6}$$

where $v_\perp$ and β are constants of integration. Integrating once more gives

$$x = v_d t + \frac{v_\perp}{\omega_c}\ sin(\omega_c t + \beta), \quad y = -\frac{v_\perp}{\omega_c}\ cos(\omega_c t + \beta) \tag{3.7}$$

where $v_d = c\frac{E_y}{B_0}$ is the *drift velocity.* The orbit is visualized as follows. Its projection onto the (x, y)-plane is a circle of radius $\frac{v_\perp}{\omega_c}$ (*Larmor radius* or *gyro-radius*) whose center (guiding center) moves along the x-direction with the drift velocity v_d. Along the z-direction, the particle is accelerated according to (3.4). If $\mathbf{E}\|\mathbf{B}$ ($E_y = 0$), $v_d = 0$.

Alternatively, (3.3) can be solved by transforming as

$$\mathbf{v} = \frac{c}{B^2}\mathbf{E} \times \mathbf{B} + \mathbf{v}', \tag{3.8}$$

$$m_e\frac{d\mathbf{v}'}{dt} = \frac{-e\mathbf{E}\cdot\mathbf{B}}{B^2}\mathbf{B} - \frac{e}{c}\mathbf{v}' \times \mathbf{B}. \tag{3.9}$$

If we ignore the motion parallel with $\mathbf{B}$, the orbit on the perpendicular plane is a circular motion represented by $\mathbf{v}'$ plus the $E \times B$ drift of the gyration center represented by

$$\mathbf{v}_d = \frac{c}{B^2}\mathbf{E} \times \mathbf{B}. \tag{3.10}$$

Note that the $E \times B$ drift is independent of the sign of charge.

Exercise. Show that the radius of curvature vector is obtained as

$$\vec{\rho} = \frac{cm}{eB^2}\left(\mathbf{v} - \frac{c}{B^2}\mathbf{E}\times\mathbf{B}\right)\times\mathbf{B}.$$

- **Constant force (F) and Constant magnetic field**

A positive charge in this case is governed by

$$m\frac{d\mathbf{v}}{dt} = \mathbf{F} + \frac{e}{c}\,\mathbf{v}\times\mathbf{B}.$$

The parallel motion is trivial: $m\frac{dv_\parallel}{dt} = F_\parallel$, so we assume $\mathbf{F}\perp\mathbf{B}$. One can easily prove that the motion is the gyration around $\mathbf{B}$ and the drift with the drift velocity

$$\mathbf{v}_d = \frac{c}{eB^2}\mathbf{F}\times\mathbf{B}.$$

3.3. Equation for guiding center

Orbits of charged particles in a space with a time-varying or inhomogeneous magnetic field being imbedded can be approximated as circular motion about the zero order magnetic field (i.e. about the guiding center) plus the drift motion of the guiding center. Mathematically, this picture is a perturbational view, valid when spatial and temporal variation of the magnetic field is small over the Larmor radius or the cyclotron period, respectively. If other forces such as electric or gravitational forces act on the particle, the guiding center moves in accordance with the equation of guiding center, which we derive in this section. The curvature of the magnetic field line and/or the inhomogeneity of the magnetic field can also give rise to drift motion of the particle. For simplicity we assume that only an inhomogeneous magnetic force is acting on a charged particle:

$$m\frac{d\mathbf{v}}{dt} = \frac{e}{c}\mathbf{v}\times\mathbf{B}. \tag{3.11}$$

Consulting Fig. 3.1, the position vector of the guiding center, we have $\mathbf{R}(t) = \mathbf{r}(t) + \vec{\rho}(t)$, where $\mathbf{r}$ is the particle position and

$$\vec{\rho} = \frac{mc}{e}\frac{\mathbf{v}\times\mathbf{B}}{B^2} \quad (for\ +charge) \tag{3.12}$$

is the radius of curvature vector. After an infinitesimal time dt, the corresponding change of the particle position is $d\mathbf{r}$, and the radius of curvature vector changes to $\vec{\rho}\,'$, giving the relation

$$d\mathbf{R} = d\mathbf{r} + \vec{\rho}\,' - \vec{\rho}. \tag{3.13}$$

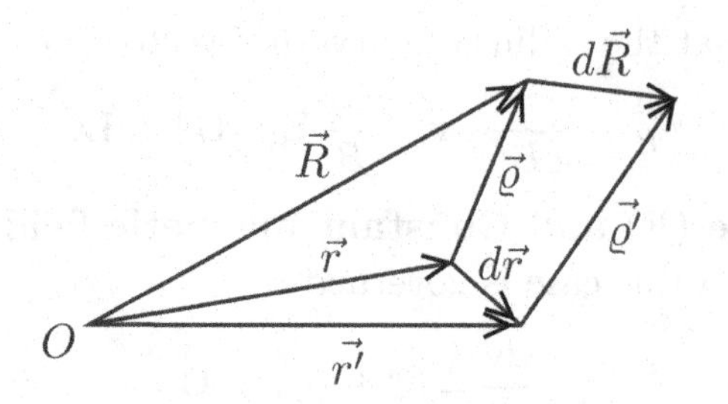

Fig. 3.1. $\vec{\rho} + d\mathbf{R} = d\mathbf{r} + \vec{\rho}'$.

From (3.12), we obtain

$$\vec{\rho}' - \vec{\rho} = d\vec{\rho} = \frac{mc}{e}\left(\frac{d\mathbf{v}\times\mathbf{B} + \mathbf{v}\times d\mathbf{B}}{B^2} - 2\frac{dB}{B^3}\mathbf{v}\times\mathbf{B}\right). \tag{3.14}$$

Using (3.11) for $d\mathbf{v}$, we obtain

$$d\mathbf{R} = \frac{\mathbf{B}}{B^2}\mathbf{v}\cdot\mathbf{B}\,dt + \frac{mc}{e}\left(\frac{\mathbf{v}\times d\mathbf{B}}{B^2} - 2\,\frac{dB}{B^3}\,\mathbf{v}\times\mathbf{B}\right). \tag{3.15}$$

Here $\mathbf{B}$ takes the value at the particle position, so we write $\mathbf{B} = \mathbf{B}(\mathbf{r}(t))$; but not explicitly depending upon t. Then

$$d\mathbf{B} = dB_i = \frac{\partial B_i}{\partial r_j}\frac{dr_j}{dt}\,dt = v_j\,\frac{\partial B_i}{\partial r_j}dt = \mathbf{v}\cdot\nabla\mathbf{B}\,dt.$$

Thus (3.15) gives the drift velocity of the guiding center,

$$\mathbf{v}_d = \frac{d\mathbf{R}}{dt} = \mathbf{v}_\parallel + \frac{mc}{e}\left(\frac{\mathbf{v}\times(\mathbf{v}\cdot\nabla)\mathbf{B}}{B^2} - 2\,\frac{\mathbf{v}\times\mathbf{B}}{B^3}\,(\mathbf{v}\cdot\nabla)B\right) \tag{3.16}$$

where $v_\parallel$ is the velocity component parallel to the magnetic field. (3.16) is the equation for the drift motion of the guiding center. We shall consider a few special cases of $\mathbf{B}$.

i) $\mathbf{B} = \hat{\mathbf{z}}B(x)$ ($\mathbf{B}$ has a gradient in x-direction). In this case we have

$$\mathbf{v}_d = \mathbf{v}_\parallel - \frac{mc}{e}\frac{v_x}{B^2}\frac{dB}{dx}(\hat{\mathbf{x}}v_y - \hat{\mathbf{y}}v_x).$$

Here the perpendicular velocity components $v_{x,y}$ are time-oscillating functions with the gyrofrequency (cyclotron frequency) ω_c. Therefore the drift velocity has a meaning when it is averaged over the period (T). We have $\langle v_x v_y\rangle = 0$ and $\langle v_x^2\rangle = \frac{1}{2}v_\perp^2$. Here $\langle\cdots\rangle = \frac{1}{T}\int_0^T(\cdots)dt$. So the above equation gives

$$\mathbf{v}_d = \mathbf{v}_\parallel + \hat{\mathbf{y}}\frac{mc}{eB^2}\frac{v_\perp^2}{2}\frac{dB}{dx}. \tag{3.17}$$

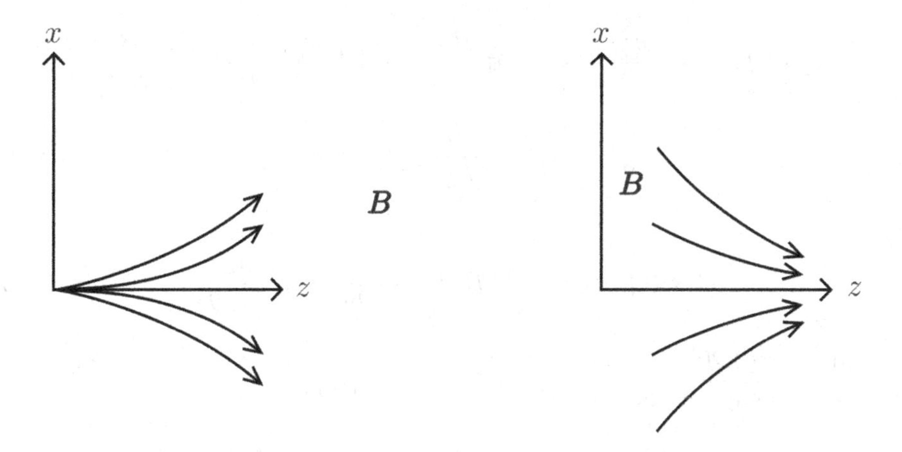

Fig. 3.2. Converging or diverging magnetic field.

(3.17) indicates that the guiding center drifts in the direction of $\mathbf{B} \times \nabla B$, and is called ∇B-drift.

ii) **B**-field line slightly curved. As shown in Fig. 3.2, **B** vector has a small B_x component in addition to the main component B_z ($B_z \gg B_x$).

We write

$$\mathbf{B} = \hat{\mathbf{x}} B_x(z) + \hat{\mathbf{z}} B_z(x, y).$$

Note that $\nabla \cdot \mathbf{B} = 0$. Introduce the differential operator $\mathbf{Q} = \mathbf{v}\,(\mathbf{v} \cdot \nabla) = v_i v_j \frac{\partial}{\partial x_j}$. Then ∇B-drift term takes a convenient form amenable to easy averaging

$$\mathbf{v}_d = \frac{mc}{eB^2}\left[\mathbf{Q} \times \mathbf{B} + \frac{2}{B}(\mathbf{B} \times \mathbf{Q})B\right]. \tag{3.18}$$

Here we have $\nabla = \hat{\mathbf{z}}\frac{\partial}{\partial z} + \nabla_\perp$, with $\nabla_\perp = \hat{\mathbf{x}}\frac{\partial}{\partial x} + \hat{\mathbf{y}}\frac{\partial}{\partial y}$, and

$$\mathbf{Q} = (\hat{\mathbf{z}} v_\parallel + \mathbf{v}_\perp)\left(v_\parallel \frac{\partial}{\partial z} + \mathbf{v}_\perp \cdot \nabla_\perp\right)$$

which consists of quadratic terms of particle velocity components. $\mathbf{v}_\perp$ is a time-oscillating function expressed by sine or cosine functions with frequency ω_c. Thus we take an average over the period, by putting $\mathbf{v}_\perp = v_{\perp 0}(\hat{\mathbf{x}} cos\ \omega_c t + \hat{\mathbf{y}} sin\ \omega_c t)$, we obtain

$$\langle \mathbf{Q} \rangle = \hat{\mathbf{z}}\ v_\parallel^2 \frac{\partial}{\partial z} + \frac{v_{\perp 0}^2}{2}\left(\hat{\mathbf{x}}\frac{\partial}{\partial x} + \hat{\mathbf{y}}\frac{\partial}{\partial y}\right)$$

$$\langle \mathbf{Q} \rangle \times \mathbf{B} = \hat{\mathbf{x}} \left(\frac{v_{\perp 0}^2}{2} \frac{\partial B_z}{\partial y} - v_\parallel^2 \frac{\partial B_y}{\partial z} \right) - \hat{\mathbf{y}} \left(\frac{v_{\perp 0}^2}{2} \frac{\partial B_z}{\partial x} - v_\parallel^2 \frac{\partial B_x}{\partial z} \right)$$
$$+ \hat{\mathbf{z}} \frac{v_{\perp 0}^2}{2} \left(\frac{\partial B_y}{\partial x} - \frac{\partial B_x}{\partial y} \right) \tag{3.19}$$

$$\mathbf{B} \times \langle \mathbf{Q} \rangle \, B = \frac{v_{\perp 0}^2}{2} \, B_z \left(-\hat{\mathbf{x}} \frac{\partial B_z}{\partial y} + \hat{\mathbf{y}} \frac{\partial B_z}{\partial x} \right). \tag{3.20}$$

In obtaining (3.20), we put $B \simeq B_z$. Also we included the component B_y which disappears at the end. Using (3.19) and (3.20) in (3.18) gives

$$\mathbf{v}_d = \frac{cm}{eB^2} \left[-\hat{\mathbf{x}} \left(\frac{v_{\perp 0}^2}{2} \frac{\partial B_z}{\partial y} + v_\parallel^2 \frac{\partial B_y}{\partial z} \right) + \hat{\mathbf{y}} \left(\frac{v_{\perp 0}^2}{2} \frac{\partial B_z}{\partial x} + v_\parallel^2 \frac{\partial B_x}{\partial z} \right) \right.$$
$$\left. + \hat{\mathbf{z}} \frac{v_{\perp 0}^2}{2} \left(\frac{\partial B_y}{\partial x} - \frac{\partial B_x}{\partial y} \right) \right]. \tag{3.21}$$

Since $\mathbf{B}$ satisfies $\nabla \times \mathbf{B} = 0$, we have the relations, $\frac{\partial B_y}{\partial x} = \frac{\partial B_x}{\partial y}$, $\frac{\partial B_y}{\partial z} = \frac{\partial B_z}{\partial y}$, and $\frac{\partial B_x}{\partial z} = \frac{\partial B_z}{\partial x}$. Putting these relations into the above equation gives

$$\mathbf{v}_d = \frac{cm}{eB^3} B_z \left(\frac{v_{\perp 0}^2}{2} + v_\parallel^2 \right) \left[-\hat{\mathbf{x}} \, \frac{\partial B_z}{\partial y} + \hat{\mathbf{y}} \, \frac{\partial B_z}{\partial x} \right]$$
$$= \left(\frac{v_{\perp 0}^2}{2} + v_\parallel^2 \right) \frac{\hat{\mathbf{z}} B_z \times \nabla B_z}{B^2 \omega_c} = \left(\frac{v_{\perp 0}^2}{2} + v_\parallel^2 \right) \frac{\mathbf{B} \times \nabla B_z}{B^2 \omega_c}. \tag{3.22}$$

The first term in (3.22) reproduces the previous result ∇B-drift. We give in the following an alternative physical interpretation of (3.22).

3.4. Magnetic moment and ∇B drift

Magnetic moment (μ) is defined by the product of the current and the area enclosed by the loop along which the current is flowing,

$$\mu = \frac{1}{c} \frac{e \, \omega_c}{2\pi} \, \pi r_L^2 = \frac{m v_\perp^2 / 2}{B}, \quad r_L = \frac{v_\perp}{\omega_c}, \quad \textit{Larmor radius} \tag{3.23}$$

where the factor $\frac{1}{c}$ enters in our electrostatic unit system. In (3.22) the ∇B_z-drift reads

$$\mathbf{V}_{\nabla B_z} = \frac{v_{\perp 0}^2 / 2}{B^2 \omega_c} \, \mathbf{B} \times \nabla B_z = \frac{c}{eB^2} \mathbf{B} \times \mu \nabla B_z. \tag{3.24}$$

Suppose a constant force $\mathbf{F} \perp \mathbf{B}$ acts upon a positive charge, in addition to the centripetal magnetic force $\frac{e}{c}\mathbf{v} \times \mathbf{B}$. Then the drift due to the force is

$$\mathbf{V}_F = \frac{c}{eB^2}\mathbf{F} \times \mathbf{B}. \tag{3.25}$$

Comparison of (3.24) and (3.25) indicates that ∇B_z gives rise to an equivalent force

$$\mathbf{F} = -\mu \nabla B_z \tag{3.26}$$

acting on a charge with a magnetic moment μ.

3.5. Force acting on a charge having a magnetic moment

In this Section, we show that an electron with a magnetic moment μ is exerted by a force (3.26) due to a magnetic field whose line of force diverges as shown in Fig. 3.2. Write $\nabla \cdot \mathbf{B} = 0$ in cylindrical coordinate, assuming the azimuthal symmetry,

$$\frac{1}{r}\frac{\partial}{\partial r}(rB_r) + \frac{\partial B_z}{\partial z} = 0. \tag{3.27}$$

Integrate this equation with respect to r from 0 to ρ, the Larmor radius:

$$\rho B_r(\rho) = -\int_0^\rho r\frac{\partial B_z}{\partial z}dr.$$

Since the main component of $\mathbf{B}$ is B_z and the line of force is slowly diverging, we can put $\frac{\partial B_z}{\partial z} \simeq \left.\frac{\partial B}{\partial z}\right|_{r=0}$, which can be taken out of the integral, to obtain

$$B_r(\rho) = -\frac{\rho}{2}\frac{\partial B}{\partial z}.$$

This radial component of the magnetic field together with the azimuthal velocity component of the gyrating motion $v_\theta = v_\perp$ gives rise to the Lorentz force $\mathbf{v}_\theta \times \mathbf{B}_r$ in the z-direction,

$$F_z = \frac{e}{c}v_\perp B_r = -\mu\frac{\partial B}{\partial z} \quad (per \quad \rho = v_\perp/\omega_c). \tag{3.28}$$

Defining magnetic moment vector,

$$\overrightarrow{\mu} = \frac{\mathbf{B}}{B^2}\frac{mv_\perp^2}{2} \tag{3.29}$$

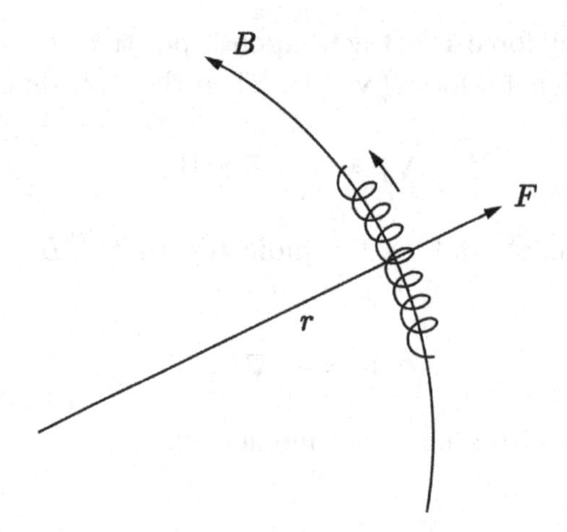

Fig. 3.3. Centrifugal force felt by a particle moving along a curved field line.

we can write (3.28) in the form

$$F_z = -\vec{\mu} \cdot \nabla B. \tag{3.30}$$

3.6. Curvature drift

The second term in (3.22) is called the curvature drift, because it is given rise to by the centripetal force sliding along the curved line of force:

$$v_\parallel^2 \frac{\mathbf{B} \times \nabla B_z}{B^2 \omega_c} : \quad curvature\ drift. \tag{3.31}$$

In Fig. 3.3, the magnetic line of force is expressed by polar coordinate as $\mathbf{B} = \hat{\theta}\ B(r)$. $\nabla \times \mathbf{B} = 0$ gives $\frac{1}{r}\frac{\partial}{\partial r}(rB) = 0$. So we have $B \propto 1/r$, and $\nabla B/B = -\mathbf{r}/r^2$. Then the second term of (3.22) is

$$\mathbf{V}_{curv} = \frac{c}{eB^2} \frac{mv_\parallel^2}{r} \hat{\mathbf{r}} \times \mathbf{B}. \tag{3.32}$$

Comparison of this equation with (3.25) indicates that the force responsible for the curvature drift is the force $\mathbf{F} = \frac{mv_\parallel^2}{r} \hat{\mathbf{r}}$, i.e. the centripetal force experienced by the particle moving along the curved line of force.

- **Magnetic moment of a charge circling around a slowly time-varying magnetic field.** This quantity is an approximate constant of motion as shown in the following. Since $\mathbf{B}(t)$ produces $\mathbf{E}(t)$ per Faraday's law, the electric force should be included in the equation of motion.

$$m\frac{d\mathbf{v}}{dt} = e\left(\mathbf{E} + \frac{1}{c}\mathbf{v} \times \mathbf{B}\right). \tag{3.33}$$

Assume $\mathbf{B}(t) = \hat{\mathbf{z}}B(t)$. Then the electric field satisfying $\nabla \cdot \mathbf{E} = 0$ and $\nabla \times \mathbf{E} = -\frac{1}{c}\frac{\partial \mathbf{B}}{\partial t}$ can be written as

$$\mathbf{E} = -\frac{1}{2c}\,\hat{\mathbf{z}} \times \mathbf{r}\,\frac{dB}{dt}. \tag{3.34}$$

$$\textit{Then,} \quad \frac{d}{dt}\frac{mv^2}{2} = -\frac{e}{2c}\frac{dB}{dt}\,\hat{\mathbf{z}} \times \mathbf{r} \cdot \frac{d\mathbf{r}}{dt}. \tag{3.35}$$

We shall consider that the particle motion takes place on the perpendicular plane, assuming $v_{\|} = 0$, and integrate (3.35) over one cycle of gyro-motion:

$$\delta\left(\frac{mv_{\perp}^2}{2}\right) = -\frac{e}{2c}\oint \frac{dB}{dt}\hat{\mathbf{z}} \cdot \mathbf{r} \times d\mathbf{r}. \tag{3.36}$$

Here we assume that B(t) is such *a slowly varying function of time that* $\frac{dB}{dt}$ *remains almost constant over one gyro-motion*, and it can be taken out of the integral. Then the integral is twice the area of Larmor circle, $2\pi\rho^2$, giving

$$\delta\left(\frac{mv_{\perp}^2}{2}\right) = \frac{e}{c}\frac{dB}{dt}\,\pi\rho^2 = \frac{mv_{\perp}^2}{2B}\frac{dB/dt}{\omega_c/2\pi}. \tag{3.37}$$

Note that

$$\delta(B) = \oint \frac{dB}{dt}\,dt \simeq \frac{dB}{dt}\frac{2\pi}{\omega_c}.$$

Therefore (3.37) is the relation $\delta(\mu B) = \mu\delta B$ which yields

$$\delta\mu = 0 \quad or \quad \mu = const.$$

This proves that μ is an *approximate constant of motion in a slowly changing magnetic field — an adiabatic invariant.* The assumption that dB/dt remains constant over several gyro-cycles is the *adiabatic assumption.* Putting it precisely, $B(t)$ changes slowly so that

$$\frac{dB}{dt} = \frac{\delta(B)}{\delta(t)} = \frac{change\ of\ B}{corresponding\ time\ interval} = constant. \tag{3.38}$$

3.7. Adiabatic invariant

Consider a one-dimensional motion in a potential $V(x, \lambda)$,

$$m\frac{d^2x}{dt^2} = -\frac{\partial V(x,\lambda)}{\partial x} \tag{3.39}$$

where λ is a parameter through which V can be varied, such as the length of an oscillating pendulum, subject to gravity. When λ is constant, we have an exact constant of integration,

$$\frac{m}{2}\left(\frac{dx}{dt}\right)^2 + V(x) = E: \ const. \tag{3.40}$$

Also the *action integral* defined as below is also a constant of motion:

$$J = \oint m \, \frac{dx}{dt} dx \tag{3.41}$$

where the integral is to be performed over the time interval between the two turning points (period).

For a simple harmonic motion with $V(x) = \frac{k}{2}x^2$ (k is the force constant),

$$J = \oint m \left(\frac{dx}{dt}\right)^2 dt = \frac{k}{2}x_0^2 \, T = ET = 2\pi \, \frac{E}{\omega} \, ,$$
$$(x_0 = amplitude, \quad \omega = \sqrt{k/m}). \tag{3.42}$$

Exercise. Carry out the integral to show

$$J = \sqrt{2m} \oint dx \sqrt{E - \frac{k}{2}x^2} = 2\pi \frac{E}{\omega}.$$

However, when λ is not constant but changes with time, (3.39) does not admit a constant of motion. Also the action integral is not a constant of motion; it is not even defined because the motion loses periodicity in the strict sense. But if $\lambda(t)$ is a *slow* enough function of t, the action integral is still an approximate constant of motion —*an adiabatic invariant.* The slow function $\lambda(t)$ changes appreciably only after many oscillations. That is, $T\frac{d\lambda}{dt} \ll \lambda$, where T is the period of oscillation defined by

$$T = \oint \frac{dx}{\dot{x}} \quad \left(\cdot = \frac{d}{dt}\right).$$

The integral $\oint$ means a round trip integral along the path between two turning points. The turning points are not fixed in the long run, but they can be assumed to be fixed in a time interval of a period. We define the average

$$\langle E \rangle = \frac{1}{T} \oint dt \left[\frac{m}{2}\dot{x}^2 + V(x, \lambda)\right]. \tag{3.43}$$

The turning points are obtained from

$$\frac{m}{2}\dot{x}^2 + V(x, \lambda) = \langle E \rangle. \tag{3.44}$$

The corresponding action integral is

$$J = \sqrt{2m} \oint dx \sqrt{\langle E \rangle - V(x, \lambda)}. \tag{3.45}$$

The proof of constancy of J goes as follows.

$$\frac{dJ}{dt} = \frac{\partial J}{\partial \langle E \rangle}\frac{d\langle E \rangle}{dt} + \frac{\partial J}{\partial \lambda}\frac{d\lambda}{dt} \tag{3.46}$$

$$\frac{\partial J}{\partial \langle E \rangle} = \sqrt{\frac{m}{2}} \oint \frac{dx}{\sqrt{\langle E \rangle - V(x, \lambda)}} = \oint \frac{dx}{\dot{x}} = T \tag{3.47}$$

$$\frac{d\langle E \rangle}{dt} = \frac{1}{T} \oint dt \left(m\dot{x}\ddot{x} + \dot{x}\frac{\partial V}{\partial x} + \dot{\lambda}\frac{\partial V}{\partial \lambda} \right). \tag{3.48}$$

The first two terms in the integral cancel. The crucial point here is that $\dot{\lambda}$ is almost constant (because λ is a slowly varying function), and can be taken out of the integral, to give

$$\frac{d\langle E \rangle}{dt} = \frac{\dot{\lambda}}{T} \oint dt \, \frac{\partial V}{\partial \lambda} \tag{3.49}$$

$$\frac{\partial J}{\partial \lambda} = -\sqrt{\frac{m}{2}} \oint dx \, \frac{\partial V/\partial \lambda}{\sqrt{\langle E \rangle - V(x, \lambda)}} = -\oint \frac{\partial V}{\partial \lambda}\frac{dx}{\dot{x}}. \tag{3.50}$$

Combining the foregoing equations yields $\frac{dJ}{dt}$=0.

- **Harmonic oscillator with slowly changing frequency.**

$$\ddot{x} + \omega^2(t)x = 0. \tag{3.51}$$

We imagine that the potential changes slowly, and so the force constant $k = m\omega^2$ is also a slowly varying function. T, the time interval between two turning points, can be defined as the period in this approximate periodic motion, and we can accordingly define $\omega = 2\pi/T$ as the frequency, albeit it is slowly changing. The solution of (3.51) can be written as

$$x(t) = A(t)sin(\omega t + \alpha(t)) \tag{3.52}$$

where A(t) and $\alpha(t)$ are slowly varying functions, $\frac{1}{A}\frac{dA}{dt} \ll \omega$, $\frac{1}{\alpha}\frac{d\alpha}{dt} \ll \omega$. The average energy taken over the fast time scale is

$$\langle E \rangle = \frac{m}{2}\langle \dot{x}^2 \rangle + \frac{1}{2}k\langle x^2 \rangle = \frac{1}{2}m\omega^2 A^2. \tag{3.53}$$

The invariant quantity is the action in (3.45):

$$J = \sqrt{2m}\oint dx \sqrt{\frac{m\omega^2 A^2}{2} - \frac{kx^2}{2}} = m\omega \oint dx \sqrt{A^2 - x^2}$$

where we regarded ω is constant and have taken out of the integral. Hence,

$$J = 2m\omega \int_{-A}^{A} dx \sqrt{A^2 - x^2} = \pi m\omega\, A^2 = 2\pi\, \frac{\langle E \rangle}{\omega} \tag{3.54}$$

where we used $\int \sqrt{A^2 - x^2}\, dx = \frac{x}{2}\sqrt{A^2 - x^2} + \frac{A^2}{2} sin^{-1}\frac{x}{A}$.

(3.54) identifies the adiabatic invariant in a slowly changing simple harmonic motion. Now we recapitulate the physical picture of the invariance of the magnetic moment in a slowly time-varying magnetic field B(t). The electric field E(t) given rise to by B(t) changes the perpendicular energy while the particle executes the cyclotron motion with a slowly changing Larmor frequency ω_c. Analogously to the simple harmonic motion considered above, we expect $\frac{m}{2}v_\perp^2 \frac{2\pi}{\omega_c}$ to be an adiabatic invariant, which is proportional to μ.

- **Charged particle motion in a time-varying magnetic field.** Here we show that the particle motion in a time-varying magnetic field can be represented by an equation of simple harmonic motion with time-varying gyro-frequency. Substituting (5.34) into (5.33) gives

$$\frac{d^2\mathbf{r}}{dt^2} + \frac{1}{2}\,\hat{\mathbf{z}} \times \mathbf{r}\,\dot{\omega}_c + \omega_c\,\hat{\mathbf{z}} \times \frac{d\mathbf{r}}{dt} = 0, \quad or \tag{3.55}$$

$$\frac{d^2x}{dt^2} - \frac{\dot{\omega}_c}{2}y - \omega_c\frac{dy}{dt} = 0, \tag{3.56}$$

$$\frac{d^2y}{dt^2} + \frac{\dot{\omega}_c}{2}x + \omega_c\frac{dx}{dt} = 0. \tag{3.57}$$

The above two equations can be written as a single equation by introducing $\xi = x + iy$:

$$\frac{d^2\xi}{dt^2} + \frac{i}{2}\dot{\omega}_c\,\xi + i\omega_c\,\frac{d\xi}{dt} = 0. \tag{3.58}$$

A change of variable

$$\xi = w(t)e^{-\frac{i}{2}\int \omega_c(t)dt} \quad gives$$

$$\frac{d^2w}{dt^2} + \frac{\omega_c^2(t)}{4}w = 0 \tag{3.59}$$

which is a simple harmonic equation with a varying frequency. Note that ξ in the above is a WKB type solution of (3.59).

Exercise. Show that the magnetic moment is given by

$$\mu = \frac{e}{2c}\frac{|\dot{\xi}|^2}{\omega_c}.$$

• Kurlsrud's proof of adiabatic invariance of the action of a slowly varying harmonic oscillator.

Reference: Kurlsrud, R. M. *Adiabatic invariant of the harmonic oscillator* Proceedings of the International School of Physics Enrico Fermi Course XXV Advanced Plasma Theory Academic Press (1964).

Kurlsrud solved the harmonic oscillator equation

$$\frac{d^2x}{dt^2} + \omega^2(t)x = 0$$

by introducing slowly varying amplitude and WKB type phase:

$$x(t) = W(t)\ sin\left(\int_0^t S(t)dt + \delta\right).$$

He expanded S and W in perturbation series in terms of the small parameter

$$\epsilon = \frac{1}{\omega^2}\frac{d\omega}{dt}$$

and worked out the double series and derived the following relation for the action $J = E/\omega$:

$$\frac{\Delta J}{J} \approx exp\left(-\frac{1}{\epsilon}f(t)\right)$$

where Δ signifies the change in a period. The important point of the above 'solution' is that the exponential function in the above expression goes to zero as $\epsilon \rightarrow 0$ when $\Delta J \sim \epsilon^n$ for any n, which means that the perturbation series with coefficients ϵ^n goes to zero for *all n*. Thus, the adiabatic invariant is invariant to all orders of ϵ^n. See §Introduction. Note that

$$\lim_{\epsilon\to 0} \frac{1}{\epsilon^n}\, exp\left(-\frac{1}{\epsilon}\right) = \lim_{\lambda\to\infty} \frac{\lambda^n}{e^{\lambda}} \to 0, \quad for\ any\ n.$$

• **Mathieu equation.** As an example, let us consider a time-varying magnetic field,

$$\mathbf{B}(t) = \hat{\mathbf{z}} B_0 (1 + \epsilon\ sin\ \Omega t) \quad (\epsilon \ll 1).$$

The induced electric field can be calculated according to (3.34). Therefore, the particle motion subject to the above time-varying magnetic field satisfies (3.59) after the transformations explained above, with the corresponding cyclotron frequency (time-varying) $\omega_c(t)$ defined as the following form:

$$\omega_c(t) = \omega_{c0}(1 + \epsilon\ sin\ \Omega t) \quad \omega_{c0} = \frac{eB_0}{mc}.$$

In this case, the equation of motion, (3.59), takes the form

$$\frac{d^2 w}{dt^2} + \frac{\omega_{c0}^2}{4} w(1 + 2\epsilon\ sin\ \Omega t) = 0$$

where we neglected the ϵ^2 term. The above equation is of the *Mathieu equation* type. If the natural frequency ω_{c0} and the modulation frequency Ω satisfy a certain resonance condition, the particle is accelerated and a parametric instability can be given rise to. In fact, the Mathieu equation is used as a model equation to study parametric instability (Chapter 9).

3.8. Multiple time perturbation analysis

The multiple scale perturbation method is a mathematical scheme which remedies the shortcoming of the regular perturbation method by its facility to remove the secularity-causing terms at every stage of iteration. In the regular perturbation analysis, the secular solutions (proportional to t) appear at the very first stage of iteration, so the solution loses its validity after a long time. That is, the regular perturbation method produces a solution which is valid only for a limited time. The multiple scale method has a device to remove the secularity-causing terms at every stage of iteration, and the perturbation solution stands free from the secularity. This method is useful in plasma physics. See Appendix B. In this section, the method is demonstrated by deriving the $\mathbf{B} \times \nabla B$-drift, as an example.

Let us consider the simple case (3.17) in which the magnetic field is assumed to be $\mathbf{B} = \hat{\mathbf{z}}B(x)$. We solve

$$\frac{dv_x}{dt} = \Omega v_y, \quad \frac{dv_y}{dt} = -\Omega v_x, \quad \textit{with} \quad \Omega(x) = \frac{eB(x)}{mc}. \tag{3.60}$$

It should be clearly understood that x is the particle position, and the magnetic field $B(x)$ is the value at the particle position x. The particle orbit consists of Larmor gyration plus the drift of the guiding center. We choose the coordinate origin as the instantaneous guiding center. Assuming that the fractional change of the magnetic field over the Larmor radius (ρ) is much smaller than unity

$$\frac{dB}{dx}\rho \ll B \tag{3.61}$$

$\Omega(x)$ in (3.60) can be Taylor expanded to express it in terms of the value at the guiding center:

$$\frac{dv_x}{dt} = (\Omega + \epsilon\Omega' x)v_y, \tag{3.62}$$

$$\frac{dv_y}{dt} = -(\Omega + \epsilon\Omega' x)v_x, \tag{3.63}$$

where $\Omega' = d\Omega/dx$ at $x = 0$ and ϵ is the indicator of smallness of the term. The perturbation analysis of the above equations is based on (3.62). We also have

$$v_x = \frac{dx}{dt}, \tag{3.64}$$

$$v_y = \frac{dy}{dt}. \tag{3.65}$$

We solve the above four equations by multiple scale perturbation method by expanding in the powers of ϵ as

$$\frac{d}{dt} = \frac{\partial}{\partial t_0} + \epsilon\frac{\partial}{\partial t_1} + \epsilon^2\frac{\partial}{\partial t_2} + \cdots \tag{3.66}$$

$$\sigma(t_0, t_1, t_2 \cdots) = \sigma^{(0)}(t_0, t_1, \cdots) + \epsilon\sigma^{(1)}(t_0, t_1, \cdots) + \epsilon^2\sigma^{(2)}(t_0, t_1, \cdots) + \cdots \tag{3.67}$$

where σ takes any of x, y, v_x, v_y. We break down the above equations order by order in the following equations:

$$v_x^{(0)} = \frac{\partial x^{(0)}}{\partial t_0} \tag{3.68}$$

$$\frac{\partial}{\partial t_0}v_x^{(0)} = \Omega v_y^{(0)} \tag{3.69}$$

$$\frac{\partial}{\partial t_0}v_x^{(1)} + \frac{\partial}{\partial t_1}v_x^{(0)} = \Omega v_y^{(1)} + \Omega' x^{(0)} v_y^{(0)} \tag{3.70}$$

$$\frac{\partial}{\partial t_0}v_y^{(0)} = -\Omega v_x^{(0)} \tag{3.71}$$

$$\frac{\partial}{\partial t_0}v_y^{(1)} + \frac{\partial}{\partial t_1}v_y^{(0)} = -\Omega v_x^{(1)} - \Omega' x^{(0)} v_x^{(0)}. \tag{3.72}$$

Combining (3.70) and (3.72) gives

$$\left(\frac{\partial^2}{\partial t_0^2} + \Omega^2\right) v_x^{(0)} = 0 \tag{3.73}$$

$$v_x^{(0)} = A(t_1) sin\ \Omega t_0 + B(t_1) cos\ \Omega t_0 \tag{3.74}$$

$$v_y^{(0)} = A(t_1) cos\ \Omega t_0 - B(t_1) sin\ \Omega t_0 \tag{3.75}$$

$$x^{(0)}(t_0, t_1) = -\frac{A(t_1)}{\Omega} cos\ \Omega t_0 + \frac{B(t_1)}{\Omega} sin\ \Omega t_0 + C(t_1). \tag{3.76}$$

(3.71) and (3.73) give

$$\left(\frac{\partial^2}{\partial t_0^2} + \Omega^2\right) v_x^{(1)} = -\frac{\partial}{\partial t_1}\frac{\partial v_x^{(0)}}{\partial t_0} - \Omega \frac{\partial}{\partial t_1} v_y^{(0)} + \Omega' \frac{\partial}{\partial t_0}(x^{(0)} v_y^{(0)}) - \Omega\Omega' x^{(0)} v_x^{(0)}, \tag{3.77}$$

$$\left(\frac{\partial^2}{\partial t_0^2} + \Omega^2\right) v_y^{(1)} = -\frac{\partial}{\partial t_1}\frac{\partial v_y^{(0)}}{\partial t_0} + \Omega \frac{\partial}{\partial t_1} v_x^{(0)} - \Omega' \frac{\partial}{\partial t_0}(x^{(0)} v_x^{(0)}) - \Omega\Omega' x^{(0)} v_y^{(0)}. \tag{3.78}$$

Using the zero order solutions in the above two equations, we have

$$RHS\ of\ (3.77) = -2\Omega\left(\frac{\partial A}{\partial t_1} cos\ \Omega t_0 + \frac{\partial B}{\partial t_1} sin\ \Omega t_0\right) - \Omega\Omega'\ CA\ sin\Omega t_0 - \Omega\Omega'\ CB\ cos\Omega t_0$$

$$+\Omega'\frac{\partial}{\partial t_0}\left(-\frac{A^2}{\Omega}cos^2\Omega t_0 - \frac{B^2}{\Omega}sin^2\Omega t_0 + \frac{2AB}{\Omega}\ sin\Omega t_0\ cos\Omega t_0\right)$$

$$-\Omega\Omega'\left[\frac{AB}{\Omega}\ sin^2\Omega t_0 - \frac{AB}{\Omega}\ cos^2\Omega t_0\right.$$

$$\left.+\left(\frac{B^2}{\Omega}-\frac{A^2}{\Omega}\right) sin\Omega t_0 cos\Omega t_0 + CA\ sin\Omega t_0 + CB\ cos\Omega t_0\right]$$

$RHS\ of$ (3.78)

$$= 2\Omega\left(\frac{\partial A}{\partial t_1}sin\ \Omega t_0 + \frac{\partial B}{\partial t_1}cos\ \Omega t_0\right)$$

$$-\Omega\Omega'\ CA\ cos\Omega t_0 + \Omega\Omega'\ CB\ sin\Omega t_0$$

$$-\Omega'\frac{\partial}{\partial t_0}\left(\frac{AB}{\Omega}sin^2\Omega t_0 - \frac{AB}{\Omega}cos^2\Omega t_0 + \frac{1}{\Omega}(B^2 - A^2)sin\Omega t_0\ cos\Omega t_0\right)$$

$$-\Omega\Omega'\left[\frac{2AB}{\Omega}\ sin\Omega t_0\ cos\Omega t_0 - \frac{A^2}{\Omega}cos^2\Omega t_0\right.$$

$$\left.-\frac{B^2}{\Omega}sin^2\Omega t_0 + CA\ cos\Omega t_0 - CB\ sin\Omega t_0\right].$$

The above equations, the right hand sides of (3.78) and (3.79), now have frequencies 0, Ω, and 2Ω, produced by mixing the frequencies due to the quadratic terms of (3.78) and (3.79). The Ω-frequency terms are the *secular terms* of (3.78) and (3.79): such terms require $v_x^{(1)}$ should be proportional to $t_0\ sin\Omega t_0$ or $t_0\ cos\Omega t_0$, which is not physically acceptable because they grow indefinitely as time increases. Multiple scale perturbation method systematically gets rid of such terms by putting the sum of all the secular terms to zero. In (3.78) we put

$$\frac{\partial A}{\partial t_1} + \Omega' BC = 0, \qquad \frac{\partial B}{\partial t_1} - \Omega' AC = 0.$$

These equations determine the slow time dependence of the coefficients, $A(t_1)$ and $B(t_1)$, but we don't have to solve them for our purpose. Once the secular terms are removed, (3.78) and (3.79) yield solutions of 0-frequency and 2Ω-frequency, of which the former is the solution that we seek for. We have from (3.79)

$$\left(\frac{\partial^2}{\partial t_0^2} + \Omega^2\right) v_y^{(1)} = \frac{\Omega'}{2}(A^2 + B^2) \tag{3.79}$$

whose particular solution is

$$v_y^{(1)} = \frac{\Omega'}{2\Omega^2}(A^2 + B^2) = \frac{\Omega'}{2\Omega^2}[v_x^{(0)2} + v_y^{(0)2}] \tag{3.80}$$

which is the ∇B-drift in the field $B_z(x)$. Note that RHS of (3.78) has no zero-frequency terms.

The multiple scale method was first invented by Frieman [Reference: E. A. Frieman, *On a new method in the theory of irreversible processes*, J. Math. Phys. **4** p. 410, 1963] for study of the Liouville equation.

3.9. Adiabatic invariant derived from Hamiltonian

Hamiltonian H is the expression for the total energy of system, expressed in terms of the generalized coordinates p and q, and time t. We have a constant of motion in a conservative force field:

$$E = H(p, q, t).$$

If H explicitly depends on t, total energy E is not a constant of motion. When H depends on t only implicitly through $p(t)$ and $q(t)$, then E is constant. This is easily seen from

$$\frac{dE}{dt} = \frac{\partial H}{\partial p}\frac{dp}{dt} + \frac{\partial H}{\partial q}\frac{dq}{dt} + \frac{\partial H}{\partial t}.$$

The first two terms on the RHS cancel owing to Hamilton's equations $\frac{\partial H}{\partial p} = \frac{dq}{dt}, \frac{\partial H}{\partial q} = -\frac{dp}{dt}$. Therefore we have $\frac{dE}{dt} = \frac{\partial H}{\partial t}$. Let $H(p, q, \lambda)$ be the Hamiltonian of the system, which depends on the parameter $\lambda(t)$. For this system, E changes according to

$$\frac{dE}{dt} = \frac{\partial H}{\partial t} = \frac{\partial H}{\partial \lambda}\frac{d\lambda}{dt}. \tag{3.81}$$

The above equation is averaged over the period (T) of the motion. In doing this, we can treat $d\lambda/dt$ as constant since λ changes only very slowly, and $d\lambda/dt$ does so.

$$\left\langle \frac{dE}{dt} \right\rangle = \frac{d\lambda}{dt}\frac{1}{T}\int_0^T \frac{\partial H}{\partial \lambda} dt. \tag{3.82}$$

Over one period, the Hamiltonian has a constant value E; $H(p, q, \lambda) = E$, constant. Thus, $p = p(q, \lambda)$ or $q = q(p, \lambda)$. Hence,

$$\frac{\partial H}{\partial \lambda} + \frac{\partial H}{\partial p}\frac{\partial p}{\partial \lambda} = 0. \tag{3.83}$$

We use Hamilton's equation $dq/dt = \partial H/\partial p$ to change dt to dq:

$$dt = \frac{dq}{\partial H/\partial p}, \quad T = \int_0^T dt = \oint \frac{dq}{\partial H/\partial p}. \tag{3.84}$$

Using (3.83) and (3.84) in (3.82) yields

$$\left\langle \frac{dE}{dt} \right\rangle = -\frac{d\lambda}{dt} \frac{\oint \frac{\partial p}{\partial \lambda} dq}{\oint \frac{dq}{\partial H/\partial p}}. \tag{3.85}$$

Also we have

$$\frac{1}{\partial H/\partial p} = \frac{1}{\partial E/\partial p} = \frac{\partial p}{\partial E}. \tag{3.86}$$

Using (3.86) in (3.85) yields

$$\oint \left(\frac{\partial p}{\partial E} \left\langle \frac{dE}{dt} \right\rangle + \frac{\partial p}{\partial \lambda} \frac{d\lambda}{dt} \right) dq = 0. \tag{3.87}$$

On the other hand, let us consider the action integral

$$I = \oint p\,(E, \lambda, q) dq. \tag{3.88}$$

We can show that the above action integral is a constant of motion by calculating dI/dt. For a loop integral, the integrated term vanishes [recall Leibnitz's rule for differentiation], and we have

$$\frac{dI}{dt} = \oint \left(\frac{\partial p}{\partial t} \right)_q dq = \oint \left(\frac{\partial p}{\partial E} \frac{dE}{dt} + \frac{\partial p}{\partial \lambda} \frac{d\lambda}{dt} \right) dq. \tag{3.89}$$

In view of (3.87), the RHS is zero, and we proved that the action integral I is constant.

- **Three adiabatic invariants.** 1) μ*−invariant*: magnetic moment is invariant in a periodic magnetic field which varies time-wise slowly on a long time scale as compared with the cyclotron period or which varies spatially with length scale much greater than the Larmor radius. 2) *longitudinal invariant*: $\int_a^b v_\parallel ds = invariant$ which is an approximate constant in a mirror type magnetic field when the magnetic field changes slowly temporally or the mirror field is not axisymmetric. In the integral, a and b are two turning points on the magnetic field line where the spiralling particle reverses its moving direction along the magnetic field line. The integral is performed along the magnetic field line, with $v_\parallel$ being the particle velocity parallel to the field line. The time scale of the change of magnetic field strength, in this case, should be long as compared with the bounce period. 3) Φ*−invariant*:

while a particle is spiralling along the field line in a mirror field it drifts across the field line due to the curvature of the field line. In Van Allen Belt, particles bounce back and forth between the turning points and simultaneously drift circumferentially, across the field lines — roughly speaking, east-west motion. The circumferential motion completes an approximate closed contour when the particle returns to the vicinity of the starting point. In this case the line integral $\oint$ *drift velocity* $\times$ *dl* along the circumferential orbit is an adiabatic invariant. By Stokes' theorem, the line integral is the same as the surface integral, which is the flux, magnetic field $\times$ the area of the equatorial plane (say). Hence, the magnetic flux Φ is adiabatic invariant.

Chapter 4

Basic equations of Vlasov–Maxwell plasma

In this Chapter, we study the Vlasov equation for the distribution function $f(\mathbf{r}, \mathbf{v}, t)$ which governs the change of the distribution of plasma particles in the electromagnetic fields $\mathbf{E}(\mathbf{r}, \mathbf{t})$ and $\mathbf{B}(\mathbf{r}, \mathbf{t})$ generated by the plasma particles themselves. The plasma current $\mathbf{J}$ and the density ρ_c are obtained from the distribution function upon appropriate integrals over velocity [see (4.8) and (4.9) below], which are the source of the plasma electromagnetic fields. Since the plasma electromagnetic fields are changed by $\mathbf{J}$ and ρ_c through the Maxwell equations, we have a coupled system of the kinetic equation and the electrodynamic equations in which the Vlasov equation and the Maxwell equations are coupled. The distribution function $f(\mathbf{r}, \mathbf{v}, t)$, and the electric and magnetic fields $\mathbf{E}(\mathbf{r}, \mathbf{t})$ and $\mathbf{B}(\mathbf{r}, \mathbf{t})$ evolve together in the course of time. This is a logically acceptable self-consistent and widely used model of a *hot* plasma in which short-range collisions between constituting particles are infrequent; long-range Coulomb collisions are prevalent. The adjective '*hot*' is appropriate since a high degree of ionization is required to be so. An interesting mathematical facet of the Vlasov equation is that it contains as a subset the solutions of a *cold* fluid equation (Euler equation); here the cold fluid equation does *not* mean the moment equation. See §Introduction. In the course of analysis of the Vlasov equation and the cold fluid equation in the Fourier space, we frequently encounter the algebraic expression $1/(\omega - kv)$ whose singularity at $\omega = kv$ poses important physical (as well as mathematical) issue. To deal with this algebraic singularity, the argument of *analytic continuation* by Landau is widely used. The issue of the singularity can be resolved by enforcing the *causal constraint* on the plasma susceptibility.

4.1. Vlasov equation

A plasma is described by the Vlasov equation which governs the time rate of change of the single particle distribution function $f_\alpha(\mathbf{r}, \mathbf{v}, t)$:

$$\frac{\partial}{\partial t} f_\alpha(\mathbf{r}, \mathbf{v}, t) + \mathbf{v} \cdot \frac{\partial f_\alpha}{\partial \mathbf{r}} + \frac{e_\alpha}{m_\alpha} \left[\mathbf{E}(\mathbf{r}, \mathbf{t}) + \frac{1}{c} \mathbf{v} \times \mathbf{B}(\mathbf{r}, t) \right] \cdot \frac{\partial f_\alpha}{\partial \mathbf{v}} = 0, \quad (4.1)$$

where the subscript α takes $e(i)$ for electron (ion) distribution function. The Vlasov equation above entails an approximation which is valid only when the number (N) of particles in the Debye sphere is much greater than unity:

$$N \lambda_{D\alpha}^3 \gg 1 \quad \left(\lambda_{D\alpha} = \sqrt{\frac{T_\alpha}{4\pi N e^2}}, \ \textit{Debye length}\right). \quad (4.2)$$

This criterion can be derived from the condition that Debye cross section (λ_D^2) is much greater than Coulomb cross section $[(e^2/T)^2]$, or $\lambda_D \gg e^2/T$. (4.1) is in fact a collisionless Boltzmann equation; we neglected the collision term. This is acceptable for sufficiently hot plasmas. If the particle collisions need to be included, we can use the Landau collision term (Chapter 1). A collisional plasma relaxes to a Maxwell–Boltzmann distribution in a characteristic time (*relaxation time*) $\tau = N\lambda_{D\alpha}^3/\omega_{p\alpha}$ where $\omega_{p\alpha} = \sqrt{4\pi N e^2/m_\alpha}$ is the plasma frequency. Use of (4.1) is therefore valid when it is applied to a dynamical change which takes place in a time short as compared to τ.

The Vlasov equation describes the behavior how the aggregate of the phase points move about in the six-dimensional phase space $(\mathbf{r}, \mathbf{v})$: In terms of the hydrodynamics terminology, the Vlasov *continuum* occupying the $(\mathbf{r}, \mathbf{v})$ phase space moves like an *incompressible* fluid, i.e.

$$\frac{Df(\mathbf{r}, \mathbf{v}, t)}{Dt} = 0 \quad (4.3)$$

$$\frac{D}{Dt} = \frac{\partial}{\partial t} + \frac{d\mathbf{r}}{dt} \cdot \frac{\partial}{\partial \mathbf{r}} + \frac{d\mathbf{v}}{dt} \cdot \frac{\partial}{\partial \mathbf{v}} \,. \quad (4.4)$$

This should be compared with the continuity equation, $\frac{d\rho}{dt} + \rho \nabla \cdot \mathbf{v} = 0$, for *compressible* fluid. The Vlasov description reflects the continuum aspect of plasma. In actuality, plasmas manifest discrete particle effects (besides the inter-particle collisions) which belong to the higher hierarchy of plasma kinetic equation. (4.1) can be derived heuristically by considering the time rate of change of the number of particles enclosed by a fixed, arbitrary volume in the phase space. In the absence of collisions among the particles, this rate is determined only by the flows entering or leaving the volume across the boundary. Here the boundary is a surface in the six-dimensional phase space.

The Vlasov fluid flux across the boundary is expressed by $(\dot{\mathbf{r}}f, \dot{\mathbf{v}}f)$ $[\cdot = d/dt]$, and we can write

$$\frac{\partial}{\partial t}\int d^3r d^3v\, f = -\int d^3v d^2\mathbf{r}\cdot\dot{\mathbf{r}}f - \int d^3r d^2\mathbf{v}\cdot\dot{\mathbf{v}}f.$$

Applying the Gauss theorem to the surface integrals on the right, we have

$$\int d^3r d^3v \left[\frac{\partial f}{\partial t} + \frac{\partial}{\partial \mathbf{r}}\cdot(\dot{\mathbf{r}}f) + \frac{\partial}{\partial \mathbf{v}}\cdot(\dot{\mathbf{v}}f)\right] = 0.$$

Using $m\dot{\mathbf{v}} = e(\mathbf{E} + \frac{1}{c}\mathbf{v}\times\mathbf{B})$ in the above gives

$$\frac{\partial f}{\partial t} + \mathbf{v}\cdot\frac{\partial f}{\partial \mathbf{r}} + \frac{e}{m}\left(\mathbf{E} + \frac{1}{c}\mathbf{v}\times\mathbf{B}\right)\cdot\frac{\partial f}{\partial \mathbf{v}} = 0 \tag{4.5}$$

which is (4.1). This derivation reflects the fluid-like behavior of the Vlasov plasma.

4.2. Maxwell equations

In (4.1), the electric and magnetic fields and the distribution functions are connected by Maxwell equations:

$$\nabla\times\mathbf{E} = -\frac{1}{c}\frac{\partial \mathbf{B}}{\partial t}, \tag{4.6}$$

$$\nabla\times\mathbf{B} = \frac{1}{c}\frac{\partial \mathbf{E}}{\partial t} + \frac{4\pi}{c}\mathbf{J}, \tag{4.7}$$

where the current $\mathbf{J}$ is generated by plasma particles:

$$\mathbf{J}(\mathbf{r},t) = \sum_\alpha e_\alpha \int d^3v\; \mathbf{v}\; f_\alpha(\mathbf{r},\mathbf{v},t) \tag{4.8}$$

and we have the Poisson equation

$$\nabla\cdot\mathbf{E}(\mathbf{r},t) = 4\pi\sum_\alpha e_\alpha \int d^3v f_\alpha(\mathbf{r},\mathbf{v},t) \equiv 4\pi\rho_c \tag{4.9}$$

where ρ_c is the charge density. $\nabla\cdot$ (4.6) gives $\frac{\partial}{\partial t}\nabla\cdot\mathbf{B} = 0$ which means $\nabla\cdot\mathbf{B} = 0$ always. $\nabla\cdot$ (4.7) yields

$$4\pi\nabla\cdot\mathbf{J} + \frac{\partial}{\partial t}\nabla\cdot\mathbf{E} = 0. \tag{4.10}$$

Using (4.9) in (4.10) gives

$$\nabla\cdot\mathbf{J} + \frac{\partial \rho_c}{\partial t} = 0. \tag{4.11}$$

This is the equation of charge conservation. (4.5)–(4.9) show how f_α and the fields $\mathbf{E}$ and $\mathbf{B}$ are coupled. Here $\mathbf{E}$ and $\mathbf{B}$ are the *self-consistent* fields, since they are generated by the plasma itself.

(4.5) is a nonlinear equation because of the last term which is a product of two unknown functions. The first step to study the Maxwell–Vlasov plasma governed by the foregoing equations is the investigation of the linear waves derived from the linearized equations. Maxwell equations are linear. Linearizing (4.5), we have the linearized Vlasov equation in the following form

$$\frac{\partial}{\partial t} f_\alpha(\mathbf{r}, \mathbf{v}, t) + \mathbf{v} \cdot \frac{\partial f_\alpha}{\partial \mathbf{r}} + \frac{e_\alpha}{m_\alpha} \left[\mathbf{E}(\mathbf{r}, t) + \frac{\mathbf{v}}{c} \times \mathbf{B}(\mathbf{r}, t) \right] \cdot \frac{\partial f_{\alpha 0}}{\partial \mathbf{v}} = 0, \tag{4.12}$$

where $f_{\alpha 0}$ is the zero order equilibrium distribution function which is usually assumed to be a Maxwellian. f_α is now the perturbation. Note that the last term vanishes ($\frac{\mathbf{v}}{c} \times \mathbf{B}(\mathbf{r}, t) \cdot \frac{\partial f_{\alpha 0}}{\partial \mathbf{v}} = 0$) if we assume $f_{\alpha 0}(v)$ to be isotropic, since $\frac{\partial f_{\alpha 0}}{\partial \mathbf{v}} \sim \mathbf{v}$ in this case. The isotropy of f_0 will be assumed in the following discussion.

Linear solutions should be thoroughly investigated for the interest in the linear equation itself and for being ready to embark on the study of nonlinear solutions. Linear solutions are valid only within a limited duration after the perturbation sets in. (4.5)–(4.9) are transformed to a homogeneous set of algebraic equations if written in Fourier amplitudes. Such a set of simultaneous homogeneous equations has nontrivial solutions only when a *dispersion relation* is satisfied. The dispersion relation is nothing but the solvability condition of the homogeneous set of the algebraic equations. Solutions of the dispersion relation determine the *normal modes* of the plasma. Assuming $\sim e^{i\mathbf{k}\cdot\mathbf{r} - i\omega t}$ solutions for all the unknown functions in the above, we have

$$(-i\omega + i\mathbf{k} \cdot \mathbf{v}) f_\alpha(\mathbf{k}, \mathbf{v}, \omega) + \frac{e_\alpha}{m_\alpha} \mathbf{E}(\mathbf{k}, \omega) \cdot \frac{\partial f_{\alpha 0}}{\partial \mathbf{v}} = 0, \tag{4.13}$$

$$\mathbf{k} \times \mathbf{E} = \frac{\omega}{c} \mathbf{B}, \tag{4.14}$$

$$\mathbf{k} \times \mathbf{B} = -\frac{\omega}{c} \mathbf{E} - \frac{4\pi i}{c} \mathbf{J}. \tag{4.15}$$

Eliminating $\mathbf{B}$ in (4.14) and (4.15) gives

$$\mathbf{k} \times (\mathbf{k} \times \mathbf{E}) + \frac{\omega^2}{c^2} \mathbf{E} = -\frac{4\pi i\omega}{c^2} \mathbf{J}. \tag{4.16}$$

Equations (4.8) and (4.13) give

$$J_i = -i\ E_j \sum_\alpha \frac{e_\alpha^2}{m_\alpha} \int d^3v\ \frac{v_i \frac{\partial f_{\alpha 0}}{\partial v_j}}{\omega - \mathbf{k}\cdot\mathbf{v}} \tag{4.17}$$

where the subscripts i and j are Cartesian indices and the summation convention is used.

4.3. Conductivity, susceptibility, dielectric permittivity

Introducing the conductivity tensor σ_{ij}, we have $J_i = \sigma_{ij}E_j$, so we find from (4.17)

$$\sigma_{ij} = -i \sum_\alpha \frac{e_\alpha^2}{m_\alpha} \int d^3v\ \frac{v_i \frac{\partial f_{\alpha 0}}{\partial v_j}}{\omega - \mathbf{k}\cdot\mathbf{v}}. \tag{4.18}$$

We define the susceptibility and the dielectric permittivity tensors:

$$\chi_{ij} = \frac{4\pi i}{\omega}\sigma_{ij}, \quad \varepsilon_{ij} = \delta_{ij} + \chi_{ij}. \tag{4.19}$$

Then we have the expression for the dielectric tensor:

$$\varepsilon_{ij} = \delta_{ij} + \sum_\alpha \frac{\omega_{p\alpha}^2}{\omega} \int d^3v\ \frac{v_i \frac{\partial f_{\alpha 0}}{\partial v_j}}{\omega - \mathbf{k}\cdot\mathbf{v}}. \tag{4.20}$$

(4.16) is written as

$$\left[n^2\left(\frac{k_i k_j}{k^2} - \delta_{ij}\right) + \varepsilon_{ij}\right] E_j = 0 \quad \left(n^2 = \frac{c^2k^2}{\omega^2},\quad \textit{refractive index}\right). \tag{4.21}$$

We have the following important relation for ε_{ij} which is valid if $f_{\alpha 0}$ is isotropic:

$$\varepsilon_{ij} = \delta_{ij}\varepsilon_T + (\varepsilon_L - \varepsilon_T)\frac{k_i k_j}{k^2} \tag{4.22}$$

$$\varepsilon_L = 1 + \Sigma\frac{\omega_{p\alpha}^2}{k^2} \int \frac{\mathbf{k}\cdot\frac{\partial f_{\alpha 0}}{\partial \mathbf{v}}}{\omega - \mathbf{k}\cdot\mathbf{v}} d^3v : \textit{longitudinal dielectric constant} \tag{4.23}$$

$$\varepsilon_T = 1 - \Sigma\frac{\omega_{p\alpha}^2}{\omega} \int \frac{f_{\alpha 0}}{\omega - \mathbf{k}\cdot\mathbf{v}} d^3v : \textit{transverse dielectric constant.} \tag{4.24}$$

Proof of (4.22) goes as follows. The integral in (4.20) can be written as

$$\int \frac{v_i \partial f_0/\partial v_j}{\omega - \mathbf{k}\cdot\mathbf{v}} d^3v = -\int f_0 \frac{\partial}{\partial v_j}\left(\frac{v_i}{\omega - \mathbf{k}\cdot\mathbf{v}}\right) d^3v$$
$$= -\delta_{ij}\int \frac{f_0 d^3v}{\omega - \mathbf{k}\cdot\mathbf{v}} - \int \frac{f_0 v_i k_j d^3v}{(\omega - \mathbf{k}\cdot\mathbf{v})^2}. \tag{4.25}$$

We use the following identity

$$\frac{1}{(\omega - \mathbf{k}\cdot\mathbf{v})^2} = \frac{k_l}{k^2}\frac{\partial}{\partial v_l}\frac{1}{\omega - \mathbf{k}\cdot\mathbf{v}}. \tag{4.26}$$

So the last term of (4.25)=

$$-\frac{k_j k_l}{k^2}\int f_0 v_i \frac{\partial}{\partial v_l}\frac{1}{\omega - \mathbf{k}\cdot\mathbf{v}} d^3v = \frac{k_j k_l}{k^2}\int \frac{f_0\delta_{il} + v_i \partial f_0/\partial v_l}{\omega - \mathbf{k}\cdot\mathbf{v}} d^3v$$
$$= \frac{k_i k_j}{k^2}\int \frac{f_0\, d^3v}{\omega - \mathbf{k}\cdot\mathbf{v}} + \frac{k_j k_l}{k^2}\int \frac{v_i \partial f_0/\partial v_l}{\omega - \mathbf{k}\cdot\mathbf{v}} d^3v. \tag{4.27}$$

Here we use $\partial f_0/\partial v_l = v_l f_0'/v$ $(f_0' = \frac{df_0}{dv})$. So the last term in (4.27) becomes

$$\frac{k_j}{k^2}\int \frac{v_i}{v}\frac{\mathbf{k}\cdot\mathbf{v}}{\omega - \mathbf{k}\cdot\mathbf{v}} f_0' d^3v$$
$$= \frac{k_j}{k^2}\int \frac{v_i}{v} f_0'\left(-1 + \frac{\omega}{\omega - \mathbf{k}\cdot\mathbf{v}}\right) d^3v = \frac{k_j\omega}{k^2}\int \frac{\partial f_0/\partial v_i}{\omega - \mathbf{k}\cdot\mathbf{v}} d^3v$$

because the -1 term vanishes upon integration (odd function integral). Integrating by parts the last term gives

$$-\frac{k_i k_j \omega}{k^2}\int \frac{f_0 d^3v}{(\omega - \mathbf{k}\cdot\mathbf{v})^2} = \frac{\omega k_i k_j}{k^4}\int \frac{k_l \partial f_0/\partial v_l}{\omega - \mathbf{k}\cdot\mathbf{v}} d^3v \tag{4.28}$$

where we used (4.26) again and integrated by parts. Collecting the above results establishes (4.22). In (4.22) we note that the dielectric tensor of an isotropic velocity distribution can be written as a linear combination of two coordinate-independent basic tensors δ_{ij} and $k_i k_j/k^2$.

Exercise. Show from (4.22) that

$$\varepsilon_L = \frac{k_i k_j}{k^2}\,\varepsilon_{ij}, \quad \varepsilon_T = \frac{1}{2}(\varepsilon_{ii} - \varepsilon_L) \tag{4.29}$$

where ε_{ii} is the trace of the tensor ε_{ij}.

Exercise. Calculate the right hand sides of (4.29) from (4.20).

4.4. Dispersion relation

(4.21) is a set of homogeneous linear simultaneous equations, and the solvability condition requires the determinant of the matrix composed by the coefficients of E_j to vanish:

$$\Lambda_{ij} \equiv n^2 \left(\frac{k_i k_j}{k^2} - \delta_{ij} \right) + \varepsilon_{ij}; \quad det\ [\Lambda_{ij}] = 0. \tag{4.30}$$

Direct expansion of the determinant gives

$$det\ [\Lambda_{ij}] = \varepsilon_L (\varepsilon_T - n^2)^2 = 0. \tag{4.31}$$

We have two modes of wave satisfying the dispersion relations:

$$\varepsilon_L = 0 \quad (electrostatic\ wave), \tag{4.32}$$

$$\varepsilon_T = n^2 \quad (electromagnetic\ wave). \tag{4.33}$$

Use of (4.22) in (4.21) gives

$$\Lambda_{ij} E_j = \varepsilon_L \frac{\mathbf{k} \cdot \mathbf{E}}{k^2} \mathbf{k} + (\varepsilon_T - n^2) \left(\mathbf{E} - \frac{\mathbf{k} \cdot \mathbf{E}}{k^2} \mathbf{k} \right) = 0. \tag{4.34}$$

If the dispersion relation (4.32) or (4.33) is substituted into the wave equation (4.21), we obtain the polarization of the electric field vector. From the above equation, we see that the $\varepsilon_L = 0$ mode satisfies the relation

$$\mathbf{E} = \hat{\mathbf{k}} (\mathbf{E} \cdot \hat{\mathbf{k}}) \quad (electrostatic\ wave,\ \nabla \times \mathbf{E} = 0). \tag{4.35}$$

Also, when $\mathbf{E}$ satisfies (4.35), (4.34) is satisfied by $\varepsilon_L = 0$. In the limit of $n \to \infty$ (or $c \gg \omega/k$), (4.34) stands meaningful only when the electric field is polarized in such a way that $\mathbf{E} = \mathbf{k}\ \mathbf{k} \cdot \mathbf{E}/k^2$. Then we have $\nabla \times \mathbf{E} = 0$ and (4.34) gives the dispersion relation $\varepsilon_L = 0$. Algebraically we obtain the electrostatic dispersion relation from the general electromagnetic dispersion relation by formally taking the limit $c \to \infty$.

The $\varepsilon_T = n^2$ mode should have the relation

$$\hat{\mathbf{k}} \cdot \mathbf{E} = 0 \quad (electromagnetic\ wave,\ \nabla \cdot \mathbf{E} = 0). \tag{4.36}$$

Electric displacement ($\mathbf{D}$) is defined by

$$\nabla \times \mathbf{B} = \frac{4\pi}{c} \mathbf{J} + \frac{1}{c} \frac{\partial \mathbf{E}}{\partial t} = \frac{1}{c} \frac{\partial \mathbf{D}}{\partial t}. \tag{4.37}$$

Electric polarization (**P**) is defined by

$$\mathbf{J} = \frac{\partial \mathbf{P}}{\partial t}. \tag{4.38}$$

So we have

$$\mathbf{D} = \mathbf{E} + 4\pi\mathbf{P}, \quad -4\pi\nabla\cdot\mathbf{P} = \nabla\cdot\mathbf{E} = 4\pi\rho_c \tag{4.39}$$

$$\rho_c = -\nabla\cdot\mathbf{P}. \quad (\textit{charge density}) \tag{4.40}$$

Equation (4.38) is characteristic of plasma: in plasma we assume that there is no magnetization current, which means that we take the magnetic permeability to be 1. The causal relationship between **D** and **E** defines **D** as the effect while **E** is regarded to be the cause. The electric displacement **D** represents the response of the medium when it is acted on by the electric field **E**. Fourier transforming (4.37) gives

$$D_i(\mathbf{k},\omega) = \varepsilon_{ij}\ (\mathbf{k},\omega)\ E_j(\mathbf{k},\omega). \tag{4.41}$$

Using (4.22) in the above equation gives

$$\mathbf{D}(\mathbf{k},\omega) = \varepsilon_T\mathbf{E}_\perp\ + \varepsilon_L\mathbf{E}_\parallel \tag{4.42}$$

where $\perp$ and $\parallel$ are referred to the direction of **k**: $\mathbf{E}_\parallel = \hat{k}\ (\hat{k}\cdot\mathbf{E})$ and $\mathbf{E}_\perp = \mathbf{E} - \hat{k}(\hat{k}\cdot\mathbf{E})$. So $\varepsilon_{T,\ L}$ determines, respectively, the contribution to the electric displacement **D** along the direction perpendicular and parallel to **k**. Depending upon whether the electric field is perpendicular or parallel to **k**, the relation between **D** and **E** reduces respectively to

$$\mathbf{D} = \varepsilon_T\mathbf{E} \quad or \quad \mathbf{D} = \varepsilon_L\mathbf{E}.$$

Dielectric tensor in anisotropic plasma. Even if no static magnetic field is present, a plasma can be anisotropic when the zeroth order distribution function has a temperature anisotropy or when the plasma is drifting. In this case the wave magnetic field term in (4.1) does not vanish. Eliminating the wave magnetic field by $\mathbf{B} = \frac{c}{\omega}\ \mathbf{k}\times\mathbf{E}$, current **J** is obtained as

$$J_i = -iE_j\ \Sigma_\alpha\ \frac{e_\alpha^2}{m_\alpha\omega}\int\frac{d^3v}{\omega - \mathbf{k}\cdot\mathbf{v}}\left[(\omega - \mathbf{k}\cdot\mathbf{v})\ v_i\ \frac{\partial f_{\alpha 0}}{\partial v_j} + v_i v_j k_l\frac{\partial f_{\alpha 0}}{\partial v_l}\right].$$

The coefficient of E_j is σ_{ij}. Then the dielectric tensor takes the form

$$\varepsilon_{ij} = \delta_{ij} + \Sigma \frac{\omega_{p\alpha}^2}{\omega^2} \int d^3v \; \frac{v_i \left[\delta_{jl}(\omega - \mathbf{k}\cdot\mathbf{v}) + v_j k_l\right] \frac{\partial f_{\alpha 0}}{\partial v_l}}{\omega - \mathbf{k}\cdot\mathbf{v}}. \tag{4.43}$$

Exercise. Show that the above dielectric tensor ε_{ij} reduces to (4.20) if $f_{\alpha 0}$ is isotropic. Hint: use $\frac{\partial f_{\alpha 0}}{\partial v_l} = \frac{v_l}{v} \frac{df_{\alpha 0}(v)}{dv}$ and discard the odd function integrals.

In this case, dispersion relation is obtained by substituting (4.43) into (4.30). However, the dielectric tensor (4.43) for anisotropic plasma cannot be expressed in terms of the basic tensors $k_i k_j / k^2$ and δ_{ij} .

4.5. Causality

Causality in electrodynamics means that the response (effect) must always follow the cause; the cause cannot be precedent to the effect. This notion is implicitly related with the condition for a function to have the Fourier transform. For a Fourier component of a wave with phasor

$$e^{ikx - i\omega t} \tag{4.44}$$

the principle of causality can be operationally incorporated by assuming that the frequency ω has an infinitesimally small *positive* imaginary part. Then the disturbance at infinitely remote past ($t = -\infty$) is nil in accordance with the causal notion that the time $t = -\infty$ is necessarily prior to the time of occurrence of any stimulus (cause) which gives rise to the onset of disturbance (effect). The small positive imaginary part (ν) of the frequency is called *adiabatic switching* on of the perturbation at $t = -\infty$. The small positive imaginary frequency ν arises mathematically if one solves the plasma equations as an initial value problem via Laplace transform, as Landau demonstrated [see next Chapter], but it can be simply invoked as an imaginary frequency in accordance with the principle of causality. Adiabatic switching supplies the first definition of causality that distinguishes the past from the future.

We mention that the phasor in the expression (4.44) is compatible with the Fourier transform convention adopted in this book: Fourier transform of a wave function $g(x,t)$ and its inversion are

$$g(x,t) = \int_{-\infty}^{\infty} \int_{-\infty}^{\infty} g(k,\omega) \; e^{ikx - i\omega t} \; \frac{dk}{2\pi} \frac{d\omega}{2\pi}, \tag{4.45}$$

$$g(k,\omega) = \int_{-\infty}^{\infty} \int_{-\infty}^{\infty} g(x,t) \; e^{-ikx + i\omega t} \; dxdt. \tag{4.46}$$

One who uses different sign in front of the time-phase in the exponential, the frequency ω should be assumed to have a negative small imaginary frequency.

In order to have an in-depth discussion of the above causality notion, let us consider a trivial example to solve a differential equation

$$\frac{df}{dt} + a\ f(t) + b = 0$$

where a and b are real constants. Let us Fourier transform the above equation to change $f(t) \to f(\omega)$. The first term gives in the transform operation

$$\int_{-\infty}^{\infty} \frac{df}{dt} e^{i\omega t} dt = \left[f(t) e^{i\omega t} \right]_{-\infty}^{\infty} - i\omega f(\omega).$$

The integrated term must vanish, so we need appropriate logic to discard it. At $t = -\infty$, we can assume $f(-\infty) = 0$, physically, because the universe was not created at the remote past. But perturbation (everything is perturbation) begins to grow afterwards to realize physical interaction. To mimic the prehistoric perturbation, we endow ω with a small (infinitesimal) *positive* imaginary part. But the primordial small imaginary part is not enough; $Im\ \omega$, the imaginary part, must be large enough so that the integrated term at $t = +\infty$ vanishes. How large $Im\ \omega$ must be depends on $f(t)$. If $f(t) \sim e^{\alpha t}$, we should have $Im\ \omega > \ \alpha$. So we have to work on a limited region of ω plane. We obtain

$$f(\omega)(\omega + ia) = -2\pi i\ b\ \delta(\omega).$$

This algebraic equation requires cautiousness in writing its solution:

$$f(\omega) = \frac{-2\pi i\ b\ \delta(\omega)}{\omega + ia} + c\ \delta(\omega + ia) \quad c : const.$$

of which inversion integral is

$$f(t) = \int_{-\infty}^{\infty} \frac{d\omega}{2\pi}\ e^{-i\omega t} \left[\frac{-2\pi i b\ \delta(\omega)}{\omega + ia} + c\ \delta(\omega + ia) \right].$$

In the preceding step, we note $(\omega + ia)\delta(\omega + ia) = 0$ identically. The value of the first integral is simply the contribution at $\omega = 0$ due to the function $\delta(\omega)$, giving $(-b/a)$. The second integral can be done by contour integration with the contour enclosing the lower ω plane by infinite semi-circle for $t > 0$ plus the real ω line integral. But the latter contour should be indented to exclude the δ function singularity at $\omega = -ia$, which is excluded by a whole infinitesimal circle surrounding the point $\omega = -ia$ and two vertical line integrals running in the opposite direction. Thus, we obtain

$$f(t) = -\frac{b}{a} + const. \times e^{-at} \quad (t > 0).$$

We consider a one-dimensional Vlasov–Poisson plasma which is governed by the following linearized equations:

$$\frac{\partial}{\partial t} g(x, v, t) + v\frac{\partial g}{\partial x} - \frac{e}{m} E(x, t) \frac{dg_0(v)}{dv} = 0 \tag{4.47}$$

$$\frac{\partial}{\partial x} E(x, t) = -4\pi e \int_{-\infty}^{\infty} dv \; g(x, v, t) \tag{4.48}$$

where $g(x, v, t)$ is the perturbed electron distribution function, $g_0(v)$ is the zero order equilibrium electron distribution function, and the ions are assumed to form the uniform neutralizing background. We Fourier transform Eqs. (4.47) and (4.48) by performing the integral $\int \int e^{-ikx+i\omega t}(\cdots)dxdt$. Let us scrutinize the time-Fourier transform of the first term of Eq. (4.47):

$$\int_{-\infty}^{\infty} \frac{\partial g(x, v, t)}{\partial t} e^{i\omega t} dt = \left[g(x, v, t)e^{i\omega t} \right]_{-\infty}^{\infty} - i\omega g(x, v, \omega).$$

At $t = \infty$, $e^{i\omega t} \to 0$ (provided $g(t)$ doesn't blow up) since ω has a small positive imaginary part. At $t = -\infty$, we have $g(x, v, t = -\infty) = 0$ according to the aforementioned causal boundary condition. Thus, Eq. (4.47) is Fourier transformed unambiguously by the prescription of endowing ω a small positive imaginary part. We obtain from Eqs. (4.47) and (4.48)

$$g(k, v, \omega) = \frac{e}{m} \frac{dg_0}{dv} \frac{i}{\omega - kv} E(k, \omega), \tag{4.49}$$

$$ikE(k, \omega) = -4\pi e \int_{-\infty}^{\infty} dv \; g(k, v, \omega). \tag{4.50}$$

Therefore the aforementioned causal principle is equivalent to the mathematical condition that Fourier transform to exist.

4.5.1. *Causality and susceptibility*

In this subsection, we calculate the dielectric function or the susceptibility in general terms to see how the principle of causality is incorporated in electrodynamics.

We have for the electric displacement vector $\mathbf{D}(t)$ and the plasma current $\mathbf{J}(t)$ the following equation:

$$\frac{1}{c}\frac{\partial \mathbf{D}}{\partial t} = \frac{1}{c}\frac{\partial \mathbf{E}}{\partial t} + \frac{4\pi}{c}\mathbf{J}. \tag{4.51}$$

Here, we consider a particular Fourier component $\mathbf{k}$ but it will not be explicitly written. $\mathbf{k}$ can be reinstated at an appropriate stage if desirable. We integrate (4.51) with the initial conditions $\mathbf{D}(t) = 0$ and $\mathbf{E}(t) = 0$ at $t = -\infty$ to get

$$\mathbf{D}(t) = \mathbf{E}(t) + 4\pi \int_{-\infty}^{t} \mathbf{J}(t')dt' = \mathbf{E}(t) + 4\pi \int_{0}^{\infty} \mathbf{J}(t-\tau)d\tau, \tag{4.52}$$

where we made a change of variable, $t' = t - \tau$. In (4.52), $\mathbf{J}$ is related with $\mathbf{E}$ through a material relation. The displacement vector $\mathbf{D}(t)$ then represents the response of the medium, or the effect caused by the input $\mathbf{E}(t)$. Introducing the Fourier integral for $\mathbf{J}(t)$,

$$\mathbf{J}(t) = \frac{1}{2\pi}\int_{-\infty}^{\infty} \mathbf{J}(\omega)e^{-i\omega t}d\omega,$$

the last term of (4.52) can be written as

$$4\pi \int_{0}^{\infty} d\tau e^{i\omega\tau} \frac{1}{2\pi}\int_{-\infty}^{\infty} \mathbf{J}(\omega)e^{-i\omega t}d\omega.$$

The $\int d\tau$-integral gives i/ω (ω has a small positive imaginary part). Writing $J_i = \sigma_{ij}E_j$, and $\chi_{ij} = \frac{4\pi i}{\omega}\sigma_{ij}$ with σ_{ij} being the conductivity tensor, the above expression becomes

$$\frac{1}{2\pi}\int_{-\infty}^{\infty} d\omega\ \chi_{ij}(\omega)E_j(\omega)e^{-i\omega t}$$

which is the inverse Fourier transform of $\chi_{ij}(\omega)E_j(\omega)$, and can be expressed as a convolution integral. Therefore, we have

$$D_i(t) = E_i(t) + \int_{-\infty}^{\infty} \chi_{ij}(t')E_j(t-t')dt'. \tag{4.53}$$

Causality demands that the kernel $\chi(t)$ in (4.53) be zero for $t < 0$. This is so because $\mathbf{D}$ at present time t should be irrelevant to the values of $\mathbf{E}$ at future times. This agrees with the basic idea of causality in natural phenomena and is vindicated by experience. Thus, we *impose* the condition,

$$\chi_{ij}(t) = 0, \quad for \quad t < 0. \tag{4.54}$$

Then, (4.53) is rewritten as

$$D_i(t) = E_i(t) + \int_0^\infty \chi_{ij}(t')E_j(t-t')dt'. \tag{4.55}$$

Fourier transforming (4.55) gives

$$D_i(\omega) = E_i(\omega) + \int_{-\infty}^\infty dt\; e^{i\omega t} \int_0^\infty \chi_{ij}(t')E_j(t-t')dt'.$$

Via a change of variable, $t - t' = t''$, $dt \to dt''$, the above equation takes the form

$$D_i(\omega) = \left(\delta_{ij} + \int_0^\infty \chi_{ij}(t')e^{i\omega t'}dt'\right) E_j(\omega). \tag{4.56}$$

This equation can be written in terms of the dielectric tensor ε_{ij} as

$$D_i(\omega) = \varepsilon_{ij}(\omega)E_j(\omega); \quad \varepsilon_{ij}(\omega) = \delta_{ij} + \chi_{ij}(\omega),$$
$$\chi_{ij}(\omega) = \int_0^\infty \chi_{ij}(t)\; e^{i\omega t}dt. \tag{4.57}$$

Clearly $\chi(\omega)$ in the above equation is analytic in the upper half ω–plane because the integrand is analytic therein and the integral is finite at the upper limit $t \to \infty$. We emphasize that the causal requirement expressed by (4.54) is an important constraint on the susceptibility function $\chi(t)$, which is defined by the inverse transform of (4.57). The constraint can be viewed as the foremost cause for the collisionless damping of plasma waves, as is discussed in the next section.

4.5.2. *Kramers–Kronig relations*

The Kramers–Kronig (K-K) relations relate the real (imaginary) part of the susceptibility function $\chi(\omega)$ to the imaginary (real) part of $\chi(\omega)$. If one knows one of the two, the other can be obtained from the known function by Hilbert transform. Usually the K-K relations are derived by the mathematical theory of the analytic continuation from the analytic region of $\chi(\omega)$, which is the upper-half of the complex ω-plane, to the real ω-axis. [See Arfken and Weber, *Mathematical methods for physicists*, Academic Press (2001).] Here we present a new and more economical way of deriving the K-K relations by directly applying the constraint on $\chi(t)$ in (4.54). One can enforce the constraint (4.54) by noting its step function nature as

$$\chi(t) = \frac{1}{2}\,[H(t)+1]\,\chi(t) \tag{4.58}$$

where $H(t)$ is the step function: H(t) = 1 for $t > 0$ and $H(t) = -1$ for $t < 0$, and the k-dependance is implicit. Clearly, the right hand side of (4.58) is equal to $\chi(t)$ for $t > 0$ and zero for $t < 0$. Fourier transform of (4.58) is written as a convolution integral

$$\chi(\omega) = \frac{1}{2}\,\frac{1}{2\pi}\int_{-\infty}^{\infty} d\omega'\chi(\omega')H(\omega-\omega') + \frac{1}{2}\chi(\omega). \tag{4.59}$$

The Fourier transform of the step function is (see Appendix A)

$$H(\omega) = 2i\;P\;\frac{1}{\omega} \tag{4.60}$$

where P stands for the symbol of the principal value. Using Eq. (4.60) in (4.59) yields

$$\chi(\omega) = \frac{i}{\pi}\;P\int_{-\infty}^{\infty}\frac{\chi(\omega')}{\omega-\omega'}d\omega'. \tag{4.61}$$

This is the implicit form of the K-K relations obtained by direct application of the causal requirement in (4.54). Writing $\chi = Re\;\chi + i\;Im\;\chi$ and separating the real and the imaginary parts, (4.61) gives the K-K relations

$$Re\;\chi(\omega) = -\frac{1}{\pi}P\int_{-\infty}^{\infty}\frac{Im\;\chi(\omega')}{\omega-\omega'}d\omega', \tag{4.62}$$

$$Im\;\chi(\omega) = \frac{1}{\pi}P\int_{-\infty}^{\infty}\frac{Re\;\chi(\omega')}{\omega-\omega'}d\omega'. \tag{4.63}$$

• **Development from (4.57).** We obtained (4.61) directly from (4.54). Here we derive (4.61) from (4.57). Let us use the Fourier transform expression for $\chi(t)$ in (4.57) and carry out the $\int dt$–integral. Then (4.57) converts to a Cauchy integral equation. This is shown as follows:

$$\begin{aligned}\chi(\omega) &= \int_0^{\infty} dt\;e^{i\omega t}\int_{-\infty}^{\infty}\chi(\omega')e^{-i\omega' t}\,\frac{d\omega'}{2\pi}\\ &= \int_{-\infty}^{\infty}\frac{d\omega'}{2\pi}\chi(\omega')\int_0^{\infty} dt\;e^{i(\omega-\omega')t}.\end{aligned}$$

Carrying out the $\int dt$-integral yields

$$\chi(\omega) = \frac{1}{2\pi i}\int_{-\infty}^{\infty}\frac{\chi(\omega')}{\omega'-\omega}d\omega' \quad (\omega_i > 0). \tag{4.64}$$

Equation (4.64) can be expressed as a contour integral with contour consisting of the real ω–axis and the infinite semi-circle winding the upper half ω–plane, because the contribution from the infinite semi-circle vanishes. Then, according to Cauchy's integral theorem, the function $\chi(\omega)$ is analytic in the upper half ω–plane. *The causal constraint in* (4.54) *endows* $\chi(\omega)$ *with the analyticity in the upper half* ω*–plane.* The Cauchy integral in (4.64) is a delta function-like relation which holds in half range Fourier integral. [cf. $\chi(\omega) = \int_{-\infty}^{\infty}\chi(\omega')\delta(\omega-\omega')d\omega'$ in the whole range integral.]

Now we let the complex variable ω in (4.64) become real by descending the point ω downward so that it lands on the real axis. This limiting process is accomplished by the following steps. We put $\omega = \omega + i\nu$ $(\nu > 0)$. Then we have

$$\frac{1}{\omega'-\omega} = \frac{1}{\omega'-\omega-i\nu} = \frac{\omega'-\omega+i\nu}{(\omega'-\omega)^2+\nu^2}$$
$$= P\,\frac{1}{\omega'-\omega} + i\pi\,\delta(\omega'-\omega)$$

where the symbol P denotes the principal value. The last expression is the result of taking the limit $\nu \to 0$. This limiting process gives

$$\chi(\omega) = \frac{1}{\pi i}P\int_{-\infty}^{\infty}\frac{\chi(\omega')}{\omega'-\omega}d\omega' \quad (\omega_i = 0)$$

which is (4.61), and is the analytic continuation of the function in (4.64) onto the domain $\omega_i = 0$.

• **Analytic continuation and indentation of the integral path.** Let us consider an analytic function $F(\omega)$, analytic in the upper half of the ω-plane with the property $lim_{\omega\to\infty} \mid F(\omega) \mid = 0$. Then, by Cauchy's integral formula, we can write

$$F(\omega) = \frac{1}{2\pi i}\int_C \frac{F(\omega')}{\omega'-\omega}d\omega'$$

where ω is an interior point of the closed contour C that consists of the real ω-axis and the infinite semi-circle in the upper half-plane. Because the contribution from the infinite semi-circle vanishes, the above integral can be written as

$$F(\omega) = \frac{1}{2\pi i}\int_{-\infty}^{\infty} \frac{F(\omega')}{\omega' - \omega} d\omega'.$$

This equation reads a complex function $F(\omega)$ of a complex variable ω. Now, we let the variable ω become real by descending the point ω downward so that it lands on the real axis. Then, the integration path in the integral, the path along the real ω-axis, gets deformed into the concave-down shape of an infinitesimal semicircle around the singular point $\omega' = \omega$. Thus, the integral is a sum of the principal value of the integral plus the contribution from the infinitesimal semicircle. The result is (4.61).

In this Section, we have shown that the causal constraint $\chi(t < 0) = 0$ in (4.54) implies the analyticity of $\chi(\omega)$ in the upper-half ω plane, and vice versa. See the summary at the end of this Chapter.

4.5.3. *Hilbert transform*

The pair of equations, (4.62) and (4.63), is Hilbert transform of each other. Any one of them can be obtained from the other by Hilbert transform. The *Hilbert inversion* is facilitated by the identity (See § 4.6.6)

$$\frac{1}{\pi^2} P \int_{-\infty}^{\infty} \frac{1}{x' - x} \frac{1}{x' - x''} dx' = \delta(x - x''), \tag{4.65}$$

which is the closure property of the Hilbert transform. Equation (4.65) can be deduced when one solves for one of the K-K relations as a given integral equation without knowing the other relation. Here, by eliminating either $Re\ \chi(\omega)$ or $Im\ \chi(\omega)$ between (4.62) and (4.63), we can ascertain that (4.65) holds. In order to invert (4.62), let us multiply both sides by $P \frac{1}{\omega - \omega''}$ and integrate $\int_{-\infty}^{\infty} d\omega(\cdots)$:

$$P \int_{-\infty}^{\infty} \frac{Re\ \chi(\omega)}{\omega - \omega''} d\omega = \frac{1}{\pi} P \int_{-\infty}^{\infty} \frac{d\omega}{\omega - \omega''} \int_{-\infty}^{\infty} \frac{Im\ \chi(\omega')}{\omega' - \omega} d\omega'.$$

If we use the formula in (4.65), the right hand side of the above equation becomes $-\pi\ Im\ \ \chi(\omega'')$, and we obtain

$$Im\ \chi(\omega) = -\frac{1}{\pi} P \int_{-\infty}^{\infty} \frac{Re\ \chi(\omega')}{\omega' - \omega} d\omega'. \tag{4.66}$$

4.6. Causality and collisionless damping of plasma wave

A wave propagating in a medium damps due to collisions between particles that take part in the oscillatory motion. This phenomenon is referred to as *dissipation.* Vlasov plasmas assume no particle-particle collisions; the collision term of the Boltzmann equation is entirely neglected. However, Landau discovered that a wave propagating in Vlasov plasma is subject to damping which is not associated with particle collisions, if a certain dynamical condition is met. This is called *Landau damping* or *collisionless damping.* Landau discovered the Landau damping when he solved Vlasov–Poisson equation as an initial value problem by Laplace transform. The mathematical theory of Landau damping will be dealt with in the next Chapter. In this Section, we show that the fundamental reason of the collisionless damping can be traced to the causal constraint in (4.54).

4.6.1. *Causality and Landau damping*

In evaluation of plasma wave dispersion relation, one often encounters the algebraic expression; the reciprocal of the Doppler-shifted frequency, $\frac{1}{\omega-kv}$. This algebraic quantity is well defined if $\omega \neq kv$ but it should be more defined to deal with the issue of the singularity at $\omega = kv$. The ambiguity at $\omega = kv$ can be removed by using the relation

$$\frac{1}{\omega - kv} = P\frac{1}{\omega - kv} + \lambda\delta(\omega - kv) \tag{4.67}$$

where the symbol P denotes the principal value and λ is an undetermined constant. Equation (4.67) is correct because of the two identities:

$$(x-a)\ P\ \frac{1}{x-a} = 1, \quad (x-a)\delta(x-a) = 0. \tag{4.68}$$

In fact, (4.67) was the mathematical motivation leading to the discovery of the Van Kampen modes of plasma wave. Here we use the causal requirement in electrodynamics to determine the constant λ, that is, (4.54):

$$\chi(t) = 0 \quad for \quad t < 0. \tag{4.69}$$

As an application, we consider the susceptibility of electron plasma (ions are assumed to be immobile) governed by the Vlasov–Poisson equation as introduced in § 4.5. We have (4.49) and (4.50):

$$g(k, v, \omega) = \frac{e}{m}\frac{dg_0}{dv}\frac{i}{\omega - kv}E(k, \omega), \tag{4.70}$$

$$ikE(k, \omega) = -4\pi e\int_{-\infty}^{\infty} dv\ g(k, v, \omega). \tag{4.71}$$

In this one-dimensional problem, one-dimensional forms of (4.39)–(4.41) will be used. The electron charge density is by definition

$$\rho_c = -e\int_{-\infty}^{\infty} dv g(k,v,\omega) = -ikP_e. \tag{4.72}$$

The *electric polarization* P_e takes the form

$$P_e = \frac{e^2}{km}\int_{-\infty}^{\infty} dv\,\frac{dg_0}{dv}\,\frac{1}{\omega - kv}E(k,\omega). \tag{4.73}$$

Then, the relation $E + 4\pi P_e = \varepsilon E = (1+\chi)E$ gives

$$\chi(k,\omega) = \frac{\omega_{pe}^2}{k}\int_{-\infty}^{\infty} dv\,\frac{dg_0}{dv}\,\frac{1}{\omega - kv} \tag{4.74}$$

where ω_{pe} is the electron plasma frequency. Landau evaluated the singular integral when ω becomes real in (4.74) along the famous *Landau contour* (see next Chapter). Here we evaluate (4.74) by satisfying the causal constraint $\chi(t<0)=0$. Let us use (4.67) in the above equation to write

$$\chi(k,\omega) = \frac{\omega_{pe}^2}{k}\int_{-\infty}^{\infty} dv\,\frac{dg_0}{dv}\left[P\frac{1}{\omega - kv} + \lambda\delta(\omega - kv)\right].$$

In order to determine the constant λ, let us invert the above equation:

$$\begin{aligned}\chi(k,t) &= \int_{-\infty}^{\infty}\chi(k,\omega)e^{-i\omega t}\frac{d\omega}{2\pi}\\ &= \frac{\omega_{pe}^2}{2\pi k}\int_{-\infty}^{\infty} dv\frac{dg_0}{dv}e^{-ikvt}\left[P\int_{-\infty}^{\infty} d\omega\,\frac{e^{-i\omega t}}{\omega} + \lambda\right].\end{aligned} \tag{4.75}$$

The principal part integral is the step function:

$$P\int_{-\infty}^{\infty}\frac{e^{-ixt}}{x}dx = \begin{cases} -i\pi & \mathit{if}\ t>0 \\ i\pi & \mathit{if}\ t<0 \end{cases}. \tag{4.76}$$

The causal requirement in (4.69) is satisfied by $\lambda = -i\pi$, and therefore the plasma susceptibility becomes

$$\chi(k,\omega) = \frac{\omega_{pe}^2}{k}\int_{-\infty}^{\infty} dv\,\frac{dg_0}{dv}\left[P\frac{1}{\omega - kv} - i\pi\delta(\omega - kv)\right]. \tag{4.77}$$

This is a direct and intuitive way of evaluating the susceptibility, which agrees with the result evaluated along the Landau contour (see Chapter 5).

The inversion of (4.77), Eq. (4.75) takes the form

$$\chi(k,t) = \begin{cases} -i\,\frac{\omega_{pe}^2}{k}\int_{-\infty}^{\infty} dv \frac{dg_0}{dv} e^{-ikvt} & if \;\; t > 0 \\ 0 & if \;\; t < 0 \end{cases} \tag{4.78}$$

in agreement with the causal requirement. We can derive the collisionless damping from the above expressions for $\chi(\omega)$ or $\chi(t)$.

Exercise. Use (4.70) and the definition of the current

$$J(k,\omega) = -e\int dvg(k,v,\omega) = \frac{\omega}{4\pi i}\chi(k,\omega)E(k,\omega)$$

to obtain (4.74).

Exercise. Use the implicit form of the K-K relations (4.61) to determine λ in (4.75)

So far we have discovered the imaginary part of $\chi(\omega)$ in (4.77). Further discussion is devoted to show that the imaginary part of $\chi(\omega)$ gives rise to damping or growing of the wave, depending upon the slope of the distribution function $g_0(v)$. The set of equations (4.70) and (4.71) has nontrivial solutions when the following solvability condition is met:

$$1 + \frac{\omega_{pe}^2}{k}\int_{-\infty}^{\infty} dv\,\frac{dg_0}{dv}\,\frac{1}{\omega - kv} = 0. \tag{4.79}$$

(4.79) can be put in the form, in view of (4.74),

$$\varepsilon(k,\omega) = 1 + \chi(k,\omega) = 0 \tag{4.80}$$

where $\chi(k,\omega)$ is now given by (4.77). (4.80) is the dispersion relation of electrostatic electron wave (Langmuir wave or electron plasma wave). When $\nabla \times \mathbf{E} = 0$, we have $\mathbf{B} = 0$ per (4.14). Then $D = E + 4\pi P_e = \varepsilon E = 0$ per (4.37) and (4.39). Dispersion relation (4.80) determines complex frequencies $\omega(k) = \omega_r(k) + i\omega_i(k)$ for real k. Solving this equation for complex ω is not an easy problem but a systematic perturbational method is available when $\omega_r \gg \omega_i$ and $Re\;\varepsilon \gg Im\;\varepsilon$, which we assume to be the case. We separate (4.80) into real and imaginary parts:

$$Re\;\varepsilon \equiv \varepsilon_r = 1 + \frac{\omega_{pe}^2}{k}\;P\int_{-\infty}^{\infty} dv\,\frac{dg_0}{dv}\,\frac{1}{\omega - kv}, \tag{4.81}$$

$$Im\;\varepsilon \equiv \varepsilon_i = -i\pi\,\frac{\omega_{pe}^2}{k}\int_{-\infty}^{\infty} dv\,\frac{dg_0}{dv}\delta(\omega - kv) = -i\pi\frac{\omega_{pe}^2}{k^2}\left[\frac{\partial g_0}{\partial v}\right]_{v=\omega/k}. \tag{4.82}$$

We write

$$\varepsilon(\omega) = \varepsilon(\omega_r + i\nu\ \omega_i) = \varepsilon_r(\omega_r + i\nu\omega_i) + i\nu\varepsilon_i(\omega_r + i\nu\omega_i) = 0$$

where ν is an indicator marking the small term. Taylor expanding in the powers of ν gives

$$\varepsilon_r(\omega_r) + i\nu\omega_i\frac{\partial\varepsilon_r}{\partial\omega_r} + i\nu\varepsilon_i(\omega_r) - \nu^2\omega_i\frac{\partial\varepsilon_i}{\partial\omega_r} + \cdots = 0.$$

Equating equal powers of ν gives

$$\varepsilon_r(\omega_r) = 0, \quad \omega_i = -\frac{\varepsilon_i(\omega_r)}{\partial\varepsilon_r/\partial\omega_r}. \tag{4.83}$$

Note that ω_r is determined from $\varepsilon_r = 0$. In (4.83), we have succeeded in solving the dispersion relation (4.80) for ω by a perturbational algebra.

Landau damping of electron plasma wave. The wave velocity of electron plasma wave is much greater than the thermal velocity of plasma: $\omega/k \gg v_{th}$. In this case, the wave velocity can be thought to be situated somewhere on the Maxwellian tail, and the integrand in (4.81) can be expanded in the powers of vk/ω. Insignificant error is introduced by taking only a few terms in the expansion. Then the real part of the dielectric constant in (4.81) becomes

$$\begin{aligned}\varepsilon_r(k,\omega) &= 1 + \frac{\omega_{pe}^2}{k\omega}\int dv\frac{\partial g_0}{\partial v}\left(1 + \frac{vk}{\omega} + \frac{v^2k^2}{\omega^2} + \frac{v^3k^3}{\omega^3} + \cdots\right)\\ &= 1 - \frac{\omega_{pe}^2}{\omega^2}\int dv g_0\left(1 + 3v^2\frac{k^2}{\omega^2} + \cdots\right)\end{aligned} \tag{4.84}$$

with g_0 one-dimensional Maxwellian,

$$g_0(v) = \left[\frac{m_e}{2\pi T_e}\right]^{\frac{1}{2}} exp\left[-\frac{m_e v^2}{2T_e}\right]. \tag{4.85}$$

[In the Langmuir wave, ion contribution is neglected which are small by the mass ratio m_e/m_i.] The average of kinetic energy $\frac{mv^2}{2}$ in one dimension is $\frac{T}{2}$, and so

$$\int_{-\infty}^{\infty} g_0(v)\ v^2 dv = \frac{T}{m} \equiv v_{th}^2. \tag{4.86}$$

Also introducing the Debye length,

$$\lambda_{De}^2 = \frac{T_\alpha}{4\pi n_0 e^2} = \frac{v_{th}^2}{\omega_{pe}^2} \tag{4.87}$$

we can write

$$\varepsilon_r(\omega, k) = 1 - \frac{\omega_{pe}^2}{\omega^2} - 3\frac{\omega_{pe}^4}{\omega^4}k^2\lambda_{De}^2. \tag{4.88}$$

$\varepsilon_r(\omega, k) = 0$ is solved by

$$\omega_r^2 - \omega_{pe}^2 - 3\frac{\omega_{pe}^4}{\omega_r^2}k^2\lambda_{De}^2 = 0. \tag{4.89}$$

If $k^2\lambda_{De}^2 \ll 1$, $\omega_r^2 \simeq \omega_{pe}^2$ is the approximate solution of (4.89). Using this, we obtain

$$\omega_r^2 = \omega_{pe}^2(1 + 3k^2\lambda_{De}^2) \quad or \quad \omega_r = \omega_{pe}\left(1 + \frac{3}{2}k^2\lambda_{De}^2\right). \tag{4.90}$$

Equations (4.82), (4.83), and (4.90) give the imaginary frequency, the Landau damping rate of electron plasma wave for a Maxwellian plasma:

$$\omega_i = -\sqrt{\frac{\pi}{8}}\frac{\omega_{pe}}{k^3\lambda_{De}^3}\, exp\left[\frac{-1}{2k^2\lambda_{De}^2} - \frac{3}{2}\right]. \tag{4.91}$$

[A more general method of obtaining (4.90) and (4.91) is presented in the next Chapter.] The imaginary frequency ω_i vanishes for $k \to 0$ and increases rapidly as k increases. The damping is comparable to ω_{pe} for $k\lambda_{De} > 0.3$ so that such waves are not observed.

4.6.2. *Application of Kramers–Kronig relations*

In this subsection, it will be shown that the real and imaginary parts of $\chi(\omega)$ in (4.77) satisfy the K-K relations, thus further supporting the view that the cause of the collisionless damping is attributed to the causal requirement. The real part of $\chi(k, \omega)$ is given by the principal value:

$$Re\ \chi(k, \omega) = -P\int_{-\infty}^{\infty} dv\ \frac{\frac{dG_0(v)}{dv}}{v - \frac{\omega}{k}} \tag{4.92}$$

where $G_0(v) = \frac{\omega_{pe}^2}{k^2}g_0(v)$. The imaginary part of $\chi(k, \omega)$ is, per (4.63),

$$\begin{aligned}
Im\ \chi(k, \omega) &= \frac{1}{\pi}P\int_{-\infty}^{\infty}\frac{Re\ \chi(\omega', k)}{\omega - \omega'}d\omega' \\
&= \frac{1}{\pi}P\int_{-\infty}^{\infty}\frac{d\omega'}{\omega' - \omega}P\int_{-\infty}^{\infty}dv\frac{\frac{dG_0(v)}{dv}}{v - \frac{\omega'}{k}} \\
&= -\frac{1}{\pi}\int_{-\infty}^{\infty}dv\frac{dG_0(v)}{dv}P\int_{-\infty}^{\infty}d\xi\ \frac{1}{\xi - v}\ \frac{1}{\xi - \omega/k}\ (\xi = \omega'/k).
\end{aligned}$$

Upon using (4.65), Hilbert inversion formula, the above expression becomes

$$Im\ \varepsilon(k,\omega) = -\pi \left[\frac{dG_0}{dv}\right]_{v=\frac{\omega}{k}} . \tag{4.93}$$

Reversely, if (4.93) is substituted into (4.62), we recover (4.92).

4.6.3. *Collisionless damping of ion acoustic wave derived from cold fluid equation*

Finally in this Chapter, we derive the plasma susceptibility by using the fluid equation, with the result being in complete agreement with the solution of the Vlasov equation. This agreement is not a mere coincidence but is due to the close kinship between the Vlasov equation and the fluid equation. We shall see in the next Chapter that the Euler equation is solved by the solution which belongs to the subset of the Vlasov equation solution. One more ingredient of the analysis is that a plasma can be considered to be a collection of noninteracting cold beams distributed over different initial velocities.

The ion acoustic wave is an electrostatic mode that is characteristic of plasma whose constituents, electrons and ions, have great disparity of mass. The restoring force of the oscillatory motion is provided by electron pressure while the inertia by the massive ions. We consider the plasma as a collection of ion beams with Boltzmann-distributed electrons in the background. An ion in the electric wave of a Fourier component with phasor $e^{ikx-i\omega t}$ is subject to the equation of motion

$$\frac{dv}{dt} = \frac{e}{m_i} E(k,\omega)\ e^{i(kx-\omega t)}. \tag{4.94}$$

Equation (4.94) is the Lagrangian equation of motion, where $x = x(t)$ is the particle position at time t. Note that the sole independent variable is t. As the zero order solution, we use the *unperturbed orbit*

$$x = x_0 + v_0 t$$

where x_0 and v_0 are the initial position and velocity, respectively, at the initial time $t = 0$. Then (4.94) with these boundary conditions is solved by

$$v(t) = v_0 - \frac{e}{m_i} E(k,\omega)\ \frac{e^{ikx_0}}{i(kv_0-\omega)} \left(1 - e^{i(kv_0-\omega)t}\right). \tag{4.95}$$

The terms v_0 and 1 in the above equation are necessary to have the initial condition satisfied. However, even without those terms, the rest of the terms in (4.95) satisfy (4.94). (4.95) is the particular solution of the differential equation (4.94). The first order homogeneous solution is

$$v(t) = \frac{e}{m_i} E(k, \omega) \, \frac{e^{ikx_0 + i(kv_0 - \omega)t}}{i(kv_0 - \omega)} \tag{4.96}$$

which corresponds to the Fourier component with phasor $e^{ikx - i\omega t}$. It is trivial to obtain Eulerian velocity corresponding to (4.96); we simply put $x_0 = x - v_0 t$ in (4.96) to get

$$v(x, t) = \frac{e}{m_i} E(k, \omega) \, \frac{e^{ikx - i\omega t}}{i(kv_0 - \omega)}. \tag{4.97}$$

We immediately recognize that (4.97) solves the Eulerian equation of motion

$$\frac{\partial v}{\partial t} + v_0 \, \frac{\partial v}{\partial x} = \frac{e}{m_i} E(k, \omega) e^{ikx - i\omega t} \tag{4.98}$$

where x and t are now independent. The above Lagrangian consideration clarifies the meaning of v_0 which is the initial velocity and v which is the perturbed velocity.

We need both electron and ion equations. The plasma at hand is considered to be a group of ion beams; each beam is characterized by the initial velocity v_0 in the background of Boltzmann-distributed electrons. Modeling a plasma as a group of beams with varying beam velocities was earlier adopted by Bohm and Gross. Reference: Bohm, D. and Gross, E. P. *Theory of plasma oscillations A. Origin of medium-like behavior*, Phys. Rev. **75**, 1851 (1949); Bohm, D. and Gross, E. P. *Theory of plasma oscillations B. Excitation and damping of oscillations* Phys. Rev. **75**, 1864 (1949).
The ion equations read

$$\frac{\partial v}{\partial t} + v_0 \, \frac{\partial v}{\partial x} = -\frac{e}{m_i} \nabla \phi \tag{4.99}$$

$$\frac{\partial n}{\partial t} + N(v_0) \, \frac{\partial v}{\partial x} + v_0 \, \frac{\partial n}{\partial x} = 0 \tag{4.100}$$

where v_0 is the zero order initial ion velocity, v is the perturbed ion velocity, $N(v_0)$ is the equilibrium ion number density of the v_0-beam, n is the perturbation of the ion number density of the v_0-beam, and ϕ is the electric potential ($\mathbf{E} = -\nabla \phi$).

We assume that the electrons are Boltzmann-distributed, forming the massless background of the plasma. Thus, the perturbed electron number density in each beam can be written as

$$n_e = N\ (\ e^{\frac{e\phi}{T_e}} - 1) \simeq \frac{e\phi}{T_e} N(v_0). \tag{4.101}$$

Then, the Poisson equation reads

$$\nabla^2 \phi = -4\pi e \sum_{v_0} \left[n(v_0) - \frac{e\phi}{T_e} N(v_0) \right]. \tag{4.102}$$

(4.99) and (4.100) yield, in terms of the Fourier amplitudes,

$$n(k, \omega, v_0) = \frac{N(v_0)}{(\omega - kv_0)^2} \frac{ek^2 \phi}{m_i}. \tag{4.103}$$

Substituting (4.103) into the Fourier transformed equation of (4.102) and integrating over the distribution of the initial velocities $f(v_0)$ by putting $N(v_0) = f(v_0)dv_0$, we obtain the dispersion relation,

$$\varepsilon(k, \omega) = 1 + \chi(k, \omega) = 1 + \frac{1}{k^2 \lambda_D^2} - \omega_{pi}^2 \int_{-\infty}^{\infty} \frac{f(v_0)dv_0}{(\omega - kv_0)^2} = 0 \tag{4.104}$$

where ω_{pi} is the ion plasma frequency and the Debye length of the plasma is defined by

$$\lambda_D^{-2} = \frac{4\pi e^2}{T_e} \int_{-\infty}^{\infty} f(v_0)dv_0.$$

Integrating by parts in (4.104), the susceptibility is found to be

$$\chi(k, \omega) = \frac{1}{k^2 \lambda_D^2} + \frac{\omega_{pi}^2}{k} \int_{-\infty}^{\infty} dv_0 \frac{\frac{df}{dv_0}}{\omega - kv_0}. \tag{4.105}$$

In order to enforce the causal requirement $\chi(k, t < 0) = 0$, we take the steps parallel to (4.74)–(4.78) by introducing (4.67) for the quantity $1/(\omega - kv_0)$ in the above integral. We can obtain the following equation after enforcing the causal requirement.

$$\chi(k, \omega) = \frac{1}{k^2 \lambda_D^2} + \frac{\omega_{pi}^2}{k} \int_{-\infty}^{\infty} dv_0 \frac{df}{dv_0} \left[P \frac{1}{\omega - kv_0} - i\pi\delta(\omega - kv_0) \right]. \tag{4.106}$$

The first term on the right side of (4.106), $(k\lambda_D)^{-2}$, comes from the electron density which we assumed to be Boltzmann-distributed, in lieu of using the electron Vlasov equation. (4.106) agrees with the result of the kinetic theory. The principal part integral yields the dispersion relation contributed by the non-resonant particles ($v_0 \neq \omega/k$) while the δ-function term gives the resonant particle ($v_0 = \omega/k$) contribution to the susceptibility. Depending upon the slope of the distribution function df/dv_0, this term yields damping or growing of the wave (Landau damping). The actual calculation of Landau damping will be carried out in Chapter 5. See also § 4.6.5.

4.6.4. *Plemelj formula as analytic continuation*

Earlier we mentioned that the adiabatic switching frequency ν (infinitesimal, > 0) is another scheme of enforcing the causality which mathematically facilitates Fourier transforms. The separation of real and imaginary parts of the expression $\frac{1}{\omega - kv}$ can be achieved by explicitly inserting ν and taking the limit $\nu \to 0$ at the end. We have

$$\frac{1}{\omega - kv + i\nu} = \frac{\omega - kv}{(\omega - kv)^2 + \nu^2} - \frac{i\nu}{(\omega - kv)^2 + \nu^2}. \tag{4.107}$$

Upon taking the limit $\nu \to 0$, the above expression becomes

$$P\,\frac{1}{\omega - kv} - i\pi\delta(\omega - kv). \tag{4.108}$$

This is the result of applying Plemelj formula [See Appendix A] to the expression $\frac{1}{\omega - kv}$. It is well known that Landau contour is equivalent to the Plemelj formula.

Two ingredients of causal requirements, that is, $\omega_i > 0$ in the wave phasor and $\chi(t > 0) = 0$, yield the same result.

4.6.5. *Useful expression for electron plasma wave Landau damping rate*

This subsection should be read in reference to § 4.6.1. When we calculated there the Landau damping rate, we assumed g_0 to be a Maxwellian. However, progress can be made with further investigation of the dispersion relation for a weakly damping wave without specifying the zero order distribution. Reference: Clemmow, P. C. and Dougherty, J. P. *Electrodynamics of particles and plasmas* Addison-Wesley (1969). For instance, consider electrostatic electron plasma wave, (4.81).

$$\varepsilon_r(k, \omega_r) = 1 - \frac{\omega_{pe}^2}{k^2} P \int_{-\infty}^{\infty} \frac{g_0'(v)\, dv}{v - \omega_r/k} = 0, \tag{4.109}$$

$$\varepsilon_i(k, \omega_r) = -\pi \frac{\omega_{pe}^2}{k^2}\, \left[g_0'\right]_{v=\omega_r/k}.$$

Then (4.83) gives

$$\omega_i = -\pi\, k\, \left[g_0'\right]_{v=\omega_r/k} \div P\int \frac{g_0'(v)dv}{(v-\omega_r/k)^2}. \tag{4.110}$$

Differentiating (4.109) with respect to k in which we consider ω_r is a function of k per $\varepsilon_r(\omega_r, k) = 0$ gives

$$\begin{aligned} P\int \frac{g_0'(v)dv}{(v-\omega_r/k)^2} &= \left(\frac{\partial\omega_r}{\partial k} - \frac{\omega_r}{k}\right)^{-1} 2P\int \frac{g_0'(v)dv}{v-\omega_r/k} \\ &= \left(\frac{\partial\omega_r}{\partial k} - \frac{\omega_r}{k}\right)^{-1} \frac{2k^2}{\omega_{pe}^2} \end{aligned}$$

$$\omega_i = \frac{\pi}{2}\frac{\omega_{pe}^2}{k^2}\left(\omega_r - k\frac{\partial\omega_r}{\partial k}\right)\, \left[g_0'\right]_{v=\omega_r/k} \simeq \frac{\pi}{2}\frac{\omega_{pe}^2}{k^2}\omega_r\, \left[g_0'\right]_{v=\omega_r/k} \tag{4.111}$$

where we used $\omega_r^2 = \omega_{pe}^2(1+3k^2\lambda_{De}^2)$, which can be obtained from fluid equation. Note that $\omega_r - k\,\partial\omega_r/\partial k = \omega_{pe}^2/\omega_r > 0$. For g_0 Maxwellian, (4.111) recovers (4.91). According to (4.111), for an instability to occur ($\omega_i > 0$), the slope of the distribution function must be positive ($g_0'(\omega_r/k) > 0$) when the phase velocity of the wave is larger than the group velocity. We can even use $\omega_r = \omega_{pe}$, the cold plasma result in (4.111).

4.6.6. *Notes added to Chapter 4*

In Section 4.5.3, we have shown that: $\chi(t<0) = 0$ implies K-K relations. Conversely, we will show that if the K-K relations hold $\chi(t<0) = 0$.

Combine (4.62) and (4.63) to write

$$\chi(\omega) = \frac{i}{\pi}P\int_{-\infty}^{\infty} \frac{\chi(\omega')}{\omega-\omega'}d\omega'. \tag{4.112}$$

Fourier inversion of the above equation is

$$\chi(t) = \frac{i}{\pi}\frac{1}{2\pi}\int_{-\infty}^{\infty} d\omega'\chi(\omega')P\int_{-\infty}^{\infty} d\omega\, \frac{e^{-i\omega t}}{\omega-\omega'}. \tag{4.113}$$

The principal part integral is $-i\pi e^{-i\omega' t}H(t)$ per (4.76). So one obtains

$$\chi(t) = \chi(t)H(t) = \begin{cases} \chi(t) & if\ \ t>0 \\ -\chi(t) & if\ \ t<0 \end{cases}. \tag{4.114}$$

The second line of the above equation indicates that $\chi(t) = 0$ if $t < 0$.

Notes on Hilbert transform. Solve the following equation for $F(x)$.

$$G(x) = \frac{1}{\pi} P \int_{-\infty}^{\infty} \frac{F(x')}{x - x'} dx'. \tag{4.115}$$

Solution: Fourier transforming the above equation gives

$$G(k) = \int_{-\infty}^{\infty} G(x) e^{-ikx} = \int_{-\infty}^{\infty} e^{-ikx} \frac{1}{\pi} P \int_{-\infty}^{\infty} \frac{F(x')}{x - x'} dx'.$$

Make a change of variable, $x - x' = x''$, $dx \to dx''$, then we have

$$G(k) = F(k) \frac{1}{\pi} P \int_{-\infty}^{\infty} dx \frac{e^{-ikx}}{x} = -iF(k)H(k).$$

Solving for $F(k)$, we have, per $1/H(k) = H(k)$ for the step function,

$$F(k) = iG(k)H(k).$$

Let us invert this equation.

$$F(x) = \frac{1}{2\pi} \int_{-\infty}^{\infty} iG(k)H(k)e^{ikx} dk$$

$$= \frac{1}{2\pi} \int_{-\infty}^{\infty} dk e^{ikx} i \int_{-\infty}^{\infty} e^{-ikx''} G(x'') dx'' \frac{i}{\pi} P \int_{-\infty}^{\infty} dx' \frac{e^{-ikx'}}{x'}.$$

$\int dk$-integral gives δ-function. The result is

$$F(x) = -\frac{1}{\pi} P \int_{-\infty}^{\infty} \frac{G(x'')}{x - x''} dx''. \tag{4.116}$$

Equations (4.115) and (4.116) are a Hilbert transform pair. Substituting (4.116) into (4.115), we have

$$G(x) = -\frac{1}{\pi^2} P \int_{-\infty}^{\infty} dx' \int_{-\infty}^{\infty} \frac{1}{x - x'} \frac{1}{x' - x''} G(x'') dx''. \tag{4.117}$$

Therefore we conclude

$$-\frac{1}{\pi^2} P \int_{-\infty}^{\infty} \frac{1}{x - x'} \frac{1}{x' - x''} dx' = \delta(x - x''). \tag{4.118}$$

• **Summary: Causality, Analyticity of $\chi(\omega)$, K-K Relations.** We have three premises in connection with the susceptibility, $\chi(\omega)$ or $\chi(t)$.

1. Causal constraint, $\chi(t<0)=0$.
2. Analyticity of $\chi(\omega)$ in the upper half ω–plane.
3. The K-K relations in $\chi(\omega)$.

The above three premises are interrelated as Arfken succinctly summarized: *Any one of the three implies the other two*, Reference: Titchmarsh, E. C. *Introduction to the theory of Fourier integrals*, Clarendon Press Oxford (1937).

The representation of $\chi(\omega)$ on $\omega_i = 0$ (4.77) is particularly important from a physical point of view because it can be considered as the outcome of the causality. The connection between Landau damping and the causal requirement was demonstrated by introducing (4.67) in $\chi(\omega)$ and requiring the causal constraint to be satisfied. The result agrees with the analytic continuation of (4.64) onto the axis $\omega_i = 0$, in accordance with the Landau contour. The plasma susceptibility obtained by satisfying the causal requirement makes it transparent that the collisionless damping in Vlasov–Poisson plasma is traced to the causality. However, it appears that we need to have the wave-particle interaction operative in order for the collisionless damping to actually occur, as will be discussed in Section 5.6.

Integral relations between D(k,t) and E(k,t). Let us evaluate $\chi(k,t)$ in (4.78), with g_0 being a Maxwellian: $g_0(v) = \sqrt{\frac{m}{2\pi T}}\, exp\,(-\frac{mv^2}{2T})$. We obtain after some algebra

$$\chi(k,t) = \begin{cases} \omega_{pe}^2\; t\; e^{-\frac{T}{2m}k^2t^2} & if \quad t>0 \\ 0 & if \quad t<0. \end{cases} \tag{4.119}$$

Using (4.119) in (4.55) yields

$$D(k,t) = E(k,t) + \omega_{pe}^2 \int_{-\infty}^{t} (t-t')\; e^{-\frac{T}{2m}k^2(t-t')^2} E(k,t')dt'. \tag{4.120}$$

This is a Voltera integral equation for $E(t)$, given the response function $D(t)$. It is useful to consider its Fourier transform.

$$\begin{aligned} D(\omega) &= E(\omega) + \omega_{pe}^2 \int_{-\infty}^{\infty} e^{i\omega t}dt \int_{-\infty}^{t} (t-t')\; e^{-\frac{T}{2m}k^2(t-t')^2} E(t')dt' \\ &= E(\omega) + \omega_{pe}^2 \int_{-\infty}^{\infty} e^{i\omega t}dt \int_{0}^{\infty} \tau\; e^{-\frac{T}{2m}k^2\tau^2} E(t-\tau)d\tau. \end{aligned}$$

Using the inverse relation $E(t-\tau) = \int_{-\infty}^{\infty} E(\omega')exp[-i\omega'(t-\tau)]d\omega'/2\pi$ in the above expression yields

$$D(\omega) = E(\omega) + \omega_{pe}^2\ E(\omega) \int_0^\infty \tau\ e^{i\omega\tau}\ e^{-\frac{T}{2m}k^2\tau^2} d\tau. \tag{4.121}$$

The integral in (4.121) can be carried out in the complex τ-plane. Integrate by parts and carry out the algebra:

$$\int_0^\infty e^{i\omega\tau}\ e^{-\frac{T}{2m}k^2\tau^2} d\tau = e^{-\frac{\omega^2}{4\alpha}} \int_{0-i\omega/2\alpha}^{\infty-i\omega/2\alpha} e^{-\alpha z^2} dz \quad (\alpha = Tk^2/2m)$$

$$= e^{-\frac{\omega^2}{4\alpha}} \left[\int_{-\omega/2\beta}^{0} i\ dy\ e^{\alpha y^2} + \int_0^\infty e^{-\alpha x^2} dx + 0 \right].$$

Thus (4.121) becomes, after collecting the terms,

$$D(\omega,k) = E(\omega,k)\left[1 + \frac{1}{k^2\lambda_D^2}\left(1 + \beta\ e^{-\beta^2}\ [-2F(\beta) + i\sqrt{\pi}\,]\right)\right], \tag{4.122}$$

where $\lambda_D = \sqrt{T/4\pi Ne^2}$, Debye length, $\beta = \omega/kv_{th}$, $mv_{th}^2/2 = T$, and the error function $F(x)$ is defined by $F(x) = \int_0^x e^{t^2} dt$. Then, the quantity in the large bracket in (4.122) is the plasma dielectric permittivity:

$$\varepsilon_L(\omega,k) = 1 + \frac{1}{k^2\lambda_D^2}\left[1 + \beta\ e^{-\beta^2}\left(-2F(\beta) + i\sqrt{\pi}\right)\right]. \tag{4.123}$$

Equation (4.123) agrees with the result obtained in terms of the *plasma dispersion function* by using the Landau contour in Chapter 5. We emphasize that (4.123) has been derived here by using only the causal constraint $\chi(t<0) = 0$ applied to the Vlasov equation with a Maxwellian zero order distribution function. From (4.123), we identify

$$Re\ \chi = \frac{1}{k^2\lambda_D^2}\left[1 - 2\beta\ e^{-\beta^2} \int_0^\beta e^{t^2} dt\right], \tag{4.124}$$

$$Im\ \chi = \frac{\sqrt{\pi}}{k^2\lambda_D^2}\beta\ e^{-\beta^2}. \tag{4.125}$$

It is instructive to show that (4.124) and (4.125) satisfy the K-K relations in (4.62) and (4.63). First, (4.62) is proven by showing

$$\begin{aligned}
1 - 2\beta\, e^{-\beta^2} & \int_0^\beta e^{t^2} dt \\
&= \frac{1}{\sqrt{\pi}} P \int_{-\infty}^{\infty} \frac{d\beta' \beta'}{\beta' - \beta} e^{-\beta'^2} = \frac{1}{\sqrt{\pi}} P \int_{-\infty}^{\infty} d\beta' \left(1 + \frac{\beta}{\beta' - \beta}\right) e^{-\beta'^2} \\
&= \frac{1}{\sqrt{\pi}} \left[\sqrt{\pi} + \beta\, P \int_{-\infty}^{\infty} d\beta'\, \frac{e^{-\beta'^2}}{\beta' - \beta}\right]. \qquad (4.126)
\end{aligned}$$

For the principal value integral, we refer to (5.86) and (5.88) in Chapter 5, and write:

$$P \int_{-\infty}^{\infty} \frac{e^{-q^2}}{q - r} dq = -2\sqrt{\pi} e^{-r^2} \int_0^r e^{w^2} dw. \qquad (4.127)$$

Using (4.127) in (4.126) furnishes the equality. The other K-K relation will be proved by verifying the equality

$$\begin{aligned}
\sqrt{\pi}\beta\, e^{-\beta^2} &= -\frac{1}{\pi} P \int_{-\infty}^{\infty} \frac{d\beta'}{\beta' - \beta} \left[1 - 2\beta'\, e^{-\beta'^2} \int_0^{\beta'} e^{w^2} dw\right] \\
&= -\frac{1}{\pi} P \int_{-\infty}^{\infty} \frac{d\beta'}{\beta' - \beta} - \frac{1}{\pi\sqrt{\pi}} P \int_{-\infty}^{\infty} \frac{d\beta'}{\beta' - \beta} \beta' P \int_{-\infty}^{\infty} \frac{e^{-q^2}}{q - \beta'} dq, \qquad (4.128)
\end{aligned}$$

where we used (4.127). If we put $\frac{\beta'}{q-\beta'} = -1 + \frac{q}{q-\beta'}$, then after cancelation, (4.128) becomes

$$\sqrt{\pi}\beta\, e^{-\beta^2} = -\frac{1}{\pi\sqrt{\pi}} \int_{-\infty}^{\infty} dq\, q\, e^{-q^2} P \int_{-\infty}^{\infty} d\beta' \frac{1}{\beta' - \beta} \frac{1}{q - \beta'}. \qquad (4.129)$$

The truth of (4.129) is readily established upon using the Poincare–Bertrand formula (4.118). Q.E.D.

In summary, we derived the plasma susceptibility and the response function D by applying only the causal constraint $\chi(t) = 0$ *for* $t < 0$, to the integral equation

$$D(t) = E(t) + \int_{-\infty}^{t} \chi(t - t') E(t') dt'. \qquad (4.130)$$

(4.123) is the equivalent representations of the familiar formula (which will be obtained in Chapter 5 by Landau contour)

$$\varepsilon_L(k,\omega) = 1 + \frac{1}{k^2\lambda_D^2}(1+\beta Z(\beta)), \tag{4.131}$$

where $Z(\beta)$, the plasma dispersion function, is

$$Z(\beta) = \frac{1}{\sqrt{\pi}}\int_{-\infty}^{\infty}\frac{e^{-q^2}}{q-\beta}dq = e^{-\beta^2}\left(i\pi - 2\int_0^{\beta}e^{w^2}dw\right). \tag{4.132}$$

[Integral without the symbol P is evaluated along the Landau contour and should be distinguished from the principal value integral.] (4.130) can be inverted to the reciprocal form

$$E(t) = \int_t^{\infty} K(t-t')D(t')dt' \tag{4.133}$$

where

$$K(t,k) = \int_{-\infty}^{\infty}\frac{d\omega}{2\pi}\frac{e^{-i\omega t}}{\varepsilon_L(\omega,k)} \tag{4.134}$$

which we call the *reciprocal kernel*. Note that the time interval $(-\infty,t)$ in (4.130) maps into the time interval (t,∞) in (4.133). The causality constraint implied in $K(t)$ is: $K(t) = 0$ for $t > 0$.

Reference: Lee, H. J. *Causality in plasma electrodynamics* J. Korean Phys. Soc. **73** (1) pp. 65–85 July (2018).

Chapter 5

Mathematical theory of Vlasov equation

In continuation to Chapter 4, the Vlasov equation as a *hyperbolic partial differential equation* will be mathematically approached to obtain its solutions. The nonlinear term $\mathbf{E}\cdot\frac{\partial f_\alpha}{\partial \mathbf{v}}$ defies simple analysis, and a perturbational scheme must be employed in Chapter 10. Here, we study the linearized equation, which has been studied thoroughly by many authors, and consequently various methods for solving the equation are known; including Fourier and Laplace transform methods. In particular, the characteristic method (§ 5.2) is useful; the method supplies a wide scope of application. Because of its usefulness, we introduce the elementary mathematical theory of the characteristic method in § 5.2. In this Chapter, we also address the *dynamical interpretation* of Landau damping. In Section 5.11 we study a nonlinear theory of Vlasov–Poisson system for a restricted wave form called BGK (Bernstein–Greene–Kruskal)wave. The BGK wave is restricted in the sense that it is a progressive wave with the variables x and t being combined in the form $\xi = x - vt$. A linear version of the BGK wave was investigated by Van Kampen (1955). Van Kampen's theory is reminiscent of the Sturm–Liouville theory of differential equation. It is interesting that an arbitrary distribution function can be expanded in terms of Van Kampen's eigen functions. Van Kampen's eigen function is closely related with beam suceptibility [Reference: H. J. Lee, *Causality in plasma electrodynamics* J. Korean Phys. Soc. **73** (1) pp. 65–85 July(2018)].

5.1. Rudimentary linear analysis

5.1.1. *Vlasov–Poisson system*

When $\nabla\times\mathbf{E} = 0$, (4.6) gives $\mathbf{B} = 0$, and so (4.5) and (4.7) become, respectively,

$$\frac{\partial f_\alpha}{\partial t} + \mathbf{v}\cdot\frac{\partial f_\alpha}{\partial \mathbf{r}} + \frac{e_\alpha}{m_\alpha}\mathbf{E}\cdot\frac{\partial f_\alpha}{\partial \mathbf{v}} = 0, \tag{5.1}$$

$$\frac{1}{c}\frac{\partial \mathbf{E}}{\partial t} + \frac{4\pi}{c}\mathbf{J} = 0. \tag{5.2}$$

(5.2) can be used instead of the Poisson equation below. $\nabla \cdot$ (5.2) and the charge conservation equation in (4.11) gives the Poisson equation ($\mathbf{E} = -\nabla\phi$)

$$\nabla \cdot \mathbf{E}(\mathbf{r}, t) = -\nabla^2\phi = 4\pi \sum_\alpha e_\alpha \int f_\alpha(\mathbf{r}, \mathbf{v}, t) d^3v. \tag{5.3}$$

In a high frequency field, the electrons are actively oscillating but the ions are more or less stationary. In this case, we can assume the latter to form a uniform neutralizing background of constant density n_0. When the electric field varies only in one coordinate,

$$\mathbf{E}(\mathbf{r}, t) = \hat{\mathbf{x}} E(x, t) = -\hat{\mathbf{x}} \frac{\partial \phi(x, t)}{\partial x}, \tag{5.4}$$

we have

$$\frac{\partial}{\partial t} f(x, \mathbf{v}, t) + v_x \frac{\partial f}{\partial x} - \frac{e}{m} E(x, t) \frac{\partial f}{\partial v_x} = 0. \tag{5.5}$$

Now we introduce the partially integrated distribution function,

$$g(x, v_x, t) = \int_{-\infty}^{\infty} \int_{-\infty}^{\infty} dv_y dv_z f(x, \mathbf{v}, t). \tag{5.6}$$

Thus the one-dimensional form of the Vlasov equation for electrons in a high frequency electrostatic field reads

$$\frac{\partial}{\partial t} g(x, v, t) + v \frac{\partial g}{\partial x} - \frac{e}{m} E(x, t) \frac{\partial g}{\partial v} = 0 \tag{5.7}$$

where $E(x, t)$ satisfies the Poisson equation

$$\frac{\partial E(x, t)}{\partial x} = 4\pi e \left[n_0 - \int_{-\infty}^{\infty} dv \; g(x, v, t) \right]. \tag{5.8}$$

We linearize the preceding equations by putting $g(x, v, t) = g_0(v) + g'(x, v, t)$ where $g_0(v)$ is the equilibrium one-dimensional distribution function (say, a one-dimensional Maxwellian distribution) and $g'(v, x, t)$ is a small perturbation. Then we have

$$\frac{\partial}{\partial t} g'(x, v, t) + v \frac{\partial g'}{\partial x} - \frac{e}{m} E(x, t) \frac{dg_0(v)}{dv} = 0 \tag{5.9}$$

$$\frac{\partial E(x, t)}{\partial x} = -4\pi e \int_{-\infty}^{\infty} dv \; g'(x, v, t)). \tag{5.10}$$

In the following, the prime on $g'(x, v, t)$ will be dropped with the understanding that g is the perturbation. (5.9) and (5.10) can be dealt with by a few different methods.

- Fourier analysis of (5.9) and (5.10) was carried out in Chapter 4 with regard to the electrodynamic causality.
- (5.9) is a first order linear differential equation in the variable t. Let us Fourier-transform (5.9) with respect to the spatial coordinate x by performing the integral $\int(...)e^{-ikx}dx$.

$$\frac{\partial}{\partial t}g(k,v,t) + ikv\ g(k,v,t) - \frac{e}{m}E(k,t)\frac{dg_0(v)}{dv} = 0. \tag{5.11}$$

In obtaining the above equation we assumed

$$g(x = \pm\infty, v, t) = 0$$

which means that there are no particles at infinity. Likewise (5.10) is Fourier-transformed with respect to x as

$$ikE(k,t) = -4\pi e\int_{-\infty}^{\infty} dv\ g(k,v,t). \tag{5.12}$$

(5.11) is a first order linear ordinary differential equation in the variable t, and is solved in the following form

$$g(k,v,t) = \frac{e}{m}\frac{dg_0(v)}{dv}\int_{t_0}^{t} E(k,t')e^{ikv(t'-t)}dt' + g(k,v,t_0)e^{ikv(t_0-t)} \tag{5.13}$$

where the last term is called the *free-streaming* term. Let us put $t_0 = -\infty$ and assume

$$g(k, v, t = -\infty) = 0.$$

The above condition means that there are no perturbations in the infinitely remote past which is often referred to as the *causality* and amounts to assuming, in our definition of Fourier transform, that ω has a small positive imaginary part. Then we have

$$g(k,v,t) = \frac{e}{m}\frac{dg_0(v)}{dv}\int_{-\infty}^{t} E(k,t')e^{ikv(t'-t)}dt'. \tag{5.14}$$

(5.14) is useful when we need explicit time-dependence of the perturbation. For further development, we now perform the Fourier transform of (5.14) with respect to the variable t.

$$g(k,v,\omega) = \frac{e}{m}\frac{dg_0(v)}{dv}\int_{-\infty}^{\infty} e^{i\omega t}dt \int_{-\infty}^{t} e^{ikv(t'-t)}E(k,t')dt'$$

$$= \frac{e}{m}\frac{dg_0(v)}{dv}\int_{-\infty}^{\infty} e^{i(\omega-kv)t}dt \int_{-\infty}^{t} e^{ikvt'}E(k,t')dt'$$

$$= \frac{e}{m}\frac{dg_0(v)}{dv}\frac{1}{i(\omega-kv)}\int_{-\infty}^{\infty}\left[\frac{d}{dt}e^{i(\omega-kv)t}\int_{-\infty}^{t} e^{ikvt'}E(k,t')dt'\right]dt.$$

The integrated term in the above is

$$\left[e^{i(\omega-kv)t}\int_{-\infty}^{t} dt' E(k,t')e^{ikvt'}\right]_{t=-\infty}^{\infty}$$

which vanishes under the assumption that ω has a small positive imaginary part. Then we have

$$g(k,v,\omega) = \frac{e}{m}\frac{dg_0(v)}{dv}\frac{i}{(\omega-kv)}E(k,\omega) \quad (cf.\ (4.49)). \tag{5.15}$$

Next, (5.12) is Fourier transformed in the form

$$ikE(k,\omega) = -4\pi e\int_{-\infty}^{\infty} g(k,v,\omega)dv. \tag{5.16}$$

Elimination of $E(k,\omega)$ in the above two equations yield the dispersion relation

$$1 + \frac{\omega_p^2}{k}\int_{-\infty}^{\infty} dv\,\frac{dg_0}{dv}\,\frac{1}{\omega-kv} = 0 \quad (cf.\ (4.79)).$$

• If the initial condition is given as $g(k,v,t=0)=0$, we have, instead of (5.14),

$$g(k,v,t) = \frac{e}{m}\frac{dg_0(v)}{dv}\int_0^t E(k,t')e^{ikv(t'-t)}dt'. \tag{5.17}$$

Laplace transform is defined by (see next section)

$$E_k(\omega) = \int_0^{\infty} dt e^{i\omega t}E_k(t).$$

Combining Laplace transforms of (5.12) and (5.17) yields

$$ikE_k(\omega) = -\omega_p^2\int_{-\infty}^{\infty} dv\frac{dg_0(v)}{dv}\int_0^{\infty} dte^{i(\omega-kv)t}\int_0^t E_k(t')e^{ikvt'}dt'.$$

Put in the above

$$e^{i(\omega-kv)t} = \frac{1}{i(\omega-kv)}\frac{d}{dt}e^{i(\omega-kv)t}$$

and integrate by parts. The integrated term vanishes at both limits of $t=\infty$ and $t=0$. Note that the $\int dv$-integral containing the factor e^{-ikvt} vanishes as $t\to\infty$, because the phase oscillates rapidly. The result is again the dispersion relation (4.79). The initial value problem is more carefully examined in the next section.

5.1.2. *Self-consistency*

The electric field $\mathbf{E}$ in (5.3) is the self-consistent field, which is generated by the plasma particles and at the same time influence particle distribution through (5.1). To see this more clearly, we calculate the potential $\phi(\mathbf{r},t)$ made at $\mathbf{r}$ by all the plasma particles of the system which consists of electrons and ions. The potential at $\mathbf{r}$ made by the particles located at $\mathbf{r}'$ is

$$\Sigma_\alpha \frac{e_\alpha}{|\mathbf{r}-\mathbf{r}'|}\int d^3v' f_\alpha(\mathbf{r}',\mathbf{v}',t).$$

The potential at $\mathbf{r}$ due to all the particles in the system is

$$\phi(\mathbf{r},t) = \Sigma_\alpha \int d^3r' \frac{e_\alpha}{|\mathbf{r}-\mathbf{r}'|}\int d^3v' f_\alpha(\mathbf{r}',\mathbf{v}',t), \tag{5.18}$$

$$\mathbf{E}(\mathbf{r},t) = -\nabla\phi = \Sigma_\alpha e_\alpha \int d^3r' \frac{\mathbf{r}-\mathbf{r}'}{|\mathbf{r}-\mathbf{r}'|^3}\int d^3v' f_\alpha(\mathbf{r}',\mathbf{v}',t), \tag{5.19}$$

where we used the relation

$$\nabla_r \frac{1}{|\mathbf{r}-\mathbf{r}'|} = -\frac{\mathbf{r}-\mathbf{r}'}{|\mathbf{r}-\mathbf{r}'|^3}.$$

Differentiating (5.19) gives (5.3):

$$\begin{aligned}\nabla\cdot\mathbf{E}(\mathbf{r},t) &= \sum_\alpha e_\alpha \int d^3r' \nabla_r \cdot \frac{\mathbf{r}-\mathbf{r}'}{|\mathbf{r}-\mathbf{r}'|^3}\int d^3v' f_\alpha(\mathbf{r}',\mathbf{v}',t) \\ &= 4\pi \sum_\alpha e_\alpha \int d^3v' f_\alpha(\mathbf{r},\mathbf{v}',t), \end{aligned} \tag{5.20}$$

$$per \quad \nabla_r \cdot \frac{\mathbf{r}-\mathbf{r}'}{|\mathbf{r}-\mathbf{r}'|^3} = -\nabla_r^2 \frac{1}{|\mathbf{r}-\mathbf{r}'|} = 4\pi\delta(\mathbf{r}-\mathbf{r}').$$

Now (5.1) takes the form

$$\frac{\partial f_\alpha}{\partial t}+\mathbf{v}\cdot\frac{\partial f_\alpha}{\partial \mathbf{r}}+\frac{e_\alpha}{m_\alpha}\frac{\partial f_\alpha}{\partial \mathbf{v}}\cdot\sum_\alpha e_\alpha\int d^3r'\frac{\mathbf{r}-\mathbf{r}'}{\mid\mathbf{r}-\mathbf{r}'\mid^3}$$

$$\times\int d^3v' f_\alpha(\mathbf{r}',\mathbf{v}',t)=0. \tag{5.21}$$

Equation (5.21) has the virtue of having gotten rid of the electric field.

Example. Solve (5.3) for **E** in reverse order. To solve

$$\nabla^2\phi(\mathbf{r},t)=-4\pi\Sigma\, e_\alpha\int d^3v f_\alpha(\mathbf{r},\mathbf{v},t)$$

recall the Green function of Laplace equation is

$$G(\mathbf{r},\mathbf{r}')=\frac{-1}{4\pi\mid\mathbf{r}-\mathbf{r}'\mid},$$

which satisfies

$$\nabla^2G(\mathbf{r},\mathbf{r}')=\delta(\mathbf{r}-\mathbf{r}').$$

Therefore we calculate:

$$\phi(\mathbf{r},t)=\int d^3r'G(\mathbf{r},\mathbf{r}')\times \textit{inhomogeneous term}\,(\mathbf{r}')$$

$$=\int\frac{d^3r'}{\mid\mathbf{r}-\mathbf{r}'\mid}\Sigma\, e_\alpha\int d^3v f_\alpha(\mathbf{r}',\mathbf{v},t)$$

which is (5.18).

5.1.3. *Convolution*

We now calculate the Fourier transform of (5.19) in which the $\int d^3r'$-integral is a *convolution.* Convolution integral of two functions is

$$(f*g)_\mathbf{r}=\int_{-\infty}^{\infty}g(\mathbf{r}')f(\mathbf{r}-\mathbf{r}')d^3r'=\int_{-\infty}^{\infty}f(\mathbf{r}')g(\mathbf{r}-\mathbf{r}')d^3r'.$$

It can be easily proven that

$$\int e^{-i\mathbf{k}\cdot\mathbf{r}}(f*g)_\mathbf{r}\,d^3r=\tilde f(\mathbf{k})\,\tilde g(\mathbf{k}).$$

In words, the Fourier transform of a convolution integral is the product of each Fourier transform. The inverse of $\tilde f(\mathbf{k})\,\tilde g(\mathbf{k})$=

$$\left(\frac{1}{2\pi}\right)^3\int e^{i\mathbf{k}\cdot\mathbf{r}}\tilde f(\mathbf{k})\,\tilde g(\mathbf{k})d^3k=(f*g)_\mathbf{r}.$$

In words, the Fourier transform (or the inverse transform) of the product of two functions is the convolution. So much for the mathematics of convolution.

Next, Fourier transforming (5.19) gives

$$\mathbf{E}(\mathbf{k},t) = \Sigma_\alpha e_\alpha \left[\int e^{-i\mathbf{k}\cdot\mathbf{r}} \frac{\mathbf{r}}{r^3} d^3r \right] \int d^3v' f_\alpha(\mathbf{k},\mathbf{v}',t).$$

The Fourier integral of [...] can be computed by introducing the spherical coordinates:

$\mathbf{r} = (r\ sin\theta cos\varphi, r\ sin\theta sin\varphi,\ r\ cos\theta)$, giving

$$\int e^{-i\mathbf{k}\cdot\mathbf{r}} \frac{\mathbf{r}}{r^3} d^3r = \int e^{-ikrcos\theta}\ \frac{\mathbf{r}}{r^3}\ r^2\ sin\theta\ d\theta\ d\varphi\ dr = -\frac{4\pi i}{k^2}\ \mathbf{k}. \tag{5.22}$$

$$Hence, \quad \mathbf{E}(\mathbf{k},t) = -\frac{4\pi i}{k^2}\ \mathbf{k}\ \Sigma_\alpha e_\alpha \int d^3v' f_\alpha(\mathbf{k},\mathbf{v}',t). \tag{5.23}$$

Exercise. Prove (5.23) by Fourier transforming (5.3).

5.2. Characteristic method

Reference: Hildebrand, F. B. *Advanced Calculus for Applications*, Prentice-Hall, Englewood Cliffs, N. J., 1976.

This is one of the most powerful methods to attack the Vlasov and Eulerian fluid equations, which belong to the class of *hyperbolic partial differential equations.* The linearized Vlasov equation is neatly integrated by applying the method of characteristics, which is well known by the name *integration along the unperturbed orbit.* The characteristic equations of the Vlasov and the Eulerian equations are identical; thus the solutions of the two equations considerably overlap. The Euler equation and the Poisson equation can be combined into a Riemann invariant form in which two *Riemann invariants* are the new unknown functions. This casts the light to investigate the Vlasov–Poisson system by the Riemannian characteristic method that is widely used in fluid dynamics.

5.2.1. *Mathematics of characteristic method*

The one-dimensional form of Vlasov equation for electrons reads

$$\frac{\partial}{\partial t} g(x,v,t) + v\frac{\partial g}{\partial x} - \frac{e}{m}E(x,t)\frac{\partial g}{\partial v} = 0. \tag{5.24}$$

This equation is mathematically a quasi-linear partial differential equation of first order. Let us consider a three-dimensional space consisting of three coordinate axes, (t, x, v). Then g is a function of these three coordinates, and we define the gradient operator in this space:

$$\nabla = \hat{\mathbf{t}}\frac{\partial}{\partial t} + \hat{\mathbf{x}}\frac{\partial}{\partial x} + \hat{\mathbf{v}}\frac{\partial}{\partial v}.$$

Also we define the vector

$$\mathbf{V} = \hat{\mathbf{t}}\ 1 + \hat{\mathbf{x}}v - \hat{\mathbf{v}}\frac{e}{m}E(x,t).$$

Then (5.24) is

$$\mathbf{V} \cdot \nabla g = 0. \tag{5.25}$$

Since ∇g is a vector normal to the surface $g = c =$ constant, (5.25) states that the vector $\mathbf{V}$ lies in the tangent plane of $g = c$. That is, at any point, the vector $\mathbf{V}$ is tangent to a curve in the integral surface $g = c$ which passes through that point. Equation (5.24) may be considered as specifying the direction of the vector $\mathbf{V}$ at a point in (t, x, v)-space. Imagine that $\mathbf{V}$ is the velocity of a particle in (t, x, v) space. If a particle moves from a given initial point in the direction of $\mathbf{V}$ and continues its motion along the direction given by $\mathbf{V}$, a space curve is drawn which we call the *characteristic curve* of the partial differential equation (5.24). Any surface which is made up from such characteristic curves contains the vector $\mathbf{V}$ on the tangent plane of that surface, and that surface (consisting of the characteristic curves) is an integral surface of (5.24). [The ordinary differential equation, $dy/dx = f(x, y)$ defines the tangential line at the point (x, y) which is tangent to the integral curve $y(x)$. Integral curve can be obtained by connecting such infinitesimal tangential segments.]

If $d\mathbf{r}$ is the difference vector between two neighboring points on a characteristic curve, $d\mathbf{r}$ is parallel to $\mathbf{V}$, or $d\mathbf{r} \times \mathbf{V} = 0$. This condition gives the equation for the characteristic curve and can be written in the form

$$\frac{dt}{1} = \frac{dx}{v} = \frac{dv}{-\frac{e}{m}E(x,t)}. \tag{5.26}$$

These relations are equivalent to two ordinary differential equations. Let the solutions of two independent equations of (5.26) be denoted by

$$u_1(x,t,v) = c_1 \quad \textit{and} \quad u_2(x,t,v) = c_2$$

with c_1 and c_2 independent constants. Then the intersection of these two surfaces for a given value of c_1 and c_2 is a characteristic curve. The intersections of the two families of the surfaces $u_1 = c_1$ and $u_2 = c_2$ generate a host of characteristic curves which form the integral surface of (5.24). Thus the general solution of (5.24) can be represented by an equation of the form

$$F(c_1, c_2) = 0 \quad or \quad F[u_1(x,t,v),\ u_2(x,t,v)] = 0. \tag{5.27}$$

Next we shall consider the linearized Vlasov equation,

$$\frac{\partial}{\partial t} g(x,v,t) + v\frac{\partial g}{\partial x} - \frac{e}{m}E(x,t)\frac{dg_0(v)}{dv} = 0. \tag{5.28}$$

We assume the solution of this equation in the form

$$u(t, x, g(t,x)) = c \tag{5.29}$$

in the sense that (5.29) determines g as a function of x and t which satisfies (5.28). It should be noted that v plays the role of a parameter only, not a variable in (5.28). Differentiating both sides of (5.29), we have

$$\frac{\partial u}{\partial t} + \frac{\partial u}{\partial g}\frac{\partial g}{\partial t} = 0, \qquad \frac{\partial u}{\partial x} + \frac{\partial u}{\partial g}\frac{\partial g}{\partial x} = 0.$$

Solving for $\partial g/\partial t$ and $\partial g/\partial x$ in the above equations and substituting the results into (5.28) give

$$\frac{\partial u}{\partial t} + v\frac{\partial u}{\partial x} + \frac{e}{m}E(x,t)\frac{dg_0(v)}{dv}\frac{\partial u}{\partial g} = 0 \tag{5.30}$$

which is of the same form as (5.24), and the relations for the characteristic curve is

$$\frac{dt}{1} = \frac{dx}{v} = \frac{dg}{\frac{e}{m}E(x,t)\frac{dg_0(v)}{dv}}. \tag{5.31}$$

In this case the perturbed distribution function g can be found by integrating the relation

$$dg = \frac{e}{m}E(x,t)\frac{dg_0(v)}{dv}\ dt \tag{5.32}$$

where x and t are related by the equation,

$$dx = v\ dt \tag{5.33}$$

where v should be regarded as a parameter (constant).

Use of (5.32) and (5.33) will be illustrated in the following. Integrating (5.33) gives

$$x = vt + c_1. \quad (c_1 = const.) \tag{5.34}$$

Integrating (5.32), we write

$$g(t) = \frac{e}{m}\frac{dg_0(v)}{dv}\int_{-\infty}^{t} E\left(x(t'), t'\right) dt' \tag{5.35}$$

where we put $g(t = -\infty) = 0$. Now we assume $E(x,t) = E(k,\omega)e^{i(kx-\omega t)}$. Then (5.35) is rewritten as

$$\begin{aligned} g(t) &= \frac{e}{m}\frac{dg_0}{dv} E(k,\omega)\int_{-\infty}^{t} e^{ik[x-v(t-t')]-i\omega t'} dt' \\ &= \frac{e}{m}\frac{dg_0}{dv} E(k,\omega)e^{ik(x-vt)}\int_{-\infty}^{t} e^{ikvt'-i\omega t'} dt' \end{aligned}$$

where we used (5.34) to write $x(t') = vt' + c_1 = vt' + x - vt$. Carrying out the integral gives

$$g(t,x,v) = \frac{e}{m}\frac{dg_0}{dv}\frac{E(k,\omega)}{i(kv-\omega)}e^{ikx-i\omega t}. \tag{5.36}$$

In (5.36), the coefficient of $e^{ikx-i\omega t}$ is the Fourier coefficient $g(k,\omega,v)$:

$$g(k,\omega,v) = \frac{e}{m}\frac{dg_0}{dv}\frac{E(k,\omega)}{i(kv-\omega)} \tag{5.37}$$

which agrees with the Fourier transform solution of (5.28).

In (5.35), we specified the integration constant as 0 at $t = -\infty$. Instead, we can set up a more general solution by introducing the second integration constant from (5.32):

$$g(t) = \frac{e}{m}\frac{dg_0}{dv}\int^{t} E\Big(x(t'), t'\Big)dt' + c_2. \tag{5.38}$$

Then the general solution of (5.28) can be written as $G(c_1, c_2) = 0$, or

$$g(x,v,t) = \frac{e}{m}\frac{dg_0}{dv}\int^{t} E\Big(x - vt + vt', t'\Big)dt' + F(x - vt). \tag{5.39}$$

We note that $F(x - vt)$ is the homogeneous solution of (5.28).

Example. Solve the following equation [O'Neil, T. (1965) *Collisionless damping of nonlinear plasma oscillations,* Phys. Fluids **8**, 2255].

$$\frac{\partial}{\partial t}f(x,v,t) + v\,\frac{\partial f}{\partial x} = \frac{e}{m}E_0 sin(k_0 x - \omega_0 t)\frac{df_0(v)}{dv}. \tag{5.40}$$

First, let us Fourier transform the above equation with respect to x to obtain

$$\frac{\partial f}{\partial t} + ikvf(k,v,t) = \frac{ie\pi}{m}E_0\frac{df_0}{dv}[e^{i\omega_0 t}\delta(k+k_0) - e^{-i\omega_0 t}\delta(k-k_0)] \tag{5.41}$$

which is an ordinary linear differential equation of first order in the variable t and is solved by

$$\begin{aligned} &f(k,v,t) \\ &\quad = \frac{\pi e}{m}E_0\frac{df_0}{dv} \times \left[\frac{\delta(k+k_0)}{kv+\omega_0}(e^{i\omega_0 t} - e^{-ikvt}) - \frac{\delta(k-k_0)}{kv-\omega_0}(e^{-i\omega_0 t} - e^{-ikvt})\right] \\ &\qquad + e^{-ikvt}f(k,v,t=0). \end{aligned}$$

Fourier inversion of the above equation gives the answer:

$$\begin{aligned} f(x,v,t) = {}& \frac{e}{m}\frac{df_0}{dv}\frac{E_0}{k_0 v - \omega_0} \\ & \times [cos(k_0(x-vt)) - cos(k_0 x - \omega_0 t)] + [f(x,v,t=0)]_{x\to x-vt}. \end{aligned}$$

Next, let us solve (5.40) with the initial condition $f(x,v,t=0) = f(v,0)$ $cos(k_0 x)$ by characteristic method. We use the characteristic relations, $x - vt = c_1$ and

$$df = \frac{e}{m}E_0\frac{df_0}{dv}sin(k_0 x - \omega_0 t)dt.$$

Integrating gives

$$f(t'=t) - f(t'=0) = \frac{eE_0}{m}\frac{df_0}{dv}\int_0^t sin[k_0(c_1 + vt') - \omega_0 t']dt'.$$

Carrying out the dt' integral, we obtain

$$\begin{aligned} f(x,t,v) - f(t'=0) = {}& \frac{eE_0}{m}\frac{df_0}{dv}\frac{1}{\omega_0 - k_0 v} \\ & \times [cos(k_0 x - \omega_0 t) - cos k_0(x - vt)] \end{aligned}$$

where

$$f(t'=0) = f(v,0)cos[k_0 x(t'=0)] = f(v,0)cos[k_0(vt'+c_1)_{t'=0}]$$
$$= f(v,0)cos[k_0(x-vt)]$$

which agrees with the Fourier transform result.

• Let us consider the particular case $E = 0$ in the linear equation (5.28) and the corresponding characteristic equation (5.31). Then the characteristic equation

$$\frac{dt}{1} = \frac{dx}{v} = \frac{dg}{0} \tag{5.42}$$

solves the equation

$$\frac{\partial}{\partial t} g(x,v,t) + v\,\frac{\partial g}{\partial x} = 0 \tag{5.43}$$

which generates the unperturbed orbits $x - vt = const. = c$. (5.42) and (5.43) simply state that $g = const. = b$ along the characteristic line $x - vt = c$. Then the general solution is $F(b,c) = 0$ with F an arbitrary function. Therefore $g = f(c) = f(x - vt)$ is the general solution, with f an arbitrary function. This trivial example is more generalized in the following.

Further remark on the general solution of Vlasov equation (5.24). The characteristic relations associated with the homogeneous equation (5.24) can be augmented:

$$\frac{dt}{1} = \frac{dx}{v} = \frac{dv}{-\frac{e}{m}E(x,t)} = \frac{dg}{0} \tag{5.44}$$

which immediately states that $g(x,v,t) = const. = c$ along the intersections of any two integral surfaces of (5.44): c is an arbitrary function of any two independent constants derived from (5.44). For example, (5.44) gives three independent constants of integral:

$$c_1(x,t) = x - vt, \quad c_2(v,t) = v + \tfrac{e}{m}\int^t E(c_1 + vt', t')dt',$$
$$c_3(x,v) = \tfrac{v^2}{2} - \tfrac{e}{m}\phi(x)$$

where $-\partial\phi(x)/\partial x = E(x)$. Then (5.24) is solved by $H(c, c_1, c_2) = 0$ with H an arbitrary function, which can be written as

$$g(x,v,t) = F_1(c_1(x,t), c_2(v,t)). \tag{5.45}$$

Likewise (5.24) is solved by

$$g(x,v,t) = F_2(c_1(x,t), c_3(x,v)) \quad or \quad F_3(c_2(v,t), c_3(x,v)) \tag{5.46}$$

with arbitrary F's.

Exercise. Show (5.45) and (5.46) satisfy (5.24) by differentiation.

Example. Find the solution of the equation

$$\frac{\partial \rho(x,t)}{\partial t} + a(\rho)\frac{\partial \rho}{\partial x} = 0. \tag{5.47}$$

The characteristic equation is

$$\frac{dt}{1} = \frac{dx}{a(\rho)} = \frac{d\rho}{0}. \tag{5.48}$$

We have $\rho(x,t) = c_1$ and $x = \int^t a[\rho(x',t')]dt' + c_2$ with c_1 and c_2 constant. Since ρ is constant, $c_2 = x - a(\rho)t$. The general solution of (5.47) is a function of c_1 and c_2:

$$\rho = f[x - t\, a(\rho(x,t))].$$

Exercise. Show by direct differentiation that the above equation satisfies (5.47).

One who is not entirely happy with the augmented characteristic relation (5.44) can take the following direct view of the characteristics. The left hand side of (5.47) (as well as (5.24)) is a directional derivative, and thus (5.47) (as well as (5.24)) states that ρ is constant along the direction parametrically represented by the curve $x = x(s)$ and $t = t(s)$ [this curve is the characteristic curve as would be turned out to be]. We have

$$\frac{\partial \rho(x,t)}{\partial t} + a(\rho)\frac{\partial \rho}{\partial x} = \frac{d}{ds}\rho(x(s),t(s)) = 0$$

$$\frac{d\rho}{ds} = \frac{\partial \rho}{\partial t}\frac{dt}{ds} + \frac{\partial \rho}{\partial x}\frac{dx}{ds}$$

$$\frac{dt}{ds} = 1, \quad \frac{dx}{ds} = a(\rho), \quad \frac{d\rho}{ds} = 0$$

which yield respectively

$$t = s \quad (t(s=0) \equiv 0), \quad x = \int^{s'=t} a\left(\rho(s')\right) ds' + x_0.$$

Because ρ is constant, the second equation gives $x = a(\rho)t + x_0$. Therefore, we obtain

$$\rho = const \equiv \rho(x_0) = \rho\left(x - t\, a(\rho)\right). \tag{5.49}$$

Clearly (5.49) is equivalent to (5.48).

5.2.2. *Vlasov equation vs fluid equation*

Let us consider the Eulerian equation of motion

$$\frac{\partial v(x,t)}{\partial t} + v\frac{\partial v}{\partial x} = -\frac{e}{m}E(x,t). \tag{5.50}$$

We assume that solution v is obtained in implicit form: $g(x,t,v(x,t)) = c$ is an integral of (5.50) in the sense that the function g determines v as a function of x and t. Then by partial differentiation

$$\frac{\partial g}{\partial t} + \frac{\partial g}{\partial v}\frac{\partial v}{\partial t} = 0, \quad \frac{\partial g}{\partial x} + \frac{\partial g}{\partial v}\frac{\partial v}{\partial x} = 0.$$

Solving for $\frac{\partial v}{\partial x}$ and $\frac{\partial v}{\partial t}$, and substituting the results into (5.50) give

$$\frac{\partial g}{\partial t} + v\frac{\partial g}{\partial x} - \frac{e}{m}E(x,t)\frac{\partial g}{\partial v} = 0 \tag{5.51}$$

which is (5.24), the Vlasov equation. We note that (5.50) and (5.51) have the same characteristic relation. Equation (5.50) determines the particle trajectory in *Eulerian coordinates.* This example shows that the characteristics of Vlasov equation are the particle trajectories. Conversely, if $g\,(x, v(x,t), t) = c$ solves (5.51), we can show that $v(x,t)$ satisfies (5.50). Therefore, the Euler equation and Vlasov equation imply each other.

Exercise. Suppose we have the equation

$$\frac{\partial x(t,v)}{\partial t} - \frac{e}{m}E(x,t)\frac{\partial x}{\partial v} = v$$

and assume that $g(t,v,x(t,v)) = c$ is an integral of the above equation. Show that this g satisfies the Vlasov equation.

Exercise. Assume a solution of (5.24) in the form $g(x,v,t(x,v)) = c$. Then (5.24) is transformed to

$$v\frac{\partial t}{\partial x} - \frac{e}{m}E(x,t)\frac{\partial t}{\partial v} = 1.$$

5.2.3. *Oscillating frame*

If a plasma is irradiated by an oscillating electric field, the zero order equilibrium distribution function is determined by

$$\frac{\partial F_\alpha(t,\mathbf{v})}{\partial t} + \frac{e_\alpha}{m_\alpha} sin(\omega_0 t)\mathbf{E_0}\cdot\frac{\partial F_\alpha}{\partial \mathbf{v}} = 0, \tag{5.52}$$

where α denotes the particle species.

For simplicity, we take $\mathbf{E_0} = \hat{\mathbf{x}}E_0$. Then we have

$$\frac{\partial G_\alpha(t,v)}{\partial t} + \frac{e_\alpha}{m_\alpha} sin(\omega_0 t) E_0 \frac{\partial G_\alpha}{\partial v} = 0$$

where G_α is the one-dimensional distribution function. The characteristic relation is

$$\frac{dt}{1} = \frac{dv}{E_0 sin(\omega_0 t) e_\alpha / m_\alpha} = \frac{dG_\alpha}{0} \tag{5.53}$$

which gives two constants of integral $G_\alpha = c_1$ and $v + \frac{e_\alpha}{m_\alpha \omega_0} E_0 cos(\omega_0 t) = c_2$. Therefore $G_\alpha(v,t) = G_\alpha(v + \frac{e_\alpha}{m_\alpha \omega_0} E_0 cos(\omega_0 t))$ is the general solution of (5.53). By defining a new velocity, $u_\alpha = v + \frac{e_\alpha}{m_\alpha \omega_0} E_0 cos(\omega_0 t)$, one goes into the *oscillating frame* [Dawson, J. and Oberman, C. (1962) *High-frequency conductivity and the emission and absorption coefficients of a fully ionized plasma*, Phys. Fluids **5** p. 517; Sanmartin, J. R. (1970) *Electrostatic plasma instabilities excited by a high frequency electric field* Phys. Fluids **13** p. 1533–1542]. If $\mathbf{E}_0$ has all the three components, the characteristic relation reads

$$\begin{aligned}\frac{dt}{1} &= \frac{dv_x}{E_{0x} sin(\omega_0 t) e_\alpha / m_\alpha} = \frac{dv_y}{E_{0y} sin(\omega_0 t) e_\alpha / m_\alpha} \\ &= \frac{dv_z}{E_{0z} sin(\omega_0 t) e_\alpha / m_\alpha} = \frac{dF_\alpha}{0}\end{aligned}$$

which gives four constants of integral: $F_\alpha = c$ and

$$v_i + \frac{e_\alpha}{m_\alpha \omega_0} E_{0i} \, cos(\omega_0 t) = c_i \quad (i = x, y, z).$$

Thus c expressed in terms of c_i is a solution of (5.52):

$$F_\alpha(\mathbf{v}, t) = F_\alpha(c_x, c_y, c_z) = F_\alpha\left(\mathbf{v} + \frac{e_\alpha}{m_\alpha \omega_0} \mathbf{E}_0 cos\ \omega_0 t\right). \tag{5.54}$$

5.3. Laplace transform method

In this Section, Laplace transform will be used to solve Vlasov–Poisson equations as an initial value problem. Preliminary mathematical discussion appears to be necessary for the subsequent development. A good mathematical reference is Hildebrand [*Advanced Calculus for Applications*, Prentice Hall, 1976].

5.3.1. *Laplace transform*

We have an identity

$$f(x) = \int_{-\infty}^{\infty} f(t)\delta(x-t)dt.$$

In this expression we use

$$\delta(x-t) = \frac{1}{2\pi}\int_{-\infty}^{\infty} e^{i(x-t)u}du$$

to obtain

$$f(x) = \frac{1}{2\pi}\int_{-\infty}^{\infty} e^{iux}\int_{-\infty}^{\infty} e^{-iut}f(t)dt\, du \quad (-\infty < x < \infty). \tag{5.55}$$

Let us express a function $e^{-cx}f(x)$ as a Fourier integral in the form of (5.55). Here we shall assume that $f(x) = 0$ for $x < 0$:

$$e^{-cx}f(x) = \frac{1}{2\pi}\lim_{R\to\infty}\int_{-R}^{R} e^{iux}\left[\int_{0}^{\infty} e^{-iut}e^{-ct}f(t)dt\right] du \tag{5.56}$$

where c is an arbitrary constant with the only requirement that the integral

$$\int_{0}^{\infty} e^{-ct}|f(t)|dt \tag{5.57}$$

exists. We rewrite (5.56):

$$f(x) = \frac{1}{2\pi}\lim_{R\to\infty}\int_{-R}^{R} e^{(iu+c)x}\left[\int_{0}^{\infty} e^{-(iu+c)t}f(t)dt\right] du. \tag{5.58}$$

Putting $iu + c = p$ gives

$$f(x) = \frac{1}{2\pi i}\lim_{R\to\infty}\int_{c-iR}^{c+iR} e^{px}\left[\int_{0}^{\infty} e^{-pt}f(t)dt\right] dp. \tag{5.59}$$

In order for the transition of (5.58) to (5.59) to be legitimate, the constant c must be large enough and $f(x)$ should be of *exponential order* so that integral (5.57) exists. We define Laplace transform of $f(t)$ as

$$F(p) = \int_{0}^{\infty} e^{-pt}f(t)dt. \tag{5.60}$$

Its inverse transform is

$$f(t) = \frac{1}{2\pi i}\lim_{R\to\infty}\int_{c-iR}^{c+iR} e^{tz}F(z)dz \tag{5.61}$$

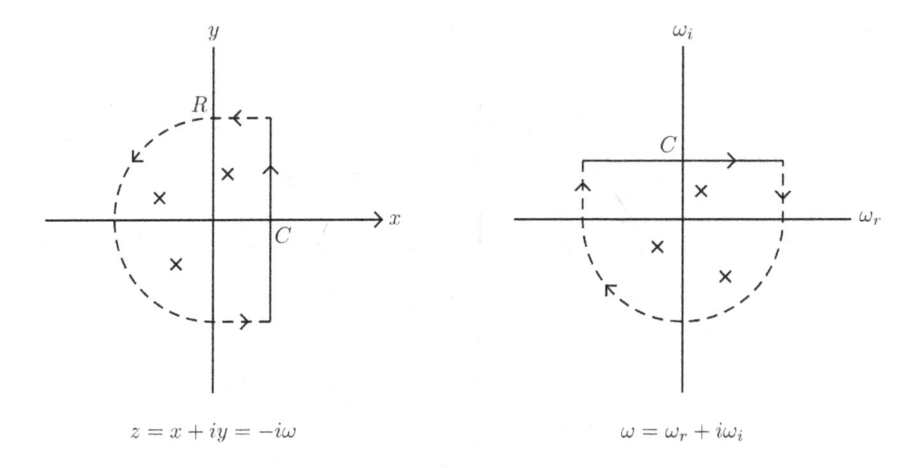

Fig. 5.1. Laplace contour with singularities marked by x.

and can be calculated by contour integral [*Bromwich integral*: see Hildebrand (1976)]. Here the integration path on the $z(= x + iy)$-plane is an infinite straight line $x = c > 0$ which is far enough off the imaginary axis so that the integral (5.57) exists. If the function $F(z)$ is analytic except at a finite number of singularities which are located on the left side of $x = c$, integral (5.61) can be performed along the contour (solid straight line and dotted curve) shown in Fig. 5.1. Here c must be large enough so that all the singularities are situated on the left side of it. If $F(z)$ is uniformly convergent along the dotted curve, the Bromwich integral can be computed by the residue theorem.

Laplace transform introduced above can be formulated in a slightly different form as favored by some plasma physicists (as adopted in this book). Introducing a complex quantity $\omega = -u + ic$ in (5.58) gives

$$f(x) = \frac{1}{2\pi} \lim_{R\to\infty} \int_{-R+ic}^{R+ic} e^{-i\omega x} \left[\int_0^\infty e^{i\omega t} f(t) dt \right] d\omega. \tag{5.62}$$

We define Laplace transform of $f(t)$:

$$F(\omega) = \int_0^\infty e^{i\omega t} f(t) dt \tag{5.63}$$

which can be remembered as 'the positive half' of the Fourier integral. Then its inversion is

$$f(t) = \frac{1}{2\pi} \lim_{R\to\infty} \int_{-R+ic}^{R+ic} e^{-i\omega t} F(\omega)\, d\omega. \tag{5.64}$$

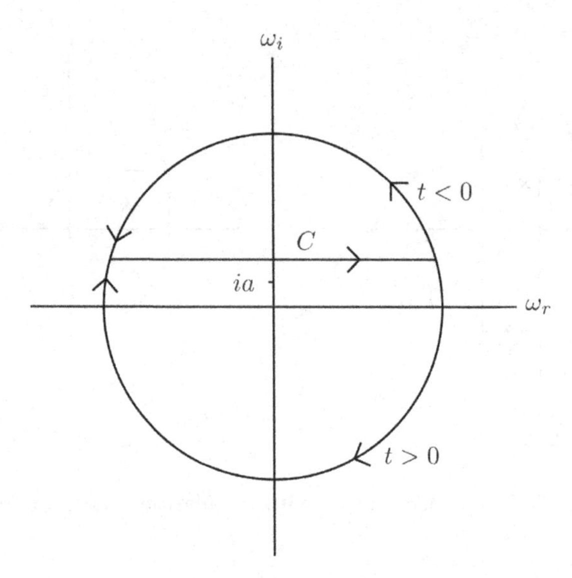

Fig. 5.2. ω plane.

Obviously (5.64) and (5.61) are obtained from each other by a change of variable $z \longleftrightarrow -i\omega$. The contour of integral (5.64) is shown in Fig. 5.2 where c must be large enough so that it runs above all the singularities of $F(\omega)$, and the contour is completed by the infinite semi-circle enclosing the lower-half ω-plane.

Remark. The straight line c runs above the uppermost singularity of $F(\omega)$ in (5.64). In the verbatim of plasma wave dynamics, the constant c in (5.64) should be larger than the growth rate of the fastest growing mode. This is tantamount to the mathematical condition for Laplace transform to exist, i.e., for the integral (5.57) to exist. The inversion integral path in (5.64) is in accord with the premise of Laplace transform: $f(t) = 0$ for $t < 0$. To see this, let us calculate $f(t)$ for $t < 0$. In this case, the inversion contour should consist of the straight path in (5.64) plus the infinite semi-circle encircling the upper-half ω-plane so that the infinite circle contribution vanishes. See Fig. 5.2. Clearly we have $f(t) = 0$ [since there are no poles inside the contour] in conformity with the assumption that we imposed on $f(t) : f(t) = 0$ for $t < 0$. If a plasma begins to be perturbed from the quiescent state ($f(t) = 0$ for $t < 0$ and $= f(t)$ for $t > 0$), the Laplace transform and the inversion formula (5.64) with c large enough is in accord with the *causality.* As a trivial example, let us consider the equation

$$\frac{df}{dt} = af(t) \quad for \quad t > 0 \quad with \quad a > 0. \tag{5.65}$$

We assume $f(t) = 0$ for $t < 0$. We know that its solution is

$$f(t) = A\ e^{at}. \quad A : const. \tag{5.66}$$

Taking Laplace transform with (5.65) gives

$$aF(\omega) = \int_0^\infty e^{i\omega t}\frac{df}{dt}dt = \lim_{t\to\infty}\ [e^{i\omega t}f(t)] - f(0) - i\omega F(\omega).$$

The limiting term should vanish in order for $F(\omega)$ to exist. To be so, we observe that ω_i should be greater than a in view of (5.66), and we have the straight line contour running above the singular point $\omega = ia$ of $F(\omega)$,

$$F(\omega) = \frac{if(0)}{\omega - ia}.$$

In view of (5.64), the working contour should encircle the lower half plane for $t > 0$. With the aid of the residue theorem,

$$f(t) = \frac{1}{2\pi}(-2\pi i)\ e^{at}\ if(0) = f(0)e^{at}$$

which is (5.66). For $t < 0$, the contour should encircle the upper half plane, and we have the result $f(t) = 0$ for $t < 0$. See Fig. 5.2.

5.3.2. *Initial value problem of Vlasov–Poisson equations*

Laplace transform of (5.11) by performing $\int_0^\infty(...)e^{i\omega t}dt$ gives

$$\begin{aligned} -i\omega g(k,v,\omega) &+ \left[g(k,v,t)e^{i\omega t}\right]_{t=\infty} - g(k,v,0) \\ &+ ikvg(k,v,\omega) - \frac{e}{m}E(k,\omega)\frac{\partial g_0}{\partial v} = 0. \end{aligned} \tag{5.67}$$

In order for the second term to vanish, we need $\omega_i > 0$. This is another aspect of the causality called adiabatic switching on. So Laplace transformed function is defined in the domain $\omega_i > 0$. The situation is the same as the case of Fourier-transformed function. Recall that, in the inverse transform, the straight line part of the Bromwich contour must be above all the singularities of $g(k,v,\omega)$ (Fig. 5.2). Keeping this in mind, we have

$$g(k,v,\omega) = \frac{i}{\omega - kv}\left[g(k,v,t=0) + \frac{e}{m}E(k,\omega)\frac{\partial g_0}{\partial v}\right]. \tag{5.68}$$

Also (5.12) gives

$$ikE(k,\omega) = -4\pi e\int_{-\infty}^{\infty} g(k,v,\omega)dv. \tag{5.69}$$

Elimination of $g(k, v, \omega)$ in the above two equations gives

$$E(k,\omega) = \left(\frac{4\pi e n_0}{k^2}\int_{-\infty}^{\infty} dv\ \frac{g(k,v,t=0)}{v-\omega/k}\right) \div \left(1 - \frac{\omega_{pe}^2}{k^2}\int_{-\infty}^{\infty} \frac{\frac{\partial g_0}{\partial v}}{v-\omega/k}dv\right) \tag{5.70}$$

whose inversion is

$$E(k,t) = \int_{-\infty+ic}^{\infty+ic} d\omega e^{-i\omega t} \times \left[\left(\frac{2en_0}{k^2}\int_{-\infty}^{\infty} dv \frac{g(k,v,t=0)}{v-\omega/k}\right) \div \left(1 - \frac{\omega_{pe}^2}{k^2}\int_{-\infty}^{\infty} \frac{\frac{\partial g_0}{\partial v}}{v-\omega/k}dv\right)\right] \tag{5.71}$$

where the integration path runs from $-\infty$ to ∞ above all the singularities of the integrand in the complex ω-plane. The numerator is analytic unless the function $g(k, v, t = 0)$ is pathological. Thus the singularities are located at the zeros of the denominator (marked by x in Fig. 5.1), i.e. the zeros of

$$\varepsilon(k,\omega) \equiv 1 - \frac{\omega_{pe}^2}{k^2}\int_{-\infty}^{\infty} dv \frac{\frac{dg_0}{dv}}{v-\omega/k} = 1 + \chi(\omega) = 0. \tag{5.72}$$

We note that (5.72) is the dispersion relation of the plasma waves obtained by Fourier transform method (See Eq. (4.79)). For inversion purpose, we use the contour in Fig. 5.1 and, we need to extend the definition of the function $\varepsilon(\omega)$ legitimately on the entire ω-plane. To do so, the function $\varepsilon(\omega)$ should be *analytically continued* onto the $\omega_i = 0$ axis and into the region $\omega_i < 0$. When it is done, the integral (5.71) can be computed by the residue theorem as

$$E(k,t) = 2\pi i\ \Sigma_j\ (Residue)_j\ of\ \left[\frac{e^{-i\omega t}\ 2en_0}{k^2\varepsilon(\omega,k)}\int_{-\infty}^{\infty} \frac{g(k,v,t=0)}{v-\omega/k}dv\right]. \tag{5.73}$$

Here the subscript j denotes each singularity scattered on the ω-plane. Originally, $\varepsilon(\omega)$ is defined only in the region $\omega_i > 0$. Then, what value should we give it on the $\omega_i = 0$ axis or in the region of $\omega_i < 0$? If a certain physical quantity that is defined in some region needs to be defined outside of that region, what is the acceptable definition of that physical quantity in the extended region? The philosophical answer to this question is that we require only the mathematical continuity of the physical function to the extended region. This is paraphrased as follows: if a physical function ceases to be defined because it is outside the originally defined region, the definition is valid in the extended region provided it is *analytically continued.* If a physical function, say, $\varepsilon(\omega)$, in its functional form, is unable to represent the physical quantity over the whole ω-plane, the whole of the analytically continued functions represents the dielectric function in the whole ω-plane, with different representations in respective domains. Here we employ a straightforward method to analytically continue $\varepsilon(\omega)$ in (5.72) from the domain $\omega_i > 0$ unto the domain $\omega_i = 0$. We write $\omega = \omega_r + \omega_i$ and

$$\frac{1}{\omega - kv} = \frac{\omega_r - kv - i\omega_i}{(\omega_r - kv)^2 + \omega_i^2} \quad (\omega_i > 0).$$

Now upon taking the limit $\omega_i \to 0$, the real part is evaluated as the principal value and the imaginary part becomes the δ -function:

$$\frac{1}{\omega - kv} = P\,\frac{1}{\omega_r - kv} - i\pi\delta(\omega_r - kv). \tag{5.74}$$

Therefore we have

$$\chi(k,\omega) = \frac{\omega_{pe}^2}{k}\int_{-\infty}^{\infty} dv \frac{dg_0}{dv}\left[P\,\frac{1}{\omega_r - kv} - i\pi\delta(\omega_r - kv)\right] \quad on \quad \omega_i = 0 \tag{5.75}$$

which is the analytic continuation onto $\omega_i = 0$ of the function $\chi(\omega)$ in (5.72) defined in the domain $\omega_i > 0$. In the domain $\omega_i < 0$, we can ascertain that the analytically continued function of $\chi(\omega)$ is

$$\chi(k,\omega) = \frac{\omega_{pe}^2}{k}\int_{-\infty}^{\infty} dv \frac{dg_0}{dv}\left[\frac{1}{\omega - kv} - 2i\pi\delta(\omega_r - kv)\right] \quad \omega_i < 0. \tag{5.76}$$

Note that (5.74) corresponds to the well-known Plemelj formula with $\omega_i \to 0^+$. The algebraic expressions for $\chi(\omega)$ in (5.75) and (5.76) are equivalent to evaluating the integral along the indented contour shown in Fig. 5.3. The deformed contour in the integration is called *Landau contour.*

Summary: The *electrostatic dielectric function* $\varepsilon(k,\omega)$, given by (5.72), is defined by three different expressions in respective regions.

$$\varepsilon(k,\omega) = 1 - \frac{\omega_{pe}^2}{k^2}\int_{-\infty}^{\infty} \frac{\frac{\partial g_0}{\partial v}}{v - \omega/k}\,dv \quad (\omega_i > 0), \tag{5.77}$$

$$\varepsilon(k,\omega) = 1 - \frac{\omega_{pe}^2}{k^2}\,P\int_{-\infty}^{\infty} \frac{\frac{\partial g_0}{\partial v}}{v - \omega/k}\,dv - i\pi\frac{\omega_{pe}^2}{k^2}\left[\frac{\partial g_0}{\partial v}\right]_{v=\omega/k} \quad (\omega_i \to 0), \tag{5.78}$$

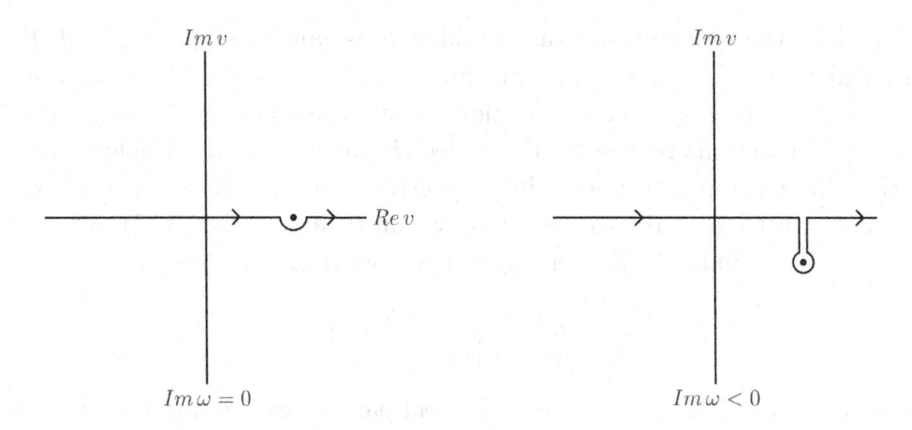

Fig. 5.3. Landau contour.

$$\varepsilon(k,\omega) = 1 - \frac{\omega_{pe}^2}{k^2}\int_{-\infty}^{\infty}\frac{\frac{\partial g_0}{\partial v}}{v-\omega/k}\,dv - 2i\pi\frac{\omega_{pe}^2}{k^2}\left[\frac{\partial g_0}{\partial v}\right]_{v=\omega/k} \quad (\omega_i < 0). \tag{5.79}$$

[To allow for the case $k < 0$, k^2 in the denominators of the last terms of (5.77) and (50) should be replaced by $k \mid k \mid$.]

Returning to (5.73), the asymptotic behavior (at large times) of the electric field depends on the poles of the integrand: poles of the numerator and zeros of the denominator $[\varepsilon(\omega,k) = 0]$. For reasonable initial perturbation $g(k,v,t = 0)$, the numerator has no poles, and the zeros of the equation $\varepsilon(k,\omega) = 0$ determine the normal modes supported by the plasma (accompanied by the *Landau damping* due to the imaginary part in (5.77)).

Use of (5.70) in (5.68) to eliminate $E(k,\omega)$ gives

$$g(k,v,\omega) = \frac{i}{\omega - kv}\left[g(k,v,t=0) + \frac{\omega_{pe}^2}{k^2}\frac{\partial g_0}{\partial v}\frac{1}{\varepsilon(\omega,k)}\int_{-\infty}^{\infty}\frac{g(k,v,t=0)}{v-\omega/k}dv\right],$$

$$g(k,v,t) = \frac{i}{2\pi}\int_{-\infty+ic}^{\infty+ic}\frac{d\omega e^{-i\omega t}}{\omega - kv}$$
$$\times\left[g(k,v,t=0) + \frac{\omega_{pe}^2}{k^2}\frac{\partial g_0}{\partial v}\frac{1}{\varepsilon(\omega,k)}\int_{-\infty}^{\infty}\frac{g(k,v,t=0)}{v-\omega/k}dv\right]. \tag{5.80}$$

In the above Laplace inversion, there are two kinds of singularities: $\omega = kv$ and the zeros of $\varepsilon(\omega,k) = 0$. The latter corresponds to the normal modes of $E(k,t)$ and Landau damps (see (75), Introduction). The pole at $\omega = kv$ gives an additional term $\sim e^{-ikvt}$, which is called *free streaming or ballistic* term. This term persists endlessly and becomes more and more oscillatory as time increases. But $\int dv$ will give almost complete cancellations, yielding no appreciable effect on macroscopic variables such as density perturbation. This term exists because the collisionless Vlasov particles carry the memory of their initial perturbations with them for all subsequent time, but their memory will be wiped out if particle collisions enter the dynamics (mathematically putting, $\omega = kv$ is no longer singular).

The imaginary part of $\varepsilon(k,\omega)$ in (5.77) is due to the *resonant particles* whose velocities are $v = \omega/k$ and which absorb the wave energy. Further evaluation of (5.77) will be carried out in the sequel.

5.4. Plasma dispersion function

In this Section, we evaluate the principal value integral in (5.78) with g_0 a Maxwellian by introducing the plasma dispersion function $Z(\zeta)$ as in (5.86) below.

5.4.1. *Dielectric function of electrostatic wave*

This subject was dealt with in § 4.3. Here we derive a convenient expression for ε applicable in Vlasov plasma for electrostatic waves. We have

$$\mathbf{E} + 4\pi\mathbf{P} = \mathbf{D} = \varepsilon\mathbf{E}, \quad \mathbf{E} = -\nabla\phi \tag{5.81}$$

$$-i\mathbf{k}\phi + 4\pi\mathbf{P} = -i\mathbf{k}\varepsilon\ \phi. \tag{5.82}$$

For electrostatic wave, $\mathbf{E} + 4\pi\mathbf{P} = 0$, so the dispersion relation is $\varepsilon = 0$. From (5.82)

$$-ik^2\phi + 4\pi\mathbf{k}\cdot\mathbf{P} = -ik^2\varepsilon\phi, \tag{5.83}$$

$$\nabla\cdot\mathbf{P} = -\frac{1}{4\pi}\nabla\cdot\mathbf{E} = -\sum_\alpha e_\alpha \int f_\alpha d^3v = i\mathbf{k}\cdot\mathbf{P}, \tag{5.84}$$

$$\varepsilon = 1 - \frac{4\pi}{k^2\phi}\sum_\alpha e_\alpha \int f_\alpha d^3v. \tag{5.85}$$

Note that f_α is proportional to ϕ in electrostatic waves, so ϕ in the denominator cancels out. Combining the linear relation (4.13) with the above equation recovers (4.23).

5.4.2. *Plasma dispersion function* $Z(\zeta)$

The velocity integrals appearing in ε_L or ε_T (Eq. (4.29)) can be systematically dealt with for Maxwellian (Eq. (4.85)) plasmas by considering the integral

$$Z(\zeta) = \frac{1}{\sqrt{\pi}}\int_{-\infty}^{\infty}\frac{e^{-x^2}}{x-\zeta}\,dx \quad : \textit{plasma dispersion function} \tag{5.86}$$

where the integral path goes along the Landau contour, indented concave-down at $x = \zeta$ when $Im(\zeta) \to 0$. Consider the principal value of $Z(\zeta)$, Z_p:

$$\frac{dZ_p}{d\zeta} = \frac{1}{\sqrt{\pi}}\int_{-\infty}^{\infty}\frac{e^{-x^2}}{(x-\zeta)^2}dx = \frac{1}{\sqrt{\pi}}\int_{-\infty}^{\infty}\frac{d}{dx}\left(\frac{-1}{x-\zeta}\right)e^{-x^2}\,dx$$

$$= \frac{-2}{\sqrt{\pi}}\int_{-\infty}^{\infty}\frac{xe^{-x^2}}{x-\zeta}\,dx$$

$$\frac{dZ_p}{d\zeta} = \frac{-2}{\sqrt{\pi}}\int_{-\infty}^{\infty}\left(1+\frac{\zeta}{x-\zeta}\right)e^{-x^2}dx = -2(1+\zeta Z_p). \tag{5.87}$$

(5.87) is a first order differential equation with the boundary condition $Z_p(0) = 0$ (an odd function integral) which is solved by

$$Z_p(\zeta) = -2e^{-\zeta^2}\int_0^{\zeta} e^{x^2}dx. \tag{5.88}$$

So the plasma dispersion function can be written as

$$Z(\zeta) = \frac{1}{\sqrt{\pi}}\int_{-\infty}^{\infty}\frac{e^{-x^2}}{x-\zeta}\,dx = e^{-\zeta^2}\left(-2\int_0^{\zeta}e^{x^2}\,dx + i\sqrt{\pi}\right). \tag{5.89}$$

The error function integral in (5.89) can be asymptotically expanded [Arfken and Weber (2001) *Mathematical methods for physicists* 5th ed. Harcourt/Academic Press] for the cases of $\zeta \gg 1$ and $\zeta \ll 1$:

$$Z(\zeta) = -\frac{1}{\zeta} - \frac{1}{2\zeta^3} - \frac{3}{4\zeta^5} - \frac{15}{8}\frac{1}{\zeta^7} - \ldots + i\sqrt{\pi}e^{-\zeta^2} \quad \zeta \gg 1, \tag{5.90}$$

$$Z(\zeta) = -2\zeta\left(1 - \frac{2}{3}\zeta^2 + \frac{4}{15}\zeta^4 - \ldots\right) + i\sqrt{\pi}e^{-\zeta^2} \quad \zeta \ll 1. \tag{5.91}$$

A useful relation is

$$\int dv\,\frac{\omega g_0}{kv-\omega} = \zeta Z(\zeta); \quad \zeta = \frac{\omega/k}{(2T/m)^{1/2}}. \tag{5.92}$$

Example. Starting from (4.23), show that

$$\varepsilon_L = 1 + 2\Sigma \frac{\omega_{p\alpha}^2}{\omega^2} \zeta_\alpha^2 \Big(1 + \zeta_\alpha Z(\zeta_\alpha)\Big) = 1 + \sum \frac{1}{k^2 \lambda_{D\alpha}^2} \Big(1 + \zeta_\alpha Z(\zeta_\alpha)\Big), \tag{5.93}$$

$$\varepsilon_T = 1 + \sum \frac{\omega_{p\alpha}^2}{\omega^2} \zeta_\alpha Z(\zeta_\alpha) \quad with \quad \zeta_\alpha = \frac{\omega/k}{(2T_\alpha/m_\alpha)^{1/2}}. \tag{5.94}$$

In (4.23), we take $\mathbf{k} = \hat{x} k_x$, and integrate $\int\int dv_y dv_z$ to write

$$\varepsilon_L = 1 + \sum \frac{\omega_{p\alpha}^2}{k_x^2} \int \frac{k_x}{\omega - k_x v_x} \frac{\partial g_{\alpha 0}}{\partial v_x} dv_x.$$

Now we suppress the subscript x with the understanding that $g_{\alpha 0}$ is one-dimensional Maxwellian. Using $dg_{\alpha 0}/dv = -\frac{m_\alpha}{T_\alpha} v g_{\alpha 0}$, we can write

$$\varepsilon_L = 1 + \Sigma \frac{\omega_{p\alpha}^2}{k^2} \frac{m_\alpha}{T_\alpha} \left(1 + \int \frac{\omega \; g_{\alpha 0}}{kv - \omega} dv\right).$$

Introducing a new integration variable, say, $q = \sqrt{\frac{m_\alpha}{2T_\alpha}}\, v$, the $\int dv$-integral is written in terms of $Z(\zeta)$, yielding the desired result. Derivation of (4.94) is left as an exercise.

Exercise. $Z(\zeta)$ satisfies the same differential equation as that satisfied by $Z_p(\zeta)$ but with different boundary condition. What is it?

Exercise. Show that $Z_p(\zeta) = -Z_p(-\zeta)$; odd function.

Exercise. Show that the real and imaginary parts of $Z(\zeta)$ are connected by the Kramers–Kronig relations.

5.5. Evaluation of the dispersion relation

Here we evaluate ε_L in (4.23) or (5.93). We investigate the dispersion relation to find the eigen frequencies and the accompanied Landau damping in a systematic way with the aid of the plasma dispersion function. We assume that the equilibrium distribution is an isotropic Maxwellian:

$$f_{\alpha 0}(v) = \left[\frac{m_\alpha}{2\pi T_\alpha}\right]^{\frac{3}{2}} exp\left[-\frac{m_\alpha v^2}{2T_\alpha}\right]. \tag{5.95}$$

5.5.1. *Cold plasma wave*

We first consider a *cold plasma*. If we take the limit $T_\alpha \to 0$ in (5.95), the Maxwellian becomes

$$f_{\alpha 0}(v) = \delta(\mathbf{v}) = \delta(v_x)\delta(v_y)\delta(v_z). \tag{5.96}$$

Remember that $\delta(x) = \lim_{n\to\infty} \frac{n}{\sqrt{\pi}} exp\,(-n^2x^2)$ is a one-dimensional δ-function. Let us put $f_0(v) = \delta(v)$, and neglect ion contribution which is small by the factor of mass ratio m_e/m_i, in (4.23):

$$\varepsilon_L(k,\omega) = 1 - \frac{\omega_{pe}^2}{k^2}\int_{-\infty}^{\infty} dv\, \frac{\frac{\partial f_0}{\partial v}}{v - \omega/k} = 0.$$

Clearly there is no singularity on the $Re(v)$-axis. Integrating by parts gives the electrostatic dispersion relation of a cold plasma,

$$\varepsilon_L = 1 - \frac{\omega_{pe}^2}{\omega^2} = 0 \quad (\textit{electrostatic, cold}). \tag{5.97}$$

In a cold plasma, it is easy to obtain by using (4.24) $\varepsilon_T = 1 - \Sigma\, \frac{\omega_{p\alpha}^2}{\omega^2} = \varepsilon_L$, so the electromagnetic dispersion relation (4.33) becomes

$$\omega^2 = \omega_{pe}^2 + c^2k^2 \quad (\textit{electromagnetic, cold}). \tag{5.98}$$

5.5.2. *Langmuir wave or electron plasma wave*

We have already obtained the dispersion relation for this case and the Landau damping rate of Langmuir wave in (4.84)–(4.91). Here we use (5.93) to get familiar with the plasma dispersion function. The wave velocity of a Langmuir wave is much greater than the electron thermal velocity, and so $\zeta_e \gg 1$. The ions are stationary in this wave, and the ion contribution to the dielectric function is negligible. Then we obtain

$$\varepsilon_L = 1 + \frac{1}{k^2\lambda_{De}^2}(1 + \zeta_e Z(\zeta_e)). \tag{5.99}$$

We use (5.90) for the asymptotic series for $Z(\zeta_e)$, and all the equations (4.84)–(4.91) are recovered.

5.5.3. *Electromagnetic wave*

To obtain the electromagnetic dispersion relation, integrate (4.24) with respect to the velocity components perpendicular to $\mathbf{k}$:

$$\varepsilon_T == 1 + \Sigma \frac{\omega_{p\alpha}^2}{k\omega}\left[P\int \frac{f_{0\alpha}}{v - \omega/k}dv - i\pi\,[f_{0\alpha}]_{v=\omega/k}\right] \tag{5.100}$$

where f_0 is one-dimensional distribution function. Clearly the ion contribution is negligible. The Landau damping term corresponds to a velocity far down on the Maxwellian tail at $v = \omega/k \simeq c$ where f_0-value is negligibly small. Expanding in the powers of vk/ω gives the dispersion relation

$$\frac{c^2k^2}{\omega^2} = 1 - \frac{\omega_{pe}^2}{\omega^2}\int_{-\infty}^{\infty} f_0(1 + kv/\omega + k^2v^2/\omega^2 + ...)dv. \tag{5.101}$$

Since $v_{th}^2 \ll c^2$, the third term can be neglected as compared to the first, giving the same result as (5.98), which is the electromagnetic wave (light wave) in a plasma. Recapitulation: The Landau damping of the electromagnetic wave in a plasma is nil because the phase velocity of the wave is greater than c, and there are no resonant particles.

Temperature Correction. The electromagnetic wave dispersion relation (4.33) reads

$$n^2 = \frac{c^2k^2}{\omega^2} = 1 + \frac{\omega_{pe}^2}{\omega^2}\zeta_e Z(\zeta_e). \tag{5.102}$$

Expanding $Z(\zeta_e)$ and neglecting the Landau damping give the temperature correction for the refractive index:

$$\omega^2 = \omega_{pe}^2 + c^2k^2 + \frac{T_e}{m_e}\frac{\omega_{pe}^2}{\omega^2}k^2 \tag{5.103}$$

$$n^2 = \frac{1 - \frac{\omega_{pe}^2}{\omega^2}}{1 + \frac{T_e}{mc^2}\frac{\omega_{pe}^2}{\omega^2}}. \tag{5.104}$$

5.5.4. *Ion acoustic wave*

Confer Section (4.6.3). Here we work out the ion acoustic wave by using (5.93). Ion acoustic wave which is a low frequency electrostatic wave is peculiar and characteristic of plasma in that the restoring force is provided by electron pressure while the inertia by ion. Both electron and ion equations should be used, and the wave velocity is such that $v_{thi} \ll \omega/k \ll v_{the}$, or $\zeta_e \ll 1$ and $\zeta_i \gg 1$. Using (5.90) for ions and (5.91) for electrons give

$$\varepsilon_L = 1 + \frac{1}{k^2\lambda_{De}^2}\left(1 - \frac{T_e}{m_i}\frac{k^2}{\omega^2} + i\sqrt{\pi}\,\zeta_e\, e^{-\zeta_e^2} + i\sqrt{\pi}\,\zeta_i\,\frac{T_e}{T_i}\, e^{-\zeta_i^2}\right).$$

Separating the real and imaginary parts, we obtain

$$\varepsilon_r = 1 + \frac{1}{k^2\lambda_{De}^2}\left(1 - \frac{T_e}{m_i}\frac{k^2}{\omega^2}\right), \tag{5.105}$$

$$\varepsilon_i = \frac{\sqrt{\pi}}{k^2\lambda_{De}^2}\left(\zeta_e\, e^{-\zeta_e^2} + \zeta_i\,\frac{T_e}{T_i}\, e^{-\zeta_i^2}\right). \tag{5.106}$$

The real part gives the dispersion relation,

$$\omega_r^2 = \frac{T_e}{m_i}\frac{k^2}{1+k^2\lambda_{De}^2}. \tag{5.107}$$

Using (4.83), we obtain the imaginary frequency, the Landau damping rate

$$\omega_i = -\sqrt{\frac{\pi}{8}}\frac{\omega_r}{(1+k^2\lambda_{De}^2)^{3/2}}\left[\sqrt{\frac{m_e}{m_i}}+\left(\frac{T_e}{T_i}\right)^{3/2} exp\left(\frac{-T_e/T_i}{2(1+k^2\lambda_{De}^2)}\right)\right]. \tag{5.108}$$

In the long wave region where the charge neutrality holds, we can set $k^2\lambda_{De}^2 \ll 1$ in (5.107) and (5.108). For future reference, we write explicitly

$$\omega_r^2 = \frac{T_e}{m_i}k^2, \quad \omega_i = -\sqrt{\frac{\pi}{8}}\omega_r\left(\frac{T_e}{T_i}\right)^{3/2} exp\left(\frac{-T_e}{2T_i}\right). \tag{5.109}$$

In (5.108) the mass ratio term $\sqrt{\frac{m_e}{m_i}}$ comes from the electron Landau damping. The ion Landau damping dominates over this, so we say that ion acoustic wave Landau damps on ions. If $T_e \simeq T_i$, the exponential term (ion term) also becomes small. Only an ion acoustic wave in a plasma of $T_e \gg T_i$ can propagate without suffering from strong Landau damping. As T_i approaches T_e, ion acoustic wave severely Landau damps. This is anticipated: in the latter case, the wave velocity ω/k lies well in the *body* of the ion distribution; if $T_e \gg T_i$, the wave velocity would be on the *tail* portion.

Exercise. If you retain one more term in the expansion of $Z(\zeta_i)$, ε_L has an extra term, $-3\frac{T_i}{m_i}k^2\omega_{pi}^2/\omega^4$. How does this term change the real and imaginary frequency? Answer:

$$\omega_r^2 = \frac{T_e}{m_i}\frac{k^2}{1+k^2\lambda_{De}^2} + 3k^2\frac{T_i}{m_i},$$

$$\omega_i = -\sqrt{\frac{\pi}{8}}\frac{\omega_r}{(1+k^2\lambda_{De}^2)^{3/2}}\left[\sqrt{\frac{m_e}{m_i}}+\left(\frac{T_e}{T_i}\right)^{3/2} exp\left(\frac{-T_e/T_i}{2[1+k^2\lambda_{De}^2]}-\frac{3}{2}\right)\right].$$

5.5.5. *Ion-acoustic instability as a two-stream instability*

The two-stream instability can arise when there is a relative motion between two species of plasma; for example, an electron beam moving through the ions. We shall consider the ions to be cold and stationary: $f_{0i} = \delta(v)$. On the other hand, the electron beam can be represented by a drifting Maxwellian,

$$f_{0e} = \left[\frac{m_e}{2\pi T_e}\right]^{\frac{1}{2}} exp\left[-\frac{m_e(v-v_0)^2}{2T_e}\right]. \tag{5.110}$$

We use these distribution functions in the dispersion relation,

$$\varepsilon_L = 1 + \sum \frac{\omega_{p\alpha}^2}{k^2} \int \frac{k\frac{\partial f_{\alpha 0}}{\partial v}}{\omega - kv} dv = 0. \tag{5.111}$$

The ion term gives after integrating by parts, $-\omega_{pi}^2/\omega^2$. Consider the electron integral,

$$\int_{-\infty}^{\infty} \frac{\partial f_{0e}/\partial v}{v-\omega/k} dv = P\int_{-\infty}^{\infty} \frac{\partial f_{0e}/\partial v}{v-\omega/k} dv + i\pi \left[f_{0e}'\right]_{v=\omega/k}. \tag{5.112}$$

The principal part becomes

$$-\frac{m_e}{T_e}\int \frac{(v-v_0)f_{e0}}{v-\omega/k} dv = -\frac{m_e}{T_e}\int \frac{uf_{e0}(u)}{u+v_0-\omega/k} du \tag{5.113}$$

where $u = v - v_0$ and $f_{e0}(u)$ is now the stationary Maxwellian. The drift velocity v_0 and the phase velocity ω/k in the denominator will be neglected because we consider they are much smaller than the thermal velocity. Then the integral (5.113) simply gives the value $-m_e/T_e$. Then the dispersion relation (5.111) reads

$$1 - \frac{\omega_{pi}^2}{\omega^2} + \frac{m_e}{T_e}\frac{\omega_{pe}^2}{k^2} - i\pi\frac{\omega_{pe}^2}{k^2}\left[f_{e0}'\right]_{u=\omega/k} = 0. \tag{5.114}$$

If we neglect the imaginary part, the dispersion relation gives the steady ion acoustic wave. The slope of the distribution function in the imaginary part of (5.114) can be different in sign from the slope of stationary distribution function ($v_0 = 0$). In the stationary plasma, the slope of the Maxwellian at the wave velocity is negative (damping); in (5.114) in which f_{e0} is a moving Maxwellian, the slope can be positive, giving rise to instability if the drift velocity is greater than the ion acoustic wave speed:

$$v_0 > \sqrt{T_e/m_i} = \omega_r/k \quad (for \quad instability). \tag{5.115}$$

The growth rate ω_i can be calculated with the aid of (4.83). We have $\partial\varepsilon_r/\partial\omega_r = 2\omega_{pi}^2/\omega_r^3$, $\varepsilon_i = -\pi(\omega_{pe}^2/k^2)(\frac{\omega_r}{k} - v_0)\sqrt{\frac{m_e}{2\pi T_2}}(-)T_e/m_e\ exp(\cdots)$. Neglecting ω_r/k as compared to v_0 and putting the exponential factor 1, we obtain (see Fig. 5.4) the growth rate

$$\omega_i = \sqrt{\frac{\pi}{8}}\sqrt{\frac{m_e}{T_e}}\,\omega_r v_0 \sim \omega_r\sqrt{\frac{\pi}{8}}\sqrt{\frac{m_e}{m_i}}. \tag{5.116}$$

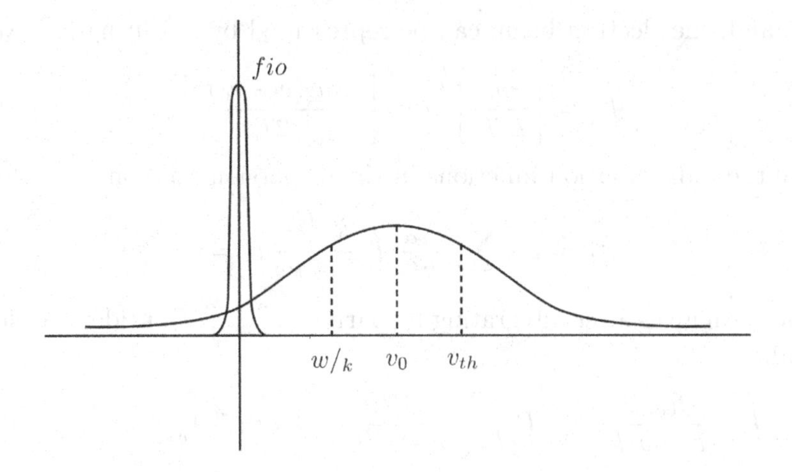

Fig. 5.4. Two stream instability of cold ions and moving Maxwellian electrons.

5.6. Physics of Landau damping

In Chapter 4, we discussed Landau damping in view of the causality. Although the causal requirement can be considered to be the foremost origin of Landau damping, we need to have the wave-particle interaction operative in order for the collisionless damping to actually occur. The wave-particle interaction results in energy exchange between the wave and the *resonant particles* whose velocities are close to the wave velocity. The orbits of such particles are periodic in the *wave frame* because they are subject to back and forth motion between two turning points. They are confined in the potential well and the particles are said to be *trapped* in the wave.

5.6.1. *Resonant particles*

Consider the equation of motion of an electron moving in electric field $E(x,t) = E_0\ sin(kx - \omega t)$,

$$\frac{dv}{dt} = -\frac{e}{m}E_0\ sin(kx - \omega t). \tag{5.117}$$

This is a *Lagrangian* description of the particle trajectory; it implicitly means that x in the above equation is the particle position at time t: $x(t)$. In this sense v is a function of t only: $v(t)$. In *Eulerian* description, the same equation is written as

$$\frac{\partial v}{\partial t} + v\frac{\partial v}{\partial x} = -\frac{e}{m}E_0\ sin(kx - \omega t) \tag{5.118}$$

where x and t are independent, and $v = v(x,t)$. The characteristic equation of (5.118) reads

$$\frac{dt}{1} = \frac{dx}{v} = \frac{dv}{-\frac{e}{m}E_0 sin(kx - \omega t)}$$

which gives $v = dx/dt$ and (5.117). The nonlinear equation (5.118) was considered in Introduction in analogy with a particle motion in a potential. Here we engage on the linearized version

$$\frac{\partial v}{\partial t} + v_0 \frac{\partial v}{\partial x} = -\frac{e}{m} E_0 \; sin(kx - \omega t) \tag{5.119}$$

where v_0 may be the initial velocity of the particle. Equation (5.119) is solved by sum of the particular solution and the homogeneous solution. The particular solution satisfies the equation itself as well as the initial conditions. The homogeneous solution satisfies the equation but is irrelevant to the initial conditions. It represents the wave expressed by the Fourier amplitude with the phasor $e^{ikx - i\omega t}$. By characteristic method, we obtain the particular solution of (5.119) in the form

$$v(x,t) = v_0 - \frac{eE_0/m}{\omega - kv_0}[cos(kx - \omega t) - cos(kx - kv_0 t)]$$
$$for \; resonant \; particles. \tag{5.120}$$

Note that v_0 in (5.120) is $v_0 = v(x,0)$, the initial velocity. If $\omega \to kv_0$, $v(x,t) = -\frac{e}{m}E_0 \; t \; sin(kx - kv_0 t)$, i.e. the particle velocity increases as $\sim t$. $\omega = kv_0$ appears to be a singular point but it is only apparent. The secular behavior $\sim t$ is manifested in the particular solution by the *resonant particles.* The motion of the *resonant particles* whose velocities are close to the wave velocity ω/k is described by (5.120), and the resonant particles exchange energy effectively with the wave and are responsible for Landau damping.

Note that

$$v(x,t) = -\frac{eE_0/m}{\omega - kv_0} cos(kx - \omega t), \quad \left(\omega \neq \frac{v_0}{k}\right)$$
$$for \; nonresonant \; ptls \tag{5.121}$$

is also a solution of (5.119) provided $\omega \neq \frac{v_0}{k}$, which can be obtained by Fourier transform and is irrelevant to initial conditions and doesn't show the $\sim t$ behavior at $\omega = kv_0$. This solution exhibits the singular behavior at $\omega = kv_0$, and is not acceptable at $\omega = kv_0$. It represents the principal part of the fraction $1/(\omega - kv_0)$. (5.121) is referred to as homogeneous solution here.

The Eulerian solution (5.120) can be obtained from the Lagrangian equation (5.117) by using the *unperturbed orbit*,

$$x = v_0 t + x_0 \tag{5.122}$$

where x_0 and v_0 are the initial position and velocity, respectively. Then (5.117) becomes

$$\frac{dv}{dt} = -\frac{e}{m} E_0 \; sin(kx_0 + kv_0 t - \omega t) \tag{5.123}$$

which is integrated to yield

$$v(t) = v_0 \; - \frac{e}{m} E_0 \left[\frac{cos(kx_0 + kv_0 t - \omega t) - cos(kx_0)}{\omega - kv_0} \right]. \tag{5.124}$$

If we go back to the *Eulerian coordinate* x, by putting $x_0 = x - v_0 t$, the above equation becomes

$$v(x,t) = v_0 \; - \frac{e}{m} E_0 \left[\frac{cos(kx - \omega t) - cos(kx - kv_0 t)}{\omega - kv_0} \right] \tag{5.125}$$

which is (5.120). The initial position x_0 serves as the *Lagrangian coordinate.* This recovery is anticipated since (5.117) and (5.118) are two alternative descriptions of the same motion. Linearization in the Eulerian equation corresponds to use of unperturbed orbit in the Lagrangian equation.

5.6.2. *Lagrangian vs Eulerian*

Instead of (5.117), we consider a Fourier form of a wave equation

$$\frac{dv}{dt} = -\frac{e}{m} E_0(k,\omega) \; e^{i(kx - \omega t)} \tag{5.126}$$

for a single Fourier component. We use the unperturbed orbit $x = x_0 + v_0 t$. Then (5.126) becomes

$$\frac{dv}{dt} = -\frac{e}{m} E(k,\omega) \; e^{ikx_0 + i(kv_0 - \omega)t} \tag{5.127}$$

whose solution satisfying the initial condition is

$$v(t) = v_0 + \frac{e}{m} E(k,\omega) \; \frac{e^{ikx_0}}{i(kv_0 - \omega)} \; [1 - e^{i(kv_0 - \omega)t}]. \tag{5.128}$$

The term 1 in the parenthesis is necessary to have the initial condition satisfied. However, without the unity, (5.128) satisfies (5.127). (5.128) is the particular solution of the differential equation (5.127). The first order homogeneous solution is

$$v(t) = -\frac{e}{m}E(k,\omega)\,\frac{e^{ikx_0+i(kv_0-\omega)t}}{i(kv_0-\omega)} \quad : \quad \textit{Lagrangian velocity.} \tag{5.129}$$

It is trivial to obtain Eulerian velocity corresponding to (5.129); simply $x_0 = x - v_0 t$ is substituted to get

$$v(x,t) = -\frac{e}{m}E(k,\omega)\,\frac{e^{ikx-i\omega t}}{i(kv_0-\omega)} \quad : \quad \textit{Eulerian velocity.} \tag{5.130}$$

(5.130) solves the Eulerian equation of motion

$$\frac{\partial v}{\partial t} + v_0\,\frac{\partial v}{\partial x} = -\frac{e}{m}E(k,\omega)e^{ikx-i\omega t}. \tag{5.131}$$

Utility of the Lagrangian solution (5.129) is evident in the following analysis to obtain the dielectric function of a plasma.

Let us consider the *first order Lagrangian coordinate* corresponding to (5.129):

$$x(t,x_0) = \frac{e}{m}E(k,\omega)\,\frac{e^{ikx_0+i(kv_0-\omega)t}}{(kv_0-\omega)^2}. \tag{5.132}$$

The Lagrangian continuity equation reads

$$n(t,x_0)J = N(x_0) \equiv N_0 \tag{5.133}$$

where $J = \frac{\partial x}{\partial x_0}$ is the Jacobian, n is the number density perturbation, $N(x_0)$ is the equilibrium initial number density. It is not easy to directly convert the Lagrangian density $n(t,x_0) = N(x_0)/J$ to the Eulerian density $n(x,t)$ because obtaining $\frac{1}{J}$ in terms of the Lagrangian coordinate is difficult. So we differentiate the expression in (5.133) to write

$$\frac{dn}{dt} + \frac{n}{J}\,\frac{dJ}{dt} = 0 \tag{5.134}$$

or its linear equation

$$\frac{dn}{dt} + \frac{N_0}{J}\,\frac{dJ}{dt} = 0. \tag{5.135}$$

We have the relations

$$\begin{aligned}\frac{1}{J}\,\frac{dJ}{dt} &= \frac{1}{J}\frac{d}{dt}\,\frac{\partial x}{\partial x_0} = \frac{1}{J}\,\frac{\partial}{\partial x_0}\,\frac{dx}{dt} = \frac{1}{J}\frac{\partial v}{\partial x_0} = \frac{\partial x_0}{\partial x}\,\frac{\partial v}{\partial x_0}\\ &= \frac{\partial v(x,t)}{\partial x}\end{aligned} \tag{5.136}$$

where we used the relation $\frac{1}{J} = \frac{\partial x_0}{\partial x}$, and the v in the last expression is the Eulerian velocity. Using (5.136) in (5.135) converts the Lagrangian equation

(5.135) to the Eulerian equation

$$\frac{\partial}{\partial t}\, n(x,t) + v_0\, \frac{\partial n}{\partial x} + N_0 \frac{\partial v}{\partial x} = 0. \tag{5.137}$$

Using (5.130) in (5.137) gives

$$n(x,t) = -i\frac{e}{m}\, k\, \frac{N_0(v_0)}{(\omega - kv_0)^2} E(k,\omega) e^{ikx - i\omega t}. \tag{5.138}$$

This is the Fourier amplitude of the density perturbation in Eulerian coordinates. See Eq. (4.103).

5.6.3. *Energy transfer in Landau damping*

Reference: D. C. Montgomery and D. A. Tidman, *Plasma Kinetic Theory*, McGraw-Hill, New York (1964); T. O'Neil, *Collisionless damping of nonlinear plasma oscillations* Phys. Fluids **3**, 2225 (1965)

We discuss the energy transfer from wave to particles or vice versa in association with Landau damping or growth. Although the foremost origin of the Landau damping can be traced to the causality, Landau damping requires wave-particle interaction for the energy transfer to actually occur. We can describe Landau damping as the effect of energy exchange between the wave and the *resonant particles*. The following derivation is due to Montgomery and Tidman (1964). The electron distribution (ions are assumed to form the uniform neutralizing background) is divided into two populations: the distribution of majority particles (f_M) and the distribution of a small number of resonant particles (f_r). The latter is nonzero only in the velocity interval (see Fig. 5.5)

$$\omega_0/k_0 - \Delta v \leq v \leq \omega_0/k_0 + \Delta v$$

where ω_0 and k_0 are referred to the following plasma wave which propagates in the plasma and interact with the plasma:

$$E(x,t) = E_0 sin(k_o x - \omega_0 t). \tag{5.139}$$

The linearized Vlasov equations for the two populations of the plasma are

$$\frac{\partial f_M}{\partial t} + v\frac{\partial f_M}{\partial x} - \frac{e}{m} E_0 sin(k_0 x - \omega_0 t)\frac{\partial f_{0M}}{\partial v} = 0, \tag{5.140}$$

$$\frac{\partial f_r}{\partial t} + v\frac{\partial f_r}{\partial x} - \frac{e}{m} E_0 sin(k_0 x - \omega_0 t)\frac{\partial f_{0r}}{\partial v} = 0, \tag{5.141}$$

$$\frac{\partial E}{\partial x} = -4\pi e \int dv\, (f_M + f_r). \tag{5.142}$$

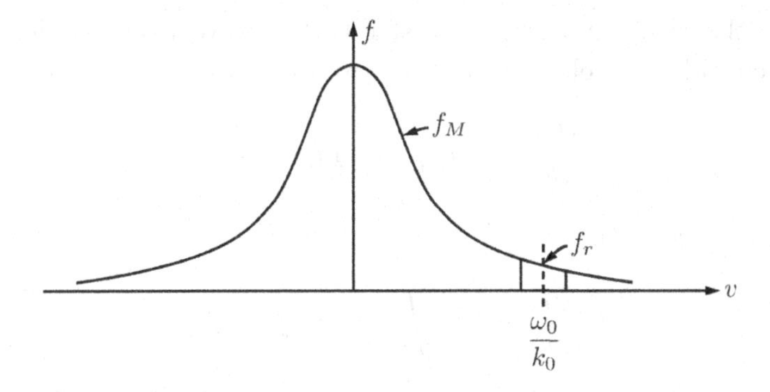

Fig. 5.5. Division of distribution, after Montgomery and Tidman (1964).

We assume in (5.142) $\int dv f_r \ll \int dv f_M$. That is, we neglect $\int dv f_r$ in the interaction of the main population with $E(x,t)$. The particles which belong to f_M are non-resonant and have an oscillatory response to the wave $E(x,t)$. Fourier-transforming (5.140) gives

$$f_M(k,\omega) = 2\pi^2 \frac{eE_0}{m}\frac{\partial f_{0M}}{\partial v} \frac{\delta(k-k_0)\delta(\omega-\omega_0) - \delta(k+k_0)\delta(\omega+\omega_0)}{\omega - kv}. \tag{5.143}$$

Inverting this,

$$f_M(x,t,v) = \frac{eE_0}{m}\frac{\partial f_{0M}}{\partial v} \frac{\cos(k_0 x - \omega_0 t)}{\omega_0 - k_0 v}. \tag{5.144}$$

Using (5.142) with neglect of f_r and (5.144) give the dispersion relation,

$$1 - \frac{\omega_{pe}^2}{k_0^2}\int_{-\infty}^{\infty} dv \frac{\partial f_{0M}/\partial v}{v - \omega_0/k_0} = 0. \tag{5.145}$$

We see that the non-resonant particles whose distribution is related with the wave according to (5.144) merely supports the wave $E(x,t)$, (5.139), under the dispersion relation (5.145).

We suppose that the wave is turned on at $t = 0$ and initially $f_r(t=0) = 0$. Using this initial condition, (5.141) is solved by characteristic method to yield

$$f_r(x,t,v) = \frac{eE_0}{m}\frac{\partial f_{0r}}{\partial v} \frac{\cos(k_0 x - \omega_0 t) - \cos(k_0 x - k_0 v t)}{\omega_0 - k_0 v}. \tag{5.146}$$

Note that (5.146) satisfies the prescribed initial condition, and (5.144) and (5.146) satisfy the same type of equations, but (5.144) is irrelevant to any boundary condition. Also note that f_r in (5.146) shows the $\sim t$ behavior if $\omega_0 \to k_0 v$ as in (5.125). The resonant particle distribution should be obtained as an initial value problem.

To see the electron energy transfer to the wave from particles or vice versa, we consider the electrostatic Maxwell equation

$$\frac{\partial E}{\partial t} = 4\pi e \int dvv\ (f_r + f_M). \tag{5.147}$$

Multiplying by $E(x,t)$,

$$E\ \frac{\partial E}{\partial t} = 4\pi e \int dvv\ E\ (f_r + f_M) \tag{5.148}$$

which expresses the rate of wave energy transfer to all the electrons. Now we substitute (5.144) and (5.146) for f_M and f_r in the right hand side and take the spatial average (denoted by $< \ >$) by integrating over a wavelength $2\pi/k_0$. Noting $<sin(k_0x - \omega_0t)\ cos(k_0x - \omega_0t)> = 0$, the f_M-term in the above vanishes upon spatial average. This means that the non-resonant electrons do not change the average electrical energy. Using

$$\langle cos(k_0x - k_0vt)sin(k_0x - \omega_0t)\rangle = \frac{1}{2}sin(k_0vt - \omega_0t),$$

(5.148) is written

$$\frac{\partial}{\partial t}\langle E^2\rangle = E_0(t)\frac{\partial E_0}{\partial t} = \frac{1}{2}\ \omega_{pe}^2 E_0^2 \int dvvf_{0r}'(v)\frac{sin(\omega_0t - k_0vt)}{\omega_0 - k_0v} \tag{5.149}$$

where we consider E_0 is a slowly varying function of t as the result of energy exchange with the electrons. Note that (5.149) is not singular at $\omega_0 = k_0v$. The integration interval in (5.149) is the narrow band centered about $v = \omega_0/k_0$, and therefore $f_{0r}'(v)$ can be taken out of the integral with $v = \omega_0/k_0$. Furthermore the value of the integral mainly comes from the region $v \simeq \omega_0/k_0$, and we can change the integration limits from $-\infty$ to ∞ without introducing much error.

$$\frac{1}{E_0}\frac{\partial E_0}{\partial t} = \frac{\omega_{pe}^2}{2}\ [f_{0r}'(v)]_{v=\omega_0/k_0} \int_{-\infty}^{\infty} dv\ v\ \frac{sin(\omega_0t - k_0vt)}{\omega_0 - k_0v} \tag{5.150}$$

where the integral $\int_{-\infty}^{\infty} dv(...)$ is equal to

$$-\frac{1}{k_0}\int_{-\infty}^{\infty} dy\ (y + \frac{\omega_0}{k_0})\frac{sin(yt)}{y} = -\frac{\omega_0}{k_0^2}\ \pi\ \ (per\ \int_{-\infty}^{\infty}\frac{sin(x)}{x}dx = \pi)$$

$$\frac{1}{E_0}\frac{\partial E_0}{\partial t} = -\frac{\pi}{2}\ \omega_{pe}^2\ \frac{\omega_0}{k_0^2}\ [f_{0r}'(v)]_{v=\omega_0/k_0} \tag{5.151}$$

which shows that the electric wave damps according to the rate given by the right hand side. Equation (5.151) agrees with the Landau damping rate (4.91).

Next, we can derive the Landau damping rate alternatively by considering the rate of change of the kinetic energy owned by the resonant particles. At early time of interaction between the resonant particles and the wave, we can assume, in (5.125), that v_0 is much greater than the second term on the right hand side of (5.125). Thus we have for the kinetic energy,

$$\frac{m_e}{2}v^2 = -eE_0v_0\frac{cos(kx-\omega t)-cos(kx-kv_0t)}{\omega-kv_0}$$

where we omitted $m_ev_0^2/2$ which is only an additive constant. Expanding the zero order distribution function in the vicinity of $v_0 = \omega/k$,

$$f_0(v) = f_0(v_0) + (v-v_0)\left[\frac{\partial f_0}{\partial v}\right]_{v=v_0} + \cdots$$

$$\int \frac{m_e}{2}v^2 f_0(v)dv = \frac{e^2E_0^2}{m_e}\int dv_0v_0\left[\frac{\partial f_0}{\partial v}\right]_{v=v_0}\frac{[cos(kx-\omega t)-cos(kx-kv_0t)]^2}{(\omega-kv_0)^2}. \tag{5.152}$$

Note that $\int \frac{m_e}{2}v^2 f_0(v_0)dv_0$ vanishes upon spatial average; this was omitted in (5.152). Also note that the integrand is not singular at $\omega = kv_0$. The dominant contribution to the integral comes from the interval around $v_0 = \omega/k$, and thus $\left[\frac{\partial f_0}{\partial v}\right]_{v_0=\omega/k}$ can be taken out of the integral, and the limits of the integration can be extended to $\pm\infty$. Using $cos(kx-\omega t)-cos(kx-kv_0t) = 2sin[(\omega-kv_0)t/2]sin[kx-(\omega+kv_0)t/2]$ and taking spatial average, we obtain

$$\int \frac{m_e}{2}v^2 f_0(v)dv = \frac{e^2E_0^2}{m_e}\left[\frac{\partial f_0}{\partial v}\right]_{v=\omega/k}\int_{-\infty}^{\infty} dv_0\ v_0\ \frac{sin^2\left(\frac{(kv_0-\omega)t}{2}\right)}{(\omega-kv_0)^2} \equiv K. \tag{5.153}$$

Using $\int_{-\infty}^{\infty} dx\ \frac{sin^2 x}{x^2} = \pi$, the kinetic energy of the resonant particles becomes

$$K = -\frac{\pi t}{2}\ \frac{e^2E_0^2}{m_e}\ \frac{\omega}{k^2}\left[\frac{\partial f_0}{\partial v}\right]_{v=\omega/k}. \tag{5.154}$$

Note that K increases linearly with t. It is readily seen that the rate of the kinetic energy increase (when the slope of f_0 is negative) dK/dt is equal to the Landau damping rate of the wave within a numerical factor. We can show from (5.151)

$$\frac{1}{8\pi}\frac{\partial}{\partial t}E_0^2 = \frac{dK}{dt}.$$

5.7. Operator method 1

Here and in the next section, we are introduced to two formal ways of writing the solution of Vlasov equation for future reference.

Linearizing (5.21) yields

$$\frac{\partial}{\partial t}f_\alpha(\mathbf{r},\mathbf{v},t) + \mathbf{v}\cdot\frac{\partial f_\alpha}{\partial \mathbf{r}} + \frac{e_\alpha}{m_\alpha}\frac{\partial f_{\alpha 0}}{\partial \mathbf{v}}$$

$$\cdot\sum_\alpha e_\alpha \int d^3r' \frac{\mathbf{r}-\mathbf{r}'}{\mid \mathbf{r}-\mathbf{r}'\mid^3}\int d^3v' f_\alpha(\mathbf{r}',\mathbf{v}',t) = 0. \tag{5.155}$$

This equation is the Vlasov–Poisson equation which we solved in various ways in the foregoing discussions. In this Section, (5.155) will be solved by another method which may be called *operator method*. For a simple equation such as (5.155), the operator method is not particularly advantageous over the Laplace transform method; rather, it may be reckoned as unnecessary artifice since the result of Laplace transform analysis is invoked eventually to identify the operator. However, the usefulness of the operator method is evident when one encounters the problem requiring double Laplace transform as in derivation of Lenard–Balescu equation (see Chapter 12). So we take this opportunity to get acquaintance with this method. Fourier transforming (5.155) with respect to the variable $\mathbf{r}$ (assuming that ions form only the neutralizing background), we have

$$(\frac{\partial}{\partial t} + i\mathbf{k}\cdot\mathbf{v})f(\mathbf{k},\mathbf{v},t) - i\frac{\omega_{pe}^2}{k^2}\mathbf{k}\cdot\frac{\partial f_0}{\partial \mathbf{v}}\int f(\mathbf{k},\mathbf{v}',t)d^3v' = 0 \tag{5.156}$$

where we used (5.23). We write (5.156) in the form

$$\left[\frac{\partial}{\partial t} + L(\mathbf{k},\mathbf{v})\right] f(\mathbf{k},\mathbf{v},t) = 0, \tag{5.157}$$

$$\text{where}\quad L(\mathbf{k},\mathbf{v})(\cdots) = \left[i\mathbf{k}\cdot\mathbf{v} - i\frac{\omega_{pe}^2}{k^2}\mathbf{k}\cdot\frac{\partial f_0}{\partial \mathbf{v}}\int d^3v\right](\cdots) \tag{5.158}$$

is an operator involving the integral operation. (5.157) is a first order linear differential equation while we regard the operator L as an algebraic quantity; thus (5.157) is solved by

$$f(\mathbf{k},\mathbf{v},t) = e^{-tL}f(\mathbf{k},\mathbf{v},0). \tag{5.159}$$

(5.159) is a formal solution of (5.157) as of an initial value problem.

The symbolic solution (5.159) is not meaningful until we identify the action of the operator L. For further development, (5.159) is Laplace-transformed:

$$f(\mathbf{k},\mathbf{v},\omega) = \frac{1}{L - i\omega} f(\mathbf{k},\mathbf{v},0). \tag{5.160}$$

Inverting (5.160) again gives

$$f(\mathbf{k},\mathbf{v},t) = \frac{1}{2\pi}\int_{-\infty+ic}^{\infty+ic} d\omega e^{-i\omega t} \frac{1}{L - i\omega} f(\mathbf{k},\mathbf{v},0). \tag{5.161}$$

(5.161) should be compared with (5.159): the operator L exponentiated in (5.159) has now transformed to algebraic operator function $1/(L-i\omega)$ in the inversion integral. Our next task is to identify the operator $(L - i\omega)^{-1}$ or specifically the quantity $\frac{1}{L-i\omega}f(\mathbf{k},\mathbf{v},0)$. To carry out this job, let us Laplace transform (5.156) to obtain

$$f(\mathbf{k},\mathbf{v},\omega) = \frac{i}{\omega - \mathbf{k}\cdot\mathbf{v}} f(\mathbf{k},\mathbf{v},0) - \frac{\frac{\omega_{pe}^2}{k^2}\mathbf{k}\cdot\frac{\partial f_0}{\partial \mathbf{v}}}{\omega - \mathbf{k}\cdot\mathbf{v}}\int_{-\infty}^{\infty} d^3v' f(\mathbf{k},\mathbf{v}',\omega). \tag{5.162}$$

Integrating (5.162) with respect to d^3v, one obtains

$$\int f(\mathbf{k},\mathbf{v},\omega)d^3v = \frac{i}{\varepsilon(k,\omega)}\int \frac{f(\mathbf{k},\mathbf{v},0)}{\omega - \mathbf{k}\cdot\mathbf{v}}d^3v \tag{5.163}$$

where ε is the dielectric function (5.72). On the other hand, (5.160) gives

$$\int f(\mathbf{k},\mathbf{v},\omega)d^3v = \int \frac{1}{L - i\omega} f(\mathbf{k},\mathbf{v},0)d^3v. \tag{5.164}$$

Comparing the above two equations, we can write

$$\frac{1}{L - i\omega} = \frac{i}{\varepsilon(k,\omega)}\frac{1}{\omega - \mathbf{k}\cdot\mathbf{v}}. \tag{5.165}$$

Using (5.165) in (5.161) yields

$$f(\mathbf{k},\mathbf{v},t) = \frac{1}{2\pi}\int_{-\infty+ic}^{\infty+ic} d\omega e^{-i\omega t}\frac{i}{\varepsilon(k,\omega)}\frac{1}{\omega - \mathbf{k}\cdot\mathbf{v}} f(\mathbf{k},\mathbf{v},0). \tag{5.166}$$

This equation should be compared with the equation (5.80). The operator method is not of great advantage in this simple example, but it is useful in a more complex situation such as in derivation of Lenard–Balescu equation, where we deal with the equation,

$$\left[\frac{\partial}{\partial t} + L_1(\mathbf{k}, \mathbf{v}_1, t) + L_2(\mathbf{k}, \mathbf{v}_2, t)\right] g(\mathbf{k}, \mathbf{v}_1, \mathbf{v}_2, t) = 0.$$

Exercise. Prove the equivalence of (5.80) and (5.166). In fact, (5.166) is more compact than (5.80), which is a meaningful result obtained in this section. The proof goes as follows. Integrate $\int_{-\infty}^{\infty}(5.80)dv$ to obtain

$$\int_{-\infty}^{\infty} dvg(k,v,t) = \frac{i}{2\pi}\int_{-\infty+ic}^{\infty+ic} d\omega e^{-i\omega t}$$
$$\times \int dv \frac{g(k,v.0)}{\omega - kv}\left[1 - \frac{\omega_p^2}{k\,\varepsilon}\int \frac{dg_0/dv}{\omega - kv}dv\right]$$
$$= \frac{i}{2\pi}\int_{-\infty+ic}^{\infty+ic} d\omega e^{-i\omega t}\,\frac{1}{\varepsilon}\int_{-\infty}^{\infty} dv \frac{g(k,v.0)}{\omega - kv} \quad (per\ (5.72))$$

from which (5.166) follows.

5.8. Operator method 2

For simplicity, we begin with the linearized one-dimensional Vlasov equation (5.28),

$$\frac{\partial}{\partial t}g(x,v,t) + v\,\frac{\partial g}{\partial x} = \frac{e}{m}\frac{\partial g_0}{\partial v}E(x,t). \tag{5.167}$$

This is a first order linear differential equation in t. Let us regard the operator $v\frac{\partial}{\partial x}$ as an algebraic quantity. Then, (5.167) is readily solved in the form, with the initial condition $g(t = -\infty) = 0$,

$$g(x,v,t) = \frac{e}{m}\frac{\partial g_0}{\partial v}\int_{-\infty}^{t} e^{(t'-t)v\frac{\partial}{\partial x}}\,E(x,t')dt'. \tag{5.168}$$

The operator $v\frac{\partial}{\partial x}$ acts on $E(x)$ through the exponential function: the exponential function is regarded as the corresponding power series. Direct substitution of (5.168) into (5.167) verifies the validity of the latter equation. It is instructive to calculate the Fourier transform:

$$g(k,v,t) = \int_{-\infty}^{\infty} dxe^{-ikx}g(x,v,t)$$
$$= \frac{e}{m}\frac{\partial g_0}{\partial v}\int_{-\infty}^{t} dt'\int_{-\infty}^{\infty} dxe^{-ikx}e^{\alpha\frac{\partial}{\partial x}}E(x,t'), \quad \alpha = (t'-t)v$$

$$= \frac{e}{m}\frac{\partial g_0}{\partial v}\int_{-\infty}^{t} dt' \int_{-\infty}^{\infty} dx\ e^{-ikx}$$
$$\times \left(1 + \alpha\frac{\partial}{\partial x} + \frac{1}{2}\alpha^2\frac{\partial^2}{\partial x^2} + \cdots\right) E(x, t'). \tag{5.169}$$

Integrating by parts yields

$$g(k, v, t) = \frac{e}{m}\frac{\partial g_0}{\partial v}\int_{-\infty}^{t} dt'\ e^{ik(t'-t)v} E(k, t'). \tag{5.170}$$

(5.170) can be obtained from (5.167) by putting $\frac{\partial}{\partial x} = ik$. One can show that Fourier-transforming (5.170) with respect to t gives

$$g(k, v, \omega) = i\ \frac{e}{m}\frac{\partial g_0}{\partial v}\frac{1}{\omega - kv}\ E(k, \omega). \tag{5.171}$$

Derivation of (5.171) is left as an exercise. Hint: Use in (5.170)

$$E(k, t') = \int_{-\infty}^{\infty} \frac{d\omega'}{2\pi} e^{-i\omega' t'} E(k, \omega')$$

and carry out the $\int dt'$-integral first.

5.9. Nyquist criterion

We have a dispersion relation for plasma waves of the form

$$\varepsilon(\omega, k) = 0. \tag{5.172}$$

We are interested in knowing whether or not the solutions of (5.172) have unstable roots ($\omega_i > 0$) without actually solving the algebraic equation $\varepsilon(\omega) = 0$ (The wave number k will be suppressed). Nyquist theorem provides means for this purpose to determine whether a dispersion relation has unstable roots in the region of interest in the ω-plane. Consider the integral in the complex ω-plane,

$$\frac{1}{2\pi i}\oint_C \frac{\varepsilon'(\omega)}{\varepsilon(\omega)}\, d\omega \quad \left(\varepsilon' = \frac{d\varepsilon}{d\omega}\right) \tag{5.173}$$

where C is a closed contour encircling the region of interest in the ω-plane in the positive (counter-clockwise) direction. To ascertain the presence of unstable roots of (5.172), let C consist of the real ω axis and the infinite semi-circle in the upper-half plane. We shall prove that integral (5.173) gives the number of roots of $\varepsilon(\omega) = 0$ within C. We assume that $\varepsilon(\omega)$ has no poles inside C. When $\varepsilon(\omega)$ has a zero at $\omega = \omega_1$, the integrand ε'/ε has a simple pole at $\omega = \omega_1$. By residue theorem, the number expressed by (5.173) is equal to

$$\lim_{\omega \to \omega_1} (\omega - \omega_1)\frac{\varepsilon'(\omega)}{\varepsilon(\omega)} = \varepsilon'(\omega_1) \lim_{\omega \to \omega_1} \frac{(\omega - \omega_1)}{\varepsilon(\omega)} = \frac{\varepsilon'(\omega_1)}{\varepsilon'(\omega_1)} = 1$$

which is the residue of the integral, and we used L'Hospital's rule in the algebra. Suppose $\varepsilon(\omega)$ has N zeros to be written in the form

$$\varepsilon(\omega) = (\omega - \omega_1)(\omega - \omega_2)...(\omega - \omega_N)g(\omega).$$

Then we have

$$\frac{\varepsilon'(\omega)}{\varepsilon(\omega)} = \frac{1}{\omega - \omega_1} + \frac{1}{\omega - \omega_2} + ... + \frac{1}{\omega - \omega_N} + \frac{g'(\omega)}{g(\omega)}$$

$$\frac{1}{2\pi i} \oint_C \frac{\varepsilon'(\omega)}{\varepsilon(\omega)} d\omega = N. \tag{5.174}$$

When $\varepsilon(\omega)$ has multiple zeros, the multiplicity should be added on the RHS of the above equation. To show this, we write

$$\varepsilon(\omega) = (\omega - \omega_0)^m g(\omega) \qquad \frac{\varepsilon'(\omega)}{\varepsilon(\omega)} = \frac{m}{\omega - \omega_0}$$

$$\frac{1}{2\pi i} \oint_C \frac{\varepsilon'(\omega)}{\varepsilon(\omega)} d\omega = m.$$

We have established the identity (5.174) (*Cauchy's integral*) with the understanding that N contains the multiplicity of the roots. Next, we introduce the polar coordinates in the ω-plane,

$$\omega = re^{i\theta}, \quad \varepsilon(\omega) = R(r,\theta)e^{i\Theta(r,\theta)}, \quad \oint_C \frac{\varepsilon'(\omega)}{\varepsilon(\omega)} d\omega = \oint_C d[\ln\ \varepsilon(\omega)].$$

Here $\ln \varepsilon(\omega) = lnR + i\Theta$. Upon introducing the polar coordinates, the contour C in the ω-plane is mapped into the contour C_ε in the (R, θ) plane (i.e. ε-plane) by the mapping functions $R(r, \theta)$ and $\Theta(r, \theta)$. Thus we have

$$\oint_C d[\ln\ \varepsilon(\omega)] = \oint_{C_\varepsilon} d\ (\ln R + i\Theta) = i \oint_{C_\varepsilon} d\Theta. \tag{5.175}$$

Note that upon completing revolutions around the closed contour, only the argument Θ changes while the modulus is intact. If the contour C_ε encircles the origin N times, the integral in (5.175) is $2\pi i N$. Now we arrive at the conclusion of the Nyquist theorem by looking at (5.174) and (5.175). If the contour C_ε in the ε plane which corresponds to the contour C in the ω plane encircles the origin of the ε plane N times, the number of the roots of $\varepsilon(\omega) = 0$ within C is N, in view of (5.174). Figures 5.6 (a), (b), and (c)

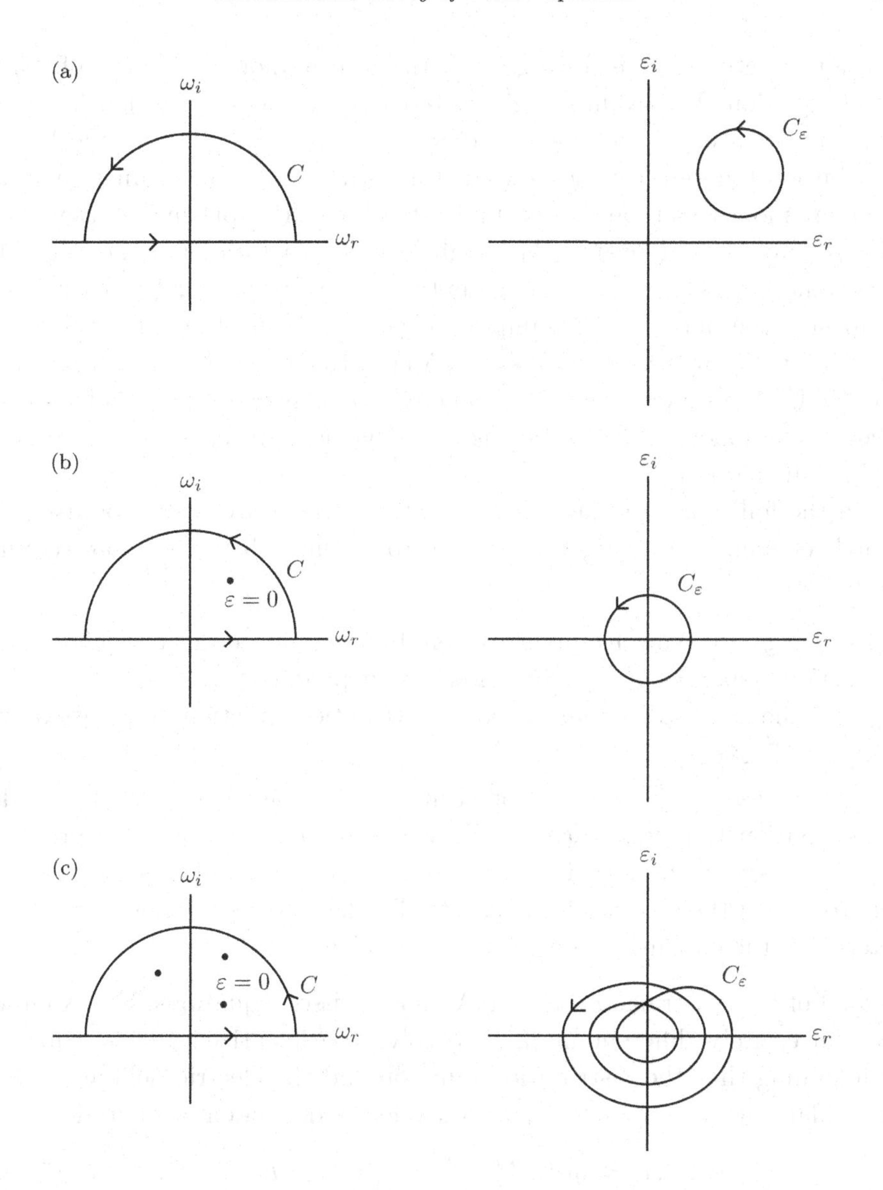

Fig. 5.6. Nyquist method.

illustrate the Nyquist criterion as to the number of the roots of $\varepsilon(\omega) = 0$, in which (a), no unstable roots; (b), 1 unstable root; (c), 3 unstable roots. Only the zeros in the upper ω plane count for instability.

5.10. Van Kampen modes

Van Kampen's approach to the solution of the Vlasov–Poisson system of equations is the third method, besides the characteristic and Laplace–Fourier transform methods. It is analogous to the Sturm–Liouville theory of differential equation. Van Kampen discovered *normal modes* or *eigen functions* of Vlasov equation, which form a complete set in the sense that the Vlasov distribution function can be expanded in terms of the eigen functions. Van Kampen modes can be better understood when a plasma is viewed as an assembly of cold beams. Although Van Kampen did not mention, the eigen functions are the plasma susceptibilities corresponding to each beam composing the plasma. Van kampen's eigen modes are the linear version of the BGK waves which we address in § 5.11.2. Reference: N. G. Van Kampen and B. U. Felderhof *Theoretical methods in plasma physics* John Wiley (1967). The following exposition is a pedagogical variation of the original Van Kampen's work.

In the following, we have three notations which are easily confusing on casual reading, and it might be helpful to distinguish them before starting discussion.

- $g(v)$ or $g_\nu(v)$: Van Kampen's eigen function in terms of which other functions (such as $g(v, t = 0)$ below) are expanded.
- $g_0(v)$: the zero order homogeneous distribution function which goes into $\eta(v) = \frac{\omega_{pe}^2}{k^2}\frac{dg_0(v)}{dv}$.
- $g(v, t = 0) \equiv g(v, 0) \equiv g_i(v)$, sometimes: the initial value of the first order distribution function, $g(x, v, t)$. In Van Kampen's initial value problem, $g(x, v, t)$ should be found in terms of $g(v, 0)$, and the latter is expanded in to obtain the solution for $g(x, v, t)$. The key step is to express $g(v, 0)$ in terms of the expansion coefficients $c(v)$ below.

Instead of Fourier transforming the Vlasov–Poisson equations, Van Kampen used an entirely different method to solve the linearized Vlasov equation by assuming that the distribution function and the electric field are simple sinusoidal wave $\sim cos(kx - \omega t)$ with a velocity-dependent amplitude:

$$g(x, t, v) = g(v)e^{ikx - i\omega t}, \quad E(x, t) = Ee^{ikx - i\omega t}. \tag{5.176}$$

The ansatz (5.176) assumes a special form: $g(x, t, v)$ is a product of a function $g(v)$ and a function $h(x, t)$, and this form suggests that the method of *separation of constant* may be applied. Here we emphasize that $g(v)$ and E are not the Fourier coefficients, but only the wave amplitudes. Later we show that the assembly of the solutions can be individualized by the index $\nu = \omega/k$, so we may write them as $g_\nu(v)$ and E_ν. The ansatz (5.176) contrasts with the wave form $e^{ikx - i\omega(k)t}$ where the dispersion relation $\omega = \omega(k)$ is obtained from the solvability condition of the linear algebraic equations consisting of the Fourier amplitudes. Substituting (5.176) into Vlasov and Poisson equations gives

$$(-i\omega + ikv)g(v) - \frac{e}{m}E\, g_0'(v) = 0 \quad \left(g_0'(v) = \frac{dg_0}{dv}\right) \tag{5.177}$$

$$ikE = -4\pi e \int_{-\infty}^{\infty} dv\, g(v). \tag{5.178}$$

Eliminating E gives

$$\left(\frac{\omega}{k} - v\right) g(v) = -\frac{\omega_{pe}^2}{k^2}\, g_0'(v) \int_{-\infty}^{\infty} dv'\, g(v'). \tag{5.179}$$

Van Kampen exercised special care when he divided (5.179) by the factor $(\omega/k - v)$. He wrote

$$g(v) = -\frac{\omega_{pe}^2}{k^2}\left(\int_{-\infty}^{\infty} dv' g(v')\right) P\, \frac{g_0'(v)}{\omega/k - v} + \lambda(k,\omega)\, \delta\left(\frac{\omega}{k} - v\right) \tag{5.180}$$

where the letter P denotes the principal value. Here we recall the formula in (4.67). The δ-function term mathematically completes the equality even when $\omega = kv$ and physically represents weakly modulated beams. In (5.180), $\lambda(k,\omega)$ can be determined by the condition that $g(v)$ is normalized $(\int_{-\infty}^{\infty} g(v)dv = 1)$:

$$\lambda(k,\omega) = 1 + \frac{\omega_{pe}^2}{k^2}\, P \int_{-\infty}^{\infty} \frac{g_0'(v)dv}{\omega/k - v}. \tag{5.181}$$

It is to be noted that the expression for λ resembles the real part of the dielectric function $\varepsilon(k,\omega)$. Now (5.180) takes the form

$$g(v) = -\frac{\omega_{pe}^2}{k^2}\, P\, \frac{g_0'(v)}{\omega/k - v} + \left[1 + \frac{\omega_{pe}^2}{k^2}\, P \int_{-\infty}^{\infty} \frac{g_0'(\xi)d\xi}{\omega/k - \xi}\right] \delta\left(\frac{\omega}{k} - v\right).$$

For a fixed wave number k, we label the Van Kampen modes by the 'eigen value' $\nu = \omega/k$:

$$g_\nu(v) = -P\, \frac{\eta(v)}{\nu - v} + \left[1 + P \int_{-\infty}^{\infty} \frac{\eta(\xi)d\xi}{\nu - \xi}\right] \delta(\nu - v) \tag{5.182}$$

with $\eta(v) = \frac{\omega_{pe}^2}{k^2}\, g_0'(v)$. The electric field corresponding to the 'elementary' wave is given by

$$E(x,t) = E\, e^{ikx - i\omega t} \quad \textit{with} \quad E = \frac{4\pi e i}{k}. \tag{5.183}$$

Integrating $\int_{-\infty}^{\infty}(5.182)dv$ gives only the identity $1 = 1$, and we have no dispersion relation; k and ω in (5.176) are independent. This was anticipated by the presence of the inhomogeneous term $\lambda(k,\omega)$ in (5.180). Equation (5.182) represents the Van Kampen modes which constitute a complete set in the sense that an arbitrary distribution function of (v,x,t) can be expanded as superposition of such modes [the completeness was proved by K. M. Case, Annals Phys. **7**, 349 (1959)]. Then the first order distribution function can be expressed by appropriate superposition of 'eigen functions' g_ν:

$$g(x,v,t) = e^{ikx}\int_{-\infty}^{\infty} d\nu e^{-ik\nu t} g_\nu(v)c(\nu) \tag{5.184}$$

where $c(\nu)$ is the expansion coefficient. The essence of Van Kampen's method is that the distribution function can be expanded in terms of the eigen functions. In terms of the initial disturbance, $g(k,v,t=0)$, we have an integral equation to determine $c(\nu)$, by substituting (5.182) into (5.184):

$$g(x,v,t=0) = e^{ikx}\int_{-\infty}^{\infty} d\nu\ c(\nu) \times \left[P\,\frac{\eta(v)}{v-\nu} + \delta(\nu - v)\left(1 + P\int_{-\infty}^{\infty}\frac{\eta(\xi)d\xi}{\nu-\xi}\right)\right]. \tag{5.185}$$

The integral involving the δ-function can be carried out to write (dropping the e^{ikx} factor)

$$g(v,t=0) \equiv g(v,0) = \eta(v)P\int_{-\infty}^{\infty} d\nu\,\frac{c(\nu)}{v-\nu} + c(v)\left(1 + P\int_{-\infty}^{\infty}\frac{\eta(\xi)d\xi}{v-\xi}\right)$$

$$or\ g_i(v) \equiv g(v,0) = \eta(v)P\int_{-\infty}^{\infty} d\nu\,\frac{c(\nu)}{v-\nu} + c(v)\left(1 + P\int_{-\infty}^{\infty}\frac{\eta(\nu)d\nu}{v-\nu}\right) \tag{5.186}$$

where the subscript i means '*initial*'. This is the integral equation that we wish to solve for $c(v)$, with $g_i(v)$ and $\eta(v)$ known. This integral equation can be solved by decomposing the functions involved in (5.186) into the corresponding half-range Fourier functions (denoted by superscripts $\pm$ in the sequel). The basic scheme of solving the integral equation is similar to the Wiener–Hopf method.

Let us consider the principal value integral

$$P\int_{-\infty}^{\infty} d\nu\,\frac{c(\nu)}{v-\nu}.$$

We introduce in the above the Fourier expression

$$c(\nu) = \int_{-\infty}^{\infty} c(\tau)e^{i\nu\tau}d\tau.$$

Then, the principal value integral becomes

$$P\int_{-\infty}^{\infty} d\nu\, \frac{c(\nu)}{v-\nu} = \int_{-\infty}^{\infty} d\tau c(\tau) P\int_{-\infty}^{\infty} d\nu\, \frac{e^{i\nu\tau}}{v-\nu}$$

where the second integral $P\int d\nu exp(i\nu t)/(v-\nu)$ yields the step function, and we write [see A. 40, Appendix]

$$P\int_{-\infty}^{\infty} d\nu\, \frac{c(\nu)}{v-\nu} = -i\pi(c^{+}(v) - c^{-}(v)) \tag{5.187}$$

where we defined:

$$c^{+}(v) = \int_{0}^{\infty} d\tau c(\tau) e^{iv\tau}, \quad c^{-}(v) = \int_{-\infty}^{0} d\tau c(\tau) e^{iv\tau}. \tag{5.188}$$

Clearly we have

$$c(v) = c^{+}(v) + c^{-}(v). \tag{5.189}$$

Likewise, we have similar expressions for other quantities in the integral equation (5.186), after decomposition into half-range Fourier integrals:

$$\eta^{+}(v) - \eta^{-}(v) = \frac{1}{\pi i} P\int_{-\infty}^{\infty} \frac{\eta(\nu)d\nu}{\nu - v} \tag{5.190}$$

$$\eta^{+}(v) + \eta^{-}(v) = \eta(v) \tag{5.191}$$

$$g_i^{+}(v) - g_i^{-}(v) = \frac{1}{\pi i} P\int_{-\infty}^{\infty} \frac{g_i(\nu)d\nu}{\nu - v} \tag{5.192}$$

$$g_i^{+}(v) + g_i^{-}(v) = g_i(v). \tag{5.193}$$

Exercise. Deduce the following relations from (5.187) and (5.189):

$$c^{+}(v) = \frac{1}{2\pi i}\left[P\int_{-\infty}^{\infty} \frac{c(\nu)d\nu}{\nu - v} + i\pi c(v)\right] = \frac{1}{2\pi i}\int_{-\infty}^{\infty} \frac{c(\nu)d\nu}{\nu - v - i\epsilon},$$

$$c^{-}(v) = \frac{-1}{2\pi i}\left[P\int_{-\infty}^{\infty} \frac{c(\nu)d\nu}{\nu - v} - i\pi c(v)\right] = \frac{-1}{2\pi i}\int_{-\infty}^{\infty} \frac{c(\nu)d\nu}{\nu - v + i\epsilon},$$

where ϵ is positive infinitesimal. Show that the above equations can be derived from Cauchy's integral formula.

Exercise. Subtracting in the above equations gives

$$c^{+}(v) - c^{-}(v) = \frac{1}{\pi i} P\int_{-\infty}^{\infty} \frac{c(\nu)d\nu}{\nu - v}.$$

Introduce the inverse Fourier relation $c(\nu) = \int_{-\infty}^{\infty} c(\tau)e^{i\nu\tau}d\tau$ in the above principal value integral. Then we have $P\int_{-\infty}^{\infty} \frac{c(\nu)d\nu}{\nu - v} = \int_{-\infty}^{\infty} d\tau c(\tau) P\int_{-\infty}^{\infty} \frac{e^{i\nu\tau}d\nu}{\nu - v} = i\pi \int_{-\infty}^{\infty} d\tau c(\tau)e^{iv\tau}H(\tau)$. Then the above equation follows.

Let us take a look at the η-equation:

$$\eta^+(v) = \frac{1}{2\pi i}\int_{-\infty}^{\infty} \frac{\eta(\nu)d\nu}{\nu - v - i\epsilon}, \quad \eta^-(v) = \frac{-1}{2\pi i}\int_{-\infty}^{\infty} \frac{\eta(\nu)d\nu}{\nu - v + i\epsilon}.$$

Changing notations $v \to \omega/k$, dummy $\nu \to v$, we have

$$\begin{aligned}\eta^+\left(\frac{\omega}{k}\right) &= \frac{1}{2\pi i}\int_{-\infty}^{\infty} \frac{\eta(v)dv}{v - \frac{\omega}{k} - i\epsilon} \\ &= \frac{1}{2\pi i}\frac{\omega_{pe}^2}{k^2}\int_{-\infty}^{\infty} \frac{g_0'(v)dv}{v - \frac{\omega}{k} - i\epsilon} = \frac{1}{2\pi i}[1 - \varepsilon(k,\omega)]. \qquad (5.194)\end{aligned}$$

So η^+ corresponds in essence to the dielectric constant ε.

Next, we use the foregoing decomposition relations to write (5.186) in the form

$$\begin{aligned}g^+(v,0) + g^-(v,0) &= [\eta^+(v) + \eta^-(v)]\ P\int_{-\infty}^{\infty} d\nu\, \frac{c(\nu)}{v - \nu} \\ &\quad + [c^+(v) + c^-(v)]\left[1 + P\int_{-\infty}^{\infty} \frac{\eta(\nu)}{v - \nu}d\nu\right] \\ &= [\eta^+(v) + \eta^-(v)](-i\pi)[c^+(v) - c^-(v)] \\ &\quad + [c^+(v) + c^-(v)][1 - i\pi(\eta^+(v) - \eta^-(v))].\end{aligned}$$

After some cancelations, we obtain

$$\begin{aligned}g^+(v,0) + g^-(v,0) &= c^+(v) + c^-(v) - 2i\pi[\eta^+(v)c^+(v) - \eta^-(v)c^-(v)] \\ &= c^+(v)[1 - 2i\pi\eta^+(v)] + c^-(v)[1 + 2i\pi\eta^-(v)].\end{aligned}$$

Separating the $\pm$ quantities yields the following relations:

$$g_i^+(v) = g^+(v,0) = c^+(v)(1 - 2\pi i\eta^+(v)), \qquad (5.195)$$

$$g_i^-(v) = g^-(v,0) = c^-(v)(1 + 2\pi i\eta^-(v)). \qquad (5.196)$$

Equations (5.195) and (5.196) connect $c(v)$ with the initial distribution $g(v,0)$:

$$c(v) = c^+(v) + c^-(v) = \frac{g^+(v,0)}{1 - 2\pi i\eta^+(v)} + \frac{g^-(v,0)}{1 + 2\pi i\eta^-(v)}. \qquad (5.197)$$

This is the desired solution of (5.186). Substituting (5.197) into (5.184) determines the $g(x, v, t)$ in terms of the initial distribution $g(x, v, 0)$.

Now what is the electric field at t in terms of $g(v, 0)$? The electric field is constructed by (5.184) and the Poisson equation

$$\frac{\partial E(x,t)}{\partial x} = -4\pi e \int_{-\infty}^{\infty} g(x,v,t)dv.$$

So using (5.182) and (5.184), and $\partial/\partial x = ik$, in the above equation gives

$$\begin{aligned}
E(x,t) &= \frac{4\pi i e}{k} e^{ikx} \int dv \int d\nu e^{-ik\nu t} c(\nu) g_\nu(v) \\
&= \frac{4\pi i e}{k} e^{ikx} \int dv \int d\nu e^{-ik\nu t} c(\nu) \\
&\quad \times \left[P\frac{\eta(v)}{v-\nu} + \delta(\nu - v)\left(1 - P\int \frac{\eta(v')dv'}{v'-\nu}\right)\right].
\end{aligned}$$

Integrals involving the principal value cancel upon $\int dv$-integration, and we are left with:

$$E(x,t) = \frac{4\pi i e}{k} e^{ikx} \int dv \int d\nu\, e^{-ik\nu t} \delta(\nu - v)\, c(\nu). \tag{5.198}$$

Using (5.197) in (5.198) gives

$$E(x,t) = \frac{4\pi i e}{k} e^{ikx} \int_{-\infty}^{\infty} d\nu\, e^{-ik\nu t} \left[\frac{g^+(\nu,0)}{1 - 2\pi i \eta^+(\nu)} + \frac{g^-(\nu,0)}{1 + 2\pi i \eta^-(\nu)} \right]. \tag{5.199}$$

[One can obtain this result more quickly by using $\int dv g_\nu(v) = 1$.] For $t > 0$, the second term has no singularities in the lower half plane, and is integrated to yield zero. Thus we finally obtain

$$\begin{aligned}
E(x,t) &= \frac{4\pi i e}{k} e^{ikx} \int_{-\infty}^{\infty} d\nu\, e^{-ik\nu t} \frac{g^+(\nu,0)}{\varepsilon(k,\nu)} \\
&= \frac{4\pi i e}{k} e^{ikx} \int_{-\infty}^{\infty} d\nu\, e^{-ik\nu t} \frac{P\int \frac{g(v,0)dv}{v-\nu} + i\pi g(\nu,0)}{2\pi i\, \varepsilon(k,\nu)}
\end{aligned}$$

where we used (5.194). Changing $\nu \to \omega/k$ in the above gives

$$E(x,t) = \frac{2e}{k^2} e^{ikx} \int d\omega e^{-i\omega t} \frac{\int \frac{g(v,0)dv}{v - \omega/k - i\epsilon}}{\varepsilon(k,\omega)}. \tag{5.200}$$

Equation (5.200) is equivalent to (5.71), if one remembers that the Landau contour should be taken in the $\int dv$-integral in (5.71). This completes the demonstration that the two methods for analysis of the linear Vlasov equation, Van Kampen's integral equation method and Landau's Laplace transform method, yield entirely equivalent results.

Next we turn to the distribution function $g(x, v, t)$ and express it as superposition of the normal modes $g_\nu(v)$. Using (5.182) in (5.184) gives

$$g(x, v, t) = e^{ikx} \int_{-\infty}^{\infty} d\nu e^{-ik\nu t} c(\nu)$$
$$\times \left[P\frac{\eta(v)}{v - \nu} + \delta(v - \nu) \left(1 - P \int_{-\infty}^{\infty} \frac{\eta(v')dv'}{v' - \nu} \right) \right]$$
$$= e^{ikx} \eta(v)\ P \int_{-\infty}^{\infty} d\nu e^{-ik\nu t} \frac{c(\nu)}{v - \nu} + k e^{ikx - ikvt} c(v)$$
$$- k e^{ikx - ikvt} c(v) P \int_{-\infty}^{\infty} \frac{\eta(v')dv'}{v' - v}. \tag{5.201}$$

Using (5.197) in the above enables one to express $g(x, v, t)$ in terms of the initial value of the perturbation. Here again the $g^-(v, 0)$-term vanishes. The middle term $(ke^{ikx-ikvt}c(v))$ in the RHS of (5.201), which is called the *free streaming term*, oscillates more and more rapidly as $t \to \infty$, and it vanishes asymptotically upon velocity integration. Its origin traces back to the term $v\ \partial f/\partial x$ in the Vlasov equation. Note that the derivation of (5.200) presented in this section utilizes only the independence of $(\pm)$ quantities from each other.

• **Physical interpretation of Van Kampen wave**. The Van Kampen wave is due to a modulated beam of particles propagating with a velocity equal to the phase velocity of the wave, as is seen from (5.176): $g(x, t, v) = g(v) exp\ (ikx - i\omega t)$. To see how this wave satisfies the Vlasov and Poisson equations we look at (5.177) again. The solution for $g(v)$ consists of two parts: the particular solution and the homogeneous solution, in which the latter solves $(\omega - kv)\ g(v) = 0$. An incorrect solution of this equation is $g(v) = 0$ because it does not cover the case $\omega - kv = 0$. The correct solution can be written as $\lambda(k, \omega) E(k, \omega) \delta(\omega - kv)$. This solution is derived from the identity (4.68), and should be added to the particular solution of (5.177), thus giving the following complete solution of (5.177):

$$(A) \quad g(v) = i\frac{e}{m} E\ P \frac{\frac{dg_0}{dv}}{\omega - kv} + \lambda E \delta(\omega - kv).$$

Substituting this equation into the Poisson equation (5.178) yields

$$(B) \quad 1 + \frac{4\pi e^2}{m} P \int \frac{g_0'}{\omega - kv} + \frac{4\pi e \lambda}{ik^2} = 0.$$

In equation (B) the first two terms give the dispersion relation of the undamped Langmuir wave. Equation (B) contains two unknown quantities, ω and λ, for a given k, and it does not yield any definite relation between ω and k. In this sense, the Van Kampen wave has no dispersion relation. Rather, ω is arbitrary for any given k, or for any given frequency ω and k, we can find λ which is connected with the number of the resonant particles whose velocity is equal to ω/k. For such a λ, we have undamped Van Kampen wave. Extrapolating this thought, it is not difficult to imagine that an appropriate superposition of the Van Kampen beams can produce Landau damping.

Interestingly, Van Kampen's eigen functions correspond to individual susceptibilities of the beams composing the plasma. Reference: Lee, H. J. *Causality in plasma electrodynamics*, J. Korean Phys. Soc. **73** (2018) 1 pp. 65-85. To elaborate this physical aspect, let us use the following electron beam equations:

$$\frac{\partial n}{\partial t} + v_0 \frac{\partial n}{\partial x} + n_0 \frac{\partial v}{\partial x} = 0,$$

$$\frac{\partial v}{\partial t} + v_0 \frac{\partial v}{\partial x} = -\frac{e}{m} E,$$

$$\frac{\partial E}{\partial x} = -4\pi e \int_{-\infty}^{\infty} n(v_0) g(v_0) dv_0,$$

where we assumed the immobile ions form the uniform neutralizing background, $g(v_0)$ is the beam distribution function, and non-subscripted quantities are perturbations. The first two equations yield, after Fourier transforming,

$$v(k,\omega) = -i\,\frac{e}{m}\,\frac{E(k,\omega)}{(\omega - kv_0)},$$

$$n(k,\omega) = -i\,\frac{e}{m}\,\frac{E(k,\omega) n_0 k}{(\omega - kv_0)^2}.$$

The plasma current due to all the beams is

$$J(k,\omega) = -e \int_{-\infty}^{\infty} (n_0 v + n v_0) g(v_0) dv_0 = i n_0 \frac{e^2}{m} \omega E \int_{-\infty}^{\infty} \frac{g(v_0)}{(\omega - kv_0)^2} dv_0.$$

Setting this quantity equal to $\omega\chi E/4\pi i$, we obtain the plasma susceptibility

$$\chi(k,\omega) = \frac{\omega_{pe}^2}{k^2} \int_{-\infty}^{\infty} \frac{\frac{dg}{dv_0}}{\frac{\omega}{k} - v_0} dv_0 \equiv \int_{-\infty}^{\infty} \widehat{\chi}(v_0) dv_0,$$

where we integrated by parts. The above expression for the plasma susceptibility is identical with the result obtained from the Vlasov equation. The integrand in the above integral, which is a function of v_0 and may be called the 'individual susceptibility', is written as

$$\frac{\omega_{pe}^2}{k^2}\frac{\frac{dg}{dv_0}}{\frac{\omega}{k}-v_0} = \frac{\omega_{pe}^2}{k^2}\frac{dg}{dv_0}\left[P\frac{1}{\frac{\omega}{k}-v_0} + \lambda\delta\left(\frac{\omega}{k}-v_0\right)\right] \equiv \widehat{\chi}(v_0).$$

$\widehat{\chi}(v_0)$ is Van Kampen's eigen function corresponding to eigen index ω/k. It is interesting that the assembly of the individual susceptibilities $\widehat{\chi}(v_0)$ corresponding to each beam can be used to expand an arbitrary distribution function of v_0.

5.11. BGK (Bernstein–Greene–Kruskal) wave

The three authors with the above initials 'BGK' succeeded in inventing exact solutions to the Vlasov–Poisson equations in the wave frame or in a form of a standing wave (progressive wave). They constructed the particle distribution functions (electron's and ion's) so that they fit the predetermined arbitrary electrostatic potential that satisfies the Poisson equation. An important ingredient in their exact solutions is a suitable population of the *trapped particles* in the given finite amplitude electrostatic potential. Their trapped particles are also called *reflected particles.* The correct distribution of electrons and ions trapped in the potential is required to have the given form of finite amplitude electrostatic wave. Their large amplitude wave is a nonlinear solution of the Vlasov and Poisson equations in the wave frame. It can be shown that the linear BGK wave can be obtained by superposition of the Van Kampen waves.

5.11.1. *Wave frame*

Consider the one-dimensional Vlasov equation,

$$\frac{\partial f(x,v,t)}{\partial t} + v\frac{\partial f}{\partial x} + \frac{e}{m}E(x,t)\frac{\partial f}{\partial v} = 0. \tag{5.202}$$

Introducing wave frame variables (denoted by prime),

$$x' = x - v_0 t, \quad v' = v - v_0, \quad t' = t,$$

where v_0 is the wave velocity, we have

$$\frac{\partial}{\partial t} = -v_0\frac{\partial}{\partial x'} + \frac{\partial}{\partial t'}, \quad \frac{\partial}{\partial x} = \frac{\partial}{\partial x'}, \quad \frac{\partial}{\partial v} = \frac{\partial}{\partial v'}.$$

Then (5.202) is written as

$$\frac{\partial f(x',v',t')}{\partial t'} + v'\frac{\partial f}{\partial x'} + \frac{e}{m}E(x',t')\frac{\partial f}{\partial v'} = 0.$$

This shows that *Vlasov equation is invariant under the Galilean transform.* If we look for a stationary wave ($\frac{\partial}{\partial t'} = 0$) in the wave frame, we solve

$$v'\frac{\partial f}{\partial x'} + \frac{e}{m}E(x')\frac{\partial f}{\partial v'} = 0. \tag{5.203}$$

This problem is equivalent to looking for a *progressive wave* in the lab frame: $f(x,v,t) = f(\xi, v)$ with $\xi = x - v_0 t$ (v_0, wave velocity). Since

$$\frac{\partial}{\partial t} = -v_0\frac{\partial}{\partial \xi}, \quad \frac{\partial}{\partial x} = \frac{\partial}{\partial \xi}$$

(5.202) is transformed to

$$(v - v_0)\,\frac{\partial f(\xi,v)}{\partial \xi} + \frac{e}{m}E(\xi)\frac{\partial f}{\partial v} = 0.$$

If we put $v - v_0 = v'$, this equation is the same as (5.203).

5.11.2. *BGK wave*

Their formulation [I. B. Bernstein, J. M. Greene, and M. D. Kruskal, *Exact nonlinear plasma oscillations* Phys. Rev. **108** p 546 (1957)] can be stated as follows: can we construct the particle distribution functions f_α (electron's and ion's) so that they fit the predetermined arbitrary electrostatic potential that satisfy the Poisson equation? They provided an affirmative answer. A steady state in one-dimensional motion is governed by the equations

$$v\frac{\partial f_\alpha}{\partial x} - \frac{e_\alpha}{m_\alpha}\frac{\partial \phi}{\partial x}\frac{\partial f_\alpha}{\partial v} = 0, \tag{5.204}$$

$$\frac{\partial^2 \phi}{\partial x^2} = 4\pi e\left(\int_{-\infty}^{\infty} f_e dv - \int_{-\infty}^{\infty} f_i dv\right). \tag{5.205}$$

The general solution of (5.204) is a function of the constant of integral (energy):

$$f_\alpha = f_\alpha(E_\alpha), \quad \textit{with} \quad E_\alpha = \frac{1}{2}m_\alpha v^2 + e_\alpha \phi. \tag{5.206}$$

By direct substitution one can verify the above relations. Note that the characteristic equations of (5.204) are

$$\frac{dx}{v} = \frac{-dv}{\frac{e_\alpha}{m_\alpha}\frac{\partial \phi}{\partial x}} = \frac{df_\alpha}{0}.$$

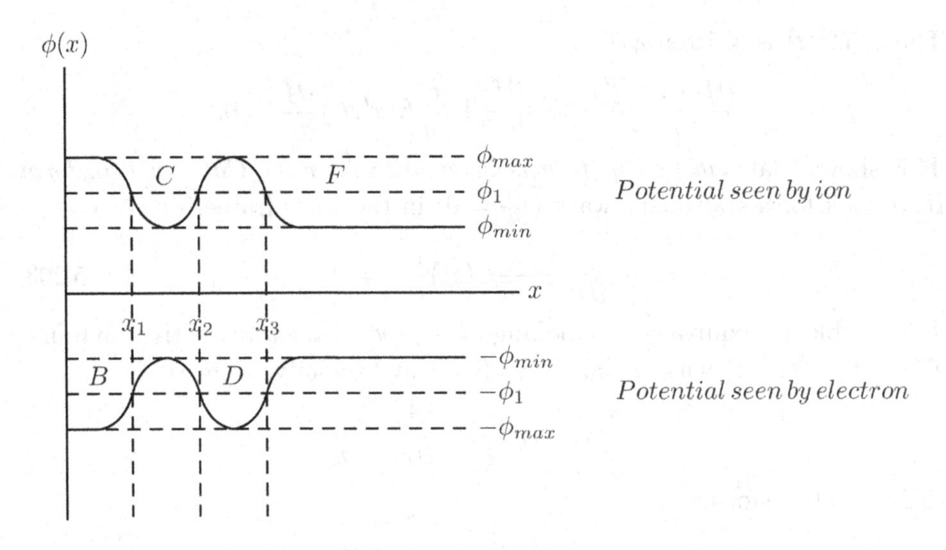

Fig. 5.7. Potential seen by charged particles.

In addition to (5.204) and (5.205), there are extra implicit conditions that have to be satisfied. Consider a potential $\phi(x)$ of the form shown in Fig. 5.7.

Ions with energy E_i such that $e\phi_{min} \leq E_i \leq e\phi_{max}$ are trapped by the potential. Thus, an ion with energy $E_i = e\phi_1$ is confined to the regions $x_1 \leq x \leq x_2$ (region C) or $x \geq x_3$ (region F), but once in C the ion cannot move into F and vice versa. [x_1, x_2, x_3 are turning points at which the ion reverses its velocity.] For these trapped particles, the distribution function must be such that it is equipartitioned between $\pm$-directions of the velocity. Ions with energies $E_i > e\phi_{max}$ can move in either direction, and the partition between the two directions of the velocity is arbitrary. The distributions of the trapped ions in regions C and F are independent since the two regions are closed to each other. Similarly, electrons of energy $E_e < -e\phi_{min}$ are trapped. For instance, an electron of energy $E_e = -e\phi_1$ can only be in regions $x \leq x_1$ (region B) or $x_2 \leq x \leq x_3$ (region D), and the trapped electrons must be equally distributed between the two directions of velocity. Electrons of energy $E_e > -e\phi_{min}$ can move unbounded, and their partition between the two directions of velocity is arbitrary.

Trapping condition of ions:

$$E_i = e\phi + \frac{1}{2}mv^2 \text{ and } E_i < e\phi_{max} \rightarrow E_i - e\phi = \frac{1}{2}mv^2 < e(\phi_{max} - \phi).$$

Trapping condition of electrons:

$$E_e = -e\phi + \frac{1}{2}mv^2 \ \ and \ \ E_e < -e\phi_{min} \rightarrow E_e + e\phi$$
$$= \frac{1}{2}mv_\cdot^2 < e(\phi - \phi_{min}).$$

Substituting (5.206) into (5.205) to change the integration variable from v to E_α, we obtain

$$\frac{d^2\phi(x)}{dx^2} = 4\pi e \int_{-e\phi}^{\infty} \frac{f_e(E_e)dE_e}{\pm\sqrt{2m_e(E_e + e\phi)}} - 4\pi e \int_{e\phi}^{\infty} \frac{f_i(E_i)dE_i}{\pm\sqrt{2m_i(E_i - e\phi)}}. \tag{5.207}$$

There are four particle distributions: trapped and untrapped electron distributions and trapped and untrapped ion distributions. If we regard any three of these distribution functions to be given, the unknown distribution function can be determined accordingly from (5.207) provided the potential is prescribed. One can look upon (5.207) as an integral equation for the distribution function of trapped electrons by writing as

$$\int_{-e\phi}^{-e\phi_{min}} dE \, \frac{f_e(E)}{\sqrt{2m_e(E + e\phi)}} = g(e\phi). \tag{5.208}$$

Note that the lower limit corresponds to $v = 0$ and the upper limit corresponds to $E_e = -e\phi_{min}$. Then (5.208) takes the form

$$g(e\phi) = \frac{1}{4\pi e}\frac{d^2\phi}{dx^2} + \int_{e\phi}^{\infty} dE \frac{f_i(E)}{\sqrt{2m_i(E - e\phi)}} - \int_{-e\phi_{min}}^{\infty} dE \frac{f_e(E)}{\sqrt{2m_e(E + e\phi)}} \tag{5.209}$$

where $f_i(E)$ is the entire ion distribution and $f_e(E)$ in (5.209) is the distribution of the untrapped electrons for $E_e > -e\phi_{min}$. (5.208) is an integral equation for the distribution of the trapped electrons which is to be determined in terms of the function $g(\phi)$. Introduce change of variables in (5.208):

$$E \rightarrow \xi \ \ via \ \ \xi = -E - e\phi_{min}; \ \ e\phi \rightarrow x \ \ via \ \ x = e\phi - e\phi_{min}.$$

Then (5.208) takes the form

$$\int_0^x \frac{f_e(\xi)d\xi}{\sqrt{x - \xi}} = \sqrt{2m_e}\, g(x) \equiv G(x) \tag{5.210}$$

which is the standard form of Voltera's integral equation (the particular form of (5.210) is known as Abel's equation). (5.210) is an integral of convolution type, and can be solved by Laplace transform [see Arfken and Weber, 3rd Ed. 1985 p. 875]. Here we use a different method [see Hildebrand, *Methods of Applied Mathematics*, Prentice-Hall (1965) p. 276] in which we divide both sides of (5.210) by $\sqrt{s-x}$, where s is a parameter, and integrate the results with respect to x over $(0, s)$. This operation leads to

$$\int_0^s \frac{G(x)dx}{\sqrt{s-x}} = \int_0^s \left(\int_0^x \frac{f_e(\xi)d\xi}{\sqrt{x-\xi}}\right) \frac{dx}{\sqrt{s-x}}.$$

If the order of integration in the right side is changed and the limits of integration are modified accordingly, the above equation becomes

$$\int_0^s \frac{G(x)dx}{\sqrt{s-x}} = \int_0^s \left(\int_\xi^s \frac{dx}{\sqrt{(x-\xi)(s-x)}}\right) f_e(\xi)d\xi.$$

The integral in the parentheses gives the constant value π. [With $x = (s-\xi)t + \xi$, this integral takes the form $\int_0^1 \frac{dt}{\sqrt{t(1-t)}}$]. Thus,

$$\int_0^x f_e(\xi)d\xi = \frac{1}{\pi}\int_0^x \frac{G(\xi)d\xi}{\sqrt{x-\xi}}.$$

By differentiating this relation, we then obtain the desired solution

$$\begin{aligned} f_e(x) &= \frac{\sqrt{2m_e}}{\pi}\frac{d}{dx}\int_0^x \frac{g(\xi)d\xi}{\sqrt{x-\xi}} \qquad (5.211) \\ &= \frac{\sqrt{2m_e}}{\pi}\left[\int_0^x \frac{\partial}{\partial x}\frac{g(\xi)}{\sqrt{x-\xi}}d\xi + \left.\frac{g(\xi)}{\sqrt{x-\xi}}\right|_{\xi=x}\right] \qquad \textit{(per Leibnitz's rule)} \\ &= \frac{\sqrt{2m_e}}{\pi}\int_0^x \left[\frac{\partial}{\partial \xi}\frac{g(\xi)}{\sqrt{x-\xi}} + \frac{\partial}{\partial x}\frac{g(\xi)}{\sqrt{x-\xi}}\right] d\xi \end{aligned}$$

where we assumed $g(0) = 0$ in accordance with (5.210). Carrying out the ξ-differentiation, the above equation becomes

$$f_e(x) = \frac{\sqrt{2m_e}}{\pi}\int_0^x \frac{1}{\sqrt{x-\xi}}\frac{dg(\xi)}{d\xi}d\xi. \qquad (5.212)$$

Here we make change of variables again: $x = -E - e\phi_{min}$, $\xi = V - e\phi_{min}$ to write

$$f_e(E) = \frac{\sqrt{2m_e}}{\pi}\int_{e\phi_{min}}^{-E} \frac{dV}{\sqrt{-E-V}}\frac{dg(V)}{dV} \quad (E < -e\phi_{min}). \qquad (5.213)$$

This is the expression for the trapped electron distribution obtained by Bernstein, *et. al.*

Exercise. Verify (5.213) by substituting it into (5.208).

In the case that the distributions of untrapped electrons, trapped electrons, and untrapped ions are specified, the foregoing procedure is readily modified to determine the distribution of trapped ions. Thus Bernstein *et. al.* showed how to construct arbitrary large amplitude wave forms of potential by using correct distributions of trapped electrons and ions in the finite amplitude potential troughs. In addition, Bernstein *et. al.* showed that in the limit of small amplitude waves the trapped portion of the distribution function becomes singular, thus recovering the δ-function term introduced by Van Kampen.

Example. BGK wave constructed with untrapped particles. We consider a simple case where each species is a cold beam of particles having the same kinetic energy. Thus we choose $E_\alpha = \frac{m_\alpha}{2} v_{\alpha 0}^2$ and

$$f_\alpha = 2N v_{\alpha 0}\, \delta \left(v^2 + \frac{2e_\alpha}{m_\alpha}\phi(x) - v_{\alpha 0}^2 \right) \tag{5.214}$$

$$with \quad v_{\alpha 0}^2 \gg \left| \frac{2e_\alpha \phi_{max}}{m_\alpha} \right| \tag{5.215}$$

which ensures that all the particles are untrapped. Note that we have chosen for the equilibrium number density $N = N_e = N_i$ to have overall charge neutrality. We designate the positive root of the argument of the δ function as $v_\alpha : \ v_\alpha = \sqrt{v_{\alpha 0}^2 - 2\frac{e_\alpha}{m_\alpha}\phi(x)}$. Using the δ-function formula $\delta[G(v)] = \delta(v - v_\alpha) \div \frac{dG}{dv} \mid_{v=v_\alpha}$, we have

$$f_\alpha = \frac{N v_{\alpha 0}}{v_\alpha(x)}\, \delta(v - v_\alpha).$$

Then the Poisson equation reads

$$\frac{d^2\phi}{dx^2} = 4\pi e N \left(\int_{-\infty}^{\infty} dv f_e(x,v) - \int_{-\infty}^{\infty} dv f_i(x,v) \right)$$

$$= 4\pi e N \left[\frac{v_{e0}}{\sqrt{v_{e0}^2 + \frac{2e}{m_e}\phi}} - \frac{v_{i0}}{\sqrt{v_{i0}^2 - \frac{2e}{m_i}\phi}} \right] \equiv -\frac{dV(\phi)}{d\phi}, \tag{5.216}$$

$$V(\phi) = -4\pi N \left[m_i v_{i0}^2 \sqrt{1 - \frac{2e\phi}{m_i v_{i0}^2}} + m_e v_{e0}^2 \sqrt{1 + \frac{2e\phi}{m_e v_{e0}^2}} \right]. \tag{5.217}$$

(5.216) describes a particle motion in a pseudo-potential $V(\phi)$ in which x plays the role of time t and ϕ the role of the coordinate x. Under the condition (5.215), we can expand the radicals in (5.216) to obtain

$$\frac{d^2\phi}{dx^2} + k_{eff}^2\,\phi = 0$$

which gives a simple harmonic motion of a wave number $k_{eff} = [4\pi Ne^2(\frac{1}{m_e v_{e0}^2} + \frac{1}{m_i v_{i0}^2})]^{1/2}$.

• **Bohm and Gross dispersion relation taking account of trapped electrons.** The number of trapped electrons at $\phi(x)$ (see Fig. 5.7) is expressed by (5.208) where $E = \frac{1}{2}mv^2 - e\phi(x)$. Let us make change of integration variable $E \to u$ per

$$E = \frac{m}{2}u^2 - e\phi_{max} \tag{5.218}$$

where u absorbed the x-dependence through $\phi(x)$. Then the number of trapped electrons at $\phi(x)$ denoted by $N(\phi)$ is

$$N(\phi) = \int_{\sqrt{\frac{2e}{m}(\phi_{max}-\phi)}}^{\sqrt{\frac{2e}{m}(\phi_{max}-\phi_{min})}} \frac{f_e(u)udu}{\sqrt{u^2 + \frac{2e}{m}(\phi - \phi_{max})}}. \tag{5.219}$$

Bohm and Gross made an ingenuous choice for f_e which results in a linear profile of $N(\phi(x))$:

$$f_e(u) = \sqrt{\frac{2e}{m}(\phi_{max} - \phi_{min}) - u^2}. \tag{5.220}$$

The integral in N can be carried out as follows [Stix, T. H. *Waves in plasmas*, American Institute of Physics, New York (1992)]. With the substitution of

$$\xi^2 = u^2 + \frac{2e}{m}(\phi - \phi_{max})$$

$N(\phi)$ becomes

$$N(\phi) = \int_0^X \sqrt{X^2 - \xi^2}, \qquad X = \sqrt{\frac{2e}{m}(\phi - \phi_{min})}. \tag{5.221}$$

On the other hand, the total number of trapped electrons in the entire trapping region is given by

$$N_1 = \int_0^Y f_e(u)du = \int_0^Y \sqrt{Y^2 - u^2}\,du,$$

$$Y = \sqrt{\frac{2e}{m}(\phi_{max} - \phi_{min})} \ (escape\ velocity). \tag{5.222}$$

Both integrals can be carried out in terms of elementary functions:

$$\int \sqrt{x^2-\xi^2}\, d\xi = \frac{\xi}{2}\sqrt{x^2-\xi^2} + \frac{x^2}{2} sin^{-1}\frac{\xi}{x}.$$

Thus we obtain

$$N(\phi) = N_1 \, \frac{\phi - \phi_{min}}{\phi_{max} - \phi_{min}}. \tag{5.223}$$

The above equation expresses the ratio of the number density of trapped electrons at $\phi(x)$ to the total number density of the trapped electrons when the trapped electron distribution function is given by (5.220).

Next, we turn to obtain the untrapped electron number density. For this purpose, we use the following *beam equations*:

$$\frac{\partial v}{\partial t} + v_0 \, \frac{\partial v}{\partial x} = -\frac{e}{m} E(x,t) \tag{5.224}$$

$$\frac{\partial n}{\partial t} + N(v_0) \, \frac{\partial v}{\partial x} + v_0 \, \frac{\partial n}{\partial x} = 0 \tag{5.225}$$

where v and n are perturbed quantities belonging to the beam whose speed is v_0. $N(v_0)$ is the equilibrium value of electron density belonging to the v_0-beam. In terms of Fourier variables, (5.224) and (5.225) are written in the form

$$(-i\omega + ikv_0)v = i\,\frac{e}{m} k\phi(k,\omega)$$

$$(-i\omega + ikv_0)n = -ikN(v_0)v$$

$$Therefore \;\; n = -\frac{e}{m} k^2 \phi \frac{N(v_0)}{(\omega - kv_0)^2}.$$

$N(v_0)$ can be expressed as $N(v_0) = f(v_0)dv_0$, where $f(v_0)$ is the distribution function of the beams which corresponds to the zero order distribution function. Thus the total electron density is obtained as the integral

$$n = -\frac{e}{m} k^2 \phi \int_{-\infty}^{\infty} \frac{f_0(v_0)dv_0}{(\omega - kv_0)^2}. \tag{5.226}$$

A part of n belongs to the untrapped particles which is expressed as

$$n_u = -\frac{e}{m} k^2 \phi \int_{-\infty}^{\frac{\omega}{k}-\Delta v} \frac{f_0(v_0)dv_0}{(\omega - kv_0)^2} - \frac{e}{m} k^2 \phi \int_{\frac{\omega}{k}+\Delta v}^{\infty} \frac{f_0(v_0)dv_0}{(\omega - kv_0)^2}. \tag{5.227}$$

Integrating by parts (5.227) gives with good approximation

$$n_u = \frac{e}{m} k\phi P \int_{-\infty}^{\infty} \frac{f_0'(v_0)dv_0}{\omega - kv_0} - \frac{2e\phi}{m} \frac{f_0(\frac{\omega}{k})}{\Delta v} \tag{5.228}$$

where P denotes the principal value. Note that $\frac{1}{2}m(\Delta v)^2$ is a mean value for the electron kinetic energy in the wave frame, below which the electron will be trapped. The minimum kinetic energy required to escape is a function of x through $\phi(x)$, so we take an average

$$\frac{1}{2}m(\Delta v)^2 = \frac{e(\phi_{max} - \phi_{min})}{2}. \tag{5.229}$$

Using (5.229) in (5.228) yields

$$n_u = \frac{e}{m} k\phi P \int_{-\infty}^{\infty} \frac{f_0'(v_0)dv_0}{\omega - kv_0} - \frac{N_2 \, \phi}{\phi_{max} - \phi_{min}} \tag{5.230}$$

where $N_2 = 2\Delta v f_0(\frac{\omega}{k})$ is the maximum possible number density of the trapped electrons [N_1 is the number density of trapped electrons, given the distribution function (5.220)]. Here the notations N_1, N_2 are the same notations used in Reference: T. H. Stix, *The theory of plasma waves*, McGraw-Hill New York (1962) § 6-7. The Poisson equation can be written in the form

$$\nabla^2\phi = 4\pi e N_1 \frac{\phi - \phi_{min}}{\phi_{max} - \phi_{min}} + 4\pi e n_u - 4\pi e n_i. \tag{5.231}$$

It is reminded that the first term in RHS of (5.231) corresponds to the trapped electrons, second term to untrapped electrons, and the third term to the ions. The constant electron term (ϕ_{min}-term) should cancel the ion terms due to charge neutrality in the zero order. Fourier transforming and using (5.230) give the dispersion relation

$$k^2 \left[1 + \frac{4\pi e^2}{mk} P \int_{-\infty}^{\infty} \frac{f_0'(v_0)dv_0}{\omega - kv_0}\right] = -4\pi e \, \frac{N_1 - N_2}{\phi_{max} - \phi_{min}}. \tag{5.232}$$

(5.232) is equivalent to the following two simultaneous equations,

$$f_1 = \frac{ieE}{m} P\left(\frac{f_0'(v)}{\omega - kv}\right) + \frac{N_1 - N_2}{2}\,\delta\left(\frac{\omega}{k} - v\right), \tag{5.233}$$

$$\nabla \cdot \mathbf{E} = ikE = -4\pi e \int_{-\infty}^{\infty} f_1 dv, \tag{5.234}$$

provided we set $E = -ik\phi = -\frac{ik}{2}(\phi_{max} - \phi_{min})$. (5.233) and (5.234) are essentially identical to the Van Kampen equations

$$f_1 = \frac{ieE}{m}\frac{\partial f_0}{\partial v}P\left(\frac{1}{\omega - kv}\right) + \lambda\delta\left(\frac{\omega}{k} - v\right) \quad cf.\ Eq.\ (5.180),$$

$$\nabla \cdot \mathbf{E} = ikE = -4\pi e\int_{-\infty}^{\infty} f_1 dv \quad cf.\ Eq.\ (5.178).$$

By appropriate choice of N_1 and N_2, the above BGK solution can be made identical to the Van Kampen mode.

Example. Solve the following integral equation for G by elementary method [L. D. Landau and E. M. Lifshitz, *Mechanics*, Pergamon (1960), § 12].

$$(A)\quad f(x) = \int_0^x \frac{dG}{d\xi}\frac{d\xi}{\sqrt{x-\xi}} \qquad [cf.\ (5.212)].$$

Divide this equation by $\sqrt{(\alpha - x)}$ and integrate with respect to x from 0 to α, where α is a parameter, to write

$$(B)\quad \int_0^\alpha \frac{f(x)dx}{\sqrt{(\alpha - x)}} = \int_0^\alpha \int_0^x \frac{dG}{d\xi}\frac{1}{\sqrt{\alpha - x}\sqrt{x-\xi}}dxd\xi.$$

Changing the order of integration, the above equation becomes

$$(C)\quad \int_0^\alpha \frac{f(x)dx}{\sqrt{(\alpha - x)}} = \int_0^\alpha \frac{dG}{d\xi}d\xi\int_\xi^\alpha \frac{dx}{\sqrt{\alpha - x}\sqrt{x-\xi}}.$$

The second integral's value is π, thus we obtain

$$(D)\quad \int_0^\alpha \frac{f(x)dx}{\sqrt{(\alpha - x)}} = \pi G(\alpha) \quad or \quad \int_0^\alpha \frac{f(\xi)d\xi}{\sqrt{(\alpha - \xi)}} = \pi G(\alpha).$$

This is Eq. (5.210).

- Proof of

$$I = \int_\xi^\alpha \frac{dx}{\sqrt{\alpha - x}\sqrt{x-\xi}} = \pi.$$

Via change of variable $x - \xi = y$,

$$I = \int_0^{\alpha-\xi} \frac{dy}{\sqrt{y}\sqrt{\alpha - \xi - y}} = \int_0^\beta \frac{dy}{\sqrt{y(\beta - y)}}.$$

This integral has the value which is independent of the constant β, as can be seen by change $\beta \longrightarrow n\beta$, $y \longrightarrow ny'$, with an arbitrary multiplying constant n. Therefore the integral I can be evaluated by

$$I = \int_0^1 \frac{dy}{\sqrt{y(1-y)}} = \pi.$$

5.12. Nonlinear prediction: what happen after Landau damping time (γ_L^{-1}) to the plasma wave?

Reference: R. Z. Sagdeev and A. A. Galeev, *Nonlinear Plasma Theory*, Benjamin, New York (1969).

Landau damping can be physically interpreted in terms of the interaction of the trapped particles with plasma wave in the framework of the linearized Vlasov equation. Since the distribution function is changed by the dynamics of the trapped particles, the Landau damping rate obtained from the linear theory is valid only for a limited time. At later stage, after a few trapping time, the distribution function f_0 must alter its original shape in the vicinity of the resonant velocity region due to the mixing of relative phase between the trapped particles, and thus the damping rate of the wave takes an oscillatorilly decreasing variation (in time), which asymptotically approaches zero (see Figs. 5.9 and 5.10). The slope of f_0 is flattened around the region of the resonant velocity (a plateau is formed). The nonlinear theory requires consideration of the quadratic term $\frac{\partial f}{\partial v}\frac{\partial \phi}{\partial x}$. If we consider a single plasma wave, the Landau damping rate is valid only at early time $t \ll \omega_b^{-1}$ (ω_b is the bounce frequency below) but it is not valid at $t \approx \omega_b$ and later. This is because the trapped particles, originally all in the same phase, disperse with respect to the phase angles. Here we might need a little elaboration about the phase θ. Linearly f is only a function of the energy W; nonlinearly, we have $f(W, \theta)$, where the angle θ which represents the extra dependency is the phase angle on the energy curve W. At $t = 0$, f is the same along the same phase plane orbit (see Fig. 5.8). The θ-symmetry is broken for the trapped particles after the bounce time. The acute θ-dependency makes f 'ragged and tattered' around the resonant velocity. To obtain a smooth picture of f, we can take time average, thus obtaining a "coarse grain" distribution function. The latter shows a plateau in the resonance region. The corresponding plasma wave becomes a BGK wave. In the following we do not attempt analytic verification of the above picture. We only present a few specific results.

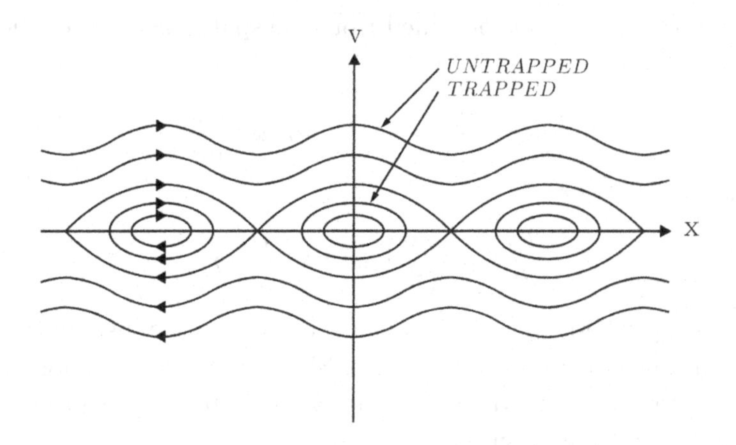

Fig. 5.8. Phase space orbits.

5.12.1. *Derivation of the nonlinear Landau damping rate*

Consider a plasma that supports the propagation of a sinusoidal electric field wave

$$E(x,t) = E_0\ sin(kx - \omega t). \tag{5.235}$$

The electron distribution (the ions are assumed to be immobile) can be divided into the main part and the part of the resonant particles (recall the discussion in Section 5.6). The main part of the distribution supports the oscillatory motion of the plasma wave and the resonant part gives rise to the Landau damping of the wave. The Landau damping rate can be obtained by equating the rate of increase of the kinetic energy of the resonant electrons with the rate of decrease of the wave energy.

It is convenient to work in the wave frame which moves with the wave velocity ω/k. Then the electric field can be written as $E_0\ sinkx$ and the rate of increase of kinetic energy can be expressed as

$$\frac{dK}{dt} = \frac{Nm}{2}\int_0^\lambda \frac{dx}{\lambda}\int_{-\infty}^{\infty} dv\left(v + \frac{\omega}{k}\right)^2 \frac{\partial f}{\partial t} \tag{5.236}$$

where N is the electron density, λ is the wavelength, and f is the electron distribution in the wave frame. The $\int dx$ integration gives the spatial average of f, so the above equation becomes

$$\frac{dK}{dt} = \frac{Nm}{2}\int_{-\infty}^{\infty} dv\left(v + \frac{\omega}{k}\right)^2 \frac{\partial \langle f\rangle}{\partial t}, \tag{5.237}$$

the quantity $\partial < f > /\partial t$ is obtained from the spatial average of the Vlasov equation

$$\frac{\partial f}{\partial t} + v\frac{\partial f}{\partial x} = \frac{e}{m} E_0 \; sin \; kx \; \frac{\partial f}{\partial v}. \tag{5.238}$$

We can write

$$\frac{\partial \langle f \rangle}{\partial t} = \left\langle \frac{e}{m} E_0 sin \; kx \; \frac{\partial f}{\partial v} \right\rangle. \tag{5.239}$$

Let $f = f_0 + f_1$, where f_0 is the unperturbed part of the distribution and f_1 is the perturbation caused by the wave. Since f_0 is spatially homogeneous, it makes no contribution to the spatial average above. f_1 can be obtained from the linearized equation

$$\frac{\partial f_1}{\partial t} + v\frac{\partial f_1}{\partial x} = \frac{e}{m} \; sin \; kx \; \frac{\partial f_0}{\partial v}. \tag{5.240}$$

The solution of (5.240) is [see Example in (5.40)]

$$f_1(x, v, t) = f_1(v, 0) cos \; (kx - kvt) - \frac{e}{m} E_0 \; \frac{\partial f_0}{\partial v} \; \frac{cos \; kx - cos \; (kx - kvt)}{kv} \tag{5.241}$$

where $f_1(v, 0)$ is the initial perturbation. Substituting (5.241) into the RHS of (5.239) gives, after the spatial averaging,

$$\frac{\partial < f >}{\partial t} = \frac{\partial}{\partial v} \left[\frac{e}{2m} f_1(v, 0) E_0 sin \; kvt + \frac{1}{2} \left(\frac{e}{m} \right)^2 \frac{\partial f_0}{\partial v} E_0^2 \; \frac{sin \; kvt}{kv} \right]. \tag{5.242}$$

Then, substituting (5.242) into (5.237) and integrating by parts give

$$\begin{aligned} \frac{dK}{dt} &= -\frac{NeE_0}{2} \int_{-\infty}^{\infty} dv f_1(v, 0) \left(v + \frac{\omega}{k} \right) sin \; kvt \\ &\quad - \frac{Ne^2 E_0^2}{2m} \int_{-\infty}^{\infty} dv \; \frac{\partial f_0}{\partial v} \left(v + \frac{\omega}{k} \right) \frac{sin \; kvt}{kv}. \end{aligned} \tag{5.243}$$

After many cycles ($t \to \infty$), the first integral phase mixes to zero and $sin \; kvt/kv \to \pi\delta(kv)$. Then (5.243) becomes

$$\frac{dK}{dt} = -\frac{\pi}{2}\omega \; \frac{\omega_p^2}{k^2} \frac{\partial f_0}{\partial v}\bigg|_{v=0} \frac{E_0^2}{4\pi} = -\frac{\pi}{2}\omega \; \frac{\omega_p^2}{k^2} \frac{\partial f_0}{\partial v}\bigg|_{v=\frac{\omega}{k}} \frac{E_0^2}{4\pi} \tag{5.244}$$

where we returned to the lab-frame in the last expression. By equating the above rate of increase of kinetic energy to the rate of decrease of wave energy,

$\frac{1}{8\pi}\frac{\partial E_0^2}{\partial t} = \gamma E_0^2/4\pi$, we recover the Landau damping rate

$$\gamma = \frac{\pi}{2}\omega\,\frac{\omega_p^2}{k^2}\,\frac{\partial f_0}{\partial v}\bigg|_{v=\frac{\omega}{k}}. \tag{5.245}$$

This formula ceases to be valid when the linear solution becomes invalid. This occurs when $\partial f_1/\partial v \simeq \partial f_0/\partial v$, that is, when the nonlinear term is no longer negligible. To look at this condition closely, we differentiate (5.241) to obtain

$$\begin{aligned}\frac{\partial f_1(x,v,t)}{\partial v} &= \frac{\partial f_1(v,0)}{\partial v} cos\ (kx-kvt) + kt\ sin\ (kx-kvt) f_1(v,0)\\ &\quad -\frac{e}{m}E_0\frac{\partial^2 f_0}{\partial v^2}\frac{cos\ kx - cos(kx-kvt)}{kv}\\ &\quad -\frac{e}{m}\frac{E_0}{k}\frac{\partial f_0}{\partial v}\frac{-ktv\ sin\ (kx-kvt) - cos\ kx + cos\ (kx-kvt)}{v^2}.\end{aligned} \tag{5.246}$$

The secular terms ($\sim t$) are responsible for $\frac{\partial f_1(x,v,t)}{\partial v}$ to grow indefinitely. In addition to the immediate one in the first line in (5.246), we have the hidden secular terms in the second and third lines. They are associated with the resonant particles with $v = 0$ (in the lab-frame $v = \omega/k$). Using L'Hospital's rule, we find a ($\sim t$) secular term in the second line and a ($\sim t^2$) secular term in the third line. Since the latter invalidates the linear solution first, we calculate the third line and obtain the condition $\partial f_0/\partial v \sim \frac{e}{m}E_0 k t^2 \partial f_0/\partial v$ or $t = \tau \equiv \sqrt{\frac{m}{eE_0 k}}$, which is the *trapping time*. After the trapping time, the linear solution is no longer valid. This is caused by the resonant particles trapped in the potential well (see Figs. 5.9 and 5.10). The trapping time is approximately the round trip period for a trapped resonant particle to bounce back and forth in the potential well, whose motion is described by $d^2x/dt^2 = \frac{e}{m}E_0 sin\ kx \simeq \frac{e}{m}E_0 kx$. This equation gives the bounce frequency $\omega_b = \sqrt{eE_0 k/m}$.

5.12.2. *Nonlinear Landau damping: characteristic method*

Reference: T. M. O'Neil, *Collisionless damping of nonlinear plasma oscillations* Phys. Fluids **3** 3225 (1965).

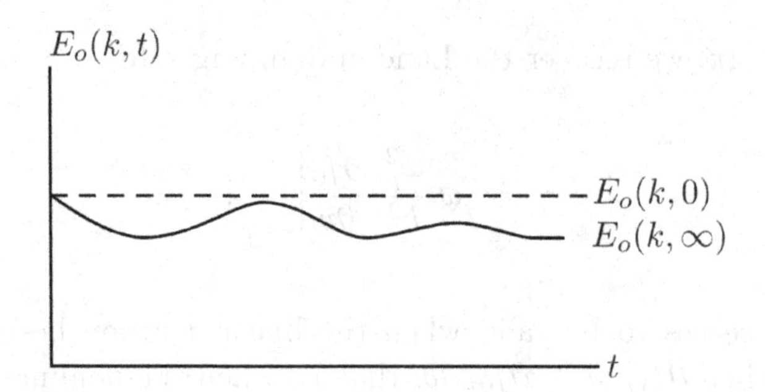

Fig. 5.9. Damping of large amplitude wave.

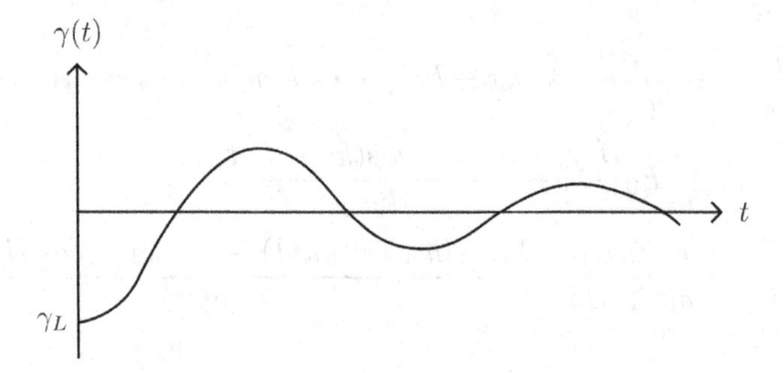

Fig. 5.10. Nonlinear change of damping rate.

Theory of nonlinear collisionless damping can be studied by considering the nonlinear Vlasov equation (5.238) in the resonant region. The characteristic equation of (5.238) is

$$\frac{dt}{1} = \frac{dx}{v} = \frac{dv}{-\frac{e}{m}E_0 sin\ kx} = \frac{df}{0}$$

which gives the orbit equation

$$\frac{d^2x}{dt^2} = -eE_0 sin\ kx. \tag{5.247}$$

The particle orbit can be found in terms of the elliptic functions which solve the above equation of motion. The phase diagram of orbits is shown in Fig. 5.8. f is a function of the constants of motion. The constants of motion can be taken as the initial phase point (x_0, v_0) from which the subsequent phase point $(x(t), v(t))$ evolves. Then, the Vlasov equation (5.238) is solved by

$$f(x,v,t) = f(x_0(x,v,t), v_0(x,v,t), 0) \tag{5.248}$$

where $f(x_0, v_0, 0)$ is the initial distribution and

$$x = x(x_0, v_0, t), \quad v = v(x_0, v_0, t) \tag{5.249}$$

give the phase point evolved from the initial phase point (x_0, v_0). To write RHS of (5.248), (5.249) should be inverted. As an example which illustrates the inversion, let us consider the approximate equation of (5.247) by considering the deeply trapped particles

$$\frac{d^2x}{dt^2} = -\frac{e}{m}E_0 kx.$$

We obtain the solution

$$x(t) = x_0 cos\ \omega(t - t_0) + \frac{v_0}{\omega} sin\ \omega(t - t_0),$$

$$v(t) = -x_0 \omega sin\ \omega(t - t_0) + v_0 cos\ \omega(t - t_0),$$

where $\omega = \sqrt{\frac{e}{m}E_0 k}$, the bounce frequency, and $x_0 = x(t = t_0)$, $v_0 = v(t = t_0)$. Inverting these, we obtain

$$x_0 = x(t) cos\ \omega(t - t_0) - \frac{v(t)}{\omega} sin\ \omega(t - t_0),$$

$$v_0 = x(t)\omega sin\ \omega(t - t_0) + v(t) cos\ \omega(t - t_0).$$

Note that $\frac{dx_0}{dt} = \frac{dv_0}{dt} = 0$. Inversion of the elliptic functions is not easy.

An initial distribution appropriate to the sinusoidal external electric field would be

$$f(x_0, v_0, 0) = f_0(v_0) + f_1(v_0, 0) cos\ kx_0. \tag{5.250}$$

In (5.250), x_0 and v_0 are

$$x_0 = x\ cos\ \omega(t - t_0) - \frac{v}{\omega} sin\ \omega(t - t_0),$$

$$v_0 = x\ \omega\ sin\ \omega(t - t_0) + v cos\ \omega(t - t_0),$$

with x, v constants of motion (initial values of x_0 and v_0 at $t_0 = t$). Note that $\partial x_0/\partial t = -v_0$, $\partial v_0/\partial t = -\partial^2 x_0/\partial t^2 = \omega^2 x_0$. To calculate the rate of increase of kinetic energy, we need $\partial f/\partial t$:

$$\begin{aligned} \frac{\partial f}{\partial t} &= \frac{\partial}{\partial x_0} f(x_0, v_0, 0)\ \frac{\partial x_0}{\partial t} + \frac{\partial}{\partial v_0} f(x_0, v_0, 0)\ \frac{\partial v_0}{\partial t} \\ &= -v_0 f_1(v_0, 0) k\ sin\ kx_0 \\ &\quad -\frac{e}{m} E_0\ sin\ kx_0 \left(\frac{\partial f_0}{\partial v_0} + \frac{\partial f_1(v_0, 0)}{\partial v_0}\ cos\ kx_0 \right). \end{aligned} \tag{5.251}$$

The rate of change of kinetic energy dK/dt is given by (5.236). O'Neil (1965) calculated the latter quantity dK/dt by using (5.251) and inverting the elliptic functions. The slow change in the amplitude E_0 [in (5.235)] of the wave due to the trapped particles can be written as

$$\frac{\partial}{\partial t}\int_{-\frac{\pi}{k}}^{\frac{\pi}{k}} \frac{E_0^2}{8\pi} sin^2 kx\, dx = -\frac{\partial}{\partial t}\frac{E_0^2}{16\pi}. \tag{5.252}$$

Equating this to dK/dt in (5.236) and using (5.251), we obtain the slow change of the wave amplitude due to interaction with the trapped particles. For times smaller than the trapping time, the wave Landau damps. For times greater than the trapping time, E_0 oscillates with time but gradually approaches a constant value. See Fig. 5.9. In Fig. 5.10, the nonlinear damping coefficient is plotted versus time. The damping coefficient begins with the linear value, which is the Landau damping rate γ_L as given by (5.245), and then oscillates with a period of approximately the trapping time. An alternative way to derive the nonlinear damping rate uses the following expression for the electric field energy,

$$W(t) = \int_0^\lambda \frac{dx}{\lambda}\frac{E^2}{4\pi}. \tag{5.253}$$

Then we have

$$\frac{dW}{dt} = \int_0^\lambda \frac{dx}{\lambda}\frac{E}{2\pi}\frac{\partial E}{\partial t}. \tag{5.254}$$

Use the Maxwell equation,

$$\frac{\partial E}{\partial t} = -4\pi en \int f(x, v, t) v dv$$

in (5.254) to obtain

$$\frac{dW}{dt} = -2ne\int_0^\lambda \frac{dx}{\lambda} E \int_{-\infty}^{\infty} f(x, v, t) v dv. \tag{5.255}$$

On the other hand, the damping rate can be written as

$$\gamma(t) = \frac{1}{2W(t)}\frac{dW(t)}{dt}. \tag{5.256}$$

Knowing $f(x, v, t)$ enables one to calculate $\gamma(t)$ with the aid of the above two equations.

Chapter 6

Kinetic theory of waves in magnetized plasma

A cosmic or fusion plasma requires consideration of the effect of a magnetic field of the zero order importance, which can be regarded as imbedded in the plasma, hence the name magnetized plasma. Plasma normal modes in a magnetized plasma have more diversity as compared to unmagnetized plasma. Vlasov equation for a magnetized plasma can be solved after exercising more labor as compared with the case without a static magnetic field therein. In isotropic plasmas, $\frac{\partial f_0(v)}{\partial \mathbf{v}} \sim \mathbf{v}$, so the magnetic force term becomes zero. However, a static magnetic field $\mathbf{B}_0$ introduces anisotropy in the plasma, and consequently, the wave solutions of Vlasov equation manifest a great complexity with regard to the dispersion relation. Algebraically, Bessel functions, which may be considered as extended sinusoidal functions, make an appearance in mathematical analysis of magnetized plasmas, both in Vlasov equation and fluid equation. This is because we need to expand the function $exp[i\ sin\ kx]$ in a Bessel series. The *Bernstein waves* are interesting because they have no direct counterparts in the fluid plasmas and make an important entry in the list of normal modes of magnetized plasma. *Cyclotron damping (or growing)*, which is the namesake of the Landau damping (growing) in magnetized plasmas, is an important mechanism for energy transfer between the plasma particles and the electric field in magnetized plasmas. *Whistler instability* is important for explaining the very low frequency (VLF) emissions in the magnetosphere.

6.1. Electrostatic wave in a magnetized plasma

To obtain solutions of Vlasov equation for a magnetized plasma requires complicated algebra which is not trivial. We first consider an electrostatic wave propagating in a magnetized plasma with $\mathbf{B_0} = \hat{\mathbf{z}}B_0$.

$$\frac{\partial}{\partial t}f_\alpha(\mathbf{r},\mathbf{v},t) + \mathbf{v}\cdot\frac{\partial f_\alpha}{\partial \mathbf{r}} + \frac{e_\alpha}{m_\alpha}\left[\mathbf{E}(\mathbf{r},t) + \frac{\mathbf{v}}{c}\times\mathbf{B}_0\right]\cdot\frac{\partial f_\alpha}{\partial \mathbf{v}} = 0 \quad (\alpha = e,\ i). \tag{6.1}$$

We linearize this equation by putting $f = f_0 + f_1$:
Zero order:

$$\mathbf{v}\cdot\frac{\partial f_{\alpha 0}}{\partial \mathbf{r}} + \frac{e_\alpha}{m_\alpha}\frac{\mathbf{v}\times\mathbf{B_0}}{c}\cdot\frac{\partial f_{\alpha 0}}{\partial \mathbf{v}} = 0. \tag{6.2}$$

First order:

$$\frac{\partial}{\partial t}f_{\alpha 1}(\mathbf{r},\mathbf{v},t) + \mathbf{v}\cdot\frac{\partial f_{\alpha 1}}{\partial \mathbf{r}} + \frac{e_\alpha}{m_\alpha}\frac{\mathbf{v}\times\mathbf{B_0}}{c}\cdot\frac{\partial f_{\alpha 1}}{\partial \mathbf{v}} = -\frac{e_\alpha}{m_\alpha}\mathbf{E}(\mathbf{r},t)\cdot\frac{\partial f_{\alpha 0}}{\partial \mathbf{v}}. \tag{6.3}$$

The static magnetic field introduces anisotropy for the plasma since the motion of charged particles along the magnetic field is different from the motion across the magnetic field. A two-temperature Maxwellian distribution which is anisotropic in the three-dimensional space but is isotropic in the plane perpendicular to $\mathbf{B_0}$ is often used for such a magnetized plasma:

$$\begin{aligned} f_{\alpha 0}(v_x, v_y, v_z) = &\left[\frac{m_\alpha}{2\pi T_{\alpha\|}}\right]^{\frac{1}{2}} exp\left[-\frac{m_\alpha v_z^2}{2T_{\alpha\|}}\right] \\ &\times \frac{m_\alpha}{2\pi T_{\alpha\perp}}\, exp\left[-\frac{m_\alpha(v_x^2+v_y^2)}{2T_{\alpha\perp}}\right]. \end{aligned} \tag{6.4}$$

It is readily seen that (6.4) satisfies (6.2).

6.1.1. *Solution by characteristic method*

We integrate (6.3) by means of the characteristic method which was used to solve the one-dimensional Vlasov–Poisson equation (5.28). The characteristic method applied to Vlasov equation is known as *integration along the unperturbed orbit.* Here 'unperturbed orbit' means the orbit that is obtained from zero order solution. The characteristic equation of (6.3) reads

$$\frac{dt}{1} = \frac{d\mathbf{r}}{\mathbf{v}} = \frac{d\mathbf{v}}{\omega_{c\alpha}\mathbf{v}\times\hat{\mathbf{z}}} = \frac{df_{\alpha 1}}{-\frac{e_\alpha}{m_\alpha}\mathbf{E}\cdot\frac{\partial f_{\alpha 0}}{\partial \mathbf{v}}} \tag{6.5}$$

where $\omega_{c\alpha} = \frac{e_\alpha B_0}{m_\alpha c}$, the cyclotron frequency.
Assuming the form

$$\mathbf{E}(\mathbf{r},t) = -i\mathbf{k}\ \phi(\mathbf{k},\omega)e^{i\mathbf{k}\cdot\mathbf{r}-i\omega t} \tag{6.6}$$

the last relation of (6.5) gives

$$f_\alpha(t,\mathbf{r},\mathbf{v}) = \frac{ie_\alpha}{m_\alpha}\,\phi\int_{-\infty}^{t} dt'\;\mathbf{k}\cdot\frac{\partial f_{\alpha 0}(v')}{\partial \mathbf{v}'}e^{i\mathbf{k}\cdot\mathbf{r}'-i\omega t'} \tag{6.7}$$

where the primed variables should be obtained by solving

$$\frac{d\mathbf{r}'(t')}{dt'} = \mathbf{v}', \quad \frac{d\mathbf{v}'(t')}{dt'} = \omega_{c\alpha}\mathbf{v}'\times\hat{\mathbf{z}}. \tag{6.8}$$

Let the integration constant of the above equations be such that at $t' = t$

$$\mathbf{r}' = \mathbf{r}, \quad \mathbf{v}' = \mathbf{v}. \tag{6.9}$$

To determine $\mathbf{r}'$, the $\mathbf{v}'$-equation in (6.8) should be solved first. In components, we have

$$\frac{dv_x'}{dt'} = \omega_{c\alpha}v_y', \quad \frac{dv_y'}{dt'} = -\omega_{c\alpha}v_x', \quad \frac{dv_z'}{dt'} = 0 \tag{6.10}$$

which are solved by, according to the boundary conditions (6.9),

$$v_x'(t') = v_\perp\ cos[\omega_{c\alpha}(t-t')+\varphi], \tag{6.11}$$

$$v_y'(t') = v_\perp\ sin[\omega_{c\alpha}(t-t')+\varphi], \tag{6.12}$$

$$v_z'(t') = v_z \equiv v_\parallel :\quad constant, \tag{6.13}$$

where $v_\perp$ and φ are two integration constants that are related with the integration constants v_x and v_y: $v_x = v_\perp cos\varphi$, $v_y = v_\perp sin\varphi$. Integrating the above three equations gives

$$x'(t') = x - \frac{v_\perp}{\omega_{c\alpha}}[sin(\omega_{c\alpha}(t-t')+\varphi) - sin\varphi], \tag{6.14}$$

$$y'(t') = y + \frac{v_\perp}{\omega_{c\alpha}}[cos(\omega_{c\alpha}(t-t')+\varphi) - cos\varphi], \tag{6.15}$$

$$z'(t') = z - v_z(t-t'). \tag{6.16}$$

Without loss of generality, we put $\mathbf{k} = \hat{\mathbf{x}}k_\perp + \hat{\mathbf{z}}k_\parallel$, where the symbols $\perp$ and $\parallel$ refer to the direction of the magnetic field. Substituting the above solutions into (6.7) and noting that $f_{\alpha 0}$ consists of the constants of motion, v_z and

$v_\perp$, and introducing a new variable $\tau = t - t'$, we write

$$f_{\alpha 1}(\mathbf{r}, t, \mathbf{v}) = \frac{ie_\alpha}{m_\alpha}\phi(\mathbf{k}, \omega)e^{i\mathbf{k}\cdot\mathbf{r}-i\omega t}\int_0^\infty d\tau\, e^{i(\omega - k_\parallel v_\parallel)\tau}$$
$$\times\left[k_\parallel\frac{\partial f_{\alpha 0}}{\partial v_\parallel} + k_\perp\frac{\partial f_{\alpha 0}}{\partial v_\perp}cos(\omega_{c\alpha}\tau + \varphi)\right]$$
$$\times\, exp\left[-i\frac{k_\perp v_\perp}{\omega_{c\alpha}}[sin(\omega_{c\alpha}\tau + \varphi) - sin\varphi]\right]. \qquad (6.17)$$

In (6.17) it will be useful to note

$$\frac{\partial f_{\alpha 0}}{\partial v'_x} = \frac{\partial f_{\alpha 0}(v_\perp, v_\parallel)}{\partial v_\perp}\frac{\partial v_\perp}{\partial v'_x} = \frac{\partial f_{\alpha 0}(v_\perp, v_\parallel)}{\partial v_\perp}\ \frac{v'_x}{v_\perp}.$$

In (6.17), the *cos*-term integral ($\equiv I$) reads

$$I = \frac{1}{2}\int_0^\infty d\tau e^{i(\omega - k_\parallel v_\parallel)\tau}[e^{i(\omega_c\tau+\varphi)} + e^{-i(\omega_c\tau+\varphi)}]e^{ia\ sin\varphi}e^{-iasin(\omega_c\tau+\varphi)}$$
$$where \quad a = \frac{k_\perp v_\perp}{\omega_c}. \qquad (6.18)$$

Using the Bessel function identity,

$$e^{ia\ sin\varphi} = \sum_{n=-\infty}^{\infty} J_n(a)e^{in\varphi}$$

$$I = \frac{1}{2}\int_0^\infty d\tau\sum_l\sum_n J_l(a)J_n(a)e^{i(l-n\pm 1)\varphi}e^{i\tau(\omega - k_\parallel v_\parallel - n\omega_c \pm \omega_c)}$$

where $\pm$–terms should be summed. Thus, we obtain

$$I = \frac{1}{2}\int_0^\infty d\tau\sum_l J_l\sum_{n'}(J_{n'+1} + J_{n'-1})\ e^{i(l-n')\varphi}\ e^{i\tau(\omega - k_\parallel v_\parallel - n'\omega_c)}.$$

Using the Bessel function identity

$$J_{n+1}(a) + J_{n-1}(a) = \frac{2n}{a}J_n(a)$$

$$I = \sum_l\sum_n J_l J_n\ \frac{n}{a}\ e^{i(l-n)\varphi}\int_0^\infty d\tau e^{i\tau(\omega - k_\parallel v_\parallel - n\omega_c)}.$$

At a later stage, we integrate over the velocity. Recall that φ is the azimuthal angle in the velocity space and only the $l = n$ terms survive in I. Thus we put $l = n$ in the above to write

$$I = \sum_{n=-\infty}^{\infty} \frac{i}{\omega - k_\parallel v_\parallel - n\omega_c} \frac{n\omega_c}{k_\perp v_\perp} J_n^2(a).$$

In (6.17), the $k_\parallel$-term can be computed similarly. We obtain the Fourier amplitude, the coefficient of the phasor $e^{i\mathbf{k}\cdot\mathbf{r}-i\omega t}$ in (6.17):

$$f_{\alpha 1}(\mathbf{k},\omega,\mathbf{v}) = -\frac{e_\alpha}{m_\alpha}\phi(\mathbf{k},\omega) \sum_{n=-\infty}^{\infty} \left[k_\parallel \frac{\partial f_{\alpha 0}}{\partial v_\parallel} + \frac{n\omega_{c\alpha}}{v_\perp}\frac{\partial f_{\alpha 0}}{\partial v_\perp}\right] \frac{[J_n(\frac{k_\perp v_\perp}{\omega_{c\alpha}})]^2}{\omega - k_\parallel v_\parallel - n\omega_{c\alpha}}. \tag{6.19}$$

When $k_\perp = 0$, only the n = 0 term is nonzero with $J_0(0) = 1$. In this parallel propagation, (6.19) is identical with the electrostatic perturbation when $B_0 = 0$, and the particles are not aware of the presence of the static magnetic field. When $k_\parallel = 0$, (6.19) shows a resonant structure at the wave frequency equal to the harmonics of the cyclotron frequency. Dispersion relation is obtained by putting (6.19) into (5.3), the Poisson equation,

$$\varepsilon_L = 1 + \sum_\alpha \frac{\omega_{p\alpha}^2}{k^2} \sum_{n=-\infty}^{\infty} \int d^3v \left[k_\parallel \frac{\partial f_{\alpha 0}}{\partial v_\parallel} + \frac{n\omega_{c\alpha}}{v_\perp}\frac{\partial f_{\alpha 0}}{\partial v_\perp}\right] \frac{[J_n(\frac{k_\perp v_\perp}{\omega_{c\alpha}})]^2}{\omega - k_\parallel v_\parallel - n\omega_{c\alpha}} = 0 \tag{6.20}$$

where $d^3v = 2\pi v_\perp dv_\perp dv_\parallel$. Here we designated the quantity in (6.20) as ε_L, the longitudinal dielectric constant defined by (4.29). This usage of the term can be justified when we have all the tensor elements of ε_{ij} of the magnetized plasma now under consideration. The entire tensor elements of ε_{ij} will be derived later. With the aid of another Bessel function identity,

$$\int_0^\infty e^{-a^2x^2} J_n^2(bx)\, x\, dx = \frac{1}{2a^2} e^{-b^2/2a^2} I_n\left(\frac{b^2}{2a^2}\right)$$

where I_n is the modified Bessel function of the nth order, the $\int dv_\perp$-integral can be carried out:

$$\varepsilon_L = 1 + \sum_\alpha \frac{\omega_{p\alpha}^2}{k^2} \sum_{n=-\infty}^{\infty} e^{-\mu_\alpha} I_n(\mu_\alpha) \int_{-\infty}^{\infty} dv_\parallel \frac{k_\parallel \frac{\partial f_{0\parallel}^\alpha}{\partial v_\parallel} - 2n\,\omega_{c\alpha}\, f_{0\parallel}^\alpha \frac{m_\alpha}{2T_{\alpha\perp}}}{\omega - k_\parallel v_\parallel - n\omega_{c\alpha}} \tag{6.21}$$

$$\mu_\alpha = \frac{k_\perp^2 T_{\alpha\perp}}{\omega_{c\alpha}^2 m_\alpha} = k_\perp^2 r_{L\alpha}^2, \qquad r_{L\alpha} = \sqrt{\frac{T_{\alpha\perp}}{m_\alpha \omega_{c\alpha}^2}} : \textit{ Larmor radius.}$$

The integral in ε_L given by (6.21) consists of two terms. The first integral, $k_\parallel$-term, is obtained, in terms of the plasma dispersion function,

$$2\sum_\alpha \frac{\omega_{p\alpha}^2}{k^2 v_{\alpha\parallel}^2} \sum_n e^{-\mu_\alpha} I_n(\mu_\alpha) \left[1 + \frac{\omega - n\omega_{c\alpha}}{k_\parallel v_{\alpha\parallel}} Z\left(\frac{\omega - n\omega_{c\alpha}}{k_\parallel v_{\alpha\parallel}}\right)\right].$$

The second integral, $\omega_{c\alpha}$–term, is obtained

$$2\sum_\alpha \frac{\omega_{p\alpha}^2}{k^2 v_{\alpha\perp}^2} \sum_n e^{-\mu_\alpha} I_n(\mu_\alpha) \frac{n\omega_{c\alpha}}{k_\parallel v_{\alpha\parallel}} Z\left(\frac{\omega - n\omega_{c\alpha}}{k_\parallel v_{\alpha\parallel}}\right)$$

where $v_\alpha^2 = 2T_\alpha/m_\alpha$. Adding the above two equations give

$$\varepsilon_L = 1 + \sum_\alpha \frac{1}{k^2 \lambda_{D\alpha}^2} \sum_{n=-\infty}^{\infty} e^{-\mu_\alpha} I_n(\mu_\alpha)$$
$$\times \left[1 + (\omega - n\omega_{c\alpha} + \frac{T_{\alpha\parallel}}{T_{\alpha\perp}} n\omega_{c\alpha}) \frac{\sqrt{m_\alpha}}{k_\parallel \sqrt{2T_{\alpha\parallel}}} Z\left(\frac{\omega - n\omega_{c\alpha}}{k_\parallel v_{\alpha\parallel}}\right)\right] \tag{6.22}$$

where $\lambda_{D\alpha}^2 = T_{\alpha\parallel}/\omega_{p\alpha}^2$, the squared Debye length.

For **parallel propagation** ($k_\perp = 0$), $\mu_\alpha = 0$ and $I_n(0) = 0$ for all n except $n = 0$. Since $I_0(0) = 1$, (6.22) reduces to

$$\varepsilon_L = 1 + \sum_\alpha \frac{\omega_{p\alpha}}{k^2} \int_{-\infty}^{\infty} dv_\parallel \frac{k_\parallel \frac{\partial f_{0\parallel}^\alpha}{\partial v_\parallel}}{\omega - k_\parallel v_\parallel}$$
$$= 1 + \sum_\alpha \frac{1}{k^2 \lambda_{D\alpha}^2} \left[1 + \frac{\omega}{k_\parallel \sqrt{2T_\alpha/m_\alpha}} Z\left(\frac{\omega}{k_\parallel \sqrt{2T_\alpha/m_\alpha}}\right)\right] \tag{6.23}$$

which is (5.93).

For **perpendicular propagation** ($k_\parallel = 0$), (6.20) reduces to

$$\varepsilon_L = 1 - 2\sum_\alpha \omega_{p\alpha}^2 \frac{e^{-\mu_\alpha}}{\mu_\alpha} \sum_{n=1}^{\infty} \frac{I_n(\mu_\alpha) n^2}{\omega^2 - n^2 \omega_{c\alpha}^2} \tag{6.24}$$

where we used the Bessel function relation $I_n(x) = I_{-n}(x)$. Putting ε_L equal to zero in (6.24), we obtain the dispersion relation of *Bernstein waves* (cyclotron harmonic waves):

$$1 - 2\sum_\alpha \omega_{p\alpha}^2 \frac{e^{-\mu_\alpha}}{\mu_\alpha} \sum_{n=1}^{\infty} \frac{I_n(\mu_\alpha)n^2}{\omega^2 - n^2\omega_{c\alpha}^2} = 0. \tag{6.25}$$

To help cross reference of (6.25), we refer to the following references: D. G. Swanson, *Plasma Waves* Academic Press New York (1989) Eq. (4.177); G. Bekefi, *Radiation Processes in Plasmas*, Wiley New York (1966) Eq. (7.32); P. C. Clemmow and J. P. Dougherty, *Electrodynamics of Particles and Plasmas*, Addison-Wesley London (1969) Eq. (9.36). The electron Bernstein waves are obtained by putting $\omega_{pi} = 0$ in (6.25). The ω *vs* $k_\perp$ curves (suitably non-dimensionalized; ω/ω_{ce} versus $\sqrt{\mu}$) are available in Fig. 7.12, Bekefi (1966). By neglecting the higher order terms of the mass ratio m_e/m_i in (6.25) we can obtain the ion Bernstein waves. The dispersion curves of the latter are available in Fig. 4.5 in Swanson (1989).

If we use the one-dimensional Maxwellian for $f_{0\|}$ with $T_\| = T_\perp$, (6.22) becomes

$$\varepsilon_L = 1 + \sum_\alpha \frac{1}{k^2\lambda_{D\alpha}^2} \sum_{n=\infty}^{\infty} e^{-\mu_\alpha} I_n(\mu_\alpha)$$
$$\times \left[1 + \frac{\omega}{\sqrt{2\pi T_\alpha/m_\alpha}} \int_{-\infty}^{\infty} \frac{e^{-m_\alpha v_\|^2/2T_\alpha}}{k_\| v_\| - \omega + n\omega_{c\alpha}} dv_\|\right] \tag{6.26}$$

where the integral path goes along the Landau contour. When the Doppler-shifted frequency $(\omega - k_\| v_\|)$ is equal to the harmonics of the cyclotron frequency, there occurs a damping. However, unlike the Landau damping, this *cyclotron damping* does not require the negative slope of $f_{0\alpha}$. Only the aforementioned resonance condition suffices to give rise to the damping. ε_L in (6.21) contains new physics. The perpendicular component of the wave vector gives rise to the *finite Larmor radius effect* which is represented by the terms $I_n(\mu_\alpha)$. This effect owes to the presence of the harmonics of the cyclotron frequency $(n\omega_{c\alpha})$ inside the velocity integral. This effect also gives rise to a new family of waves known as Bernstein waves which have no corresponding waves in cold plasmas.

We use the following asymptotic expansions for $I_n(x)$ to investigate the large and small $k_\perp r_L$ behavior of the dispersion relation (6.23):

$$I_n(x) = \left(\frac{x}{2}\right)^n \sum_{m=0}^{\infty} \frac{(x/2)^{2m}}{m!(m+n)!} \quad (x \ll 1),$$

$$I_n(x) \sim \frac{e^x}{\sqrt{2\pi x}} \quad (x \to \infty).$$

Assume the ions are immobile ($\omega_{pi} = 0$). When $k_\perp r_L \ll 1$, we have from (6.23)

$$1 - \frac{\omega_{pe}^2}{\omega^2 - \omega_{ce}^2} - \sum_{n=2}^{\infty} \frac{1}{n!} \frac{n^2 \omega_{pe}^2}{\omega^2 - n^2 \omega_{ce}^2} (k_\perp^2 r_L^2/2)^{n-1} = 0.$$

In the limit $k_\perp \to 0$, the solutions of the above equation are

$$\omega = \sqrt{\omega_{pe}^2 + \omega_{ce}^2}, \qquad \omega = n\, \omega_{ce}. \tag{6.27}$$

When $k_\perp r_L \gg 1$, we have

$$1 - \sqrt{\frac{2}{\pi}}\, \mu_e^{-3/2} \sum_{n=1}^{\infty} \frac{n^2 \omega_{pe}^2}{\omega^2 - n^2 \omega_{ce}^2} = 0.$$

In the limit of $k_\perp r_L \to \infty$, the solutions of the above dispersion relation are

$$\omega = n\, \omega_{ce}. \tag{6.28}$$

• **More about the Bernstein wave dispersion relation.** Let us consider the electron wave in (6.22) by putting $\omega_{pi} = 0$. Also $T_\perp = T_\parallel$. $\varepsilon_L = 0$ gives the dispersion relation

$$(A) \quad 1 + \lambda_D^2 (k_\perp^2 + k_\parallel^2) + \sqrt{\frac{m}{2T}} \frac{\omega}{k_\parallel} \sum_{n=-\infty}^{\infty} I_n(\mu) Z\left(\frac{\omega - n\omega_c}{k_\parallel \sqrt{\frac{2T}{m}}} \right) = 0$$

where we used the identity [see Arfken and Weber (2001)]

$$\sum_{n=-\infty}^{\infty} I_n(x) = e^x.$$

To obtain the perpendicular dispersion relation, we take the limit $k_\parallel \to 0$. Taking the first term in the asymptotic expansion of the Z-function, $Z \simeq -k_\parallel \sqrt{\frac{2T}{m}}/(\omega - n\omega_c)$. Then (A) becomes

$$(B) \quad 1 + \lambda_D^2 k_\perp^2 - e^{-\mu} I_0(\mu) - \omega e^{-\mu} \sum_{n=1}^{\infty} I_n(\mu) \left(\frac{1}{\omega - n\omega_c} + \frac{1}{\omega + n\omega_c} \right) = 0$$

where we used $I_n(x) = I_{-n}(x)$. (B) is the original Bernstein's equation for the Bernstein wave dispersion relation. [See I. B. Bernstein, *Waves in a plasma in a magnetic field* Phys. Rev. **109**, pp 10, 1958; also T. H. Stix, *Waves in Plasmas*, AIP New york (1992) Eq. (11–87), pp. 295.] The function $\lambda_D^2 k_\perp^2$ is plotted versus ω/ω_c for various values of μ in Bernstein's paper (1958) Figs. 1, 2, 3.

6.1.2. *Cyclotron resonance*

Consider the following equation of motion which describes the motion of a positively charged particle acted on by an electric wave in a magnetized plasma.

$$\frac{d\mathbf{v}}{dt} = \hat{\mathbf{x}}\ \frac{e}{m} E_0 cos(k_0 x - \omega_0 t) + \omega_c \mathbf{v} \times\ \hat{\mathbf{z}}. \tag{6.29}$$

This is a Lagrangian description, and x and $\mathbf{v}$ in the RHS are functions of t only. In the zeroth order, the particle orbit is the gyration around $\mathbf{B}_0$: $\frac{d\mathbf{v}}{dt} = \omega_c \mathbf{v} \times \hat{z}$. This is solved by $v_x = v_\perp cos(\omega_c t + \theta)$, $x = x_0 + (v_\perp/\omega_c) sin(\omega_c t + \theta)$ with $v_\perp$ and θ being constants of integration. Using this orbit, the first order velocity is obtained by the integration,

$$v_1(t) - v_1(0) = \frac{e}{m} E_0 \int_0^t cos[k_0 x(t') - \omega_0 t'] dt' \tag{6.30}$$

where $x(t') = x_0 + (v_\perp/\omega_c) sin(\omega_c t' + \theta)$. Here we choose $x_0 = x$ and $\theta = -\omega_c t$ so that $x(t') = x$ when $t' = t$. This orbit is obtained by putting $\varphi = 0$ in (6.11) and (6.14). The *cos*–integral in (6.30) can be obtained by taking the real part of the integral,

$$\int_0^t e^{ik_0 x - i\omega_0 t' + ia sin[\omega_c (t'-t)]} = \sum_n J_n(a) e^{ik_0 x - in\omega_c t} \int_0^t e^{-i\omega_0 t' + in\omega_c t'} dt'$$

where $a = k_0 v_\perp/\omega_c$. Carrying out the integral and taking the real part yield

$$v_1(t) - v_1(0) = \frac{e}{m} E_0 \sum_n J_n(a) \frac{sin(k_0 x - n\omega_c t) - sin(k_0 x - \omega_0 t)}{\omega_0 - n\omega_c}. \tag{6.31}$$

If $\omega_0 = n\omega_c$, $v_1(t)$ is not singular but increases as $\sim t\ cos(k_0 x - \omega_0 t)$ which is the familiar resonance behavior. (Physically this is unacceptable since the velocity increases indefinitely. This is due to mathematically too crude a perturbation scheme.) In this *cyclotron resonance*, the wave energy is transferred to the kinetic energy of particles, thus increasing kinetic temperature of plasma. This method of heating is known as *radio frequency heating* of plasma by cyclotron resonance.

Remark. The same problem can be described by Eulerian equation,

$$\frac{\partial \mathbf{v}}{\partial t} + \mathbf{v} \cdot \frac{\partial \mathbf{v}}{\partial \mathbf{r}} = \hat{x} \frac{e}{m} E_0 cos(k_o x - \omega_0 t) + \omega_c \mathbf{v} \times \hat{z}. \tag{6.32}$$

In the zeroth order, we have

$$\frac{\partial \mathbf{v}^0}{\partial t} = \omega_c \mathbf{v}^0 \times \hat{z}.$$

In the first order,

$$\frac{\partial \mathbf{v}^1}{\partial t} + \mathbf{v}^0 \cdot \frac{\partial \mathbf{v}^1}{\partial \mathbf{r}} = \hat{x}\frac{e}{m}E_0 cos(k_o x - \omega_0 t)$$

whose characteristic relations are

$$\frac{dt}{1} = \frac{d\mathbf{r}}{\mathbf{v}^0} = \frac{d\mathbf{v}^1}{\frac{e}{m}\hat{x}E_0 cos(k_0 x - \omega_0 t)}$$

which is the same as (6.29).

6.1.3. *Mathematical notes on the integration along the unperturbed orbit*

We have Eq. (6.5):

$$f(t, \mathbf{r}, \mathbf{v}) = \frac{-e_\alpha}{m_\alpha} \int_{-\infty}^{t} dt' \mathbf{E}(\mathbf{r}', t') \cdot \frac{\partial f_0}{\partial \mathbf{v}'} \tag{6.33}$$

which is the characteristic equation of the Vlasov equation. In this note, the Fourier transform of the above expression will be derived in a more mathematical way. Fourier transforming (6.33) gives

$$\begin{aligned} f(\omega, \mathbf{k}, \mathbf{v}) &= \frac{-e_\alpha}{m_\alpha} \int_{-\infty}^{\infty} d^3r \int_{-\infty}^{\infty} dt\, e^{-i\mathbf{k}\cdot\mathbf{r}+i\omega t} \int_{-\infty}^{t} dt' \mathbf{E}(\mathbf{r}', t') \cdot \frac{\partial f_0}{\partial \mathbf{v}'} \\ &= \frac{-e_\alpha}{m_\alpha} \int_{-\infty}^{\infty} d^3r \int_{-\infty}^{\infty} dt\, e^{-i\mathbf{k}\cdot\mathbf{r}+i\omega t} \\ &\quad \times \int_{-\infty}^{t} dt' \left(\frac{1}{2\pi}\right)^4 \int_{-\infty}^{\infty} d^3k' \int_{-\infty}^{\infty} d\omega' e^{i\mathbf{k}'\cdot\mathbf{r}'-i\omega' t'} \mathbf{E}(\mathbf{k}', \omega') \cdot \frac{\partial f_0}{\partial \mathbf{v}'}. \end{aligned} \tag{6.34}$$

We have

$$\mathbf{r}' = \mathbf{r} - \mathbf{q} \quad Eqs.\ (6.14) - (6.16) \tag{6.35}$$

where

$$q_x = \frac{v_\perp}{\omega_c}[sin(\omega_c(t - t') + \varphi) - sin\varphi],$$

$$q_y = -\frac{v_\perp}{\omega_c}[cos(\omega_c(t - t') + \varphi) - cos\varphi],$$

$$q_z = v_z(t - t'),$$

$$\mathbf{v}'(t') = -\frac{d\mathbf{q}}{dt'}.$$

Substituting (6.35) into (6.34) yields

$$f(\omega, \mathbf{k}, \mathbf{v}) = \frac{-e_\alpha}{m_\alpha} \int_{-\infty}^{\infty} d^3r \int_{-\infty}^{\infty} dt\ e^{-i(\mathbf{k}-\mathbf{k}')\cdot\mathbf{r}+i\omega t} \int_{-\infty}^{t} dt'$$
$$\times \left(\frac{1}{2\pi}\right)^4 \int_{-\infty}^{\infty} d^3k' \int_{-\infty}^{\infty} d\omega' e^{-i\mathbf{k}'\cdot\mathbf{q}-i\omega' t'}\ \mathbf{E}(\mathbf{k}', \omega') \cdot \frac{\partial f_0}{\partial \mathbf{v}'}$$
$$= \frac{-e_\alpha}{m_\alpha} \int_{-\infty}^{\infty} dt\ e^{i\omega t} \int_{-\infty}^{t} dt' \frac{1}{2\pi} \int_{-\infty}^{\infty} d\omega' e^{-i\mathbf{k}\cdot\mathbf{q}-i\omega' t'}\ \mathbf{E}(\mathbf{k}, \omega') \cdot \frac{\partial f_0}{\partial \mathbf{v}'}. \tag{6.36}$$

Introducing a new variable, $\tau = t - t'$, we have

$$f(\omega, \mathbf{k}, \mathbf{v}) = \frac{-e_\alpha}{m_\alpha} \int_{-\infty}^{\infty} dt\ e^{i\omega t} \int_{0}^{\infty} d\tau\ \frac{1}{2\pi}$$
$$\times \int_{-\infty}^{\infty} d\omega' e^{-i\mathbf{k}\cdot\mathbf{q}(\tau)} e^{-i\omega'(t-\tau)} \mathbf{E}(\mathbf{k}, \omega') \cdot \frac{\partial f_0}{\partial \mathbf{v}'}.$$

The $\int dt$-integral gives $\delta(\omega - \omega')$. Thus we finally obtain

$$f(\omega, \mathbf{k}, \mathbf{v}) = \frac{-e_\alpha}{m_\alpha}\ \mathbf{E}(\mathbf{k}, \omega) \cdot \int_{0}^{\infty} d\tau\ e^{-i\mathbf{k}\cdot\mathbf{q}(\tau)+i\omega\tau}\ \frac{\partial f_0}{\partial \mathbf{v}'} \tag{6.37}$$

which is (6.17). It can be shown that the coefficient of the phasor $e^{i\mathbf{k}\cdot\mathbf{r}-i\omega t}$ in (6.17) is the RHS of (6.37).

Example. Show by direct substitution that

$$f(t, \mathbf{r}, \mathbf{v}) = -\frac{e}{m} \int_{-\infty}^{t} dt' G(\mathbf{r}', t', \mathbf{v}') \tag{6.38}$$

satisfies

$$\frac{\partial f}{\partial t} + \mathbf{v} \cdot \frac{\partial f}{\partial \mathbf{r}} + \omega_c \mathbf{v} \times \hat{\mathbf{z}} \cdot \frac{\partial f}{\partial \mathbf{v}} = -\frac{e}{m} G(\mathbf{r}, t, \mathbf{v}) \tag{6.39}$$

where $\mathbf{r}'$ is given by (6.14)–(6.16), $\mathbf{v}'$ by (6.11)–(6.13).

Noting that $v_x = v_\perp cos\varphi,\quad v_y = v_\perp sin\varphi$, (6.14–6.16) can be written as

$$x' = x'(x, v_x, x_y, \tau) = x + \frac{v_y}{\omega_c} - \frac{1}{\omega_c}(v_x\ sin\omega_c\tau + v_y\ cos\omega_c\tau),$$

$$y' = y'(x, v_x, x_y, \tau) = y - \frac{v_x}{\omega_c} + \frac{1}{\omega_c}(v_x\ cos\omega_c\tau - v_y\ sin\omega_c\tau),$$

$$z' = z'(z, v_z, \tau) = z - v_z\tau.$$

Also we have (6.11)–(6.13):

$$v_x' = v_x'(v_x, v_y, \tau) = v_x\ cos\omega_c\tau - v_y\ sin\omega_c\tau,$$

$$v_y' = v_y'(v_x, v_y, \tau) = v_x\ sin\omega_c\tau + v_y\ cos\omega_c\tau,$$

$$v_z' = v_z.$$

Note that $\mathbf{r}'$ in (6.38) is $\mathbf{r}'(\mathbf{r}, \mathbf{v}, t)$:

$$\frac{\partial f}{\partial t} = -\frac{e}{m}G[\mathbf{r}'(t' = t), t, \mathbf{v}'(t' = t)] - \frac{e}{m}\int_{-\infty}^{t} dt' \frac{\partial}{\partial t}G(\mathbf{r}', t', \mathbf{v}').$$

$G[\mathbf{r}'(t' = t), t, \mathbf{v}'(t' = t)]$ is $G(\mathbf{r}, t, \mathbf{v})$, so it produces the right hand side of (6.39). In the last term,

$$\frac{\partial}{\partial t}G(\mathbf{r}', t', \mathbf{v}') = \frac{\partial G}{\partial \mathbf{r}'}\cdot\frac{\partial \mathbf{r}'}{\partial t} + \frac{\partial G}{\partial \mathbf{v}'}\cdot\frac{\partial \mathbf{v}'}{\partial t}.$$

Also we have

$$\mathbf{v}\cdot\frac{\partial G}{\partial \mathbf{r}} = \frac{\partial G}{\partial \mathbf{r}'}\cdot\left(\mathbf{v}\cdot\frac{\partial}{\partial \mathbf{r}}\right)\mathbf{r}' + \frac{\partial G}{\partial \mathbf{v}'}\cdot\left(\mathbf{v}\cdot\frac{\partial}{\partial \mathbf{r}}\right)\mathbf{v}'$$

$$\omega_c\mathbf{v}\times\hat{z}\cdot\frac{\partial G}{\partial \mathbf{v}} = \frac{\partial G}{\partial \mathbf{r}'}\cdot\left(\omega_c\mathbf{v}\times\hat{z}\cdot\frac{\partial}{\partial \mathbf{v}}\right)\mathbf{r}' + \frac{\partial G}{\partial \mathbf{v}'}\cdot\left(\omega_c\mathbf{v}\times\hat{z}\cdot\frac{\partial}{\partial \mathbf{v}}\right)\mathbf{v}'.$$

So the remaining terms in (6.39) are added in the form

$$\frac{\partial G}{\partial \mathbf{r}'}\cdot\frac{d\mathbf{r}'}{dt} + \frac{\partial G}{\partial \mathbf{v}'}\cdot\frac{d\mathbf{v}'}{dt}$$

$$where\quad \frac{d}{dt} = \frac{\partial}{\partial t} + \mathbf{v}\cdot\frac{\partial}{\partial \mathbf{r}} + \omega_c\mathbf{v}\times\hat{\mathbf{z}}\cdot\frac{\partial}{\partial \mathbf{v}}.$$

$\mathbf{r}'$ and $\mathbf{v}'$ are constants of motion in unprimed equations, and so $\frac{d\mathbf{r}'}{dt} = 0$ and $\frac{d\mathbf{v}'}{dt} = 0$ which can be also directly proven by using the foregoing relations. This completes the proof.

6.2. Dielectric tensor for electromagnetic wave in magnetized plasma

In this section, the general form of the dielectric tensor ε_{ij} for electromagnetic wave (Section 4.3) will be derived by solving the Vlasov equation

$$\frac{\partial}{\partial t} f_{\alpha 1}(\mathbf{r}, \mathbf{v}, t) + \mathbf{v} \cdot \frac{\partial f_{\alpha 1}}{\partial \mathbf{r}} + \frac{e_\alpha}{m_\alpha} \frac{\mathbf{v} \times \mathbf{B_0}}{c} \cdot \frac{\partial f_{\alpha 1}}{\partial \mathbf{v}}$$

$$= -\frac{e_\alpha}{m_\alpha}\left(\mathbf{E}(\mathbf{r}, t) + \frac{1}{c}\mathbf{v} \times \mathbf{B}(\mathbf{r}, t)\right) \cdot \frac{\partial f_{\alpha 0}}{\partial \mathbf{v}} \tag{6.40}$$

where the zero order distribution $f_{\alpha 0}(\mathbf{v})$ is given by (6.4), the two-temperature Maxwellian. The above equation includes the magnetic field perturbation while (6.3) does not. The algebra is quite complicated, but we derive the well-documented result by employing the mathematical method used in the previous section.

6.2.1. *Whistler wave*

Before embarking on the most general polarization (the direction of **E** with respect to **k**) of the electromagnetic wave, we take up, as a special polarization, the *whistler wave.* The whistler mode is a typical electromagnetic wave propagating in magnetized plasmas, and is important in ionosphere and magnetosphere research. Whistler mode is a *circularly polarized* electromagnetic wave that propagates along the direction of the static magnetic field ($\mathbf{B}_0 = \hat{\mathbf{z}} B_0$). The electric and magnetic fields of a whistler mode can be represented by

$$\mathbf{E}(z, t) = E_1[\hat{\mathbf{x}}\ sin(kz - \omega t) + \hat{\mathbf{y}}\ cos(kz - \omega t)], \quad (E_1\ is\ real) \tag{6.41}$$

$$\mathbf{B}(z, t) = B_1[\hat{\mathbf{x}}\ cos(kz - \omega t) - \hat{\mathbf{y}}\ sin(kz - \omega t)]. \tag{6.42}$$

We have $B_1 = -\frac{ck}{\omega} E_1$ per Maxwell equation $\nabla \times \mathbf{E} = -\frac{1}{c}\frac{\partial \mathbf{B}}{\partial t}$. Note $\mathbf{E} \cdot \mathbf{B} = 0$. Alternatively we can write, putting $\omega t \to \omega t - \pi/2$,

$$\mathbf{E}(z, t) = E_1[\hat{\mathbf{x}}\ cos(kz - \omega t) - \hat{\mathbf{y}} sin(kz - \omega t)], \tag{6.43}$$

$$\mathbf{B}(z, t) = -B_1[\hat{\mathbf{x}}\ sin(kz - \omega t) + \hat{\mathbf{y}}\ cos(kz - \omega t)]. \tag{6.44}$$

The tip of the electric vector draws a circle on the (x, y) plane in the counter clockwise direction with the wave vector $\mathbf{k} = \hat{\mathbf{z}} k \parallel \mathbf{B}_0$ (wave of this polarization is called *R-wave*). The electric vector in (6.43) can be obtained from

$$\mathbf{E} = Re\ (\hat{\mathbf{x}} + i\hat{\mathbf{y}})E_1\ e^{ikz-i\omega t}. \quad (take\ real\ part) \tag{6.45}$$

Before starting kinetic analysis, we work out the whistler wave dispersion relation from the ionic theory. The equation of motion for electrons (ions are assumed immobile) reads

$$\frac{d\mathbf{v}}{dt} = -\frac{e}{m}\mathbf{E} - \frac{e}{mc}\mathbf{v} \times \mathbf{B}_0. \tag{6.46}$$

In the dipole approximation, since $E(z,t)$ does not change much over the wavelengths or wave periods, the equation of motion is solved by assuming $E \sim e^{-i\omega t}$. In components,

$$-i\omega v_x = -\frac{e}{m}E_x - \omega_{ce}v_y, \tag{6.47}$$

$$-i\omega v_y = -\frac{e}{m}E_y + \omega_{ce}v_x. \quad \omega_{ce} = \frac{eB_0}{mc} > 0 \tag{6.48}$$

Solving the above equations,

$$v_x = \frac{-ieE_1/m}{\omega - \omega_{ce}}, \quad v_y = \frac{eE_1/m}{\omega - \omega_{ce}}. \tag{6.49}$$

Thus the current is obtained as

$$\mathbf{J} = -eN\mathbf{v} = (\hat{\mathbf{x}} + i\hat{\mathbf{y}})\frac{iE_1}{\omega - \omega_{ce}}\frac{Ne^2}{m}. \quad (N:\ electron\ number\ density) \tag{6.50}$$

On the other hand, (4.16) with $(\mathbf{k} \cdot \mathbf{E} = 0)$ gives

$$(1 - n^2)\ \mathbf{E} = -\frac{4\pi i}{\omega}\mathbf{J}. \quad (n^2 = c^2k^2/\omega^2) \tag{6.51}$$

Substituting (6.45) and (6.50) into (6.51) yields the whistler dispersion relation,

$$n^2 = 1 - \frac{\omega_{pe}^2}{\omega(\omega - \omega_{ce})}. \tag{6.52}$$

Exercise. When the electric vector is polarized as

$$\mathbf{E} = Re\ (\hat{\mathbf{x}} - i\hat{\mathbf{y}})E_1\ e^{ikz-i\omega t}, \tag{6.53}$$

determine the sense of rotation and repeat the above algebra to derive the dispersion relation of the *L-wave*,

$$n^2 = 1 - \frac{\omega_{pe}^2}{\omega(\omega + \omega_{ce})}. \tag{6.54}$$

Exercise. Derive the R-wave dispersion relation including the ion current.

$$\frac{c^2k^2}{\omega^2} = 1 - \frac{\omega_{pi}^2}{\omega^2}\frac{\omega}{\omega+\omega_{ci}} - \frac{\omega_{pe}^2}{\omega^2}\frac{\omega}{\omega-\omega_{ce}} : \quad \textit{Answer.} \tag{6.55}$$

Kinetic analysis. Integrating (6.40) gives

$$f_\alpha(t,\mathbf{r},\mathbf{v}) = -\frac{e_\alpha}{m_\alpha}\int_{-\infty}^{t} dt' \ (\mathbf{E}' + \frac{1}{c}\mathbf{v}' \times \mathbf{B}') \cdot \frac{\partial f_{\alpha 0}(\mathbf{v}')}{\partial \mathbf{v}'} \tag{6.56}$$

where the subscript 1 was omitted, and the primed quantities are functions of t'. The wave magnetic field corresponding to the electric field (6.45) is

$$\mathbf{B}(z,t) = Re\ B_1(i\hat{\mathbf{x}} - \hat{\mathbf{y}})e^{ikz-i\omega t} \quad (B_1 = -\frac{ck}{\omega}E_1). \tag{6.57}$$

Equations (6.45) and (6.57) give

$$\begin{aligned}\mathbf{E}' + \frac{1}{c}\mathbf{v}' \times \mathbf{B}' = &\left[\hat{\mathbf{x}}\ (1 - \frac{kv_z'}{\omega}) + i\hat{\mathbf{y}}\ (1 - \frac{kv_z'}{\omega}) + \hat{\mathbf{z}}\ \frac{k}{\omega}\ (v_x' + iv_y')\right] \\ &\times E_1\ e^{ikz'-i\omega t'}.\end{aligned} \tag{6.58}$$

When (6.58) is substituted into (6.56), we use the orbits (6.11)–(6.16):

$$v_x'(t') = v_\perp\ cos[\omega_{c\alpha}(t-t') + \varphi], \tag{6.59}$$

$$v_y'(t') = v_\perp\ sin[\omega_{c\alpha}(t-t') + \varphi]; \tag{6.60}$$

$$v_z'(t') = v_z \equiv v_\parallel : \quad costant, \tag{6.61}$$

$$z'(t') = z - v_z(t-t'). \tag{6.62}$$

Using (6.4) and (6.58) gives

$$\begin{aligned}f_\alpha(t,\mathbf{r},\mathbf{v}) = &-\frac{e_\alpha}{m_\alpha}E_1\left[\frac{k}{\omega}\frac{\partial f_{\alpha 0}}{\partial v_\parallel} + \left(1 - \frac{kv_\parallel}{\omega}\right)\frac{1}{v_\perp}\frac{\partial f_{\alpha 0}}{\partial v_\perp}\right] \\ &\times \int_{-\infty}^{t} dt'\ (v_x' + iv_y')e^{ikz'-i\omega t'}.\end{aligned}$$

Substituting the orbits into the time integral, we obtain

$$\int_{-\infty}^{t} dt'(...) = \frac{iv_\perp e^{i\varphi}}{\omega + \omega_{c\alpha} - kv_\parallel}e^{ikz-i\omega t}.$$

Note that we have no Bessel function expansions in this parallel propagation. Thus the current is obtained as

$$\mathbf{J} = -iE_1 e^{ikz-i\omega t}\Sigma_\alpha \frac{e_\alpha^2}{m_\alpha}\int v_\perp dv_\perp dv_\parallel d\varphi\ (\hat{\mathbf{x}}v_x + \hat{\mathbf{y}}v_y + \hat{\mathbf{z}}v_\parallel)\frac{v_\perp e^{i\varphi}}{\omega + \omega_{c\alpha} - kv_\parallel}$$
$$\times \left[\frac{k}{\omega}\frac{\partial f_{\alpha 0}}{\partial v_\parallel} + \left(1 - \frac{kv_\parallel}{\omega}\right)\frac{1}{v_\perp}\frac{\partial f_{\alpha 0}}{\partial v_\perp}\right]. \tag{6.63}$$

Note that the z-component of the current vanishes because of the $\int d\varphi$-integral. Putting $v_x = v_\perp cos\varphi$ and $v_y = v_\perp sin\varphi$, the $\int d\varphi$-integral can be carried out:

$$\mathbf{J} = -i\pi E_1\ (\hat{\mathbf{x}} + i\hat{\mathbf{y}})e^{ikz-i\omega t}\Sigma_\alpha \frac{e_\alpha^2}{m_\alpha}\int \frac{v_\perp^3 dv_\perp dv_\parallel}{\omega + \omega_{c\alpha} - kv_\parallel}$$
$$\times \left[\frac{k}{\omega}\frac{\partial f_{\alpha 0}}{\partial v_\parallel} + \left(1 - \frac{kv_\parallel}{\omega}\right)\frac{1}{v_\perp}\frac{\partial f_{\alpha 0}}{\partial v_\perp}\right]. \tag{6.64}$$

Substituting (6.64) into (6.51) yields the whistler dispersion relation,

$$\frac{c^2k^2}{\omega^2} = 1 + \pi\sum_\alpha \frac{\omega_{p\alpha}^2}{\omega}\int \frac{v_\perp^3 dv_\perp dv_\parallel}{\omega \pm \omega_{c\alpha} - kv_\parallel}\left[\frac{k}{\omega}\frac{\partial f_{\alpha 0}}{\partial v_\parallel} + \left(1 - \frac{kv_\parallel}{\omega}\right)\frac{1}{v_\perp}\frac{\partial f_{\alpha 0}}{\partial v_\perp}\right] \tag{6.65}$$

where $\pm$ signs correspond respectively to R $(+)$ and L$(-)$ wave.

Cold plasma limit. The equilibrium distribution function $f_{\alpha 0}$ is given by (6.4). Consider the zero temperature limit. The first term in the integral ($\sim \partial f_0/\partial v_\parallel$) vanishes. The $kv_\parallel/\omega$-term also vanishes. Integrating by parts with respect to $v_\perp$ and using $\int_0^\infty v_\perp dv_\perp f_{\alpha 0}(v_\perp) = 1/2\pi$, the cold plasma limit of (6.65) is obtained to be

$$1 - \frac{c^2k^2}{\omega^2} - \Sigma_\alpha \frac{\omega_{p\alpha}^2}{\omega(\omega \pm \omega_{c\alpha})} = 0 \tag{6.66}$$

which agrees with (6.55). Let us consider the R-wave:

$$1 - \frac{c^2k^2}{\omega^2} - \frac{\omega_{pi}^2}{\omega(\omega + \omega_{ci})} - \frac{\omega_{pe}^2}{\omega(\omega - \mid \omega_{ce}\mid)} = 0.$$

Recall that $\omega_{c\alpha}$ contains sign. In the low frequency domain, $\omega \ll \omega_{ci}$, binomial expansion of the above gives

$$1 - \frac{k^2c^2}{\omega^2} - \frac{1}{\omega}\left[\frac{\omega_{pi}^2}{\omega_{ci}} - \frac{\omega_{pe}^2}{|\omega_{ce}|}\right] + \frac{\omega_{pi}^2}{\omega_{ci}^2} + \frac{\omega_{pe}^2}{\omega_{ce}^2} = 0.$$

Note that the square bracket term is zero. Then the above equation becomes

$$1 - \frac{k^2c^2}{\omega^2} + \frac{c^2}{v_A^2} = 0 \quad \left(v_A = \sqrt{\frac{B_0^2}{4\pi N m_i}},\ \textit{Alfven velocity}\right) \tag{6.67}$$

which is the dispersion relation of the *Alfven wave.*

Example. Solve (6.46) without using the dipole approximation and modify dispersion relation (6.52). Take the x-component of (6.46),

$$\frac{dv_x}{dt} = -\frac{e}{m}E_x - \omega_{ce}v_y.$$

Integrating this,

$$\begin{aligned} v_x(t) &= \int_{-\infty}^{t}\left[-\frac{e}{m}E_x(t') - \omega_{ce}v_y(t')\right]dt' \\ &= \int_{-\infty}^{t}\left[-\frac{e}{m}\tilde{E}_x - \omega_{ce}\tilde{v}_y\right]e^{ikz(t')-i\omega t'}dt' \end{aligned}$$

where the tilded quantities are Fourier coefficients. Since v_z is a constant of motion, we have $z(t') = v_{\parallel}t' + z - v_{\parallel}t$. Using this, the integral can be done:

$$v_x(z,t) = i\ e^{ikz-i\omega t}\ \frac{\frac{e}{m}\tilde{E}_x + \omega_{ce}\tilde{v}_y}{kv_{\parallel} - \omega}. \tag{6.68}$$

Likewise we obtain

$$v_y(z,t) = i\ e^{ikz-i\omega t}\ \frac{\frac{e}{m}\tilde{E}_y - \omega_{ce}\tilde{v}_x}{kv_{\parallel} - \omega}. \tag{6.69}$$

Without the exponential factors, the above equations give the Fourier coefficients, $\tilde{v}_{x,\ y}$. Solving for these, we obtain

$$\tilde{v}_x = -\frac{e}{m}\ \frac{i\omega'\tilde{E}_x - \omega_{ce}\tilde{E}_y}{\omega'^2 - \omega_{ce}^2}, \tag{6.70}$$

$$\tilde{v}_y = +\frac{e}{m}\ \frac{-i\omega'\tilde{E}_y + \omega_{ce}\tilde{E}_x}{\omega'^2 - \omega_{ce}^2}, \tag{6.71}$$

where $\omega' = \omega - kv_\parallel$ is the Doppler-shifted frequency. Using these, the current **J** can be obtained. Then (6.51) can be written in components,

$$(1-n^2)\tilde{E}_x = \frac{-i\omega_{pe}^2}{\omega\,(\omega'^2 - \omega_{ce}^2)}(i\omega'\tilde{E}_x + \omega_{ce}\tilde{E}_y), \tag{6.72}$$

$$(1-n^2)\tilde{E}_y = \frac{-i\omega_{pe}^2}{\omega\,(\omega'^2 - \omega_{ce}^2)}(i\omega'\tilde{E}_y - \omega_{ce}\tilde{E}_x). \tag{6.73}$$

These are two simultaneous equations for the quantities $\tilde{E}_{x,\,y}$. Nontrivial solutions are obtained when the determinant of the coefficients are zero, giving

$$\left[1 - n^2 - \frac{\omega'\omega_{pe}^2}{\omega(\omega'^2 - \omega_{ce}^2)}\right]^2 = \left[\frac{\omega_{ce}\omega_{pe}^2}{\omega(\omega'^2 - \omega_{ce}^2)}\right]^2. \tag{6.74}$$

Equation (6.74) shows that the electromagnetic wave propagating along the direction of the static magnetic field has always two modes which correspond to the $\pm$ signs when the square root is taken in the above. We obtain

$$1 - n^2 = \frac{\omega_{pe}^2}{\omega(\omega - kv_\parallel \pm \omega_{ce})} \tag{6.75}$$

which are R (+sign) and L (−sign) waves. Note that the Doppler-shifted frequency enters in the dynamics.

If (6.46) is cast into the form of the corresponding Eulerian equation, we should solve

$$\frac{\partial \mathbf{v}}{\partial t} + v_\parallel \frac{\partial \mathbf{v}}{\partial z} = -\frac{e}{m}\mathbf{E} - \frac{e}{mc}\mathbf{v}\times\mathbf{B}_0. \tag{6.76}$$

This equation yields identical results, in an easier way. Clearly the Doppler-shifted frequency emerges from the left hand side of (6.76).

Exercise. In the kinetic dispersion relation (6.65), put

$$f_{\alpha 0} = f_{\alpha 0}(v_\parallel)\delta(v_\perp) \tag{6.77}$$

where $f_{\alpha 0}(v_\parallel)$ is the one-dimensional Maxwellian. Obtain the corresponding dispersion relation.

6.2.2. *Dielectric tensor of magnetized plasma: general polarization*

In this Section, the dielectric tensor ε_{ij} of electromagnetic plasma waves propagating in general direction with general polarization (**k** not necessarily parallel to $\mathbf{B}_0$ as in whistler) is obtained by solving (6.40). The starting equation is (6.40) or (6.56) with the zeroth order distribution function (6.4). In terms of Fourier coefficients, we have $\mathbf{B} = \frac{c}{\omega}\mathbf{k} \times \mathbf{E}$. Using this in (6.56) gives

$$f_\alpha(t, \mathbf{r}, \mathbf{v}) = -\frac{e_\alpha}{m_\alpha}\int_{-\infty}^{t} dt' \left[E_j \frac{\partial f_{\alpha 0}}{\partial v_j'}(1 - \frac{k_l v_l'}{\omega}) + \frac{k_l}{\omega}\frac{\partial f_{\alpha 0}}{\partial v_l'} v_j' E_j\right] e^{i\mathbf{k}\cdot\mathbf{r}' - i\omega t'} \tag{6.78}$$

where the primed quantities are functions of t', and E_j is the Fourier coefficient. Therefore the current becomes

$$J_i = -E_j \sum_\alpha \frac{e_\alpha^2}{m_\alpha}\int d^3v v_i \int_{-\infty}^{t} dt' \left[\frac{\partial f_{\alpha 0}}{\partial v_j'}(1 - \frac{k_l v_l'}{\omega}) + \frac{k_l}{\omega}\frac{\partial f_{\alpha 0}}{\partial v_l'} v_j'\right] e^{i\mathbf{k}\cdot\mathbf{r}' - i\omega t'}. \tag{6.79}$$

Without loss of generality, we set $\mathbf{k} = \hat{\mathbf{x}}k_\perp + \hat{\mathbf{z}}k_\parallel$. Then the array in the square bracket above has components

$$[\ \cdots\ ]_x = \frac{U}{\omega v_\perp} v_x', \quad [\ \cdots\ ]_y = \frac{U}{\omega v_\perp} v_y',$$

where

$$U(v_\perp, v_\parallel) = (\omega - k_\parallel v_\parallel)\frac{\partial f_{\alpha 0}}{\partial v_\perp} + k_\parallel v_\perp \frac{\partial f_{\alpha 0}}{\partial v_\parallel},$$

$$[\ \cdots\ ]_z = \frac{\partial f_{\alpha 0}}{\partial v_\parallel} + \Gamma\, v_x', \quad \Gamma = \frac{k_\perp v_\parallel}{\omega}\left(\frac{1}{v_\perp}\frac{\partial f_{\alpha 0}}{\partial v_\perp} - \frac{1}{v_\parallel}\frac{\partial f_{\alpha 0}}{\partial v_\parallel}\right). \tag{6.80}$$

Furthermore we make a change of variable, $t - t' = \tau$. Then we have

$$J_i = -E_j\ \Sigma_\alpha \frac{e_\alpha^2}{m_\alpha}\int d^3v\ v_i\ e^{-i\omega t}\int_0^\infty d\tau\ [\ \cdots\ ]_j\, e^{ik_\perp x' + ik_\parallel z' + i\omega\tau}. \tag{6.81}$$

The orbit to be used here is

$$v_x'(\tau) = v_\perp\ cos(\omega_{c\alpha}\tau + \varphi), \quad v_y'(\tau) = v_\perp\ sin(\omega_{c\alpha}\tau + \varphi), \quad v_z'(\tau) = v_\parallel : const.$$

$$x'(\tau) = x - \frac{v_\perp}{\omega_{c\alpha}}[sin(\omega_{c\alpha}\tau + \varphi) - sin\varphi], \quad z'(\tau) = z - v_\parallel\tau.$$

To evaluate (6.81), we need three kinds of integral:

$$(I_1,\ I_{vx},\ I_{vy}) \equiv \int_0^\infty (1,\ v'_x,\ v'_y)\ e^{ik_\perp x' + ik_\parallel z' + i\omega\tau}\ d\tau.$$

These are calculated by means of the type of algebra used in (6.17).

$$I_1 = ie^{ik_\perp x + ik_\parallel z}\ \Sigma_n \Sigma_l J_n(a_\alpha) J_l(a_\alpha)\ \frac{e^{i(l-n)\varphi}}{\omega - k_\parallel v_\parallel - n\omega_{c\alpha}}, \tag{6.82}$$

$$I_{vx} = iv_\perp e^{ik_\perp x + ik_\parallel z}\ \Sigma_n \Sigma_l \frac{n}{a_\alpha} J_n(a_\alpha) J_l(a_\alpha)\ \frac{e^{i(l-n)\varphi}}{\omega - k_\parallel v_\parallel - n\omega_{c\alpha}}, \tag{6.83}$$

$$I_{vy} = v_\perp e^{ik_\perp x + ik_\parallel z}\ \Sigma_n \Sigma_l J'_n(a_\alpha) J_l(a_\alpha)\ \frac{e^{i(l-n)\varphi}}{\omega - k_\parallel v_\parallel - n\omega_{c\alpha}}, \tag{6.84}$$

where

$$a_\alpha = \frac{k_\perp v_\perp}{\omega_{c\alpha}}, \quad J'_n(a_\alpha) = \frac{dJ_n(a_\alpha)}{da_\alpha}.$$

We also used

$$J_{n+1}(a) + J_{n-1}(a) = \frac{2n}{a} J_n(a), \quad J_{n+1}(a) - J_{n-1}(a) = 2J'_n(a). \tag{6.85}$$

Now we are ready to derive the conductivity tensor σ_{ij} from (6.81). The coefficient of $E_j\ e^{ik_\perp x + ik_\parallel z - i\omega t}$ is the tensor element of σ_{ij}, corresponding to the term $v_i[\cdots]_j$. We have

$$\sigma_{xx} = -i\ \Sigma \frac{e_\alpha^2}{\omega\ m_\alpha} \int d^3v v_x U \Sigma_n \Sigma_l \frac{n}{a_\alpha} J_n J_l \frac{e^{i(n-l)\varphi}}{\omega - k_\parallel v_\parallel - n\omega_{c\alpha}}.$$

Putting $v_x = v_\perp(e^{i\varphi} + e^{-i\varphi})/2$, the $\int d\varphi$ integral yields two terms $\delta_{l,\ n-1}$ and $\delta_{l,\ n+1}$. Combining these terms by (6.85) gives

$$\sigma_{xx} = -i\ \Sigma_\alpha \frac{e_\alpha^2}{\omega\ m_\alpha}\ \Sigma_n \int d^3v\ v_\perp\ \frac{U\ [\frac{n}{a_\alpha} J_n]^2}{\omega - k_\parallel v_\parallel - n\omega_{c\alpha}}. \tag{6.86}$$

Likewise we obtain

$$\sigma_{xy} = -\ \Sigma_\alpha \frac{e_\alpha^2}{\omega\ m_\alpha}\ \Sigma_n \int d^3v\ v_\perp\ \frac{U\ \frac{n}{a_\alpha} J_n\ J'_n}{\omega - k_\parallel v_\parallel - n\omega_{c\alpha}}, \tag{6.87}$$

$$\sigma_{yx} = -\sigma_{xy}. \tag{6.88}$$

We have, using (6.80),

$$\sigma_{xz} = -\Sigma_\alpha \frac{e_\alpha^2}{m_\alpha} \Sigma_n \Sigma_l \int d^3v v_x \ i \ J_n J_l \left[\frac{\partial f_{\alpha 0}}{\partial v_\parallel} + \Gamma v_\perp \frac{n}{a_\alpha} \right] \frac{e^{i(l-n)\varphi}}{\omega - k_\parallel v_\parallel - n\omega_{c\alpha}}$$

$$= -i \ \Sigma_\alpha \frac{e_\alpha^2}{m_\alpha \omega} \Sigma_n \int d^3v v_\perp \frac{\frac{n}{a_\alpha} W \ J_n^2}{\omega - k_\parallel v_\parallel - n\omega_{c\alpha}} \tag{6.89}$$

where

$$W(v_\perp, v_\parallel) = (\omega - n\omega_{c\alpha}) \frac{\partial f_{\alpha 0}}{\partial v_\parallel} + n\omega_{c\alpha} \frac{v_\parallel}{v_\perp} \frac{\partial f_{\alpha 0}}{\partial v_\perp}, \tag{6.90}$$

$$\sigma_{zx} = -i \ \Sigma_\alpha \ \frac{e_\alpha^2}{\omega m_\alpha} \ \Sigma_n \int d^3v \ v_\parallel \ \frac{\frac{n}{a_\alpha} \ U \ J_n^2}{\omega - k_\parallel v_\parallel - n\omega_{c\alpha}}, \tag{6.91}$$

$$\sigma_{yy} = -\Sigma \frac{e_\alpha^2}{\omega m_\alpha} \int d^3v v_y U \Sigma_n \Sigma_l J_n' J_l \frac{e^{i(n-l)\varphi}}{\omega - k_\parallel v_\parallel - n\omega_{c\alpha}}$$

$$= -i \ \Sigma \frac{e_\alpha^2}{\omega m_\alpha} \ \Sigma_n \int d^3v \ v_\perp \frac{U \ (J_n')^2}{\omega - k_\parallel v_\parallel - n\omega_{c\alpha}}, \tag{6.92}$$

$$\sigma_{zz} = -\Sigma_\alpha \frac{e_\alpha^2}{m_\alpha} \Sigma_n \Sigma_l \int d^3v v_\parallel \ i \ J_n J_l \left[\frac{\partial f_{\alpha 0}}{\partial v_\parallel} + \Gamma v_\perp \frac{n}{a_\alpha} \right] \frac{e^{i(l-n)\varphi}}{\omega - k_\parallel v_\parallel - n\omega_{c\alpha}}$$

$$= -i \ \Sigma_\alpha \ \frac{e_\alpha^2}{m_\alpha \omega} \ \Sigma_n \int d^3v \ v_\parallel \ \frac{W \ J_n^2}{\omega - k_\parallel v_\parallel - n\omega_{c\alpha}}. \tag{6.93}$$

We have, using (6.80),

$$\sigma_{yz} = -\Sigma_\alpha \frac{e_\alpha^2}{m_\alpha} \Sigma_n \Sigma_l \int d^3v \ v_y \ i \ J_n J_l \left[\frac{\partial f_{\alpha 0}}{\partial v_\parallel} + \Gamma v_\perp \frac{n}{a_\alpha} \right] \frac{e^{i(l-n)\varphi}}{\omega - k_\parallel v_\parallel - n\omega_{c\alpha}}$$

$$= \ \Sigma_\alpha \frac{e_\alpha^2}{m_\alpha \omega} \ \Sigma_n \int d^3v \ v_\perp \ \frac{W \ J_n J_n'}{\omega - k_\parallel v_\parallel - n\omega_{c\alpha}}, \tag{6.94}$$

$$\sigma_{zy} = - \ \Sigma_\alpha \ \frac{e_\alpha^2}{\omega m_\alpha} \ \Sigma_n \int d^3v \ v_\parallel \ \frac{U \ J_n J_n'}{\omega - k_\parallel v_\parallel - n\omega_{c\alpha}}. \tag{6.95}$$

Equations (6.86)–(6.95) give all the nine elements of the conductivity tensor. The above results are collected to give the following expression for the dielectric tensor

$$\varepsilon_{ij} = \delta_{ij} + \sum_\alpha \frac{\omega_{p\alpha}^2}{\omega^2} \sum_{n=-\infty}^{\infty} \int d^3v \frac{S_{ij}}{\omega - k_\parallel v_\parallel - n\omega_{c\alpha}} \tag{6.96}$$

where the tensor elements S_{ij} are given by

$$\mathbf{S_{ij}} = \begin{pmatrix} v_\perp U \left(\frac{nJ_n}{a_\alpha}\right)^2 & -iv_\perp U \frac{n}{a_\alpha} J_n J_n' & v_\perp W \frac{n}{a_\alpha} J_n^2 \\ iv_\perp U \frac{n}{a_\alpha} J_n J_n' & v_\perp U J_n'^2 & iv_\perp W J_n J_n' \\ v_\parallel U \frac{n}{a_\alpha} J_n^2 & -iv_\parallel U J_n J_n' & v_\parallel W J_n^2 \end{pmatrix}.$$

Exercise. Show that $v_\perp W$ in S_{13} and S_{23} can be replaced by $v_\parallel U$, so that the tensor S_{ij} satisfies Onsager's relation (see later page).

Exercise. Equation (6.96) for ε_{ij} can also be put into the following form:

$$\varepsilon_{ij} = (1 - \sum_\alpha \frac{\omega_{p\alpha}^2}{\omega^2})\delta_{ij} + \sum_\alpha \frac{\omega_{p\alpha}^2}{\omega^2} \sum_{n=-\infty}^{\infty} \int d^3v \frac{k_\parallel \frac{\partial f_{\alpha 0}}{\partial v_\parallel} + \frac{n\omega_{c\alpha}}{v_\perp}\frac{\partial f_{\alpha 0}}{\partial v_\perp}}{\omega - k_\parallel v_\parallel - n\omega_{c\alpha}} S_{ij}', \tag{6.97}$$

where S_{ij}' are given by

$$\mathbf{S_{ij}'} = \begin{pmatrix} \left(\frac{n\omega_{c\alpha}}{k_\perp} J_n\right)^2 & iv_\perp \frac{n\omega_{c\alpha}}{k_\perp} J_n J_n' & \frac{n\omega_{c\alpha}}{k_\perp} v_\parallel J_n^2 \\ -iv_\perp \frac{n\omega_{c\alpha}}{k_\perp} J_n J_n' & (v_\perp J_n')^2 & -iv_\perp v_\parallel J_n J_n' \\ \frac{n\omega_{c\alpha}}{k_\perp} v_\parallel J_n^2 & iv_\parallel v_\perp J_n J_n' & (v_\parallel J_n)^2 \end{pmatrix}.$$

Hint: The following would be useful.

$$\int_0^\infty e^{-a^2x^2} J_n^2(bx)\, x\, dx = \frac{1}{2a^2} e^{-\lambda}\, I_n(\lambda) \quad (\lambda = b^2/2a^2),$$

$$\sum_{n=-\infty}^{\infty} n^2\, I_n(\lambda) = \lambda\, e^\lambda,$$

$$\sum_{n=\infty}^{\infty} n^2 \int d^3v v_\perp \frac{J_n^2(a)}{a^2} \frac{\partial f_0}{\partial v_\perp} = -1.$$

The dispersion relation is obtained by substituting (6.97) into Maxwell's equation (4.21):

$$\left[n^2(\frac{k_i k_j}{k^2} - \delta_{ij}) + \varepsilon_{ij}\right] E_j = 0 \quad (n^2 = \frac{c^2k^2}{\omega^2}, \quad \textit{refractive index}). \tag{6.98}$$

Specializing without loss of generality as $\mathbf{k} = (k\ sin\theta, 0, k\ cos\theta)$, the determinant of the coefficients of E_j can be written as, after a long algebra,

$$An^4 + Bn^2 + C = 0$$

$$A = k_i k_j \varepsilon_{ij}/k^2, \quad C = det(\varepsilon_{ij})$$

$$B = \varepsilon_{xz}^2 - \varepsilon_{xx}\varepsilon_{zz} + 2sin\theta\ cos\theta\ (\varepsilon_{xy}\varepsilon_{yz} - \varepsilon_{yy}\varepsilon_{zx})$$
$$-cos^2\theta(\varepsilon_{yy}\varepsilon_{zz} - \varepsilon_{yz}^2) - sin^2\theta\ (\varepsilon_{xy}^2 + \varepsilon_{xx}\varepsilon_{yy}).$$

It can be shown that $|E_\parallel| \gg |E_\perp|$ provided $A = 0$, where the symbols $\parallel$ and $\perp$ refer to the direction of the static magnetic field. In this sense of approximation, the electrostatic dispersion relation in a magnetized plasma reads $A = 0$. It is to be noted that the purely longitudinal wave in the strict sense ($|E_\perp| = 0$) exists only when $\theta = 0$.

Exercise. Using the tensor elements of ε_{ij}, show that Eq. (6.20) is indeed the electrostatic dielectric constant because it is equal to $\varepsilon_L = k_i k_j \varepsilon_{ij}/k^2$.

$$\varepsilon_L = \frac{1}{k^2}(k_x^2\varepsilon_{xx} + 2k_x k_z \varepsilon_{xz} + k_z^2 \varepsilon_{zz})$$
$$= 1 + \sum_\alpha \frac{\omega_{p\alpha}^2}{k^2\omega^2} \sum_n \int \frac{d^3v}{\omega - k_\parallel v_\parallel - n\omega_{c\alpha}}$$
$$\times \left(\frac{n^2\omega_{c\alpha}^2}{v_\perp} J_n^2\ U + \frac{2k_\parallel v_\parallel n\omega_{c\alpha}}{v_\perp} J_n^2\ U + k_\parallel^2 v_\parallel J_n^2\ W \right)$$

which reduces to (6.20).

6.2.3. *Alternative integration*

We have the Vlasov equation (6.40) which we write here again

$$\frac{\partial}{\partial t} f_\alpha(\mathbf{r}, \mathbf{v}, t) + \mathbf{v} \cdot \frac{\partial f_\alpha}{\partial \mathbf{r}} + \frac{e_\alpha}{m_\alpha} \frac{\mathbf{v} \times \mathbf{B_0}}{c} \cdot \frac{\partial f_\alpha}{\partial \mathbf{v}} = -\frac{e_\alpha}{m_\alpha}$$
$$\times \left(\mathbf{E}(\mathbf{r}, t) + \frac{1}{c}\mathbf{v} \times \mathbf{B}(\mathbf{r}, t) \right) \cdot \frac{\partial f_{\alpha 0}}{\partial \mathbf{v}} \tag{6.99}$$

which we Fourier-transform with respect to the variables $\mathbf{r}$ and t. Putting $f_\alpha \sim e^{i\mathbf{k}\cdot\mathbf{r} - i\omega t}$, we have

$$(-i\omega + i\mathbf{k} \cdot \mathbf{v}) f(\mathbf{k}, \mathbf{v}, \omega) + \omega_c \mathbf{v} \times \hat{\mathbf{z}} \cdot \frac{\partial f}{\partial \mathbf{v}} = -\frac{e}{m}$$
$$\times \left(\mathbf{E}(\mathbf{k}, \omega) + \frac{1}{c}\mathbf{v} \times \mathbf{B}(\mathbf{k}, \omega) \right) \cdot \frac{\partial f_0}{\partial \mathbf{v}} \tag{6.100}$$

where we omitted the subscript α denoting the species, which will be reinstated at the final result. Introducing the cylindrical coordinates in the velocity space

$$\mathbf{v} = \hat{\mathbf{v}}_\perp v_\perp + \hat{\mathbf{z}} v_z = \hat{\mathbf{x}} v_\perp cos\varphi + \hat{\mathbf{y}} v_\perp sin\varphi + \hat{\mathbf{z}} v_z,$$

where z-direction is the direction of the static magnetic field $\mathbf{B}_0$, we have

$$\frac{\partial f}{\partial \mathbf{v}} = \hat{\mathbf{v}}_\perp \frac{\partial f}{\partial v_\perp} + \frac{\hat{\varphi}}{v_\perp}\frac{\partial f}{\partial \varphi} + \hat{\mathbf{z}} \frac{\partial f}{\partial v_z}.$$

Then (6.100) takes the form

$$\frac{\partial f}{\partial \varphi} + \frac{i(\omega - \mathbf{k}\cdot\mathbf{v})}{\omega_c} f = \frac{e}{m\omega_c}\left(\mathbf{E}(\mathbf{k},\omega) + \frac{1}{c}\mathbf{v}\times\mathbf{B}(\mathbf{k},\omega)\right)\cdot\frac{\partial f_0}{\partial \mathbf{v}}. \tag{6.101}$$

This equation is an ordinary first order differential equation with respect to the angular coordinate φ. Without loss of generality, we set $\mathbf{k} = \hat{\mathbf{x}} k_x + \hat{\mathbf{z}} k_z$. Multiplying the integration factor, the above equation can be written as

$$\frac{d}{d\varphi}\left(f_\alpha\, e^{-i\Psi_\alpha(\varphi)}\right) = \frac{e_\alpha}{m_\alpha \omega_{c\alpha}} e^{-i\Psi_\alpha(\varphi)}\, Q_\alpha(\varphi) \tag{6.102}$$

$$\Psi_\alpha(\varphi) = \frac{-(\omega - k_z v_z)\varphi + k_x v_\perp\ sin\varphi}{\omega_{c\alpha}}$$

$$Q_\alpha(\varphi) = \left(\mathbf{E}(\mathbf{k},\omega) + \frac{1}{c}\mathbf{v}\times\mathbf{B}(\mathbf{k},\omega)\right)\cdot\frac{\partial f_{\alpha 0}}{\partial \mathbf{v}}.$$

Integrating (6.102) from $\varphi = \varphi_0$ to φ gives

$$\left[f_\alpha\, e^{-i\Psi_\alpha(\varphi)}\right]_{\varphi_0}^{\varphi} = \frac{e_\alpha}{m_\alpha\omega_{c\alpha}} \int_{\varphi_0}^{\varphi} e^{-i\Psi_\alpha(\varphi')}\, Q_\alpha(\varphi')d\varphi'. \tag{6.103}$$

The exponential term $exp(-i\Psi_\alpha(\varphi))$ vanishes at $\varphi = +\infty$ $(\varphi = -\infty)$ when $\omega_{c\alpha} > 0$ $(\omega_{c\alpha} < 0)$ if we assume ω has a small positive imaginary part (causality; as we have done throughout this book). Thus we obtain

$$f_\alpha(\varphi) = \frac{e_\alpha}{m_\alpha\omega_{c\alpha}} e^{i\Psi_\alpha(\varphi)} \int_{\pm\infty}^{\varphi} e^{-i\Psi_\alpha(\varphi')} Q_\alpha(\varphi')d\varphi' \tag{6.104}$$

where $+\infty(-\infty)$ corresponds to ion (electron) equation.

Excercise. Use (6.104) for electrostatic wave to show that the result agrees with (6.19).

6.2.4. *Cold plasma limit of the kinetic dielectric tensor*

Since the Larmor radius $v_\perp/\omega_{c\alpha} \to 0$, in the cold plasma, we first should take the limit $a_\alpha \to 0$ in the matrix S_{ij}. Use can be made of: $J_n(a) \to \delta_{n,0}$, $\frac{n}{a}J_n(a) = \frac{1}{2}[J_{n+1}(a) + J_{n-1}(a)] \to \frac{1}{2}(\delta_{n,1} + \delta_{n,-1})$, $J_n'(a) = \frac{1}{2}[J_{n-1}(a) - J_{n+1}(a)] \to \frac{1}{2}(\delta_{n,1} - \delta_{n,-1})$, $[\frac{nJ_n(a)}{a}]^2 \to \frac{1}{4}(\delta_{n,1} + \delta_{n,-1})$, $\frac{n}{a}J_n(a)J_n'(a) \to \frac{1}{4}(\delta_{n,1} - \delta_{n,-1})$, $\frac{n}{a}J_n(a)J_n(a) \to 0$, $J_n(a)J_n'(a) \to 0$.

Thus we have

$$\mathbf{S_{ij}} = \begin{pmatrix} v_\perp U\ \frac{1}{4}(\delta_{n,1} + \delta_{n,-1}) & iv_\perp U\ \frac{1}{4}(\delta_{n,1} - \delta_{n,-1}) & 0 \\ iv_\perp U\ \frac{1}{4}(\delta_{n,-1} - \delta_{n,1}) & v_\perp U\ \frac{1}{4}(\delta_{n,-1} + \delta_{n,1}) & 0 \\ 0 & 0 & v_\parallel W\delta_{n,0} \end{pmatrix},$$

$$U(v_\perp, v_\parallel) = \left(\omega - k_\parallel v_\parallel\right)\frac{\partial f_{\alpha 0}}{\partial v_\perp} + k_\parallel v_\perp \frac{\partial f_{\alpha 0}}{\partial v_\parallel},$$

$$W(v_\perp, v_\parallel) = (\omega - n\omega_{c\alpha})\frac{\partial f_{\alpha 0}}{\partial v_\parallel} + n\omega_{c\alpha}\frac{v_\parallel}{v_\perp}\frac{\partial f_{\alpha 0}}{\partial v_\perp}.$$

This form of S_{ij} is also valid for the parallel propagation, $\mathbf{k} = k\mathbf{B}_0/B_0$. The operations of taking the limit $a \to 0$ and $\int d^3v$ with the delta function distribution are commutative. Next the tensor elements ε_{ij} are evaluated by using the delta function distribution in cylindrical coordinates,

$$f_0(\mathbf{v}) = \delta(\mathbf{v}) = \frac{1}{v_\perp}\delta(v_\perp)\delta(\varphi)\delta(v_\parallel).$$

Using the above equations in (6.96) gives

$$\varepsilon_{11} = 1 + \frac{1}{4}\Sigma\,\frac{\omega_{p\alpha}^2}{\omega^2}\int d^3v\left[(\omega - k_\parallel v_\parallel)\frac{\partial f_{0\alpha}}{\partial v_\perp} + k_\parallel v_\perp\frac{\partial f_{0\alpha}}{\partial v_\parallel}\right]$$
$$\times\left(\frac{1}{\omega - k_\parallel v_\parallel - \omega_{c\alpha}} + \frac{1}{\omega - k_\parallel v_\parallel + \omega_{c\alpha}}\right)$$

where $d^3v = v_\perp dv_\perp dv_\parallel d\varphi$. The second integral vanishes. Integrating by parts in the first integral, one obtains

$$\varepsilon_{\mathbf{ij}} = \begin{pmatrix} 1 - \Sigma\frac{\omega_{p\alpha}^2}{\omega^2 - \omega_{c\alpha}^2} & -i\,\Sigma\,\frac{\omega_{c\alpha}}{\omega}\,\frac{\omega_{p\alpha}^2}{\omega^2 - \omega_{c\alpha}^2} & 0 \\ i\,\Sigma\,\frac{\omega_{c\alpha}}{\omega}\,\frac{\omega_{p\alpha}^2}{\omega^2 - \omega_{c\alpha}^2} & 1 - \Sigma\frac{\omega_{p\alpha}^2}{\omega^2 - \omega_{c\alpha}^2} & 0 \\ 0 & 0 & 1 - \Sigma\,\frac{\omega_{p\alpha}^2}{\omega^2} \end{pmatrix}.$$

6.2.5. *Symmetry of dielectric tensor: Onsager's principle*

Onsager's principle states as follows: the dielectric tensor $\varepsilon_{ij}(\omega, \mathbf{k})$ in an isotropic plasma from which a magnetic field is absent has symmetry such that

$$\varepsilon_{ij}(-\omega, -\mathbf{k}) = \varepsilon^*_{ij}(\omega, \mathbf{k}) \quad (* \textit{ denotes the complex conjugate}). \tag{6.105}$$

Onsager's principle asserts that $\varepsilon_{ij}(\omega, \mathbf{k}, \mathbf{B})$ in the presence of a magnetic field has the following symmetry

$$\varepsilon_{ij}(\omega, \mathbf{k}, \mathbf{B}) = \varepsilon_{ji}(\omega, -\mathbf{k}, -\mathbf{B}). \tag{6.106}$$

That is, the dielectric tensor is symmetrical under the simultaneous change of the signs of the magnetic field and the wave vector. If the plasma is invariant under spatial inversion, we have

$$\varepsilon_{ij}(\omega, \mathbf{k}, \mathbf{B}) = \varepsilon_{ji}(\omega, \mathbf{k}, -\mathbf{B}). \tag{6.107}$$

6.3. Energy of electromagnetic wave in plasma

References: L. D. Landau and E. M. Lifshitz, *Electrodynamics of continuous media*, Pergamon Press Oxford (1966) § 61; G. Bekefi, *Radiation processes in plasmas*, Wiley New York (1966); B. B. Kadomtsev, *Plasma turbulence*, Academic Press London (1965)

Plasmas are dispersive media whose dielectric permittivity $\varepsilon(\mathbf{k}, \omega)$ depends on the wave frequency and the wave vector. Dispersion of a wave in general accompanies a dissipation of energy, which means that a dispersive medium absorbs energy from the wave. The imaginary part of $\varepsilon(\mathbf{k}, \omega)$ determines the absorption of the wave energy by the medium. Here we recall that the real and the imaginary parts of $\varepsilon(\mathbf{k}, \omega)$ are connected by the Kramers–Kronig relations. Thus, the relations may be useful in calculating the energy dissipation. We have the Poynting theorem ((6.110) below)

$$-\nabla \cdot \frac{c}{4\pi}(\mathbf{E} \times \mathbf{B}) = \frac{1}{4\pi}\left[\mathbf{E} \cdot \frac{\partial \mathbf{D}}{\partial t} + \mathbf{B} \cdot \frac{\partial \mathbf{B}}{\partial t}\right].$$

By averaging with respect to time, the RHS yield the energy production rate in the medium, i.e. heat generated per unit time and volume. For a monochromatic wave of single frequency, this time average vanishes, and the entropy is conserved. Here we quote Landau and Lifshitz (1966): "the strict periodicity results in no steady accumulation of electromagnetic energy." To have correspondence with real irreversible process, we must consider a multi-chromatic wave or a wave packet. We consider a simple form of quasi-chromatic wave with a slowly changing amplitude in the form (6.114) below. Mathematically, we have a two-time scale problem in calculating the electromagnetic field energy in plasma.

6.3.1. *Field energy of a dispersive wave*

In this section, we consider how an external force (currents) changes the electromagnetic energy of a plasma. To calculate the work done by the external current, we write

$$\nabla \times \mathbf{B} = \frac{1}{c}\frac{\partial \mathbf{D}}{\partial t} + \frac{4\pi}{c}\mathbf{J}_{ext} \quad (cf.\ Eq.\ (4.37)), \tag{6.108}$$

$$\nabla \times \mathbf{E} = -\frac{1}{c}\frac{\partial \mathbf{B}}{\partial t}. \tag{6.109}$$

The scalar product of (6.108) with $\mathbf{E}$ gives the work performed by the external current. Also forming the divergence of the Poynting vector $\frac{c}{4\pi}\mathbf{E}\times\mathbf{B}$, we obtain

$$\frac{\partial W}{\partial t} \equiv \frac{1}{4\pi}\left[\mathbf{E}\cdot\frac{\partial \mathbf{D}}{\partial t} + \mathbf{B}\cdot\frac{\partial \mathbf{B}}{\partial t}\right] = -\nabla\cdot\frac{c}{4\pi}(\mathbf{E}\times\mathbf{B}) - \mathbf{E}\cdot\mathbf{J}_{ext}. \tag{6.110}$$

The above equation defines the *field energy* W. When $\mathbf{J}_{ext} = 0$, $\int W d^3r = const$, that is, the total field energy W contained in the whole space does not change with time. Note that the $\mathbf{E}\cdot\frac{\partial \mathbf{D}}{\partial t}$ term is the sum of the electric wave energy plus the particle kinetic energy that enters through the polarization current. This quantity has a meaning only when it is averaged over the period of the wave.

For simplicity we shall assume that the response vector $\mathbf{D}$ is directed along the direction of the input field $\mathbf{E}$. So we consider

$$D(t) = E(t) + \int_0^\infty \chi(\tau)E(t-\tau)d\tau. \tag{6.111}$$

Fourier-transforming gives

$$D(\omega) = \left(1 + \int_0^\infty dt\chi(t)e^{i\omega t}\right)E(\omega) \equiv \varepsilon(\omega)E(\omega). \tag{6.112}$$

From the knowledge of the dielectric permeability $\varepsilon(\omega)$, we can determine $D(t)$ by inverting the above equation. Differentiating $D(t) = \int D(\omega)e^{-i\omega t}d\omega/2\pi$ gives

$$\frac{\partial D}{\partial t} = \frac{-i}{2\pi}\int_{-\infty}^{\infty} \omega\ \varepsilon(\omega)E(\omega)e^{-i\omega t}d\omega. \tag{6.113}$$

Here we assume that $E(t)$ in (6.111) is quasi-monochromatic, i.e. it oscillates with a certain frequency ω_0 with a slowly changing amplitude:

$$E(t) = E_0(\nu t)e^{-i\omega_0 t} \tag{6.114}$$

where νt indicates the slow time with $\nu \ll \omega_0$. It should be emphasized that $E_0(\nu t)$ varies much more slowly than the phase $e^{-i\omega_0 t}$ ($\frac{\partial E_0}{\partial t}\frac{1}{E_0} \ll \omega_0$). The slow modulation of the amplitude occurs in reality due to interaction between waves or due to change of the medium property.

If E_0 is a strict constant, we have

$$E(\omega) = 2\pi E_0 \delta(\omega - \omega_0).$$

Thus (6.112) gives $D(\omega) = 2\pi E_0 \delta(\omega - \omega_0)\varepsilon(\omega_0)$. In this case, the response function $D(\omega)$ is contributed by $\varepsilon(\omega_0)$, solely at the frequency $\omega = \omega_0$, sharp. However, for a slowly modulated electric field, $D(\omega)$ is expected to be contributed by the values of $\varepsilon(\omega)$ over some narrow range of ω centered about ω_0. Let us put the slowly modulated amplitude as

$$E_0(\nu t) = E_0 e^{i\nu t}$$

where E_0 is constant. Then we have

$$E(\omega) = E_0 \int_{-\infty}^{\infty} e^{it(\omega - \omega_0 + \nu)} dt. \tag{6.115}$$

Therefore

$$\begin{aligned}\frac{\partial D}{\partial t} &= -i\,\frac{E_0}{2\pi}\int_{-\infty}^{\infty} d\omega\; \omega\; \varepsilon(\omega) e^{-i\omega t} \int_{-\infty}^{\infty} dt' e^{it'(\omega - \omega_0 + \nu)} \\ &= -\frac{E_0}{2\pi}\int_{-\infty}^{\infty} d\omega\; \varepsilon(\omega)\; e^{-i\omega t} \int_{-\infty}^{\infty} dt' e^{it'(-\omega_0+\nu)} \frac{\partial}{\partial t'} e^{i\omega t'}.\end{aligned}$$

Integrating by parts gives

$$\begin{aligned}\frac{\partial D}{\partial t} &= iE_0\,(-\omega_0 + \nu)\int_{-\infty}^{\infty} d\omega \varepsilon(\omega)\; e^{-i\omega t}\delta(\omega - \omega_0 + \nu) \\ &= iE_0\,(-\omega_0 + \nu)\;\varepsilon(\omega_0 - \nu)\; e^{-i(\omega_0 - \nu)t}.\end{aligned} \tag{6.116}$$

Expand $\varepsilon(\omega_0 - \nu) = \varepsilon(\omega_0) - \nu\,\frac{\partial \varepsilon}{\partial \omega_0} + \cdots$, then (6.116) can be arranged into the form

$$\frac{\partial D}{\partial t} = -i\omega_0 \varepsilon(\omega_0) E_0(t) e^{-i\omega_0 t} + e^{-i\omega_0 t}\,\frac{\partial}{\partial \omega_0}(\omega_0 \varepsilon(\omega_0))\,\frac{\partial}{\partial t} E_0(t) \tag{6.117}$$

where we put $i\nu E_0 e^{i\nu t} = \partial E_0(t)/\partial t$, and neglected ν^2–term, and $\nu \ll \omega_0$.

Equation (6.117) can be interpreted operationally as follows. In the equation, $D(\omega) = \varepsilon(\omega)E(\omega)$, ω contains Fourier frequency ω_0 and a slow time variation $\partial/\partial t_0$, i.e.

$$\omega = \omega_0 + i\,\frac{\partial}{\partial t_0} \quad \textit{with} \quad \omega_0 \gg \frac{\partial}{\partial t_0} \tag{6.118}$$

where the slow time variable t_0 is the time variable of the amplitude E_0. Then we have

$$\varepsilon(\omega) = \varepsilon\left(\omega_0 + i\frac{\partial}{\partial t_0}\right) = \varepsilon(\omega_0) + \frac{\partial \varepsilon}{\partial \omega_0} i \frac{\partial}{\partial t_0}$$

$$\begin{aligned}\frac{\partial D}{\partial t} &= -i\omega D(\omega) = \left(-i\omega_0 + \frac{\partial}{\partial t_0}\right)\left(\varepsilon(\omega_0) + i\frac{\partial \varepsilon}{\partial \omega_0}\frac{\partial}{\partial t_0}\right) E_0(t_0) e^{-i\omega_0 t} \\ &= e^{-i\omega_0 t}\left(-i\omega_0\varepsilon(\omega_0) + \omega_0 \frac{\partial \varepsilon}{\partial \omega_0}\frac{\partial}{\partial t_0} + \varepsilon(\omega_0)\frac{\partial}{\partial t_0}\right) E_0(t_0)\end{aligned} \tag{6.119}$$

which agrees with (6.117). In (6.119), we neglected $\partial^2/\partial t_0^2$.

For a quasi-monochromatic plane wave, we now calculate the time average (denoted by $<\cdots>$) of $\mathbf{E}\cdot\frac{\partial \mathbf{D}}{\partial t}$ in (6.110).

$$\begin{aligned}\mathbf{E}\cdot\frac{\partial \mathbf{D}}{\partial t} &\to \frac{1}{4}(\mathbf{E}+\mathbf{E}^*)\cdot\left(\frac{\partial \mathbf{D}}{\partial t} + \frac{\partial \mathbf{D}^*}{\partial t}\right) \\ &= \frac{1}{4}\left(\mathbf{E}\cdot\frac{\partial \mathbf{D}}{\partial t} + \mathbf{E}^*\cdot\frac{\partial \mathbf{D}^*}{\partial t} + \mathbf{E}^*\cdot\frac{\partial \mathbf{D}}{\partial t} + \mathbf{E}\cdot\frac{\partial \mathbf{D}^*}{\partial t}\right).\end{aligned}$$

The first two terms in the above equation vanish upon time-averaging, since they vary as $e^{\pm 2i\omega_0 t}$. Using (6.114) and (6.117) yields for the electric energy (W_E)

$$\begin{aligned}\frac{d\,W_E}{dt} &= \frac{1}{4\pi}\left\langle \mathbf{E}\cdot\frac{\partial \mathbf{D}}{\partial t}\right\rangle = \frac{1}{16\pi}\left[\frac{\partial}{\partial \omega}(\omega\varepsilon(\omega))\right]_{\omega=\omega_0}\left(\mathbf{E}_0\cdot\frac{\partial \mathbf{E}_0^*}{\partial t} + \mathbf{E}_0^*\cdot\frac{\partial \mathbf{E}_0}{\partial t}\right) \\ &= \frac{1}{16\pi}\left[\frac{\partial}{\partial \omega}(\omega\varepsilon(\omega))\right]_{\omega=\omega_0}\frac{\partial}{\partial t}|E_0|^2 \\ &= \frac{1}{16\pi}\left[\frac{\partial}{\partial \omega}(\omega\varepsilon(\omega))\right]_{\omega=\omega_0}\frac{\partial}{\partial t}|E|^2.\end{aligned} \tag{6.120}$$

In obtaining (6.120), we assumed $\varepsilon(\omega)$ is real (*transparent region*). It is seen that only the terms involving $\frac{\partial}{\partial\omega}(\omega\varepsilon)$ survive the time-averaging. From the above equation, it is natural to define the electric energy as

$$W_E = \frac{1}{16\pi}\left[\frac{\partial}{\partial\omega}(\omega\varepsilon(\omega))\right]_{\omega=\omega_0}|E|^2. \tag{6.121}$$

The magnetic field energy is simply

$$W_B = \frac{1}{16\pi}|B|^2 \tag{6.122}$$

so that the field energy W in (6.110) is $W = W_E + W_B$. In W_E, the particle kinetic energy is included through the polarization current.

In general, the displacement vector is connected to the electric field by the dielectric tensor: $D_i(\omega) = \varepsilon_{ij}(\omega)E_j(\omega)$ (neglecting the spatial dispersion). In this case, we have

$$\begin{aligned}\frac{\partial D_i}{\partial t} &= -i\omega D_i(\omega) = \left(-i\omega_0 + \frac{\partial}{\partial t_0}\right)\left[\varepsilon_{ij}(\omega_0) + i\,\frac{\partial\varepsilon_{ij}}{\partial\omega_0}\,\frac{\partial}{\partial t_0}\right]E_{j0}(t_0)e^{-i\omega_0 t}\\ &= -i\omega_0\varepsilon_{ij}(\omega_0)E_{j0}(t_0)e^{-i\omega_0 t} + \left[\frac{\partial}{\partial\omega}(\omega\varepsilon_{ij}(\omega))\right]_{\omega=\omega_0}\frac{\partial E_{j0}}{\partial t_0}e^{-i\omega_0 t}.\end{aligned} \tag{6.123}$$

Thus we have

$$\begin{aligned}\left\langle \mathbf{E}\cdot\frac{\partial\mathbf{D}}{\partial t}\right\rangle &= \frac{1}{4}\,i\,\omega_0\varepsilon^*_{ij}(\omega_0)E_{i0}E^*_{j0} - \frac{1}{4}\,i\,\omega_0\varepsilon_{ij}(\omega_0)E^*_{i0}E_{j0}\\ &\quad + \frac{1}{4}\left[\frac{\partial}{\partial\omega}(\omega\varepsilon_{ij}(\omega))\right]_{\omega=\omega_0}E^*_{i0}\frac{\partial E_{j0}}{\partial t_0}\\ &\quad + \frac{1}{4}\left[\frac{\partial}{\partial\omega}(\omega\varepsilon^*_{ij}(\omega))\right]_{\omega=\omega_0}E_{i0}\frac{\partial E^*_{j0}}{\partial t_0}.\end{aligned} \tag{6.124}$$

• **Hermitian and Anti-Hermitian Tensor.** For further development of (6.124), the dielectric tensor, which is not real in general, is separated into non-dissipative and dissipative parts; they are respectively termed Hermitian (H) and anti-Hermitian (A) part. The anti-Hermitian part is the relevant generalization of the imaginary part in tensors or matrices. They are defined as

$$\varepsilon^H_{ij} = \frac{1}{2}(\varepsilon_{ij} + \varepsilon^*_{ji}), \quad \varepsilon^A_{ij} = \frac{1}{2i}(\varepsilon_{ij} - \varepsilon^*_{ji}). \tag{6.125}$$

Note that ε_{ij}^H, ε_{ij}^A, themselves are complex quantities in general. From the above definition, we can derive

$$\varepsilon_{ij} = \varepsilon_{ij}^H + i\,\varepsilon_{ij}^A, \quad \varepsilon_{ji}^* = \varepsilon_{ij}^H - i\,\varepsilon_{ij}^A. \tag{6.126}$$

The matrix ε_{ji}^*, the *adjoint matrix* of ε_{ij}, is obtained by taking the complex conjugate after transposing ε_{ij}. If $\varepsilon_{ji}^* = \varepsilon_{ij}$, ε_{ij} is self-adjoint or the medium is *loss-free* or *non-dissipative.* As can be seen from (6.126), the dielectric tensor is loss free if $\varepsilon_{ij}^H \gg \varepsilon_{ij}^A$. From (6.125) one can show that

$$\varepsilon_{ij}^H = \varepsilon_{ji}^{H*}, \quad \varepsilon_{ij}^A = \varepsilon_{ji}^{A*}.$$

That is, ε_{ij}^H, ε_{ij}^A themselves are self-adjoint but ε_{ij} is not self-adjoint in general. In **transparent region** of $(\omega, \mathbf{k})$, $\varepsilon_{ij}^H \gg \varepsilon_{ij}^A$ (the medium is loss-free), the waves are only weakly damped. Dissipation of waves arises from the anti-Hermitian part ε_{ij}^A or the imaginary frequency ω_i. In the transparent region, the imaginary part of the dispersion relation determinant (Eq. (4.31)) is much less than the real part ($|Im\,\Lambda| \ll Re\,|\Lambda|$). Therefore, the dispersion relation is approximately written as

$$Re\ \Lambda(\omega_r, \mathbf{k}) = 0 \quad (in\ \ transparent\ \ region).$$

The damping rate can be obtained from the formula (4.83) in the form

$$damping\ rate = -\frac{Im\ \Lambda(\omega_r)}{\partial Re\ \Lambda(\omega_r)/\partial \omega_r}.$$

Going back to (6.124), the first two terms add to

$$\begin{aligned}\frac{i}{4}\omega_0 E_{j0} E_{i0}^* (\varepsilon_{ji}^* - \varepsilon_{ij}) &= \frac{i}{4}\omega_0 E_{j0} E_{i0}^* (-2i) \varepsilon_{ij}^A \\ &= \frac{\omega_0}{2} E_{i0} E_{jo}^* \varepsilon_{ji}^A \quad (per\ i \longleftrightarrow j).\end{aligned}$$

In adding up the last two terms in (6.124), we assume $\varepsilon_{ij}^H \gg \varepsilon_{ij}^A$. Also by interchanging the dummy indices, $i \longleftrightarrow j$, when necessary, one obtains for the sum of the last two terms in (6.124)

$$\frac{1}{4}\frac{\partial}{\partial t}\left[E_{j0}^* \frac{\partial}{\partial \omega}(\omega \varepsilon_{ji}^H) E_{i0}\right].$$

Then averaging (6.110) leads to the energy equation

$$\frac{dW}{dt} + \left\langle \nabla \cdot \frac{c}{4\pi}\mathbf{E} \times \mathbf{B} \right\rangle = -\frac{\omega_0}{8\pi} E_i E_j^* \varepsilon_{ji}^A \tag{6.127}$$

where the wave energy W is defined by

$$W = \frac{1}{16\pi}\left(|B|^2 + E_j^* \left[\frac{\partial}{\partial\omega}(\omega\varepsilon_{ji}^H(\omega))\right]_{\omega=\omega_0} E_i\right). \tag{6.128}$$

Note that W consists of three components: the magnetic field energy $|B|^2$, the electric field energy and coherent (oscillating) particle kinetic energy which are both contained in the second term of W. Using $\varepsilon_{ij} = \delta_{ij} + \frac{4\pi i}{\omega}\sigma_{ij}$, we have $E_j^* \frac{\partial}{\partial\omega}(\omega\varepsilon_{ji}^H)E_i = E_j E_j^* + 4\pi i E_i E_j^* \sigma_{ij}^H$. The first term is the electric field energy while the second term is seen to be the particle mechanical energy. In (6.127), note that plasma wave energy W may either decrease or increase depending upon the anti-Hermitian part ε_{ji}^A. For a *negative energy wave*, W in (6.128) can be negative if the particles have some source of non-equilibrium kinetic energy in the unperturbed state (as in a beam).

Remark. For a stationary cold plasma, the dielectric constant is $\varepsilon(\omega) = 1 - \omega_p^2/\omega^2$ for plasma oscillation, and $\frac{\partial}{\partial\omega}(\omega\varepsilon(\omega)) \equiv U = 1 + \frac{\omega_p^2}{\omega^2} > 0$ (positive energy wave). If the plasma is moving with a drifting velocity v_0, the corresponding wave has

$$\varepsilon(\omega) = 1 - \frac{\omega_p^2}{(\omega - kv_0)^2}.$$

Show that $U < 0$ (negative energy wave) for $kv_0 > \omega_p$.

Exercise. In non-absorbing isotropic medium, use (4.22) in (6.128) for ε_{ji}^H to show that

$$W = \frac{1}{16\pi}\frac{1}{k^2}\left[|\mathbf{k}\cdot\mathbf{E}|^2 \frac{\partial}{\partial\omega}\omega\varepsilon_L(\omega,k) + |\mathbf{k}\times\mathbf{E}|^2 \frac{\partial}{\partial\omega}\omega\left(\varepsilon_T(\omega,k) - \frac{c^2k^2}{\omega^2}\right)\right]. \tag{6.129}$$

$$Hint: \ use \ \ |\mathbf{k}\times\mathbf{E}|^2 = k^2\,|\mathbf{E}|^2 - (\mathbf{k}\cdot\mathbf{E})(\mathbf{k}\cdot\mathbf{E}^*).$$

6.3.2. k-*dependance of* D, *spatial dispersion*

Electric displacement vector $\mathbf{D}(\mathbf{r},t)$ is defined by [see Eq. (4.52)]

$$\mathbf{D}(\mathbf{r},t) = \mathbf{E}(\mathbf{r},t) + 4\pi\int_{-\infty}^{t} \mathbf{J}(\mathbf{r},t')dt'. \tag{6.130}$$

This form of equation suggests the name *electric displacement*. When the medium is homogeneous in time and space, the conductivity tensor σ_{ij}

depends on $(t-t')$ and $(\mathbf{r}-\mathbf{r}')$, and consequently, the current is obtained by

$$J_i(\mathbf{r},t) = \int_{-\infty}^{t} dt' \int_{-\infty}^{\infty} d^3r' \sigma_{ij}(t-t', \mathbf{r}-\mathbf{r}') E_j(t', \mathbf{r}'). \tag{6.131}$$

This relation is called *material relation.* It would be instructive to show that the Fourier transform of (6.131) leads to

$$J_i(\mathbf{k},\omega) = \sigma_{ij}(\mathbf{k},\omega) E_j(\mathbf{k},\omega). \tag{6.132}$$

The proof goes as follows. Via change of variable, $t-t' = \tau$, (6.131) becomes

$$J_i(\mathbf{r},t) = \int_{0}^{\infty} d\tau \int_{-\infty}^{\infty} d^3r' \sigma_{ij}(\tau, \mathbf{r}-\mathbf{r}') E_j(t-\tau, \mathbf{r}'). \tag{6.133}$$

Here $\sigma_{ij}(t<0) = 0$: this is the causal requirement (see Chapter 4). So the $\int d\tau$–integral can be performed from $-\infty$ to $+\infty$,

$$\begin{aligned} J_i(\mathbf{r},t) &= \int_{-\infty}^{\infty} d\tau \int_{-\infty}^{\infty} d^3r' \sigma_{ij}(\tau, \mathbf{r}-\mathbf{r}') E_j(t-\tau, \mathbf{r}') \\ &= \int_{-\infty}^{\infty} d\tau \int_{-\infty}^{\infty} d^3r' \sigma_{ij}(t-\tau, \mathbf{r}-\mathbf{r}') E_j(\tau, \mathbf{r}'). \end{aligned}$$

The second line expression is valid because of the property of the convolution integral. The convolution integral is the inverse Fourier transform of the product of the two Fourier transformed functions:

$$J_i(\mathbf{r},t) = \left(\frac{1}{2\pi}\right)^4 \int e^{i\mathbf{k}\cdot\mathbf{r}-i\omega t} \sigma_{ij}(\mathbf{k},\omega) E_j(\mathbf{k},\omega) d^3k d\omega$$

from which (6.132) follows.

If the causal requirement $[\sigma_{ij}(t<0) = 0]$ is not used, the alternative proof of (6.132) can be done in the following way. By direct Fourier transform, we write

$$\begin{aligned} J_i(\mathbf{k},\omega) &= \int_{-\infty}^{\infty} d^3r e^{-i\mathbf{k}\cdot\mathbf{r}} \int_{-\infty}^{\infty} dt e^{i\omega t} \int_{-\infty}^{t} dt' \\ &\quad \times \int_{-\infty}^{\infty} d^3r' \sigma_{ij}(t-t', \mathbf{r}-\mathbf{r}') E_j(t', \mathbf{r}'). \end{aligned} \tag{6.134}$$

The $\int d^3r'$–integral is a straightforward convolution integral, so (6.134) can be written as

$$\begin{aligned} J_i(\mathbf{k},\omega) &= \int_{-\infty}^{\infty} d^3r e^{-i\mathbf{k}\cdot\mathbf{r}} \int_{-\infty}^{\infty} dt e^{i\omega t} \int_{-\infty}^{t} dt' \\ &\quad \times \left(\frac{1}{2\pi}\right)^3 \int_{-\infty}^{\infty} e^{i\mathbf{k}'\cdot\mathbf{r}} \sigma_{ij}(t-t', \mathbf{k}') E_j(t', \mathbf{k}') d^3k'. \end{aligned} \tag{6.135}$$

The $\int d^3r$–integral can be done to yield

$$\begin{aligned}
J_i(\mathbf{k},\omega) &= \int_{-\infty}^{\infty} dte^{i\omega t}\int_{-\infty}^{t} dt'\sigma_{ij}(t-t',\mathbf{k})E_j(t',\mathbf{k}) \\
&= \int_{-\infty}^{\infty} dte^{i\omega t}\int_{0}^{\infty} dt''\sigma_{ij}(t'',\mathbf{k})E_j(t-t'',\mathbf{k}) \\
&= \int_{-\infty}^{\infty} dte^{i\omega t}\int_{0}^{\infty} dt''\int_{-\infty}^{\infty}\sigma_{ij}(\omega',\mathbf{k})e^{-i\omega' t''}\frac{d\omega'}{2\pi} \\
&\quad\times\int_{-\infty}^{\infty} E_j(\omega'',\mathbf{k})e^{-i\omega''(t-t'')}\frac{d\omega''}{2\pi}. \qquad (6.136)
\end{aligned}$$

After $\int dt$–integral, we obtain

$$J_i(\mathbf{k},\omega) = \int_0^{\infty} dt''\int_{-\infty}^{\infty}\sigma_{ij}(\omega',\mathbf{k})e^{-i\omega' t''}\frac{d\omega'}{2\pi}E_j(\omega,\mathbf{k})e^{i\omega t''}. \qquad (6.137)$$

In the above, we have

$$\int_0^{\infty} dt'' e^{i(\omega-\omega')t''} = \frac{i}{\omega-\omega'}.$$

So (6.137) becomes

$$\begin{aligned}
J_i(\mathbf{k},\omega) &= E_j(\omega,\mathbf{k})\,\frac{i}{2\pi}\int_{-\infty}^{\infty}\sigma_{ij}(\omega',\mathbf{k})\frac{d\omega'}{\omega-\omega'} \\
&= E_j(\omega,\mathbf{k})\,\frac{i}{2\pi}\oint_C\sigma_{ij}(\omega',\mathbf{k})\frac{d\omega'}{\omega-\omega'}
\end{aligned}$$

where the closed contour is the infinite semicircle winding the upper half ω–plane. Because $\sigma_{ij}(\omega)$ is analytic in the upper half ω–plane, the contour integral gives the value $-2\pi i\sigma_{ij}(\omega)$, and thus we have

$$J_i(\mathbf{k},\omega) = E_j(\omega,\mathbf{k})\;\sigma_{ij}(\omega,\mathbf{k}).$$

The above two derivations of (6.132) imply that the causality condition is equivalent to the analyticity of $\sigma(\omega)$ in the upper ω-plane.

Next, let us consider the following expression that is to be used in (6.130):

$$J_i(\mathbf{r},t') = \left(\frac{1}{2\pi}\right)^4\int e^{i\mathbf{k}\cdot\mathbf{r}-i\omega t'}J_i(\mathbf{k},\omega)d^3k d\omega.$$

It is left as an exercise to show

$$4\pi\int_{-\infty}^{t} dt' J_i(\mathbf{r},t') = \int_{-\infty}^{\infty} d^3r'\int_{-\infty}^{\infty} dt'\chi_{ij}(t-t',\mathbf{r}-\mathbf{r}')E_j(t',\mathbf{r}') \qquad (6.138)$$

where $\chi_{ij}(\mathbf{k},\omega) = \frac{4\pi i}{\omega}\sigma_{ij}(\mathbf{k},\omega)$ and (6.132) would be useful. Therefore, (6.130) can be written as

$$D_i(\mathbf{r},t) = \int_{-\infty}^{\infty} d^3r' \int_{-\infty}^{\infty} dt' \varepsilon_{ij}(t-t',\mathbf{r}-\mathbf{r}')E_j(t',\mathbf{r}'), \tag{6.139}$$

$$\varepsilon_{ij}(t-t',\mathbf{r}-\mathbf{r}') = \delta_{ij}\delta(\mathbf{r}-\mathbf{r}')\delta(t-t') + \chi_{ij}(t-t',\mathbf{r}-\mathbf{r}'). \tag{6.140}$$

The above equation gives the Fourier transform of $\varepsilon_{ij}(\mathbf{r},t)$ in the form

$$\varepsilon_{ij}(\mathbf{k},\omega) = \delta_{ij} + \chi_{ij}(\mathbf{k},\omega) \tag{6.141}$$

because the Fourier transform of the delta function is unity. (6.139) is a convolution integral. Its Fourier transform is the product of each Fourier transform:

$$D_i(\mathbf{k},\omega) = \varepsilon_{ij}(\mathbf{k},\omega)E_j(\mathbf{k},\omega). \tag{6.142}$$

Exercise. Prove the following relations:

$$\varepsilon_{ij}^H(\omega,\mathbf{k}) = \varepsilon_{ji}^H(-\omega,-\mathbf{k}), \quad \varepsilon_{ij}^A(\omega,\mathbf{k}) = -\varepsilon_{ji}^A(-\omega,-\mathbf{k}),$$
$$\varepsilon_{ij}(-\omega,-\mathbf{k}) = \varepsilon_{ij}^*(\omega,\mathbf{k}). \tag{6.143}$$

6.3.3. *Field energy of electromagnetic wave with slow temporal and spatial modulation*

In this Section, we generalize the analysis of the field energy of almost monochromatic wave with a slow temporal modulation (§ 6.3.1) so as to obtain electromagnetic wave energy with spatial modulation included. The key quantity in the development is $\partial\mathbf{D}(\mathbf{r},t)/\partial t$. We consider an electric field:

$$\mathbf{E}(\mathbf{r},t) = \mathbf{E}_0(\mathbf{r},t)e^{i\mathbf{k}_0\cdot\mathbf{r}-i\omega_0 t} \tag{6.144}$$

where the amplitude $\mathbf{E}_0$ is a slow function of $\mathbf{r}$ and t. The analysis is facilitated by putting

$$\mathbf{E}_0(\mathbf{r},t) = \mathbf{A}e^{-i\mathbf{q}\cdot\mathbf{r}+i\nu t} \tag{6.145}$$

with $q \ll k_0$ and $\nu \ll \omega_0$ and $\mathbf{A}$ a constant vector. Then we have

$$\frac{\partial E_{0i}}{\partial t} = i\nu E_{0i}, \quad \frac{\partial E_{0i}}{\partial r_j} = -iq_j E_{0i}. \tag{6.146}$$

Fourier transforming (6.144) gives

$$\mathbf{E}(\mathbf{k},\omega) = \mathbf{A}\ (2\pi)^4\ \delta(\omega - \omega_0 + \nu)\ \delta(\mathbf{k} - \mathbf{k}_0 + \mathbf{q}). \tag{6.147}$$

We work with the expression [see (6.113)]

$$\frac{\partial D_i(\mathbf{r},t)}{\partial t} = -\frac{i}{(2\pi)^4}\int_{-\infty}^{\infty} d\omega \int_{-\infty}^{\infty} d^3k\ e^{i\mathbf{k}\cdot\mathbf{r}-i\omega t}\omega\ \varepsilon_{ij}(\omega,\mathbf{k})E_j(\omega,\mathbf{k}). \tag{6.148}$$

Using (6.147) in (6.148) gives

$$\begin{aligned}\frac{\partial D_i(\mathbf{r},t)}{\partial t} &= -i\ E_{0j}\ e^{i\mathbf{k}_0\cdot\mathbf{r}-i\omega_0 t}\ (\omega_0-\nu)\ \varepsilon_{ij}(\omega_0-\nu,\ \mathbf{k}_0-\mathbf{q})\\ &= -i\ E_{0j}\ e^{i\mathbf{k}_0\cdot\mathbf{r}-i\omega_0 t}\left[\omega_0\varepsilon_{ij}(0) - \nu\frac{\partial}{\partial\omega_0}(\omega_0\varepsilon_{ij}(0)) - q_l\omega_0\frac{\partial\varepsilon_{ij}(0)}{\partial k_{0l}}\right]\end{aligned} \tag{6.149}$$

where we expanded ε_{ij}, and the argument $(0) \equiv (\mathbf{k}_0,\ \omega_0)$. Using the relations (6.146) in the above gives

$$\begin{aligned}\frac{\partial D_i(\mathbf{r},t)}{\partial t} = \Bigg[&-i\omega_0\varepsilon_{ij}(\mathbf{k}_0,\omega_0)E_{0j} + \frac{\partial}{\partial\omega_0}(\omega_0\varepsilon_{ij}(\mathbf{k}_0,\omega_0))\frac{\partial E_{0j}}{\partial t}\\ &-\omega_0\frac{\partial\varepsilon_{ij}(\mathbf{k}_0,\omega_0)}{\partial k_{0l}}\frac{\partial E_{0j}}{\partial r_l}\Bigg] \times\ e^{i\mathbf{k}_0\cdot\mathbf{r}-i\omega_0 t}.\end{aligned} \tag{6.150}$$

Compare the above equation with (6.123). The last term in (6.150) is a new term derived from the spatial variation. With the inclusion of the spatial modulation, averaging (6.110) by the analogous way that led to (6.127) amends the energy equation in the form

$$\frac{dW}{dt} + \left\langle\nabla\cdot\frac{c}{4\pi}\mathbf{E}\times\mathbf{B}\right\rangle = -\frac{\omega_0}{8\pi}E_iE_j^*\varepsilon_{ji}^A - \frac{\omega_0}{8\pi}\frac{\partial E_i^*}{\partial r_l}\frac{\partial\varepsilon_{ij}^H}{\partial k_l}E_j. \tag{6.151}$$

$$W = \frac{1}{16\pi}\left(|B|^2 + E_j^*\left[\frac{\partial}{\partial\omega}(\omega\varepsilon_{ji}^H(\omega))\right]_{\omega=\omega_0}E_i\right). \tag{6.152}$$

The new term represents the mechanical energy flux due to particles flowing coherently with the wave.

6.3.4. *Compact form of wave energy equation*

• **Alternative energy equation.** We rederive (6.151) and (6.152) by the technique of *derivative expansion*, and cast them into a compact form. In

Fourier representation, Maxwell equations are

$$\mathbf{k} \times \mathbf{E} = \frac{\omega}{c}\,\mathbf{B}, \tag{6.153}$$

$$\mathbf{k} \times \mathbf{B} = -\frac{\omega}{c}\mathbf{D} = -\frac{\omega}{c}\,\varepsilon_{ij}E_j. \tag{6.154}$$

Here we consider the field components, $\mathbf{k}$, ω are all complex quantities: $\mathbf{k} = \mathbf{k}_r + i\mathbf{k}_i$, $\omega = \omega_r + i\omega_i$. Performing (6.153)$\cdot$ $\mathbf{B}^* +$ (6.154)$^* \cdot \mathbf{E}$ (* denotes the complex conjugate) yields

$$\mathbf{E} \cdot \mathbf{k}^* \times \mathbf{B}^* + \mathbf{B}^* \cdot \mathbf{k} \times \mathbf{E} = \frac{\omega}{c}B^2 - \frac{\omega^*}{c}E_i\varepsilon_{ij}^*E_j^*. \tag{6.155}$$

Taking the complex conjugate of the above equation gives

$$\mathbf{E}^* \cdot \mathbf{k} \times \mathbf{B} + \mathbf{B} \cdot \mathbf{k}^* \times \mathbf{E}^* = \frac{\omega^*}{c}B^2 - \frac{\omega}{c}E_i^*\varepsilon_{ij}E_j. \tag{6.156}$$

Subtracting (6.156) from (6.155) yields

$$2i\;\mathbf{k}_i \cdot (\mathbf{E} \times \mathbf{B}^* + \mathbf{E}^* \times \mathbf{B}) = 2i\omega_i B^2 + \frac{1}{c}E_i^*(\omega\varepsilon_{ij} - \omega^*\varepsilon_{ji}^*)E_j. \tag{6.157}$$

In the last term we interchanged the dummy indices. We assume $\omega_r \gg \omega_i$, $k_r \gg k_i$, and Taylor expand

$$\omega\varepsilon_{ij}(\omega, \mathbf{k}) = \omega_r\varepsilon_{ij}(\omega_r, \mathbf{k}_r) + i\;\mathbf{k}_i \cdot \frac{\partial}{\partial \mathbf{k}_r}(\omega_r\varepsilon_{ij}) + i\omega_i\frac{\partial}{\partial \omega_r}(\omega_r\varepsilon_{ij}) \tag{6.158}$$

where we understand that ω_r is independent of $\mathbf{k}_r$. Taking the complex conjugate of the above equation and interchanging the indices $i \leftrightarrow j$ yields

$$\omega^*\varepsilon_{ji}^*(\omega, \mathbf{k}) = \omega_r\varepsilon_{ji}^*(\omega_r, \mathbf{k}_r) - i\;\mathbf{k}_i \cdot \frac{\partial}{\partial \mathbf{k}_r}(\omega_r\varepsilon_{ji}^*) - i\omega_i\frac{\partial}{\partial \omega_r}(\omega_r\varepsilon_{ji}^*). \tag{6.159}$$

Subtraction gives, using (6.125),

$$\omega\varepsilon_{ij} - \omega^*\varepsilon_{ji}^* = 2i\omega_r\varepsilon_{ij}^A + 2i\mathbf{k}_i \cdot \frac{\partial}{\partial \mathbf{k}_r}(\omega_r\varepsilon_{ij}^H) + 2i\omega_i\frac{\partial}{\partial \omega_r}(\omega_r\varepsilon_{ij}^H). \tag{6.160}$$

Using (6.160) in (6.157) yields

$$\mathbf{k}_i \cdot \left[\mathbf{E} \times \mathbf{B}^* + \mathbf{E}^* \times \mathbf{B} - \frac{1}{c}E_i^*\frac{\partial}{\partial \mathbf{k}_r}(\omega_r\varepsilon_{ij}^H)E_j\right]$$
$$= \omega_i\left[B^2 + \frac{1}{c}E_i^*\frac{\partial}{\partial \omega_r}(\omega_r\varepsilon_{ij}^H)E_j\right] + \frac{1}{c}E_i^*\omega_r\varepsilon_{ij}^A E_j. \tag{6.161}$$

The physical meaning of each term of the energy equation (6.161) goes as follows. The first term on the LHS is the time-averaged energy flux (Poynting vector). The second term represents the non-electromagnetic energy flux due to particles flowing coherently with the wave. In a cold plasma, ε_{ij} is independent of $\mathbf{k}$, and this term vanishes. In a purely electrostatic wave, the electromagnetic energy flux (the first term) vanishes, and all the energy flow is contained in the second term. On the RHS of (6.161), the first term is just the time average of the magnetic field energy. The second term is the wave energy which is the sum of electric field energy and the oscillating particle kinetic energy. The last term in (6.161) represents the dissipation of the electric field energy.

(6.161) can be converted to the differential form of energy equation which applies to a single wave of the form $\mathbf{E}(\mathbf{r},t)e^{i\mathbf{k}\cdot\mathbf{r}-i\omega t}$ where the modulus is slowly varying as compared to the phase in $\mathbf{r}$ and t. We have the correspondence

$$\mathbf{k}_i \to -\frac{\partial}{\partial \mathbf{r}}, \quad \mathbf{k}_r \to \mathbf{k}, \quad \omega_i \to \frac{\partial}{\partial t}, \quad \omega_r \to \omega.$$

Then (6.161) takes the form

$$\frac{\partial W}{\partial t} + \omega E_i^* \varepsilon_{ij}^A E_j = c\,\frac{\partial}{\partial \mathbf{r}} \cdot (\mathbf{E}\times\mathbf{B}^* + \mathbf{E}^*\times\mathbf{B}) - \omega\frac{\partial}{\partial \mathbf{r}} \cdot E_i^* \frac{\partial \varepsilon_{ij}^H}{\partial \mathbf{k}} E_j, \tag{6.162}$$

$$\begin{aligned} W &= |\mathbf{B}|^2 + \frac{\partial}{\partial\omega}(\omega\varepsilon_{ij}^H)\, E_i^* E_j \\ &= \frac{c^2}{\omega^2}k^2|\mathbf{E}|^2 - \frac{c^2}{\omega^2}(\mathbf{k}\cdot\mathbf{E})(\mathbf{k}\cdot\mathbf{E}^*) + \frac{\partial}{\partial\omega}(\omega\varepsilon_{ij}^H)\, E_i^* E_j. \end{aligned}$$

Here we define:

$$G = \omega\varepsilon_{ij}^H E_i^* E_j - \frac{c^2}{\omega}k^2|\mathbf{E}|^2 + \frac{c^2}{\omega}(\mathbf{k}\cdot\mathbf{E})(\mathbf{k}\cdot\mathbf{E}^*); \quad W = \frac{\partial G}{\partial \omega}. \tag{6.163}$$

Using $\mathbf{B} = \frac{c}{\omega}\mathbf{k}\times\mathbf{E}$ in the RHS of (6.162), we obtain

$$\begin{aligned} RHS\ of\ (6.162) &= \frac{\partial}{\partial \mathbf{r}} \cdot \frac{\partial G}{\partial \mathbf{k}} \\ &= -\frac{\partial}{\partial \mathbf{r}} \cdot \left[\frac{c^2}{\omega}[-2|\mathbf{E}|^2\mathbf{k} + (\mathbf{k}\cdot\mathbf{E})\mathbf{E}^* + (\mathbf{k}\cdot\mathbf{E}^*)\mathbf{E}] + E_i^* E_j \frac{\partial\,(\omega\varepsilon_{ij}^H)}{\partial \mathbf{k}}\right]. \end{aligned}$$

Therefore (6.162) can be written as

$$\frac{\partial W}{\partial t} + \frac{\partial}{\partial \mathbf{r}} \cdot \mathbf{S} = -\omega \varepsilon_{ij}^{A} E_i^* E_j; \quad \mathbf{S} = \frac{\partial G}{\partial \mathbf{k}}$$

$$or \quad \frac{\partial}{\partial t}\frac{\partial G}{\partial \omega} + \frac{\partial}{\partial \mathbf{r}} \cdot \frac{\partial G}{\partial \mathbf{k}} = -\omega \varepsilon_{ij}^{A} E_i^* E_j. \tag{6.164}$$

[This form of energy equation would be obtained alternatively by calculating the Poynting vector term in (6.151).]

The wave equation in Fourier space in isotropic plasma can be written as

$$\left[\frac{c^2k^2}{\omega^2}\left(\frac{k_i k_j}{k^2} - \delta_{ij}\right) + \varepsilon_{ij}\right] E_j \equiv \Lambda_{ij}\; E_j = 0.$$

In the transparent region ($Re\ \varepsilon_{ij} \gg Im\ \varepsilon_{ij}$ or $\varepsilon_{ij}^{H} \gg \varepsilon_{ij}^{A}$), the quantity G in (6.163) is found to be

$$G = \Lambda_{ij} E_i^* E_j = \omega \varepsilon_{ij} E_i^* E_j - \frac{c^2}{\omega} k^2 |\mathbf{E}|^2 + \frac{c^2}{\omega}(\mathbf{k}\cdot\mathbf{E})(\mathbf{k}\cdot\mathbf{E}^*). \tag{6.165}$$

Let us introduce the polarization vector $\hat{\mathbf{a}}$:

$$\mathbf{E}(\mathbf{k},\omega) = \hat{\mathbf{a}}|\mathbf{E}(\mathbf{k},\omega)| = \hat{\mathbf{a}}E(\mathbf{k},\omega). \tag{6.166}$$

It is the unit vector directed along the direction of the electric field. In general, it is a complex quantity normalized so that $\hat{\mathbf{a}}\cdot\hat{\mathbf{a}}^* = 1$.

Using (6.166) in (6.165) gives

$$G = \left[\omega \varepsilon_{ij} a_i^* a_j - \frac{c^2}{\omega} k^2 + \frac{c^2}{\omega}|\mathbf{k}\cdot\hat{\mathbf{a}}|^2\right] |E|^2 \equiv G_1 |E|^2.$$

Now the RHS of (6.164) is $|E|^2$ times a small imaginary part of G_1 (in the transparent region). Thus, (6.164) can be written as

$$\frac{\partial I_k}{\partial t} + \frac{\partial \omega(\mathbf{k})}{\partial \mathbf{k}} \cdot \frac{\partial I_k}{\partial \mathbf{r}} = 2\gamma_k I_k, \tag{6.167}$$

$$I_k = |E_k|^2, \quad \frac{\partial \omega(\mathbf{k})}{\partial \mathbf{k}} = \frac{\frac{\partial G_1}{\partial \mathbf{k}}}{\frac{\partial G_1}{\partial \omega}}, \quad 2\gamma_k = -\frac{Im\ G_1}{\frac{\partial G_1}{\partial \omega}}. \tag{6.168}$$

where I_k is the spectral distribution of the electric field energy.

Reference: B. B. Kadomtev, *Cooperative effects in plasmas*, Reviews of plasma physics **22**, Kluer Axademic/Consultants Bureau New York (2001).

6.3.5. *Ponderomotive force*

A charged particle in a high frequency inhomogeneous electric field experiences, in addition to a high frequency oscillatory force, a zero frequency (d.c.) *ponderomotive force* (radiation pressure) which can be represented by $-\nabla\Phi$ (Φ is the ponderomotive force potential). It is a nonlinear force that survives the time average in a sinusoidal force field. Use of the ponderomotive force makes it easier to analyze parametric instabilities and wave-particle interactions in general. The ponderomotive force causes a slow variation of the density of the medium in which the high frequency electromagnetic wave is propagated (electric striction).

We consider an electron in external high frequency electric field with inhomogeneous amplitude $\mathbf{E}(\mathbf{r},t) = \mathbf{E}(\mathbf{r}) sin\ \omega_0 t$. The magnetic field accompanying this electric field is $\mathbf{B}(\mathbf{r},t) = \mathbf{B}(\mathbf{r}) cos\ \omega_0 t$, with $\mathbf{B}(\mathbf{r}) = \frac{c}{\omega_0}\nabla \times \mathbf{E}(\mathbf{r})$. In Eulerian description, the equation of motion reads

$$\frac{\partial}{\partial t}\mathbf{v}(\mathbf{r},t) + \mathbf{v}\cdot\nabla\mathbf{v} = \frac{e}{m}\left[\mathbf{E}(\mathbf{r},t) + \frac{1}{c}\mathbf{v}\times\mathbf{B}(\mathbf{r},t)\right]. \tag{6.169}$$

In the zero order linear approximation with respect to the external field $\mathbf{E}(\mathbf{r}) sin\ \omega_0 t$, we have

$$\frac{\partial}{\partial t}\mathbf{v}_0(\mathbf{r},t) = \frac{e}{m}\mathbf{E}(\mathbf{r}) sin\ \omega_0 t,$$

$$\mathbf{v}_0(\mathbf{r},t) = \frac{-e}{m\omega_0}\mathbf{E}(\mathbf{r}) cos\ \omega_0 t : \quad high\ frequency.$$

Note that the magnetic field term is of higher order ($\frac{v}{c} \ll 1$). In the order of $|\mathbf{E}(\mathbf{r})|^2$, we have

$$\frac{\partial}{\partial t}\mathbf{v}_1(\mathbf{r},t) = -\mathbf{v}_0\cdot\nabla\mathbf{v}_0 + \frac{e}{mc}\mathbf{v}_0\times\mathbf{B}(\mathbf{r}) cos\ \omega_0 t.$$

Here the nonlinear terms on RHS produce the zero frequency component. The nonlinear terms have rapidly changing phases, and only the average taken over a period has physical meaning. The ponderomotive force is the force still remaining after averaging:

$$\begin{aligned}\mathbf{F}_p &= \frac{\omega_0}{2\pi}\int_0^{\frac{2\pi}{\omega_0}} dt\ m\frac{\partial}{\partial t}\mathbf{v}_1(\mathbf{r},t) = \frac{-e^2}{2m\omega_0^2}(\mathbf{E}\cdot\nabla\mathbf{E} + \mathbf{E}\times\nabla\times\mathbf{E}) \\ &= \frac{-e^2}{4m\omega_0^2}\nabla E^2(\mathbf{r}).\end{aligned} \tag{6.170}$$

The steady d.c. ponderomotive force repels electrons and ions (regardless of the charge sign) from the region of a strong field to the region of weaker field.

It would be instructive to solve the same problem of finding the average force due to the inhomogeneous high frequency electric field by means of Lagrangian equation:

$$\ddot{\mathbf{r}} = \frac{e}{m}\left[\mathbf{E}(\mathbf{r}) sin\ \omega_0 t + \frac{1}{c}\ \dot{\mathbf{r}} \times \mathbf{B}(\mathbf{r}) cos\omega_0 t\right] \quad \left(\cdot\ = \frac{d}{dt}\right) \tag{6.171}$$

where $\mathbf{r}$ is the particle position. The oscillation starts at some position $\mathbf{r}_0$. This *oscillation center* slowly drifts (like a guiding center) due to the inhomogeneity of $\mathbf{E}(\mathbf{r})$. So we can write the particle position as

$$\mathbf{r}(t) = \mathbf{r}_0(t) + \mathbf{r}_1(t),$$

where $\mathbf{r}_1$ is the oscillating displacement vector of frequency ω_0 and the oscillation center $\mathbf{r}_0(t)$ is a slow function of t, much slower than ω_0. Assuming that the electric field does not carry the particle too far from the oscillation center in a period, we can expand the electric and magnetic fields about the oscillation center,

$$\mathbf{E}(\mathbf{r}) = \mathbf{E}(\mathbf{r}_0) + \mathbf{r}_1 \cdot \nabla_0 \mathbf{E}(\mathbf{r}_0) + \cdots,$$

$$\mathbf{B}(\mathbf{r}) = \mathbf{B}(\mathbf{r}_0) + \mathbf{r}_1 \cdot \nabla_0 \mathbf{B}(\mathbf{r}_0) + \cdots.$$

Then we have

$$\begin{aligned}\ddot{\mathbf{r}}_0 + \ddot{\mathbf{r}}_1 = \frac{e}{m}\Big[&\mathbf{E}(\mathbf{r}_0) sin\omega_0 t + \mathbf{r}_1 \cdot \nabla_0 \mathbf{E}(\mathbf{r}_0) sin\omega_0 t + \frac{1}{c}\dot{\mathbf{r}}_0 \times \mathbf{B}(\mathbf{r}_0) cos\omega_0 t \\ &+ \frac{1}{c}\dot{\mathbf{r}}_1 \times \mathbf{B}(\mathbf{r}_0) cos\omega_0 t + \frac{1}{c}\dot{\mathbf{r}}_0 \times (\mathbf{r}_1 \cdot \nabla_0)\mathbf{B}(\mathbf{r}_0) cos\omega_0 t \\ &+ \frac{1}{c}\dot{\mathbf{r}}_1 \times (\mathbf{r}_1 \cdot \nabla_0)\mathbf{B}(\mathbf{r}_0) cos\omega_0 t\Big].\end{aligned} \tag{6.172}$$

In the zero order, the two large terms are balanced:

$$\ddot{\mathbf{r}}_1 = \frac{e}{m}\mathbf{E}(\mathbf{r}_0) sin\ \omega_0 t, \quad \mathbf{r}_1 = \frac{-e}{m\omega_0^2}\mathbf{E}(\mathbf{r}_0) sin\omega_0 t.$$

Clearly the last two terms in (6.172) are of higher order and neglected. Taking the time average of the remaining terms, we obtain for the time average of $(m\ddot{\mathbf{r}}_0)$ the same result as (6.170). This ponderomotive force causes *electric striction* which gives rise to variation of the density of the medium when an intense high frequency electromagnetic wave propagates in it.

Equation (6.170) is the force acting on an electron, and the force density $(\mathbf{f}_p)$ in high frequency inhomogeneous electric field may be defined in terms of electron number density (n),

$$\mathbf{f}_p = n\mathbf{F}_p = -\frac{ne^2}{4m\omega^2}\nabla E^2 = -\frac{\omega_p^2}{\omega^2}\ \nabla \frac{E^2}{16\pi}$$

where $E^2/16\pi$ is recognized as the mean electric pressure.

The preceding derivation of ponderomotive force indicates that the origin of the ponderomotive force is the low frequency remnant resulted in by beating of two high frequency quantities in the terms of $\mathbf{v}\cdot\nabla\mathbf{v}$ and $\mathbf{v}\times\mathbf{B}$. The low frequency contribution is obtained from the average taken over the fast oscillations.

Note in the above that there are two distinctive time scales: the fast change $\omega_0\mathbf{v}_1$ and the slow change $\partial v_1/\partial t$. We show in the following that the ponderomotive force is a dynamical consequence of two distinctive time scales involved in the dynamics. Let us consider the electron equation of motion

$$\frac{\partial\mathbf{v}}{\partial t}+\nabla\,\frac{v^2}{2}-\mathbf{v}\times(\nabla\times\mathbf{v})=\frac{e}{m}(\mathbf{E}+\frac{1}{c}\mathbf{v}\times\mathbf{B}) \tag{6.173}$$

where we used the identity $\mathbf{v}\cdot\nabla\mathbf{v}=\nabla\,\frac{v^2}{2}-\mathbf{v}\times(\nabla\times\mathbf{v})$. Suppose that the plasma field consists of two oscillations of very different time scales: $\mathbf{v}=\mathbf{v}_f+\mathbf{v}_s$ where the subscripts f and s respectively mean the fast and slow fields. [To match with earlier analysis, the subscript f corresponds to 0 and s to 1.] We also have Maxwell's equation

$$\frac{\partial\mathbf{B}}{\partial t}=-c\nabla\times\mathbf{E}. \tag{6.174}$$

For the time derivative of the fast quantity $\mathbf{v}_f$, or $\frac{\partial\mathbf{v}_f}{\partial t}$, we can write

$$\frac{\partial\mathbf{v}_f}{\partial t}=\frac{e}{m}\mathbf{E}_f \tag{6.175}$$

because all the high frequency beat terms v_f^2, v_fB_f, generate only the second harmonics of high frequency or zero frequency, and do not contribute to $\frac{\partial\mathbf{v}_f}{\partial t}$ (in other words they are non-resonant). The above two equations give

$$\nabla\times\mathbf{v}_f=-\frac{e}{mc}\,\mathbf{B}_f. \tag{6.176}$$

The change of the slow field, $\partial\mathbf{v}_s/\partial t$, can be obtained by averaging (6.173) over many periods of fast oscillation; or by inspection, retaining the nonlinear beat terms of two fast oscillations. Since we have $-\mathbf{v}_f\times(\nabla\times\mathbf{v}_f)=\frac{e}{mc}\mathbf{v}_f\times\mathbf{B}_f$, we are left with

$$\frac{\partial\mathbf{v}_s}{\partial t}+\frac{1}{2}\nabla v_f^2=\frac{e}{m}\mathbf{E}_s. \tag{6.177}$$

When the external electric field is a high frequency wave $\sim sin\ \omega_0 t$ as in (6.169), put $\mathbf{E}_s=0$ and recover (6.170) upon averaging. Note that terms like v_fv_s, v_sv_s are non-resonant and contribute very little to $\frac{\partial\mathbf{v}_s}{\partial t}$. (6.177) is valid both for electromagnetic and electrostatic fields [if electrostatic, $\mathbf{B} = 0$ and $\nabla \times \mathbf{v} = 0$]. The ion ponderomotive force is smaller than the electron ponderomotive force by the factor of mass ratio m_e/m_i.

Exercise. In terms of the dielectric constant, $\varepsilon = 1 - \frac{\omega_p^2}{\omega^2}$, the force density of ponderomotive force is the average of the divergence of the Maxwell stress tensor:

$$f_k = \frac{\varepsilon - 1}{16\pi}\frac{\partial}{\partial x_k}\langle E^2\rangle = \frac{1}{8\pi}\frac{\partial}{\partial x_i}\left[\varepsilon\langle E_i E_k\rangle + \langle B_i B_k\rangle - \frac{\delta_{ik}}{2}(\langle E^2\rangle + \langle B^2\rangle)\right].$$

Hint: use $\nabla \times \mathbf{E} = -\frac{1}{c}\frac{\partial \mathbf{B}}{\partial t}$, $\nabla \times \mathbf{B} = \frac{\partial}{\partial t}(\varepsilon \mathbf{E})$, $\nabla \cdot \mathbf{B} = 0$, $\nabla \cdot (\varepsilon \mathbf{E}) = 0$.

6.4. Kinetic theory of magnetohydrodynamic waves

Reference: S. I. Braginskii and A. P. Kazantsev, *Magnetohydrodynamic waves in a rarefied plasma* (work done in 1957; the Journal of this article is unknown to the author.)

In this Section we derive MHD waves from the Vlasov equation. MHD waves look somewhat detached from the kinetic waves derived from the Vlasov–Maxwell equations in a magnetized plasma. In fact, MHD waves are a subset of the general kinetic waves whose frequency range is restricted to the low frequency $\omega \ll \omega_{ci}$, the ion cyclotron frequency. To appreciate this structure of plasma physics, it may be worthwhile to derive the MHD waves from the Vlasov equation. This can be done by calculating the low frequency limit in the result obtained in § 6.2; but the calculation involves power series expansion of the plasma dispersion function and the asymptotic expansion of Bessel functions, thus working with the double series renders the algebra quite cumbersome. Here we present Braginskii and Kazantsev's derivation which is based upon the principle of harmonic balance.

We begin with the linearized Vlasov equation governing a magnetized plasma

$$\frac{\partial}{\partial t}f_\alpha(\mathbf{r}, \mathbf{v}, t) + \mathbf{v} \cdot \frac{\partial f_\alpha}{\partial \mathbf{r}} + \frac{e_\alpha}{m_\alpha}\frac{\mathbf{v} \times \mathbf{B_0}}{c} \cdot \frac{\partial f_\alpha}{\partial \mathbf{v}}$$
$$= -\frac{e_\alpha}{m_\alpha}\left(\mathbf{E}(\mathbf{r}, t) + \frac{1}{c}\mathbf{v} \times \mathbf{B}(\mathbf{r}, t)\right) \cdot \frac{\partial F_\alpha}{\partial \mathbf{v}}.$$

In fully electromagnetic analysis, the wave magnetic field term should be retained, but for MHD wave that term is neglected assuming that $v/c \ll 1$. Recall that the equation was integrated along the unperturbed orbit in § 6.2; then $\mathbf{v}$ was the particle orbit. MHD is a slow process and v can be assumed to be $\ll c$. So the wave magnetic field term can be dropped out in the slow MHD wave dynamics. Neglecting that term and introducing Fourier variables, we write

$$-i\omega f_\alpha(\mathbf{k},\omega,\mathbf{v}) + i\mathbf{k}\cdot\mathbf{v}f_\alpha + \omega_{c\alpha}\mathbf{v}\times\hat{\mathbf{z}}\cdot\frac{\partial f_\alpha}{\partial \mathbf{v}} = -\frac{e_\alpha}{m_\alpha}\mathbf{E}(\mathbf{k},\omega)\cdot\frac{\partial F_\alpha}{\partial \mathbf{v}} \tag{6.178}$$

where F_α is zero order equilibrium distribution function. Introducing the cylindrical coordinate in the velocity space, the above equation becomes

$$(-i\omega + ik_x v_\perp cos\varphi + ik_z v_z)f_\alpha - \omega_{c\alpha}\frac{\partial f_\alpha}{\partial \varphi}$$
$$= \frac{e_\alpha}{T_\alpha}F_\alpha\,(v_\perp E_x cos\varphi + v_\perp E_y sin\varphi + E_z v_z). \tag{6.179}$$

Let us expand f in a series of harmonic functions:

$$f = \sum_{m=-\infty}^{\infty} f_m(v_\perp, v_z)e^{im\varphi}.$$

Substituting this series into (6.179) and equating the same harmonic terms give

$$m=0:\ (-i\omega + ik_z v_z)f_0 + \frac{i}{2}k_x v_\perp(f_1 + f_{-1}) = \frac{e}{T}FE_z v_z,$$

$$m=1:\ (-i\omega + ik_z v_z - i\omega_c)f_1 + \frac{i}{2}k_x v_\perp(f_0 + f_2) = \frac{e}{T}\frac{v_\perp}{2}(E_x - iE_y),$$

$$m=-1:\ (-i\omega + ik_z v_z + i\omega_c)f_{-1} + \frac{i}{2}k_x v_\perp(f_0 + f_{-2}) = \frac{e}{T}\frac{v_\perp}{2}(E_x + iE_y),$$

$$m=2:\ (-i\omega + ik_z v_z - 2i\omega_c)f_2 + \frac{i}{2}k_x v_\perp(f_1 + f_3) = 0,$$

$$m=-2:\ (-i\omega + ik_z v_z + 2i\omega_c)f_{-2} + \frac{i}{2}k_x v_\perp(f_{-1} + f_{-3}) = 0,$$

where we dropped the subscript α (which will be reinstated later). We discard $f_{\pm 3}$ to truncate the infinite set of equations. To investigate low frequency wave, higher harmonics may be ignored. We assume F_α to be Maxwellian and are interested in the frequency range

$$\omega_c \gg \omega,\ k_z v_z.$$

Then, we obtain

$$f_2 = \frac{k_x v_\perp}{4\omega_c}f_1,\quad f_{-2} = -\frac{k_x v_\perp}{4\omega_c}f_{-1},$$

$$\left(\omega - k_z v_z + \omega_c - \frac{k_x^2 v_\perp^2}{8\omega_c}\right)f_1 - \frac{1}{2}k_x v_\perp f_0 = \frac{ie}{2T}v_\perp F(E_x - iE_y), \tag{6.180}$$

$$\left(\omega - k_z v_z - \omega_c + \frac{k_x^2 v_\perp^2}{8\omega_c}\right) f_{-1} - \frac{1}{2} k_x v_\perp f_0 = \frac{ie}{2T} v_\perp F(E_x + iE_y), \quad (6.181)$$

$$(-i\omega + ik_z v_z) f_0 + \frac{i}{2} k_x v_\perp (f_1 + f_{-1}) = \frac{e}{T} F E_z v_z. \quad (6.182)$$

We have the above three equations for three unknowns f_0, f_1, and f_{-1}. Let

$$f_x = \frac{1}{2}(f_1 + f_{-1}), \quad f_y = \frac{i}{2}(f_1 - f_{-1}). \quad (6.183)$$

(6.180)–(6.182) gives

$$(\omega - k_z v_z) f_x - \left(\omega_c - \frac{k_x^2 v_\perp^2}{8\omega_c}\right) i\, f_y - \frac{k_x v_\perp}{2} f_0 = \frac{ie}{2T} v_\perp F E_x, \quad (6.184)$$

$$\left(\omega_c - \frac{k_x^2 v_\perp^2}{8\omega_c}\right) f_x - (\omega - k_z v_z) i f_y = \frac{e}{2T} v_\perp F E_y, \quad (6.185)$$

$$(\omega - k_z v_z) f_0 + k_x v_\perp f_x = \frac{ie}{T} F E_z v_z. \quad (6.186)$$

Let us introduce symbols

$$\omega' \equiv \omega - k_z v_z, \quad \omega_c' \equiv \omega_c - \frac{k_x^2 v_\perp^2}{8\omega_c}.$$

We assume $\frac{k_x^2 v_\perp^2}{\omega_c} \ll 1$. Then, we have $\omega_c' \simeq \omega_c$. (6.184)–(6.186) yield

$$f_0 = \frac{eF}{T}\left[-i\frac{k_x v_\perp^2}{2\omega_c'^2} E_x + \frac{k_x v_\perp^2}{2\omega' \omega_c'} E_y + i\frac{v_z}{\omega'} E_z\right], \quad (6.187)$$

$$f_x = \frac{eF}{T}\left[-\frac{i}{2}\frac{\omega'}{\omega_c'^2} v_\perp E_x + \frac{v_\perp}{2\omega_c'} E_y - i\frac{k_x v_\perp v_z}{2\omega_c'^2} E_z\right], \quad (6.188)$$

$$f_y = \frac{eF}{T}\left[-\frac{1}{2}\frac{v_\perp}{\omega_c'} E_x - \frac{i v_\perp}{2\omega_c'^2}\left(\omega' - \frac{k_x^2 v_\perp^2}{2\omega'}\right) E_y - \frac{k_x v_\perp v_z}{2\omega' \omega_c'} E_z\right]. \quad (6.189)$$

In (6.187), we neglected $\omega'^2/\omega_c^2 \ll 1$. In the above we put $\omega_c' \to \omega_c$. The above equations are used to calculate

$$\begin{aligned} f_\alpha &= f_0 + f_1 e^{i\varphi} + f_{-1} e^{-i\varphi} + f_2 e^{i\varphi} + f_{-2} e^{-2i\varphi} \\ &= f_0 + 2 f_x cos\varphi + 2 f_y sin\varphi + f_2 e^{i\varphi} + f_{-2} e^{-2i\varphi} \end{aligned}$$

$$\mathbf{J} = \sum_\alpha e_\alpha \int_0^{2\pi} d\varphi \int_{-\infty}^{\infty} dv_z \int_0^{\infty} v_\perp dv_\perp \mathbf{v} f_\alpha$$

$$J_x = 2\pi \sum_\alpha e_\alpha \int dv_z \int dv_\perp v_\perp^2 f_x = \sigma_{xx}E_x + \sigma_{xy}E_y + \sigma_{xz}E_z, \quad (6.190)$$

$$J_y = 2\pi \sum_\alpha e_\alpha \int dv_z \int dv_\perp v_\perp^2 f_y = \sigma_{yx}E_x + \sigma_{yy}E_y + \sigma_{yz}E_z, \quad (6.191)$$

$$J_z = 2\pi \sum_\alpha e_\alpha \int dv_z v_z \int dv_\perp v_\perp f_0 = \sigma_{zx}E_x + \sigma_{zy}E_y + \sigma_{zz}E_z, \quad (6.192)$$

$$\sigma_{xx} = -i\pi \sum_\alpha \frac{e_\alpha^2}{T_\alpha \omega_{c\alpha}^2} \int dv_z \omega' \int dv_\perp v_\perp^3 F_\alpha,$$

$$\sigma_{xy} = -\sigma_{yx} = \pi \sum_\alpha \frac{e_\alpha^2}{T_\alpha \omega_{c\alpha}} \int dv_z \int dv_\perp v_\perp^3 F_\alpha,$$

$$\sigma_{xz} = 0 = \sigma_{zx},$$

$$\sigma_{yy} = -i\pi \sum_\alpha \frac{e_\alpha^2}{T_\alpha \omega_{c\alpha}^2} \int dv_z \int dv_\perp v_\perp^3 \left(\omega' - \frac{k_x^2 v_\perp^2}{2\omega'}\right) F_\alpha,$$

$$\sigma_{yz} = -\pi \sum_\alpha \frac{e_\alpha^2 k_x}{T_\alpha \omega_{c\alpha}} \int dv_z \frac{v_z}{\omega'} \int dv_\perp v_\perp^3 F_\alpha = -\sigma_{zy},$$

$$\sigma_{zz} = 2i\pi \sum_\alpha \frac{e_\alpha^2}{T_\alpha} \int dv_z \frac{v_z^2}{\omega'} \int dv_\perp v_\perp F_\alpha.$$

Velocity integrals require the following formulas; let

$$I(n) \equiv \int_0^\infty e^{-\frac{m}{2T}x^2} x^n dx,$$

$$I(0) = \frac{1}{2}\sqrt{\frac{2\pi T}{m}}, \quad I(1) = \frac{T}{m}, \quad I(2) = \frac{T}{2m}\sqrt{\frac{2\pi T}{m}},$$

$$I(3) = 2\left(\frac{T}{m}\right)^2, \quad I(4) = \frac{3}{2}\left(\frac{T}{m}\right)^2\sqrt{\frac{2\pi T}{m}}, \quad I(5) = \left(\frac{2T}{m}\right)^3.$$

Integration over $dv_\perp$: let us assume

$$F(v_\perp, v_z) = F_0(v_z)\frac{m}{2\pi T}e^{-\frac{mv_\perp^2}{2T}},$$

with $F_0(v_z)$ a one-dimensional Maxwellian.

$$\sigma_{xx} = -i \sum_\alpha \frac{e_\alpha^2}{m_\alpha \omega_{c\alpha}^2} \int dv_z \omega' F_0(v_z),$$

$$\sigma_{xy} = \sum_\alpha \frac{e_\alpha^2}{m_\alpha \omega_{c\alpha}} = 0 = \sigma_{yx},$$

$$\sigma_{yy} = -i \sum_\alpha \frac{e_\alpha^2}{m_\alpha \omega_{c\alpha}^2} \int dv_z F_0(v_z) \left(\omega' - 2 \frac{T_\alpha}{m_\alpha} \frac{k_x^2}{\omega'} \right),$$

$$\sigma_{yz} = - \sum_\alpha \frac{e_\alpha^2 k_x}{m_\alpha \omega_{c\alpha}} \int dv_z F_0(v_z) \frac{v_z}{\omega'} = -\sigma_{zy},$$

$$\sigma_{zz} = i \sum_\alpha \frac{e_\alpha^2}{T_\alpha} \int dv_z F_0(v_z) \frac{v_z^2}{\omega'}.$$

Integration over dv_z:

$$F_0(v_z) = \sqrt{\frac{m}{2\pi T}} e^{-\frac{mv_z^2}{2T}},$$

$$Let \quad A(n) = \int dv_z \, \frac{v_z^n F_0(v_z)}{\omega - k_z v_z}, \quad \zeta = \sqrt{\frac{m}{2T}} \frac{\omega}{k_z},$$

$$A(0) = -\frac{1}{k_z} \sqrt{\frac{m}{2T}} Z(\zeta), \quad A(1) = -\frac{1}{k_z} (1 + \zeta Z(\zeta)),$$

$$A(2) = -\frac{\omega}{k_z^2} (1 + \zeta Z(\zeta)),$$

$$\sigma_{xx} = -i\omega \sum_\alpha \frac{e_\alpha^2}{m_\alpha \omega_{c\alpha}^2},$$

$$\sigma_{xy} = \sigma_{yx} = \sigma_{xz} = \sigma_{zx} = 0,$$

$$\sigma_{yy} = -i \sum_\alpha \frac{e_\alpha^2}{m_\alpha \omega_{c\alpha}^2} \left(\omega + \frac{k_x^2}{k_z} \sqrt{\frac{2T_\alpha}{m_\alpha}} Z(\zeta_\alpha) \right),$$

$$\sigma_{yz} = \frac{k_x}{k_z} \sum_\alpha \frac{e_\alpha^2}{m_\alpha \omega_{c\alpha}} (1 + \zeta_\alpha Z(\zeta_\alpha)) = -\sigma_{zy},$$

$$\sigma_{zz} = -i \, \frac{\omega}{k_z^2} \sum_\alpha \frac{e_\alpha^2}{T_\alpha} (1 + \zeta_\alpha Z(\zeta_\alpha)).$$

Let us use the relation

$$\frac{4\pi i}{\omega} \sigma_{ij} = \varepsilon_{ij}.$$

Then the following elements of ε_{ij} are obtained.

- $$\varepsilon_{xx} = \sum_\alpha \frac{\omega_{p\alpha}^2}{\omega_{c\alpha}^2} \simeq \frac{\omega_{pi}^2}{\omega_{ci}^2},$$
- $$\varepsilon_{xy} = \varepsilon_{yx} = \varepsilon_{xz} = \varepsilon_{zx} = 0,$$
- $$\varepsilon_{yy} = \sum_\alpha \frac{\omega_{p\alpha}^2}{\omega_{c\alpha}^2}\left(1 + \frac{k_x^2}{k_z^2}\frac{k_z}{\omega}\sqrt{\frac{2T_\alpha}{m_\alpha}}Z(\zeta_\alpha)\right),$$
- $$\varepsilon_{yz} = -\varepsilon_{zy} = i\frac{k_x}{k_z}\sum_\alpha \frac{\omega_{p\alpha}^2}{\omega\omega_{c\alpha}}(1 + \zeta_\alpha Z(\zeta_\alpha)),$$
- $$\varepsilon_{zz} = \frac{1}{k_z^2}\sum_\alpha \frac{4\pi N e_\alpha^2}{T_\alpha}(1 + \zeta_\alpha Z(\zeta_\alpha)).$$

Dispersion relation is obtained from the relation

$$k^2\mathbf{E} - \mathbf{k}\,\mathbf{k}\cdot\mathbf{E} = \frac{4\pi i\omega}{c^2}\mathbf{J} = \frac{\omega^2}{c^2}\varepsilon_{ij}E_j$$

which is written in components

$$\left(k_z^2 - \frac{\omega^2}{c^2}\varepsilon_{xx}\right)E_x - k_x k_z E_z = 0,$$

$$\left(k^2 - \frac{\omega^2}{c^2}\varepsilon_{yy}\right)E_y - \frac{\omega^2}{c^2}\varepsilon_{yz}E_z = 0,$$

$$-k_x k_z E_x - \frac{\omega^2}{c^2}\varepsilon_{zy}E_y + \left(k_x^2 - \frac{\omega^2}{c^2}\varepsilon_{zz}\right)E_z = 0.$$

The above three equations give the dispersion relation

$$\left(k_z^2 - \frac{\omega^2}{c^2}\varepsilon_{xx}\right)\left[\left(\varepsilon_{yy} - \frac{c^2k^2}{\omega^2}\right)\varepsilon_{zz} + \varepsilon_{yz}^2\right] = 0$$

where we neglected

$$\frac{c^2k_x^2}{\omega^2} \ll \varepsilon_{zz}.$$

The first factor gives the Alfven wave

$$\omega^2 = v_A^2 k_z^2. \tag{6.193}$$

Putting the second bracket to zero yields the dispersion relation of magnetosonic waves in the frequency region such that

$$\sqrt{\frac{T_i}{m_i}} \ll \frac{\omega}{k} \ll \sqrt{\frac{T_e}{m_e}}. \tag{6.194}$$

Thus, we need to evaluate the dielectric tensor elements for the frequencies in the region of (6.194). According to (6.19), we have $\zeta_i \gg 1$ and $\zeta_e \ll 1$. The asymptotic series for the plasma dispersion function $Z(\zeta)$ are as follows:

$$Z(\zeta_i) \simeq -\frac{1}{\zeta_i} - \frac{1}{2}\frac{1}{\zeta_i^3} \quad (\zeta_i \gg 1), \quad Z(\zeta_e) \simeq -2\zeta_e\left(1 - \frac{2}{3}\zeta_e^2\right) \quad (\zeta_e \ll 1),$$

which yields

- $$\varepsilon_{yy} = \frac{\omega_{pi}^2}{\omega_{ci}^2}\left(1 - \frac{k_x^2}{\omega^2}\frac{2T_i}{m_i}\right) - 2\frac{\omega_{pe}^2}{\omega_{ce}^2}\frac{k_x^2}{k_z^2} \simeq \frac{\omega_{pi}^2}{\omega_{ci}^2} \quad \left(\frac{m_e}{m_i}\frac{k_x^2}{k_z^2} \ll 1\right),$$

- $$\varepsilon_{yz} \simeq -i\frac{k_x}{k_z}\frac{\omega_{pi}^2}{\omega\omega_{ci}},$$
- $$\varepsilon_{zz} \simeq -\frac{\omega_{pi}^2}{\omega^2} + \frac{4\pi N e^2}{k_z^2 T_e}.$$

Using the above values for the dielectric tensor elements in the second bracket gives the dispersion relation in the form

$$(\omega^2 - k^2 v_A^2)(\omega^2 - k_z^2 v_s^2) - \omega^2 k_x^2 v_s^2 = 0 \tag{6.195}$$

where we used the relations

$$\frac{\omega_{pi}^2}{\omega_{ci}^2} = \frac{c^2}{v_A^2}, \quad \frac{T_e}{m_i} = v_s^2 \quad (\textit{sound velocity}).$$

This completes kinetic derivation of MHD waves. In this work, the authors avoided the use of Bessel functions; instead, they used simple sinusoidal harmonics. The mathematical idea underlying this work is that the Bessel functions are kind of extended sinusoidal functions.

Chapter 7

Wave

7.1. Dispersive waves

A dispersion relation means a certain functional relationship between ω and $\mathbf{k}$, which are, respectively, the Fourier variables conjugate to t and $\mathbf{r}$ in ordinary space. The mathematical description of the general wave analysis is based on the idea that a physical disturbance in a medium can be represented by Fourier integral in the form $A(x,t) = \int A(k,\omega)exp\ [ikx - i\omega(k)t]dxdt$. Without knowing the specific form of $\omega(k)$, we can extract significant general information from this integral. Also a dispersion relation itself contains information concerning the propagation of the waves and the properties of the medium supporting the waves. We first sort out a few salient general features which can be extracted from the relation $\omega = \omega(k)$.

i) ω is real for real k.
This implies that the energy of the wave is conserved and the propagation is not inhibited.

ii) ω is complex or purely imaginary for real k.
The energy of the wave is not conserved. Usually this means that dissipative mechanism is at work in the medium and the wave becomes damped as it propagates (see § 6.3.1). But if there is some mechanism of free energy liberation from the medium, nonreal ω implies instability.

iii) k is complex for real ω.
This is a conservative system in which a number of modes propagate with possible exchange among them (mode conversion). A mode with a complex wave number would thus lose its energy to (or absorb energy from) other modes at appropriate rate to conserve the total energy. This situation occurs, for example, for certain modes of electromagnetic waves in plasmas.

iv) $k = 0$ for real ω.
This case occurs when k changes from real values to imaginary with the corresponding ω varying continuously. The particular frequency yielding zero k is called the *cutoff frequency*. The wave evanesces when k becomes purely imaginary. What actually happens is that the wave is mostly reflected at cutoff where the phase velocity of the wave is infinite.

v) k is infinite for real ω.
This is where resonances take place. Since the phase velocity is nearly zero, the wave can spend a long enough time at one locality to be absorbed by the medium.

7.2. Asymptotic evaluation of Fourier integral and group velocity

Actual signals propagating in a plasma are not monochromatic but have finite spread of frequencies or wavelengths. Such disturbances or wave packets can be synthesized as Fourier integral, and we find, in the asymptotic evaluation of the integral for large time, that the group velocity emerges to be an important physical variable in its asymptotic evaluation for large time. For simplicity, we assume one-dimensional propagation and represent the wave packet as a Fourier integral in the form

$$f(x,t) = \int_{-\infty}^{\infty} F(k) e^{i[kx-\omega(k)t]} dk. \tag{7.1}$$

In this Section, (7.1) will be evaluated asymptotically by using the method of *steepest descent*; in doing so the concept of the phase and the group velocities emerges naturally. Before embarking on evaluation of integral (7.1), it is instructive to consider a wave packet superposed about a well-defined central wave number k_0 and the corresponding frequency $\omega(k_0)$ with a finite spread of wave number Δk:

$$f(x,t) = \frac{1}{\Delta k} \int_{k_0-\Delta k}^{k_0+\Delta k} F(k) e^{i[kx-\omega(k)t]} dk. \tag{7.2}$$

The factor $\frac{1}{\Delta k}$ is introduced so that f will remain non-zero as $\Delta k \to 0$. Expanding $\omega(k)$ in the vicinity of k_0 as a power series,

$$\omega(k) = \omega(k_0) + (k - k_0) \left(\frac{\partial \omega}{\partial k}\right)_{k_0} \cdot$$

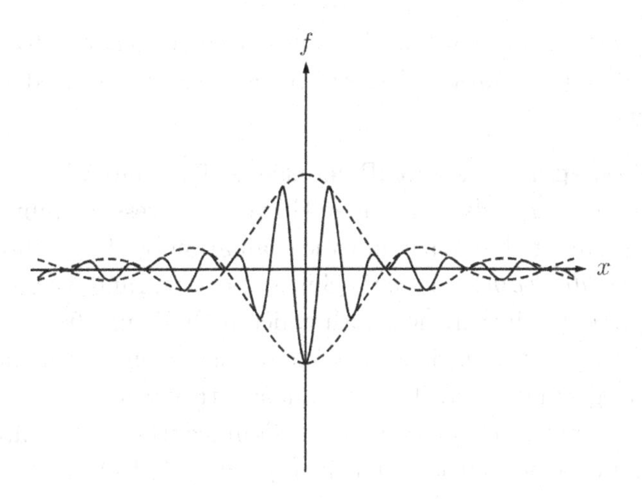

Fig. 7.1. Wave packet.

(7.2) becomes

$$f(x,t) = \frac{F(k_0)}{\Delta k} e^{i[k_0 x - \omega(k_0)t]} \int_{k_0-\Delta k}^{k_0+\Delta k} e^{i(k-k_0)x - i(k-k_0)(\frac{\partial\omega}{\partial k})_{k_0} t}\, dk$$

$$= 2F(k_0)\frac{sin[\Delta k(x - \omega_0' t)]}{\Delta k(x - \omega_0' t)} e^{i(k_0 x - \omega_0 t)} \tag{7.3}$$

where $\omega_0 = \omega(k_0), \omega_0' = (\frac{\partial\omega}{\partial k})_{k_0}$. This is a wave train whose wave number and frequency are respectively the central wave number k_0 and the corresponding frequency ω_0 of the constituent Fourier wave trains, and having its amplitude modulated by a slowly varying envelope ($sin[\cdots]term$) to form a system of wave packets (see Fig. 7.1). The packets travel with the group velocity $(\frac{\partial\omega}{\partial k})_{k_0}$ while the wave train contained inside the envelope propagates with the phase velocity ω_0/k_0. The amplitude of the modulation is a varying function. The maximum packet is the one centered about the position $x = \omega_0' t$ where the amplitude is $2F(k_0)$.

Going back to (7.1), we write the phase as $t\ \varphi(k)$, where

$$\varphi(k) = i\left[\frac{kx}{t} - \omega(k)\right]. \tag{7.4}$$

Then (7.1) becomes

$$f(x,t) = \int_{-\infty}^{\infty} F(k) e^{t\varphi(k)} dk. \tag{7.5}$$

We evaluate integral (7.5) when t becomes large. This asymptotic evaluation will reveal the hidden characteristics of the Fourier-synthesized wave, or wave packet in general.

Method of steepest descent [Reference: Arfken and Weber (2001); E. J. Lighthill, *Waves in Fluids* Cambridge University Press, Cambridge, (1978)]. Integral (7.5) is amenable to asymptotic evaluation by the method of steepest descent (or *saddle point method*). Since the integrand is analytic in the complex k-plane, the integration path which is the k-axis from $-\infty$ to ∞ can be deformed to a path which we can choose as we like with the end points fixed. This is legitimate according to Cauchy's theorem. We assume here that the ends of the path do not contribute significantly to the integral. When t is large, the exponential function in the integrand of (7.5) has a rapidly oscillating phase which makes almost exact cancellations in the integral. However, this is not true when there are points of stationary phase where $\varphi'(k) \equiv d\varphi/dk = 0$. Let us write the phase function which we assume to be analytic as

$$\varphi(z) = u(x,y) + iv(x,y) \quad z \equiv k = k_r + ik_i, \quad (x = k_r, \ \ y = k_i)$$

where we designated the complex k-plane as the more friendly z-plane. Most of the contribution to f in (7.5) comes from the section centered about the point where u is largest (local maximum). [The local maximum of u is also at the same time the point of stationary phase.] The idea of the method of steepest descent is to deform the integration path so that the region of large u is compressed into as short a path as possible. At the stationary phase point $z = z_0$, $\varphi'(z_0) = 0$, which implies $\frac{\partial u}{\partial x} = \frac{\partial v}{\partial x} = 0$ and $\frac{\partial u}{\partial y} = \frac{\partial v}{\partial y} = 0$ at $z_0 = (x_0, y_0)$. This is the necessary condition that $u(x,y)$ has a local maximum or a local minimum at z_0. Since $\varphi(z)$ is analytic, we have from the Cauchy–Riemann condition,

$$\frac{\partial^2 u}{\partial x^2} + \frac{\partial^2 u}{\partial y^2} = 0. \tag{7.6}$$

Thus $\frac{\partial^2 u}{\partial x^2} > 0$ implies $\frac{\partial^2 u}{\partial y^2} < 0$. In words, $u(x,y)$ is maximum at $z = z_0$ along the x-coordinate, but minimum along the y-coordinate. And vice versa. Therefore the stationary phase point z_0 is a saddle point when we view it in the z-plane. This is the consequence of the analyticity of $\varphi(z)$. By Taylor expansion in the vicinity of $z = z_0$, we have

$$\varphi(z) = \varphi(z_0) + \frac{1}{2}\varphi''(z_0)(z - z_0)^2.$$

We write

$$\varphi''(z_0) = \varrho e^{i\vartheta}, \quad z - z_0 = \gamma e^{i\phi}, \quad \varphi(z) = \varphi(z_0) + \frac{1}{2}\varrho\, \gamma^2 e^{i\vartheta + 2i\phi} \tag{7.7}$$

which gives

$$u(x,y) = u(x_0, y_0) + \frac{1}{2}\varrho\gamma^2\ cos(\vartheta + 2\phi),$$

$$v(x,y) = v(x_0, y_0) + \frac{1}{2}\varrho\gamma^2\ sin(\vartheta + 2\phi).$$

Furthermore, we impose the additional requirement that the imaginary part v is constant while traversing the segment of the deformed contour centered about the saddle point. If v is not constant, the rapidly oscillating exponential factor will make the integral vanishingly small, but it is not actually nil. This inaccuracy will be more and more augmented as the time t becomes large. Thus we put $sin(\vartheta + 2\phi) = 0$, which gives $cos\ (\vartheta + 2\phi) = \pm 1$ where the upper sign should be rejected since $u(x_0, y_0)$ in this case does not correspond to the maximum u. So our deformed path centered about the saddle point is described by

$$u(x,y) = u(x_0, y_0) - \frac{1}{2}\varrho\gamma^2, \tag{7.8}$$

$$v = v(x_0, y_0) = const. \tag{7.9}$$

[When u and v are harmonic, satisfying (7.6), the curves $u(x,y) = const.$ and the curves $v(x,y) = const.$ are orthogonal. Since ∇u is perpendicular to $u = const$, ∇u and the curves $v = const.$ are parallel. Thus the integration path $v = const.$ is the path of steepest descent.]

Now (7.7) becomes (reinstating k in place of z)

$$\varphi(k) = \varphi(k_0) - \frac{1}{2}\varrho\gamma^2$$

$$dk = dz = e^{i\phi} d\gamma$$

where the angle ϕ is undetermined until we specify the deformed contour. Since we will choose the path of the steepest descent as a straight line (shortest), ϕ is constant. So we have

$$f(x,t) = F(k_0) \int e^{t\varphi(k)} dk = F(k_0) e^{i\phi} \int e^{t[\varphi(k_0) - \frac{\varrho}{e}\gamma^2]} d\gamma. \tag{7.10}$$

In (7.10), the limits of the integration can be taken from $-\infty$ to $+\infty$ because the integral has non-zero contribution only from the segment of the steepest descent. Then the $d\gamma$-integral gives $\sqrt{2\pi/t\varrho}$. Recalling $\varrho = |\varphi''(k_0)|$, (7.10) becomes

$$f(x,t) = F(k_0)\sqrt{\frac{2\pi}{t}}e^{i\phi}|\varphi''(k_0)|^{-1/2}e^{t\varphi(k_0)}. \tag{7.11}$$

The above formula is quite general but has a room to be more concrete. The angle ϕ must be determined case by case according to the problem under consideration.

Determination of the angle ϕ. Going back to (7.4) and (7.5), it is more convenient to define

$$\Psi(k) = \frac{kx}{t} - \omega(k): \quad f(x,t) = \int_{-\infty}^{\infty} F(k)e^{it\Psi(k)}dk. \tag{7.12}$$

Suppose $\Psi'(k) > 0$ everywhere (this may occur for large x/t). We slightly raise the path by an amount $i\varepsilon(\varepsilon > 0)$ above the real k-axis. Then along the raised path, Ψ has the value

$$\Psi(k+i\varepsilon) = \Psi(k) + i\varepsilon\Psi'(k) \quad (k,\ real).$$

The positive imaginary part of the phase function Ψ, which we write $\delta = \varepsilon\Psi'(k)$, makes integral (7.12) exponentially small for large t.

When $\Psi'(k) < 0$ everywhere (this is the case where $x = 0$ and $d\omega/dk > 0$), integral (7.12) is again vanishingly small ($\sim e^{-t\delta}$), as can be seen by lowering the path by an amount $\varepsilon = \delta/\Psi'(k)$. Note that $\Psi'(k)$ changes sign upon crossing the stationary phase point $k = k_0$ where $\Psi'(k_0) = 0$. Therefore the raised (lowered) path on one side of k_0 should be connected to the lowered (raised) path on the other side of k_0 (see Fig. 7.2). Now we have to consider the second derivative term in the expansion of $\Psi(k)$ near k_0:

$$\Psi(k) = \Psi(k_0) + \frac{1}{2}\Psi''(k_0)(k-k_0)^2 \quad (k = k_r + ik_i)$$

which gives

$$Im[\Psi(k)] = k_i(k_r - k_0)\Psi''(k_0).$$

The contours of $Im[\Psi(k)] = \delta$ are the hyperbolas (see Fig. 7.2),

$$k_i = \frac{\delta}{\Psi''(k_0)(k_r - k_0)}. \tag{7.13}$$

It is immediately seen that the shortest straight line which connects the two branches of the hyperbola (two contours above and below) and passes through $k = k_0$ is either $\phi = \pi/4$ or $-\pi/4$.

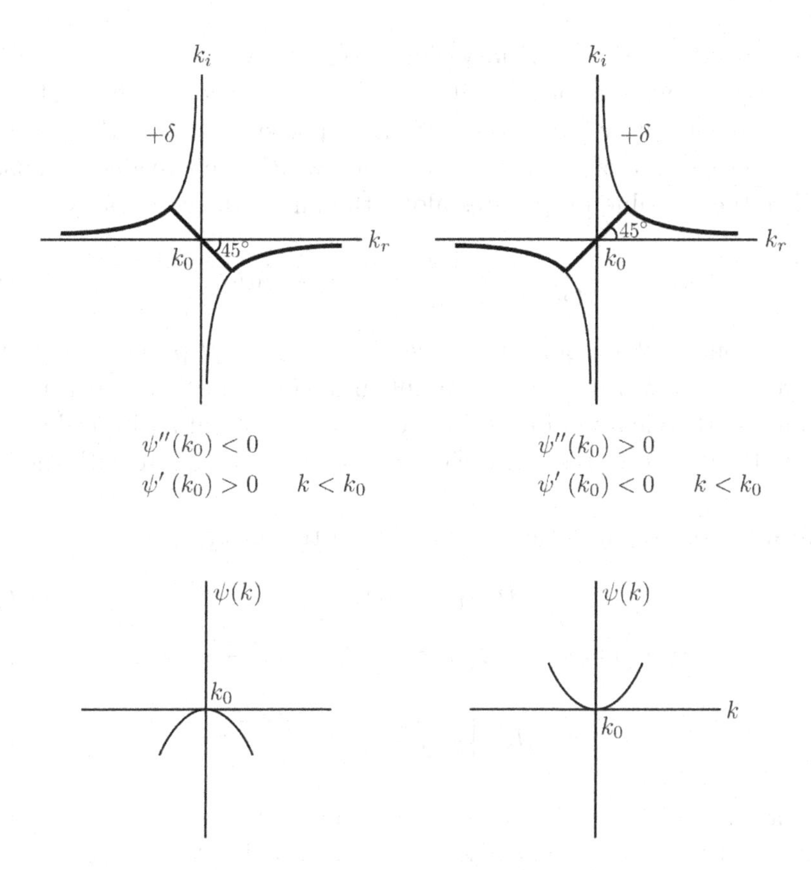

Fig. 7.2. Path of steepest descent.

Now $f(x,t)$ in (7.11) is determined without ambiguity, and can be written as a single equation for both cases in Fig. 7.2:

$$f(x,t) = F(k_0)\left[\frac{2\pi}{t|\omega''(k_0)|}\right]^{1/2} e^{i\left(k_0x-\omega_0t+\frac{\pi}{4}\frac{\omega''(k_0)}{|\omega''(k_0)|}\right)} \tag{7.14}$$

where $\omega_0 = \omega(k_0)$ and k_0 is the wave number for which

$$\omega'(k_0) = \frac{x}{t}. \tag{7.15}$$

The preceding method of evaluating integral (7.12) fails if $\Psi''(k_0) = 0$ for the stationary phase point k_0. Readers who are interested in the asymptotic evaluation of the integral when $\Psi''(k_0) = 0$ are referred to the book by Lighthill (1978), *Waves in Fluids*. If there are more than one stationary phase point satisfying (7.15), the asymptotic value of the integral should be a sum of the terms each of which is similar to the right hand side of (7.14).

Let us note that the stationary phase point k_0, which is defined by (7.15), is the same for two sets of (x, t) if $\frac{x_1}{t_1} = \frac{x_2}{t_2} = c$ (*const.*). In fact, along the straight line defined by $\frac{x}{t} = c$, the stationary phase point k_0 is the same. The straight line is a line of characteristics along which the k_0-value is constant. Therefore the k_0-value propagates along the line with the velocity

$$\frac{x_2 - x_1}{t_2 - t_1} = \frac{x_2}{t_2} = \frac{x_1}{t_1} = \omega'(k_0). \tag{7.16}$$

So the stationary phase point k_0 for which $\omega'(k_0) = x/t$ propagates with the group velocity $\omega'(k_0)$ in (7.16). The integral $f(x, t)$ in (7.5) has a non-zero value only in the vicinity of k_0, so it is quite obvious that a physical quantity associated with a wave packet, such as energy, will propagate with the group velocity.

We write the asymptotic result (7.14) in the form,

$$f(x, t) = Ae^{i\theta(x,t)}, \tag{7.17}$$

$$\theta(x, t) = k_0 x - \omega(k_0)t = t\ [k_0\omega'(k_0) - \omega(k_0)], \tag{7.18}$$

$$A = F(k_0) \left[\frac{2\pi}{t|\omega''(k_0)|}\right] e^{\frac{i\pi}{4}\frac{\omega''(k_0)}{|\omega''(k_0)|}}. \tag{7.19}$$

Since the value of k_0 can be determined from (7.15) at any point (x, t), it is a function of (x, t). Consequently, $\omega(k_0)$ is also a function of (x, t) through the dispersion relation. Therefore the amplitude A is also a function of (x, t). Equation (7.17) indicates that $f(x, t)$ is represented by a 'monochromatic' wave whose amplitude, wave number, and frequency are all slowly varying functions of (x, t). This result is interesting since the Fourier superposition ends up with a monochromatic plane wave expression (albeit with varying parameters) after going through all the constructive and destructive interferences. The last paragraph is quoted from Lighthill (1978). In conclusion, we can state that the group velocity of a wave packet constructed by Fourier synthesis (7.1) is the velocity of the constructive interference maximum of the wave. When $F(k)$ in (7.1) is a slowly varying function, the maximum at each instant of time occurs at the point of stationary phase which moves with the group velocity $\omega'(k_0)$. This result also fully justifies (6.114) that we assumed for the consideration of the field energy.

Differentiation of (7.18) with respect to x and t gives, respectively,

$$\frac{\partial\theta}{\partial x} = k + [x - \omega'(k)t]\ \frac{\partial k}{\partial x}, \tag{7.20}$$

$$\frac{\partial \theta}{\partial t} = -\omega(k) + [x - \omega'(k)t]\,\frac{\partial k}{\partial t}, \tag{7.21}$$

where we omitted the subscript '0'. Since k is the stationary phase point, the above equations reduce to

$$\frac{\partial \theta}{\partial x} = k(x,t), \tag{7.22}$$

$$\frac{\partial \theta}{\partial t} = -\omega(x,t). \tag{7.23}$$

Therefore the slowly varying phase function $\theta(x,t)$ is the eikonal whose spatial and temporal derivatives give, respectively, the local wave number and the local frequency in a nonuniform wave train $\sim e^{ik(x,t)x - i\omega(x,t)t}$. The above two equations yield

$$\frac{\partial k}{\partial t} + \frac{\partial \omega}{\partial x} = 0 \quad or \quad \frac{\partial k}{\partial t} + \omega'(k)\frac{\partial k}{\partial x} = 0. \tag{7.24}$$

Using the dispersion relation, $k = k(\omega)$, we also have

$$\frac{\partial \omega}{\partial t} + \omega'(k)\,\frac{\partial \omega}{\partial x} = 0. \tag{7.25}$$

Equations (7.24) and (7.25) have the solutions in the form

$$k(x,t) = k(x - \omega'(k)t), \tag{7.26}$$

$$\omega(x,t) = \omega(x - \omega'(k)t). \tag{7.27}$$

The above equations indicate that the wave number k and the frequency ω remain constant along the *line of characteristics* defined by

$$\frac{dx}{dt} = \omega'(k).$$

7.3. Kinematics of wave, group velocity in three-dimension

In the preceding discussion, we showed that the group velocity of a wave packet that is formed of Fourier integral is the velocity of the moving stationary phase point in the (x,t)-plane. In fact, the concept of the group velocity can be established independently of Fourier integral, even valid in inhomogeneous media and for nonlinear waves, for both of which the Fourier superposition is not appropriate. In this Section, we derive the group velocity independently of the Fourier integral representation of a wave packet, extending also to three-dimensional propagation. First, we assume that a wave quantity $f(x,t)$ is written as

$$f(\mathbf{x},t) = A(\mathbf{x},t)e^{i\theta(\mathbf{x},t)}. \tag{7.28}$$

Compare this expression with (7.17). For a wave propagating in an inhomogeneous medium whose property is slowly changing (the fractional change of the property in one local wavelength is much smaller than unity), the amplitude A can be assumed to be a slowly varying function of $\mathbf{x}$ and t. Over a small region of space and short interval of time, the phase (eikonal) $\theta(\mathbf{x},t)$ may be expanded in Taylor series in the vicinity of $\mathbf{x} = 0$ and $t = 0$.

$$\theta(\mathbf{x},t) = \theta_0 + \mathbf{x}\cdot\nabla\theta + t\,\frac{\partial\theta}{\partial t}. \tag{7.29}$$

Over a small region of space and short duration of time, the wave (7.28) can be regarded as a plane wave. Therefore, we can define the wave vector and the frequency at each point as

$$\mathbf{k} = \nabla\theta, \quad \omega = -\frac{\partial\theta}{\partial t}. \tag{7.30}$$

Compare (7.30) with (7.22) and (7.23). Combining the two equations in (7.30) gives

$$\frac{\partial\mathbf{k}}{\partial t} = -\nabla\omega. \tag{7.31}$$

In a homogeneous medium, the frequency of a wave depends on $\mathbf{k}$ only: $\omega = \omega(\mathbf{k})$. Therefore (7.31) is a differential equation for $\mathbf{k}$:

$$\frac{\partial\mathbf{k}}{\partial t} + \frac{\partial\omega}{\partial\mathbf{k}}\cdot\nabla\mathbf{k} = 0.$$

By solving the differential equation, we obtain the temporal change of $\mathbf{k}(\mathbf{x},t)$. In one-dimensional case, (7.31) takes a simple form

$$\frac{\partial k}{\partial t} + \frac{\partial\omega}{\partial k}\frac{\partial k}{\partial x} = 0$$

which is (7.24) and indicates that the wave number k is merely transported with the group velocity. In an inhomogeneous medium, the dispersion relation takes the form:

$$\omega = \omega\big(\mathbf{k}(\mathbf{x},t),\mathbf{x},t\big). \tag{7.32}$$

Substitution of (7.32) into (7.31) yields

$$\frac{\partial k_i}{\partial t} + \frac{\partial\omega}{\partial k_j}\frac{\partial k_j}{\partial x_i} + \frac{\partial\omega}{\partial x_i} = 0.$$

Upon using $\frac{\partial k_i}{\partial x_j} = \frac{\partial k_j}{\partial x_i}$, which is $\nabla \times \mathbf{k} = 0$, the above equation is written as

$$\frac{\partial \mathbf{k}}{\partial t} + \mathbf{v}_g \cdot \nabla \mathbf{k} = -\nabla \omega \tag{7.33}$$

where $\mathbf{v}_g = \partial\omega/\partial\mathbf{k}$ is the three-dimensional group velocity. (7.33) is also a differential equation for $\mathbf{k}$ because ω is a function of $(\mathbf{x}, t)$ per (7.32). The above equation is written in the characteristic form as

$$\frac{d\mathbf{k}}{dt} = -\nabla \omega \tag{7.34}$$

where the total derivative $d/dt = \partial/\partial t + \mathbf{v}_g \cdot \nabla$ is the directional derivative taken along the characteristic line

$$\frac{d\mathbf{x}}{dt} = \mathbf{v}_g = \frac{\partial \omega}{\partial \mathbf{k}}. \tag{7.35}$$

When the medium is homogeneous, $\partial\omega/\partial\mathbf{x} = 0$, and $\partial\omega/\partial\mathbf{k}$ is a constant vector. Then the characteristics are straight lines in $(\mathbf{x}, t)$ space, and each value of $\mathbf{k}$ propagates with the corresponding constant group velocity $\mathbf{v}_g$. In an inhomogeneous medium, the characteristics are not straight lines, and also the value of $\mathbf{k}$ varies as they propagate along the characteristics.

The total derivative of frequency ω taken along the characteristics is

$$\frac{d\omega}{dt} = \frac{\partial \omega}{\partial t} + \frac{\partial \omega}{\partial \mathbf{k}} \cdot \frac{d\mathbf{k}}{dt} + \frac{\partial \omega}{\partial \mathbf{x}} \cdot \frac{d\mathbf{x}}{dt}$$

where the last two terms add to zero owing to (7.34) and (7.35), yielding

$$\frac{d\omega}{dt} = \frac{\partial \omega}{\partial t}. \tag{7.36}$$

Thus the frequency is constant along each characteristics in a steady state ($\partial\omega/\partial t = 0$). It is interesting that (7.34) and (7.35) are formally identical with Hamilton's equations in mechanics

$$\dot{\mathbf{p}} = -\frac{\partial H}{\partial \mathbf{x}}, \quad \dot{\mathbf{x}} = \frac{\partial H}{\partial \mathbf{p}}$$

with the correspondence, momentum $\mathbf{p} \to$ wave vector $\mathbf{k}$ and Hamiltonian (energy)$\to$ frequency ω. The second equation of (7.30) gives, upon introducing the dispersion relation,

$$\frac{\partial \theta}{\partial t} + \omega\left(\frac{\partial \theta}{\partial \mathbf{x}}, \mathbf{x}, t\right) = 0. \tag{7.37}$$

This is the Hamilton–Jacobi equation in which the eikonal $\theta(\mathbf{x}, t)$ plays the role of the action of the particle. In a sound or electromagnetic wave whose dispersion relation is of the form $\omega = ck$, (7.37) takes the form

$$(\nabla\theta)^2 - \frac{1}{c^2}\left(\frac{\partial\theta}{\partial t}\right)^2 = 0. \tag{7.38}$$

This is the basic equation of geometrical acoustics or optics. The characteristics defined by (7.35) are often called '*rays*' in geometrical acoustics.

7.4. Wave energy

Let us consider how the wave energy propagates. Here we use the Fourier integral result obtained in (7.17)–(7.19). The wave energy contained between two planes $x = x_1$ and $x = x_2$ is proportional to

$$Q(t) = \int_{x_1}^{x_2} |A|^2 dx = 2\pi \int_{x_1}^{x_2} \frac{F(k_0)F^*(k_0)}{t|\omega''(k_0)|} dx.$$

In this integral, k_0 is given by the relation, $\omega'(k_0) = x/t$. Here the two planes $x = x_1$ and $x = x_2$ move with the group velocities $\omega'(k_{01})$ and $\omega'(k_{02})$, respectively. It is advantageous to change the integration variable from x to k_0, using the relation

$$dx = t\,\omega''(k_0)dk_0.$$

Then we have

$$Q(t) = \pm\, 2\pi \int_{k_{01}}^{k_{02}} F(k_0)F^*(k_0)dk_0$$

where $\pm$ signs correspond to $\omega'' > 0$ and $\omega'' < 0$, respectively, and k_{01} and k_{02} are defined by

$$x_1 = t\,\omega'(k_{01}), \quad x_2 = t\,\omega'(k_{02}).$$

Since k_{01} and k_{02} are fixed wave numbers, $Q(t)$ is a constant. Therefore we have

$$\frac{dQ}{dt} = \frac{d}{dt}\int_{x_1(t)}^{x_2(t)} |A|^2 dx = 0$$

which gives

$$\int_{x_1}^{x_2} \frac{\partial |A|^2}{\partial t} dx + |A|^2_{x_2}\omega'(k_{02}) - |A|^2_{x_1}\omega'(k_{01}) = 0.$$

We have thus proven that the total amount of energy between any pair of planes moving with respective group velocities remains constant. The above integral relation can be converted to a differential equation by taking the limit $x_2 - x_1 \to 0$:

$$\frac{\partial |A|^2}{\partial t} + \frac{\partial}{\partial x}\left(\omega'(k_0)|A|^2\right) = 0. \tag{7.39}$$

This equation is a conservation equation relating the energy density $|A|^2$ and the energy flux $\omega'(k_0)|A|^2$. This equation shows that the group velocity $\omega'(k_0)$ is the propagation velocity of the energy. The three-dimensional extension of the energy equation takes the form

$$\frac{\partial |A|^2}{\partial t} + \nabla \cdot \left(\mathbf{v}_g |A|^2\right) = 0. \tag{7.40}$$

The three-dimensional energy equation (7.40) (as well as (7.39)) can be derived without resorting to Fourier integral but directly from the quasi-monochromatic wave expression (7.28). [See §6.3.4 and Reference: B. B. Kadomtsev, *Cooperative effects in plasmas*, Reviews of plasma physics **22** Kluer Academic/Consultants Bureau New York (2001) equation (5.113), where the electromagnetic field energy can be expressed in such a form.]

7.5. Nonlinear waves; distortion of wave profile

We consider a model equation

$$\frac{\partial \rho}{\partial t} \pm c_0 \frac{\partial \rho}{\partial x} = 0. \quad (c_0 = const.) \tag{7.41}$$

A more general wave equation

$$\frac{\partial^2 \rho}{\partial t^2} - c_0^2 \frac{\partial^2 \rho}{\partial x^2} = 0$$

can be factored into

$$\left(\frac{\partial \rho}{\partial t} + c_0 \frac{\partial \rho}{\partial x}\right)\left(\frac{\partial \rho}{\partial t} - c_0 \frac{\partial \rho}{\partial x}\right) = 0$$

which reduces to the model equation (7.41). The solution to (7.41) is $\rho(x,t) = f(x \mp c_0 t)$, where $f(x)$ is an arbitrary function which is required only to satisfy the prescribed initial conditions. The wave motion governed by (7.41) is a parallel translation of the initial shape $f(x)$ along the x-direction with the constant velocity $\pm c_0$. Now we consider a little more complicated equation

$$\frac{\partial \rho}{\partial t} + c(\rho)\,\frac{\partial \rho}{\partial x} = 0 \tag{7.42}$$

where the velocity c is not constant but a function of ρ. The characteristic equation associated with (7.42) is

$$\frac{dt}{1} = \frac{dx}{c(\rho)} = \frac{d\rho}{0}$$

which states that $\rho(x,t) = const.$ along the characteristic line:

$$\rho(x,t) = const: \quad \frac{dx}{dt} = c(\rho). \tag{7.43}$$

This statement can be put into mathematical form

$$\rho(x,t) = f\big(x - c(\rho(x,t))t\big). \tag{7.44}$$

By direct substitution one can show that (7.44) satisfies (7.42). The proof goes as follows.

$$\frac{\partial \rho}{\partial t} = f' \cdot \left[-c - t\,\frac{dc}{d\rho}\frac{\partial \rho}{\partial t}\right], \quad \frac{\partial \rho}{\partial x} = f' \cdot \left[1 - t\,\frac{dc}{d\rho}\frac{\partial \rho}{\partial x}\right]$$

where f' is the derivative with respect to the argument. Therefore we have

$$\frac{\partial \rho}{\partial t} + c\frac{\partial \rho}{\partial x} = \left(\frac{\partial \rho}{\partial t} + c\frac{\partial \rho}{\partial x}\right)\left(-\frac{dc}{d\rho}\right)\,tf'$$

$$\textit{or} \quad \left(\frac{\partial \rho}{\partial t} + c\frac{\partial \rho}{\partial x}\right)\left(1 + \frac{dc}{d\rho}\,tf'\right) = 0.$$

The second factor is not zero for an arbitrary t, and we arrive at the equality given by (7.42). It is observed from (7.44) that ρ is constant when the argument of f is constant ($\equiv \xi$):

$$\rho = f(\xi) = const. \quad when \quad x = \xi + c(f(\xi))t. \tag{7.45}$$

Note that $f(\xi)$ is just $\rho(x,0)$, the initial distribution of density ρ. We can summarize (7.45) by saying that the initial value of ρ corresponding to the initial position ξ is conserved along the straight line on (x,t) plane given by the second equation in (7.45). Equation (7.45) that paraphrases (7.44) constitute an equivalent representation of the implicit relation (7.44). The straight lines on the (x,t) plane in (7.45) are the characteristics of the *hyperbolic* partial differential equation (7.42). According to (7.45), the initial shape $\rho(x,0) = f(x)$ is translated along the positive x-direction with velocity $c(\rho)$. In contrast to (7.41), the translation velocity c is not constant but depends

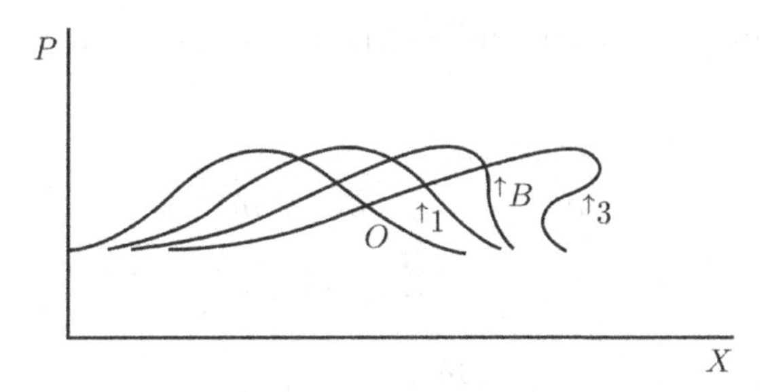

Fig. 7.3. Steepening of wave form, after Whitham (1974).

on ρ. Therefore the initial shape of ρ will be distorted. This distortion of the wave shape is due to the nonlinearity of (7.42). When $dc/d\rho > 0$, ρ of higher value propagates faster than ρ of smaller value. ρ of a higher value catches up the "runners" of smaller ρ ahead. In Fig. 7.3 [After G. B. Whitham, *Linear and Nonlinear Waves* (John Wiley, New York, 1974) Chapter 2], the distortion of original wave shape $\rho(x, 0)$ is shown for different times when $dc/d\rho > 0$. It is seen that the wave profile ultimately breaks to give a triple-valued solution (at $t = t_B$). The breaking starts at $t = t_B$ where the profile of ρ develops a vertical slope. In most physical problems $\rho(x,t)$ should be single-valued. Therefore, when breaking occurs, (7.42) is no longer valid to describe the physical process under consideration (shock or discontinuity is formed). A specific numerical example of the implicit relation (7.44) is given in the reference [R. K. Dodd, J. C. Eilbeck, J. D. Gibbon and H. C. Morris *Solitons and Nonlinear wave equations* (Academic Press, New York, 1982)], Dodd *et al.* (1982) in which the wave steepening is graphically shown.

7.6. Inclusion of dispersion; steepening vs dispersion

Reference: B. B. Kadomtsev, *Cooperative effects in plasmas*, Reviews of Plasma Physics Vol 22 Ed. V. D. Shafranov, Consultant Bureau New York (2001).

If a dispersion term is added in (7.42), that term can counter-balance the wave steepening to form a stationary wave profile; a solitary pulse (or soliton) in the form of a progressive wave. The dispersion term is characterized by $\sim \partial^3/\partial x^3$ [see (7.55) below]. As a specific example, we consider ion acoustic wave in a plasma. The ion acoustic wave is associated with the ions which provide the inertia of the medium supporting the wave, on the background of Boltzmann-distributed electrons (massless electrons are assumed). The electron pressure provides the restoring force which is necessary for the oscillation. The ion and electron equations, for one-dimensional motion, are

$$\frac{dv}{dt} = \frac{\partial v}{\partial t} + v\,\frac{\partial v}{\partial x} = \frac{e}{m_i}E, \tag{7.46}$$

$$\frac{\partial n_i}{\partial t} + \frac{\partial}{\partial x}(n_i v) = 0, \tag{7.47}$$

$$T_e\,\frac{\partial n_e}{\partial x} = -eEn_e, \tag{7.48}$$

$$\frac{\partial E}{\partial x} = 4\pi e(n_i - n_e). \tag{7.49}$$

In (7.48), we neglected the electron inertial force because the electrons can be considered to have infinite mobility during the much slower ion oscillations. Also the electrons have enough time to attain the equilibrium temperature (T_e) during the slow ion motion. The above equations are linearized with respect to the equilibrium state, $v = 0$, $n_e = n_i = n_0 = const$. Denoting the perturbed quantities by prime, we have from (7.48) and (7.49), assuming the phasor in the form $\sim e^{ikx-i\omega t}$,

$$ikT_e n_e' = -eEn_0,$$

$$ikE = 4\pi e(n_i' - n_e').$$

The above two equations yield, after eliminating E,

$$n_i' = (1 + k^2\lambda_e^2)n_e' \quad \lambda_e = \sqrt{\frac{T_e}{4\pi n_0 e^2}}. \tag{7.50}$$

(7.47) gives

$$-i\omega n_i' + n_0\,\frac{\partial v}{\partial x} = 0. \tag{7.51}$$

(7.48) gives

$$E = -\frac{T_e}{e}\,\frac{1}{1 + k^2\lambda_e^2}\,\frac{1}{v_{ph}}\,\frac{\partial v}{\partial x} \tag{7.52}$$

where $v_{ph} = \omega/k$ is the phase velocity and we used (7.50). Now (7.46) becomes upon using (7.52)

$$\frac{dv}{dt} + \frac{c_0^2}{v_{ph}(1 + k^2\lambda_e^2)}\,\frac{\partial v}{\partial x} = 0 \tag{7.53}$$

where $c_0 = \sqrt{T_e/m_i}$ is the sound velocity. This is the equation of nonlinear ion acoustic wave which was derived by calculating the driving force (the electric field term) linearly. In the linear approximation, $dv/dt = \partial v/\partial t = -i\omega v$, and (7.53) yields the linear dispersion relation.

$$\omega^2 = \frac{c_0^2 k^2}{1 + k^2\lambda_e^2}.$$

When $k^2\lambda_e^2 \ll 1$, the ion acoustic wave has no dispersion: $\omega/k = c_0 = const.$ We take into account weak dispersion effects. For ion acoustic waves propagating in the positive x-direction, the phase velocity is equal to $v_{ph} = c_0(1 + k^2\lambda_e^2)^{-1/2} \doteq c_0(1 - \frac{1}{2}k^2\lambda_e^2)$. Substituting this value into (7.53), we obtain

$$\frac{\partial v}{\partial t} + v\frac{\partial v}{\partial x} + c_0\left(1 - \frac{1}{2}k^2\lambda_e^2\right)\frac{\partial v}{\partial x} = 0. \tag{7.54}$$

Here we introduce a new frame of reference moving with velocity $c_0 : v = u - c_0$. Also we reinstate $\partial/\partial x$ by putting $k^2 = -\partial^2/\partial x^2$. Then (7.54) becomes

$$\frac{\partial u}{\partial t} + u\,\frac{\partial u}{\partial x} + \frac{1}{2}\,c_0\lambda_e^2\frac{\partial^3 u}{\partial x^3} = 0. \tag{7.55}$$

Korteweg–de Vries equation. (7.55) is the famous Korteweg–de Vries equation which is dealt with by many authors in the literature. The title of the original paper written by the two authors is interesting: D. J. Korteweg and G. de Vries, *On the change of form of long waves advancing in a rectangular canal, and on a new type of long stationary waves*, Phil. Mag., **39** (1895) p. 422-443. Now we investigate what effect is brought about by the dispersion term. We seek the solution of (7.55) in the form of progressive wave,

$$u(x,t) = u(\xi) \quad where \quad \xi = x - ct.$$

Then we have

$$-c\frac{du}{d\xi} + u\frac{du}{d\xi} + \mu\frac{d^3u}{d\xi^3} = 0 \quad (\mu = \frac{1}{2}c_0\lambda_e^2).$$

Integrating this equation gives

$$\mu\,\frac{d^2u}{d\xi^2} + \frac{u^2}{2} - cu = const. \equiv 0$$

$$or \quad \mu\,\frac{d^2u}{d\xi^2} = cu - \frac{u^2}{2} \equiv -\frac{d}{du}\phi(u) \tag{7.56}$$

$$where \quad \phi(u) = -\frac{c}{2}u^2 + \frac{u^3}{6} + c' \quad (c' = const.). \tag{7.57}$$

(7.56) should be compared with Newton's equation for a particle of mass μ subject to a force of potential $\phi(x)$: $\mu\ \frac{d^2x}{dt^2} = -\frac{d\phi}{dx}$ with the correspondence $u \rightarrow x$ and $\xi \rightarrow t$. (7.56) allows for an energy integral:

$$\frac{\mu}{2}\left(\frac{du}{d\xi}\right)^2 + \phi(u) = const. \equiv E. \tag{7.58}$$

Letting the constant E ('energy') absorb the constant c' in (7.57), we write the 'potential' in the form

$$\phi(u) = -\frac{c}{2}u^2 + \frac{u^3}{6}. \tag{7.59}$$

The potential $\phi(u)$ is plotted in Fig. 7.4.

The function ϕ has a maximum at $u = 0$, a minimum at $u = 2c$, and $\phi = 0$ at $u = 0$ and $3c$. Depending upon the values of E, different kinds of solutions emerge. When the energy E takes the value somewhere between 0 and ϕ_{min} (corresponding to $u = 2c$), the particle executes asymmetric oscillation in (u, ξ)-plane. This is because the right slope of the potential well is steeper than the left slope. A more interesting solution emerges when $E = 0$. In this case the particle starts to move near the point $u = 0$ (where $\phi = 0$), rolls down to the bottom of the well, and climbs up the hill until to reach the point $u = 3c$ which is the turning point. The turned-around particle takes the backward course to reach the point $u = 0$, taking infinite time. Thus the particle orbit has only one crest ($u = 3c$) in this case. When $E = 0$, (7.58) gives

$$\sqrt{\frac{c}{\mu}}\,\xi = \int \frac{du}{u\sqrt{1-\frac{u}{3c}}} \equiv I. \tag{7.60}$$

To calculate the integral I, we put

$$\sqrt{1-\frac{u}{3c}} = \psi, \quad u = 3c(1-\psi^2).$$

Then we have

$$I = -2\int \frac{d\psi}{1-\psi^2} = -\int d\psi\left(\frac{1}{1-\psi} + \frac{1}{1+\psi}\right).$$

Note that $1 - u/3c > 0$ and also $u/3c > 0$. Therefore $\psi < 1$. Thus we obtain

$$I = ln\frac{1-\psi}{1+\psi},$$

$$\psi = \frac{1-e^{\sqrt{\frac{c}{\mu}}\xi}}{1+e^{\sqrt{\frac{c}{\mu}}\xi}}, \quad u = \frac{3c}{cosh^2\left(\frac{1}{2}\sqrt{\frac{c}{\mu}}\,\xi\right)}. \tag{7.61}$$

A plot u vs ξ is shown in Fig. 7.4.

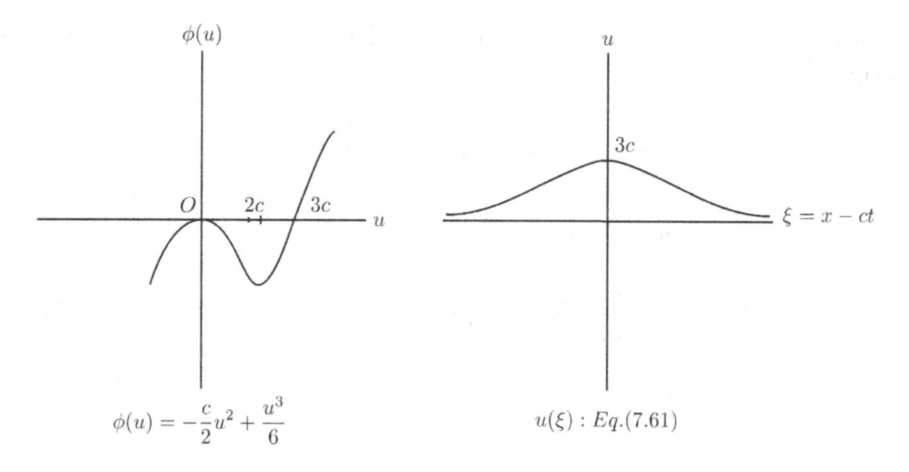

Fig. 7.4. Particle motion in a given 'potential'.

It is suggestive to write (7.61) in the form

$$u = \frac{u_0}{cosh^2 \frac{\xi}{\Delta}}$$

with $u_0 = 3c$: soliton amplitude (height), and $\Delta = 2\sqrt{\frac{\mu}{c}}$: soliton thickness. Since c denotes the soliton speed, a soliton which is moving faster has taller height. Also we have $u_0\Delta^2 = 12\mu = 6c_0\lambda_e^2 = const$; the product of the square of the thickness and the amplitude is constant. If we have a *positive dispersion*, the LHS of (7.56) should have the negative sign (change $\mu \to -\mu$). In this case, we still have the soliton solution in the form of (7.61) provided the sign of c is changed. Thus, the shape of the soliton is upside-down, and moves toward the negative direction.

Boussinesq equation. Equation (7.53) can be converted to a differential equation by putting $k = -i\frac{\partial}{\partial x}$ and $\omega = i\frac{\partial}{\partial t}$ without making the weak dispersion approximation, $k^2\lambda_e^2 \ll 1$. In this case, we obtain

$$\left(\frac{\partial^2}{\partial t^2} - \lambda_e^2 \frac{\partial^2}{\partial x^2}\frac{\partial^2}{\partial t^2} - c_0^2 \frac{\partial^2}{\partial x^2}\right) v = \left(1 - \lambda_e^2 \frac{\partial^2}{\partial x^2}\right) \frac{\partial^2}{\partial t \partial x}\frac{v^2}{2}. \quad (7.62)$$

The operator acting on v in the left hand side of (7.62) is the Boussinesq wave operator. This equation describes the waves propagating in both $\pm x$ direction while the KdV equation (7.55) is relevant to the wave propagating in only $+x$ direction.

Remark. Equation (7.48) can be integrated upon introducing the electric potential $\mathbf{E} = -\nabla\phi$:

$$n_e = n_0 \, e^{\frac{e\phi}{T_e}} \quad or \quad \phi = \frac{T_e}{e} \, ln \frac{n_e}{n_0}. \tag{7.63}$$

Then the Poisson equation reads

$$\frac{\partial^2 \phi}{\partial x^2} = 4\pi e \left(n_0 \, e^{\frac{e\phi}{T_e}} - n_i \right). \tag{7.64}$$

Now we have the ion continuity equation

$$\frac{\partial n_i}{\partial t} + \frac{\partial}{\partial x}(n_i v) = 0 \tag{7.65}$$

and the ion equation of motion

$$\frac{\partial v}{\partial t} + v \frac{\partial v}{\partial x} = -\frac{e}{m_i} \frac{\partial \phi}{\partial x}. \tag{7.66}$$

The above three equations are fully nonlinear and include the dispersion effect (i.e. effect of charge separation) for the ion acoustic wave. Equations (7.64)–(7.66) have been analyzed by various perturbation methods. In Appendix C, it is shown that the method of *strained coordinates* derives the KdV equation from (7.64)–(7.66).

Note. If we assume the charge neutrality, putting $n_i \simeq n_e$ and using (7.46) and (7.63) give

$$\frac{\partial v}{\partial t} + v \frac{\partial v}{\partial x} = -\frac{T_e}{m_i} \frac{1}{n_i} \frac{\partial n_i}{\partial x}. \tag{7.67}$$

(7.65) and (7.67) are the equations of ideal gas dynamics for an adiabatic law of $\gamma = 1$ (γ is the ratio of specific heats). The wave governed by (7.65) and (7.67) steepens its own wave profile and is developed into a shock wave. This can be rigorously proven by using *Riemann invariants.* Reference: H. J. Lee, *Nonlinear analysis of the cold fluid-Poisson plasma by using the characteristic method*, J. Korean Phys. Soc. **69** (2016) p. 1191–1211.

Sagdeev potential. We look for an exact nonlinear solution of the ion acoustic wave equations (7.64)–(7.66) under given boundary conditions. The solution is sought in the form of a progressive wave. Write $\phi(x, t) = \phi(\xi)$ with $\xi = x - ut$. Then, $\partial/\partial x = \partial/\partial\xi$ and $\partial/\partial t = -u \, \partial/\partial\xi$, and (7.56) gives

$$-u\,\frac{\partial n_i}{\partial \xi}+\frac{\partial}{\partial \xi}(n_i v)=0\longrightarrow n_i(v-u)=const.\equiv c_1.$$

The boundary conditions are: $n_i=n_0; v=0$ at $\xi\to\infty$, which gives

$$n_i(v-u)=-n_0 u=c_1. \tag{7.68}$$

From (7.66) we obtain

$$-u\,\frac{\partial v}{\partial \xi}+v\frac{\partial v}{\partial \xi}=-\frac{e}{m_i}\frac{\partial \phi}{\partial \xi}\longrightarrow (v-u)^2+2\,\frac{e\phi}{m_i}=u^2 \tag{7.69}$$

where we used the boundary condition, $\phi=v=0$ at $\xi=\infty$. Eliminating v from (7.68) and (7.69), we obtain

$$n_i=\frac{n_0 u}{\sqrt{u^2-\frac{2e\phi}{m_i}}}. \tag{7.70}$$

Substituting (7.70) into (7.64), we can write

$$\frac{d^2\phi}{d\xi^2}=4\pi e n_0\left[e^{\frac{e\phi}{T_e}}-\frac{u}{\sqrt{u^2-\frac{2e\phi}{m_i}}}\right]\equiv-\frac{dV(\phi)}{d\phi}. \tag{7.71}$$

Equation (7.71) is again Newton's equation of a particle moving in a potential $V(\phi)$, which is called the *Sagdeev potential* and takes the form

$$V(\phi)=4\pi e n_0\left[-\frac{T_e}{e}e^{\frac{e\phi}{T_e}}-\frac{m_i u^2}{e}\sqrt{1-\frac{2e\phi}{m_i u^2}}\right]+C.$$

$$(C:\ integration\ const.)$$

Let us determine C so that $V(\phi=0)=0$. Thus we obtain

$$V(\phi)=-4\pi n_0\left[T_e\,e^{\frac{e\phi}{T_e}}+m_i u^2\sqrt{1-\frac{2e\phi}{m_i u^2}}-T_e-m_i u^2\right]. \tag{7.72}$$

Note that $V(\phi)$ in (7.72) satisfies $dV/d\phi=0$ at $\phi=0$. So it is possible to have the shape of the potential *well* as shown in Fig. 7.5. [A shape of potential *hill* is also possible.] A particle situated at the origin $V=\phi=0$ which is the point of unstable equilibrium draws a 'trajectory' of a solitary pulse. When there are dissipations, the particle loses energy while it bounces back and forth in the potential well. After each cycle, the amplitude of oscillation decreases, and these ever diminishing oscillations repeat several times until the particle

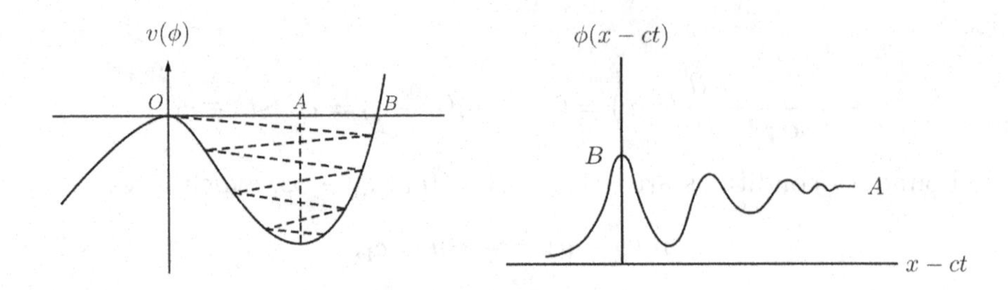

Fig. 7.5. Potential in the presence of dissipation.

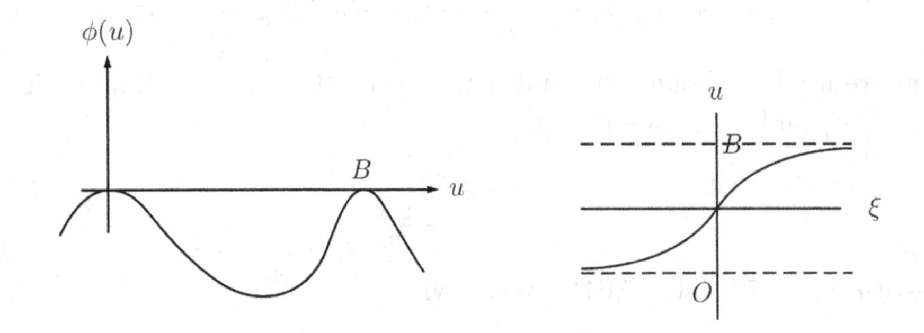

Fig. 7.6. Shock-like transition.

settles down at the bottom of the well. The plot ϕ vs ξ is shown in Fig. 7.5. It represents the quasi-shock structure which is called the *ion acoustic shock*.

When the potential has two maxima as shown in Fig. 7.6, the corresponding 'motion' is is described by a hyperbolic tangent function. The shape of the curve is a *shock-like transition*.

Nonlinear Schrödinger equation. The nonlinear Schrödinger equation describes amplitude modulation of a monochromatic wave in a nonlinear medium (see Fig. 7.1).

The amplitude of a wave can be modulated by interactions among different Fourier components. Parametric interaction among waves in plasma often results in amplitude-modulated wave of normal mode. Here we derive heuristically the nonlinear Schrödinger equation by *derivative expansion* technique applied to a *nonlinear dispersion relation*. In a weakly nonlinear system, the dispersion relation can be amplitude-dependent in the form

$$\omega = \omega(\mathbf{k}, |A|^2). \tag{7.73}$$

Such an amplitude-dependent dispersion relation emerges in nonlinear optics as well as in plasmas when the dielectric permeability of the medium depends on the electric field of the wave propagated in it. Suppose that the wave function $\psi(\mathbf{r}, t)$ takes the form

$$\psi(\mathbf{r}, t) = A(\mathbf{r}, t)e^{i\mathbf{k}_0\cdot\mathbf{r} - i\omega_o(\mathbf{k}_0)t} \tag{7.74}$$

where ω_0 corresponds to the eigen frequency of the plane harmonic wave with wave vector $\mathbf{k}_0$. We assume that the wave propagates along the x-direction: $\mathbf{k}_0 = \hat{\mathbf{x}}k_0$. We also assume that, the wave amplitude $A(\mathbf{r}, t)$ depends not only on x but also on the perpendicular coordinates. So we rewrite (7.74) as

$$\psi(\mathbf{r}, t) = A(x, \mathbf{r}_\perp, t)e^{i\mathbf{k}_0\cdot\mathbf{r} - i\omega_o(\mathbf{k}_0)t}. \tag{7.75}$$

We regard ω and k in the nonlinear dispersion relation (7.73) as corresponding to the operators: $\omega \to i\frac{\partial}{\partial t}$, and $\mathbf{k} \to -i\nabla$. Differentiating (7.75) gives

$$i\frac{\partial\psi}{\partial t} = \left[\omega_0 A + i\left(\frac{\partial}{\partial t}\right)_\epsilon A\right] e^{-i\omega_0 t + ik_0 x}$$

$$-i\nabla\psi = (\hat{\mathbf{x}}k_0 A - i\nabla_\epsilon A)e^{-i\omega_0 t + ik_0 x}$$

where $(\frac{\partial}{\partial t})_\epsilon$ and ∇_ϵ denote the slow changes of the amplitude $A(\mathbf{r}, t)$. Explicitly, we have the correspondence:

$$\omega = \omega_0 + i\left(\frac{\partial}{\partial t}\right)_\epsilon, \quad \mathbf{k} = \hat{\mathbf{x}}k_0 - i\nabla_\epsilon.$$

And the slow changes can be found from the nonlinear dispersion relation (7.73) after Taylor expansion of $\omega(\mathbf{k}, |A|^2)$ around the wave number k_0 and the frequency ω_0. We assume that the wave propagates in an isotropic medium, so that the dispersion relation reads

$$\omega = \omega(k, |A|^2) \tag{7.76}$$

where we assume that the wave vector $\mathbf{k}$ is predominantly along the $\hat{\mathbf{x}}$-direction:

$$\begin{aligned} k &= \sqrt{k_x^2 + k_\perp^2} = k_x\sqrt{1 + \frac{k_\perp^2}{k_x^2}} \simeq k_x\left(1 + \frac{k_\perp^2}{2k_x^2}\right) = k_x + \frac{1}{2}\frac{k_\perp^2}{k_x} \\ &\simeq k_0 + \left(k_x - k_0 + \frac{1}{2}\frac{k_\perp^2}{k_0}\right). \end{aligned} \tag{7.77}$$

Note that the quantity in the parenthesis is much smaller than k_0. Substituting (7.77) into (7.76) and expanding the right hand side of (7.76) in a Taylor series, we obtain

$$\omega = \omega_0 + \left(\frac{\partial \omega}{\partial k}\right)_{k_0} \left(k_x - k_0 + \frac{1}{2}\frac{k_\perp^2}{k_0}\right) + \frac{1}{2}\left(\frac{\partial^2 \omega}{\partial k^2}\right)_{k_0} (k_x - k_0)^2$$
$$+ \left(\frac{\partial \omega}{\partial |A|^2}\right)_{k_0} |A|^2 + \cdots \tag{7.78}$$

where $\omega_0 = \omega(k_0, |A|^2 = 0)$. Using the relations, $\omega - \omega_0 = i(\frac{\partial}{\partial t})_\epsilon$, $k_x - k_0 = -i(\frac{\partial}{\partial x})_\epsilon$, and $k_\perp^2 = -\nabla_\perp^2$ in (7.78), we obtain the equation for the slow change of the amplitude $A(\mathbf{r}, t)$:

$$i\left[\frac{\partial}{\partial t} + \left(\frac{\partial \omega}{\partial k}\right)_{k_0} \frac{\partial}{\partial x}\right] A + \frac{1}{2}\left(\frac{\partial^2 \omega}{\partial k^2}\right)_{k_0} \frac{\partial^2 A}{\partial x^2}$$
$$+\frac{1}{2}\left(\frac{\partial \omega}{\partial k}\right)_{k_0} \frac{\nabla_\perp^2 A}{k_0} - \left(\frac{\partial \omega}{\partial |A|^2}\right)_{k_0} |A|^2 A = 0. \tag{7.79}$$

Equation (7.79) is called *nonlinear parabolic equation* [Reference: Karpman, V. I. *Non-linear waves in dispersive media*, Pergamon, 1975]. If we neglect the variation of A in the perpendicular direction ($\nabla_\perp^2 A = 0$), (7.79) is the nonlinear Schrödinger equation.

7.7. Inverse scattering method

This is a mathematical method to find the exact solution of the KdV equation. The references are supplied at the end of this section. The KdV equation is put into the form

$$u_t - 6uu_x + u_{xxx} = 0 \tag{7.80}$$

where the subscript t or x denotes differentiation: $(\cdots)_t = \partial(\cdots)/\partial t = \partial_t(\cdots)$. We introduce a new function ψ via the transformation

$$u = \frac{\psi_{xx}}{\psi} + \lambda(t), \quad \textit{or} \quad \psi_{xx} + (\lambda - u)\psi = 0. \tag{7.81}$$

This is a Schrödinger equation with potential u and eigenvalue λ. With the transformation (7.81), the KdV equation (7.80) can be put into the form

$$\frac{d\lambda}{dt}\psi^2 + (\psi Q_x - \psi_x Q)_x = 0 \tag{7.82}$$

where

$$Q = \psi_t + u_x\psi - 2(u + 2\lambda)\psi_x$$
$$= \psi_t + \psi_{xxx} - 3(u + \lambda)\psi_x \quad [\textit{per } (7.84) \textit{ below}]. \tag{7.83}$$

(7.82) is an important development toward obtaining solutions of (7.80) and requires quite an ingenuity to arrive at the expression. We note that Q in (7.82) satisfies the Schrödinger equation in the form (7.81) if $\lambda_t = 0$. Q may be determined as in (7.83) once (7.82) is postulated. Here we only check the correctness of (7.82) by going through the following algebra.

Differentiating the Schrödinger equation (7.81),

$$\psi_{xxx} + (\lambda - u)\psi_x - u_x\psi = 0, \tag{7.84}$$

$$\psi_{xxt} + (\lambda_t - 6uu_x + u_{xxx})\psi + (\lambda - u)\psi_t = 0. \tag{7.85}$$

In (7.85) we used (7.80) for u_t. Since $\psi_{xxt} = (\psi_t)_{xx}$, (7.85) can be written as

$$(\partial_{xx} - u + \lambda)\,\psi_t + (u_{xxx} - 6uu_x)\psi + \lambda_t\psi = 0. \tag{7.86}$$

We have an identity

$$u_{xxx}\psi = (\partial_{xx} - u + \lambda)\,(u_x\psi) - 2u_{xx}\psi_x \tag{7.87}$$

which can be proven by direct differentiation. Using (7.87) in (7.86) gives

$$(\partial_{xx} - u + \lambda)\,(\psi_t + u_x\psi) - 2(3uu_x\psi + u_{xx}\psi_x) + \lambda_t\psi = 0. \tag{7.88}$$

To change $u_{xx}\psi_x$, consider

$$(u\psi_x)_{xx} = (u_x\psi_x + u\psi_{xx})_x = u_{xx}\psi_x + 2u_x\psi_{xx} + u\psi_{xxx}. \tag{7.89}$$

Using (7.81) and (7.84) in the above yields

$$\begin{aligned} u_{xx}\psi_x &= (u\psi_x)_{xx} - u(u - \lambda)\psi_x + 2\lambda u_x\psi - 3uu_x\psi \\ &= (\partial_{xx} - u + \lambda)\,(u\psi_x) + 2\lambda u_x\psi - 3uu_x\psi. \end{aligned}$$

Therefore,

$$\begin{aligned} u_{xx}\psi_x + 3uu_x\psi &= (\partial_{xx} - u + \lambda)\,(u\psi_x) + 2\lambda[\psi_{xxx} + (\lambda - u)\psi_x] \quad [per\ (7.84)] \\ &= (\partial_{xx} - u + \lambda)\,(u + 2\lambda)\psi_x. \end{aligned} \tag{7.90}$$

Using (7.90) in (7.88) yields

$$(\partial_{xx} - u + \lambda)\,Q = -\lambda_t\psi. \tag{7.91}$$

The above equation is multiplied by ψ to obtain

$$-\lambda_t\psi^2 = \psi(\partial_{xx} - u + \lambda)Q = \psi Q_{xx} + Q(\lambda - u)\psi$$
$$= \psi Q_{xx} - Q\psi_{xx} = (\psi Q_x - \psi_x Q)_x$$

which is (7.82). Although we have successfully checked the key equation (7.82), we wonder how the original authors were able to discover it. To ease this curiosity, readers are advised to consult more references on the soliton theory, for example: A. C. Scott, F. Y. F. Chu, and D. W. McLaughlin (1973) *The soliton: A new concept in applied science* Proc. IEEE **61** (1973) p. 1443–1483, in which also Gel'fand–Levitan equation (later) is also derived.

Going back to the time-independent Schrödinger equation (7.81),

$$\frac{\partial^2\psi}{\partial x^2} + [\lambda - u(x,t)]\psi = 0. \qquad (7.81)$$

Note that ψ and λ depend parametrically on t. Here we assume that the solutions $u(x,t)$ of the KdV equation are smooth, bounded, and tend to zero as $|x| \to \infty$. Then (7.81) will have a finite number of bound states with energies $\lambda_n = -\kappa_n^2 \quad (n = 1, 2, \cdots, N)$ and a continuous spectrum for positive $\lambda = k^2$. At fixed t, the eigen functions ψ_n associated with the bound states λ_n can be written as

$$\psi_n(x,t) = C_n(t)\; e^{-\kappa_n x} \quad as \quad x \to +\infty \tag{7.92}$$

since $u \to 0$ as $x \to \infty$. The bound state wave function ψ_n can be normalized: $\int_{-\infty}^{\infty} \psi_n^2 dx = 1$.

The boundary conditions for the continuum states are defined as

$$\psi = e^{-ikx} + R(k,t)e^{ikx} \quad as \quad x \to +\infty, \tag{7.93}$$

$$\psi = T(k,t)e^{-ikx} \quad as \quad x \to -\infty, \tag{7.94}$$

which are scattering solutions of (7.81) with R and T being the reflection and transmission coefficients, respectively ($|R|^2 + |T|^2 = 1$).

According to the inverse scattering technique of quantum mechanics, we can find the potential $u(x,t)$ in the Schrödinger equation (7.81) from a given initial potential $u(x,0)$ and the scattering data by the following three-step procedure.

i) Direct problem: Solve the time-independent Schrödinger equation (7.81) for a given initial potential $u(x,0)$, and obtain the discrete eigenvalues λ_n $(n = 1, 2, ...N)$ and scattering data, the reflection and the transmission coefficients $R(k,0)$ and $T(k,0)$ as $|x| \to \infty$ at $t = 0$.

ii) Evolution of the scattering data: Use (7.82) and (7.83), and the asymptotic boundary conditions (7.92)–(7.94) to determine the time evolution of the scattering data.
iii) Inverse problem: Find the potential $u(x,t)$ from the scattering data at an arbitrary time t with the aid of Gel'fand–Levitan equation (which is a linear integral equation) (see below).

[Solving nonlinear KdV equation is reduced to the problem of solving two linear equations, Schrödinger equation in step i) and the linear Gel'fand–Levitan integral equation in step iii).]

Now we first proceed to step ii) in which (7.82) plays the crucial role. For discrete eigenvalues, integrating (7.82) over x gives

$$\frac{d\lambda_n}{dt}\int_{-\infty}^{\infty}\psi_n^2 dx + [\psi_n Q_{nx} - \psi_{nx}Q_n]_{-\infty}^{+\infty} = 0.$$

The bracketed term vanishes and we have

$$\frac{d\lambda_n}{dt} = 0 \quad or \quad \kappa_n(t) = \kappa_n(0). \tag{7.95}$$

The discrete wave number κ_n is constant, independent of t. For the continuous spectrum, the wave function is not normalized, and we cannot deduce $\lambda_t = 0$ as is done above. But we may take λ to be a constant by choosing a specific wave with a fixed value of k and we have in either case

$$(\partial_{xx} + \lambda - u)Q = 0, \quad or \quad (\psi Q_x - \psi_x Q)_x = 0. \tag{7.96}$$

(7.96) indicates that Q satisfies the same Schrödinger equation as ψ does.

Now we integrate (7.82) (with $\lambda_t = 0$) twice with respect to x to obtain, for discrete eigen functions,

$$\frac{Q_n}{\psi_n} = F_n(t) + D_n(t)\int^x \frac{dx}{\psi_n^2} \tag{7.97}$$

where D and F are integration constants. Since $1/\psi_n^2$ diverges exponentially as $x \to \infty$, we must set $D_n = 0$. Then we have per (7.83)

$$F_n\psi_n = \frac{\partial\psi_n}{\partial t} + \frac{\partial^3\psi_n}{\partial x^3} - 3(u+\lambda_n)\frac{\partial\psi_n}{\partial x}. \tag{7.98}$$

Multiply the above equation by ψ_n and integrate with respect to x to write

$$F_n\int_{-\infty}^{\infty}\psi_n^2 dx = \frac{1}{2}\frac{\partial}{\partial t}\int_{-\infty}^{\infty}\psi_n^2 dx + \int_{-\infty}^{\infty}dx\left[\psi_n\frac{\partial^3\psi_n}{\partial x^3} - 3(u+\lambda_n)\psi_n\frac{\partial\psi_n}{\partial x}\right]. \tag{7.99}$$

The wave function is normalized to unity, $\int_{-\infty}^{\infty} \psi_n^2 dx = 1$, so

$$F_n = \int_{-\infty}^{\infty} dx \left[\psi_n \frac{\partial^3 \psi_n}{\partial x^3} - 6\lambda_n \psi_n \frac{\partial \psi_n}{\partial x} - 3 \frac{\partial \psi_n}{\partial x} \frac{\partial^2 \psi_n}{\partial x^2} \right]$$

$$= \left[\psi_n \frac{\partial^2 \psi_n}{\partial x^2} - 2(\frac{\partial \psi_n}{\partial x})^2 - 3\lambda_n \psi_n^2 \right]_{-\infty}^{+\infty} = 0$$

where we used (7.81) and the property that ψ and its derivatives vanish at infinity because ψ is a bound state. $F_n = 0$ means

$$Q_n = \frac{\partial \psi_n}{\partial t} + \frac{\partial^3 \psi_n}{\partial x^3} - 3(u + \lambda_n) \frac{\partial \psi_n}{\partial x} = 0 \quad with \ \lambda_n = -\kappa_n^2. \tag{7.100}$$

Substituting (7.92) into (7.100) gives

$$\frac{dC_n}{dt} - 4\kappa_n^3 C_n(t) = 0 \tag{7.101}$$

which is integrated to yield

$$C_n(t) = C_n(0) e^{4\kappa_n^3 t}. \tag{7.102}$$

For the continuum state $\lambda = k^2$, $\psi(x,t)$ evolves according to

$$Q = \frac{\partial \psi}{\partial t} + \frac{\partial^3 \psi}{\partial x^3} - 3(u + k^2) \frac{\partial \psi}{\partial x} = F(t)\psi + D(t)\psi \int^x \frac{dx}{\psi^2}. \tag{7.103}$$

Here we exploit the fact that Q and ψ satisfy the same Schrödinger equation, and so they should be different at most by some x-independent factor. Thus we can discard the last term by putting $D(t) = 0$. It follows from the asymptotic behavior as $x \to -\infty$ (per (7.94) and note $u \simeq 0$)

$$\frac{\partial T(k,t)}{\partial t} + \left[4ik^3 - F(t)\right] T(k,t) = 0. \tag{7.104}$$

For $x \to +\infty$, where the asymptotic behavior is given by (7.93) and $u \simeq 0$, we obtain two relations owing to the independence of e^{ikx} and e^{-ikx},

$$F(t) = 4ik^3 \tag{7.105}$$

$$\frac{\partial R(k,t)}{\partial t} - 8ik^3 R(k,t) = 0 \quad or \quad R(k,t) = R(k,0) e^{8ik^3 t} \tag{7.106}$$

which in turn gives for (7.104)

$$\frac{\partial T(k,t)}{\partial t} = 0 \quad or \quad T(k,t) = T(k,0). \tag{7.107}$$

The foregoing results obtained in step ii) determine the scattering data at an arbitrary time t in terms of given initial data, which are summarized as follows.

$$\kappa_n(t) = \kappa_n(0), \tag{7.108}$$

$$C_n(t) = C_n(0)e^{4\kappa_n^3 t}, \tag{7.109}$$

$$T(k,t) = T(k,0), \tag{7.110}$$

$$R(k,t) = R(k,0)e^{8ik^3 t}. \tag{7.111}$$

According to the inverse scattering theory of quantum mechanics, knowledge of $\kappa_n(t)$ and $C_n(t)$ $(n = 1, 2, ...N)$ and $R(k,t)$ is sufficient to construct the potential $u(x,t)$ from its initial value $u(x,0)$. The potential $u(x,t)$, the required solution of the KdV equation (7.80), is then given by the relation

$$u(x,t) = -2\frac{d}{dx}K(x,x,t). \tag{7.112}$$

Here the function $K(x,x,t)$ is provided by the solution to the Gel'fand–Levitan integral equation

$$K(x,y,t) + B(x+y,t) + \int_x^\infty B(y+z,t)K(x,z,t)dz = 0 \tag{7.113}$$

with the kernel B given by

$$B(\xi,t) = \sum_{n=1}^{N} C_n^2(t)e^{-\kappa_n \xi} + \frac{1}{2\pi}\int_{-\infty}^{\infty} R(k,t)e^{ik\xi}dk. \tag{7.114}$$

The introduction of $\kappa_n(t)$, $C_n(t)$, and $R(k,t)$ into (7.114) determines $B(\xi,t)$:

$$B(\xi,t) = \sum_{n=1}^{N} C_n^2(0)e^{8\kappa_n^3 t - \kappa_n \xi} + \frac{1}{2\pi}\int_{-\infty}^{\infty} R(k,0)e^{i8k^3 t + ik\xi}dk. \tag{7.115}$$

Thus the solution $K(x,y,t)$ for (7.113) determines the desired solution or potential $u(x,t)$ per (7.112). However, before we address the Gel'fand–Levitan integral equation (7.113) to determine $K(x,y,t)$ (step iii), we must find $C_n(0)$ and $R(k,0)$ by solving the direct scattering problem of the Schrödinger equation taking the initial function $u(x,0)$ as the potential (step i: direct problem).

To proceed with a specific example, let us consider an initial potential

$$u(x,0) \equiv u_0(x) = -2sech^2 x. \quad \text{(see Fig. 7.7)} \tag{7.116}$$

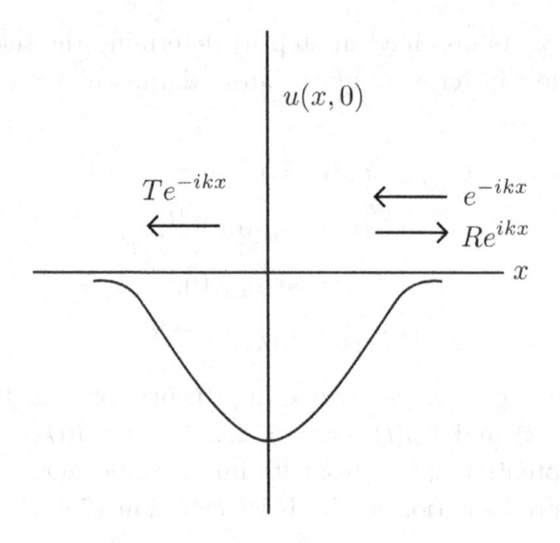

Fig. 7.7. Initial shape of potential.

Clearly the shape of the potential indicates *reflectionless* ($R(k,0)=0$), as is also shown below. We first consider the continuum states by solving

$$\left(\frac{d^2}{dx^2}+k^2+2sech^2x\right)\psi=0 \tag{7.117}$$

subject to the asymptotic boundary conditions (7.93) and (7.94). The general solution is [see Dodd *et al.* 1982]

$$\psi(x)=A(tanh\ x+ik)e^{-ikx}+B(tanh\ x-ik)e^{ikx} \tag{7.118}$$

as direct substitution proves. Let us apply the asymptotic boundary conditions (7.93) and (7.94).

$$x\to-\infty\ :\ \ \psi=A(-1+ik)e^{-ikx}+B(-1-ik)e^{ikx}=Te^{-ikx},$$

$$x\to+\infty\ :\ \ \psi=A(1+ik)e^{-ikx}+B(1-ik)e^{ikx}=e^{-ikx}+Re^{ikx}.$$

The above two equations give $B=0,\ R=0$ (*reflectionless*), $A=\frac{1}{1+ik}$, $T=\frac{ik-1}{ik+1}$, and $|T|^2=1$.

To obtain the discrete spectrum, we should solve

$$\left(\frac{d^2}{dx^2}-\kappa_n^2+2sech^2x\right)\psi=0. \tag{7.119}$$

Putting $k=i\kappa$ in (7.118), we can write the general solution in the form

$$\psi(x)=A'(tanh\ x-\kappa)e^{\kappa x}+B'(tanh\ x+\kappa)e^{-\kappa x}.$$

Since $\psi = 0$ as $|x| \to \infty$, we set $B' = 0$ (for ψ to be zero at $x = -\infty$) and $\kappa = 1$ (to satisfy the condition at $x = +\infty$). Therefore we obtain

$$\psi(x) = A'(tanh\ x - 1)e^x = -A'\ sech\ x. \tag{7.120}$$

We have only one bound state corresponding to $\kappa = 1$ for the potential (7.116). Normalizing the wave function (7.120) gives

$$\psi_{\kappa=1}(x) = \frac{1}{\sqrt{2}}\ sech\ x = \frac{\sqrt{2}}{e^x + e^{-x}} \sim \sqrt{2}\ e^{-x} \quad as \quad x \to \infty.$$

Therefore we have from (7.92) $C_{n=1}(0) = \sqrt{2}$, and

$$C_{n=1}(t) = \sqrt{2}\ e^{4t} \quad per\ (7.102), \tag{7.121}$$

$$B(x,t) = 2\ e^{8t-x}. \tag{7.122}$$

Then the Gel'fand–Levitan equation (7.113) becomes

$$K(x,y,t) + 2e^{8t-x-y} + 2\int_x^\infty e^{8t-y-z}K(x,z,t)dz = 0. \tag{7.123}$$

We seek the solution of (7.123) in the form $K(x,y,t) = F(x,t)e^{-y}$ to obtain

$$F + 2e^{8t-x} + 2Fe^{8t}\int_x^\infty e^{-2z}dz = 0, \quad or \quad F(x,t) = \frac{-2e^{8t-x}}{1+e^{8t-2x}}.$$

$$Therefore \quad K(x,x,t) = F(x,t)e^{-x} = \frac{-2\ e^{8t-2x}}{1+e^{8t-2x}}.$$

Thus the final solution is

$$u(x,t) = -2\frac{dK(x,x,t)}{dx} = \frac{\partial}{\partial x}\frac{4e^{8t-2x}}{1+e^{8t-2x}} = \frac{-8e^{8t-2x}}{(1+e^{8t-2x})^2}$$

$$= \frac{-8e^{-2\vartheta}}{(1+e^{-2\vartheta})^2} = -2\ \frac{4}{e^{2\vartheta}+2+e^{-2\vartheta}} = -2sech^2\vartheta \quad (\vartheta = x - 4t)$$

which is the single soliton solution of amplitude -2 and speed 4. The above solution, comparing with (7.61), suggests that the KdV equation is solved by a progressive wave form of $u(x - v_0 t)$ with the wave speed $v_0 = 4$.

Addendum Here we present the steps leading to the Gel'fand–Levitan integral equation (7.113). We begin with the Schrödinger equation in the form

$$-\frac{d^2\psi}{dx^2} + u(x,t)\psi = E(t)\psi \tag{7.124}$$

where $\psi = \psi(x, E(t), t)$. For fixed t, (7.124) determines ψ for a 'particle' in the potential $u(x,t)$. Since the solutions $u(x,t)$ of the KdV equation are smooth, bounded, and tend to zero as $x \to \pm\infty$, (7.124) will have at most a finite number of negative energy bound states ($E = -\kappa_n^2$, $n = 1, 2, \cdots N$) and a continuous spectrum for positive E ($E = k^2$, k *real*). We write (7.124) in the form

$$-\frac{d^2\psi}{dx^2} + u(x,t)\psi = k^2\psi \tag{7.125}$$

where k is real for continuous spectrum and $k = i\kappa$ for bound states. We introduce a solution $\phi_1(x,k)$ of (7.125) which satisfies the boundary condition

$$\phi_1(x,k) = e^{ikx}, \quad as \quad x \to +\infty. \tag{7.126}$$

The Schrödinger equation (7.125) together with the boundary condition (7.126) is equivalent to the Volterra integral equation

$$\phi_1(x,k) = e^{ikx} - \frac{1}{k}\int_x^\infty sin[k(x-x')]u(x')\phi_1(x',k)dx'. \tag{7.127}$$

[For derivation of (7.127), see Examples later.] Similarly, we define another solution satisfying the boundary condition

$$\phi_2(x,k) = e^{-ikx}, \quad as \quad x \to -\infty. \tag{7.128}$$

(7.125) together with the boundary condition (7.128) is equivalent to the Volterra integral equation

$$\phi_2(x,k) = e^{-ikx} + \frac{1}{k}\int_{-\infty}^x sin[k(x-x')]u(x')\phi_2(x',k)dx'. \tag{7.129}$$

For fixed t, we define the scattering solution of (7.125) as shown in Fig. 7.7 by the boundary conditions

$$\begin{aligned}\psi(x,k,t) &= e^{-ikx} + R(k,t)e^{ikx}, \; as\; x \to \infty \\ &= T(k,t)e^{-ikx}, \; as\; x \to -\infty\end{aligned} \tag{7.130}$$

and the bound state solutions by the boundary conditions

$$\begin{aligned}\psi_n(x,\kappa_n(t),t) &= e^{-\kappa_n(t)x}, \quad as \quad x \to \infty \\ &= c_n(\kappa_n(t),t)e^{\kappa_n(t)x}, \quad as \quad x \to -\infty.\end{aligned} \tag{7.131}$$

It should be noted that the wave function in (7.130) is related to ϕ_1 and ϕ_2 by

$$\begin{aligned}\psi(x,k) &= T(k)\phi_2(x,k) \\ &= \phi_1(x,-k) + R(k)\phi_1(x,k).\end{aligned} \tag{7.132}$$

(7.132) is the consistency relation with upper and lower line each satisfying the prescribed boundary conditions at $x = \pm\infty$, respectively.

The integral representations (7.127) and (7.128) are instrumental in writing the following expressions for the solutions of (7.125) (*Jost solution*):

$$\phi_1(x,k) = e^{ikx} + \int_x^\infty K(x,z)e^{ikz}dz, \tag{7.133}$$

$$\phi_2(x,k) = e^{-ikx} + \int_{-\infty}^x K(x,z)e^{-ikz}dz, \tag{7.134}$$

where the kernel $K(x,z)$, which is connected to the potential $u(x)$, is found to be $K(x,z) = 0$ for $z < x$ and

$$\lim_{z\to\pm\infty} K(x,z) = 0, \quad \lim_{z\to\pm\infty} \frac{\partial K(x,z)}{\partial z} = 0. \tag{7.135}$$

Next, we substitute (7.133) into (7.125) to obtain

$$\begin{aligned} &e^{ikx}\left[-k^2 - \frac{dK(x,x)}{dx} - ikK(x,x) - \frac{\partial K(x,z)}{\partial x}|_{z=x} - u + k^2\right] \\ &\quad + \int_x^\infty \left[\frac{\partial^2 K(x,z)}{\partial x^2} - uK(x,z) + k^2 K(x,z)\right] e^{ikz}dz = 0. \end{aligned} \tag{7.136}$$

The last term in (7.136) is twice integrated by parts to be written as

$$\begin{aligned} k^2\int_x^\infty K(x,z)e^{ikz}dz &= ikK(x,x)e^{ikx} - \frac{\partial K(x,z)}{\partial z}|_{z=x}e^{ikx} \\ &\quad - \int_x^\infty \frac{\partial^2 K(x,z)}{\partial z^2}e^{ikz}dz. \end{aligned} \tag{7.137}$$

Substituting (7.137) into (7.136) yields

$$\begin{aligned} &e^{ikx}\left[-\frac{dK(x,x)}{dx} - \frac{\partial K(x,z)}{\partial x}|_{z=x} - \frac{\partial K(x,z)}{\partial z}|_{z=x} - u\right] \\ &\quad + \int_x^\infty dz e^{ikz}\left[\frac{\partial^2 K(x,z)}{\partial x^2} - \frac{\partial^2 K(x,z)}{\partial z^2} - uK(x,z)\right] = 0. \end{aligned}$$

The quantities in the large brackets should be zero separately. The first three terms in the upper line add to $-2\frac{dK(x,x)}{dx}$ per $\frac{dk(x,z)}{dx} = \frac{\partial K(x,z)}{\partial x} + \frac{\partial K}{\partial z}\frac{\partial z}{\partial x}$.

yielding

$$u(x,t) = -2\,\frac{d}{dx}K(x,x) \tag{7.138}$$

$$\frac{\partial^2 K(x,z)}{\partial x^2} - \frac{\partial^2 K(x,z)}{\partial z^2} = uK(x,z). \tag{7.139}$$

Now we substitute $\phi_1(x, \pm k)$ from (7.133) into (7.132) to write

$$T(k)\phi_2(x,k) = e^{-ikx} + \int_x^\infty K(x,z)e^{-ikz}dz$$
$$+R(k)\left[e^{ikx} + \int_x^\infty K(x,z)e^{ikz}dz\right]. \tag{7.140}$$

Multiplying (7.140) by $\frac{1}{2\pi}e^{iky}$ where $y > x$, and integrating with respect to k from $-\infty$ to ∞ gives

$$\frac{1}{2\pi}\int_{-\infty}^{\infty} T(k)\phi_2(x,k)e^{iky}dk$$
$$= \frac{1}{2\pi}\int_{-\infty}^{\infty} e^{ik(y-x)}dk + \int_x^\infty dzK(x,z)\frac{1}{2\pi}\int_{-\infty}^{\infty} e^{ik(y-z)}dk$$
$$+\frac{1}{2\pi}\int_{-\infty}^{\infty} R(k)e^{ik(y+x)}dk + \int_x^\infty dzK(x,z)\frac{1}{2\pi}\int_{-\infty}^{\infty} R(k)e^{ik(y+z)}dk. \tag{7.141}$$

The first term in the RHS of (7.141) vanishes because it is $\delta(y-x)$ with $y \neq x$. Defining the Fourier transform of the reflection coefficient $R(k)$,

$$B(x) = \frac{1}{2\pi}\int_{-\infty}^{\infty} R(k)e^{ikx}dk. \tag{7.142}$$

(7.141) can be put into the form

$$\frac{1}{2\pi}\int_{-\infty}^{\infty} T(k)\phi_2(x,k)e^{iky}dk$$
$$= K(x,y) + B(x+y) + \int_x^\infty K(x,z)B(y+z)dz. \tag{7.143}$$

$B(k)$ in (7.142) is a known quantity if $R(k)$ is measured. The LHS of (7.143) involves the unknown quantity K (through ϕ_2), and (7.143) is an integral equation for K which connects the potential $u(x)$. In order to determine K, we further calculate the LHS of (7.143). $T(k)$ can be expressed as the ratio of two Wronskians (W),

$$T(k) = \frac{W[\phi_1(x,-k),\phi_1(x,k)]}{W[\phi_2(x,k),\phi_1(x,k)]} \tag{7.144}$$

where $W[a,b] = a\frac{\partial b}{\partial x} - b\frac{\partial a}{\partial x}$. The LHS of (7.143) can be contour-integrated by picking the residues at the simple poles of $T(k)$ associated with the bound states. Without going through the details, we write the result of that contribution:

$$\frac{1}{2\pi}\int_{-\infty}^{\infty} T(k)\phi_2(x,k)e^{iky}dk = -c_\kappa^2 e^{-\kappa(x+y)} - \int_x^{\infty} c_\kappa^2 K(x,z)e^{-\kappa(y+z)}dz. \tag{7.145}$$

The RHS should include contributions from every bound state of the potential. Therefore the RHS of (7.145) should be written as a summation over the contributions from all the bound states. We combine (7.145) with (7.143) and express the time-dependance explicitly to write the integral equation for $K(x,y,t)$ in the form

$$K(x,y,t) + B(x+y,t) + \int_x^{\infty} K(x,z,t)B(y+z,t)dz = 0 \tag{7.146}$$

where $B(x,t)$ is now defined by

$$B(x,t) = \sum_{n=1}^{N} c_n^2(t)e^{-\kappa_n x} + \frac{1}{2\pi}\int_{-\infty}^{\infty} R(k,t)e^{ikx}dk. \tag{7.147}$$

(7.146) with $B(x,t)$ given by (7.147) is known as the Gel'fand–Levitan-Marchenko integral equation. The solution $K(x,y,t)$ of this equation enables one to determine the potential $u(x,t)$ by the relation

$$u(x,t) = -2\frac{d}{dx}K(x,x,t). \tag{7.148}$$

Example. Solve by Fourier transform

$$\frac{d^2\psi}{dx^2} - k_0^2\psi = f(x).$$

Let us Fourier transform both sides to obtain

$$\psi(k) = -\frac{f(k)}{k^2+k_0^2}.$$

We have from the Table (Dwight, H. B. *Tables of integrals and other mathematical data*, MacMillan New York 1967, 859.001)

$$\int_{-\infty}^{\infty} \frac{e^{ikx}}{k^2 + k_0^2} \frac{dk}{2\pi} = \int_{-\infty}^{\infty} \frac{coskx}{k^2 + k_0^2} \frac{dk}{2\pi} = \frac{1}{2\pi} \frac{\pi}{k_0} e^{-k_0|x|}.$$

Let us put

$$F(k) = \frac{1}{k^2 + k_0^2}; \quad F(x) = \frac{1}{2\pi} \frac{\pi}{k_0} e^{-k_0|x|}.$$

Therefore we can write $\psi(k) = -F(k)f(k)$, whose inversion is the convolution integral

$$\psi(x) = -(F * f)_x = -\int_{-\infty}^{\infty} F(x - x')f(x')dx'$$
$$= -\frac{1}{2k_0} \int_{-\infty}^{\infty} e^{-k_0|x-x'|} f(x')dx'.$$

Example. Solve by Fourier transform

$$\frac{d^2q}{dt^2} + \omega_1^2 q(t) = f(t). \tag{7.149}$$

This is the harmonic oscillator model of the polarization vector. Fourier transform of the above equation gives

$$q(\omega) = -\frac{f(\omega)}{\omega^2 - \omega_1^2} \equiv -F(\omega)f(\omega). \tag{7.150}$$

We need the inversion of $F(\omega)$:

$$F(t) = \int_{-\infty}^{\infty} \frac{1}{\omega^2 - \omega_1^2} e^{-i\omega t} \frac{d\omega}{2\pi} = \frac{1}{4\pi\omega_1} \int_{-\infty}^{\infty} d\omega \left[\frac{e^{-i\omega t}}{\omega - \omega_1 + i\epsilon} - \frac{e^{-i\omega t}}{\omega + \omega_1 + i\epsilon}\right]$$

where we introduced a small positive imaginary part of ω to take care of the singular points located on the real axis of ω. Using the Plemelj formula in the above, we obtain

$$F(t) = \frac{1}{4\pi\omega_1} \left[P \int_{-\infty}^{\infty} d\omega \frac{e^{-i\omega t}}{\omega - \omega_1} - i\pi e^{-i\omega_1 t} - P \int_{-\infty}^{\infty} d\omega \frac{e^{-i\omega t}}{\omega + \omega_1} + i\pi e^{i\omega_1 t}\right].$$

According to Appendix (A.40), we have (by putting $x \to -t$)

$$P\int_{-\infty}^{\infty} \frac{e^{-i\omega t}}{\omega - \omega_1} d\omega = -i\pi e^{-i\omega_1 t}(t > 0); \;\; i\pi e^{-i\omega_1 t}(t < 0),$$

$$P\int_{-\infty}^{\infty} \frac{e^{-i\omega t}}{\omega + \omega_1} d\omega = -i\pi e^{i\omega_1 t}(t > 0); \;\; i\pi e^{i\omega_1 t}(t < 0).$$

Therefore we obtain

$$F(t) = -\frac{1}{\omega_1} \; sin \; \omega_1 t \; (t > 0); \;\; F(t) = 0 \; (t < 0).$$

Inverting (7.150) yields

$$q(t) = -(F * f)_t = -\int_{-\infty}^{\infty} F(t - t')f(t')dt'.$$

Since $F(t - t') = 0$ for $t' > t$, we have

$$q(t) = \frac{1}{\omega_1} \int_{-\infty}^{t} \; sin \; \omega_1(t - t')f(t')dt'. \tag{7.151}$$

This result conforms to the causality requirement.

Example. Solve

$$\frac{d^2\psi}{dx^2} + k_0^2\psi = f(x) \tag{7.152}$$

with the boundary condition $\psi(x) = e^{ik_0 x}$ at $x = \infty$.

i) Use of the Wronskian.

The homogeneous solution consists of ($cos \; k_0 x$, $sin \; k_0 x$), and the particular solution which is proportional to f is obtained

$$\psi(x) = \int_{\infty}^{x} dx' \; f(x') \frac{cos k_0 x' \; sin k_0 x - sin \; k_0 x' cos \; k_0 x}{Wronskian[cos k_0 x', \; sin k_0 x']} + e^{ik_0 x}. \tag{7.153}$$

The Wronskian is k_0. Thus we obtain

$$\psi(x) = \frac{1}{k_0} \int_{\infty}^{x} dx' \; f(x') \; sin k_0(x - x') + e^{ik_0 x}. \tag{7.154}$$

ii) Fourier transform method.

Fourier transforming (7.152) gives

$$\psi(k) = -f(k)F(k) \tag{7.155}$$

$$where \;\; F(k) = \frac{1}{k^2 - k_0^2} = -\frac{1}{2k_0}\left(\frac{1}{k + k_0} - \frac{1}{k - k_0}\right). \tag{7.156}$$

Inverting the above equation gives

$$F(x) = \frac{1}{2k_0}\int_{-\infty}^{\infty}\frac{dk}{2\pi}\left[\frac{e^{ikx}}{k-k_0} - \frac{e^{ikx}}{k+k_0}\right]. \tag{7.157}$$

$k = k_0, k = -k_0$ is the singularity in the respective integral. The integrals are written as the sum of two terms (see Eq. (4.67)):

$$F(x) = \frac{1}{4\pi k_0}\left[P\int_{-\infty}^{\infty} dk \frac{e^{ikx}}{k-k_0} + \lambda e^{ik_0 x}\right]$$
$$-\frac{1}{4\pi k_0}\left[P\int_{-\infty}^{\infty} dk \frac{e^{ikx}}{k+k_0} + \lambda e^{-ik_0 x}\right]. \tag{7.158}$$

The principal part integral is a step function, and the above equation becomes

$$F(x) = \frac{1}{4\pi k_0}\left[\lambda e^{ik_0 x} + i\pi e^{ik_0 x} \ (x>0) \quad or \quad -i\pi e^{ik_0 x} \ (x<0)\right]$$
$$-\frac{1}{4\pi k_0}\left[\lambda e^{-ik_0 x} + i\pi e^{-ik_0 x} \ (x>0) \ \ or \ \ -i\pi e^{-ik_0 x} \ \ (x<0)\right]. \tag{7.159}$$

The constant λ will be determined according to the boundary conditions.
a) We assume that $F(x) = 0$ for $x > 0$. This boundary condition can be satisfied by $\lambda = -i\pi$. Then we obtain

$$F(x) = \frac{1}{k_0}\ sin k_0 x \quad (x<0). \tag{7.160}$$

b) We assume that $F(x) = 0$ for $x < 0$. This boundary condition can be satisfied by $\lambda = i\pi$. Then we obtain

$$F(x) = -\frac{1}{k_0}\ sin k_0 x \quad (x>0). \tag{7.161}$$

Inversion of (7.155) is the convolution integral:

$$\psi(x) = -\int_{-\infty}^{\infty} F(x')f(x-x')dx'. \tag{7.162}$$

a) With the aid of the function $F(x)$ in (7.160), the solution to (7.152) is written as

$$\psi(x) = -\int_{-\infty}^{0} F(x')f(x-x')dx'$$
$$= \int_{\infty}^{x} F(x-x')f(x')dx' = \frac{1}{k_0}\int_{\infty}^{x} sin k_0(x-x')f(x')dx'. \tag{7.163}$$

The general solution is obtained by adding the homogeneous solution $e^{\pm ik_0x}$ to this solution, which is appropriate to enforce the boundary condition in (7.126): $\psi = e^{ik_0x}$ as $x \to \infty$. Thus, we write

$$\psi(x, k_0) = e^{ik_0x} + \frac{1}{k_0}\int_{\infty}^{x} sink_0(x - x')f(x')dx'. \tag{7.164}$$

b) With the aid of the function $F(x)$ in (7.161), the solution to (7.152) is written as

$$\psi(x) = -\int_0^{\infty} F(x')f(x - x')dx' = \int_{-\infty}^{x} \frac{1}{k_0} sink_0(x - x')f(x')dx'. \tag{7.165}$$

Enforcing the boundary condition in (7.128), $\psi = e^{-ik_0x}$ as $x \to -\infty$, we write

$$\psi(x, k_0) = e^{-ik_0x} + \frac{1}{k_0}\int_{-\infty}^{x} sink_0(x - x')f(x')dx'. \tag{7.166}$$

• **Two-soliton solution.** Two-soliton solution corresponds to the initial potential

$$u(x, 0) = -6\ sech^2 x. \tag{7.167}$$

Note that here the amplitude and width of the wave do not match with the solitary wave solution. We have to solve

$$\psi_{xx} + [\lambda - u(x, 0)]\,\psi = 0. \tag{7.168}$$

The potential $u(x, 0)$ is again reflectionless. The scattering data for this potential are given by $\kappa_1 = 1$, $\kappa_2 = 2$, and R = 0, where we solved (7.168) with $\lambda = -\kappa^2$. The details for obtaining these eigen values are omitted but readers can refer to [P. L. Bhatnagar, *Nonlinear waves in one-dimensional dispersive media*, Oxford University Press, Oxford (1979)]. Solving the Gel'fand–Levitan equation, we can obtain the following two-soliton solution

$$u(x, t) = -12\,\frac{3 + 4cosh(2x - 8t) + cosh(4x - 64t)}{[3cosh(x - 28t) + cosh(3x - 36t)]^2}. \tag{7.169}$$

For large t, (7.169) tends to the superposition of two solitons of the form

$$u(x, t) = -12sech^2(x - 4t + \delta_1) \tag{7.170}$$

and

$$u(x, t) = -48sech^2[2(x - 16t) + \delta_2). \tag{7.171}$$

Exercise. Show that (7.169) agrees with the initial condition (7.167).

References for the inverse scattering method:

Bhatnagar, P. L. *Nonlinear Waves in One-dimensional Dispersive Media*, Oxford University Press Oxford 1979.

Debnath, L. *Nonlinear Partial Differential Equations for Scientists and Engineers*, Birkhauser, Boston, 1997.

Davidson, R. C. *Methods in Nonlinear Plasma Theory*, Academic Press New York, 1972.

Jeffrey A. and Kawahara, T. *Asymptotic Methods in Nonlinear Wave Theory*, Pitman Advanced Publishing Program, Boston, 1982.

Scott, A. C. Chu, F. Y. F. and McLaughlin, D. W. The soliton: A new concept in applied science, *Proc. IEEE* **61** (1973) pp. 1443–1483.

Chapter 8

Two-stream instability

A plasma with components of different streaming velocity (beams) tends to be unstable if a certain condition is met. In comparison, a moving plasma with a uniform velocity as a whole would not be different from a stationary plasma, as is evident from the Galilean point of view. A plasma consisting of two groups of electrons streaming with different speeds can be unstable because of the growing Langmuir oscillation (in the background of immobile ions). Similarly, a plasma with two ion beams in the background of hot Maxwellian electrons can be unstable on the growing ion acoustic wave perturbation (because in this case ion wave Landau damping can be weak) — *ion acoustic instability*.

Plasmas with beams draw attention because such plasmas can be used to make a traveling wave tube or an amplifier in the laboratory. Solar wind is also a streaming plasma appearing in nature. Two-stream instabilities can be used to heat plasma by injecting a particle beam into the plasma. Streaming of charged particles is also given rise to by currents induced in a plasma by external electromagnetic fields. In this case, two-stream instabilities arise as the result of a beam-plasma interaction. Analytically, the streaming velocity $\mathbf{V}_0$ is the zero order quantity upon which the small perturbation $\mathbf{v}$ should be superposed. The term $\mathbf{k} \cdot \mathbf{V}_0$ that appears in the dispersion relation significantly modifies the dispersion relation of stationary plasma wave.

The study of the dispersion relations of beam plasma wave also stimulated the effort to establish the criterion of two different kinds of instability: convective vs absolute instability. A useful reference to this subject is:

R. J. Briggs, *Electron-Stream Interaction with Plasmas* MIT Press, Cambridge, Massachusetts, (1964)

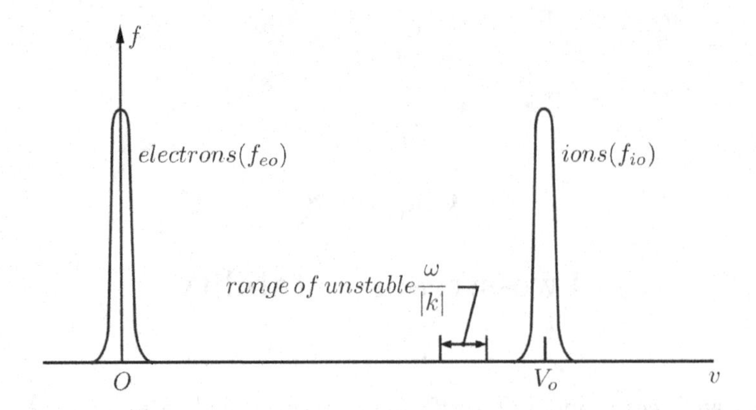

Fig. 8.1. Distributions for a two-stream instability.

8.1. Analysis of two-stream instability

A plasma can be regarded as an assembly of beams individualized by different initial velocities. The plasma susceptibility can be obtained by summing up all the *cold* individual beam susceptibilities. Combining this result with the Poisson equation yields the usual dispersion relation, which agrees with that obtained from the Vlasov equation. (see § 5.10). To obtain beam instability, we need at least two beams of different streaming speed. This requires a hump-like component superposed on a smoothly varying velocity distribution function; for example, *a bump-in-tail instability* [N. A. Krall and A. W. Trivelpiece, *Principles of plasma physics*, McGraw-Hill New York, (1973)]. Also see Fig. 8.1. In terms of the distribution function, a cold beam can be represented by a delta function. Figure 8.1 shows the distribution function for a cold ion beam streaming at speed V_0 through a cold electron 'beam' (in fact, the figure represents a stationary group of electrons).

The delta function distributions serve as the zero order distribution functions when we obtain the kinetic dielectric function of the beam plasma (cf. Eq. (4.23), for exmple). For a warm beam with thermal spread, we can use the distribution function of the form (see Fig. 8.2)

$$f_0 = \frac{N_1}{N}\left(\frac{m}{2\pi T_1}\right)^{3/2} exp\left(-\frac{mv^2}{2T_1}\right) + \frac{N_2}{N}\left(\frac{m}{2\pi T_2}\right)^{1/2} exp\left(-\frac{m(v_z - V_0)^2}{2T_2}\right)\delta(v_x)\delta(v_y).$$

$$(N = N_1 + N_2)$$

In the following, we suppose that a plasma consists of several *cold* streaming species characterized by zero order velocities V_{jo} and zero order constant number densities N_{j0}, where $j = 1, 2, \ldots$ is the index designating each

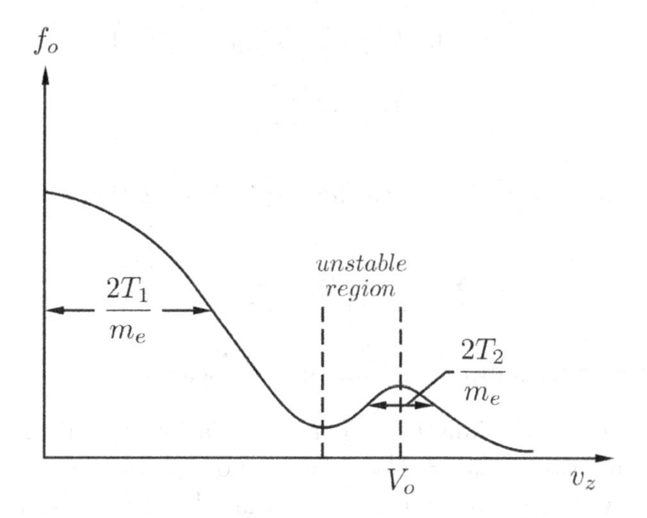

Fig. 8.2. Distribution for 'bump in tail' instability.

streaming species. We assume that $\Sigma_j\ e_j\ N_{j0} = 0$ (charge neutrality), and there is no electric field in the unperturbed state. Each beam individualized by the index $'j'$ is governed by the cold fluid equations,

$$\frac{\partial \mathbf{v}_j}{\partial t} + \mathbf{v}_j \cdot \nabla \mathbf{v}_j = \frac{e_j}{m_j}\mathbf{E}, \tag{8.1}$$

$$\frac{\partial n_j}{\partial t} + \nabla \cdot (n_j \mathbf{v}_j) = 0, \tag{8.2}$$

$$\nabla \cdot \mathbf{E} = 4\pi \sum_j e_j n_j. \tag{8.3}$$

Linearizing the above equations by writing $\mathbf{v}_j = \mathbf{V}_{j0} + \mathbf{v}'_j$, $n_j = N_{j0} + n'_j$ where the primed quantities are the perturbations, we obtain

$$\frac{\partial \mathbf{v}'_j}{\partial t} + \mathbf{V}_{j0} \cdot \nabla \mathbf{v}'_j = \frac{e_j}{m_j}\mathbf{E}, \tag{8.4}$$

$$\frac{\partial n'_j}{\partial t} + \mathbf{V}_{j0} \cdot \nabla n'_j + N_{j0} \nabla \cdot \mathbf{v}'_j = 0, \tag{8.5}$$

$$\nabla \cdot \mathbf{E} = 4\pi \sum_j e_j n'_j. \tag{8.6}$$

If the first order quantities vary as $e^{i\mathbf{k}\cdot\mathbf{r} - i\omega t}$, the linearized equations take the form

$$-i(\omega - \mathbf{k} \cdot \mathbf{V}_{j0})\mathbf{v}'_j = \frac{e_j}{m_j}\mathbf{E}, \tag{8.7}$$

$$-i(\omega - \mathbf{k} \cdot \mathbf{V}_{j0})n'_j + iN_{j0}\mathbf{k} \cdot \mathbf{v}'_j = 0 \tag{8.8}$$

$$i\mathbf{k} \cdot \mathbf{E} = 4\pi \sum_j e_j n'_j. \tag{8.9}$$

Solving for n'_j and using (8.9) give the dispersion relation

$$\sum_j \frac{\omega_{pj}^2}{(\omega - \mathbf{k} \cdot \mathbf{V}_{j0})^2} = 1 \tag{8.10}$$

where ω_{pj} is the plasma frequency of the j-th beam.

Exercise. Obtain the kinetic dispersion relation when the zero order distribution functions $F_{i0} = \delta(\mathbf{v} - \mathbf{V}_0)$ and $F_{e0} = \delta(\mathbf{v})$ are perturbed electrostatically. Use (4.23) and ascertain that the result agrees with (8.10).

We consider only two streams of particles with $\mathbf{k} \| \mathbf{V}_{j0}$ (j = 1,2). These may be a stream of electrons and a stream of ions with different velocities. Or these may be two streams of electrons, limiting to the waves of high frequencies in which the ions form stationary background. Then the dispersion relation (8.10) becomes

$$\frac{\omega_{p1}^2}{(\omega - kV_1)^2} + \frac{\omega_{p2}^2}{(\omega - kV_2)^2} = 1. \tag{8.11}$$

Since this is a quartic equation in ω, ω cannot be obtained as a simple function of k. If we introduce the phase velocity $v_\phi = \omega/k$, the above equation can be written as

$$k^2 = \frac{\omega_{p1}^2}{(v_\phi - V_1)^2} + \frac{\omega_{p2}^2}{(v_\phi - V_2)^2} \equiv F(v_\phi). \tag{8.12}$$

This equation has four roots for v_ϕ, possibly including two pairs of complex conjugates. The general form of $F(v_\phi)$ is shown in Fig. 8.3(a),(b).

In Fig. 8.3(a), there are four real roots of (8.12), corresponding to four propagating waves. In Fig. 8.3(b), there are only two real roots; the remaining two solutions are a pair of complex conjugates, one of which corresponds to a growing wave in time. The instability is called the two-stream instability. We can find the value k_c [for $k > k_c$ ($k < k_c$), we have stable (unstable) stream] from the condition $dF/dv_\phi = 0$, which gives

$$v_\phi = \frac{V_2\, \omega_{p1}^{2/3} + V_1\, \omega_{p2}^{2/3}}{\omega_{p1}^{2/3} + \omega_{p2}^{2/3}}. \tag{8.13}$$

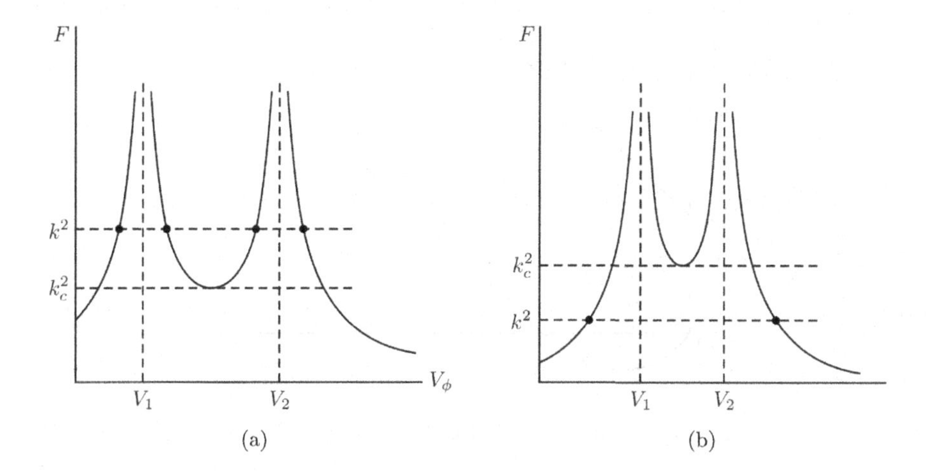

Fig. 8.3. Two-stream instability (a) stable (b) unstable.

Using (8.13) in (8.12) yields

$$k_c^2 = \frac{\omega_{p1}^{2/3} + \omega_{p2}^{2/3}}{(V_1 - V_2)^2}. \tag{8.14}$$

We can see from (8.14) that large (small) k_c corresponds to the case that the two streaming velocities are narrowly (widely) separated. Analysis of (8.12) requires numerical work, but there is a special case for which the dispersion relation can be solved analytically. The quartic equation (8.11) (in ω) becomes a quadratic equation in ω^2 if

$$V_1 = -V_2 \equiv V, \quad \omega_{p1}^2 = \omega_{p2}^2 \equiv \omega_p^2.$$

This includes the case where two identical electron beams travel in opposite direction. In this case, (8.11) becomes

$$\frac{\omega_p^2}{(\omega - kV)^2} + \frac{\omega_p^2}{(\omega + kV)^2} = 1 \tag{8.15}$$

whose solution can be written as

$$\omega^2 = \omega_p^2 + k^2V^2 \pm \omega_p(\omega_p^2 + 4k^2V^2)^{1/2}. \tag{8.16}$$

The plot ω^2 vs k^2 is shown in Fig. 8.4. Note the resemblance with the surface polariton dispersion curve (see Eq. (13.59)).

$\oplus$ curve indicates that real ω corresponds to real k, thus all waves satisfying the dispersion relation are stable. This case is represented by the four dots in Fig. 8.3(a). $\ominus$ curve shows that for $k < k_c$, we can have an instability. This curve corresponds to the two dots in Fig. 8.3(b).

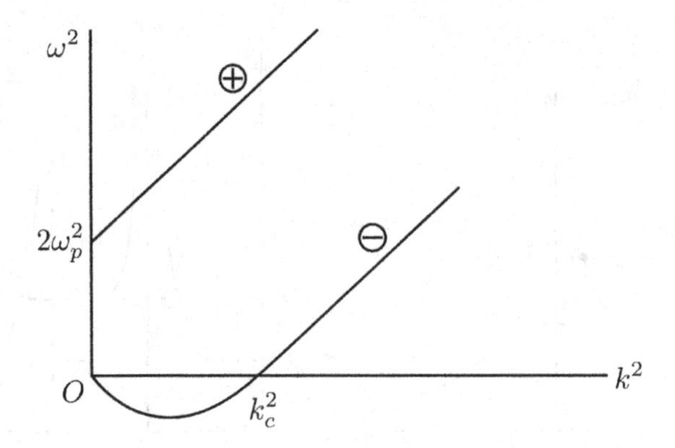

Fig. 8.4. Plot of Eq. (8.16).

8.2. Absolute instability and convective instability I

The dispersion relation $D(k,\omega) = 0$ is, in general, an algebraic equation connecting a complex $k = k_r + ik_i$ to a complex $\omega = \omega_r + i\omega_i$. With the phasor $e^{ikx-i\omega t}$, a wave is said to be unstable (damped) if we have $\omega_i > 0 (\omega_i < 0)$ for some real wave number k. One also speaks of amplifying (evanescent) wave if $k_i < 0$ ($k_i > 0$) for some real ω. This classification does not apply when ω and k are both complex at the same time. In particular, when $\omega_i > 0$, there are two kinds of instability: absolute and convective instability. In convective instability, the instability propagates away from a fixed point of observation so that the disturbance at a fixed point in the medium may decay with time (as in a traveling wave). On the other hand, an absolute instability grows with time at every point in space (as in a standing wave). See Fig. 8.5.

We approach this problem through an example of specific model equations

$$\frac{\partial \rho_1}{\partial t} + c_1 \frac{\partial \rho_1}{\partial x} = a\rho_2, \tag{8.17}$$

$$\frac{\partial \rho_2}{\partial t} - c_2 \frac{\partial \rho_2}{\partial x} = a\rho_1. \tag{8.18}$$

The above equations describe two oppositely propagating waves which are coupled (with 'a' the coupling constant). The corresponding dispersion relation is obtained from $(-i\omega + ikc_1)\rho_1 = a\rho_2, (-i\omega - ikc_2)\rho_2 = a\rho_1$, or

$$(kc_1 - \omega)(kc_2 + \omega) = a^2. \tag{8.19}$$

In uncoupled state, ρ_1 is constant along the characteristic line $\frac{dx}{dt} = c_1$, and ρ_2 is constant along the characteristic line $\frac{dx}{dt} = -c_2$. Let us recall that

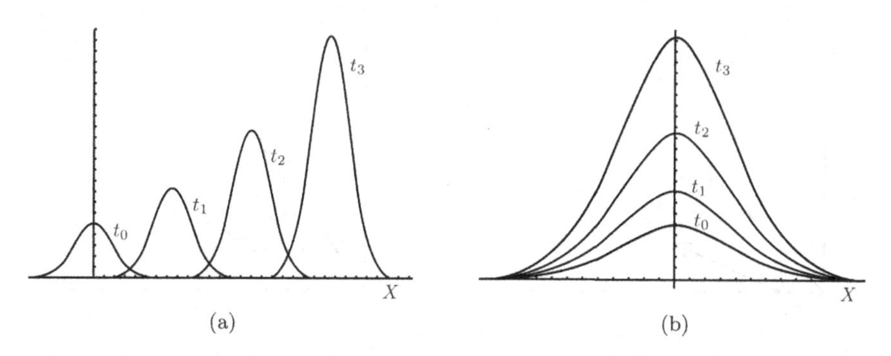

Fig. 8.5. Convective (a) and absolute (b) instability at four equidistant times (after M. Brambilla, *Kinetic theory of plasma waves* p. 43 Clarendon Oxford 1998).

a characteristic line is a dividing line of two *different natures* of solutions which cannot be connected through Taylor expansion across it. This property of the characteristic line defines the *region of influence* for a perturbation propagating on (x, t)-plane. In Fig. 8.6(a), we drew two characteristic lines emanating from the origin on the (x, t)-plane. Note that the signs of the slope of the characteristic lines are opposite. In this case, the region of influence is the wedge-shaped region comprising the t-axis. A moving detector of the disturbance situated at P initially, which is subject to the perturbation, cannot escape from the region of disturbance since the path of the moving detector as shown in the figure is always inside the region of influence. If the disturbance at P grows, it is an *absolute instability*. In Fig. 8.6(b), we drew two characteristic lines emanating from the origin with the same sign of the slopes. A stationary detector situated at the point P will be eventually out of the region of influence, in this case. The perturbation at the point P is convected away, and the instability is a *convective instability*.

In Fig. 8.7, we plotted two model dispersion relations,

$$(kc_1 - \omega)(kc_2 + \omega) = \pm a^2 \tag{8.20}$$

where the + sign corresponds to (8.19). Firstly, it should be noted that the asymptotes of the hyperbolas are $\frac{\omega}{k} = c_1$ and $\frac{\omega}{k} = -c_2$ which coincide with the slopes of the characteristics, respectively. In general we have

$$\lim_{k \to \infty} \frac{\omega}{k} = V, \textit{ the slope of characteristics.} \tag{8.21}$$

Figure 8.7 corresponds to the case that the characteristics run with slopes of different signs, and therefore the unstable solutions of (8.19), if there are any, should indicate an absolute instability. The $\oplus$ curve shows that there is a section of real k (on the k-axis in the diagram) to which no real ω

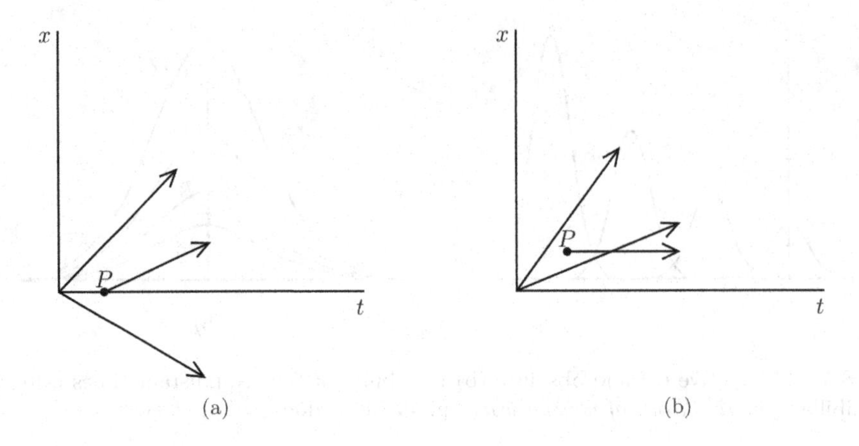

Fig. 8.6. Absolute (a) and convective (b) instability.

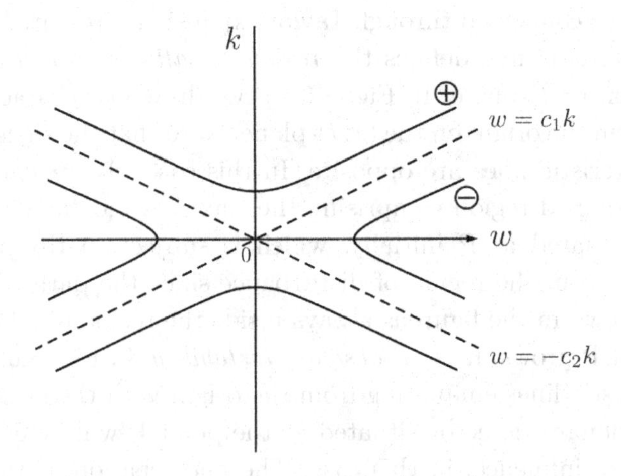

Fig. 8.7. Dispersion relation for two beams traveling in the opposite directions.

is corresponding. Solving the quadrature, one finds that ω is complex for $-\frac{2a}{c_1+c_2} < k < \frac{2a}{c_1+c_2}$. Clearly we can have an absolute instability. The $\ominus$ curve shows that there is a section of real ω (on the ω-axis in the diagram) to which no real k is corresponding. Solving the quadrature, one finds that k is complex for $-\frac{2a\sqrt{c_1c_2}}{c_1+c_2} < \omega < \frac{2a\sqrt{c_1c_2}}{c_1+c_2}$. The complex k means that the wave can be amplifying or evanescent in this case.

In Fig. 8.8, we plotted another set of model dispersion relations

$$(kc_1 - \omega)(kc_2 - \omega) = \pm a^2.$$

The curves correspond to the characteristics running in the same direction. The $\oplus$ curve in Fig. 8.8 shows that ω is real for all real k ($-\infty < k < \infty$). Also

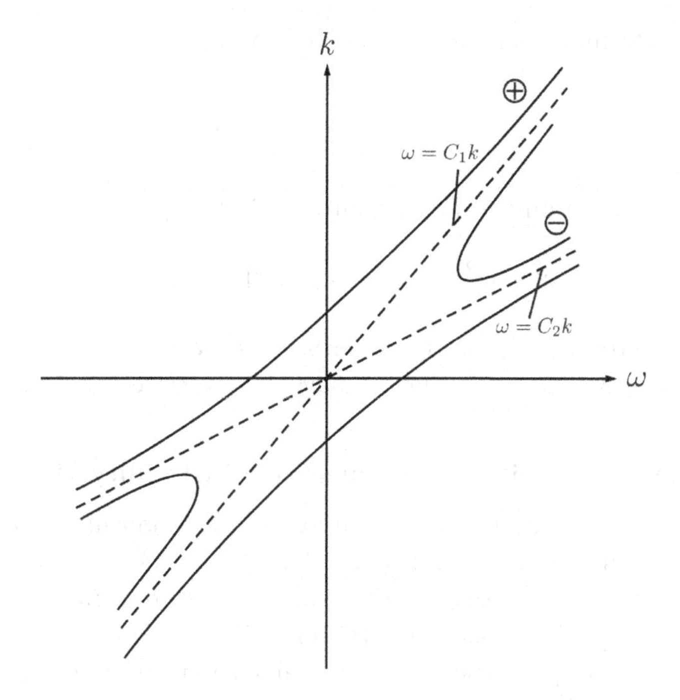

Fig. 8.8. Dispersion relation for two beams traveling in the same direction.

k is real for all real ω ($-\infty < \omega < \infty$). There is no instability nor damping; no amplification nor evanescence in this case. The $\ominus$ curve indicates that there is a segment of real ω, $-\frac{2a\sqrt{c_1c_2}}{|c_1-c_2|} < \omega < \frac{2a\sqrt{c_1c_2}}{|c_1-c_2|}$ which corresponds to complex k (no instability can occur but a mode conversion is possible.). Also there is a section of real k, $-\frac{2a}{|c_1-c_2|} < k < \frac{2a}{|c_1-c_2|}$ which corresponds to complex ω. In this range of k, we have a convective instability.

Remark. To find the characteristics of the system of equations (8.17) and (8.18), we eliminate ρ_1 or ρ_2 to get

$$-c_1c_2\rho_{xx} + (c_1 - c_2)\rho_{xt} + \rho_{tt} - a^2\rho = 0 \tag{8.22}$$

where ρ stands for ρ_1 or ρ_2, and the subscript x or t denotes the partial derivative: $\rho_x = \frac{\partial \rho}{\partial x}$, $\rho_{xx} = \frac{\partial^2 \rho}{\partial x^2}$, etc.. We seek the solution of the above equation in the form $\rho(x,t) = \rho(\xi)$ where $\xi = x - c_1t$. Then (8.22) is solved by $\rho(\xi) = 0$: in words, $\rho = 0$ along the characteristics $\frac{dx}{dt} = c_1$ or $x - c_1t = \xi = constant$. Equation (8.22) also permits solution in the form $\rho(x,t) = \rho(\eta)$ where $\eta = x + c_2t$. Then (8.22) is solved by $\rho(\eta) = 0$: in words, $\rho = 0$ along the characteristics $\frac{dx}{dt} = -c_2$ or $x + c_2t = \eta = constant$. Clearly $\frac{dx}{dt} = c_1$ and $\frac{dx}{dt} = -c_2$ are two families of the characteristic lines of (8.22).

Exercise. Determine the nature of instability for

$$1 - \frac{\omega_p^2}{\omega^2} - \frac{\omega_b^2}{(\omega - kv_b)^2} = 0.$$

Solution: This equation can be cast into the form

$$(\omega^2 - \omega_p^2)[(\omega - kv_b)^2 - \omega_b^2] = \omega_b^2 \omega_p^2.$$

Draw the asymptotes of the hyperbolas, which are the two characteristic lines. State why we conclude that we have a convective instability.

8.3. Absolute instability and convective instability II

In this Section, we derive the criterion of distinguishing the absolute and convective instabilities by Laplace–Fourier inversion of the dispersion relation. Reference: R. J. Briggs, *Electron-Stream Interaction with Plasmas* MIT Press, Cambridge, Massachusetts, (1964).

If a plasma is perturbed by external current (source) the Maxwell equation can be written in the form

$$\Lambda_{ij}(\mathbf{k}, \omega) E_j(\mathbf{k}, \omega) = S_i(\mathbf{k}, \omega) \tag{8.23}$$

where $S_i(\mathbf{k}, \omega)$ is the Laplace–Fourier transform of the source, and

$$\Lambda_{ij}(\mathbf{k}, \omega) = \frac{c^2 k^2}{\omega^2} \left[\frac{k_i k_j}{k^2} - \delta_{ij} \right] + \varepsilon_{ij}, \quad \varepsilon_{ij} : \textit{dielectric tensor.}$$

Inverting (8.23) gives

$$\mathbf{E}(\mathbf{k}, \omega) = \mathbf{\Lambda}^{-1} \cdot \mathbf{S} = \frac{\mathbf{\Lambda}_{adj} \cdot \mathbf{S}}{|\Lambda_{ij}|}$$

where $\mathbf{\Lambda}_{adj}$ is the adjoint matrix of Λ_{ij}. Let us write $\mathbf{\Lambda}_{adj} \cdot \mathbf{S} = \mathbf{Q}(\mathbf{k}, \omega)$ and $|\Lambda_{ij}| = D(\mathbf{k}, \omega)$.

Fourier–Laplace inversion of the above equation yields

$$\mathbf{E}(\mathbf{r}, t) = \int_{-\infty + i\sigma}^{\infty + i\sigma} \frac{d\omega}{2\pi} e^{-i\omega t} \int_{-\infty}^{\infty} \frac{d^3 k}{(2\pi)^3} e^{i\mathbf{k} \cdot \mathbf{r}} \frac{\mathbf{Q}(\mathbf{k}, \omega)}{D(\mathbf{k}, \omega)} \tag{8.24}$$

where the relation $D(\mathbf{k}, \omega) = 0$ is the dispersion relation. To investigate the time-asymptotic behavior of $\mathbf{E}(\mathbf{r}, t)$, it is sufficient to consider the Green function

$$G(\mathbf{r},t) = \int_{-\infty+i\sigma}^{\infty+i\sigma} \frac{d\omega}{2\pi} e^{-i\omega t} \int_{-\infty}^{\infty} \frac{d^3k}{(2\pi)^3} e^{i\mathbf{k}\cdot\mathbf{r}} \frac{1}{D(\mathbf{k},\omega)}. \tag{8.25}$$

That $G(\mathbf{r},t)$ in (8.25) is indeed the Green function is seen from the following equations:

$$G(\omega,\mathbf{r}) = \int_{-\infty}^{\infty} \frac{d^3k}{(2\pi)^3} \frac{1}{D(\mathbf{k},\omega)} e^{i\mathbf{k}\cdot\mathbf{r}},$$

$$G(\omega,\mathbf{k}) = \int_{-\infty}^{\infty} d^3r' G(\omega,\mathbf{r}') e^{-i\mathbf{k}\cdot\mathbf{r}'} = \frac{1}{D(\omega,\mathbf{k})},$$

$$\mathbf{E}(\omega,\mathbf{r}) = \int_{-\infty}^{\infty} d^3r' \mathbf{Q}(\omega,\mathbf{r}') G(\omega,\mathbf{r}-\mathbf{r}').$$

The last equation yields (8.24) after Laplace inversion. We consider the one-dimensional version of (8.25):

$$G(z,t) = \int_{-\infty}^{\infty} \frac{dk}{2\pi} e^{ikz} \int_{-\infty+i\sigma}^{\infty+i\sigma} \frac{d\omega}{2\pi} e^{-i\omega t} \frac{1}{D(k,\omega)} \tag{8.26}$$

$$\textit{or} \quad G(z,t) = \int_{-\infty+i\sigma}^{\infty+i\sigma} \frac{d\omega}{2\pi} e^{-i\omega t} F(z,\omega) \tag{8.27}$$

$$\textit{where} \quad F(z,\omega) = \int_{-\infty}^{\infty} \frac{dk}{2\pi} e^{ikz} \frac{1}{D(k,\omega)}. \tag{8.28}$$

In the following, we shall assume that $z > 0$ (waves are propagating $+z$ direction). (8.26) involves two contour integrations whose contours are shown in Fig. 8.9(a), (b).

In Landau's treatment of the initial value problem of the Vlasov equation, the wave number k is assumed to be real, and it was not necessary to perform k-inversion to investigate the time-asymptotic behavior of the perturbation; but now in the problem of more general instability analysis which includes the spatial amplification, we must consider k to be complex, and it adds extra complexity. In the k-integration, the contour must encircle the upper-half k-plane since $z > 0$. In the Laplace contour, L must run above all the singularities of the integrand. It should be kept in mind that the two contour integrals are interwoven; ω and k are not independent in $D(\omega,k)$.

Let us consider the ω-integration first in (8.26). The singularities of the integrand are given by

$$D(k = k_r, \omega) = 0 \tag{8.29}$$

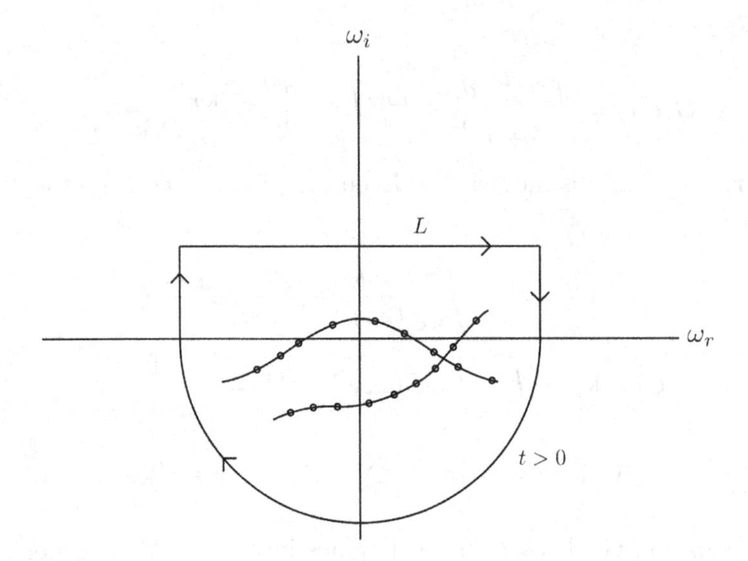

Fig. 8.9. (a): L is the Laplace contour which is above all the branches of $D(\omega, k) = 0$. The zeros are marked by o.

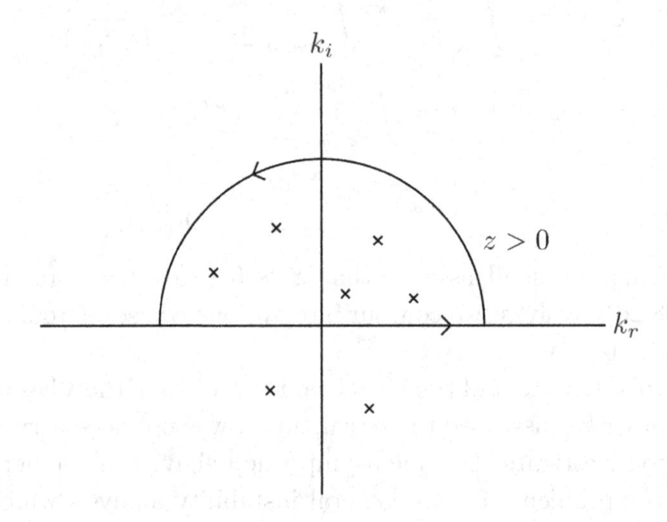

Fig. 8.9. (b): Fourier contour enclosing the upper k-plane for $z > 0$. $k(\omega_L)$, k's corresponding to ω on the Laplace contour L, are marked by x.

where k_r is the real part of k; this must be so because the k-integration is to be taken along the real k axis in (8.26). (8.29) is written in equivalent form

$$\omega = \omega(k_r) = \omega_r(k_r) + i\omega_i(k_r) \tag{8.30}$$

with obvious subscripts. (8.30) is the map of the real k-axis into the complex ω-plane through the mapping function $D(\omega, k) = 0$. In Fig. 8.10(a), possible branches of the resultant mapping are shown. These branches are the normal modes obtained from the dispersion relation (8.29) as usually. Causality demands that the line L in Fig. 8.10(a) should be above all the branches. In (8.28), the singularities of the integrand are given by

$$D(k, \omega_L) = 0 \tag{8.31}$$

where ω_L denotes ω on the Laplace contour line L. (8.31) is written in equivalent form

$$k(\omega_L) = k_r(\omega_{Lr} + i\omega_{Li}) + ik_i(\omega_{Lr} + i\omega_{Li}). \tag{8.32}$$

Clearly, $D(k_r, \omega_L) \neq 0$, and the $k's$ satisfying (8.31) are off the real k-axis as shown in Fig. 8.10(b).

(8.26), (8.30), and (8.32) specify the time and spatial evolution of the Green function $G(z,t)$. Time-asymptotic response of $G(z,t)$ can be elicited without going into the details of the integration. In a stable plasma, all the branches of $\omega(k_r)$ (see Fig. 8.10(a),(b)) have $\omega_i(k_r) < 0$ for all k_r. Then the Laplace contour L can be lowered below the ω_r axis, and hence

$$\lim_{t\to\infty} G(z,t) \to 0, \quad (for \;\; all \;\; z).$$

Instability arises when a branch $\omega(k_r)$ or a portion of it has $\omega_i(k_r) > 0$. Every singularity can be accounted for by the deformed contour L' (see Fig. 8.10(a)).

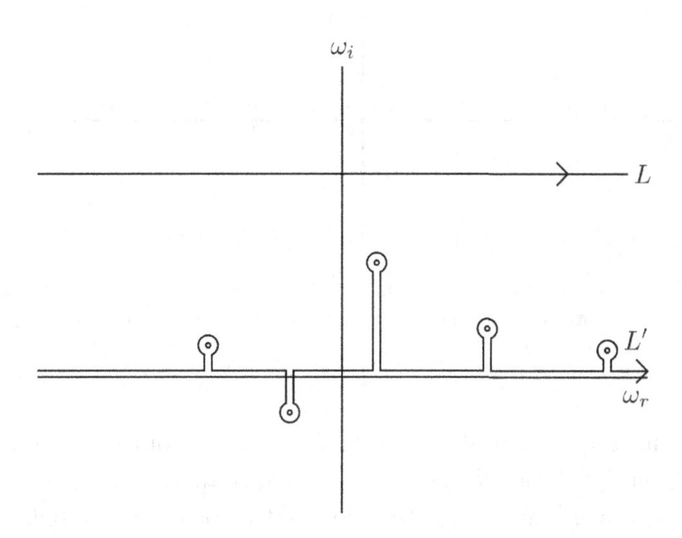

Fig. 8.10. (a): The original contour L is deformed to the contour L' for asymptotic evaluation of the Laplace inversion. o represents a point on $\omega(k_r)$ or $D(\omega, k_r) = 0$.

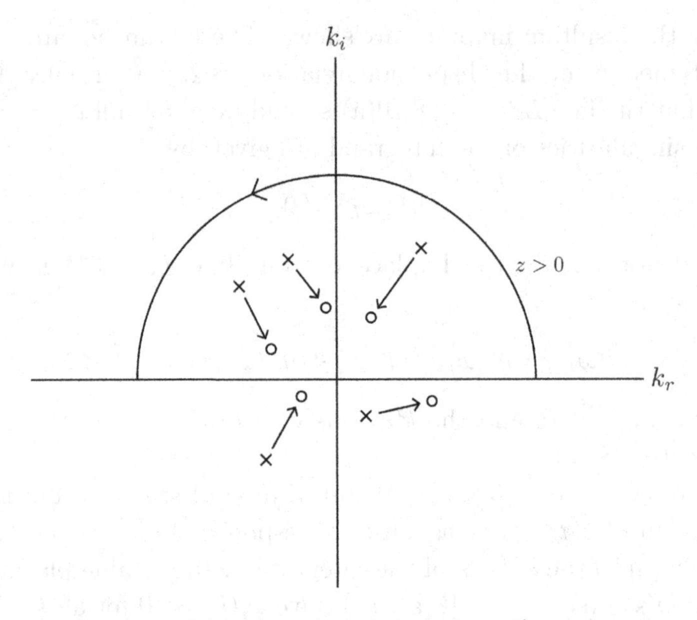

Fig. 8.10. (b): Migration of $k(\omega_L)$ as L is deformed to L'. ω_L is each point on L in Fig. 8.10(a). o : $k(\omega'_L)$ x: $k(\omega_L)$.

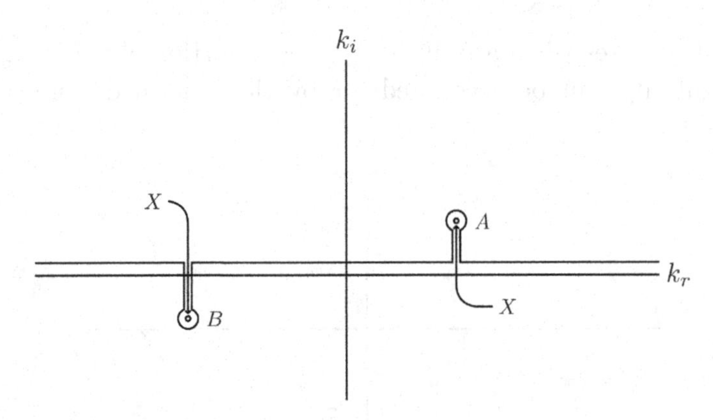

Fig. 8.10. (c): Crossing the real k-axis in the course of migration and the analytic continuation by deforming the Fourier contour. The pole A(B) corresponds to an evanescent (amplifying) wave.

Deforming Laplace contour from L to L' accompanies relocation of the k-plane singularities $k(\omega_L) \rightarrow k(\omega_{L'})$. Laplace inversion along the deformed contour L' should go together with k-inversion integral in which the singularities are now $k(\omega_{L'})$. In Fig. 8.11(b), migrations of k as the Laplace contour L is deformed to L' are indicated by arrows. Performing the

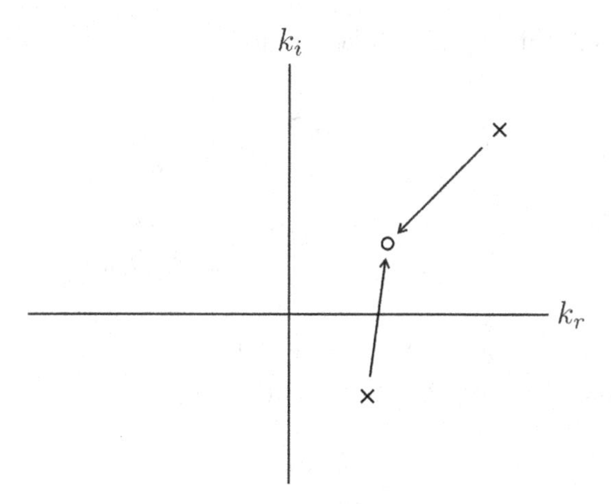

Fig. 8.11. (a): Merging of two poles at o whose coordinate is (k_c, ω_c) in the text. Originally the two poles were separated by the real k-axis (Fig. 8.9b). The absolute instability is given rise to.

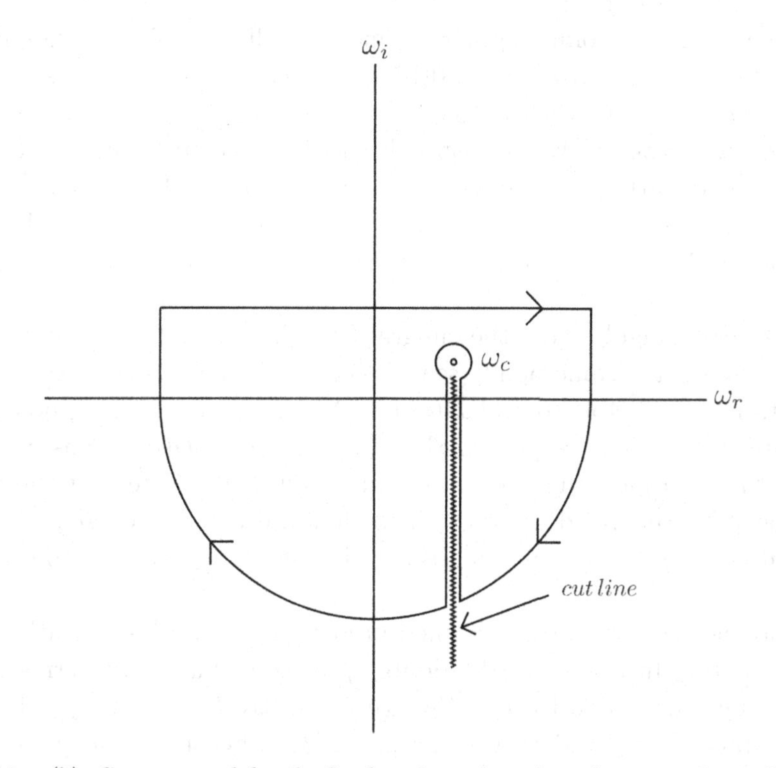

Fig. 8.11. (b): Contour used for the Laplace inversion when the two poles in Fig. 8.11 (a) merge at (k_c, ω_c). The merging point becomes a branch point and a branch cut must be introduced.

k-integration, (8.26) and (8.28) take the forms

$$F(z,\omega) = i \sum_n \frac{e^{ik_n z}}{[\partial D/\partial k]_{k_n}}, \tag{8.33}$$

$$G(z,t) = \int_{L'} \frac{d\omega}{2\pi} e^{-i\omega t}\, i \sum_n \frac{e^{ik_n z}}{[\partial D/\partial k]_{k_n}}, \tag{8.34}$$

where k_n is the n-th root of $D(k_n, \omega_{L'}) = 0$.

We see from (8.34) that the point satisfying

$$\frac{\partial D(k,\omega)}{\partial k} = 0 \tag{8.35}$$

is a singularity in the ω-integration. Since $D(k,\omega)$ in (8.35) simultaneously satisfies $D(k,\omega) = 0$, the singular point is a saddle point of the dispersion curve in the (ω, k)-plane.

How this saddle point singularity emerges will be explained shortly, but now we pay attention to Fig. 8.10(b), (c) where the singularities move in the k-plane and may reach or try to cross the real k-axis for some ω. In this case, the path of the k-integral is modified as shown in Fig. 8.11(c). This indented path gives analytic continuation of the function $F(z,\omega)$ in (8.28), and these values of ω are not singularities in the integral (8.34). The pole A(B) in Fig. 8.10(c) corresponds to an *evanescent (amplifying)* wave.

However, a singularity of the integral $F(z,\omega)$ occurs, if two moving $k(\omega_L)$ migrate toward the same point (see Fig. 8.11(a)): the two poles collide and merge to form a double root of $D(\omega, \mathbf{k})$. Originally, the merging poles were separated by the real k-axis. At the merging point, $D(k,\omega)$ has a double root at the corresponding ω ($\equiv \omega_c$), and (8.35) holds. Note that the values of ω_c must be independent of z, since they are determined solely by the function $D(k,\omega)$. In this case an absolute instability is given rise to provided $Im\ \omega_c > 0$.

A double root can also be formed in merging of two k's migrating from the *same side*. In this case, the double pole contributes two terms to the integral, but they cancel each other (use directly (8.28)). Only when the merging point is pinched by two merging k's, k_c becomes a singular point. In this case, the corresponding ω_c is a branch point, and ω-integral should be performed along the contour as shown in Fig. 8.11(b).

When a double root is formed in $D(k,\omega)$ at $\omega = \omega_c$ and $k = k_c$, we expand the dispersion function near $\omega = \omega_c$ and $k = k_c$:

$$D(\omega, k) \simeq (\omega - \omega_c) \left.\frac{\partial D}{\partial \omega}\right|_{\omega_c, k_c} + \frac{1}{2}(k - k_c)^2 \left.\frac{\partial^2 D}{\partial k^2}\right|_{\omega_c, k_c} = 0 \tag{8.36}$$

which yields in turn

$$k - k_c = \pm\sqrt{2}\, i \left[(\omega - \omega_c) \left.\frac{\partial D}{\partial \omega}\right|_c \div \left.\frac{\partial^2 D}{\partial k^2}\right|_c \right]^{\frac{1}{2}} \tag{8.37}$$

where the subscript 'c' stands for (ω_c, k_c). We also have

$$\frac{\partial D}{\partial k} \simeq \left.\frac{\partial^2 D}{\partial k^2}\right|_c (k - k_c). \tag{8.38}$$

Using (8.37) in (8.38) gives

$$\frac{\partial D}{\partial k} = \pm\sqrt{2}\, i\, \sqrt{\omega - \omega_c} \left[\frac{\partial^2 D}{\partial k^2} \frac{\partial D}{\partial \omega} \right]_c^{\frac{1}{2}}. \tag{8.39}$$

We take a single term corresponding to the upper-most singularity in the ω-plane which dominates the time-asymptotic behavior of $G(z,t)$. Using (8.39) in (8.33) gives

$$F(z, \omega) = \frac{\pm e^{ik_c z}}{\left[2\, \frac{\partial^2 D}{\partial k^2} \frac{\partial D}{\partial \omega} \right]_c^{\frac{1}{2}}} \frac{1}{\sqrt{\omega - \omega_c}}. \tag{8.40}$$

Here the $\pm$ sign can be determined from the trajectories of two merging poles. We write the Laplace inversion of (8.40):

$$G(t, z) = \left[2\, \frac{\partial^2 D}{\partial k^2} \frac{\partial D}{\partial \omega} \right]_c^{-\frac{1}{2}} e^{ik_c z - i\omega_c t} \int_{L'} \frac{d\omega}{2\pi} \frac{e^{-i(\omega - \omega_c)t}}{\sqrt{\omega - \omega_c}}. \tag{8.41}$$

Using the contour in Fig. 8.11(b), the Laplace integral can be evaluated to give

$$G(t, z) = \left[2\pi i\, \frac{\partial^2 D}{\partial k^2} \frac{\partial D}{\partial \omega} \right]_c^{-\frac{1}{2}} \frac{e^{ik_c z - i\omega_c t}}{\sqrt{t}} \tag{8.42}$$

where the double root of $D(k,\omega) = 0$ occurs for $k = k_c$ and $\omega = \omega_c$. When the imaginary part of ω_c is > 0, $G(t,z)$ in (8.42) grows indefinitely in time regardless of z, which is absolute instability.

Exercise. Prove (8.42). [See F. B. Hildebrand, *Advanced calculus for applications*, Prentice-Hall (1976) p. 627 Equation (14)].

Now we summarize the results obtained in the forgoing discussion. Let us suppose that the dispersion relation $D(\omega, k) = 0$ has unstable roots $(Im\ \omega > 0)$ for real k. Then the system is unstable. What determines the instability as convective or absolute? If the dispersion relation $D(\omega, k) = 0$ has a double root for $k = k_c$ and for $\omega = \omega_c$ with $\omega_{ci} > 0$ $[\partial D/\partial k_c = 0$ at $\omega = \omega_c]$, we have an absolute instability. Otherwise, the instability is convective. The condition $\partial D/\partial k_c = 0$ at $\omega = \omega_c$ is equivalent to the condition $\partial \omega_c/\partial k_c = 0$, i.e. the group velocity is zero. The wave energy grows in time at every point while the wave is spreading out without propagation. Readers are advised to confer the article in Introduction from equation (37) on.

Chapter 9

Parametric instability

Parametric instabilities are nonlinear interactions involving waves and particles (*wave-particle interaction*) and can be explained in terms of the nonlinearities represented by the quadratic terms in the governing equations. In the fluid equations, we have the nonlinear term $\mathbf{v} \cdot \nabla \mathbf{v}$ in the equation of motion and the nonlinear current $en\mathbf{v}$ in the continuity equation. In the Vlasov equation, we have $\mathbf{E} \cdot \frac{\partial f}{\partial \mathbf{v}}$. Parametric instabilities basically involve three waves and are conveniently regarded as scattering event in which the energy and momentum conservations among the waves involved are satisfied. Then the conservation relations become the matching conditions, which we may call the resonance conditions to enhance the interactive oscillations. Although parametric instabilities start from the nonlinear interactions, the actual analysis becomes a linear one, once the resonance conditions are satisfied with the driving wave (pump) being regarded as the zero order quantity. Thus, the literature on this subject compiled a vast amount of papers. We begin with the simple case (§ 9.1) where the electrostatic plasma normal modes (Langmuir wave and the ion acoustic wave) are coupled via a driving electric field in the form of a sinusoidal time variation. This case of a sinusoidal driving electric field is a special one in which a propagating wave is approximated by assuming an infinite wavelength ($k \simeq 0$) *(dipole approximation)*, and a two-time scale perturbational analysis can be employed since the physical resonance condition eliminates the secular terms in the perturbation series. In the case that a dipole approximation is not appropriate, the propagating nature of the three waves involved should be fully respected, with the resonance conditions:

$$\omega_1 = \omega_2 + \omega_3, \quad \mathbf{k}_1 = \mathbf{k}_2 + \mathbf{k}_3$$

where the frequency (wave vector) relation signifies the energy (momentum) conservation.

In the field of parametric instability, we have another *nonlinear Landau damping* that should be regarded as unconnected with the nonlinear Landau damping discussed in Chapter 5. The latter signifies the repetitive damping over multiple cycles of a plasma wave due to Landau damping. The former in parametric instabilities involves a beat wave with the resonance conditions in the form

$$\omega_1 - \omega_2 = (\mathbf{k}_1 - \mathbf{k}_2) \cdot \mathbf{v}$$

where subscript 1(2) denotes the primary (scattered) wave; thus the beat wave damps on the particle distribution. The beat wave nonlinear Landau damping is an important mechanism representing nonlinear wave-particle interaction and expands the scope of the *parametric decay instability* by including a quasi-mode.

9.1. Coupled mode parametric instability

If a plasma is irradiated by a high frequency oscillating electric field (pump), the pump of frequency ω_0 can couple two plasma normal modes and can give rise to a parametric instability as will be shown below. We consider two electrostatic plasma waves which are the high frequency electron wave (Langmuir wave) and the low frequency ion acoustic wave. We easily see that, by mixing of two waves through the nonlinear quadratic terms, the pump coupled with the low (high) frequency wave produces a high (low) frequency daughter wave. Therefore we have a coupled system of wave equations that may give rise to an instability. This kind of interaction is called the *parametric decay instability.* This terminology is understood from the resonance condition satisfied among the three waves, $\omega_0 = \omega_{low} + \omega_{high}$, which suggests that the external electromagnetic wave (pump) of frequency ω_0 is split into two plasma normal modes via the nonlinear coupling. We shall investigate this instability by using the two descriptions of plasma: fluid and kinetic equations. First, we use the fluid equations and perform a perturbational analysis by assuming a 'weak coupling'. We shall see that the dominant coupling terms are derived from the resonant terms. We have

$$\frac{\partial n_\alpha}{\partial t} + \nabla \cdot (n_\alpha \mathbf{v}_\alpha) = 0 \quad (\alpha = e,\ i), \tag{9.1}$$

$$m_\alpha n_\alpha \left(\frac{\partial \mathbf{v}_\alpha}{\partial t} + \mathbf{v}_\alpha \cdot \nabla \mathbf{v}_\alpha \right) = -\nabla p_\alpha + e_\alpha n_\alpha (\mathbf{E} + \mathbf{E}_0 sin\omega_0 t) - m_\alpha n_\alpha \nu_\alpha \mathbf{v}_\alpha, \tag{9.2}$$

$$\nabla \cdot \mathbf{E} = 4\pi \sum_\alpha e_\alpha n_\alpha. \tag{9.3}$$

We assume that the equilibrium state is represented by uniform density N, and the zero order velocity is the particle quiver velocity $\mathbf{v}_{0\alpha}$ as given by

$$\mathbf{v}_{0\alpha}(t) = -\frac{e_\alpha}{m_\alpha \omega_0}\mathbf{E}_0 cos\ \omega_0 t. \tag{9.4}$$

If the above equations are linearized with respect to the perturbed quantities $n'_\alpha(\mathbf{r},t) = n_\alpha - N$ and $\mathbf{v}'_\alpha = \mathbf{v}_\alpha - \mathbf{v}_{0\alpha}(t)$, then we obtain

$$\left(\frac{\partial}{\partial t} + \mathbf{v}_{0\alpha}\cdot\nabla\right)^2 n'_\alpha + \nu_\alpha\left(\frac{\partial}{\partial t} + \mathbf{v}_{0\alpha}\cdot\nabla\right) n'_\alpha$$

$$-\frac{\gamma_\alpha T_\alpha}{m_\alpha}\nabla^2 n'_\alpha + \omega_{p\alpha}^2(n'_\alpha - n'_\beta) = 0 \tag{9.5}$$

where $\omega_{p\alpha} = (4\pi N e^2/m_\alpha)^{1/2}$ and the subscript β becomes e (i) if α is i (e). In obtaining the above equation we used the equation of state $\nabla p_\alpha = \gamma_\alpha T_\alpha \nabla n_\alpha$.

Exercise. Derive (9.5). In the absence of the pump ($\mathbf{v}_{\alpha\mathbf{0}} = 0$), the electron and the ion equations are only weakly coupled through the Poisson equation: the coupling disappears if we put the mass ratio $m_e/m_i \to 0$. Note that the zero order velocity $\mathbf{v}_{\alpha\mathbf{0}}$ changes the differential operator $\partial/\partial t \to \partial/\partial t + \mathbf{v}_{\alpha\mathbf{0}}\cdot\nabla$, as is evident from the linearized continuity equation, $(\partial/\partial t + \mathbf{v}_{\alpha\mathbf{0}}\cdot\nabla)n_\alpha + N_\alpha\nabla\cdot\mathbf{v}_\alpha = 0$.

The above equation is Fourier transformed with respect to the spatial coordinate: $n'_\alpha(\mathbf{r},t) = n'_\alpha(t)e^{i\mathbf{k}\cdot\mathbf{r}}$. Also we transform as

$$n'_\alpha(t) = f_\alpha(t)e^{-ia_\alpha sin\omega_0 t};$$

where

$$a_\alpha = \frac{-e_\alpha \mathbf{k}\cdot\mathbf{E}_0}{m_\alpha\omega_0^2}$$

is the ratio of particle excursion length in the external field to the wavelength of the oscillation. This parameter will be assumed to be $\ll 1$ (weak pump). Then, (9.5) takes the form

$$D_e f_e \equiv \left(\frac{d^2}{dt^2} + \nu_e\frac{d}{dt} + \omega_e^2\right) f_e = \omega_{pe}^2 f_i e^{iasin\omega_0 t}, \tag{9.6}$$

$$D_i f_i \equiv \left(\frac{d^2}{dt^2} + \nu_i\frac{d}{dt} + \omega_i^2\right) f_i = \omega_{pi}^2 f_e e^{-iasin\omega_0 t}, \tag{9.7}$$

$$where \quad a = a_e - a_i \simeq e\mathbf{k}\cdot\mathbf{E}_0/m_e\omega_0^2, \quad \omega_\alpha^2 = \omega_{p\alpha}^2 + k^2\gamma_\alpha T_\alpha/m_\alpha. \tag{9.8}$$

Equations (9.6) and (9.7) are the basic equations for parametric decay instability induced by homogeneous oscillating electric field (here the pump is assumed to have infinite wavelength: dipole approximation).

Case of no applied external field, $a = 0$. Eliminating f_e or f_i in (9.6) and (9.7) gives

$$\left(\frac{d^2}{dt^2} + \nu_i \frac{d}{dt} + \omega_i^2\right)\left(\frac{d^2}{dt^2} + \nu_e \frac{d}{dt} + \omega_e^2\right) f_e = \omega_{pe}^2 \omega_{pi}^2 f_e.$$

Neglecting the damping terms, the dispersion relation of natural oscillations is

$$(\omega^2 - \omega_e^2)(\omega^2 - \omega_i^2) = \omega_{pe}^2 \omega_{pi}^2.$$

Approximate solutions of the above quadratic equation (in ω^2) can be obtained, keeping only the first order terms in $\omega_i^2/\omega_e^2 (\simeq m_e/m_i)$:

$$\omega^2 = \omega_e^2 = \omega_{pe}^2 + k^2 \frac{\gamma_e T_e}{m_e} \quad (high\ frequency\ wave) \tag{9.9}$$

$$\omega^2 = \omega_i^2 - \omega_{pe}^2 \omega_{pi}^2 / \omega_e^2 = \omega_{pi}^2 \left[1 + \frac{k^2 \gamma_i T_i}{m_i \omega_{pi}^2} - \frac{\omega_{pe}^2}{\omega_{pe}^2 + k^2 \gamma_e T_e / m_e}\right]. \tag{9.10}$$

The above expression becomes

$$\omega^2 = k^2 \frac{\gamma_i T_i + \gamma_e T_e}{m_i} \quad (low\ frequency\ wave) \tag{9.11}$$

when the electron Debye length (λ_e) is such that $k^2 \lambda_e^2 \ll 1$. This analysis shows that the natural oscillations of plasma in electrostatic approximation are the two waves, Langmuir wave (electron plasma wave) and ion acoustic wave. The ion acoustic wave is severely Landau-damped unless $T_e \gg T_i$, so as a propagating mode of plasma wave it is characterized by the dispersion relation

$$\omega^2 = k^2 \frac{\gamma_e T_e}{m_i}. \tag{9.12}$$

We have two methods to solve the coupled equations (9.6) and (9.7): (1) directly solve the differential equations and (2) use Fourier transform.

9.1.1. *Multiple time perturbation analysis*

This perturbation method is used to solve directly the coupled equations (9.6) and (9.7). Eliminating f_i gives

$$L\ f_e = 0 \tag{9.13}$$

where

$$L = D_i D_e - \omega_{pi}^2\omega_{pe}^2 - ia\omega_0\left(- \omega_0 sin\ \omega_0 t + \nu_i\ cos\ \omega_0 t + 2cos\ \omega_0 t\ \frac{d}{dt}\right)D_e$$
$$- a^2\omega_0^2\ cos^2\omega_0 t D_e$$

$$a = a_e - a_i \approx e\mathbf{k}\cdot\mathbf{E}_0/m_e\omega_0^2. \tag{9.14}$$

Let $a = \epsilon \ll 1$ be the expansion parameter in the multiple time perturbation series for the solution of (9.13). Then we write

$$f_e = f_0 + \epsilon f_1 + \epsilon^2 f_2 + \cdots$$
$$t = t_0 + \epsilon t_1 + \epsilon^2 t_2 + \cdots$$
$$\frac{\partial}{\partial t} = \frac{\partial}{\partial t_0} + \epsilon\frac{\partial}{\partial t_1} + \epsilon^2\frac{\partial}{\partial t_2} + \cdots. \tag{9.15}$$

The zero order equation of (9.13) is

$$\left(D_i^{(0)}D_e^{(0)} - \omega_{pi}^2\omega_{pi}^2\right)f_0 = 0 \tag{9.16}$$
$$with\quad D_\alpha^{(0)} = \frac{\partial^2}{\partial t_0^2} + \omega_\alpha^2$$

where we assumed that the collision frequency ν_α is a quantity of $O(\epsilon)$. (9.16) determines two normal modes of electrostatic perturbation in the absence of the pump:

$$f_0 = A_2(t_1)e^{i\omega_2 t_0} + A_{-2}(t_1)e^{-i\omega_2 t_0} + A_3(t_1)e^{i\omega_3 t_0} + A_{-3}(t_1)e^{-i\omega_3 t_0} \tag{9.17}$$

where ω_2 is given by (9.9) (high frequency) and ω_3 is given by (9.11) (low frequency). The first order equation of (9.13) is

$$\left(D_i^{(0)}D_e^{(0)} - \omega_{pi}^2\omega_{pi}^2\right)f_1 = -L_1 f_0, \tag{9.18}$$

$$L_1 = 4\frac{\partial^4}{\partial t_0^3 t_1} + (\nu_i + \nu_e)\frac{\partial^3}{\partial t_0^3} + 2(\omega_e^2 + \omega_i^2)\frac{\partial^2}{\partial t_0\partial t_1} + (\nu_i\omega_e^2 + \nu_e\omega_i^2)\frac{\partial}{\partial t_0}$$
$$+ ia\omega_0\left(\omega_0 sin\ \omega_0 t - 2cos\ \omega_0 t\ \frac{\partial}{\partial t_0}\right)\left(\frac{\partial^2}{\partial t_0^2} + \omega_e^2\right). \tag{9.19}$$

To facilitate calculation of $L_1 f_0$, we consider a typical term, $A(t_1)\ e^{i\omega t_0}$, operated by L_1. We obtain

$$L_1 A(t_1) e^{i\omega t_0} = \left(F_0(\omega) A + F_1(\omega) \frac{\partial A}{\partial t_1} \right) e^{i\omega t_0} + F_+(\omega) A e^{i(\omega+\omega_0)t_0}$$

$$+ F_-(\omega) A e^{i(\omega-\omega_0)t_0}, \tag{9.20}$$

$$F_0(\omega) = i\omega\big(\nu_i \omega_e^2 + \nu_e \omega_i^2 - (\nu_e + \nu_i)\omega^2\big), \tag{9.21}$$

$$F_1(\omega) = 2i\omega(\omega_e^2 + \omega_i^2 - 2\omega^2), \tag{9.22}$$

$$F_\pm(\omega) = a\omega_0(\omega_e^2 - \omega^2)(\omega \pm \frac{\omega_0}{2}). \tag{9.23}$$

Then (9.18) can be written as

$$(D_i^{(0)} D_e^{(0)} - \omega_{pi}^2 \omega_{pi}^2) f_1$$

$$= -\sum_j \left[\left(F_0(\omega_j) A_j + F_1(\omega_j) \frac{\partial A_j}{\partial t_1} \right) e^{i\omega_j t_0} + F_+(\omega_j) A_j e^{i(\omega_j + \omega_0)t_0} \right.$$

$$\left. + F_-(\omega_j) A_j e^{i(\omega_j - \omega_0)t_0} \right] \tag{9.24}$$

where the summation index j runs over ± 2 and ± 3. Also note $\omega_{-j} = -\omega_j$ and $F_\pm$ are the mixing terms of the pump and a natural mode.

The secularity causing terms in (9.24) are those terms oscillating with frequencies $\pm\omega_2$ or $\pm\omega_3$, which are the frequencies of the homogeneous solution (recall § 3.8). In addition to these terms, there are also resonant terms which cause the secularity by satisfying the following resonant conditions:

$$-\omega_0 + \omega_2 + \omega_3 = \Delta' : \quad \textit{Stokes coupling},$$

$$-\omega_0 + \omega_2 - \omega_3 = \Delta'' : \quad \textit{anti-Stokes coupling}.$$

Strict satisfaction of resonant conditions means $\Delta' = \Delta'' = 0$, but we allowed for a small frequency mismatch. All these secularity causing terms in (9.24) should be removed for each frequency. For example: if we take $j = 2$, the terms $F_+(\pm\omega_3)$ are the resonant terms. So we write

$$\left[F_1(\omega_2) \frac{\partial A_2}{\partial t_1} + F_0(\omega_2) A_2 \right] e^{i\omega_2 t_0} + F_+(-\omega_3) A_{-3} e^{i(\omega_0 - \omega_3)t_0}$$

$$+ F_+(\omega_3) A_3 e^{i(\omega_0 + \omega_3)t_0} = 0.$$

Canceling the exponential factors yields (9.25) below, and the rest of the equations are likewise obtained as below.

$$F_1(\omega_2)\,\frac{\partial A_2}{\partial t_1} + F_0(\omega_2)A_2 + F_+(-\omega_3)A_{-3}e^{-i\Delta' t_1} + F_+(\omega_3)A_3e^{-i\Delta'' t_1} = 0, \tag{9.25}$$

$$F_1(-\omega_2)\,\frac{\partial A_{-2}}{\partial t_1} + F_0(-\omega_2)A_{-2} + F_-(\omega_3)A_3e^{i\Delta' t_1} + F_-(-\omega_3)A_{-3}e^{i\Delta'' t_1} = 0, \tag{9.26}$$

$$F_1(\omega_3)\,\frac{\partial A_3}{\partial t_1} + F_0(\omega_3)A_3 + F_+(-\omega_2)A_{-2}e^{-i\Delta' t_1} + F_-(\omega_2)A_2e^{i\Delta'' t_1} = 0, \tag{9.27}$$

$$F_1(-\omega_3)\frac{\partial A_{-3}}{\partial t_1} + F_0(-\omega_3)A_{-3} + F_-(\omega_2)A_2e^{i\Delta' t_1} + F_+(-\omega_2)A_{-2}e^{-i\Delta'' t_1} = 0. \tag{9.28}$$

It would be useful to know that $\omega_2^2 \approx \omega_e^2 \approx \omega_{pe}^2, \quad \omega_i^2 \ll \omega_e^2$. Using the definitions for F's, the above expressions are simplified to yield

$$\frac{\partial A_2}{\partial t_1} + \Gamma_e A_2 + \alpha\left(A_{-3}e^{-i\Delta' t_1} + A_3e^{-i\Delta'' t_1}\right) = 0, \tag{9.29}$$

$$\frac{\partial A_{-2}}{\partial t_1} + \Gamma_e A_{-2} + \alpha\left(A_3e^{i\Delta' t_1} + A_{-3}e^{i\Delta'' t_1}\right) = 0, \tag{9.30}$$

$$\frac{\partial A_3}{\partial t_1} + \Gamma_i A_3 + \beta\left(A_{-2}e^{-i\Delta' t_1} - A_2e^{i\Delta'' t_1}\right) = 0, \tag{9.31}$$

$$\frac{\partial A_{-3}}{\partial t_1} + \Gamma_i A_{-3} + \beta\left(A_2e^{i\Delta' t_1} - A_{-2}e^{-i\Delta'' t_1}\right) = 0, \tag{9.32}$$

$$\Gamma_{e,i} = \nu_{e,i}/2, \quad \alpha = ia\omega_0^2\omega_e^2/4\omega_2^3,$$
$$\beta = a\omega_0(\omega_2 - \omega_0/2)\omega_{pi}^2\omega_{pe}^2/(2i\omega_3\omega_2^2\omega_e^2)$$

where α and β are the coupling constants. Note that $'a'$ in α and β is the pump intensity.

• Summary: (+)–subscripted wave and (−)–subscripted wave propagate in opposite directions. (9.29) describes slow change (first order) of ω_2– wave due to the coupling of the pump and the low frequency waves $\omega_{\pm 3}$. (9.31) describes slow change of ω_3– wave due to the coupling of the pump and the high frequency waves $\omega_{\pm 2}$. We have the four equations (9.29)–(9.32) for the 4 unknown functions $A_{\pm 2}$ and $A_{\pm 3}$ which describe the resonant interactions among the four waves when the Stokes and anti-Stokes resonant conditions are met; they include all the possible couplings among the waves propagating both of $\pm$ directions, with both resonance conditions of Stokes and anti-Stokes being met. Now we consider more restricted couplings.

• If we exclude the low frequency wave propagating in the negative direction, put $A_3 = 0$ to write

$$\frac{\partial A_2}{\partial t_1} + \Gamma_e A_2 + \alpha A_{-3} e^{-i\Delta' t_1} = 0, \tag{9.33}$$

$$\frac{\partial A_{-2}}{\partial t_1} + \Gamma_e A_{-2} + \alpha A_{-3} e^{i\Delta'' t_1} = 0, \tag{9.34}$$

$$\frac{\partial A_{-3}}{\partial t_1} + \Gamma_i A_{-3} + \beta\left(A_2 e^{i\Delta' t_1} - A_{-2} e^{-i\Delta'' t_1}\right) = 0. \tag{9.35}$$

• When the anti-Stokes coupling is excluded, we have

$$\frac{\partial A_2}{\partial t_1} + \Gamma_e A_2 + \alpha A_{-3} e^{-i\Delta' t_1} = 0, \tag{9.36}$$

$$\frac{\partial A_{-3}}{\partial t_1} + \Gamma_i A_{-3} + \beta A_2 e^{i\Delta' t_1} = 0. \tag{9.37}$$

(9.36) and (9.37) are a subset of (9.33)–(9.35) as they are obtained by excluding the anti-Stokes coupling terms in the latter equations. (9.33)–(9.35) yield two kinds of instability: *parametric decay instability* and *oscillating two-stream instability* (OTS). The parametric decay instability can be also obtained from (9.36) and (9.37) in exactly identical form. (9.35) is suggestive of the characteristics of the OTS instability. Putting $\Delta' = \Delta'' = 0$ in (9.35) gives the β-term as $\sin\, \omega_3 t$, a standing wave form which grows in the fashion of *purely growing.* This non-propagating instability is also called the *modulational instability.* It is a result of a four-wave interaction involving the pump, two side band waves ($\omega_{\pm 2}$), and the low frequency wave ω_{-3}, while the parametric decay instability is a three-wave interaction.

For future reference, we derive the dispersion relations from the above two sets of equations.

• Let us eliminate $A_{\pm 2}$ in (9.33)–(9.35):

$$\frac{\partial^3 A'_{-3}}{\partial t_1^3} + \frac{\partial^2 A'_{-3}}{\partial t_1^2}[2\gamma + i(\Delta'' - \Delta')]$$
$$+ (\gamma - i\Delta')(\gamma + i\Delta'')\frac{\partial A'_{-3}}{\partial t_1} - i\alpha\beta(\Delta' + \Delta'')A'_{-3} = 0 \tag{9.38}$$

where $A'_{-3} = A_{-3}e^{\Gamma_i t}$, $\gamma = \Gamma_e - \Gamma_i$. Since $\frac{\partial}{\partial t_1} = \frac{\partial}{\partial t} - \frac{\partial}{\partial t_0} = -i\omega + i\omega_3$, putting $\frac{\partial}{\partial t_1} = i(\omega_3 - \omega) + \Gamma_i$ gives the dispersion relation

$$[i(\omega - \omega_3) - \Gamma_i][-\omega^2 + \Gamma_e^2 - 2i\omega\Gamma_e + \delta^2] - 2i\delta\alpha\beta = 0 \tag{9.39}$$

where $\delta = \omega_0 - \omega_2$, and we used $\Delta' - \Delta'' = 2\omega_3$, $\Delta' + \Delta'' = -2\delta$, $\Delta'\Delta'' = \delta^2 - \omega_3^2$. (9.39) can be cast into a slightly different form by multiplying both sides by $i(\omega + \omega_3 + i\Gamma_i)$:

$$\begin{aligned}(\omega^2 + 2i\omega\Gamma_i - \Gamma_i^2 - \omega_3^2)(\omega^2 + 2i\omega\Gamma_e - \Gamma_e^2 - \delta^2)\\ = -2\alpha\beta\delta(\omega + \omega_3 + i\Gamma_i) \approx -4\alpha\beta\delta\ \omega_3\end{aligned} \tag{9.40}$$

where we used $\omega + \omega_3 \approx 2\omega_3$. The coupling coefficient takes the value

$$\alpha\beta = \frac{a^2\omega_{pi}^2\omega_{pe}^2}{16\omega_2\omega_3} = \frac{e^2k^2E_0^2}{16m_im_e\omega_2\omega_3}. \tag{9.41}$$

If Γ_i is neglected as compared to ω_3, (9.40) is the dispersion relation for the OTS instability derived in Nishikawa I [Nishikawa, K. (1968) *Parametric excitation of coupled waves I. General formulation*, *J. Phys. Soc. Japan*, **24**, Eq. 31; our ω_3 is ω_1 therein, and ω_2 stands for the same high frequency].

• Next, we eliminate A_2 from (9.36) and (9.37) to derive the dispersion relation of the parametric decay instability:

$$\frac{\partial^2 A_{-3}}{\partial t_1^2} + (\Gamma_e + \Gamma_i - i\Delta')\frac{\partial A_{-3}}{\partial t_1} + \Gamma_i(\Gamma_e - i\Delta')A_{-3} = \alpha\ \beta\ A_{-3}. \tag{9.42}$$

Putting $\frac{\partial}{\partial t_1} = i(\omega_3 - \omega)$ in the above equation gives the dispersion relation in the form

$$\begin{aligned}&(\omega - \omega_3)^2 + (\omega - \omega_3)\ (i\Gamma_i + i\Gamma_e + \Delta') - \Gamma_i\Gamma_e + i\Gamma_i\Delta' + \alpha\ \beta\\ &= (\omega - \omega_3 + i\Gamma_i)(\omega - \omega_3 + \Delta' + i\Gamma_e) + \alpha\ \beta\\ &= (\omega - \omega_3 + i\Gamma_i)(\omega - \omega_0 + \omega_2 + i\Gamma_e) + \alpha\ \beta = 0.\end{aligned} \tag{9.43}$$

(9.43) agrees with Nishkawa's dispersion relation [Nishikawa I (1968) Eq. 16; note that Nishikawa's $\Delta = -\Delta'$].

Exercise. Explain the physical context of the relation $\frac{\partial}{\partial t_1} = i(\omega_3 - \omega)$ contained in (9.39) and (9.43).

9.1.2. *Fourier transform derivation of the parametric dispersion relation*

Instead of using a direct perturbational method (in § 9.1.1), we analyze (9.6)–(9.7) by Fourier transform to derive (9.40) and (9.43). We use the Bessel function identity in (9.6) and (9.7)

$$e^{iasin\omega_0 t} = \sum_p J_p(a) e^{ip\omega_0 t}$$

and obtain Fourier transformed equations

$$D_e(\omega) f_e(\omega) + \omega_{pe}^2 \sum_{p=-\infty}^{\infty} J_p(a) f_i(\omega + p\omega_0) = 0, \tag{9.44}$$

$$D_i(\omega) f_i(\omega) + \omega_{pi}^2 \sum_{p=-\infty}^{\infty} J_p(-a) f_e(\omega + p\omega_0) = 0, \tag{9.45}$$

$$D_\alpha(\omega) = \omega^2 + i\ \nu_\alpha \omega - \omega_\alpha^2. \tag{9.46}$$

We consider $a \ll 1$ (weak pump), so we retain only $p = 0,\ \pm 1$ terms in the Bessel series:

$$D_e(\omega) f_e(\omega) + \omega_{pe}^2 \left[f_i(\omega) + J_1(a) f_i(\omega + \omega_0) + J_{-1}(a) f_i(\omega - \omega_0) \right] = 0, \tag{9.47}$$

$$D_i(\omega) f_i(\omega) + \omega_{pi}^2 \left[f_e(\omega) + J_1(-a) f_e(\omega + \omega_0) + J_{-1}(-a) f_e(\omega - \omega_0) \right] = 0, \tag{9.48}$$

where we used $J_0(\pm a) = 1$. By shifting, we can write

$$\begin{aligned} &D_e(\omega + \omega_0) f_e(\omega + \omega_0) \\ &\quad + \omega_{pe}^2 \left[f_i(\omega + \omega_0) + J_1(a) f_i(\omega + 2\omega_0) + J_{-1}(a) f_i(\omega) \right] = 0, \end{aligned} \tag{9.49}$$

$$\begin{aligned} &D_e(\omega - \omega_0) f_e(\omega - \omega_0) \\ &\quad + \omega_{pe}^2 \left[f_i(\omega - \omega_0) + J_1(a) f_i(\omega) + J_{-1}(a) f_i(\omega - 2\omega_0) \right] = 0. \end{aligned} \tag{9.50}$$

In the above two equations, we consider the frequency ω is in the region of the low frequency near the ion acoustic wave frequency. In the absence of the pump, $f_i(\omega)$ is the ion density perturbation. Thus, $f_i(\omega \pm \omega_0)$, $f_i(\omega \pm 2\omega_0)$, etc. which are perturbations in the high frequency field, are small as compared to $f_i(\omega)$. Neglecting those small terms, the preceding equations yield the dispersion relation

$$D_i(\omega) - \frac{\omega_{pe}^2\omega_{pi}^2}{D_e(\omega)} = \omega_{pe}^2\omega_{pi}^2 J_1^2(a)\left[\frac{1}{D_e(\omega-\omega_0)} + \frac{1}{D_e(\omega+\omega_0)}\right]. \quad (9.51)$$

Since ω is of low frequency, $D_e(\omega) \approx -\omega_e^2$. And we have the relation

$$\begin{aligned} D_i(\omega) - \frac{\omega_{pe}^2\omega_{pi}^2}{D_e(\omega)} &\simeq \omega^2 + i\nu_i\omega - \omega_i^2 + \frac{\omega_{pe}^2\omega_{pi}^2}{\omega_e^2} \\ &\simeq \omega^2 + i\nu_i\omega - \omega_{pi}^2 + \frac{\omega_{pe}^2\omega_{pi}^2}{\omega_e^2} \\ &\simeq \omega^2 + i\nu_i\omega - \frac{k^2\gamma_e T_e}{m_i}, \end{aligned}$$

Then, the dispersion relation takes the form

$$\omega^2 + i\omega\nu_i - \frac{k^2\gamma_e T_e}{m_i} = \frac{1}{4}\, a^2\, \omega_{pe}^2\omega_{pi}^2\left[\frac{1}{D_e(\omega-\omega_0)} + \frac{1}{D_e(\omega+\omega_0)}\right] \quad (9.52)$$

where we used $J_1(a) = a/2$. In (9.52), $D_e(\omega\mp\omega_0)$ corresponds to, respectively, the Stokes (anti-Stokes) coupling. Retaining only $D_e(\omega-\omega_0)$ term recovers (9.43). Keeping both terms recovers (9.40). To show this, we need further simplification of (9.52) to filter out small terms by employing the *resonance approximation*. The approximation is based on $\omega \approx \omega_3 \ll \omega_0 \approx \omega_e = \omega_2$. The LHS of (9.52) is, putting $\omega_3^2 = k^2\gamma_e T_e/m_i$,

$$LHS = \omega^2 - \omega_3^2 + i\omega\nu_i = (\omega+\omega_3)(\omega-\omega_3) + 2i\omega\Gamma_i \approx 2\omega_3(\omega - \omega_3 + i\Gamma_i), \quad (9.53)$$

$$D_e(\omega-\omega_0) \approx -2\omega_2(\omega - \omega_0 + \omega_2 + i\Gamma_e), \quad (9.54)$$

$$D_e(\omega+\omega_0) \approx 2\omega_2(\omega + \omega_0 - \omega_2 + i\Gamma_e). \quad (9.55)$$

Using the above relations in (9.52) recovers (9.40) and (9.43). It is noteworthy that the multiple time perturbation scheme automatically takes care of the resonance approximations.

So far we derived the dispersion relation of the coupled mode parametric instability in two ways with the identical results. Next, the growth rates and thresholds of the instabilities will be calculated. Let us write (9.40) or (9.52) in the form

$$\left(\omega^2 + i\omega\nu_i - k^2\frac{T_e}{m_i}\right)\left[\left(\omega + \frac{i\nu_e}{2}\right)^2 - \delta^2\right] + \frac{\delta}{\omega_0}\omega_{pi}^2|\mathbf{k}\cdot\mathbf{v}_0|^2 = 0 \quad (9.56)$$

where $\delta = \omega_0 - \omega_e$, $\mathbf{v}_0 = \frac{e\mathbf{E}_0}{m_e\omega_0}$ is the electron quiver velocity in the pump field. The form of the dispersion relation (9.56) is similar to the dispersion relation of two stream instability (hence the name OTS instability). When v_0 is small, (9.56) relates with the growth of the ion acoustic wave. The solution can be sought by writing $\omega = kv_s + i\gamma$ ($v_s = \sqrt{\gamma_e T_e/m_i}$, the speed of the ion acoustic wave, γ is the growth rate). In this case, the instability is called a *decay instability.* However, when v_0 is large, ω can become as large as kv_s. Then (9.56) becomes

$$\omega^2(\omega^2 - \delta^2) + \frac{\delta}{\omega_0}\omega_{pi}^2|\mathbf{k}\cdot\mathbf{v}_0|^2 = 0.$$

This mode is entirely independent of the ion acoustic wave. Furthermore it disappears when $v_0 = 0$ (while the ion acoustic wave doesn't). Such a mode which disappears if the pump is turned off is called a *quasi mode.* Returning to (9.56), it yields two different solutions: see § 9.1.3 and § 9.1.4.

Reference: Nishikawa, K. (1968). *Parametric excitation of coupled waves II. Parametric plasmon-photon interaction*, J. Phys. Soc. Japan **24**, p. 1152–1158.

9.1.3. *Purely growing instability (oscillating two stream instability)*

When $\delta = \omega_0 - \omega_e < 0$, (9.56) permits $\omega = i\gamma$ solution, as can be seen from

$$(\gamma^2 + k^2v_s^2)\left[\left(\gamma + \frac{\nu_e}{2}\right)^2 + \delta^2\right] + \frac{\delta}{\omega_0}\omega_{pi}^2|\mathbf{k}\cdot\mathbf{v}_0|^2 = 0. \qquad (9.57)$$

Exercise. Show that in the regime of strong growth ($\gamma \gg kv_s$) the maximum growth rate is

$$\gamma_{max} = \left[\frac{m_e\omega_0|\mathbf{k}\cdot\mathbf{v}_0|^2}{2m_i}\right]^{1/3}$$

which is obtained at $\gamma = -\delta$. Neglect $k^2v_s^2 \ll \gamma^2$ and obtain the optimum γ per $\frac{\partial}{\partial\delta}(9.57) = 0$.

Exercise. Show that in the regime of weak growth the maximum growth rate is

$$\gamma_{max} = \frac{\omega_0}{2}\frac{v_0^2}{u^2} - \frac{\nu_e}{2} \qquad (u = \sqrt{\gamma_e T_e/m_e})$$

which is obtained at $\delta = -\frac{\omega_0}{2}\frac{v_0^2}{u^2}$. The threshold is obtained from setting $\gamma = 0$ which gives

$$\left(\frac{v_0}{u}\right)^2 \geq \frac{\nu_e}{\omega_0}. \qquad (9.58)$$

9.1.4. *Parametric decay instability when* $\delta > 0$

Put $\omega = kv_s + i\gamma$ and $\delta = kv_s(> 0)$ and assume $kv_s \gg \gamma$ in (9.56) to obtain

$$4\gamma^2 + 2\gamma(\nu_i + \nu_e) + \nu_i\nu_e - \frac{\omega_{pi}^2 k^2 v_0^2}{\omega_0 k v_s} = 0. \tag{9.59}$$

The growth rate is found by solving the above quadratic equation. The threshold for the instability is simply obtained by setting $\gamma = 0$:

$$\frac{v_0^2}{u^2} = \frac{\nu_e\nu_i}{\omega_0 k v_s}. \tag{9.60}$$

Note that the threshold for the decay instability depends on the product of the two collision frequencies.

The parametric decay instability can be analyzed by retaining only the Stokes coupling in (9.52),

$$\omega^2 + i\omega\nu_i - k^2 v_s^2 = \frac{\omega_{pi}^2 k^2 v_0^2}{D(\omega - \omega_0)} \tag{9.61}$$

with the result (9.59). It is interesting that (9.56) which have contributions from both of the Stokes and anti-Stokes components contains the solution of (9.61) as a subset. One can verify that (9.61) yields the growth rate given by (9.59) [use (9.54), put $\delta = kv_s$, and neglect $\gamma \ll kv_s$].

Addendum: Nishikawa's solution of Equation (9.52).

Reference: Nishikawa, K. (1968). *Parametric excitation of coupled waves I. General formulation*, J. Phys. Soc. Japan **24**, p. 916–922.
(9.52) is written in the form, in the dipole approximation,

$$(A1): \ \omega^2 + i\omega\nu_i - \frac{T_e k^2}{m_i}$$
$$= \omega_{pi}^2 k^2 v_0^2 \left(\frac{1}{D(\omega - \omega_0, k)} + \frac{1}{D(\omega + \omega_0, k)}\right)\left(v_0^2 = \frac{1}{4}\frac{e^2 E_0^2}{m_e^2 \omega_0^2}\right).$$

i) When $\omega = \sqrt{\frac{T_e}{m_i}}\, k \left(\sqrt{\frac{T_e}{m_i}}\, k = v_s k \gg \text{frequency shift in } \omega\right)$.

$$\omega^2 + i\omega\nu_i - \frac{T_e k^2}{m_i} = (\omega + \omega_1)(\omega - \omega_1) + i\omega\nu_i \simeq 2\omega_1(\omega - \omega_1 + i\Gamma_1)$$

where we changed notation: $\omega_1 = \sqrt{\frac{T_e k^2}{m_i}}$ is our earlier ω_3, and $\Gamma_1 = \nu_i/2$. In this case, the anti-Stokes coupling can be neglected.

$$D(\omega - \omega_0, k) \simeq -2\omega_e(\omega - \omega_0 + \omega_e + i\Gamma_2) \quad (\omega \ll \omega_0, \quad \Gamma_2 = \nu_e/2).$$

Using the above two equations and introducing the frequency shift $\Delta = \omega_0 - \omega_1 - \omega_e$, Equation (A1) is written in the form

$$(A2): (\omega - \omega_1 + i\Gamma_1)(\omega - \omega_1 + i\Gamma_2 - \Delta) + \frac{\omega_{pi}^2 k^2 v_0^2}{4\omega_1 \omega_e} = 0.$$

Put $\omega = x + iy$ in (A2) and separate the real and imaginary parts:

$$\textit{real part}: \quad (x - \omega_1)(x - \omega_1 - \Delta) - (y + \Gamma_1)(y + \Gamma_2) + \frac{\omega_{pi}^2 k^2 v_0^2}{4\omega_1 \omega_e} = 0.$$

$$\textit{Imaginary part}: \quad (x - \omega_1)(2y + \Gamma_1 + \Gamma_2) = \Delta(y + \Gamma_1).$$

Eliminating x in the above two equations gives

$$(y + \Gamma_1)(y + \Gamma_2)\left(1 + \frac{\Delta^2}{(2y + \Gamma_1 + \Gamma_2)^2}\right) = \frac{\omega_{pi}^2 k^2 v_0^2}{4\omega_1 \omega_e}.$$

The threshold exists since LHS is greater than zero. The threshold is obtained by putting $y = 0$:

$$\frac{\omega_{pi}^2 k^2 v_0^2}{4\omega_1 \omega_e} = \Gamma_1 \Gamma_2 \left(1 + \frac{\Delta^2}{(\Gamma_1 + \Gamma_2)^2}\right).$$

Optimum threshold is obtained when $\Delta = 0$:

$$\frac{v_0^2}{u^2} = \frac{\nu_i \nu_e}{k v_s \omega_{pe}} \quad \left(u^2 = \frac{T_e}{m_e}\right).$$

ii) When $v_s k \ll \omega_e$, the anti-Stokes coupling should be included. Also, $\omega_0 \simeq \omega_e$. Using the resonance approximation, we have

$$D(\omega + \omega_0, k) = 2\omega_e(\omega + \delta + i\Gamma_2),$$

$$D(\omega - \omega_0, k) = -2\omega_e(\omega - \delta + i\Gamma_2),$$

where $\delta = \omega_0 - \omega_e$. Then (A1) takes the form

$$(A3): \omega^2 + i\omega\nu_i - \frac{T_e k^2}{m_i} = \frac{\omega_{pi}^2 k^2 v_0^2}{2\omega_0}\left[\frac{-2\delta}{(\omega + i\Gamma_2)^2 - \delta^2}\right].$$

It is readily seen that $\omega = iy$ (purely growing) solution is possible. Putting $\omega = iy$ in the above gives

$$[\delta^2 + (y + \Gamma_2)^2]\left[y^2 + 2y\Gamma_1 + \frac{T_e}{m_i}k^2\right] + \delta\,\frac{k^2 v_0^2 \omega_{pi}^2}{\omega_0} = 0.$$

Note that $\delta < 0$ is necessary for this equation to be valid. The threshold is obtained from the above equation by putting $y = 0$:

$$(A4): \quad v_0^2 = -\frac{\omega_0}{\delta}\,\frac{\delta^2 + \Gamma_2^2}{\omega_{pi}^2}\,\frac{T_e}{m_i}.$$

Note that this is independent of Γ_1. The optimum threshold can be obtained from the condition $\partial v_0^2/\partial\delta = 0$: We obtain the threshold for the purely growing instability

$$(A5): \quad \frac{v_0^2}{u^2} = \frac{\nu_e}{\omega_0} \quad (when\ \delta = \Gamma_2).$$

Equation (A3) also allows for $\omega = x + iy$ solution. Separating (A3) into real and imaginary parts after putting $\omega = x + iy$, we obtain

$$(x^2 - y^2 - 2y\Gamma_1 - \omega_1^2)F(x,y) = \delta^2 - x^2 + (y + \Gamma_2)^2$$

$$(y + \Gamma_1)F(x,y) = y + \Gamma_2$$

where

$$F(x,y) = \frac{\omega_0}{\omega_{pi}^2 k^2 v_0^2 \delta}\,[(x+\delta)^2 + (y+\Gamma_2)^2][(x-\delta)^2 + (y+\Gamma_2)^2].$$

We observe that a growing solution ($y > 0$) is possible only when $F(x,y) > 0$ i. e. $\delta > 0$. We eliminate $F(x,y)$ and x^2 to have an equation for the growth rate y. Then putting $y = 0$ in that equation gives the equation for the threshold:

$$\omega_{pi}^2 k^2 v_0^2 = \frac{\Gamma_1\Gamma_2\omega_0}{\delta}\left[4\delta^2 + \frac{(\Gamma_2^2 + 2\Gamma_1\Gamma_2 + \omega_1^2 - \delta^2)^2}{(\Gamma_1 + \Gamma_2)^2}\right].$$

We take $\delta = \omega_1$ and assume that $\omega_1 \gg \Gamma_1,\ \Gamma_2$. Then we have $\omega_{pi}^2 k^2 v_0^2 = 4\Gamma_1\Gamma_2\omega_0\omega_1$. Or

$$v_0^2 = \frac{4\Gamma_1\Gamma_2\omega_0\omega_1}{k^2\omega_{pi}^2} = u^2\,\frac{\nu_e\nu_i}{\omega_0 k v_s}.$$

This is the threshold for the parametric decay instability obtained earlier.

9.2. Physical derivation of coupled wave equations

The coupled equations in the preceding section were derived mathematically. For the sake of more transparent physical presentation of the parametric interactions, we explicitly exhibit the coupling terms incorporated in the governing plasma equations; this helps to identify what the important nonlinear mechanism is in the parametric instability.

Parametric decay instability occurs when a high frequency electromagnetic wave (represented by the quiver velocity (9.4)) decays into a high frequency electron wave (Langmuir wave) and a low frequency ion acoustic wave. We use:

• High frequency electron equation of motion:

$$\frac{\partial \mathbf{v}_e}{\partial t} = -\frac{e}{m_e}\mathbf{E} - \frac{\gamma_e T_e}{N m_e}\nabla n_e - \nu_e \mathbf{v}_e. \tag{9.62}$$

• Low frequency electron equation of motion: (the low frequency quantities will be denoted by δ)

$$-\frac{e}{m_e}\delta\mathbf{E} - \frac{T_e}{N m_e}\nabla\delta n_e = \mathbf{v}_0 \cdot \nabla \mathbf{v}_e \tag{9.63}$$

where $\mathbf{v}_0$ has been given by (9.4). In (9.63), the electron inertia is neglected, and the last term, the nonlinear term, is the electron ponderomotive force which will give rise to a low frequency beat by mixing two high frequency waves.

• Low frequency ion equation of motion:

$$\frac{\partial}{\partial t}\delta\mathbf{v}_i = \frac{e}{m_i}\delta\mathbf{E} - \nu_i \delta\mathbf{v}_i. \tag{9.64}$$

• High frequency electron continuity equation:

$$\frac{\partial n_e}{\partial t} + N\nabla \cdot \mathbf{v}_e = 0. \tag{9.65}$$

• Low frequency ion continuity equation:

$$\frac{\partial\, \delta n_i}{\partial t} + N\nabla \cdot \delta\mathbf{v}_i = 0. \tag{9.66}$$

Equations (9.62) to (9.66) should be supplemented by Maxwell equation.

$$\frac{\partial \mathbf{E}}{\partial t} = 4\pi e(N\mathbf{v}_e + \delta n_e \mathbf{v}_0): \quad \textit{high frequency}. \tag{9.67}$$

In (9.67) $\mathbf{v}_0 \delta n_e$, the nonlinear electron current, is the important nonlinearity, together with the ponderomotive force in (9.63), which excites the decay instability. Both of the nonlinearities are electron's. (9.62), (9.65), and (9.67) are arranged in the form

$$\left(\frac{\partial^2}{\partial t^2} + \nu_e \frac{\partial}{\partial t} + \omega_{pe}^2 - \frac{\gamma_e T_e}{m_e}\nabla\nabla\cdot\right)\mathbf{v}_e = -\omega_{pe}^2 \frac{\delta n_e}{N}\mathbf{v}_0. \tag{9.68}$$

The low frequency equations, (9.63) and (9.64) are combined to give

$$-\frac{m_i}{m_e}\left(\frac{\partial}{\partial t}\,\delta\mathbf{v}_i + \nu_i \delta\mathbf{v}_i\right) - \frac{T_e}{N m_e}\nabla\delta n_e = \mathbf{v}_0 \cdot \nabla \mathbf{v}_e. \tag{9.69}$$

Perform $\nabla\cdot$ on the above equation and use (9.66) in which we put $\delta n_i \simeq \delta n_e$ (this charge neutrality condition is justified since we are considering the low frequency wave) to get

$$\left(\frac{\partial^2}{\partial t^2} + \nu_i \frac{\partial}{\partial t} - \frac{T_e}{m_i}\nabla^2\right)\frac{\delta n_e}{N} = \frac{m_e}{m_i}\nabla \cdot (\mathbf{v}_0 \cdot \nabla \mathbf{v}_e). \tag{9.70}$$

Equations (9.68) and (9.70) are the two coupled equations describing parametric interactions. The RHS of (9.68) represents the nonlinear current which excites the scattered field, and the RHS of (9.70) is the ponderomotive force which gives rise to, in turn, the electron density perturbation δn_e.

Dipole approximation is not used. In (9.68) and (9.70), the electron quiver velocity $\mathbf{v}_0$ (given in (9.4)) is written in the dipole approximation. In this section we remove this restriction to derive the parametric equations. Consider a large amplitude plane polarized electromagnetic pump wave is propagated in the plasma:

$$\mathbf{E}_p(\mathbf{r}, t) = \mathbf{E}_0 e^{i\mathbf{k}_0\cdot\mathbf{r} - i\omega_0 t} + \mathbf{E}_0^* e^{-i\mathbf{k}_0\cdot\mathbf{r} + i\omega_0 t},$$

$$(* = c.c. \quad complex\ conjugate). \tag{9.71}$$

The particle quiver velocity

$$\mathbf{v}_{0\alpha}(\mathbf{r}, t) = \frac{ie_\alpha}{m_\alpha \omega_0}\mathbf{E}_0\ e^{i(\mathbf{k}_0\cdot\mathbf{r} - \omega_0 t)} + c.c. \tag{9.72}$$

represents the equilibrium electron velocity in the plasma under consideration. The frequency ω_0 is so high that only the electrons oscillate with the high quiver velocity in the pump field with the ions forming a stationary background (also note that the ion nonlinearities are much smaller than the electron's by the mass ratio m_e/m_i). The electric vector is polarized so as to be $\mathbf{k}_0 \cdot \mathbf{E}_0 = 0$. And the pump wave satisfies the usual electromagnetic dispersion relation,

$$\omega_0^2 = \omega_{pe}^2 + c^2 k_0^2. \tag{9.73}$$

When the pump frequency is near to the plasma frequency, the dipole approximation ($\mathbf{k}_0 \simeq 0$) is valid. When the pump is a propagating wave as given by (9.72), (9.70) should be modified to have an extra term:

$$\left(\frac{\partial^2}{\partial t^2} + \nu_i \frac{\partial}{\partial t} - \frac{T_e}{m_i}\nabla^2\right)\frac{\delta n_e}{N} = \frac{m_e}{m_i}\nabla \cdot [\mathbf{v}_0 \cdot \nabla \mathbf{v}_e + \mathbf{v}_e \cdot \nabla \mathbf{v}_0]. \tag{9.74}$$

Then, the parametric equations for the pump field (9.72) are (9.68) and (9.74). Fourier-transforming (9.68) yields

$$(-\omega^2 - i\omega\nu_e + \omega_{pe}^2)\mathbf{v}_e(\mathbf{k},\omega) + \frac{\gamma_e T_e}{m_e}\mathbf{k}\cdot\mathbf{v}_e(\mathbf{k},\omega)\mathbf{k}$$
$$= -\frac{\omega_{pe}^2}{N}[\mathbf{v}_0\delta n_e(\omega-\omega_0,\mathbf{k}-\mathbf{k}_0) + \mathbf{v}_0^*\delta n_e(\omega+\omega_0,\mathbf{k}+\mathbf{k}_0)] \quad (9.75)$$

where

$$\mathbf{v}_0 = \frac{-ie\mathbf{E}_0}{m_e\omega_0} \quad (9.76)$$

[the *amplitude* of the quiver velocity (9.76) will be written without the argument. The quiver velocity itself (see (9.72)) is also written sometimes without the argument if there is no danger of confusion. One can recognize that $\mathbf{v}_0$ is the amplitude if $\mathbf{v}_0^*$ appears simultaneously.]

Fourier-transforming (9.74) yields

$$\left(-\omega^2 - i\omega\nu_i + \frac{T_e}{m_i}k^2\right)\frac{\delta n_e(\omega,\mathbf{k})}{N}$$
$$= \frac{m_e}{m_i}\times[-\mathbf{v}_0\cdot(\mathbf{k}-\mathbf{k}_0)\ \ \mathbf{k}\cdot\mathbf{v}_e(\omega-\omega_0,\mathbf{k}-\mathbf{k}_0)$$
$$-\mathbf{v}_0^*\cdot(\mathbf{k}+\mathbf{k}_0)\ \mathbf{k}\cdot\mathbf{v}_e(\omega+\omega_0,\mathbf{k}+\mathbf{k}_0) - \mathbf{v}_0\cdot\mathbf{k}\ \ \mathbf{k}_0\cdot\mathbf{v}_e(\omega-\omega_0,\mathbf{k}-\mathbf{k}_0)$$
$$+\mathbf{v}_0^*\ \cdot\mathbf{k}\ \ \mathbf{k}_0\cdot\mathbf{v}_e(\omega+\omega_0,\mathbf{k}+\mathbf{k}_0)]. \quad (9.77)$$

Putting $\mathbf{k}_0\cdot\mathbf{v}_0 = 0$ simplifies the above equation,

$$\left(\omega^2 + i\omega\nu_i - \frac{T_e}{m_i}k^2\right)\frac{\delta n_e(\omega,\mathbf{k})}{N} = \frac{m_e}{m_i}[(\mathbf{k}\cdot\mathbf{v}_0)\ (\mathbf{k}+\mathbf{k}_0)\cdot\mathbf{v}_e(\omega-\omega_0,\mathbf{k}-\mathbf{k}_0)$$
$$+(\mathbf{k}\cdot\mathbf{v}_0^*)\ (\mathbf{k}-\mathbf{k}_0)\cdot\mathbf{v}_e(\omega+\omega_0,\mathbf{k}+\mathbf{k}_0)]. \quad (9.78)$$

In (9.75), we can set $\mathbf{v}_e \| \mathbf{k}$ since we are considering electrostatic waves. Then we have

$$\mathbf{v}_e(\omega-\omega_0,\mathbf{k}-\mathbf{k}_0) = \frac{\omega_{pe}^2}{N}\frac{\mathbf{v}_0^*\ \delta n_e(\omega,\mathbf{k})}{D(\omega-\omega_0,\mathbf{k}-\mathbf{k}_0)}, \quad (9.79)$$

$$\mathbf{v}_e(\omega+\omega_0,\mathbf{k}+\mathbf{k}_0) = \frac{\omega_{pe}^2}{N}\frac{\mathbf{v}_0\ \delta n_e(\omega,\mathbf{k})}{D(\omega+\omega_0,\mathbf{k}+\mathbf{k}_0)}, \quad (9.80)$$

$$D(\omega,\mathbf{k}) = \omega^2 + i\omega\nu_e - \omega_{pe}^2 - \frac{\gamma_e T_e}{m_e}k^2, \quad (9.81)$$

where we neglected the off-resonance terms, $\delta n_e(\omega \pm 2\omega_0, \mathbf{k} \pm 2\mathbf{k}_0)$. Then (9.75) and (9.78) give, with the aid of above equations, the following dispersion relation,

$$\omega^2 + i\omega\nu_i - \frac{T_e k^2}{m_i}$$
$$= \omega_{pi}^2(\mathbf{k}\cdot\mathbf{v}_0)(\mathbf{k}\cdot\mathbf{v}_0^*)\left[\frac{1}{D(\omega-\omega_0, \mathbf{k}-\mathbf{k}_0)} + \frac{1}{D(\omega+\omega_0, \mathbf{k}+\mathbf{k}_0)}\right]. \tag{9.82}$$

Exercise. Derive (9.77).

• **Wave vector resonance condition.** The foregoing steps leading to the dispersion relation (9.82) requires an intensive caution in the algebra of Fourier transform, so an easier and straight way is desirable in hindsight. For this purpose, we stipulate from the beginning the resonance conditions among the central wave (k_0, ω_0) and the two sidebands:

$$k' = k_0 - k, \quad \omega' = \omega_0 - \omega \quad : \textit{Stokes} \tag{9.83}$$

$$k'' = k_0 + k, \quad \omega'' = \omega_0 + \omega \quad : \textit{anti-Stokes} \tag{9.84}$$

where we understand that ω and k belong to the low frequency wave. For simplicity, we shall assume $\nabla = \hat{\mathbf{x}}\frac{\partial}{\partial x}$, $\mathbf{v}_e = \hat{\mathbf{x}} v_e$, and $\mathbf{v}_0 = \hat{\mathbf{x}} v_0$ to write (9.68) and (9.70) in the form

$$\left(\frac{\partial^2}{\partial t^2} + \nu_e\frac{\partial}{\partial t} + \omega_{pe}^2 - \frac{\gamma_e T_e}{m_e}\frac{\partial^2}{\partial x^2}\right) v_e = -\omega_{pe}^2\frac{\delta n_e}{N} v_0, \tag{9.85}$$

$$\left(\frac{\partial^2}{\partial t^2} + \nu_i\frac{\partial}{\partial t} - \frac{T_e}{m_i}\frac{\partial^2}{\partial x^2}\right)\frac{\delta n_e}{N} = \frac{m_e}{m_i}\frac{\partial}{\partial x}\left(v_0\frac{\partial}{\partial x} v_e\right). \tag{9.86}$$

We write

$$v_0 = \tilde{v}_0 e^{ik_0x - i\omega_0 t} + \tilde{v}_0^* e^{-ik_0x + i\omega_0 t} \tag{9.87}$$

$$v_e = v_e' e^{ik'x - i\omega' t} + v_e'^* e^{-ik'x + i\omega' t} + v_e'' e^{ik''x - i\omega'' t} + v_e''^* e^{-ik''x + i\omega'' t} \tag{9.88}$$

$$\frac{\delta n_e}{N} = n(x,t) = \tilde{n} e^{ikx - i\omega t} + \tilde{n}^* e^{-ikx + i\omega t}. \tag{9.89}$$

Substituting (9.88) and (9.89) into (9.85) yields, after discarding all the terms which do not satisfy the resonance conditions,

$$(\omega'^2 + i\omega'\nu_e - \omega_{pe}^2 - k'^2 u^2) v_e' = \omega_{pe}^2 \tilde{v}_0 \tilde{n}^* \quad (u^2 = \gamma_e T_e/m_e), \tag{9.90}$$

$$(\omega''^2 + i\omega''\nu_e - \omega_{pe}^2 - k''^2 u^2)v_e'' = \omega_{pe}^2 \tilde{v}_0 \tilde{n}, \tag{9.91}$$

$$(\omega^2 + i\omega\nu_i - v_s^2 k^2)\tilde{n} = \frac{m_e}{m_i} k^2 (\tilde{v}_0 v_e'^* + \tilde{v}_0^* v_e''). \quad (v_s^2 = T_e/m_i). \tag{9.92}$$

We can show that the above three equations agree with (9.82).

9.3. Parametric decay instability and nonlinear Landau damping

The threshold for the parametric decay instability can be derived not only from the dispersion relation (9.56) but also from consideration of power flow among the two interacting waves: the low frequency wave $n(\omega)$ and the scattered wave (lower side band wave) $v'(\omega')$. We begin with (9.92) where we neglect the anti-Stokes component:

$$\tilde{n} = \frac{m_e}{m_i} \frac{k^2 \tilde{v}_0 v_e'^*}{\omega^2 + i\nu_i\omega - v_s^2 k^2}. \tag{9.93}$$

This is the low frequency ion density perturbation determined as a beat mode of the two high frequency waves with frequencies ω_0 and ω'. In the low frequency region, the quasi-neutrality holds, and therefore, this can be put equal to the electron density perturbation, which couples the pump to drive a high frequency nonlinear *electron current*:

$$J_e' = -eN\tilde{v}_0\tilde{n}^* = -\frac{m_e}{m_i} \frac{eNk^2|\tilde{v}_0|^2 v_e'}{\omega^2 - i\nu_i\omega - v_s^2 k^2}. \tag{9.94}$$

The dissipation rate of the ω'-wave energy is obtained by

$$\frac{\partial}{\partial t}|E'|^2 = -4\pi(J_e' E'^* + J'^* E')$$
$$= 4\pi e N k^2 |\tilde{v}_0|^2 \frac{m_e}{m_i}\left[\frac{v_e' E'^*}{\omega^2 - i\nu_i\omega - v_s^2 k^2} + \frac{v_e'^* E'}{\omega^2 + i\nu_i\omega - v_s^2 k^2}\right].$$

Here we assume $\nu_i \gg \omega,\ kv_s$, and keep only the $\nu_i\omega$-term in the denominator and use $v_e' = \frac{-ie}{m_e}\frac{E'}{\omega'}$ to obtain

$$\frac{\partial}{\partial t}|E'|^2 = 4\pi e N k^2 |\tilde{v}_0|^2 \frac{m_e}{m_i}|E'|^2 \frac{e}{m_e\omega'}\frac{2}{\nu_i\omega}.$$

Equating the above expression to the collisional dissipation rate of the ω'-wave, i.e. $\nu_e|E'|^2$, we obtain the threshold of the parametric decay instability as

$$\frac{\tilde{v}_0^2}{u^2} = \frac{1}{2}\frac{\nu_e\nu_i}{kv_s\omega_0} \tag{9.95}$$

where we put $\omega \simeq kv_s$, $\omega' \simeq \omega_{pe} \simeq \omega_0$. Except for the factor $\frac{1}{2}$, (9.95) is the same as (9.60).

When the frequency ν_i dominates over ω and kv_s, the ion acoustic wave loses its identity as a normal mode; it is a degenerated virtual mode due to heavy damping. When the plasma has $T_e \sim T_i$, the ion wave is heavily Landau-damped on ions if the wave-particle resonant condition,

$$\omega_0 - \omega' = (k_0 - k') \times \textit{ ion thermal velocity}$$

is satisfied. When we identify ν_i as the Landau damping rate of the ion acoustic wave, the parametric decay process described above is called *nonlinear Landau damping.* The beat wave energy is channeled to the scattered wave to give rise to a parametric decay instability.

9.4. More on oscillating two stream instability (OTSI)

The purely growing wave is a zero-frequency wave. This is an extended parametric decay instability which arises when both Stokes and anti-Stokes couplings are included. At the onset of the OTSI, the 'dispersion relation' is obtained by putting $\omega = 0$ in (9.52):

$$-\frac{T_e}{m_i}k^2 = \omega_{pi}^2 k^2 v_0^2 \left[\frac{1}{\omega_0^2 - i\omega_0\nu_e - \omega_{ek}^2} + \frac{1}{\omega_0^2 + i\omega_0\nu_e - \omega_{ek}^2}\right] = \frac{\omega_{pi}^2 k^2 v_0^2}{\omega_0}\,\frac{\delta}{\delta^2 + \frac{\nu_e^2}{4}}$$

which gives

$$v_0^2 = \frac{T_e}{m_i}\frac{\omega_0}{\omega_{pi}^2}\,\frac{\delta^2 + \frac{\nu_e^2}{4}}{\delta}.$$

The right hand side has the minimum value at $\delta = \nu_e/2$, and the corresponding v_0 is

$$\frac{v_0^2}{u^2} = \frac{\nu_e}{\omega_0}. \quad (\textit{see } (9.58))$$

At the onset of OTSI, the scattered waves (ω', ω'') form a standing wave. In (9.88) we substitute the following expressions (obtained per $\omega = k_0 = 0$), introducing a phase α,

$$v_e' = |v_e|e^{-i\alpha}e^{-ikx-i\omega_0 t} + c.c. = 2|v_e|\ cos(kx + \omega_0 t + \alpha),$$
$$v_e'' = |v_e|e^{-i\alpha}e^{ikx-i\omega_0 t} + c.c. = 2|v_e|\ cos(kx - \omega_0 t - \alpha).$$

So we can represent the scattered field as a standing wave

$$v_e = |v_e| coskx\ cos(\omega_0 t + \alpha). \tag{9.96}$$

The corresponding electric field is

$$E_s = -\frac{m_e}{e}\frac{\partial v_e}{\partial t} = |E_s|\ coskx\ sin(\omega_0 t + \alpha) \tag{9.97}$$

with

$$|E_s| = \frac{m_e}{e}\omega_0 |v_e|. \quad (s:\ scattered)$$

The low frequency equations are

$$-\frac{e}{m_e}\delta E - \frac{T_e}{N m_e}\frac{\partial \delta n_e}{\partial x} = v_0\frac{\partial v_e}{\partial x}, \quad (cf.\ (9.63)) \tag{9.98}$$

$$\frac{\partial}{\partial t}\delta v_i = \frac{e}{m_i}\delta E - \frac{T_i}{N m_i}\frac{\partial}{\partial x}\delta n_i. \quad (cf.\ (9.64)) \tag{9.99}$$

In (9.99) we neglected the ion collision frequency ν_i which is not essential in OTSI and included T_i although it is not essential. Eliminating δE in the above two equations, we get

$$\frac{\partial}{\partial t}\delta v_i + \frac{T_i + T_e}{N m_i}\frac{\partial}{\partial x}\delta n_e = -\frac{m_e}{m_i}\frac{\partial}{\partial x}(v_0 v_e). \tag{9.100}$$

We put $\partial/\partial t = 0$ in (9.100) because we consider a zero frequency wave to obtain

$$\delta n_e = -\frac{N m_e}{T_e + T_i}\langle v_0 v_e\rangle \tag{9.101}$$

where $\langle\cdots\rangle$ is the time average and $\langle v_0 v_e\rangle$ is the ponderomotive force. Note that to pick up the zero frequency component from the RHS we have to take time average. For a pump of the form

$$E_0 = |E_0| sin\ \omega_0 t, \quad or \quad v_0 = \xi\ \omega_0\ cos\ \omega_0 t$$

where $\xi = \frac{e|E_0|}{m_e\omega_0^2}$ is the amplitude of the electron oscillation, the time average is

$$\langle v_0 v_e\rangle = \frac{\xi|E_s|}{2}\,\frac{e}{m_e}\,cos\;kx\;cos\;\alpha \tag{9.102}$$

where we used (9.96). The current responsible for the excitation of the scattered wave is

$$J = -e\delta n_e\; v_0 = -e\omega_0\xi\; cos\omega_0 t\;\delta n_e.$$

The power per unit volume delivered by this current to the standing wave is given by

$$\langle E_s\; J\rangle = \langle E_s\; cos\; kx\; sin(\omega_0 t + \alpha)\; J\rangle$$

$$= \frac{1}{2}\;|E_s|^2\;\xi^2\frac{e^2 N\omega_0}{T_e + T_i}\langle cos^2 kx\; cos^2\omega_0 t\rangle\; sin\alpha\; cos\;\alpha$$

where the time and space average give factor (1/2), respectively, and we put $\alpha = \pi/4$, corresponding to the maximum power, to get

$$\langle E_s J\rangle = \frac{1}{16}|E_s|^2\;\xi^2\;\frac{Ne^2\omega_0}{T_e + T_i}.$$

At the threshold, this power must be equal to the power of the scattered wave dissipated by collision, $\frac{\nu_e}{16\pi}|E_s|^2$, giving in terms of the Debye lengths,

$$\xi^2 = 4\frac{\nu_e}{\omega_{pe}}\;(\lambda_e^2 + \lambda_i^2) \quad or \quad |E_0|^2 = 16\pi N\;\frac{\nu_2}{\omega_{pe}}(T_e + T_i). \tag{9.103}$$

which can be put into the form (use $|v_0| = \xi\omega_0$)

$$\frac{|v_0|^2}{u^2} = 4\;\frac{\nu_e}{\omega_{pe}}. \quad (cf.\;(9.58))$$

The factor 4 discrepancy from (9.58) arises because of our use of *sin* and *cos* representation instead of the exponential functions.

9.5. Two-time scale analysis of parametric interaction

The parametric interaction between the electron and ion waves is basically a phenomenon involving two time scales of great difference; slow and fast oscillations. Mode couplings among the plasma normal modes make the amplitudes of the participating waves change slowly while the phases of the waves oscillate rapidly. Without the coupled mode interaction, the Fourier amplitudes of the normal modes remain constant in time and space. Naturally, analysis of parametric instability is facilitated by employing two time perturbation methods which clearly delineate the different time scales characterizing the dynamics of mode coupling. This section complements the outright perturbation scheme presented in the earlier section. Here we shall assume $k_0 = 0$ (dipole approximation) for the pump,

$$v_0(t) = \tilde{v}_0 e^{-i\omega_0 t} + \tilde{v}_0^* e^{i\omega_0 t}.$$

Then the x coordinate in (9.85) and (9.86) can be Fourier transformed away by replacing $\partial/\partial x = ik$ with the understanding that k now is the wave number of the low frequency wave. Then the two equations read

$$\left(\frac{\partial^2}{\partial t^2} + \nu_e \frac{\partial}{\partial t} + \omega_{pe}^2 + u^2 k^2\right) v_e(k,t) = -\omega_{pe}^2 n(k,t) v_0(t), \qquad (9.104)$$

$$\left(\frac{\partial^2}{\partial t^2} + \nu_i \frac{\partial}{\partial t} + v_s^2 k^2\right) n(k,t) = -\frac{m_e}{m_i} k^2 v_0(t) v_e(k,t). \qquad (9.105)$$

Let us write $t = t_0 + \epsilon t_1$ where t_0 is the fast time which scales the rapidly oscillating phases and t_1 is the slow time which scales the slowly changing amplitudes. Accordingly the time derivative is written

$$\frac{\partial}{\partial t} = \frac{\partial}{\partial t_0} + \epsilon \frac{\partial}{\partial t_1}. \qquad (9.106)$$

Also we write

$$n(k,t) = \tilde{n}(k,t_1) e^{-i\omega t_0} + \tilde{n}^*(k,t_1) e^{+i\omega t_0}. \qquad (9.107)$$

Because of the coupling in (9.104), there will appear quasi plasma waves whose frequencies are up and down shifted:

$$v_e' = A'(t_1) e^{-i\omega' t_0} + A'^*(t_1) e^{i\omega' t_0}, \qquad (9.108)$$

$$v_e'' = A''(t_1) e^{-i\omega'' t_0} + A''^*(t_1) e^{i\omega'' t_0}, \qquad (9.109)$$

$$\textit{where} \quad \omega' = \omega_0 - \omega, \quad \omega'' = \omega_0 + \omega. \qquad (9.110)$$

Using (9.106) in (9.104), the quasi waves should satisfy

$$\left[\frac{\partial^2}{\partial t_0^2} + \omega_{ek}^2 + \epsilon\left(2\frac{\partial}{\partial t_0}\frac{\partial}{\partial t_1} + \nu_e \frac{\partial}{\partial t_0}\right)\right] A'(t_1) e^{-i\omega' t_0} = -\omega_{pe}^2 \tilde{n}^* v_0 e^{-i(\omega_0 - \omega)t_0} \qquad (9.111)$$

$$\left[\frac{\partial^2}{\partial t_0^2} + \omega_{ek}^2 + \epsilon\left(2\frac{\partial}{\partial t_0}\frac{\partial}{\partial t_1} + \nu_e \frac{\partial}{\partial t_0}\right)\right] A''(t_1) e^{-i\omega'' t_0} = -\omega_{pe}^2 \tilde{n} v_0 e^{-i(\omega_0 + \omega)t_0} \qquad (9.112)$$

$$\textit{where} \quad \omega_{ek}^2 = \omega_{pe}^2 + u^2 k^2 \qquad (9.113)$$

and all the non-resonant terms are discarded in the RHS. In the ϵ^0 order we have

$$\omega'^2 = \omega_{ek}^2, \quad \omega''^2 = \omega_{ek}^2. \tag{9.114}$$

In view of (9.110), the two equations in (9.114) can't be both right. So we relax the resonance conditions by introducing the frequency mismatches,

$$-\omega_0 + \omega + \omega' = \Delta', \tag{9.115}$$

$$-\omega_0 - \omega + \omega'' = \Delta''. \tag{9.116}$$

Then in ϵ^1 order, we obtain

$$\left(\frac{\partial}{\partial t_1} + \Gamma_e\right) A' = -\frac{i}{2}\frac{\omega_{pe}^2}{\omega'}\tilde{v}_0\tilde{n}^* e^{i\Delta' t_1} \tag{9.117}$$

$$\left(\frac{\partial}{\partial t_1} + \Gamma_e\right) A'' = -\frac{i}{2}\frac{\omega_{pe}^2}{\omega''}\tilde{v}_0\tilde{n} e^{i\Delta'' t_1} \tag{9.118}$$

where $\Gamma_e = \nu_e/2$. Likewise, (9.105) gives the linear dispersion relation

$$\omega^2 = v_s^2 k^2 \tag{9.119}$$

and the equation for the amplitude

$$\left(\frac{\partial}{\partial t_1} + \Gamma_i\right) \tilde{n} = -\frac{i}{2}\frac{m_e}{m_i}\frac{k^2}{\omega}(\tilde{v}_0 A'^* e^{i\Delta' t_1} + \tilde{v}_0^* A'' e^{-i\Delta'' t_1}). \quad (\Gamma_i = \nu_i/2) \tag{9.120}$$

(9.117)–(9.120) completely agree with the earlier mode coupling equations (9.33)–(9.35); but the coupling coefficients are more explicit.

Exercise. Verify that $A_{-3} \equiv \tilde{n}\, e^{\Gamma_i t}$ satisfies (9.38).

Exercise. Verify that A_{-3} satisfies (9.42) if the anti-Stokes coupling is discarded.

9.6. Various parametric processes

Reference: W. L. Kruer *The physics of laser plasma interactions*, Addison-Wesley 1988.

9.6.1. *Stimulated Raman scattering*

A large amplitude electromagnetic wave (light wave) disturbs a plasma to give rise to an electron density perturbation which in turn scatters off the incident wave into another electromagnetic wave of a different frequency. To investigate this process, we use the following equations.

$$\frac{\partial \mathbf{v}}{\partial t} + \mathbf{v} \cdot \nabla \mathbf{v} = -\frac{e}{m_e}\left[\mathbf{E} + \frac{1}{c}\mathbf{v} \times \mathbf{B}\right] - \frac{\nabla p}{N m_e} - \nu_e \mathbf{v}_e. \tag{9.121}$$

The electron velocity $\mathbf{v}$ can be separated into two components: a component in the light field ($\mathbf{v}_L$)and a component in the plasma wave field ($\mathbf{v}_e$). In the light field, the magnetic force can be neglected for non-relativistic motion and the second term in (9.121) can be neglected as compared to the first term.

$$\frac{|\partial v/\partial t|}{|\mathbf{v} \cdot \nabla \mathbf{v}|} = \frac{\omega v}{k v^2} = \frac{c}{v} \gg 1.$$

So we use

$$\frac{\partial \mathbf{v}_L}{\partial t} = \frac{-e}{m_e}\mathbf{E}_L. \quad (\textit{light wave}) \tag{9.122}$$

In the electrostatic plasma wave equation, the ponderomotive force, the beat of two light waves (the pump wave and the scattered light wave) should be included:

$$\frac{\partial \mathbf{v}_e}{\partial t} + \mathbf{v}_L \cdot \nabla \mathbf{v}_L = \frac{e}{m_e}\nabla\phi - \frac{\nabla p}{N m_e} - \nu_e \mathbf{v}_e - \frac{e}{c m_e}\mathbf{v}_L \times \mathbf{B}. \tag{9.123}$$

Use the vector identity, $\mathbf{v}_L \cdot \nabla \mathbf{v}_L = (\nabla \times \mathbf{v}_L) \times \mathbf{v}_L + \nabla v_L^2/2$. Note that (9.122) gives

$$\frac{\partial}{\partial t}\nabla \times \mathbf{v}_L = -\frac{e}{m_e}\nabla \times \mathbf{E}_L. \tag{9.124}$$

Here, the Maxwell equation

$$\nabla \times \mathbf{E}_L = -\frac{1}{c}\frac{\partial \mathbf{B}}{\partial t} \tag{9.125}$$

gives $\nabla \times \mathbf{v}_L = \frac{e\mathbf{B}}{m_e c}$. So we have a cancellation in (9.123):

$$\frac{\partial \mathbf{v}_e}{\partial t} = -\nabla\frac{v_L^2}{2} + \frac{e}{m_e}\nabla\phi - \frac{\nabla p}{N m_e} - \nu_e \mathbf{v}_e \quad (\textit{plasma wave}) \tag{9.126}$$

where $\mathbf{v}_L = \mathbf{v}_0 + \mathbf{v}_L'$ ($'$ denotes the scattered light wave). The other necessary equations are

$$\frac{\partial n_e}{\partial t} + N\nabla \cdot \mathbf{v}_e = 0, \tag{9.127}$$

$$\nabla^2\phi = 4\pi e n_e. \tag{9.128}$$

Taking divergence of (9.126) and putting $p = \gamma_e T_e n_e$ yield

$$\left(\frac{\partial^2}{\partial t^2} + \nu_e \frac{\partial}{\partial t} + \omega_{pe}^2 - \frac{\gamma_e T_e}{m_e}\nabla^2\right) n_e = N\nabla^2(\mathbf{v}_0 \cdot \mathbf{v}'_L). \tag{9.129}$$

The equations for the scattered light wave are (9.122), (9.125), and

$$\nabla \times \mathbf{B} = \frac{1}{c}\frac{\partial \mathbf{E}}{\partial t} + \frac{4\pi}{c}\mathbf{J}, \tag{9.130}$$

$$\mathbf{J} = -e(N\mathbf{v}'_L + n_e \mathbf{v}_0). \tag{9.131}$$

Taking curl of (9.125) yields ($\nabla \cdot \mathbf{E}'_L = 0$)

$$-\nabla^2 \mathbf{E}'_L = -\frac{1}{c^2}\frac{\partial^2 \mathbf{E}'_L}{\partial t^2} - \frac{4\pi e}{c^2}\left[N\frac{\partial \mathbf{v}'_L}{\partial t} + \frac{\partial}{\partial t}(n_e \mathbf{v}_0)\right].$$

Use of (9.122) in the above equation gives

$$\left(c^2\nabla^2 - \frac{\partial^2}{\partial t^2} - \omega_{pe}^2\right)\mathbf{v}'_L = \frac{4\pi e^2}{m_e}(n_e \mathbf{v}_0). \tag{9.132}$$

Note that the important nonlinearities are the ponderomotive force in the electron equation (9.123) and the nonlinear current in (9.131) for the scattered field. (9.129) and (9.132) are the coupled equations for the Raman scattering.

9.6.2. *Stimulated Brillouin scattering*

Similarly to Raman scattering, a large amplitude light wave perturbs a plasma to excite a density fluctuation associated with an ion acoustic wave, which in turn scatters off the incident wave into another light wave of different frequency. Since the low frequency ion motion is interconnected with the low frequency electron motion through the Poisson equation, the low frequency equations of both electrons and ions should be used. Electron equation is given by (9.126) in which we can neglect the electron inertia:

$$-\frac{m_e}{2}\nabla v_L^2 + e\nabla\phi - \frac{\gamma_e T_e}{N}\nabla n_e = 0. \tag{9.133}$$

Ion equations are:

$$\left(\frac{\partial}{\partial t} + \nu_i\right)\mathbf{v}_i = -\frac{e}{m_i}\nabla\phi, \tag{9.134}$$

$$\frac{\partial n_i}{\partial t} + N\nabla \cdot \mathbf{v}_i = 0. \tag{9.135}$$

The above two equations give

$$\left(\frac{\partial}{\partial t} + \nu_i\right)\frac{\partial n_i}{\partial t} = \frac{eN}{m_i}\nabla^2\phi. \tag{9.136}$$

Instead of the Poisson equation, one can use the charge neutrality which is valid in a low frequency region, $n_e \simeq n_i$. So n_i in the above equation can be replaced by n_e. Taking divergence of (9.133) and eliminating $\nabla^2\phi$ yield

$$\frac{\partial^2 n_e}{\partial t^2} + \nu_i\frac{\partial n_e}{\partial t} - \frac{\gamma_e T_e}{m_i}\nabla^2 n_e = N\,\frac{m_e}{m_i}\nabla^2(\mathbf{v}_0\cdot\mathbf{v}_L'). \tag{9.137}$$

Equations (9.132) and (9.137) are the coupled equations to describe the Brillouin process.

9.6.3. *2 ω_{pe} instability*

Here we consider an instability in which a light wave decays into two electron plasma waves. The frequency of the light wave is approximately equal to twice of the plasma frequency, hence the name. We need only the electron equation which we write in the form

$$\frac{\partial \mathbf{v}}{\partial t} + \mathbf{v}\cdot\nabla\mathbf{v} = -\frac{e}{m_e}(\mathbf{E}_0 - \nabla\phi) - \frac{c}{m_e}\mathbf{v}\times\mathbf{B} - \frac{\gamma_e T_e}{N m_e}\nabla n_e - \nu_e\mathbf{v}_e' \tag{9.138}$$

where

$$\mathbf{v} = \mathbf{v}_0 + \mathbf{v}_e'$$

with $\mathbf{v}_0$ the light wave field and $\mathbf{v}_e'$ the plasma wave field ($\nabla\times\mathbf{v}_e' = 0$). As discussed earlier, we have

$$\frac{\partial \mathbf{v}_0}{\partial t} = -\frac{e}{m_e}\mathbf{E}_0, \tag{9.139}$$

$$\frac{\partial \mathbf{v}_e'}{\partial t} + \frac{1}{2}\nabla v^2 = \frac{e}{m_e}\nabla\phi - \frac{\gamma_e T_e}{N m_e}\nabla n_e - \nu_e\mathbf{v}_e'. \tag{9.140}$$

Maxwell equation in the electrostatic form is

$$\frac{\partial}{\partial t}\nabla\phi + 4\pi e(n_e\mathbf{v}_0 + N\mathbf{v}_e') = 0 \tag{9.141}$$

$$\nabla^2\phi = 4\pi e n_e. \tag{9.142}$$

The above two equations give

$$\frac{\partial n_e}{\partial t} + N\nabla\cdot\mathbf{v}_e' + \mathbf{v}_0\cdot\nabla n_e = 0 \tag{9.143}$$

which is the continuity equation [in this way we can clearly see that the nonlinear term stems from the nonlinear current in (9.141)]. Taking divergence of (9.140) and using the preceding equations yield

$$-\left(\frac{\partial}{\partial t}+\nu_e\right)\left(\frac{\partial n_e}{\partial t}+\mathbf{v}_0\cdot\nabla n_e\right)+N\nabla^2\mathbf{v}_0\cdot\mathbf{v}_e'=\omega_{pe}^2 n_e-\frac{\gamma_e T_e}{m_e}\nabla^2 n_e. \tag{9.144}$$

Equations (9.143) and (9.144) are the two coupled equations for $2\omega_{pe}$ instability.

9.7. Unified theory of parametric instability

In the preceding sections, parametric instabilities were considered case by case. In a unified picture, a large amplitude pump wave beat with low frequency waves to excite Stokes and anti-Stokes daughter waves ((9.146) below). The latter beat with the pump again to give rise to a low frequency ponderomotive force that enhances the low frequency wave ((9.156) below). This input and feedback mechanism of parametric instabilities can be integrated to establish a general formula, comprising the various cases. When the pump is a large amplitude electromagnetic wave, the parametric interaction is often called *stimulated scattering*. The nonlinearities responsible for the instabilities are the ponderomotive force in electron equation of motion (ion ponderomotive force is negligible by the mass ratio) and the nonlinear electron current (ion nonlinear current is also negligible by the mass ratio). For the scattered field (side band), Maxwell equations read

$$\nabla\times\mathbf{E}_s=-\frac{1}{c}\frac{\partial\mathbf{B}_s}{\partial t}, \tag{9.145}$$

$$\nabla\times\mathbf{B}_s=\frac{1}{c}\frac{\partial\mathbf{E}_s}{\partial t}+\frac{4\pi}{c}\mathbf{J}_s-\frac{4\pi e}{c}\mathbf{v}_0\delta n_e, \tag{9.146}$$

where $\mathbf{J}_s(=-eN\mathbf{v}_s)$ is the linear current contributing to the scattered field and the last term is the electron nonlinear current responsible for the scattered field, with δn_e the electron density fluctuation. Eliminating $\mathbf{B}_s$ in (9.145) and (9.146) and Fourier transforming give

$$\mathbf{k}\times(\mathbf{k}\times\mathbf{E}_s)+\frac{\omega^2}{c^2}\mathbf{E}_s(\mathbf{k},\omega)+\frac{4\pi i\omega}{c^2}\mathbf{J}_s(\mathbf{k},\omega)=\mathbf{Q} \tag{9.147}$$

$$where\quad \mathbf{Q}=\frac{4\pi e i\omega}{c^2}\int \mathbf{v}_0(\mathbf{r},t)\delta n_e(\mathbf{r},t)\,e^{-i\mathbf{k}\cdot\mathbf{r}+i\omega t}d^3r\,dt. \tag{9.148}$$

The pump obeys the equation

$$\frac{\partial \mathbf{v}_0}{\partial t} = -\frac{e}{m_e}\mathbf{E}_0\, e^{i\mathbf{k}_0\cdot\mathbf{r}-i\omega_0 t} + c.c.$$

$$Therefore \quad \mathbf{v}_0(\mathbf{r},t) = \frac{-ie}{m_e\omega_0}\mathbf{E}_0 e^{i\mathbf{k}_0\cdot\mathbf{r}-i\omega_0 t} + \frac{ie}{m_e\omega_0}\mathbf{E}_0^* e^{-i\mathbf{k}_0\cdot\mathbf{r}+i\omega_0 t}. \tag{9.149}$$

The linear current $\mathbf{J}_s$ can be written in terms of $\mathbf{E}_s$: $J_{si} = \sigma_{ij}E_{sj}$. We here consider only an isotropic plasma (no magnetic field, no beam). Then (9.147) is written as, using $\varepsilon_{ij} = \delta_{ij} + \frac{4\pi i}{\omega}\sigma_{ij} = \delta_{ij}\varepsilon_T + (\varepsilon_L - \varepsilon_T)\frac{k_i k_j}{k^2}$,

$$\left[\left(\frac{\omega^2}{c^2}\varepsilon_T - k^2\right)\delta_{ij} + \left(k^2 + (\varepsilon_L - \varepsilon_T)\frac{\omega^2}{c^2}\right)\frac{k_i k_j}{k^2}\right] E_{sj} = Q_i. \tag{9.150}$$

The inverse matrix of (9.150) is obtained as [the determinant is given by (4.31)]

$$Inverse\ of\ [\ ]_{ij} = \frac{\delta_{ij}}{\beta} - \frac{\alpha}{\beta(\alpha+\beta)}\frac{k_i k_j}{k^2} = \frac{\delta_{ij}}{\beta} - \frac{k_i k_j}{k^2}\left(\frac{1}{\beta} - \frac{1}{\alpha+\beta}\right)$$

$$\alpha = k^2 + (\varepsilon_L - \varepsilon_T)\frac{\omega^2}{c^2}, \quad \beta = \frac{\omega^2}{c^2}\varepsilon_T - k^2.$$

Thus, we write

$$E_{si}(\mathbf{k},\omega) = c^2\,\Lambda_{ij}Q_j \tag{9.151}$$

$$where \quad \Lambda_{ij} = \frac{\delta_{ij} - \frac{k_i k_j}{k^2}}{\omega^2\varepsilon_T - c^2k^2} + \frac{k_i k_j}{k^2\omega^2\varepsilon_L}. \tag{9.152}$$

Exercise. Show that (9.151) satisfies (9.150).

Carrying out the Fourier integral in (9.148) we have

$$\mathbf{Q} = \frac{4\pi i e\omega}{c^2}\left[\mathbf{v}_0\delta n_e(\omega-\omega_0, \mathbf{k}-\mathbf{k}_0) + \mathbf{v}_0^*\delta n_e(\omega+\omega_0, \mathbf{k}+\mathbf{k}_0)\right] \tag{9.153}$$

$$where \quad \mathbf{v}_0 = \frac{-ie}{m_e\omega_0}\mathbf{E}_0.$$

From (9.151) and (9.153) we can write

$$E_{si}(\mathbf{k},\omega) = \Lambda_{ij}(\mathbf{k},\omega)4\pi i e\omega[v_{0j}\delta n_e(\omega-\omega_0) + v_{0j}^*\delta n_e(\omega+\omega_0)].$$

Shifting gives

$$\begin{aligned}\mathbf{E}^- &\equiv \mathbf{E}_s(\omega-\omega_0, \mathbf{k}-\mathbf{k}_0)\\ &= 4\pi i e(\omega-\omega_0)\,\Lambda_{ij}(\omega-\omega_0, \mathbf{k}-\mathbf{k}_0)v_{0j}^*\delta n_e(\omega,\mathbf{k}),\end{aligned} \tag{9.154}$$

$$\mathbf{E}^+ \equiv \mathbf{E}_s(\omega + \omega_0, \mathbf{k} + \mathbf{k}_0)$$
$$= 4\pi ie(\omega + \omega_0)\ \Lambda_{ij}(\omega + \omega_0, \mathbf{k} + \mathbf{k}_0) v_{0j} \delta n_e(\omega, \mathbf{k}), \qquad (9.155)$$

where we neglected $\delta n_e(\omega \pm 2\omega_0, \mathbf{k} \pm 2\mathbf{k}_0)$. In the above, $\mathbf{E}^\pm$ are, respectively, anti-Stokes and Stokes components of the scattered field. So far we obtained the scattered field components in terms of the electron density perturbation δn_e. We calculate δn_e by means of fluid equations,

$$\left(\frac{\partial}{\partial t} + \nu_e\right) \mathbf{v}_e = \frac{e}{m_e} \nabla\phi - \frac{\gamma_e T_e}{N m_e} \nabla \delta n_e \ - \nabla(\mathbf{v}_0 \cdot \mathbf{v}_s), \qquad (9.156)$$

$$\frac{\partial \delta n_e}{\partial t} + N \nabla \cdot \mathbf{v}_e = 0, \qquad (9.157)$$

$$-\nabla^2 \phi = 4\pi e(\delta n_i - \delta n_e), \qquad (9.158)$$

$$\frac{\partial \mathbf{v}_i}{\partial t} = -\frac{e}{m_i} \nabla\phi, \qquad (9.159)$$

$$\frac{\partial \delta n_i}{\partial t} + N \nabla \cdot \mathbf{v}_i = 0. \qquad (9.160)$$

If δn_e is a high frequency perturbation ($\omega \gg \omega_{pi}$), the ion perturbation can be neglected ($\delta n_i = 0$). If the frequency ω is sufficiently small, the ion response to the electric force $\nabla\phi$ should be retained, and the Poisson equation should be used. (9.158)–(9.160) are Fourier-transformed and combined to give

$$-k^2 \phi = \frac{4\pi e}{1 - \frac{\omega_{pi}^2}{\omega^2}} \delta n_e(\mathbf{k}, \omega).$$

Then (9.156) and (9.157) yield

$$\left[\omega(\omega + i\nu_e) - \frac{\omega_{pe}^2}{1 - \frac{\omega_{pi}^2}{\omega^2}} - \frac{\gamma_e T_e k^2}{m_e}\right] \delta n_e(\mathbf{k}, \omega)$$
$$= -N \int e^{-i\mathbf{k}\cdot\mathbf{r} + i\omega t} \nabla^2 (\mathbf{v}_s \cdot \mathbf{v}_0(\mathbf{r}, t))\ d^3r\ dt$$
$$= N k^2 \Big[\mathbf{v}_0 \cdot \mathbf{v}_s(\omega - \omega_0, \mathbf{k} - \mathbf{k}_0) + \ \mathbf{v}_0^* \cdot \mathbf{v}_s(\omega + \omega_0, \mathbf{k} + \mathbf{k}_0)\Big] \qquad (9.161)$$

$$\textit{where} \quad \mathbf{v}_s = -\frac{ie}{m_e \omega} \mathbf{E}_s. \qquad (9.162)$$

(9.162) is valid because the scattered wave has the wave velocity much greater than the electron thermal velocity.

Exercise. Show that the linear part of (9.161) gives the Langmuir wave and the ion acoustic wave, respectively, according as $\omega^2 \gg \omega_{pi}^2$ or $\omega^2 \ll \omega_{pi}^2$.

From (9.154), (9.155), and (9.161), we obtain the dispersion relation

$$\omega\left(\omega + i\nu_e\right) - \frac{\omega_{pe}^2}{1 - \frac{\omega_{pi}^2}{\omega^2}} - \frac{\gamma_e T_e k^2}{m_e}$$
$$= k^2\omega_{pe}^2\left[\frac{|\mathbf{k}_+ \times \mathbf{v}_0|^2}{k_+^2\left(\omega_+^2\varepsilon_T^+ - c^2k_+^2\right)} + \frac{|\mathbf{k}_+ \cdot \mathbf{v}_0|^2}{k_+^2\omega_+^2\varepsilon_L^+} + \frac{|\mathbf{k}_- \times \mathbf{v}_0|^2}{k_-^2\left(\omega_-^2\varepsilon_T^- - c^2k_-^2\right)} + \frac{|\mathbf{k}_- \cdot \mathbf{v}_0|^2}{k_-^2\omega_-^2\varepsilon_L^-}\right] \tag{9.163}$$

where $\omega_\pm = \omega \pm \omega_0$, $\mathbf{k}_\pm = \mathbf{k} \pm \mathbf{k}_0$, $\varepsilon_{L,T}^\pm = \varepsilon_{L,T}(\omega_\pm, \mathbf{k}_\pm)$, $|\mathbf{k} \times \mathbf{v}|^2 = k^2|\mathbf{v}|^2 - (\mathbf{k} \cdot \mathbf{v})(\mathbf{k} \cdot \mathbf{v}^*)$.

• **Electrostatic dispersion relation.** When the scattered wave can be approximated as an electrostatic mode, (9.147) reduces to, per $\mathbf{k} \times \mathbf{E}_s = 0$,

$$\varepsilon_{ij}E_{sj} = \frac{c^2}{\omega^2}\,Q_i.$$

Using

$$\varepsilon_{ij} = \varepsilon_L\frac{k_ik_j}{k^2} + \varepsilon_T\left(\delta_{ij} - \frac{k_ik_j}{k^2}\right), \qquad E_{si} = \frac{k_i}{k}E_s$$

we have

$$\varepsilon_{ij}E_{sj} = \varepsilon_L\frac{k_ik_j}{k^2}E_{sj} = \frac{c^2}{\omega^2}\,Q_i.$$

Performing $\mathbf{k}\cdot$ into both sides gives

$$E_{sj} = \frac{c^2}{\varepsilon_L\omega^2k^2}k_jk_iQ_i.$$

Comparing this equation with (9.151), we identify as

$$\Lambda_{ij} = \frac{c^2}{\varepsilon_L\omega^2k^2}\,k_ik_j$$

in this electrostatic case. This identification is directly obtained by taking the limit $c \to \infty$ in (9.152). The corresponding electrostatic parametric dispersion relation takes the form of (9.163) with $|\mathbf{k}_\pm \times \mathbf{v}_0|^2$–terms discarded.

References: Drake, J. F., Kaw, P. K., Lee, Y. C., Schmidt, G., Liu, C. S., and Rosenbluth, M. N. (1974). *Parametric instabilities of electromagnetic waves in plasmas*, Phys. Fluids **17**, p. 778; L. Stenflo, *Stimulated scattering of large amplitude waves in the ionosphere* Physica Scripta **T30** pp. 166–169

(1990); L. Stenflo and P. K. Shukla *Theory of stimulated scattering of large-amplitude waves* J. Plasma Phys. **64** part 4 pp. 353–357 (2000).

9.8. Parametric instability by time-oscillating pump: kinetic theory

In this Section, we derive the kinetic theory version of the parametric dispersion relation, (9.52). That is a kinetic theory of the parametric decay instability by sinusoidal high frequency electric field (dipole approximation for the pump).

Reference: Sanmartin, J. R. *Electrostatic plasma instabilities excited by a high frequency electric field*, Phys. Fluids, **13**, p. 1533–1542 (1970).
We use the Vlasov equation in the form

$$\frac{\partial f_\alpha}{\partial t} + \mathbf{v}\cdot\frac{\partial f_\alpha}{\partial \mathbf{r}} + \frac{e_\alpha}{m_\alpha}\left(\mathbf{E}_0 sin\omega_0 t - \frac{\partial \phi(\mathbf{r},t)}{\partial \mathbf{r}}\right)\cdot\frac{\partial f_\alpha}{\partial \mathbf{v}} = 0. \tag{9.164}$$

Linearizing this by putting $f_\alpha = F_\alpha(\mathbf{v},t) + f'_\alpha(\mathbf{r},\mathbf{v},t)$, we have

$$\frac{\partial F_\alpha}{\partial t} + \frac{e_\alpha}{m_\alpha} sin\omega_0 t\, \mathbf{E}_0\cdot\frac{\partial F_\alpha}{\partial \mathbf{v}} = 0 \tag{9.165}$$

$$\frac{\partial f'_\alpha}{\partial t} + \mathbf{v}\cdot\frac{\partial f'_\alpha}{\partial \mathbf{r}} + \frac{e_\alpha}{m_\alpha}\mathbf{E}_0 sin\omega_0 t\cdot\frac{\partial f'_\alpha}{\partial \mathbf{v}} - \frac{e_\alpha}{m_\alpha}\frac{\partial \phi(\mathbf{r},t)}{\partial \mathbf{r}}\cdot\frac{\partial F_\alpha}{\partial \mathbf{v}} = 0 \tag{9.166}$$

$$\nabla^2\phi = -4\pi\sum_\alpha e_\alpha\int f'_\alpha d^3v. \tag{9.167}$$

In the following, we shall omit the prime on f'. The zero order equation (9.165) is solved in the following form by characteristic method (see § 5.2.3.)

$$F_\alpha(\mathbf{v},t) = F_\alpha(\mathbf{u}_\alpha), \quad \mathbf{u}_\alpha = \mathbf{v} + \frac{e_\alpha\mathbf{E}_0}{m_\alpha\omega_0}cos\omega_0 t. \tag{9.168}$$

Note that $\mathbf{u}_\alpha$ is the constant of the motion of the characteristic equation of (9.165).

Exercise. Verify that a Maxwellian

$$F(v,t) = \sqrt{\frac{m}{2\pi T}}\; exp\left[-\frac{m}{2T}\,(v + \frac{eE_0}{m\omega_0}cos\omega_0 t)^2\right]$$

satisfies $\frac{\partial F}{\partial t} + \frac{e}{m}E_0\; sin\omega_0 t\; \frac{\partial F}{\partial v} = 0$.

Therefore, it is advantageous to transform old variable set $(t, \mathbf{r}, \mathbf{v}) \rightarrow$ new variable set $(t', \mathbf{r}, \mathbf{u}(\mathbf{v}, t))$ with $t = t'$ in (9.166): we have

$$\frac{\partial}{\partial t} = \frac{\partial}{\partial t'} + \frac{\partial u_j}{\partial t}\frac{\partial}{\partial u_j} = \frac{\partial}{\partial t'} - \frac{e_\alpha}{m_\alpha} sin\omega_0 t\ E_{0j}\frac{\partial}{\partial u_j}, \quad \frac{\partial}{\partial v_j} = \frac{\partial}{\partial u_j}. \tag{9.169}$$

Also Fourier-transforming with respect to $\mathbf{r}$, one obtains

$$\frac{\partial}{\partial t} f_\alpha(\mathbf{k}, \mathbf{u}, t) + i\mathbf{k}\cdot\mathbf{u} f_\alpha + i\omega_0 a_\alpha cos\omega_0 t\ f_\alpha - i\phi\frac{e_\alpha}{m_\alpha}\mathbf{k}\cdot\frac{\partial F_\alpha}{\partial \mathbf{u}} = 0, \tag{9.170}$$

$$where \quad a_\alpha = -\frac{e_\alpha}{m_\alpha \omega_0^2}\mathbf{k}\cdot\mathbf{E}_0. \tag{9.171}$$

We further transform as $f_\alpha = g_\alpha e^{-ia_\alpha sin\omega_0 t}$ to obtain

$$\frac{\partial}{\partial t} g_\alpha(\mathbf{k}, \mathbf{u}, t) + i\mathbf{k}\cdot\mathbf{u}\ g_\alpha - i\frac{e_\alpha}{m_\alpha}\ \phi\ \mathbf{k}\cdot\frac{\partial F_\alpha}{\partial \mathbf{u}} e^{ia_\alpha sin\omega_0 t} = 0 \tag{9.172}$$

$$k^2\phi = 4\pi\sum_\alpha e_\alpha e^{-ia_\alpha sin\omega_0 t}\int g_\alpha\ d^3u. \tag{9.173}$$

Substituting (9.173) into (9.172) for ϕ, we obtain

$$\frac{\partial g_e}{\partial t} + i\mathbf{k}\cdot\mathbf{u}\ g_e - i\frac{\omega_{pe}^2}{k^2}\mathbf{k}\cdot\frac{\partial F_e}{\partial \mathbf{u}}\int g_e d^3u = -i\frac{\omega_{pe}^2}{k^2}\mathbf{k}\cdot\frac{\partial F_e}{\partial \mathbf{u}} e^{iasin\omega_0 t}\int g_i d^3u, \tag{9.174}$$

$$\frac{\partial g_i}{\partial t} + i\mathbf{k}\cdot\mathbf{u}\ g_i - i\frac{\omega_{pi}^2}{k^2}\mathbf{k}\cdot\frac{\partial F_i}{\partial \mathbf{u}}\int g_i d^3u = -i\frac{\omega_{pi}^2}{k^2}\mathbf{k}\cdot\frac{\partial F_i}{\partial \mathbf{u}} e^{-iasin\omega_0 t}\int g_e d^3u, \tag{9.175}$$

where 'a' is defined in (9.8). Kinetic theory yields integro-differential equations. The above coupled equations should be compared with (9.6) and (9.7). Fourier transform the above equations with respect to t to obtain

$$\begin{aligned}(\omega - \mathbf{k}\cdot\mathbf{u})g_e(\omega) &+ \frac{\omega_{pe}^2}{k^2}\mathbf{k}\cdot\frac{\partial F_e}{\partial \mathbf{u}}\int g_e(\omega) d^3u \\ &= \frac{\omega_{pe}^2}{k^2}\mathbf{k}\cdot\frac{\partial F_e}{\partial \mathbf{u}}\Sigma_p J_p(a)\int g_i(\omega + p\omega_0) d^3u,\end{aligned} \tag{9.176}$$

$$\begin{aligned}(\omega - \mathbf{k}\cdot\mathbf{u})g_i(\omega) &+ \frac{\omega_{pi}^2}{k^2}\mathbf{k}\cdot\frac{\partial F_i}{\partial \mathbf{u}}\int g_i(\omega) d^3u \\ &= \frac{\omega_{pi}^2}{k^2}\mathbf{k}\cdot\frac{\partial F_i}{\partial \mathbf{u}}\Sigma_p J_p(-a)\int g_e(\omega + p\omega_0) d^3u,\end{aligned} \tag{9.177}$$

where $g_\alpha(\omega) = g_\alpha(\omega, \mathbf{k}, \mathbf{u})$.

Divide the above equations by $(\omega - \mathbf{k}\cdot\mathbf{u})$ and integrate over $\int d^3u$ to write

$$\left[1 + \frac{1}{\chi_e(\omega)}\right] G_e(\omega) = \Sigma_p J_p(a)\ G_i(\omega + p\omega_0), \tag{9.178}$$

$$\left[1 + \frac{1}{\chi_i(\omega)}\right] G_i(\omega) = \Sigma_p J_p(-a)\ G_e(\omega + p\omega_0), \tag{9.179}$$

$$\chi_\alpha(\omega) = \frac{\omega_{p\alpha}^2}{k^2} \int d^3u \frac{\mathbf{k}\cdot\frac{\partial F_\alpha}{\partial \mathbf{u}}}{\omega - \mathbf{k}\cdot\mathbf{u}}, \quad \textit{susceptibility},$$

$$G_\alpha(\omega) = \int d^3u\ g_\alpha(\mathbf{k}, \mathbf{u}, \omega)$$

and the $\mathbf{k}$–dependance was suppressed. For a weak pump, we can take three terms in the Bessel series: $p = 0, \ \pm 1$. We write

$$\left[1 + \frac{1}{\chi_e(\omega)}\right] G_e(\omega) = G_i(\omega) + J_1(a)G_i(\omega + \omega_0) + J_{-1}(a)G_i(\omega - \omega_0), \tag{9.180}$$

$$\left[1 + \frac{1}{\chi_i(\omega)}\right] G_i(\omega) = G_e(\omega) + J_1(-a)G_e(\omega + \omega_0) + J_{-1}(-a)G_e(\omega - \omega_0). \tag{9.181}$$

The above two equations correspond to (9.47) and (9.48). The parametric dispersion relation can be obtained in an analogous way. Shifting in (9.180) gives

$$\left[1 + \frac{1}{\chi_e(\omega + \omega_0)}\right] G_e(\omega + \omega_0)$$
$$= G_i(\omega + \omega_0) + J_1(a)G_i(\omega + 2\omega_0) + J_{-1}(a)G_i(\omega), \tag{9.182}$$

$$\left[1 + \frac{1}{\chi_e(\omega - \omega_0)}\right] G_e(\omega - \omega_0)$$
$$= G_i(\omega - \omega_0) + J_1(a)G_i(\omega) + J_{-1}(a)G_i(\omega - 2\omega_0). \tag{9.183}$$

Eliminate $G_e(\omega \pm \omega_0)$ in (9.181) by using (9.182) and (9.183). In doing so, neglect $G_i(\omega \pm \omega_0), \quad G_i(\omega \pm 2\omega_0)$ as off-resonant to obtain

$$1 + \frac{1}{\chi_i(\omega)} - \frac{\chi_e(\omega)}{1 + \chi_e(\omega)} = J_1^2(a)\left[\frac{\chi_e(\omega + \omega_0)}{1 + \chi_e(\omega + \omega_0)} + \frac{\chi_e(\omega - \omega_0)}{1 + \chi_e(\omega - \omega_0)}\right]. \tag{9.184}$$

The linear dispersion relation is obtained by putting $a = 0$ in the above:

$$1 + \chi_i(\mathbf{k}, \omega) + \chi_e(\mathbf{k}, \omega) = 0. \tag{9.185}$$

In (9.184), we approximate:

$$\chi_e(\omega) = \frac{-\omega_{pe}^2}{\omega^2 - \gamma_e T_e k^2/m_e}, \quad \chi_i = -\frac{\omega_{pi}^2}{\omega^2}$$

$$\frac{1}{\chi_i(\omega, \mathbf{k})} + \frac{1}{1 + \chi_e(\omega, \mathbf{k})} = J_1^2(a)\omega_{pe}^2\omega_{pi}^2 \left[\frac{1}{D_e(\omega + \omega_0, \mathbf{k})} + \frac{1}{D_e(\omega - \omega_0, \mathbf{k})}\right] \tag{9.186}$$

where D_e is defined in (9.46). In the LHS, neglect ω^2 as compared to ω_{pe}^2, $T_e k^2/m_e$ and $k^2\lambda_e^2 \ll 1$, then the LHS of the above equation becomes $\omega^2 - k^2\gamma_e T_e/m_i$, yielding the identical form as (9.52).

(9.186) agrees with Equation (17) in Sanmartin (1970). (9.184) was obtained by Silin (1965): V. P. Silin, *Parametric resonance in a plasma* Sov. Phys. JETP **21**, pp. 1127, (1965): JETP **24**, pp. 1242, (1967).

9.9. Nonlinear Landau damping: kinetic theory of parametric decay instability

The parametric decay instability is given rise to by wave-particle interaction when the low frequency beat wave (denoted by subscript 3 below) of two high frequency waves satisfy the Landau resonance conditions. See § 9.3 for its dipole approximation version. In this Section, we study the kinetic process of the nonlinear Landau damping by using the Vlasov equation. This treatment will make the physical process of the decay instability more clear. We consider three waves with frequencies and wave vectors $(\mathbf{k}_j, \omega_j)$ $(j = 1, 2, 3)$ in which the high frequency waves are designated by $j = 1, 2$ and the low frequency wave is subscripted by $'3'$. We assume that the three waves satisfy the resonance conditions,

$$\omega_1 - \omega_2 = \omega_3, \quad \mathbf{k}_1 - \mathbf{k}_2 = \mathbf{k}_3. \tag{9.187}$$

The Vlasov equations corresponding to each frequency are

$$\frac{\partial f_1}{\partial t} + \mathbf{v} \cdot \frac{\partial f_1}{\partial \mathbf{r}} - \frac{e}{m_e}\mathbf{E}_1 \cdot \frac{\partial F_{0e}}{\partial \mathbf{v}} = 0, \tag{9.188}$$

$$\frac{\partial f_2}{\partial t} + \mathbf{v} \cdot \frac{\partial f_2}{\partial \mathbf{r}} - \frac{e}{m_e} \mathbf{E}_2 \cdot \frac{\partial F_{0e}}{\partial \mathbf{v}} = 0, \tag{9.189}$$

$$\frac{\partial f_{3e}}{\partial t} + \mathbf{v} \cdot \frac{\partial f_{3e}}{\partial \mathbf{r}} - \frac{e}{m_e} \mathbf{E}_3 \cdot \frac{\partial F_{0e}}{\partial \mathbf{v}} - \frac{e}{m_e} \left(\mathbf{E}_1 \cdot \frac{\partial f_2}{\partial \mathbf{v}} + \mathbf{E}_2 \cdot \frac{\partial f_1}{\partial \mathbf{v}} \right) = 0, \tag{9.190}$$

$$\frac{\partial f_{3i}}{\partial t} + \mathbf{v} \cdot \frac{\partial f_{3i}}{\partial \mathbf{r}} + \frac{e}{m_i} \mathbf{E}_3 \cdot \frac{\partial F_{0i}}{\partial \mathbf{v}} = 0. \tag{9.191}$$

Note that the high frequency ion equation is not used (ions are stationary in a high frequency field); all the important nonlinearities are contained in the electron equations; two high frequency waves beat to produce the low frequency perturbation f_{3e}; ion nonlinearities are small by the mass ratio and can be neglected in (9.191). $F_{0e,i}$ are the respective zero order distribution functions. We also have Maxwell equations

$$\frac{\partial \mathbf{E}_j}{\partial t} = -4\pi \mathbf{J}_j \quad (j = 1, 2, 3), \tag{9.192}$$

$$\mathbf{J}_1 = -en_0 \int \mathbf{v} f_1 d^3v - en_3 \int \mathbf{v} f_2 d^3v, \tag{9.193}$$

$$\mathbf{J}_2 = -en_0 \int \mathbf{v} f_2 d^3v - en_3 \int \mathbf{v} f_1 d^3v, \tag{9.194}$$

$$\mathbf{J}_3 = en_0 \int \mathbf{v}(f_{3i} - f_{3e}) d^3v, \tag{9.195}$$

where

$$n_3 = \int f_{3i}\, d^3v \tag{9.196}$$

is the low frequency number density fluctuation; by the charge neutrality in the low frequency region, we have $n_e \simeq n_i \equiv n_3$. In the above governing equations, the nonlinear terms are: 1) $\mathbf{E} \cdot \frac{\partial f}{\partial \mathbf{v}}$ in the low frequency electron equation in (9.190), which corresponds to the low frequency ponderomotive force in fluid equation; 2) the nonlinear currents in the two high frequency electron currents $\mathbf{J}_1$ and $\mathbf{J}_2$, the last terms of (9.193) and (9.194). Let us introduce the Fourier coefficients by putting

$$f_j(\mathbf{r}, \mathbf{v}, t) = f_j(\mathbf{v}) e^{i\mathbf{k}\cdot\mathbf{r} - i\omega_j t} + f_j^*(\mathbf{v}) e^{-i\mathbf{k}\cdot\mathbf{r} + i\omega_j t} \quad (j = 1, 2, 3)$$

where

$$f_j(\mathbf{v}) = \frac{ie}{m_e} \frac{\mathbf{E}_j \cdot \frac{d}{d\mathbf{v}} F_{0e}}{\omega_j - \mathbf{k}_j \cdot \mathbf{v}} \quad (j = 1, 2), \tag{9.197}$$

$$f_{3e}(\mathbf{v}) = \frac{ie}{m_e} \frac{\mathbf{E}_3 \cdot \frac{d}{d\mathbf{v}} F_{0e} + \mathbf{E}_1 \cdot \frac{d}{d\mathbf{v}} f_2^* + \mathbf{E}_2^* \cdot \frac{d}{d\mathbf{v}} f_1}{\omega_3 - \mathbf{k}_3 \cdot \mathbf{v}}, \tag{9.198}$$

$$f_{3i}(\mathbf{v}) = -\frac{ie}{m_i} \frac{\mathbf{E}_3 \cdot \frac{d}{d\mathbf{v}} F_{0i}}{\omega_3 - \mathbf{k}_3 \cdot \mathbf{v}}. \tag{9.199}$$

Equating $\int f_{3e} d^3v = \int f_{3i} d^3v$ (charge neutrality) gives $\mathbf{E}_3$ in terms of the beat terms. The result is substituted into (9.199) to obtain

$$\begin{aligned} n_3 = &-\frac{e^2}{m_i m_e} \int d^3v \frac{\hat{\mathbf{k}}_3 \cdot \frac{d}{d\mathbf{v}} F_{0i}}{\omega_3 - \mathbf{k}_3 \cdot \mathbf{v}} \int d^3v \frac{\mathbf{E}_1 \cdot \frac{d}{d\mathbf{v}} f_2^* + \mathbf{E}_2^* \cdot \frac{d}{d\mathbf{v}} f_1}{\omega_3 - \mathbf{k}_3 \cdot \mathbf{v}} \\ &\div \left[\frac{ie}{m_e} \int d^3v \frac{\hat{\mathbf{k}}_3 \cdot \frac{d}{d\mathbf{v}} F_{0e}}{\omega_3 - \mathbf{k}_3 \cdot \mathbf{v}} + \frac{ie}{m_i} \int d^3v \frac{\hat{\mathbf{k}}_3 \cdot \frac{d}{d\mathbf{v}} F_{0i}}{\omega_3 - \mathbf{k}_3 \cdot \mathbf{v}} \right] \end{aligned} \tag{9.200}$$

where $\hat{\mathbf{k}}_3 = \mathbf{k}_3/k_3$, and we used $\mathbf{E}_3 = \hat{\mathbf{k}}_3 E_3$. (9.200) gives the low frequency number density perturbation generated as the two high frequency waves beat. Using the one-dimensional Maxwellian

$$F_{0\alpha} = \sqrt{\frac{m_\alpha}{2\pi T_\alpha}} e^{-\frac{m_\alpha v^2}{2T_\alpha}} \quad (\alpha = e,\ i)$$

the velocity integrals in (9.200) can be calculated. Assume $\frac{\omega_3}{k_3} \ll v_{eth}$ (electron thermal velocity) and expand $1/(\omega_3 - k_3 v)$ in the powers of $\omega_3/k_3 v$ to obtain

$$\int dv \frac{\hat{\mathbf{k}}_3 \cdot \frac{d}{d\mathbf{v}} F_{0e}}{\omega_3 - \mathbf{k}_3 \cdot \mathbf{v}} = \frac{m_e}{T_e k_3}. \tag{9.201}$$

The ion integral in the denominator of (9.200) should be integrated along the Landau contour since the wave-particle resonance condition can be satisfied. So we write

$$\begin{aligned} &\int dv \frac{\hat{\mathbf{k}}_3 \cdot \frac{d}{d\mathbf{v}} F_{0i}}{\omega_3 - \mathbf{k}_3 \cdot \mathbf{v}} = -\frac{m_i}{T_i k_3} W \\ &\textit{with} \quad W = \int_{-\infty}^{\infty} \frac{v F_{oi}}{\frac{\omega_3}{k_3} - v} dv. \end{aligned} \tag{9.202}$$

Note that $\omega_3/k_3 \sim v_{thi}$ (ion thermal velocity), which is the wave-particle resonance condition. The latter may be rewritten in view of (9.187) as

$$\omega_1 - \omega_2 = (k_1 - k_2) \cdot v_{thi}.$$

Now, in (9.200), we have, per (9.197) and (9.198) (assume all the **k**'s are parallel), the integral

$$\int \frac{E_1 \frac{\partial f_2^*}{\partial v} + E_2^* \frac{\partial f_1}{\partial v}}{\omega_3 - k_3 v} dv = \frac{-ie}{m_e} E_1 E_2^* k_3 \int dv \, \frac{F'_{0e}}{(\omega_3 - k_3 v)^2} \left(\frac{1}{\omega_1 - k_1 v} - \frac{1}{\omega_2 - k_2 v} \right)$$

$$= \frac{ie}{T_e \omega_{pe}^2} E_1 E_2^* \tag{9.203}$$

where we put $\omega_1 \simeq \omega_2 \simeq \omega_{pe}$, and the prime denotes v-differentiation.

Exercise. Prove (9.203). Assume $\frac{\omega_3}{k_3} \ll v_{eth} \ll \frac{\omega_1}{k_1}, \frac{\omega_2}{k_2}$ and expand the velocity-dependent functions in appropriate power series.

Collecting the foregoing results, we obtain

$$n_3 = \frac{e^2}{m_e T_i \omega_{pe}^2} \frac{W}{1 - \frac{T_e}{T_i} W} E_1 E_2^*. \tag{9.204}$$

The dispersion relation of the linear low frequency wave is obtained by putting the denominator equal to zero:

$$1 - \frac{T_e}{T_i} W \equiv \varepsilon(k_3, \omega_3) = 0$$

from which we can write

$$\varepsilon_r = 1 - \frac{T_e}{T_i} W_r \quad (\textit{real part}), \tag{9.205}$$

$$\varepsilon_i = -\frac{T_e}{T_i} W_i \quad (\textit{imaginary part}), \tag{9.206}$$

$$W_r = \frac{k_3}{\omega_3} \int dv \; F_{0i} \; v(1 + \frac{k_3 v}{\omega_3} + \cdots) = \frac{T_i}{m_i} \frac{k_3^2}{\omega_3^2}. \tag{9.207}$$

$$\textit{Therefore,} \quad \varepsilon_r = 1 - \frac{T_e}{m_i} \frac{k_3^2}{\omega_3^2}. \tag{9.208}$$

To calculate W_i, we write

$$W = -\sqrt{\frac{m_i}{2\pi T_i}} \int \frac{v e^{-\frac{m_i v^2}{2T_i}}}{v - \frac{\omega_3}{k_3}} dv \tag{9.209}$$

from which we obtain by picking up the half-residue at the Landau pole

$$W_i = \pi i(-)\sqrt{\frac{m_i}{2\pi T_i}} \left[v e^{-\frac{m_i v^2}{2T_i}} \right]_{v=\frac{\omega_3}{k_3}} = -\pi i \sqrt{\frac{m_i}{2\pi T_i}} \frac{\omega_3}{k_3} \, e^{-T_e/2T_i}. \tag{9.210}$$

The Landau damping rate (γ) of the low frequency ion acoustic wave is obtained from the well known formula

$$\gamma = -\frac{\varepsilon_i}{\partial\varepsilon_r/\partial\omega_3} = i\sqrt{\frac{\pi}{8}}\,\omega_3\left(\frac{T_e}{T_i}\right)^{3/2} e^{-T_e/2T_i}. \tag{9.211}$$

In terms of γ, we have

$$W_i = \frac{T_i}{T_e}\frac{2\gamma}{\omega_3}. \tag{9.212}$$

Using (9.207) and (9.212) in (9.204) gives

$$n_3 = \frac{(k_3^2 + i2\gamma\omega_3\frac{m_i}{T_e})e^2 E_1 E_2^*}{m_e m_i \omega_{pe}^2\,(\omega_3^2 - v_s^2 k_3^2 - 2i\gamma\omega_3)}. \tag{9.213}$$

(9.213) should be compared with (9.93). It can be ascertained that the imaginary part in the numerator cancels out in the calculation of the dissipation rate of the ω_2-wave energy. The result for the threshold is same as (9.95) with the replacement of $\nu_i \to 2\gamma$. (See Exercise below.)

Exercise. Show that

$$n_3 = \frac{i\omega_3 - 2\gamma}{2\gamma}\cdot\frac{e^2 E_1 E_2^*}{T_e m_e \omega_{pe}^2},$$

$$\frac{\partial}{\partial t}|E_2|^2 = \frac{\omega_3}{\gamma}\cdot\frac{e^2|E_1|^2|E_2|^2}{T_e\, m_e\, \omega_1}.$$

$$\mathit{Threshold:}\quad |E_1|^2 = \frac{4\pi n_0\, T_e\, \gamma\, \nu_e}{\omega_3\, \omega_{pe}} \quad \mathit{or} \quad \frac{|v_1|^2}{u^2} = \frac{\gamma\nu_e}{\omega_1 k_3 v_s} \quad (\mathit{see}\ (9.95)) \tag{9.214}$$

where ν_e is the damping rate (collisional+ Landau) of the wave energy of 2-wave: $\frac{\partial}{\partial t}|E_2|^2 = \nu_e|E_2|^2$. Use $\frac{\partial E_2}{\partial t} = -4\pi J_2$ where J_2 is the beat current of 1 and 3-wave.

Reference: Y.-Y. Kuo and J. A. Fejer (1972). *Spectral-line structures of saturated parametric instabilities*, Phys. Rev. Lett. **29** No. 25 pp. 1667–1670; H. J. Lee, *Kinetic theory of coherent resonant interactions of electrostatic plasma waves* J. Korean Phys. Soc. **7** pp. 256–265 (1984).

9.10. Nonlinear dispersion relation for parametric decay instability

The content of the last section can be reformulated in terms of the nonlinear dispersion relations, which make the parametric couplings more transparent.

The nonlinear dispersion relation can be converted to differential equations for the wave amplitude. It is a two-time analysis. Fourier transforming (9.192) and dotting $\mathbf{k}_1$ into the result give

$$-i\omega_1 \mathbf{k}_1 \cdot \mathbf{E}_1 = i\omega_{pe}^2 \left[\int d^3v \, \frac{\mathbf{k}_1 \cdot \mathbf{v} \, \mathbf{E}_1 \cdot \frac{d}{d\mathbf{v}} F_{0e}}{\omega_1 - \mathbf{k}_1 \cdot \mathbf{v}} + \frac{n_3}{n_0} \int d^3v \, \frac{\mathbf{k}_1 \cdot \mathbf{v} \, \mathbf{E}_2 \cdot \frac{d}{d\mathbf{v}} F_{0e}}{\omega_2 - \mathbf{k}_2 \cdot \mathbf{v}} \right].$$

Let us use

$$\frac{\mathbf{k}_1 \cdot \mathbf{v}}{\omega_1 - \mathbf{k}_1 \cdot \mathbf{v}} = -1 + \frac{\omega_1}{\omega_1 - \mathbf{k}_1 \cdot \mathbf{v}}$$

and integrate by parts the first integral to obtain

$$\left(1 - \omega_{pe}^2 \int \frac{F_{0e} \, d^3v}{(\omega_1 - \mathbf{k}_1 \cdot \mathbf{v})^2} \right) \mathbf{k}_1 \cdot \mathbf{E}_1 = -\frac{\omega_{pe}^2}{\omega_1} \frac{n_3}{n_0} \int d^3v \, \frac{\mathbf{k}_1 \cdot \mathbf{v} \, \mathbf{E}_2 \cdot \frac{d}{d\mathbf{v}} F_{0e}}{\omega_2 - \mathbf{k}_2 \cdot \mathbf{v}}.$$

The integral in the right hand side is equal to

$$\begin{aligned}
&-\frac{m_e}{T_e} \frac{E_2}{k_2} \int d^3v \, \frac{\mathbf{k}_1 \cdot \mathbf{v} \, \mathbf{k}_2 \cdot \mathbf{v} \, F_{0e}}{\omega_2 - \mathbf{k}_2 \cdot \mathbf{v}} \\
&\quad = -\frac{m_e}{T_e} \frac{\omega_2 E_2}{k_2} \int d^3v \, F_{0e} \frac{\mathbf{k}_1 \cdot \mathbf{v}}{\omega_2 - \mathbf{k}_2 \cdot \mathbf{v}} \\
&\quad = -\frac{m_e}{T_e} \frac{E_2}{k_2} \int d^3v \, F_{0e} \, \mathbf{k}_1 \cdot \mathbf{v} \, \left(1 + \frac{\mathbf{k}_2 \cdot \mathbf{v}}{\omega_2} + \cdots \right) \\
&\quad = -\frac{m_e}{T_e \omega_2} \frac{E_2}{k_2} \int d^3v \, F_{0e} \, (k_{1x} k_{2x} v_x^2 + k_{1y} k_{2y} v_y^2 + k_{1z} k_{2z} v_z^2), \\
&\quad \quad (\textit{cross terms vanished}) \\
&\quad = -\frac{m_e}{T_e \omega_2} \frac{E_2}{k_2} \mathbf{k}_1 \cdot \mathbf{k}_2 \int d^3v \, F_{0e} \, v_x^2 = -\frac{1}{\omega_2} \mathbf{k}_1 \cdot \mathbf{E}_2.
\end{aligned}$$

We also have

$$n_3 = \int f_{3i} d^3v = \frac{ie \, W}{T_i k_3} E_3. \tag{9.215}$$

Collecting the above results, we finally can write

$$\varepsilon(k_1, \omega_1) \mathbf{k}_1 \cdot \mathbf{E}_1 = \frac{ieW}{k_3 T_i} \, (\mathbf{k}_1 \cdot \mathbf{E}_2) \, E_3 \tag{9.216}$$

$$\textit{where} \quad \varepsilon(k_1, \omega_1) = 1 - \omega_{pe}^2 \int \frac{F_{0e} d^3v}{(\omega_1 - \mathbf{k}_1 \cdot \mathbf{v})^2} = 1 - \frac{\omega_{pe}^2}{\omega_1^2} (1 + 3k_1^2 \lambda_e^2) \tag{9.217}$$

is the dielectric function of ω_1-wave. Likewise we obtain

$$\varepsilon(k_2,\omega_2)\mathbf{k}_2\cdot\mathbf{E}_2 = -\frac{ieW^*}{k_3T_i}\,(\mathbf{k}_2\cdot\mathbf{E}_1)\,E_3^* \tag{9.218}$$

$$where \quad \varepsilon(k_2,\omega_2) = 1-\omega_{pe}^2\int\frac{F_{0e}d^3v}{(\omega_2-\mathbf{k}_2\cdot\mathbf{v})^2} = 1-\frac{\omega_{pe}^2}{\omega_2^2}(1+3k_2^2\lambda_e^2) \tag{9.219}$$

is the dielectric function of ω_2-wave. To obtain the ω_3-equation, it is more convenient to use the Poisson equation

$$i\,\mathbf{k}_3\cdot\mathbf{E}_3 = 4\pi e\int(f_{3i}-f_{3e})d^3v.$$

Using (9.201) and (9.202) and carrying out the integral of (9.198), one obtains

$$\varepsilon(k_3,\omega_3)E_3 = \frac{i}{4\pi n_0 e}\left[(\mathbf{k}_2\cdot\mathbf{E}_1)\mathbf{E}_2^*\cdot\hat{\mathbf{k}_3}-(\mathbf{k}_1\cdot\mathbf{E}_2^*)\mathbf{E}_1\cdot\hat{\mathbf{k}_3}\right] \tag{9.220}$$

$$where \quad \varepsilon(k_3,\omega_3) = 1-\frac{T_e}{T_i}W \quad (\lambda_e^2k_3^2\ll 1). \tag{9.221}$$

This is the low frequency dielectric function as we defined earlier (see the equation below (9.204)). (9.216), (9.218), and (9.220) are the three coupled equations for the three waves connected by the resonance condition (9.187). It is immediately seen that no couplings take place if $\mathbf{k}_1\perp\mathbf{k}_2$, so we assume all the $\mathbf{k}$'s to be parallel. Then we have

$$\varepsilon(k_1,\omega_1)E_1 = \frac{ieW}{k_3T_i}E_2E_3, \tag{9.222}$$

$$\varepsilon(k_2,\omega_2)E_2 = -\frac{ieW^*}{k_3T_i}E_1E_3^*, \tag{9.223}$$

$$\varepsilon(k_3,\omega_3)E_3 = -\frac{ik_3}{4\pi n_0 e}E_1E_2^*. \tag{9.224}$$

Eliminating E_2 gives

$$1-\frac{T_e}{T_i}W = \frac{W}{4\pi n_0T_i}\,\frac{|E_1|^2}{\varepsilon(k_2,\omega_2)}. \tag{9.225}$$

Exercise. Use (9.212) and show that the nonlinear dispersion relation (9.225) can be cast into the form

$$\omega_3^2+2i\gamma\omega_3-\frac{T_e}{m_i}k_3^2 = \frac{k_3^2\omega_{pe}^2}{4\pi n_0m_i}\,\frac{|E_1|^2}{\omega_2^2-\omega_{pe}^2-3k_2^2\frac{T_e}{m_e}} \tag{9.226}$$

which is equivalent to (9.61).

References: H. J. Lee, *Kinetic theory of coherent resonant interactions of electrostatic plasma waves*, J. Korean Phys. Soc. **17** pp. 256–265 (1984); H. J. Lee, *Time evolution of coherently resonant waves in a collisionless homogeneous plasma*, J. Korean Phys. Soc. **7** pp. 71–80 (1982).

9.11. Parametric instability due to an oscillating electric field in a magnetized plasma: kinetic theory

One method will be presented to derive the dispersion relation of the electrostatic parametric instability due to an oscillating electric field in a magnetized plasma by using the Vlasov equation. We only attempt to derive the results corresponding to (9.172) and (9.173). Then the remaining job can be carried out analogously to the analysis in Section 9.8. The Vlasov equation in a magnetized plasma reads ($\mathbf{B}_0 = \hat{\mathbf{z}}B_0$)

$$\frac{\partial f_\alpha}{\partial t} + \mathbf{v}\cdot\frac{\partial f_\alpha}{\partial \mathbf{r}} + \frac{e_\alpha}{m_\alpha}\left(\mathbf{E}_0 sin\omega_0 t + \frac{1}{c}\mathbf{v}\times\mathbf{B}_0 - \frac{\partial\phi(\mathbf{r},t)}{\partial\mathbf{r}}\right)\cdot\frac{\partial f_\alpha}{\partial\mathbf{v}} = 0. \tag{9.227}$$

Linearizing, after putting $f_\alpha = F_\alpha(\mathbf{v},t) + f'_\alpha(\mathbf{r},\mathbf{v},t)$, gives

$$\frac{\partial F_\alpha}{\partial t} + \frac{e_\alpha}{m_\alpha}\left(\mathbf{E}_0 sin\omega_0 t + \frac{1}{c}\mathbf{v}\times\mathbf{B}_0\right)\cdot\frac{\partial F_\alpha}{\partial\mathbf{v}} = 0 \tag{9.228}$$

$$\frac{\partial f'_\alpha}{\partial t} + \mathbf{v}\cdot\frac{\partial f'_\alpha}{\partial \mathbf{r}} + \frac{e_\alpha}{m_\alpha}\left(\mathbf{E}_0 sin\omega_0 t + \frac{1}{c}\mathbf{v}\times\mathbf{B}_0\right)\cdot\frac{\partial f'_\alpha}{\partial\mathbf{v}} - \frac{e_\alpha}{m_\alpha}\frac{\partial\phi(\mathbf{r},t)}{\partial\mathbf{r}}\cdot\frac{\partial F_\alpha}{\partial\mathbf{v}} = 0. \tag{9.229}$$

The characteristic equation of (9.228) is

$$\frac{dt}{1} = \frac{dv_x}{\frac{e_\alpha}{m_\alpha} sin\omega_0 t\ E_{0x} + \omega_{c\alpha}v_y} = \frac{dv_y}{\frac{e_\alpha}{m_\alpha} sin\omega_0 t\ E_{0y} - \omega_{c\alpha}v_x} = \frac{dv_z}{\frac{e_\alpha}{m_\alpha} E_{0z} sin\omega_0 t} = \frac{dF_\alpha}{0}.$$

Or, writing separately, we have

$$\frac{dv_x}{dt} = \frac{e_\alpha}{m_\alpha}E_{0x} sin\omega_0 t + \omega_{c\alpha}v_y, \tag{9.230}$$

$$\frac{dv_y}{dt} = \frac{e_\alpha}{m_\alpha}E_{0y} sin\omega_0 t - \omega_{c\alpha}v_x, \tag{9.231}$$

$$\frac{dv_z}{dt} = \frac{e_\alpha}{m_\alpha}E_{0z} sin\omega_0 t. \tag{9.232}$$

Then F_α is a function of the constants of motion of the above equations.

Now (9.230), (9.231), and (9.232) will be solved. Eliminating v_y gives

$$\frac{d^2 v_x}{dt^2} + \omega_{c\alpha}^2 v_x = \frac{e_\alpha}{m_\alpha}(\omega_0 E_{0x} cos\omega_0 t + \omega_{c\alpha} E_{0y} sin\omega_0 t), (\omega_{c\alpha} = \frac{e_\alpha B_0}{c\, m_\alpha}) \tag{9.233}$$

whose solution consists of the particular solution and the homogeneous solution, as is obtained below. Also the rest of the equations (9.231) and (9.232) can be solved straightly, giving

$$v_x = u_x sin\ \omega_{c\alpha} t + u_y cos\ \omega_{c\alpha} t - \frac{e_\alpha/m_\alpha}{\omega_0^2 - \omega_{c\alpha}^2}(\omega_0 E_{0x} cos\ \omega_0 t + \omega_{c\alpha} E_{0y} sin\ \omega_0 t), \tag{9.234}$$

$$v_y = u_x cos\ \omega_{c\alpha} t - u_y sin\ \omega_{c\alpha} t - \frac{e_\alpha/m_\alpha}{\omega_0^2 - \omega_{c\alpha}^2}(\omega_0 E_{0y} cos\ \omega_0 t - \omega_{c\alpha} E_{0x} sin\ \omega_0 t), \tag{9.235}$$

$$v_z = -\frac{e_\alpha}{m_\alpha \omega_0} E_{0z} cos\omega_0 t + u_z, \tag{9.236}$$

where u_x, u_y, u_z are the constants of motion of integration. Then, the general solution of (9.228) can be written as

$$F_\alpha(\mathbf{v}, t) = F_\alpha(u_x, u_y, u_z). \tag{9.237}$$

Solving for the constants of integration, u_x, u_y, u_z, we have

$$\begin{aligned} u_x &= v_x sin\ \omega_{c\alpha} t + v_y cos\ \omega_{c\alpha} t \\ &+ \frac{e_\alpha/m_\alpha}{\omega_0^2 - \omega_{c\alpha}^2}\big[\omega_0\ cos\omega_0 t(E_{0x} sin\ \omega_{c\alpha} t + E_{0y} cos\ \omega_{c\alpha} t) \\ &- \omega_{c\alpha} sin\omega_0 t(E_{0x} cos\omega_{c\alpha} t - E_{0y} sin\omega_{c\alpha} t)\big], \end{aligned} \tag{9.238}$$

$$\begin{aligned} u_y &= v_x cos\ \omega_{c\alpha} t - v_y sin\ \omega_{c\alpha} t \\ &+ \frac{e_\alpha/m_\alpha}{\omega_0^2 - \omega_{c\alpha}^2}\big[\omega_0\ cos\omega_0 t(E_{0x} cos\ \omega_{c\alpha} t - E_{0y} sin\ \omega_{c\alpha} t) \\ &+ \omega_{c\alpha} sin\omega_0 t(E_{0x} sin\omega_{c\alpha} t + E_{0y} cos\omega_{c\alpha} t)\big], \end{aligned} \tag{9.239}$$

$$u_z = v_z + \frac{e_\alpha}{m_\alpha \omega_0} E_{0z} cos\ \omega_0 t. \tag{9.240}$$

Exercise. Prove that $F_\alpha(\mathbf{v}, t) = F_\alpha(u_x(t), u_y(t), u_z(t))$ satisfies (9.228).

Thus the t-dependence of $F_\alpha(\mathbf{v}, t)$ is only implicit, through $\mathbf{u}(t)$.

Next, we make a change of variable in (9.229): $f(\mathbf{r},\mathbf{v},t) \to f(\mathbf{r},\mathbf{u}(\mathbf{v},t),t)$ by introducing $\mathbf{u}$ via (9.238)–(9.240). In (9.229), we change

$$\frac{\partial f}{\partial t} \to \frac{\partial f}{\partial t} + \frac{\partial f}{\partial \mathbf{u}} \cdot \frac{\partial \mathbf{u}}{\partial t}, \quad \frac{\partial f}{\partial \mathbf{v}} = \frac{\partial f}{\partial \mathbf{u}}.$$

Then, among the terms in (9.229), we have

$$\frac{\partial f_\alpha}{\partial \mathbf{u}} \cdot \frac{\partial \mathbf{u}}{\partial t} + \frac{e_\alpha}{m_\alpha}\left(\mathbf{E}_0 sin\omega_0 t + \frac{1}{c}\mathbf{v} \times \mathbf{B}_0\right) \cdot \frac{\partial f_\alpha}{\partial \mathbf{v}} = 0.$$

This equation holds because it is nothing but (9.228) with F being replaced by f. Therefore, we are left with

$$\frac{\partial f_\alpha(\mathbf{u})}{\partial t} + i\mathbf{k}\cdot\mathbf{v}f_\alpha - i\frac{e_\alpha}{m_\alpha}\phi(\mathbf{k},t)\mathbf{k}\cdot\frac{\partial F_\alpha}{\partial \mathbf{v}} = 0 \tag{9.241}$$

where we omitted prime on f_α, and $\mathbf{v}$ should be written in terms of $\mathbf{u}$. It can be easily shown by direct calculation that

$$\mathbf{k}\cdot\frac{\partial F_\alpha}{\partial \mathbf{v}} = \mathbf{q}_\alpha(t)\cdot\frac{\partial F_\alpha}{\partial \mathbf{u}}$$

$$where\ \mathbf{q}_\alpha = (k_x sin\omega_{c\alpha}t + k_y\ cos\omega_{c\alpha}t,\quad k_x cos\omega_{c\alpha}t - k_y\ sin\omega_{c\alpha}t,\quad k_z) \tag{9.242}$$

which is the rotation of the wave vector $\mathbf{k}$ by the rotation matrix

$$\mathbf{R}_\alpha = \begin{pmatrix} sin\ \omega_{c\alpha}t & cos\ \omega_{c\alpha}t & 0 \\ cos\ \omega_{c\alpha}t & -sin\ \omega_{c\alpha}t & 0 \\ 0 & 0 & 1 \end{pmatrix}.$$

Using (9.234)–(9.236) in (9.241) gives

$$\frac{\partial}{\partial t}f_\alpha(\mathbf{q}_\alpha,t,\mathbf{u}) + i\mathbf{q}_\alpha\cdot\mathbf{u}\ f_\alpha - \frac{ie_\alpha}{m_\alpha}\phi(\mathbf{k},t)\mathbf{q}_\alpha\cdot\frac{\partial F_\alpha}{\partial \mathbf{u}} +$$

$$-i\ f_\alpha\left[\beta_\alpha\omega_0 cos\ \omega_0 t\ \mathbf{k}_\perp\cdot\mathbf{E}_{0\perp} + \beta_\alpha\omega_{c\alpha}sin\ \omega_0 t\ \hat{\mathbf{z}}\cdot(\mathbf{k}\times\mathbf{E}_0)\right.$$

$$\left.+\frac{e_\alpha}{m_\alpha\omega_0}k_z E_{0z}\ cos\ \omega_0 t\right] = 0 \tag{9.243}$$

$$\beta_\alpha = \frac{e_\alpha}{m_\alpha}\frac{1}{\omega_0^2 - \omega_{c\alpha}^2}.$$

The symbol $\perp$ is referred to the direction of $\mathbf{B}_0$. We further transform

$$f_\alpha = g_\alpha e^{i\xi_\alpha(t)}$$

$$with \quad \xi_\alpha(t) = \left(\beta_\alpha \mathbf{k}_\perp \cdot \mathbf{E}_{0\perp} + \frac{e_\alpha}{m_\alpha \omega_0^2} k_z E_{oz}\right) sin\ \omega_0 t$$
$$-\beta_\alpha \frac{\omega_{c\alpha}}{\omega_0} \hat{\mathbf{z}} \cdot (\mathbf{k} \times \mathbf{E}_0) cos\ \omega_0 t. \tag{9.244}$$

Then we have

$$\frac{\partial}{\partial t} g_\alpha(\mathbf{k}, \mathbf{u}, t) + i\mathbf{q}_\alpha(\mathbf{k}, t) \cdot \mathbf{u} g_\alpha - \frac{ie_\alpha}{m_\alpha} \phi\ \mathbf{q}_\alpha \cdot \frac{\partial F_\alpha}{\partial \mathbf{u}} e^{-i\xi_\alpha(t)} = 0. \tag{9.245}$$

The Poisson equation takes the form (the Jacobian is 1)

$$k^2\ \phi(\mathbf{k}, t) = 4\pi \sum_\alpha e_\alpha\ e^{i\xi_\alpha(t)} \int g_\alpha d^3u. \tag{9.246}$$

(9.245) and (9.246) are the main results of this Section and their further analysis is left for readers. (9.245) and (9.246) are identical in the formal structure with (9.172) and (9.173). The analysis of (9.245) and (9.246) may proceed as is done in Section 9.8 but the details would be considerably more complex because we have still the time-dependent coefficient $\mathbf{q}_\alpha$, and ξ_α is composed of three terms as compared with a single term in the case of unmagnetized plasma. The three terms in ξ_α represent the relative strength of the pump with respect to the direction of the magnetic field. Note that in the case of unmagnetized plasma ($\omega_{c\alpha} = 0$) we have $\xi_\alpha = -a_\alpha$, and (9.245) and (9.246) reduce to (9.172) and (9.173), respectively. Tzoar (1969) dealt with the same problem by an entirely different transformation:

Tzoar, N. (1969) *Parametric excitations in plasma in a magnetic field*, Phys. Rev. **178**, pp. 356–362.

Also a useful reference for the parametric instability in magnetized plasmas is:

P. K. Kaw *Parametric excitation of electromagnetic waves in magnetized plasmas*, Advances in Plasma Physics Vol. 6, Ed. by A Simon and W. B. Thompson, Wiley New York (1976).

9.12. Compact form of the dispersion relation of parametric instability in a magnetized plasma

This Section extends § 9.7 by considering a magnetized plasma with a static magnetic field $\mathbf{B}_0 = \hat{\mathbf{z}} B_0$. The mathematical steps are analogous to § 9.7, and a compact form of the dispersion relation of parametric instability in a magnetized plasma is derived by using fluid equations. We consider a high frequency propagating electromagnetic pump wave to be coupled with electrostatic plasma normal modes. The latter can be the high frequency Langmuir wave or the low frequency ion acoustic wave. This is the generalization for a magnetized plasma of the work by Drake, *et al.* [Drake, J. F., Kaw, P. K., Lee, Y. C., Schmidt, G., Liu, C. S., and Rosenbluth, M. N. (1974), *Parametric instabilities of electromagnetic waves in plasmas, Phys. Fluids*, **17**, 778]. We hope to integrate the various parametric processes in a magnetized plasma in a compact form. Let us include the magnetic force terms in the two fluid equations in (9.156)–(9.160).

$$\frac{\partial \mathbf{v}_e}{\partial t} = \frac{e}{m_e}\nabla\phi - \frac{e}{m_e c}\mathbf{v}_e \times \mathbf{B}_0 - \frac{\gamma_e T_e}{N m_e}\nabla\delta n_e - \nabla(\mathbf{v}_0 \cdot \mathbf{v}_s), \tag{9.247}$$

$$\frac{\partial \mathbf{v}_i}{\partial t} = -\frac{e}{m_i}\nabla\phi + \frac{e}{m_i c}\mathbf{v}_i \times \mathbf{B}_0, \tag{9.248}$$

$$\frac{\partial \delta n_e}{\partial t} + N\nabla\cdot\mathbf{v}_e = 0, \tag{9.249}$$

$$\frac{\partial \delta n_i}{\partial t} + N\nabla\cdot\mathbf{v}_i = 0, \tag{9.250}$$

$$-\nabla^2\phi = 4\pi e(\delta n_i - \delta n_e). \tag{9.251}$$

The last term in (9.247) is the nonlinear force (ponderomotive force), which is important for the parametric interaction under consideration. Another important nonlinearity is the nonlinear current that must be included in the equation for $\mathbf{v}_s$, the scattered field, which is (9.262) below. These two nonlinear terms are responsible for the parametric instabilities that we intend to integrate in a compact form. Using (9.247) and (9.249), δn_e can be determined in terms of $(\phi - \frac{m_e}{e}\mathbf{v}_0 \cdot \mathbf{v}_s)$. The general formal solution can be written in terms of the electron and ion susceptibilities, $\chi_{e,i}$, which give the linear connections in terms of Fourier variables

$$\chi_e\phi = \frac{4\pi e}{k^2}\delta n_e, \tag{9.252}$$

$$\chi_i\phi = -\frac{4\pi e}{k^2}\delta n_i. \tag{9.253}$$

Then δn_e is obtained from (9.247) and (9.249) in the form

$$\delta n_e = \frac{\chi_e\, k^2}{4\pi e}\left(\phi - \frac{m_e}{e}\mathbf{v}_0 \cdot \mathbf{v}_s\right). \tag{9.254}$$

Equations (9.251), (9.252), and (9.253) give $(1+\chi_e+\chi_i)\phi = 0$, so the linear dispersion relation reads

$$1 + \chi_e + \chi_i \equiv \varepsilon_L = 0 \tag{9.255}$$

where ε_L is the dielectric constant. From (9.252) and (9.255), we can write

$$\phi = -\frac{4\pi e}{k^2}\frac{\delta n_e}{1+\chi_i}. \tag{9.256}$$

Using (9.256) in (9.254) yields

$$\delta n_e(\mathbf{k},\omega) = -\frac{Nk^2}{\omega_{pe}^2}\frac{\chi_e(1+\chi_i)}{\varepsilon_L}\left[\mathbf{v}_0 \cdot \mathbf{v}_s^*\right]_{(\mathbf{k},\omega)}. \tag{9.257}$$

Equation (9.257) is general in the sense that $\chi_{e,i}$ can be calculated kinetically as well as fluidly.

- **Susceptibility.** We have

$$\frac{\partial \mathbf{P}}{\partial t} = \mathbf{J}, \quad J_i = \sigma_{ij}E_j = -i\omega P_i$$

where $\mathbf{P}$ is polarization. The susceptibility is defined by $\chi_{ij} = \frac{4\pi i}{\omega}\sigma_{ij}$. Therefore, we have the relation $\mathbf{P}_\alpha = \frac{\chi_\alpha}{4\pi}\,\mathbf{E}$, and therefore

$$-i\omega\nabla\cdot\mathbf{P}_\alpha = \nabla\cdot\mathbf{J}_\alpha = i\omega\ \rho_\alpha \ (\textit{charge conservation}),$$
$$\frac{\chi_\alpha}{4\pi}\nabla\cdot\mathbf{E} = -e_\alpha\delta n_\alpha \quad (\alpha = e.\ i)$$

from which (9.252)–(9.253) follow.

When one uses the fluid equations above, we have

$$\chi_i = \frac{-\ \eta_i\omega_{pi}^2}{\omega^2-\omega_{ci}^2}, \tag{9.258}$$

$$\chi_e = \frac{-\eta_e\omega_{pe}^2}{\omega^2-\omega_{ce}^2-\eta_e\,\frac{\gamma_e T_e k^2}{m_e}}, \tag{9.259}$$

$$\textit{where} \quad \eta_\alpha = 1 - \frac{\omega_{c\alpha}^2}{\omega^2}\frac{k_z^2}{k^2} \quad (\alpha = e,\ i) \tag{9.260}$$

with $\omega_{p\alpha}$ and $\omega_{c\alpha}$ plasma and cyclotron frequency, respectively. The correctness of (9.257) can be directly checked from the governing equations (9.247)–(9.251). One can solve the vector equation (9.247) by Fourier transforming and resolving the resulting equation into the three directions: $\mathbf{k}$, $\hat{\mathbf{z}}$, and $\mathbf{k}\times\hat{\mathbf{z}}$. Then we end up with the expression

$$\delta n_e = -\frac{N}{\omega_{pe}^2 \eta_e} \frac{\chi_e(1+\chi_i)}{\varepsilon_L} \mathbf{k}\cdot\left(\mathbf{k} + \frac{\omega_{ce}}{\omega}\mathbf{k}\times\hat{\mathbf{z}} - k_z\frac{\omega_{ce}^2}{\omega^2}\hat{\mathbf{z}}\right)[\mathbf{v}_0\cdot\mathbf{v}_s^*]_{(\mathbf{k},\omega)} \tag{9.261}$$

where we used (9.258)–(9.260). It is immediately seen that (9.261) reduces to (9.257) in unmagnetized plasma.

The electrostatic normal modes in a magnetized plasma are obtained from the linear dispersion relation $\varepsilon_L = 1+\chi_i+\chi_e = 0$ with (9.258)–(9.260). Also the linear dispersion relation can be put into the form

$$0 = \frac{\varepsilon_L}{(1+\chi_i)\chi_e} = \frac{1}{1+\chi_i} + \frac{1}{\chi_e}. \quad (cf.\ (9.255))$$

Exercise. Verify the following relations in unmagnetized plasma.

$$\frac{1}{1+\chi_i} + \frac{1}{\chi_e} = \begin{cases} -\dfrac{1}{\omega_{pi}^2}\left(\omega^2 - \dfrac{\gamma_e T_e k^2}{m_i}\right) & (\omega \ll \omega_{pi}) \\ -\dfrac{1}{\omega_{pe}^2}\left(\omega^2 - \omega_{pe}^2 - \dfrac{\gamma_e T_e k^2}{m_e}\right) & (\omega \gg \omega_{pi}) \end{cases}.$$

Next we consider the scattered wave (side band) which we represent by primed quantities, $\mathbf{k}', \omega'$. We first deal with the Stokes coupling with the resonance conditions

$$\mathbf{k}' = \mathbf{k}_0 - \mathbf{k}, \quad \omega' = \omega_0 - \omega.$$

The Maxwell equations in Fourier variables are arranged in the form (cf. (9.146))

$$\mathbf{k}'\times(\mathbf{k}'\times\mathbf{E}') + \frac{\omega'^2}{c^2}\mathbf{E}' + \frac{4\pi i\omega'}{c^2}\mathbf{J}' = \frac{4\pi i e\omega'}{c^2}\,[\tilde{\mathbf{v}}_0\delta n_e^*]_{(\mathbf{k}',\omega')} \tag{9.262}$$

where the last term is the electron nonlinear current (tilded quantity is a complex Fourier amplitude and * denotes the complex conjugate of a Fourier amplitude). The linear current

$$\mathbf{J}' = -eN\mathbf{v}' \tag{9.263}$$

is calculated from the cold electron equation of motion

$$-i\,\omega'\mathbf{v}' = -\omega_{ce}\mathbf{v}'\times\hat{\mathbf{z}} - \frac{e}{m_e}\mathbf{E}' \tag{9.264}$$

where we neglected the wave magnetic force $\mathbf{v}' \times \mathbf{B}'$ assuming $v'^2/c^2 \ll 1$. Using (9.263) for $\mathbf{J}'$ and (9.264) for $\mathbf{E}'$, (9.262) is written as

$$\left(\frac{\omega'^2 - \omega_{pe}^2}{c^2} - k'^2\right)\mathbf{v}' + \mathbf{k}'\,\mathbf{k}'\cdot\mathbf{v}' + i\,\frac{\omega_{ce}}{\omega'}\left(\frac{\omega'^2}{c^2} - k'^2\right)\mathbf{v}' \times \hat{\mathbf{z}}$$

$$+ i\,\frac{\omega_{ce}}{\omega'}\mathbf{k}'\;\mathbf{v}'\cdot\hat{\mathbf{z}}\times\mathbf{k}' = \frac{\omega_{pe}^2}{Nc^2}\;\tilde{\mathbf{v}}_0\delta n_e^* \equiv \mathbf{Q}. \tag{9.265}$$

Exercise. Consider unmagnetized plasma ($\omega_{ce} = 0$). Solve the above equation for $\mathbf{v}'$ in terms of $\mathbf{Q}$. Substitute the result into (9.257) and obtain the parametric dispersion relation

$$\frac{1}{k^2}\left(\frac{1}{\chi_e(\mathbf{k},\omega)} + \frac{1}{1+\chi_i(\mathbf{k},\omega)}\right) = \frac{1}{k'^2}\left[\frac{|\tilde{\mathbf{v}}_0\cdot\mathbf{k}'|^2}{\omega_{pe}^2 - \omega'^2} + \frac{|\mathbf{k}'\times\tilde{\mathbf{v}}_0|^2}{c^2k'^2 + \omega_{pe}^2 - \omega'^2}\right]. \tag{9.266}$$

The contribution from the anti-Stokes component can be easily included. Upon inclusion of that component in the above equation, we recover Eq. (13) in Drake *et al.* (1974). Compare (9.266) with (9.163).

The above equation (9.265) can be solved for $\mathbf{v}$ in terms of the source term $\mathbf{Q}$ by resolving the equation into the three directions, $\mathbf{k}$, $\hat{\mathbf{z}}$, and $\mathbf{k}\times\hat{\mathbf{z}}$. [Primes will be omitted for a while.] After a long algebra, we obtain

$$\mathbf{v} = \frac{1}{\lambda^2(1 - A_3^2)}\left(\mathbf{F}(\mathbf{Q}) - \;\mathbf{F}(\mathbf{k})\;\frac{\frac{\omega_{ce}}{\omega}\,G(\mathbf{Q}) + \mathbf{k}\cdot\mathbf{F}(\mathbf{Q})}{A_2 + \mathbf{k}\cdot\mathbf{F}(\mathbf{k})}\right) \tag{9.267}$$

$$\textit{where}\quad \mathbf{F}(\mathbf{k}) = \mathbf{k} - iA_3\;\mathbf{k}\times\hat{\mathbf{z}} - A_3^2(\hat{\mathbf{z}}\cdot\mathbf{k})\;\hat{\mathbf{z}}, \tag{9.268}$$

$$\mathbf{F}(\mathbf{Q}) = \mathbf{Q} - iA_3\;\mathbf{Q}\times\hat{\mathbf{z}} - A_3^2(\hat{\mathbf{z}}\cdot\mathbf{Q})\;\hat{\mathbf{z}}, \tag{9.269}$$

$$G(\mathbf{Q}) = i\;\mathbf{Q}\cdot\hat{\mathbf{z}}\times\mathbf{k} - A_3\;(\hat{\mathbf{z}}\times\mathbf{k})\cdot(\hat{\mathbf{z}}\times\mathbf{Q}), \tag{9.270}$$

$$\lambda^2 = \frac{\omega^2 - \omega_{pe}^2}{c^2} - k^2, \tag{9.271}$$

$$A_2 = \lambda^2 - \frac{\omega_{ce}}{\omega}\left(\frac{\omega^2}{c^2} - k_z^2\right)A_3, \tag{9.272}$$

$$A_3 = \frac{\omega_{ce}}{\omega}\;\frac{1}{\lambda^2}\;\left(\frac{\omega^2}{c^2} - k^2\right). \tag{9.273}$$

One can check the correctness of (9.267) by substituting it into (9.265) and ascertaining many cancellations. After combining (9.257) and (9.267) we obtain the desired dispersion relation of the parametric instability in the intermediate form

$$\frac{1}{k^2}\left(\frac{1}{\chi_e(\omega,\mathbf{k})}+\frac{1}{1+\chi_i(\omega,\mathbf{k})}\right)=-\frac{1}{\omega_{pe}^2}\tilde{\mathbf{v}}_0\cdot\tilde{\mathbf{v}}_s^* \tag{9.274}$$

where (9.267) should be substituted for $\mathbf{v}_s$ with $\mathbf{Q}$ now standing for $\frac{\omega_{pe}^2}{c^2}$. [$\delta n_e$ has been canceled on both sides.] On the RHS of (9.274), the scattered wave quantities depend on $\mathbf{k}'$ and ω'— the Stokes component. Using Eq. (9.267) gives

$$\tilde{\mathbf{v}}_0\cdot\tilde{\mathbf{v}}_s^*=\frac{1}{\lambda^2(1-A_3^2)}\left[|\tilde{\mathbf{v}}_0|^2-A_3^2|\tilde{\mathbf{v}}_0\cdot\hat{\mathbf{z}}|^2\right.$$
$$\left.-\frac{\tilde{\mathbf{v}}_0\cdot\mathbf{F}^*(\mathbf{k})}{A_2+\mathbf{k}\cdot\mathbf{F}(\mathbf{k})}\left(\mathbf{k}\cdot\mathbf{F}^*(\tilde{\mathbf{v}}_0)+\frac{\omega_{ce}}{\omega}G^*(\tilde{\mathbf{v}}_0)\right)\right] \tag{9.275}$$

to which the factor ω_{pe}^2/c^2 should be attached. Note that $\mathbf{k}\cdot\mathbf{F}(\mathbf{k})$ is real. We calculate:

$$\begin{aligned}
&\mathbf{k}\cdot\mathbf{F}^*(\tilde{\mathbf{v}}_0)\ \ \tilde{\mathbf{v}}_0\cdot\mathbf{F}^*(\mathbf{k})\\
&\quad=\big(\mathbf{k}\cdot(\tilde{\mathbf{v}}_0^*+iA_3\tilde{\mathbf{v}}_0^*\times\hat{\mathbf{z}}-A_3^2\hat{\mathbf{z}}\cdot\tilde{\mathbf{v}}_0^*\hat{\mathbf{z}})\big)\big(\tilde{\mathbf{v}}_0\cdot(\mathbf{k}+iA_3\mathbf{k}\times\hat{\mathbf{z}}-A_3^2\hat{\mathbf{z}}\cdot\mathbf{k}\,\hat{\mathbf{z}})\big)\\
&\quad=\big(\tilde{\mathbf{v}}_0^*\cdot(\mathbf{k}-iA_3\mathbf{k}\times\hat{\mathbf{z}}-A_3^2\hat{\mathbf{z}}\cdot\mathbf{k}\,\hat{\mathbf{z}})\big)\big(\tilde{\mathbf{v}}_0\cdot(\mathbf{k}+iA_3\mathbf{k}\times\hat{\mathbf{z}}-A_3^2\hat{\mathbf{z}}\cdot\mathbf{k}\,\hat{\mathbf{z}})\big)\\
&\quad=|\tilde{\mathbf{v}}_0\cdot(\mathbf{k}+iA_3\mathbf{k}\times\hat{\mathbf{z}}-A_3^2\hat{\mathbf{z}}\cdot\mathbf{k}\,\hat{\mathbf{z}})|^{\,2},\\
&G^*(\tilde{\mathbf{v}}_0)\tilde{\mathbf{v}}_0\cdot\mathbf{F}^*(\mathbf{k})\\
&\quad=\big(-\tilde{\mathbf{v}}_0^*\cdot(A_3\mathbf{k}+i\hat{\mathbf{z}}\times\mathbf{k}-A_3\hat{\mathbf{z}}\cdot\mathbf{k}\,\hat{\mathbf{z}})\big)\big(\tilde{\mathbf{v}}_0\cdot(\mathbf{k}-iA_3\hat{\mathbf{z}}\times\mathbf{k}-A_3^2\,\hat{\mathbf{z}}\cdot\mathbf{k}\,\hat{\mathbf{z}})\big).
\end{aligned} \tag{9.276}$$

In (9.274) and (9.275), we see that the denominator $\lambda^2(1-A_3^2)=0$ gives the linear dispersion relation of the side band wave. $\lambda^2=0$ yields the usual electromagnetic wave $\omega^2=\omega_{pe}^2+c^2k^2$, and the relation $A_3=\pm1$ can be rearranged in the form

$$\frac{c^2k^2}{\omega^2}=1-\frac{\omega_{pe}^2}{\omega(\omega\mp\omega_{ce})} \tag{9.277}$$

where $+(-)$ sign corresponds to the L (R, whistler) wave. We put $A_3=\pm1$ in (9.276). This approximation may be permissible for a weak pump. Thus we have

$$G^*(\mathbf{v}_0)\tilde{\mathbf{v}}_0 \cdot \mathbf{F}^*(\mathbf{k}) = \mp|\tilde{\mathbf{v}}_0 \cdot (\mathbf{k} \pm i\mathbf{k} \times \hat{\mathbf{z}} - \hat{\mathbf{z}} \cdot \mathbf{k}\, \hat{\mathbf{z}})|^2 \tag{9.278}$$

where the upper (lower) signs correspond to the R(L) wave. Collecting the foregoing results, the dispersion relation can be written as

$$\frac{1}{k^2}\left(\frac{1}{\chi_e(\mathbf{k},\omega)} + \frac{1}{1+\chi_i(\mathbf{k},\omega)}\right) = -\frac{1}{c^2}\frac{1}{\lambda^2(1-A_3^2)}\Bigg[|\tilde{\mathbf{v}}_0|^2 - A_3|\hat{\mathbf{z}}\cdot\tilde{\mathbf{v}}_0|^2 - \frac{|\tilde{\mathbf{v}}_0 \cdot (\mathbf{k}' + iA_3\mathbf{k}' \times \hat{\mathbf{z}} - A_3^2\hat{\mathbf{z}} \cdot \mathbf{k}'\, \hat{\mathbf{z}})|^2 \mp \frac{\omega_{ce}}{\omega'}|\tilde{\mathbf{v}}_0 \cdot (\mathbf{k}' \pm i\mathbf{k}' \times \hat{\mathbf{z}} - \hat{\mathbf{z}} \cdot \mathbf{k}'\, \hat{\mathbf{z}})|^2}{A_2 - A_3^2 k_z'^2 + k'^2}\Bigg] \tag{9.279}$$

where λ, A_2, A_3 are functions of primed variables, ω', $\mathbf{k}'$. Equation (9.279) is the final result of the electromagnetic parametric dispersion relation when the waves are coupled through the Stokes coupling. The contribution from the anti-Stokes coupling is easily added to the RHS by writing additional terms with the replacement of $(') \to ('')$ in the Stokes terms, which implies the resonance conditions $\mathbf{k}'' = \mathbf{k}_0 + \mathbf{k}$ and $\omega'' = \omega_0 + \omega$.

Next we show that (9.279) recovers the earlier result for unmagnetized plasma in the limit of $\omega_{ce} = 0$. The right hand side of (9.279) reduces, upon putting $\omega_{ce} = 0$, to

$$-\frac{1}{\omega'^2 - \omega_{pe}^2 - c^2k'^2}\left[|\tilde{\mathbf{v}}_0|^2 - \frac{c^2|\tilde{\mathbf{v}}_0 \cdot \mathbf{k}'|^2}{\omega'^2 - \omega_{pe}^2}\right].$$

We use the identity in the above

$$|\tilde{\mathbf{v}}_0|^2 = \frac{|\tilde{\mathbf{v}}_0 \cdot \mathbf{k}'|^2 + |\mathbf{k}' \times \tilde{\mathbf{v}}_0|^2}{k'^2} \tag{9.280}$$

to obtain the dispersion relation

$$\frac{1}{k^2}\left(\frac{1}{\chi_e(\mathbf{k},\omega)} + \frac{1}{1+\chi_i(\mathbf{k},\omega)}\right) = \frac{1}{k'^2}\left[\frac{|\tilde{\mathbf{v}}_0 \cdot \mathbf{k}'|^2}{\omega_{pe}^2 - \omega'^2} + \frac{|\mathbf{k}' \times \tilde{\mathbf{v}}_0|^2}{c^2k'^2 + \omega_{pe}^2 - \omega'^2}\right] \tag{9.281}$$

for unmagnetized plasma, which agrees with (9.163) [see also Eq. (13) in Drake *et al.*].

Next we consider the electrostatic limit ($c \to \infty$) of Eq. (9.279). From (9.271)–(9.273), we obtain

$$\lambda^2(1 - A_3^2) \to -k'^2(1 - \frac{\omega_{ce}^2}{\omega'^2}),$$

$$A_3 \to \frac{\omega_{ce}}{\omega'},$$

$$A_2 - A_3^2 k_z'^2 + k'^2 \to \frac{\omega'^2 - \omega_{pe}^2}{c^2} - \frac{\omega_{ce}}{\omega'}\left(\frac{\omega'^2}{c^2} - k_z'^2\right) A_3 - A_3^2 k_z'^2$$
$$= \frac{\omega'^2 - \omega_{pe}^2 - \omega_{ce}^2}{c^2}.$$

We see that c^2 cancels and the electrostatic dispersion relation is found to be

$$\frac{1}{k^2}\left(\frac{1}{\chi_e(\mathbf{k},\omega)} + \frac{1}{1+\chi_i(\mathbf{k},\omega)}\right) = \frac{(1 - \frac{\omega_{ce}^2}{\omega'^2})^{-1}}{k'^2(\omega_{pe}^2 + \omega_{ce}^2 - \omega'^2)}$$
$$\times\left[\left|\tilde{\mathbf{v}}_0 \cdot \left(\mathbf{k}' + i\frac{\omega_{ce}}{\omega'}\mathbf{k}' \times \hat{\mathbf{z}} - \frac{\omega_{ce}^2}{\omega'^2}\,\hat{\mathbf{z}} \cdot \mathbf{k}'\,\hat{\mathbf{z}}\right)\right|^2 \mp \frac{\omega_{ce}}{\omega'}|\tilde{\mathbf{v}}_0 \cdot (\mathbf{k}' \pm i\mathbf{k}' \times \hat{\mathbf{z}} - \hat{\mathbf{z}} \cdot \mathbf{k}'\,\hat{\mathbf{z}})|^2\right] \tag{9.282}$$

where the upper and the lower signs corresponds to two cyclotron waves of different helicity. If $\omega_{ce} = 0$, the above equation reduces to the electrostatic limit of (9.281). The singularity at $\omega' = \omega_{ce}$ (upper signs) or $\omega' = -\omega_{ce}$ (lower signs) is only apparent since the numerator also vanishes in this case.

Algebraic remark: alternative expressions

- (9.264) is solved for $\mathbf{v}$ in the form

$$\mathbf{v} = \frac{-ie}{\omega m_e}\,\frac{1}{1 - \omega_{ce}^2/\omega^2}\left(\mathbf{E} - i\frac{\omega_{ce}}{\omega}\mathbf{E} \times \hat{\mathbf{z}} - \frac{\omega_{ce}^2}{\omega^2}\hat{\mathbf{z}}E_z\right). \tag{9.283}$$

- (9.262) can be written in terms of $\mathbf{E}$:

$$A_1'\mathbf{E} + c^2\mathbf{k} \cdot \mathbf{E}\,\mathbf{k} + i\,A_2'\mathbf{E} \times \hat{\mathbf{z}} + A_3'\hat{\mathbf{z}}E_z = 4\pi\,ie\omega[\mathbf{v}_0\delta n_e^*] \equiv \mathbf{Q}' \tag{9.284}$$

which is solved for $\mathbf{E}$ in the form

$$\mathbf{E} = \frac{A_1'}{A_1'^2 - A_2'^2}\left[\mathbf{F}'(\mathbf{Q}') - \frac{c^2}{S}\mathbf{F}'(\mathbf{k})\mathbf{Q}' \cdot \mathbf{F}'^*(\mathbf{k})\right], \tag{9.285}$$

$$\mathbf{F}'(\mathbf{Q}') \equiv \mathbf{Q}' - i\frac{A_2'}{A_i'}\mathbf{Q}' \times \hat{\mathbf{z}} - \hat{\mathbf{z}}Q_z'\,\frac{A_3' + A_2'^2/A_1'}{A_1' + A_3'},$$

$$\mathbf{F}'(\mathbf{k}) = \mathbf{k} - i\frac{A_2'}{A_i'}\mathbf{Q}' \times \hat{\mathbf{z}} - \hat{\mathbf{z}}Q_z' \frac{A_3' + A_2'^2/A_1'}{A_1' + A_3'},$$

$$\mathbf{F}'^*(\mathbf{k}) = \mathbf{k} + i\frac{A_2'}{A_i'}\mathbf{Q}' \times \hat{\mathbf{z}} - \hat{\mathbf{z}}Q_z' \frac{A_3' + A_2'^2/A_1'}{A_1' + A_3'}, \tag{9.286}$$

$$A_1' = \omega^2 - c^2k^2 - \frac{\omega^2\omega_{pe}^2}{\omega^2 - \omega_{ce}^2}, \tag{9.287}$$

$$A_2' = \frac{\omega\omega_{ce}\omega_{pe}^2}{\omega^2 - \omega_{ce}^2}, \quad A_3' = \frac{\omega_{ce}^2\omega_{pe}^2}{\omega^2 - \omega_{ce}^2}, \tag{9.288}$$

$$S = A_1' - \frac{A_2'^2}{A_1'} + c^2\left(k^2 - k_z^2\frac{A_3' + A_2'^2/A_1'}{A_1' + A_3'}\right). \tag{9.289}$$

In the above we inverted the wave equation (9.262) in a magnetized cold plasma. One can show the equivalence of (9.267) and (9.285) at the expense of a long algebra.

- Whistler.

When $\mathbf{k} \cdot \mathbf{E} = 0$ in (9.262), we have more manageable expressions. (9.262)–(9.264) yield, after eliminating $\mathbf{E}'$,

$$a_1\mathbf{v} + ia_2\mathbf{v} \times \hat{\mathbf{z}} = \mathbf{q}, \tag{9.290}$$

$$a_1 = \omega^2 - c^2k^2 - \omega_{pe}^2, \quad a_2 = \frac{\omega_{ce}}{\omega}(\omega^2 - c^2k^2), \quad \mathbf{q} = \frac{\omega_{pe}^2}{N}[\mathbf{v}_0\delta n_e^*]. \tag{9.291}$$

Solving (9.290) gives

$$\mathbf{v} = \frac{a_1}{a_1^2 - a_2^2}\left[\mathbf{q} - i\frac{a_2}{a_1}\mathbf{q} \times \hat{\mathbf{z}} - \frac{a_2^2}{a_1^2}\hat{\mathbf{z}}\, q_z\right]. \tag{9.292}$$

Solving (9.262) for $\mathbf{E}$ with $\mathbf{k} \cdot \mathbf{E} = 0$ gives

$$\mathbf{E} = \frac{A_1'}{A_1'^2 - A_2'^2}\,\mathbf{F}'(\mathbf{Q}'), \quad \mathbf{Q}' = 4\pi ie\omega[\mathbf{v}_0\delta n_e^*] \tag{9.293}$$

where A_1', A_2' A_3' are given by (9.287)–(9.288). Equivalence of (9.292) and (9.293) can be proven via (9.264) or (9.283). The above inversions are useful for deriving the compact forms.

- **Electrostatic Stokes side band.**

Put $\mathbf{k}' \times \mathbf{E}' = 0$ in (9.262):

$$\mathbf{E}' + \frac{4\pi i}{\omega'}\mathbf{J}' = \frac{4\pi i e}{\omega'} \; [\tilde{\mathbf{v}}_0 \delta n_e^*]_{(\mathbf{k}',\omega')} \,. \tag{9.294}$$

Also we have

$$\mathbf{J}' = -eN\mathbf{v}', \tag{9.295}$$

$$-i\;\omega'\mathbf{v}' = -\omega_{ce}\mathbf{v}' \times \hat{\mathbf{z}} - \frac{e}{m_e}\mathbf{E}'. \tag{9.296}$$

Solving the above three equations yield

$$\mathbf{v} = \frac{(\omega^2 - \omega_p^2)\mathbf{Q} - i\omega\omega_c\mathbf{Q} \times \hat{\mathbf{z}} - \frac{\omega^2\omega_c^2}{\omega^2 - \omega_p^2}\hat{\mathbf{z}}Q_z}{(\omega^2 - \omega_p^2)^2 - \omega^2\omega_c^2}, \tag{9.297}$$

$$\mathbf{E} = \frac{4\pi i e N}{\omega}\left[\frac{\frac{\omega^2}{\omega_p^2}(\omega^2 - \omega_p^2 - \omega_c^2)\mathbf{Q} - i\omega\omega_c\mathbf{Q} \times \hat{\mathbf{z}} - \frac{\omega^2\omega_c^2}{\omega^2 - \omega_p^2}\hat{\mathbf{z}}Q_z}{(\omega^2 - \omega_p^2)^2 - \omega^2\omega_c^2}\right], \tag{9.298}$$

$$where \quad \mathbf{Q} = \frac{\omega_{pe}^2}{N}[\mathbf{v}_0 \delta n_e^*].$$

Thus we have checked a few special cases.

Additional references to this Chapter: L. Stenflo J. Plasma Phys. **61** 129; L. Stenflo *Resonant three-wave interactions in plasmas* Physica Scripta **T50** pp. 15–19 (1994); L. Stenflo and P. K. Shukla *Theory of stimulated scattering of large-amplitude waves* J. Plasma Phys. **64** part 4 pp. 353–357 (2000).

Chapter 10

Nonlinear theory of Vlasov equation

The Vlasov equation contains the nonlinear term $\mathbf{E} \cdot \frac{\partial f}{\partial \mathbf{v}}$ from which various nonlinear effects are derived. One step higher development over the linearized Vlasov equation is the *quasilinear theory* which replaces the Vlasov–Poisson (or Vlasov–Maxwell) system of equations by a set of two linear equations. The quasilinear theory is in essence two-time or two-space scale analysis of the distribution function. The nonlinear term in the Vlasov equation has been studied under the classification called *wave-particle interaction* and *wave-wave interaction.* Basically, the mathematical scheme to deal with the nonlinearity is iteration with linear solutions.

Reference: R. Z. Sagdeev and A. A. Galeev *Nonlinear plasma theory*, Benjamin New York (1969).

10.1. Quasilinear theory

In the linear theory, the zero order distribution function which is assumed to be independent of $\mathbf{r}$ and t determine the growth or damping rates of the excited wave. As the wave grows, the zero order distribution function will be affected by the wave. This was discussed in conjunction with the trapped particle dynamics in earlier chapter. However, the linear theory itself doesn't take account of the influence of the growing wave on the zero order distribution function. The quasilinear theory remedies this shortcoming of the linear Vlasov equation by allowing a slow time variation for the "zero order" distribution function. The zero order distribution function f_0 is the spatially homogeneous part extracted from $f(\mathbf{r}, \mathbf{v}, t)$ (the extraction can be done by picking up the $\mathbf{k} = 0$ component from the Fourier series) and furthermore, f_0 is endowed with slow time variation.

We start with the Vlasov equation for the electron distribution function $f(\mathbf{r}, \mathbf{v}, t)$ (ions are assumed to form a uniform background of immobile positive charges)

$$\frac{\partial}{\partial t} f(\mathbf{r}, \mathbf{v}, t) + \mathbf{v} \cdot \frac{\partial f}{\partial \mathbf{r}} - \frac{e}{m} \mathbf{E}(\mathbf{r}, t) \cdot \frac{\partial f}{\partial \mathbf{v}} = 0 \tag{10.1}$$

and the Poisson equation

$$\nabla \cdot \mathbf{E} = 4\pi e N \left(1 - \int d^3 v \; f(\mathbf{r}, \mathbf{v}, t)\right). \tag{10.2}$$

Let us Fourier transform (10.1) and (10.2) with respect to $\mathbf{r}$. Transforming the product term proceeds as follows.

$$I \equiv \int d^3 r e^{-i\mathbf{k}\cdot\mathbf{r}} \mathbf{E}(\mathbf{r}) f(\mathbf{r}).$$

We substitute into the integral I the expression of the inverse transform

$$\mathbf{E}(\mathbf{r}) = \left(\frac{1}{2\pi}\right)^3 \int d^3 k' e^{i\mathbf{k}'\cdot\mathbf{r}} \mathbf{E}(\mathbf{k}').$$

Then we can write

$$\begin{aligned}
I &= \int d^3 r e^{-i\mathbf{k}\cdot\mathbf{r}} f(\mathbf{r}) \left(\frac{1}{2\pi}\right)^3 \int d^3 k' e^{i\mathbf{k}'\cdot\mathbf{r}} \mathbf{E}(\mathbf{k}') \\
&= \left(\frac{1}{2\pi}\right)^3 \int d^3 k' \mathbf{E}(\mathbf{k}') \int d^3 r e^{-i(\mathbf{k}-\mathbf{k}')\cdot\mathbf{r}} f(\mathbf{r}) \\
&= \left(\frac{1}{2\pi}\right)^3 \int d^3 k' \mathbf{E}(\mathbf{k}') f(\mathbf{k}-\mathbf{k}') = \left(\frac{1}{2\pi}\right)^3 \int d^3 k' \mathbf{E}(\mathbf{k}-\mathbf{k}') f(\mathbf{k}').
\end{aligned}$$

The last equality is due to the property of the convolution integral. If one wishes to change the above Fourier integral to Fourier series, put $\left(\frac{1}{2\pi}\right)^3 \int d^3 k' \to \sum_{\mathbf{k}'}$. Then (10.1) yields

$$\frac{\partial}{\partial t} f(\mathbf{k}, \mathbf{v}, t) + i\mathbf{k} \cdot \mathbf{v} f - \frac{e}{m} \sum_{\mathbf{k}'} \mathbf{E}(\mathbf{k}', t) \cdot \frac{\partial}{\partial \mathbf{v}} f(\mathbf{k}-\mathbf{k}', \mathbf{v}, t) = 0. \tag{10.3}$$

In the above equation, we separate the equation corresponding to $\mathbf{k} = 0$ and $\mathbf{k} \neq 0$:

$$\mathbf{k} = 0: \quad \frac{\partial}{\partial t} f(\mathbf{k}=0, \mathbf{v}, t) = \frac{e}{m} \sum_{\mathbf{k}'} \mathbf{E}(-\mathbf{k}', t) \cdot \frac{\partial}{\partial \mathbf{v}} f(\mathbf{k}', \mathbf{v}, t), \quad (A)$$

$$\begin{aligned}
\mathbf{k} \neq 0: &\frac{\partial}{\partial t} f(\mathbf{k}, \mathbf{v}, t) + i\mathbf{k} \cdot \mathbf{v} f - \frac{e}{m} \mathbf{E}(\mathbf{k}, t) \cdot \frac{\partial f(\mathbf{k}=0, \mathbf{v}, t)}{\partial \mathbf{v}} \\
&= \frac{e}{m} {\sum_{\mathbf{k}'}}' \mathbf{E}(\mathbf{k}', t) \cdot \frac{\partial}{\partial \mathbf{v}} f(\mathbf{k}-\mathbf{k}', \mathbf{v}, t), \quad (B)
\end{aligned}$$

where the prime on $\sum$ denotes that the $\mathbf{k}'=\mathbf{k}$ term is excluded in the $\sum$.

Let us assume in Eq. (B) $f(\mathbf{k}=0,\mathbf{v},t) \gg f(\mathbf{k},\mathbf{v},t)$ and neglect the nonlinear term in the RHS:

$$\frac{\partial}{\partial t}f(\mathbf{k},\mathbf{v},t) + i\mathbf{k}\cdot\mathbf{v}f - \frac{e}{m}\mathbf{E}(\mathbf{k},t)\cdot\frac{\partial f(\mathbf{k}=0,\mathbf{v},t)}{\partial \mathbf{v}} = 0. \quad (C)$$

Neglect of the RHS of (B), the 'collision term', appears to be an analogous approximation to the Vlasov equation which was also made 'collisionless' by neglecting the higher order correlation term. In the higher order theory, RHS of (B) should be included to develop the coupled mode equations. This subject is dealt with in the following. Equation (C) is refinement over the Vlasov equation in as much as we take account of the time dependance of $f(\mathbf{k}=0,\mathbf{v},t)$ which we assume to be slowly varying. To express this idea, we shall write the function $f(\mathbf{k}=0,\mathbf{v},t)$ as $f(\mathbf{k}=0,\mathbf{v},\epsilon t)$, with ϵt meaning the slow time. Thus we have two quasilinear equations (C) and (A) which we write in the form

$$\frac{\partial}{\partial t}f(\mathbf{k},\mathbf{v},t) + i\mathbf{k}\cdot\mathbf{v}f - \frac{e}{m}\mathbf{E}(\mathbf{k},t)\cdot\frac{\partial f(\mathbf{k}=0,\mathbf{v},\epsilon t)}{\partial \mathbf{v}} = 0, \quad (10.4)$$

$$\frac{\partial}{\partial t}f(\mathbf{k}=0,\mathbf{v},\epsilon t) = \frac{e}{m}\sum_{\mathbf{k}}\mathbf{E}(-\mathbf{k},t)\cdot\frac{\partial}{\partial \mathbf{v}}f(\mathbf{k},\mathbf{v},t). \quad (10.5)$$

Together with (10.4) and (10.5), we have the Poisson equation expanded in Fourier series:

$$\nabla\cdot\sum_{k}\mathbf{E}(\mathbf{k},t)e^{i\mathbf{k}\cdot\mathbf{r}} = 4\pi eN\left[1-\int d^3v\sum_{k}f(\mathbf{k},\mathbf{v},t)e^{i\mathbf{k}\cdot\mathbf{r}}\right].$$

The $\mathbf{k}=0$ term on the RHS cancels 1, and we have

$$i\mathbf{k}\cdot\mathbf{E}(\mathbf{k},t) = -4\pi Ne\int f(\mathbf{k},\mathbf{v},t)d^3v. \quad (10.6)$$

Equations (10.4)–(10.6) are the self-consistent quasilinear equations. Clearly (10.4) and (10.5) are an improvement over the linearized Vlasov equation. The slow function $f(\mathbf{k}=0,\mathbf{v},\epsilon t)$ corresponds to the zero order equilibrium distribution function F_0 in the linearized equation, which we assumed to be unvarying in time. In the quasilinear theory, the corresponding quantity changes slowly of which time variation is determined according to the evolution of the system, $\mathbf{E}(\mathbf{k},t)$ and f($\mathbf{k}$,$\mathbf{v}$,t), as expressed by (10.5).

For simplicity we go to the one-dimensional case: take $\mathbf{k}=\hat{\mathbf{x}}k$, say, and integrate (10.4) with respect to the two velocity components perpendicular to $\mathbf{k}$, to obtain (note $\mathbf{E_k} \parallel \mathbf{k}$)

$$\frac{\partial f_k}{\partial t} + ikvf_k + \frac{e}{m} ik\phi_k \frac{\partial f_0(v, \epsilon t)}{\partial v} = 0, \tag{10.7}$$

$$k^2 \phi_k(t) = -4\pi e N \int dv f_k(v,t), \tag{10.8}$$

$$\frac{\partial}{\partial t} f_0(v, \epsilon t) = \frac{e}{m} \sum_k ik\phi_{-k} \frac{\partial f_k}{\partial v}, \tag{10.9}$$

where we changed notations slightly in an obvious way and introduced potential $E_k = -ik\phi_k$. The first order linear equation (10.7) can be solved to obtain

$$f_k(v,t) = -\frac{e}{m} ik \int_0^t dt' e^{ikv(t'-t)} \phi_k(t') \frac{\partial}{\partial v} f_0(v, \epsilon t'). \tag{10.10}$$

Using (10.10) in (10.8) gives

$$\begin{aligned} \phi_k(t) &= ik \frac{\omega_{p0}^2}{k^2} \int_0^t dt' \phi_k(t') \int_{-\infty}^{\infty} dv e^{ikv(t'-t)} \frac{\partial}{\partial v} f_0(v, \epsilon t') \\ &= \omega_{p0}^2 \int_0^t dt' \phi_k(t')(t'-t) \int_{-\infty}^{\infty} dv e^{ikv(t'-t)} f_0(v, \epsilon t') \end{aligned}$$

from which we obtain

$$\begin{aligned} \frac{d^2 \phi_k}{dt^2} &= -\left(\omega_{p0}^2 \int_{-\infty}^{\infty} f_0(\epsilon t, v) dv \right) \phi_k(t) \\ &\quad - ik\omega_{p0}^2 \int_0^t dt' \phi_k(t') \int_{-\infty}^{\infty} dv e^{ikv(t'-t)} v^2 \frac{\partial f_0}{\partial v} \end{aligned} \tag{10.11}$$

where $\omega_{p0}^2 = 4\pi N e^2/m$. The last term vanishes asymptotically as $t \to \infty$ since the exponential factor oscillates rapidly to yield almost complete cancellation in the integral. The quantity in the round parenthesis defines a slowly changing plasma frequency $\omega_p^2(\epsilon t)$. We write

$$\frac{d^2 \phi_k}{dt^2} + \omega_p^2(\epsilon t) \phi_k(t) = 0. \tag{10.12}$$

The asymptotic result, (10.12), clearly reveals the two-time nature inherited from (10.7) and suggests using the method of two-time perturbation analysis or WKB approximation to deal with (10.7)–(10.9). Here we adopt the latter approach.

We review the relevant physics implied by (10.7). The linear theory indicates that the plasma is stable if $f_0(v, \epsilon t)$ is a single-humped function of v. If there is a small second hump in the distribution function at velocities somewhat larger than the thermal velocity, a weak instability occurs with $\omega_k = \omega_k^r + i\gamma_k$ with $\gamma_k \ll \omega_k^r$, as given by (4.111). Clearly the wave will not keep on growing without limit. The physical mechanism of the instability is understood by the role of the resonant particles. Particles with velocities a little less (greater) than the phase velocity of a wave are accelerated (decelerated) on the average. The sign of $\partial f_0/\partial v$ at the wave velocity determines whether the wave would grow or damp, and the double humped distribution in Fig. (8.2) predicts instability. This physical process, which accelerates the slower particles and decelerates the faster particles changes the shape of the zero order distribution $f_0(v)$ to form a plateau, and $f_0(v)$ should not be considered as time-independent.

Equations (10.7) and (10.8) are solved by introducing the WKB phasor in an analogous manner as the linearized Vlasov equation is solved:

$$f_k(v,t) = \widetilde{f}_k(v)e^{\int_0^t dt'[-i\omega_k(t')+\gamma_k(t')]}, \tag{10.13}$$

$$\phi_k(t) = \widetilde{\phi}_k e^{\int_0^t dt'[-i\omega_k(t')+\gamma_k(t')]}. \tag{10.14}$$

Using (10.13) in (10.7) gives

$$f_k(v,t) = \frac{e}{m}(-ik)\phi_k \frac{\frac{\partial}{\partial v} f_0(v,t)}{ikv - i\omega_k(t) + \gamma_k(t)}. \tag{10.15}$$

Using (10.15) in (10.8) yields

$$\frac{\omega_{p0}^2}{k}\int dv \, \frac{\frac{\partial}{\partial v} f_0(v,t)}{kv - \omega_k(t) - i\gamma_k(t)} = 1. \tag{10.16}$$

(10.16) is solved for γ_k in the form (cf. Chapter 5)

$$\gamma_k = \frac{\pi}{2}\frac{\omega_k\omega_{pe}^2}{k^2}\left[\frac{\partial f_0}{\partial v}\right]_{v=\omega_k/k} . \quad (cf. \ (4.111)) \tag{10.17}$$

On the other hand, substituting (10.15) into (10.9) gives

$$\frac{\partial f_0}{\partial t} = \frac{e^2}{m^2}\sum_k | E_k |^2 \frac{\partial}{\partial v}\left[\frac{\gamma_k}{(kv-\omega_k)^2+\gamma_k^2}\frac{\partial}{\partial v} f_0(v)\right] \tag{10.18}$$

where we used the reality condition $\omega_{-k} = -\omega_k$. When $\omega_k/\gamma_k \gg 1$, we have

$$\frac{\gamma_k}{(kv-\omega_k)^2+\gamma_k^2} \simeq \pi\delta(kv-\omega_k).$$

The Fourier series in (10.18) can be converted to Fourier integral by changing $\sum_k \to \frac{1}{2\pi}\int \, dk$:

$$\frac{\partial f_0}{\partial t} = \frac{e^2}{2m^2}\frac{\partial}{\partial v}\int_{-\infty}^{\infty} dk \mid E_k \mid^2 \delta(kv-\omega_k)\frac{\partial f_0}{\partial v}.$$

Carrying out the integration, we obtain

$$\frac{\partial}{\partial t} f_0(v,t) = \frac{e^2}{2m^2}\frac{\partial}{\partial v}\left[\frac{1}{\mid v \mid} \mid E_k \mid^2 \frac{\partial f_0(v,t)}{\partial v}\right] \tag{10.19}$$

with $v = \omega_k/k$.

(10.14) gives the relation, $\frac{\partial}{\partial t} \mid E_k \mid^2 = 2\gamma_k \mid E_k \mid^2$, which can be written in the form

$$\frac{\partial}{\partial t} \mid E_k \mid^2 = \frac{4\pi^2 e^2}{m\omega_{pe}} v^2 \frac{\partial f_0(v,t)}{\partial v} \mid E_k \mid^2 . \tag{10.20}$$

(10.19) and (10.20) are the main results of the quasilinear theory.

Example. Use (10.10) to derive the quasilinear equation (10.19).

Solution: Let us write the potential in the WKB form

$$\phi_k(t) = \phi_k(0) exp\left(\int_0^t [-i\omega_k + \gamma_k(\epsilon t')]dt'\right). \tag{10.21}$$

Use (10.10) and (10.21) in (10.9) and the reality condition $\omega_{-k} = -\omega_k$, $\gamma_{-k} = \gamma_k$ to get

$$\frac{\partial}{\partial t} f_0(v,\epsilon t) = \frac{e}{m}\sum_k ik\phi_{-k}(0)e^{\int_0^t [i\omega_k + \gamma_k(\epsilon t')]dt'}$$
$$\times\left(-\frac{e}{m} ik \int_0^t dt' e^{ikv(t'-t)}\phi_k(0) e^{\int_0^{t'}[-i\omega_k+\gamma_k(\epsilon t'')]dt''}\frac{\partial}{\partial v} f_0(v,\epsilon t')\right). \tag{10.22}$$

Putting $\mid \phi_k(t) \mid^2 = \phi_k(0)\phi_{-k}(0)e^{2\int_0^t \gamma_k(\epsilon t'')dt''}$ in (10.22), we obtain

$$\frac{\partial}{\partial t} f_0(v,\epsilon t) = \frac{\partial}{\partial v}\int_0^t dt' \left(\frac{e}{m}\right)^2 \sum_k k^2 \mid \phi_k(t) \mid^2 e^{i(kv-\omega_k)(t'-t)}$$
$$\times e^{\int_t^{t'} \gamma_k(\epsilon t'')dt''}\frac{\partial}{\partial v} f_0(v,\epsilon t').$$

In the above integral, the functions of slow time variable ϵt can be taken out of the integral, assigning them the values at time t:

$$\frac{\partial}{\partial t} f_0(v,\epsilon t) = \left(\frac{e}{m}\right)^2 \frac{\partial}{\partial v}\left[\sum_k k^2 |\phi_k(t)|^2 \left(\int_0^t dt' e^{[i(kv-\omega_k)+\gamma_k(t)](t'-t)}\right)\frac{\partial}{\partial v} f_0(v,\epsilon t)\right].$$

Carrying out the $\int dt'$-integral, we obtain

$$\frac{\partial}{\partial t} f_0(v, \epsilon t) = \left(\frac{e}{m}\right)^2 \frac{\partial}{\partial v}\left[\sum_k k^2 \mid \phi_k(t) \mid^2 \frac{1 - e^{-i(kv-\omega_k)t-\gamma_k t}}{i(kv - \omega_k) + \gamma_k} \frac{\partial}{\partial v} f_0(v, \epsilon t)\right]. \tag{10.23}$$

The exponential term vanishes asymptotically. Using the familiar relation

$$\frac{1}{i(kv - \omega_k) + \gamma_k} = \frac{P}{i(kv - \omega_k) + \gamma_k} + \pi\ \delta(kv - \omega_k)$$

(10.23) can be written as

$$\frac{\partial f_0}{\partial t} = \frac{\partial}{\partial v} D(v) \frac{\partial f_0}{\partial v} \tag{10.24}$$

$$D(v) = \left(\frac{e}{m}\right)^2 \frac{\partial}{\partial v} \sum_k k^2 \mid \phi_k(t) \mid^2 \left[\frac{\gamma_k}{(kv - \omega_k)^2 + \gamma_k^2} + \pi\ \delta(kv - \omega_k)\right] \tag{10.25}$$

where neglecting the non-resonant part yields (10.19) and we used the reality condition $\omega_{-k} = -\omega_k$ and γ_k is given by (10.17). To complement (10.25) we have

$$\frac{\partial}{\partial t} \mid \phi_k \mid^2 = 2\gamma_k(t) \mid \phi_k \mid^2 . \tag{10.26}$$

• **Extraction of f_0 from f by spatial averaging.** In the preceding section, we focused on the mathematical formulation of the quasilinear theory only with the physical motivation as to how we can remedy the limitation of the linearized Vlasov equation in view of the trapped particle dynamics. The physical motivation can be strengthened by the development presented in this Section. The same quasilinear equations can be derived more physically by the process of spatially averaging the interacting terms over the plasma volume. We start from the nonlinear equation (10.1) and split up $f(\mathbf{r}, \mathbf{v}, t)$ into two parts:

$$f(\mathbf{r}, \mathbf{v}, t) = f_0(\mathbf{v}, t) + f_1(\mathbf{r}, \mathbf{v}, t), \tag{10.27}$$

$$\textit{where}\ \ f_0(\mathbf{v}, t) = \langle f(\mathbf{r}, \mathbf{v}, t)\rangle_r \tag{10.28}$$

is the spatially averaged distribution function, regarded as a slowly varying function of t. The waves are represented by $f_1(\mathbf{r}, \mathbf{v}, t)$, and we have $\langle f_1(\mathbf{r}, \mathbf{v}, t)\rangle_r = 0$. With this separation, we assume $f_1 \ll f_0$. Averaging (10.1) over space, we can write

$$\frac{\partial}{\partial t}\langle f(\mathbf{r},\mathbf{v},t)\rangle_r + \mathbf{v}\cdot\left\langle\frac{\partial f}{\partial \mathbf{r}}\right\rangle_r - \frac{e}{m}\left\langle\mathbf{E}\cdot\frac{\partial f}{\partial \mathbf{v}}\right\rangle_r = 0. \tag{10.29}$$

In the second term, we have typically

$$\frac{v_x}{V}\int\frac{\partial f}{\partial x}dxdydz = \frac{v_x}{V}\int f(x,y,z)\Big|_{x=boundary} dydz$$

where V is the volume of the plasma, and we put the boundary at ∞. Since there are no particles at the boundary this integral vanishes. The third term in (10.29) is simplified by using the assumption $\langle \mathbf{E}\rangle_r = 0$, which is true for the ensemble of randomly phased plasma waves.

$$\begin{aligned}\left\langle\mathbf{E}\cdot\frac{\partial f}{\partial \mathbf{v}}\right\rangle_r &= \left\langle\mathbf{E}\cdot\frac{\partial f_0}{\partial \mathbf{v}}\right\rangle_r + \left\langle\mathbf{E}\cdot\frac{\partial f_1}{\partial \mathbf{v}}\right\rangle_r \\ &= \langle\mathbf{E}\rangle_r\cdot\frac{\partial f_0}{\partial \mathbf{v}} + \left\langle\mathbf{E}\cdot\frac{\partial f_1}{\partial \mathbf{v}}\right\rangle_r = \left\langle\mathbf{E}\cdot\frac{\partial f_1}{\partial \mathbf{v}}\right\rangle_r.\end{aligned}$$

Then (10.29) becomes

$$\frac{\partial f_0}{\partial t} = \frac{e}{m}\left\langle\mathbf{E}\cdot\frac{\partial f_1}{\partial \mathbf{v}}\right\rangle_r. \tag{10.30}$$

The equation for f_1 is obtained by linearizing (10.1) ($f_1 \ll f_0$) in the form

$$\frac{\partial f_1}{\partial t} + \mathbf{v}\cdot\frac{\partial f_1}{\partial \mathbf{r}} - \frac{e}{m}\mathbf{E}(\mathbf{r},t)\cdot\frac{\partial f_0(\mathbf{v},\epsilon t)}{\partial \mathbf{v}} = 0. \tag{10.31}$$

Together with the Poisson equation

$$\nabla\cdot\mathbf{E} = -4\pi e\int d^3v f_1(\mathbf{r},\mathbf{v},t), \tag{10.32}$$

(10.31) and (10.32) are formally identical with the linearized Vlasov and Poisson equations. We obtain the normal mode Langmuir wave frequencies

$$\omega(k,\epsilon t) = \omega_{pe}(\epsilon t)\left(1+\frac{3}{2}k^2\lambda_e^2\right) + i\frac{\pi}{2}\frac{\omega_{pe}^3}{k^2}\left[\frac{\partial g(u,\epsilon t)}{\partial u}\right]_{\frac{\omega}{k}} \tag{10.33}$$

where ϵt is the slow time, expressing the slow change of $f_0(v,\epsilon t)$. Note that the Langmuir frequency, the real part (ω_r) as well as the imaginary part (ω_i), is now slowly changing in t [cf. (4.91)]. For simplicity, we assume the one-dimensional propagation; $\mathbf{k} = k\hat{\mathbf{x}}$. We consider that the linear waves form a spectrum of waves with phasor $exp[ikx - i\omega(k,\epsilon t)t]$ with $\omega(k,\epsilon t)$ as given by (10.33). The first order electric field can be expressed as spatial

Fourier representation (in the following we suppress the slow time indicator ϵt):

$$E(x,t) = \int_{-\infty}^{\infty} \frac{dk}{2\pi} E(k) e^{ikx - i\omega(k)t}, \tag{10.34}$$

$$f_1(v,x,t) = \int_{-\infty}^{\infty} \frac{dk}{2\pi} f_1(v,k) e^{ikx - i\omega(k)t}, \tag{10.35}$$

$$\begin{aligned}
\langle E f_1 \rangle_x &= \frac{1}{L} \int dx \int \frac{dk'}{2\pi} E(k') e^{ik'x - i\omega(k')t} \int \frac{dk}{2\pi} f_1(v,k) e^{ikx - i\omega(k)t} \\
&= \frac{1}{2\pi L} \int dk \int dk' E(k') f_1(v,k) e^{-i[\omega(k)+\omega(k')]t} \delta(k+k') \\
&= \frac{1}{2\pi L} \int dk \int dk' E(-k) f_1(v,k) e^{-i[\omega(k)+\omega(-k)]t}.
\end{aligned} \tag{10.36}$$

Substituting the linear relation

$$f_1(v,k) = i \frac{e}{m} E(k) \frac{\partial f_0/\partial v}{\omega - kv} \tag{10.37}$$

into (10.36) gives

$$\langle E f_1 \rangle_x = \frac{i}{2\pi L} \frac{e}{m} \int dk E(-k) E(k) \frac{\partial f_0/\partial v}{\omega - kv} e^{-i[\omega(k)+\omega(-k)]t}. \tag{10.38}$$

Since $E(x,t)$ in (10.34) is a real quantity, we have the following parity relations:

$$E(-k) = E^*(k), \quad \omega_r(-k) = -\omega_r(k), \quad \omega(-k) = \omega_i(k) \tag{10.39}$$

where the asterisk denotes the complex conjugate. The real (imaginary) part of ω is an odd (even) function of k. Consequently, we have

$$E(k)E(-k) = \mid E(k) \mid^2, \quad \omega(k) + \omega(-k) = 2i\omega_i(k)$$

and (10.38) reduces to

$$\langle E f_1 \rangle_x = \frac{i}{2\pi L} \frac{e}{m} \int dk \mid E(k) \mid^2 \frac{\partial f_0/\partial v}{\omega - kv} e^{2i\omega_i(k)t}. \tag{10.40}$$

Therefore the time evolution of $f_0(v,t)$, (10.30), takes the form

$$\frac{\partial f_0}{\partial t} = \frac{i}{2\pi L} \frac{e^2}{m^2} \frac{\partial}{\partial v} \int dk \mid E(k) \mid^2 \frac{\partial f_0/\partial v}{\omega - kv} e^{2i\omega_i(k)t}. \tag{10.41}$$

In the preceding presentation of the quasilinear equations, we saw that the zero order distribution function f_0 (see Eq. (10.28)) as defined by spatial average of the exact distribution function allows a complete separation of the rapidly varying distribution function $f_1(\mathbf{r}, \mathbf{v}, t)$ from the slowly changing zero order $f_0(\mathbf{v}, \epsilon t)$. The quasilinear equations were derived by calculating the spatial average of the product terms, each of which is rapidly varying [see (10.30)].

References: D. R. Nicholson *Introduction to plasma theory* Wiley New York (1983); P. M. Bellan *Fundamentals of plasma physics* Cambridge Univ. Press, Cambridge (2006).

• **Extraction of f_0 from f by averaging over the random phases.** In this Section we show that the separation of f_1 and f_0 can be achieved alternatively by taking average over the random phases of the fluctuating waves. We start from (10.1) and (10.2) where we write the electric field $\mathbf{E}$ as a wave packet:

$$\mathbf{E}(\mathbf{r}, t) = \sum_k \mathbf{E}_k(t) e^{i\mathbf{k}\cdot\mathbf{r} - i\omega(\mathbf{k})t} \tag{10.42}$$

where $\omega(\mathbf{k})$ is determined by the linear dispersion relation. (10.42) is a two-time representation of $\mathbf{E}(\mathbf{r}, t)$, in which the complex amplitude is slowly varying while the phase is rapidly oscillating. The slow change of the amplitude is caused by Landau damping or the growth (γ), depending upon the shape of the initial distribution function. The slow change of the amplitude implies $\gamma(\mathbf{k}) \ll \omega(\mathbf{k})$. Thus the slow amplitude can be represented as

$$\mathbf{E}_k(t) = \hat{\mathbf{a}}(\mathbf{k}) E_k(t) e^{i\theta(\mathbf{k})}, \quad (\hat{\mathbf{a}}(\mathbf{k})\ \textit{polarization vector};\ \ E_k(t)\ \ \textit{real})$$

where we assume that $\theta(\mathbf{k})$ corresponding to different $\mathbf{k}$ are random and independent. This means that the amplitude vanishes when averaged over the phase:

$$\langle \mathbf{E}_k(t) \rangle = \frac{\int_0^{2\pi} d\theta(\mathbf{k}) \mathbf{E}_k(t)}{\int_0^{2\pi} d\theta(\mathbf{k})} = 0.$$

To show this, it is sufficient to calculate

$$\int_0^{2\pi} d\theta(\mathbf{k}) [cos\theta(\mathbf{k})\ \ or\ \ sin\theta(\mathbf{k})] = 0.$$

It follows from this assumption of random phases that

$$\langle \mathbf{E}_k(t) \cdot \mathbf{E}^*_{k'}(t) \rangle = \mid \mathbf{E}_k(t) \mid^2 \delta(\mathbf{k} - \mathbf{k}').$$

We plan to work on the quantity which is the deviation from the mean value:

$$f_1(\mathbf{r}, \mathbf{v}, t) = f(\mathbf{r}, \mathbf{v}, t) - f_0(\mathbf{v}, t) \tag{10.43}$$

where $f_0 \equiv \langle f \rangle$ is the average of $f(\mathbf{r}, \mathbf{v}, t)$ over the phases. Accordingly, we have $\langle f_1(\mathbf{r}, \mathbf{v}, t) \rangle = 0$. Substituting $f = f_0 + f_1$ into (10.1) and taking average over the ensemble of phases, we obtain

$$\frac{\partial f_0}{\partial t} + \mathbf{v} \cdot \frac{\partial f_0}{\partial \mathbf{r}} - \frac{e}{m} \left\langle \mathbf{E}(\mathbf{r}, t) \cdot \frac{\partial f_1}{\partial \mathbf{v}} \right\rangle = 0. \tag{10.44}$$

We consider the plasma to be uniform in the original state, and the averaged distribution function f_0 will be assumed to be independent of spatial coordinate. Then the above equation becomes

$$\frac{\partial f_0}{\partial t} - \frac{e}{m} \left\langle \mathbf{E}(\mathbf{r}, t) \cdot \frac{\partial f_1}{\partial \mathbf{v}} \right\rangle = 0. \tag{10.45}$$

Subtracting (10.44) from (10.1) gives

$$\begin{aligned}
&\frac{\partial}{\partial t} f_1(\mathbf{r}, \mathbf{v}, t) + \mathbf{v} \cdot \frac{\partial f_1}{\partial \mathbf{r}} - \frac{e}{m} \mathbf{E}(\mathbf{r}, t) \cdot \frac{\partial f_0}{\partial \mathbf{v}} \\
&\quad - \frac{e}{m} \mathbf{E}(\mathbf{r}, t) \cdot \frac{\partial f_1}{\partial \mathbf{v}} - \frac{e}{m} \left\langle \mathbf{E}(\mathbf{r}, t) \cdot \frac{\partial f_1}{\partial \mathbf{v}} \right\rangle = 0.
\end{aligned}$$

In the *quasilinear approximation*, the last two terms in the above are neglected which account for nonlinear wave-wave interactions. Then, we have

$$\frac{\partial}{\partial t} f_1(\mathbf{r}, \mathbf{v}, t) + \mathbf{v} \cdot \frac{\partial f_1}{\partial \mathbf{r}} - \frac{e}{m} \mathbf{E}(\mathbf{r}, t) \cdot \frac{\partial f_0}{\partial \mathbf{v}} = 0. \tag{10.46}$$

In the two quasilinear equations, (10.45) and (10.46), which are completely identical to (10.30) and (10.31), the average is taken over the random phases in contrast with the spatial average as indicated in (10.30). However, in view of the correlation relationship between the random phased electric fields (the equation above (10.43)), it is reasonably clear that the spatial average (see (10.36)) and the average over the phases would produce the same result. The subsequent development of (10.45) and (10.46) is entirely parallel to the foregoing presentation. Our further development of (10.45) and (10.46) contains complexity ensuing from the removal of one dimensionality. In addition, we shall consider, in accordance with the wave packet form of electric field (10.42), the wave packet form of f_1 is:

$$f_1(\mathbf{r}, \mathbf{v}, t) = \sum_k f_1(\mathbf{k}, \mathbf{v}, \epsilon t) e^{i\mathbf{k}\cdot\mathbf{r} - i\omega(\mathbf{k})t} \tag{10.47}$$

where $f_1(\mathbf{k}, \mathbf{v}, \epsilon t)$ is a slowly varying function of t. Substituting (10.47) into (10.46) gives, for the Fourier amplitude,

$$\frac{\partial f_k(\epsilon t)}{\partial t} + i(\mathbf{k}\cdot\mathbf{v} - \omega_k) f_k + i\frac{e}{m}\phi_k(t)\mathbf{k}\cdot\frac{\partial f_0(\epsilon t, \mathbf{v})}{\partial \mathbf{v}} = 0 \quad (cf.\ (10.7)) \tag{10.48}$$

where and in the following the subscript $'1'$ is omitted. Keeping in mind that the first term in (10.48) is due to Landau damping or growth, we make the transformation $f_k(\epsilon t, \mathbf{v}) = g_k(\mathbf{v}) e^{\gamma_k t}$. Then (10.48) becomes

$$(i\mathbf{k}\cdot\mathbf{v} - i\omega_k + \gamma_k) g_k + i\frac{e}{m}\phi_k(t)\ e^{-\gamma_k t}\mathbf{k}\cdot\frac{\partial f_0(\epsilon t, \mathbf{v})}{\partial \mathbf{v}} = 0$$

$$or \quad (i\mathbf{k}\cdot\mathbf{v} - i\omega_k + \gamma_k) f_k(\mathbf{v}, \epsilon t) + i\frac{e}{m}\phi_k(t)\ \mathbf{k}\cdot\frac{\partial f_0(\epsilon t, \mathbf{v})}{\partial \mathbf{v}} = 0. \tag{10.49}$$

We instated the slow time derivative in (10.48) as the imaginary part of frequency ω_k in (10.49). We have the Poisson equation in the form

$$k^2\phi_k(t) = -4\pi e N \int d^3v f_k(\mathbf{v}, t). \tag{10.50}$$

Using (10.49) in (10.50) gives the linear dispersion relation

$$1 + \frac{\omega_{pe}^2}{k^2}\int d^3v\ \frac{\mathbf{k}\cdot\frac{\partial f_0}{\partial \mathbf{v}}}{\omega_k - \mathbf{k}\cdot\mathbf{v} + i\gamma_k} = 0.$$

Next, let us develop (10.45). Substituting $\mathbf{E}(\mathbf{r}, t)$ and $f(\mathbf{r}, \mathbf{v}, t)$ in their Fourier representation and using (10.49), we can write (10.45) in the form

$$\frac{\partial f_0}{\partial t} = \frac{e}{m}\left\langle \sum_{k'} \mathbf{E}_{k'}(t) e^{i\mathbf{k}'\cdot\mathbf{r} - i\omega_{k'}t}\cdot\frac{\partial}{\partial \mathbf{v}}\sum_k f_k(t) e^{i\mathbf{k}\cdot\mathbf{r} - i\omega_k t}\right\rangle$$

$$= \frac{e}{m}\left\langle \sum_{k'} -i\mathbf{k}'\phi_{k'}(t) e^{i\mathbf{k}'\cdot\mathbf{r} - i\omega_{k'}t}\cdot\frac{\partial}{\partial \mathbf{v}}\sum_k e^{i\mathbf{k}\cdot\mathbf{r} - i\omega_k t}\ \frac{-i\frac{e}{m}\phi_k\mathbf{k}\cdot\frac{\partial f_0}{\partial \mathbf{v}}}{i\mathbf{k}\cdot\mathbf{v} - i\omega_k + \gamma_k}\right\rangle.$$

The average taken over random phases gives $\delta(\mathbf{k} + \mathbf{k}')$. Note that $\omega_k + \omega_{-k} = 0$, $\phi_k\phi_{-k} = |\phi_k|^2$. Thus the above equation reduces to

$$\frac{\partial f_0}{\partial t} = \left(\frac{e}{m}\right)^2 \sum_k |\phi_k|^2\,\mathbf{k}\cdot\frac{\partial}{\partial \mathbf{v}}\frac{\mathbf{k}\cdot\frac{\partial f_0}{\partial \mathbf{v}}}{i\mathbf{k}\cdot\mathbf{v} - i\omega_k + \gamma_k}. \quad (cf.\ (10.23)) \tag{10.51}$$

$\gamma_k = \gamma_k(\epsilon t)$, a slowly varying function of t. In linear theory where the distribution function is constant, the dependence of E_k^2 on t is simply exponential, $E_k^2 \sim e^{2\gamma_k t}$, but in quasilinear theory in which the distribution has a weak time dependence, we have

$$\frac{dE_k^2}{dt} = 2\gamma_k(t)E_k^2.$$

[The original author of the material presented in this subsection is unknown to the author of this book, who transported the analysis with gratitude.]

10.2. Nonlinear theory of Vlasov–Poisson plasma

In this section, the nonlinear term that we discarded in the linear and quasilinear theory of the Vlasov equation is accounted for. The Vlasov equation is expanded in a perturbation series with the nonlinear term iterated by the linear solution. In terms of Fourier coefficients, the nonlinear term produces quadratic and triadic terms, and even higher order terms. The series solution is investigated under certain resonance conditions. Then, the Vlasov–Poisson system (or Vlasov–Maxwell system) can be reduced to a form from which important physical consequence can be elicited. The resonance conditions are two kinds of which products are: nonlinear Landau damping or growth (*wave-particle interaction*) and mode coupling (*wave-wave-interaction*).

Reference: R. Z. Sagdeev and A. A. Galeev *Nonlinear plasma theory,* Benjamin New York (1969).

10.2.1. *Straight expansion of Vlasov–Poisson equation in powers of wave amplitude*

We start from the Vlasov equation for the distribution function $f_\alpha(\mathbf{r}, \mathbf{v}, t)$

$$\frac{\partial}{\partial t} f_\alpha(\mathbf{r}, \mathbf{v}, t) + \mathbf{v} \cdot \frac{\partial f_\alpha}{\partial \mathbf{r}} + \frac{e_\alpha}{m_\alpha}\mathbf{E}(\mathbf{r}, t) \cdot \frac{\partial f_\alpha}{\partial \mathbf{v}} = 0 \tag{10.52}$$

and the Poisson equation

$$\nabla \cdot \mathbf{E} = 4\pi \sum_\alpha \int d^3v\, f_\alpha(\mathbf{r}, \mathbf{v}, t), \quad (\mathbf{E} = -\nabla\phi). \tag{10.53}$$

Expand f in the series

$$f_\alpha = f_{\alpha 0}(\mathbf{v}) + f_\alpha^{(1)}(\mathbf{r}, \mathbf{v}, t) + f_\alpha^{(2)}(\mathbf{r}, \mathbf{v}, t) + f_\alpha^{(3)}(\mathbf{r}, \mathbf{v}, t) + \cdots$$

and solve (10.52) by successive approximation.

$$\frac{\partial}{\partial t} f_\alpha^{(n)}(\mathbf{r},\mathbf{v},t) + \mathbf{v}\cdot\frac{\partial f_\alpha^{(n)}}{\partial \mathbf{r}} + \frac{e_\alpha}{m_\alpha}\mathbf{E}(\mathbf{r},t)\cdot\frac{\partial f_\alpha^{(n-1)}}{\partial \mathbf{v}} = 0 \quad (n = 1,2,3,\ldots f^{(0)} = f_0). \tag{10.54}$$

The first order equation gives

$$f_\alpha^{(1)}(\mathbf{k},\mathbf{v},\omega) = -\frac{e_\alpha}{m_\alpha}\frac{\mathbf{k}\cdot\frac{\partial f_{\alpha 0}}{\partial \mathbf{v}}}{\omega - \mathbf{k}\cdot\mathbf{v}}\,\phi(\mathbf{k},\omega). \tag{10.55}$$

The second order equation gives

$$f_\alpha^{(2)}(\mathbf{k},\mathbf{v},\omega) = -\frac{e_\alpha}{m_\alpha}\sum_{k',\omega'}\frac{\mathbf{k}'\cdot\frac{\partial f_\alpha^{(1)}(\mathbf{k}'',\omega'',\mathbf{v})}{\partial \mathbf{v}}}{\omega - \mathbf{k}\cdot\mathbf{v}}\,\phi(\mathbf{k}',\omega')$$

$$(\mathbf{k}'' = \mathbf{k} - \mathbf{k}', \quad \omega'' = \omega - \omega').$$

Using (10.55) in the above gives [for algebraic details, see **Exercise** at the end of § 10.2.3.]

$$f_\alpha^{(2)}(\mathbf{k},\mathbf{v},\omega) = -\frac{(e_\alpha/m_\alpha)^2}{\omega - \mathbf{k}\cdot\mathbf{v}}\,\frac{1}{2}\sum_{k',\omega'}\left[\mathbf{k}'\cdot\frac{\partial}{\partial \mathbf{v}}\frac{\mathbf{k}''\cdot\frac{\partial f_{\alpha 0}}{\partial \mathbf{v}}}{\omega'' - \mathbf{k}''\cdot\mathbf{v}} + term\ ('\leftrightarrow'')\right]$$

$$\times\,\phi(\mathbf{k}',\omega')\phi(\mathbf{k}'',\omega''). \tag{10.56}$$

Substituting the above results into (10.53) yields the equation for ϕ in the form

$$\varepsilon^{(1)}(\mathbf{k},\omega)\phi(\mathbf{k},\omega) + \sum_{k=k'+k'',\ \omega=\omega'+\omega''}\varepsilon^{(2)}(\mathbf{k}',\omega',\mathbf{k}'',\omega'')\phi(\mathbf{k}',\omega')\phi(\mathbf{k}'',\omega'')$$

$$+ \sum_{k=k'+k''+k''',\ \omega=\omega'+\omega''+\omega'''}\varepsilon^{(3)}(\mathbf{k}',\omega',\mathbf{k}'',\omega'',\mathbf{k}''',\omega''')$$

$$\times\,\phi(\mathbf{k}',\omega')\phi(\mathbf{k}'',\omega'')\phi(\mathbf{k}''',\omega''') + \cdots = 0 \tag{10.57}$$

where

$$\varepsilon^{(1)}(\mathbf{k},\omega) = 1 + \Sigma_\alpha\,\frac{\omega_{p\alpha}^2}{k^2}\int d^3v\,\frac{\mathbf{k}\cdot\frac{\partial f_{\alpha 0}}{\partial \mathbf{v}}}{\omega - \mathbf{k}\cdot\mathbf{v}}, \tag{10.58}$$

$$\varepsilon^{(2)}(\mathbf{k}', \omega', \mathbf{k}'', \omega'') = -\frac{1}{2} \sum_{\alpha} \frac{e_\alpha}{m_\alpha} \frac{\omega_{p\alpha}^2}{k^2} \int \frac{d^3 v}{\omega - \mathbf{k} \cdot \mathbf{v}}$$

$$\times \left[\mathbf{k}' \cdot \frac{\partial}{\partial \mathbf{v}} \frac{\mathbf{k}'' \cdot \frac{\partial f_{\alpha 0}}{\partial \mathbf{v}}}{\omega'' - \mathbf{k}'' \cdot \mathbf{v}} + \mathbf{k}'' \cdot \frac{\partial}{\partial \mathbf{v}} \frac{\mathbf{k}' \cdot \frac{\partial f_{\alpha 0}}{\partial \mathbf{v}}}{\omega' - \mathbf{k}' \cdot \mathbf{v}} \right]. \tag{10.59}$$

(10.57) is solved by expanding in powers of the wave amplitude. The lowest order solution is obtained from

$$\varepsilon^{(1)}(\mathbf{k}, \omega)\phi^{(1)}(\mathbf{k}, \omega) = 0 \tag{10.60}$$

whose solution we write in the form

$$\phi^{(1)}(\mathbf{k}, \omega) = \phi_k \delta(\omega - \omega(\mathbf{k})) \tag{10.61}$$

where $\omega(\mathbf{k})$ is the root of $\varepsilon^{(1)}(\mathbf{k}, \omega) = 0$; the normal mode frequency of the linear solution. Note in (10.60) that $\phi^{(1)}(\mathbf{k}, \omega) = 0$ if $\varepsilon^{(1)}(\mathbf{k}, \omega) \neq 0$ but $\phi^{(1)}(\mathbf{k}, \omega)$ has a value if $\varepsilon^{(1)}(\mathbf{k}, \omega) = 0$. The second order equation for $\phi(\mathbf{k}, \omega)$ in (10.57) reads

$$\varepsilon^{(1)}(\mathbf{k}, \omega)\phi^{(2)}(\mathbf{k}, \omega)$$

$$+ \sum_{k=k'+k'',\ \omega=\omega'+\omega''} \varepsilon^{(2)}(\mathbf{k}', \omega', \mathbf{k}'', \omega'')\phi^{(1)}(\mathbf{k}', \omega')\phi^{(1)}(\mathbf{k}'', \omega'') = 0. \tag{10.62}$$

Using (10.61) in (10.62) gives

$$\phi^{(2)}(\mathbf{k}, \omega) = - \sum_{k=k'+k''} \frac{\varepsilon^{(2)}(\mathbf{k}', \omega', \mathbf{k}'', \omega'')}{\varepsilon^{(1)}(\mathbf{k}, \omega)} \phi_{k'} \phi_{k''} \delta(\omega - \omega(\mathbf{k}') - \omega(\mathbf{k}'')). \tag{10.63}$$

To develop the equation for $|\ \phi(\mathbf{k}, t)\ |^2$, the wave kinetic equation, we multiply (10.57) by $\phi^*(\mathbf{k}, \Omega) e^{-i\omega t} e^{i\Omega t}$ and integrate over $\int d\Omega d\omega$:

$$\int d\omega \int d\Omega\ Eq.\ (10.57) \times \phi^*(\mathbf{k}, \Omega) e^{-i\omega t} e^{i\Omega t}. \tag{10.64}$$

We take in (10.64) the imaginary part only. [Why? Because it yields the desired kinetic equation. The real part produces its own relation but is of no concern. Also note

$$\mid \phi(\mathbf{k},t) \mid^2 = \left(\frac{1}{2\pi}\right)^2 \int e^{-i\omega t}\phi(\mathbf{k},\omega)d\omega \int e^{i\Omega t}\phi^*(\mathbf{k},\Omega)d\Omega \Big].$$

The first term in (10.57) gives the integral

$$I = \int d\omega \int d\Omega\ \varepsilon^{(1)}(\mathbf{k},\omega)\phi(\mathbf{k},\omega)\phi^*(\mathbf{k},\Omega)e^{-i\omega t}e^{i\Omega t} = \phi^*(\mathbf{k},t)\ J \qquad (10.65)$$

where

$$J = \int d\omega e^{-i\omega t}\varepsilon^{(1)}(\mathbf{k},\omega)\phi(\mathbf{k},\omega) = \int d\omega e^{-i\omega t}[\varepsilon_k'(\omega) + i\varepsilon_k''(\omega)]\phi(\mathbf{k},\omega),$$

where we separated $\varepsilon^{(1)} = \varepsilon_k'(\omega) + i\varepsilon_k''(\omega)$ into real and imaginary parts. Our purpose is to express J explicitly in terms of $\phi(\mathbf{k},t)$. Since $\phi(\mathbf{k},\omega)$ is peaked around $\omega(\mathbf{k}) \equiv \omega_k$, we can write

$$\phi(\mathbf{k},t) = \int d\omega e^{-i\omega t}\phi(\mathbf{k},\omega) \simeq \Delta\omega e^{-i\omega_k t}\ \phi_k(\omega_k),$$

$$\int d\omega\varepsilon_k''(\omega)e^{-i\omega t}\phi(\mathbf{k},\omega) \simeq \Delta\omega\varepsilon_k''(\omega_k)e^{-i\omega_k t}\ \phi_k(\omega_k) = \varepsilon_k''(\omega_k)\phi(\mathbf{k},t),$$

therefore

$$J = i\ \varepsilon_k''(\omega_k)\phi(\mathbf{k},t) + \int d\omega \left(\varepsilon'(\mathbf{k},\omega_k) + (\omega-\omega_k)\left[\frac{\partial\varepsilon'}{\partial\omega}\right]_{\omega_k}\right)\phi(\mathbf{k},\omega)e^{-i\omega t}$$

where we Taylor-expanded $\varepsilon'(\omega)$ in the neighborhood of $\omega = \omega_k$. Note $\varepsilon'(\mathbf{k},\omega_k) = 0$.

$$\begin{aligned} Therefore\ J &= i\ \varepsilon_k''(\omega_k)\phi(\mathbf{k},t) + \left[\frac{\partial\varepsilon'}{\partial\omega}\right]_{\omega_k}\int d\omega(\omega-\omega_k)\phi(\mathbf{k},\omega)e^{-i\omega t} \\ &= i\ \varepsilon_k''(\omega_k)\phi(\mathbf{k},t) + \left[\frac{\partial\varepsilon'}{\partial\omega}\right]_{\omega_k}\left(-\omega_k\phi(\mathbf{k},t) + \int d\omega\omega\phi(\mathbf{k},\omega)e^{-i\omega t}\right) \\ &= i\ \varepsilon_k''(\omega_k)\phi(\mathbf{k},t) + \left[\frac{\partial\varepsilon'}{\partial\omega}\right]_{\omega_k}\left(-\omega_k\phi(\mathbf{k},t) + i\ \frac{\partial\phi(\mathbf{k},t)}{\partial t}\right). \end{aligned}$$

Using the above in (10.65) gives for the imaginary part

$$Im\ I = \varepsilon_k''(\omega_k) \mid \phi(\mathbf{k},t) \mid^2 + \left[\frac{\partial\varepsilon'}{\partial\omega}\right]_{\omega_k}\frac{1}{2}\frac{\partial}{\partial t} \mid \phi(\mathbf{k},t) \mid^2. \qquad (10.66)$$

Next, we consider the quadratic term of (10.57) in the integral (10.64). The quadratic term is expressed as follows by putting it in terms of the products of $\phi's$:

$$\sum \varepsilon^{(2)}(\mathbf{k}',\omega',\mathbf{k}'',\omega'')[\phi^{(1)}(\mathbf{k}',\omega')\phi^{(1)}(\mathbf{k}'',\omega'') + \phi^{(2)}(\mathbf{k}',\omega')\phi^{(2)}(\mathbf{k}'',\omega'') + \phi^{(1)}(\mathbf{k}',\omega')\phi^{(2)}(\mathbf{k}'',\omega'') + \phi^{(2)}(\mathbf{k}',\omega')\phi^{(1)}(\mathbf{k}'',\omega'') + \cdots]. \tag{10.67}$$

The above expression is multiplied by $\phi^{(1)*}(\mathbf{k})\delta(\omega-\Omega)e^{i\Omega t}$ and is integrated over $d\Omega$. Here we invoke the random phase approximation which states that the ensemble average takes the form: $\langle\phi^{(1)}(\mathbf{k})\phi^{(1)*}(\mathbf{k}')\rangle = \mid \phi^{(1)}(\mathbf{k}) \mid^2 \delta_{k,k'}$. Clearly the odd power terms of $\phi^{(1)}(\mathbf{k})$ will be ensemble averaged to zero, and thus the first and the second terms in the RHS of (10.67) will vanish. We also neglect terms that are higher than fourth order in $\phi^{(1)}(\mathbf{k})$. Using (10.63) for the remaining last two terms in (10.67), we obtain the kinetic equation in the following form [Sagdeev and Galeev (1969), III-9]

$$\begin{aligned}
\left[\frac{\partial\varepsilon'}{\partial\omega}\right]_{\omega_k} \frac{1}{2}\frac{\partial}{\partial t} \mid \phi(\mathbf{k},t) \mid^2 &= -\varepsilon''_k(\omega_k) \mid \phi(\mathbf{k},t) \mid^2 \\
&+ 2Im \sum_{\mathbf{k}=\mathbf{k}'+\mathbf{k}''} \frac{\mid \varepsilon^{(2)}(\mathbf{k}',\mathbf{k}'',\omega_{k'},\omega_{k''}) \mid^2}{\varepsilon^{(1)*}(\mathbf{k}'+\mathbf{k}'',\ \omega_{k'}+\omega_{k''})} \mid \phi_{k'} \mid^2 \mid \phi_{k''} \mid^2 \\
&+ Im \sum_{\mathbf{k}'} \left[4\, \frac{\varepsilon^{(2)}(\mathbf{k}',\mathbf{k}-\mathbf{k}',\omega_{k'},\omega_k-\omega_{k'})\varepsilon^{(2)}(\mathbf{k},-\mathbf{k}',\omega_k,-\omega_{k'})}{\varepsilon^{(1)}(\mathbf{k}-\mathbf{k}',\omega_k-\omega_{k'})} \right. \\
&\left. -3\,\varepsilon^{(3)}(\mathbf{k}',-\mathbf{k}',\mathbf{k},\omega_{k'},-\omega_{k'},\omega_k) \right] \times \mid \phi_{k'} \mid^2 \mid \phi_k \mid^2 .
\end{aligned} \tag{10.68}$$

The first term of the RHS of (10.68) is the quasilinear growth rate since it is

$$\omega_i(\mathbf{k}) = -\frac{\varepsilon''_k(\omega_k)}{\partial\varepsilon'_k/\partial\omega_k}.$$

The second term describes the rate at which $\phi_{k'}$ and $\phi_{k''}$ merge to ϕ_k through wave-wave interaction. The third term determines the rate at which ϕ_k and $\phi_{k'}$ generate $\phi_{k''}$. Furthermore, this term contains a contribution to wave-particle interaction (nonlinear Landau damping) as will be shown in the following. The resonance condition for a wave-wave interaction to occur (in the second term) depends on the type of the dispersion curve. In the case of electron plasma oscillation, the dispersion curve is non-decay type.

• **Decay type and Non-decay type.** Let $\overrightarrow{\kappa} \equiv (\mathbf{k}, \omega)$ represent the four-vector as indicated. For $\overrightarrow{\kappa_1}$ and $\overrightarrow{\kappa_2}$ which satisfy $\varepsilon(\overrightarrow{\kappa_1}) = 0$ and $\varepsilon(\overrightarrow{\kappa_2}) = 0$, respectively, the dispersion curve is of non-decay type, if $\varepsilon(\overrightarrow{\kappa_1} + \overrightarrow{\kappa_2}) = 0$ cannot be satisfied. If $\varepsilon(\overrightarrow{\kappa_1} + \overrightarrow{\kappa_2}) = 0$ can be satisfied, the wave is decay type. The phase velocity decreases (increases) with k in the non-decay (decay) type wave. The resonance condition $\mathbf{k} = \mathbf{k}' + \mathbf{k}''$ can be satisfied among three waves belonging to a branch of the dispersion curve, if the dispersion curve is of decay type. If the dispersion curve is of non-decay type, the resonance condition can be satisfied by involving a third component belonging to a different branch (for instance, two Langmuir waves and an ion acoustic wave). In the linear portion of the ion acoustic wave spectrum, the wave can decay. The change of wave energy (LHS of (10.68)) of a Langmuir wave is determined by the quasilinear growth rate and nonlinear Landau damping in the third term, as is discussed below.

10.2.2. *Nonlinear Landau damping*

To evaluate the nonlinear Landau damping term in (10.68), we identify as follows. $\mathbf{k} \equiv \mathbf{k}_1$: a high frequency Langmuir wave, $\mathbf{k}' \equiv \mathbf{k}_2$: another high frequency Langmuir wave, $\mathbf{k}'' = \mathbf{k} - \mathbf{k}' = \mathbf{k}_1 - \mathbf{k}_2 \equiv \mathbf{k}_3$: low frequency wave, which should be an ion acoustic wave because Langmuir wave is not decayable. Therefore the third term for a single triad is

$$\frac{\varepsilon^{(2)}(\mathbf{k}_2, \omega_2, \mathbf{k}_3, \omega_3)\varepsilon^{(2)}(\mathbf{k}_1, \omega_1, -\mathbf{k}_2, -\omega_2)}{\varepsilon^{(1)}(\mathbf{k}_3, \omega_3)} \mid \phi_{k_1} \mid^2 \mid \phi_{k_2} \mid^2 \tag{10.69}$$

where $\varepsilon^{(2)}$ are electrons and $\varepsilon^{(1)}$ consists of both electron and ion contributions. To see how the third term came about more clearly, let us go back to the nonlinear dispersion relation in (10.57), and write the three-wave interaction in the form

$$\begin{aligned}\varepsilon^{(1)}(\mathbf{k}_1, \omega_1)\phi(\mathbf{k}_1, \omega_1) = -\varepsilon^{(2)}(\mathbf{k}_2, \omega_2, \mathbf{k}_3, \omega_3)[\phi^{(1)}(\mathbf{k}_2, \omega_2)\phi^{(2)}(\mathbf{k}_3, \omega_3) \\ + \phi^{(2)}(\mathbf{k}_2, \omega_2)\phi^{(1)}(\mathbf{k}_3, \omega_3)].\end{aligned} \tag{10.70}$$

The second term in the above is nonresonant and need not be considered. Using

$$\phi^{(2)}(\mathbf{k}_3, \omega_3) = -\frac{\varepsilon^{(2)}(\mathbf{k}_1, \omega_1, -\mathbf{k}_2, -\omega_2)}{\varepsilon^{(1)}(\mathbf{k}_3, \omega_3)}\phi_{k_1}\phi_{-k_2}$$

in (10.70) gives (10.69).

Now let us calculate the dielectric functions, i.e., the integrals in (10.58) and (10.59). Let us use the inequalities

$$v_i \ll \frac{\omega_3}{k_3} \ll v_e \ll \frac{\omega_1}{k_1}, \frac{\omega_2}{k_2} \quad (v_{e,i} : \textit{thermal velocity})$$

and neglect small terms. Then we obtain

$$\begin{aligned}
&\varepsilon^{(2)}(\mathbf{k}_1, \omega_1, -\mathbf{k}_2, -\omega_2) \\
&= \frac{1}{2}\frac{\omega_{pe}^2}{k_3^2}\frac{e}{m_e}\int \frac{d^3v}{\omega_3 - \mathbf{k}_3\cdot\mathbf{v}}\left[\mathbf{k}_1\cdot\frac{\partial}{\partial\mathbf{v}}\frac{-\mathbf{k}_2\cdot\frac{\partial f_{e0}}{\partial\mathbf{v}}}{-\omega_2 + \mathbf{k}_2\cdot\mathbf{v}} - \mathbf{k}_2\cdot\frac{\partial}{\partial\mathbf{v}}\frac{\mathbf{k}_1\cdot\frac{\partial f_{e0}}{\partial\mathbf{v}}}{\omega_1 - \mathbf{k}_1\cdot\mathbf{v}}\right] \\
&= \frac{1}{2}\frac{\omega_{pe}^2}{k_3^2}\frac{e}{m_e}\int \frac{d^3v}{\omega_3 - \mathbf{k}_3\cdot\mathbf{v}}\left[\frac{\mathbf{k}_1\cdot\mathbf{k}_2\mathbf{k}_2\cdot\frac{\partial f_{e0}}{\partial\mathbf{v}}}{(\omega_2 - \mathbf{k}_2\cdot\mathbf{v})^2} + \frac{\mathbf{k}_1\cdot\frac{\partial}{\partial\mathbf{v}}\mathbf{k}_2\cdot\frac{\partial f_{e0}}{\partial\mathbf{v}}}{\omega_2 - \mathbf{k}_2\cdot\mathbf{v}}\right. \\
&\qquad \left. - \frac{\mathbf{k}_1\cdot\mathbf{k}_2\mathbf{k}_1\cdot\frac{\partial f_{e0}}{\partial\mathbf{v}}}{(\omega_1 - \mathbf{k}_1\cdot\mathbf{v})^2} - \frac{\mathbf{k}_2\cdot\frac{\partial}{\partial\mathbf{v}}\mathbf{k}_1\cdot\frac{\partial f_{e0}}{\partial\mathbf{v}}}{\omega_1 - \mathbf{k}_1\cdot\mathbf{v}}\right]
\end{aligned} \tag{10.71}$$

where we used $\mathbf{k}_1\cdot\frac{\partial}{\partial\mathbf{v}}(\mathbf{k}_2\cdot\mathbf{v}) = \mathbf{k}_1\cdot\mathbf{k}_2$. The second and the fourth terms in the above cancel since $(\mathbf{k}_2\cdot\mathbf{v})$ and $(\mathbf{k}_1\cdot\mathbf{v})$ in the denominators can be neglected as compared to $\omega_1 \simeq \omega_2 \simeq \omega_{pe}$. Also neglect ω_3 as compared to $(\mathbf{k}_3\cdot\mathbf{v})$, then the remaining terms give

$$\varepsilon^{(2)}(\mathbf{k}_1, \omega_1, -\mathbf{k}_2, -\omega_2) \simeq -\frac{1}{2}\frac{e}{T_e}\frac{\mathbf{k}_1\cdot\mathbf{k}_2}{k_3^2}, \tag{10.72}$$

$$\begin{aligned}
\varepsilon^{(2)}(\mathbf{k}_2, \omega_2, \mathbf{k}_3, \omega_3) &= \frac{1}{2}\frac{\omega_{pe}^2}{k_1^2}\frac{e}{m_e}\int \frac{d^3v}{\omega_1 - \mathbf{k}_1\cdot\mathbf{v}} \\
&\quad \times\left[\mathbf{k}_2\cdot\frac{\partial}{\partial\mathbf{v}}\frac{\mathbf{k}_3\cdot\frac{\partial f_{e0}}{\partial\mathbf{v}}}{\omega_3 - \mathbf{k}_3\cdot\mathbf{v}} + \mathbf{k}_3\cdot\frac{\partial}{\partial\mathbf{v}}\frac{\mathbf{k}_2\cdot\frac{\partial f_{e0}}{\partial\mathbf{v}}}{\omega_2 - \mathbf{k}_2\cdot\mathbf{v}}\right] \\
&= \frac{1}{2}\frac{\omega_{pe}^2}{k_1^2}\frac{e}{m_e}\int d^3v \\
&\quad \times\left[\frac{\mathbf{k}_1\cdot\mathbf{k}_2}{(\omega_1 - \mathbf{k}_1\cdot\mathbf{v})^2}\frac{\mathbf{k}_3\cdot\frac{\partial f_{e0}}{\partial\mathbf{v}}}{\omega_3 - \mathbf{k}_3\cdot\mathbf{v}} + \frac{\mathbf{k}_1\cdot\mathbf{k}_3}{(\omega_1 - \mathbf{k}_1\cdot\mathbf{v})^2}\frac{\mathbf{k}_2\cdot\frac{\partial f_{e0}}{\partial\mathbf{v}}}{\omega_2 - \mathbf{k}_2\cdot\mathbf{v}}\right]
\end{aligned}$$

where we integrated by parts. In the above, $(\mathbf{k}_2 \cdot \mathbf{v})$ and $(\mathbf{k}_1 \cdot \mathbf{v})$ in the denominators can be neglected as compared to $\omega_1 \simeq \omega_2 \simeq \omega_{pe}$. The second term vanishes because it is an odd function integral. Also neglect ω_3 as compared to $\mathbf{k}_3 \cdot \mathbf{v}$. Then we obtain

$$\varepsilon^{(2)}(\mathbf{k}_2, \omega_2, \mathbf{k}_3, \omega_3) \simeq -\frac{1}{2}\frac{e}{T_e}\frac{\mathbf{k}_1 \cdot \mathbf{k}_2}{k_1^2}, \tag{10.73}$$

$$\varepsilon^{(1)}(\mathbf{k}_3, \omega_3) = 1 + \frac{\omega_{pi}^2}{k_3^2}\int d^3v \frac{\mathbf{k}_3 \cdot \frac{\partial f_{i0}}{\partial \mathbf{v}}}{\omega_3 - \mathbf{k}_3 \cdot \mathbf{v}} + \frac{\omega_{pe}^2}{k_3^2}\int d^3v \frac{\mathbf{k}_3 \cdot \frac{\partial f_{e0}}{\partial \mathbf{v}}}{\omega_3 - \mathbf{k}_3 \cdot \mathbf{v}}. \tag{10.74}$$

(10.74) agrees with the definition of the longitudinal dielectric permeability. In the last term (electron term), ω_3 can be neglected, and it becomes $1/\lambda_e^2 k_3^2$. The ion term integral is calculated along the Landau contour because the ions can resonate with the low frequency ion wave. We write

$$\varepsilon^{(1)}(\mathbf{k}_3, \omega_3) = 1 + \frac{1}{\lambda_e^2 k_3^2} - \frac{W}{k_3^2 \lambda_i^2}, \tag{10.75}$$

$$W = \int d^3v \, \frac{v_z f_{io}(v^2)}{\frac{\omega_3}{k_3} - v_z} = \int dv_z \, \frac{v_z f_{io}(v_z)}{\frac{\omega_3}{k_3} - v_z}, \tag{10.76}$$

where $f_{io}(v_z) = \sqrt{\frac{m_i}{2\pi T_i}} exp[-\frac{m_i v_z^2}{2T_i}]$, one-dimensional Maxwellian, and we take $\mathbf{k}_3$ to be parallel to the z-direction without loss of generality. The function W is related with the plasma dispersion function $Z(\zeta)$.

Exercise. Show that

$$W = -1 - \zeta Z(\zeta); \quad Z(\zeta) = \frac{1}{\sqrt{\pi}}\int_{-\infty}^{\infty} \frac{e^{-x^2}}{x - \zeta}\, dx, \quad \zeta = \frac{\omega_3/k_3}{\sqrt{2T_i/m_i}}.$$

[Sagdeev and Galeev (1969) introduced a similar function $W_S(\zeta) = -\frac{i}{\sqrt{\pi}} Z(\zeta)$. Our function W in (10.76) is equal to $-1 - i\sqrt{\pi}\,\zeta\, W_S(\zeta)$.]

The real part of $W (\equiv W_r)$ is the principal value of the integral, and can be evaluated by expanding the denominator in the powers of the small quantity $k_3 v_z/\omega_3$. One obtains

$$W_r = \frac{T_i}{m_i}\frac{k_3^2}{\omega_3^2}. \tag{10.77}$$

Exercise. Obtain the imaginary part of W:

$$W_i = -\sqrt{\frac{\pi}{2}}\left(\frac{T_e}{T_i}\right)^{\frac{1}{2}} e^{-\frac{T_e}{2T_i}}. \tag{10.78}$$

Solution: W_i is the negative of the imaginary part of the integral

$$\int dv \; \frac{v \; f_{i0}(v)}{v - \frac{\omega + i\nu}{k}} : \quad \nu \rightarrow 0^+.$$

[ω has a small imaginary part in accord with the causality.] Use

$$\frac{1}{v - u \mp i\nu} = P \; \frac{1}{v - u} \pm i\pi\delta(v - u)$$

$$W_i = -\pi\delta(v - \frac{\omega_3}{k_3}) \; v \; f_{i0}(v); \quad \frac{\omega_3}{k_3} = \sqrt{\frac{T_e}{m_i}}$$

which gives (10.78).

Thus, finally, the kinetic equation (10.68) in which the dominant nonlinear interaction is due to the nonlinear Landau damping is obtained by using (10.72), (10.73), and (10.75) in the form

$$\frac{\partial}{\partial t} \left[k^2 \mid \phi(\mathbf{k}, t) \mid^2 \right] = Im \sum_{k'} \frac{(\mathbf{k} \cdot \mathbf{k}')^2 \omega_{pe}}{4\pi N T_i}$$

$$\times \left(\frac{T_i}{T_e} \right)^2 \mid \phi(\mathbf{k}, t) \mid^2 \mid \phi(\mathbf{k}', t) \mid^2 \; \frac{1}{\frac{T_i}{T_e} - W}, \tag{10.79}$$

$$\zeta = \frac{\omega_k - \omega_{k'}}{\mid \mathbf{k} - \mathbf{k}' \mid \; \sqrt{2T_i/m_i}}, \qquad N : \textit{plasma number density}$$

where we used $\partial\varepsilon'/\partial\omega_k = 2/\omega_{pe}$. [in equation III-13 in Sagdeev and Galeev's monograph, the factor $(T_i/T_e)^2$ is missing and there is a sign difference in front of i.]

Using (10.77) in (10.75) gives the real and imaginary parts of $\varepsilon^{(1)}(\omega_3, k_3)$:

$$\varepsilon'(\omega_3, k_3) = 1 + \frac{1}{\lambda_e^2 k_3^2} - \frac{\omega_{pi}^2}{\omega_3^2}; \quad \varepsilon'' = -\frac{W_i}{k_3^2 \lambda_i^2}. \tag{10.80}$$

We obtain the low frequency ion acoustic wave dispersion relation from $\varepsilon' = 0$:

$$\omega_3^2 = k_3^2 \; \frac{T_e}{m_i} \quad or \quad W_r = \frac{T_i}{T_e} \tag{10.81}$$

where we neglected $\lambda_e^2 k_3^2 \ll 1$. Ion Landau damping rate is calculated by

$$\gamma_3 = -\frac{\varepsilon''}{\partial\varepsilon'/\partial\omega_3} = \frac{\omega_3}{2} \frac{T_e}{T_i} W_i = -\sqrt{\frac{\pi}{8}} \; \omega_3 \left(\frac{T_e}{T_i} \right)^{\frac{3}{2}} \; e^{-\frac{T_e}{2T_i}}. \tag{10.82}$$

One can show

$$Im\left[\frac{1}{\frac{T_i}{T_e}-W}\right]=\frac{1}{W_i}=\frac{1}{2}\frac{\omega_3}{\gamma_3}\frac{T_e}{T_i}. \tag{10.83}$$

Using the foregoing relations, (10.79) takes the form for single triad of interacting waves

$$\frac{\partial}{\partial t}\left[k_1^2\mid\phi(\mathbf{k}_1,t)\mid^2\right]=\frac{1}{2}\frac{(\mathbf{k}_1\cdot\mathbf{k}_2)^2\omega_{pe}}{4\pi NT_e}\frac{\omega_3}{\gamma_3}\mid\phi(\mathbf{k}_1,t)\mid^2\mid\phi(\mathbf{k}_2,t)\mid^2. \tag{10.84}$$

This is the rate of energy transfer to the wave k_1 by resonant interaction of k_2 and k_3 waves through nonlinear Landau damping.

10.2.3. *Derivative expansion*

We derived differential equation (10.68) from the algebraic equation (10.57) by Fourier inversion. In the course of Fourier inversion, ω acted like the operator $i\frac{\partial}{\partial t}$. Converting nonlinear dispersion relation to a differential equation with respect to t can be accomplished more directly by the correspondence $\omega\to\omega_k+i\frac{\partial}{\partial t}$. Let us consider the equation

$$\varepsilon(\omega,k)\phi(k,\omega)=R \tag{10.85}$$

where R represents the nonlinear term and $\varepsilon(\omega,k)$ is regarded as an operator. The Fourier representation $\phi(k,\omega)$ is considered to be a slow function of t due to the nonlinear term in R: $\phi(k,\omega)\to\phi(k,\omega,\epsilon t)$. Taylor expansion gives

$$\varepsilon=\varepsilon'\left(\omega_k+i\frac{\partial}{\partial t}\right)+i\varepsilon''=\varepsilon'(\omega_k)+\frac{\partial\varepsilon'}{\partial\omega_k}i\frac{\partial}{\partial t}+i\varepsilon'';\quad \varepsilon'(\omega_k)=0.$$

$$\textit{Therefore}\quad \frac{\partial\phi(k,\omega,\epsilon t)}{\partial t}+\frac{\varepsilon''}{\partial\varepsilon'/\partial\omega_k}\phi(k,\omega,\epsilon t)=\frac{R}{i\partial\varepsilon'/\partial\omega_k}. \tag{10.86}$$

Let us elaborate this. The linear solution can be written as $\phi(k,\omega,\epsilon t)=A(\epsilon t)\delta(\omega-\omega_k)$ with ω_k the normal mode frequency satisfying $\varepsilon(\omega_k)=0$. Expanding $\varepsilon(\omega)$ near ω_k as before, we have

$$\left(-\omega_k\frac{\partial\varepsilon'}{\partial\omega_k}+i\varepsilon''(\omega_k)\right)\phi(k,\omega,\epsilon t)+\frac{\partial\varepsilon'}{\partial\omega_k}\omega\int\phi(k,t)e^{i\omega t}dt=R$$

where we put $\varepsilon'(\omega_k) = 0$. The following integral is evaluated as

$$\begin{aligned}
\omega \int \phi(k,t)e^{i\omega t}dt &= \omega \int dt e^{i\omega t}\frac{1}{2\pi}\int d\omega' \phi(k,\omega')e^{-i\omega' t}\\
&= -i\int dt \frac{\partial}{\partial t}e^{i\omega t}\frac{1}{2\pi}\int d\omega' \phi(k,\omega')e^{-i\omega' t}\\
&= i\int dt e^{i\omega t}\frac{1}{2\pi}\int d\omega' \frac{\partial}{\partial t}\Big(\phi(k,\omega'\epsilon t)e^{-i\omega' t}\Big)\\
&= i\int dt e^{i\omega t}\frac{1}{2\pi}\int d\omega' \Big(\frac{\partial}{\partial t}\phi(k,\omega',\epsilon t) - i\omega'\phi(k,\omega',\epsilon t)\Big)e^{-i\omega' t}\\
&= i\int d\omega' \delta(\omega-\omega')\Big(\frac{\partial}{\partial t}\phi(k,\omega',\epsilon t) - i\omega'\phi(k,\omega',\epsilon t)\Big)\\
&= i\Big(\frac{\partial}{\partial t}\phi(k,\omega,\epsilon t) - i\omega\phi(k,\omega,\epsilon t)\Big).
\end{aligned}$$

Therefore, we have

$$\Big(-\omega_k\frac{\partial\varepsilon'}{\partial\omega_k} + i\varepsilon''(\omega_k)\Big)\phi(k,\omega,\epsilon t) + i\frac{\partial\varepsilon'}{\partial\omega_k}\Big(\frac{\partial}{\partial t}\phi(k,\omega,\epsilon t) - i\omega\phi(k,\omega,\epsilon t)\Big) = R.$$

In the above, the first term and the last term cancel because $\phi(k,\omega) \sim \delta(\omega - \omega_k)$. So we finally obtain (10.86) above.

Exercise. Carry out the algebra leading to (10.56).

$$\begin{aligned}
&\phi = \phi^{(1)}(\mathbf{r},t) + \phi^{(2)}(\mathbf{r},t) + \phi^{(3)}(\mathbf{r},t) + \cdots,\\
&\nabla^2(\phi^{(1)} + \phi^{(2)} + \phi^{(3)} + \cdots) + 4\pi\sum_\alpha e_\alpha \int d^3v(f_\alpha^{(1)} + f_\alpha^{(2)} + f_\alpha^{(3)} + \cdots) = 0.
\end{aligned} \tag{10.87}$$

We break down (10.52) and (10.53) order by order:

$$\frac{\partial}{\partial t}f_\alpha^{(1)} + \mathbf{v}\cdot\frac{\partial f_\alpha^{(1)}}{\partial \mathbf{r}} - \frac{e_\alpha}{m_\alpha}\nabla\phi^{(1)}(\mathbf{r},t)\cdot\frac{\partial f_{\alpha 0}}{\partial \mathbf{v}} = 0, \tag{10.88}$$

$$\nabla^2\phi^{(1)} + 4\pi\Sigma_\alpha\, e_\alpha\int d^3v\, f_\alpha^{(1)} = 0, \tag{10.89}$$

$$\frac{\partial}{\partial t} f_\alpha^{(2)} + \mathbf{v} \cdot \frac{\partial f_\alpha^{(2)}}{\partial \mathbf{r}} - \frac{e_\alpha}{m_\alpha} \left[\nabla \phi^{(2)} \cdot \frac{\partial f_{\alpha 0}}{\partial \mathbf{v}} + \nabla \phi^{(1)} \cdot \frac{\partial f_\alpha^{(1)}}{\partial \mathbf{v}} \right] = 0, \tag{10.90}$$

$$\nabla^2 \phi^{(2)} + 4\pi \Sigma_\alpha \, e_\alpha \int d^3 v \; f_\alpha^{(2)} = 0, \tag{10.91}$$

$$\frac{\partial}{\partial t} f_\alpha^{(3)} + \mathbf{v} \cdot \frac{\partial f_\alpha^{(3)}}{\partial \mathbf{r}} - \frac{e_\alpha}{m_\alpha} \left[\nabla \phi^{(3)} \cdot \frac{\partial f_{\alpha 0}}{\partial \mathbf{v}} + \nabla \phi^{(1)} \cdot \frac{\partial f_\alpha^{(2)}}{\partial \mathbf{v}} + \nabla \phi^{(2)} \cdot \frac{\partial f_\alpha^{(1)}}{\partial \mathbf{v}} \right] = 0, \tag{10.92}$$

$$\nabla^2 \phi^{(3)} + 4\pi \Sigma_\alpha \, e_\alpha \int d^3 v \; f_\alpha^{(3)} = 0. \tag{10.93}$$

$$(10.88): \quad f_\alpha^{(1)}(\mathbf{k}, \mathbf{v}, \omega) = -\frac{e_\alpha}{m_\alpha} \frac{\mathbf{k} \cdot \frac{\partial f_{\alpha 0}}{\partial \mathbf{v}}}{\omega - \mathbf{k} \cdot \mathbf{v}} \phi^{(1)}(\mathbf{k}, \omega), \tag{10.94}$$

$$(10.89): \quad \varepsilon^{(1)}(\mathbf{k}, \omega) \; \phi^{(1)}(\mathbf{k}, \omega) = 0, \tag{10.95}$$

$$\varepsilon^{(1)}(\mathbf{k}, \omega) = 1 + \Sigma_\alpha \frac{\omega_{p\alpha}^2}{k^2} \int d^3 v \; \frac{\mathbf{k} \cdot \frac{\partial f_{\alpha 0}}{\partial \mathbf{v}}}{\omega - \mathbf{k} \cdot \mathbf{v}} \tag{10.96}$$

$$(10.90): \quad f_\alpha^{(2)}(\mathbf{k}, \mathbf{v}, \omega) = -\frac{e_\alpha}{m_\alpha} \frac{\mathbf{k} \cdot \frac{\partial f_{\alpha 0}}{\partial \mathbf{v}}}{\omega - \mathbf{k} \cdot \mathbf{v}} \phi^{(2)}(\mathbf{k}, \omega)$$

$$-\frac{e_\alpha}{m_\alpha} \sum_{k', \omega'} \frac{\mathbf{k}' \cdot \frac{\partial f_\alpha^{(1)}(\mathbf{k}'', \omega'', \mathbf{v})}{\partial \mathbf{v}}}{\omega - \mathbf{k} \cdot \mathbf{v}} \phi^{(1)}(\mathbf{k}', \omega') \tag{10.97}$$

$$\textit{where} \quad \mathbf{k}'' = \mathbf{k} - \mathbf{k}', \quad \omega'' = \omega - \omega',$$

$$(10.91): \quad \varepsilon^{(1)}(\mathbf{k}, \omega) \phi^{(2)}(\mathbf{k}, \omega) = -\sum_{k', \omega'} \int d^3 v \frac{\mathbf{k}' \cdot \frac{\partial f_\alpha^{(1)}(\mathbf{k}'', \omega'', \mathbf{v})}{\partial \mathbf{v}}}{\omega - \mathbf{k} \cdot \mathbf{v}} \phi^{(1)}(\mathbf{k}', \omega'). \tag{10.98}$$

Use (10.94) in (10.98) to obtain

$$\varepsilon^{(1)}(\mathbf{k}, \omega) \phi^{(2)}(\mathbf{k}, \omega) = -\sum_{k', \omega'} \varepsilon^{(2)}(\mathbf{k}', \omega', \mathbf{k}'', \omega'') \phi^{(1)}(\mathbf{k}', \omega') \phi^{(1)}(\mathbf{k}'', \omega''), \tag{10.99}$$

$$\varepsilon^{(2)}(\mathbf{k}', \omega', \mathbf{k}'', \omega'') = -\sum_\alpha \frac{e_\alpha}{m_\alpha} \frac{\omega_{p\alpha}^2}{k^2} \int \frac{d^3 v}{\omega - \mathbf{k} \cdot \mathbf{v}} \, \mathbf{k}' \cdot \frac{\partial}{\partial \mathbf{v}} \frac{\mathbf{k}'' \cdot \frac{\partial f_{\alpha 0}}{\partial \mathbf{v}}}{\omega'' - \mathbf{k}'' \cdot \mathbf{v}}. \tag{10.100}$$

The expression for $\varepsilon^{(2)}$ can be symmetrized by adding the term obtained by interchanging (′) and (″) and dividing by 2 (the term obtained by interchanging is the same as the original term):

$$\varepsilon^{(2)}(\mathbf{k}', \omega', \mathbf{k}'', \omega'') = -\frac{1}{2}\sum_{\alpha} \frac{e_\alpha}{m_\alpha}\frac{\omega_{p\alpha}^2}{k^2}\int \frac{d^3v}{\omega - \mathbf{k}\cdot\mathbf{v}} \times \left[\mathbf{k}'\cdot\frac{\partial}{\partial \mathbf{v}}\frac{\mathbf{k}''\cdot\frac{\partial f_{\alpha 0}}{\partial \mathbf{v}}}{\omega'' - \mathbf{k}''\cdot\mathbf{v}} + \mathbf{k}''\cdot\frac{\partial}{\partial \mathbf{v}}\frac{\mathbf{k}'\cdot\frac{\partial f_{\alpha 0}}{\partial \mathbf{v}}}{\omega' - \mathbf{k}'\cdot\mathbf{v}}\right]. \tag{10.101}$$

Now (10.99) can be written in the form (10.62). The perturbation series in (10.87) can be written, upon using (10.95) and (10.62), in the form (10.56).

10.2.4. *Coupled mode equations: wave-wave interaction*

A plasma can be populated by various collective oscillations excited by instability and spontaneous noise, in addition to particles. In such a *turbulent plasma*, the interaction between the various components of waves plays an important role. Anomalous diffusion of a plasma across a magnetic field can take place in such a turbulent state of plasma. *Langmuir paradox* is a phenomenon that electrons are Maxwell–Boltzmann distributed even in a very rarefied plasma. This can be explained by strong scattering of electrons by collective oscillations in turbulent plasma. Dynamical equation for describing wave-wave interactions in such an aggregate of interacting oscillations is useful to study turbulent plasma. This term is applied to a plasma which has a large number of collective degrees of wave oscillation, given rise to spontaneously or by instability. In (10.68) the second term on the RHS represents mathematically the dynamical effect of the interaction between the waves. Instead of exploring the second term, we study this category of wave-wave interaction by means of the basic fluid equations. Fluid equations are advantageous if we focus on the wave-wave interaction because the resonant particle dynamics are automatically skimmed.

In this Section, coupled mode equations are derived for a plasma whose wave constituents are the high frequency Langmuir waves and the low frequency ion acoustic waves. The constituent waves beat each other to produce dynamical effects for the turbulent plasma. The fluid equations will be used for each component of the waves with superscript 'l' for Langmuir wave and 's' for ion acoustic wave (subscripts 'e' and 'i' denote electron and ion quantities). Readers should also consult Chapter 9.

$$\mathbf{v}_e = \mathbf{v}_e^l + \mathbf{v}_e^s, \quad \mathbf{v}_i = \mathbf{v}_i^s + \mathbf{v}_i^l.$$

Here the superscripts 'l' and 's' can be also read as high and low frequency, respectively. The ions in the high frequency field are stationary ($\mathbf{v}_i^l = 0$), so $\mathbf{v}_i = \mathbf{v}_i^s$. The electrons in the low frequency ion acoustic wave field will be assumed to form Boltzmann distribution which is static (electron inertia is neglected), so we put $\mathbf{v}_e^s = 0$.

$$\mathbf{v}_e = \mathbf{v}_e^l, \quad \mathbf{v}_i = \mathbf{v}_i^s, \tag{10.102}$$

$$\mathbf{E} = -\nabla\phi, \quad \phi = \phi^l + \phi^s, \quad n_i = n_0 + n_i^s, \tag{10.103}$$

$$n_e = n_0 + n_e^l + n_e^s, \quad n_e^s = \frac{en_o}{T_e}\phi^s, \tag{10.104}$$

$$\frac{\partial \mathbf{v}_e^l}{\partial t} = -\frac{e}{m_e}\mathbf{E}^l - \frac{3T_e}{n_0 m_e}\nabla n_e^l - \mathbf{v}_e^l \cdot \nabla \mathbf{v}_e^l. \tag{10.105}$$

The last term of (10.105) is written considering the possibility that two Langmuir wave components beat to produce the third Langmuir wave component. But a Langmuir wave spectrum is not of decay type, and the last term is put to zero. The high (Langmuir) and low (ion acoustic), electron and ion equations are summarized as follows:

$$\frac{\partial \mathbf{v}_e^l}{\partial t} = -\frac{e}{m_e}\mathbf{E}^l - \frac{3T_e}{n_0 m_e}\nabla n_e^l, \tag{10.106}$$

$$\frac{\partial n_e^l}{\partial t} + n_0 \nabla \cdot \mathbf{v}_e^l + \nabla \cdot (n_e^s \mathbf{v}_e^l) = 0, \tag{10.107}$$

$$\nabla \cdot \mathbf{E}^l = -\nabla^2 \phi^l = -4\pi e n_e^l, \tag{10.108}$$

$$-\frac{e}{m_e}\mathbf{E}^s - \frac{T_e}{n_0 m_e}\nabla n_e^s = \mathbf{v}_e^l \cdot \nabla \mathbf{v}_e^l, \tag{10.109}$$

$$\frac{\partial n_e^s}{\partial t} + n_0 \nabla \cdot \mathbf{v}_e^s = 0, \tag{10.110}$$

$$\nabla \cdot \mathbf{E}^s = 4\pi e(n_i^s - n_e^s), \tag{10.111}$$

$$\frac{\partial \mathbf{v}_i^s}{\partial t} = \frac{e}{m_i}\mathbf{E}^s, \tag{10.112}$$

$$\frac{\partial n_i^s}{\partial t} + n_0 \nabla \cdot \mathbf{v}_i^s = 0, \tag{10.113}$$

$$\frac{\partial \mathbf{E}^l}{\partial t} = 4\pi e \mathbf{v}_e^l (n_0 + n_e^s), \tag{10.114}$$

$$\frac{\partial \mathbf{E}^s}{\partial t} = 4\pi e n_0 (\mathbf{v}_e^s - \mathbf{v}_i^s), \tag{10.115}$$

$$n_e^s = \frac{e n_0}{T_e}\,\phi^s. \tag{10.116}$$

Not all the equations are independent. (10.114) is equivalent to (10.108) upon using (10.107). (10.115) is equivalent to (10.111) upon using (10.110) and (10.113). (10.116) will be used in lieu of (10.110). One can use (10.114) and (10.115) in lieu of the Poisson equations (10.108) and (10.111). Combining (10.106) and (10.107) yields

$$\left(\frac{\partial^2}{\partial t^2} + \omega_{pe}^2 - \frac{3T_e}{m_e}\nabla^2\right) n_e^l = -\nabla\cdot\frac{\partial}{\partial t}(n_e^s \mathbf{v}_e^l) = -\nabla\cdot\left(n_e^s \frac{\partial \mathbf{v}_e^l}{\partial t}\right)$$

where we neglected the time derivative of the low frequency quantity. In the above, use (10.106) with neglect of the thermal correction term and (10.116) to write

$$\left(\frac{\partial^2}{\partial t^2} + \omega_{pe}^2 - \frac{3T_e}{m_e}\nabla^2\right) n_e^l = \frac{n_0 e^2}{T_e m_e}\nabla\cdot(\mathbf{E}^l\phi^s).$$

Performing Fourier transform, $\times \int e^{-i\mathbf{k}\cdot\mathbf{r}} d^3r$, gives

$$\left(\frac{\partial^2}{\partial t^2} + \omega_{pe}^2 + \frac{3T_e}{m_e}k^2\right) n_e^l(\mathbf{k},t) = \frac{n_0 e^2}{T_e m_e} i\mathbf{k}\cdot\int e^{-i\mathbf{k}\cdot\mathbf{r}}\mathbf{E}^l(\mathbf{r},t)\phi^s(\mathbf{r},t)d^3r$$

$$= \frac{n_0 e^2}{T_e m_e} i\mathbf{k}\cdot\int d^3r e^{-i\mathbf{k}\cdot\mathbf{r}}\left[\int \mathbf{E}^l(\mathbf{k}',t)e^{i\mathbf{k}'\cdot\mathbf{r}}\frac{d^3k'}{(2\pi)^3}\right]\left[\int \phi^s(\mathbf{k}'',t)e^{i\mathbf{k}''\cdot\mathbf{r}}\frac{d^3k''}{(2\pi)^3}\right].$$

Integral $\int d^3r$ gives $(2\pi)^3\delta(\mathbf{k}-\mathbf{k}'-\mathbf{k}'')$. Using, in the above, the following relation

$$n_e^l(\mathbf{k},t) = -\frac{k^2}{4\pi e}\phi^l(\mathbf{k}.t),$$

we obtain

$$\left(\frac{\partial^2}{\partial t^2} + \omega_{pe}^2 + \frac{3T_e}{m_e}k^2\right)\phi^l(\mathbf{k},t)$$

$$= -\frac{e\,\omega_{pe}^2}{T_e}\int\frac{d^3k'}{(2\pi)^3}\int d^3k''\frac{\mathbf{k}\cdot\mathbf{k}'}{k^2}\phi^l(\mathbf{k}',t)\phi^s(\mathbf{k}'',t)\delta(\mathbf{k}-\mathbf{k}'-\mathbf{k}''). \tag{10.117}$$

Note the appearance of the wave vector matching condition, $\delta(\mathbf{k}-\mathbf{k}'-\mathbf{k}'')$. Taking divergence of (10.109) and using (10.112) and (10.113) and the charge neutrality $n_i^s \simeq n_e^s$ (acceptable in the low frequency region) yield

$$\left(\frac{\partial^2}{\partial t^2} - \frac{T_e}{m_i}\nabla^2\right) n_e^s = \frac{m_e}{m_i} n_0 \nabla \cdot (\mathbf{v}_e^l \cdot \nabla \mathbf{v}_e^l). \tag{10.118}$$

Fourier transforming (10.118) yields

$$\left(\frac{\partial^2}{\partial t^2} + \frac{T_e}{m_i}k^2\right) n_e^s(\mathbf{k},t) = \frac{m_e}{m_i} n_0 i\mathbf{k} \cdot \int d^3 r e^{-i\mathbf{k}\cdot\mathbf{r}} \mathbf{v}_e^l(\mathbf{r},t) \cdot \nabla \mathbf{v}_e^l(\mathbf{r},t)$$

$$= \frac{m_e}{m_i} n_0 i\mathbf{k} \cdot \int d^3 r e^{-i\mathbf{k}\cdot\mathbf{r}} \int \mathbf{v}_e^l(\mathbf{k}',t) e^{i\mathbf{k}'\mathbf{r}} \frac{d^3k'}{(2\pi)^3} \int \frac{d^3k''}{(2\pi)^3} \cdot i\mathbf{k}'' \mathbf{v}_e^l(\mathbf{k}'',t) e^{i\mathbf{k}''\mathbf{r}}$$

$$\left(\frac{\partial^2}{\partial t^2} + \frac{T_e}{m_i}k^2\right) \phi^s(\mathbf{k},t)$$

$$= -\frac{m_e}{m_i}\frac{T_e}{e} \int \frac{d^3k'}{(2\pi)^3} \int d^3k'' \delta(\mathbf{k}-\mathbf{k}'-\mathbf{k}'') \mathbf{k}'' \cdot \mathbf{v}_e^l(\mathbf{k}',t) \mathbf{k} \cdot \mathbf{v}_e^l(\mathbf{k}'',t) \tag{10.119}$$

where we used (10.116). (10.117) and (10.119) are the set of the mode coupling equations in a turbulent plasma whose wave constituents are Langmuir and ion acoustic waves. (10.117) and (10.119) will be further reduced to standard forms of coupled mode equations. Let us first consider the linear solutions of (10.117) and (10.119), with neglect of the RHS. We write from (10.117)

$$\left(\frac{\partial^2}{\partial t^2} + \omega_e^2(\mathbf{k})\right) \phi^l(\mathbf{k},t) = 0, \quad \omega_e^2(\mathbf{k}) = \omega_{pe}^2 + \frac{3T_e}{m_e}k^2$$

whose solution can be written as

$$\phi^l(\mathbf{k},t) = A(\mathbf{k}) e^{-i\omega_e^+(\mathbf{k})t} + B(\mathbf{k}) e^{-i\omega_e^-(\mathbf{k})t}, \quad \omega_e^{\pm}(\mathbf{k}) = \pm\sqrt{\omega_{pe}^2 + \frac{3T_e}{m_e}k^2}. \tag{10.120}$$

Because of the nonlinear driving term on the RHS, the amplitudes $A(\mathbf{k})$ and $B(\mathbf{k})$ become slowly varying, so we write the nonlinear solution in the form

$$\phi^l(\mathbf{k},t) = A(\mathbf{k},t) e^{-i\omega_e^+(\mathbf{k})t} + B(\mathbf{k},t) e^{-i\omega_e^-(\mathbf{k})t}. \tag{10.121}$$

Implicitly, (10.121) is the two-time scale solution: the amplitude is slowly changing while the phase changes rapidly.

Since $\phi(\mathbf{r},t)$ is a real quantity, we have

$$\phi(\mathbf{r},t) = \int_{-\infty}^{\infty} d^3k e^{i\mathbf{k}\cdot\mathbf{r}} \phi(\mathbf{k},t) = \int_{-\infty}^{\infty} d^3k e^{-i\mathbf{k}\cdot\mathbf{r}} \phi^*(\mathbf{k},t)$$
$$= \int_{-\infty}^{\infty} d^3k e^{i\mathbf{k}\cdot\mathbf{r}} \phi^*(-\mathbf{k},t)$$

$$\phi(\mathbf{k},t) = \phi^*(-\mathbf{k},t), \quad \phi(\mathbf{k},0) = \phi^*(-\mathbf{k},0).$$

$$\textit{Therefore}, \quad A(\mathbf{k}) = A^*(-\mathbf{k}), \quad B(\mathbf{k}) = B^*(-\mathbf{k}).$$

Likewise, the ion sound wave solution is written as

$$\phi^s(\mathbf{k},t) = S^+(\mathbf{k},t)e^{-i\omega_s^+(\mathbf{k})t} + S^-(\mathbf{k},t)e^{-i\omega_s^-(\mathbf{k})t}, \quad \omega_s^{\pm}(k) = \pm\sqrt{\frac{T_e}{m_i}}\,k. \tag{10.122}$$

Neglecting the thermal correction ($k^2\lambda_e^2 \ll 1$), we have

$$\frac{\partial}{\partial t}\mathbf{v}_e^l(\mathbf{k},t) = i\,\frac{e}{m_e}\mathbf{k}\phi^l(\mathbf{k},t).$$

Using (10.121) in the above equation gives

$$\mathbf{v}_e^l(\mathbf{k},t) = -\frac{e}{m_e}\mathbf{k}\left[\frac{A(\mathbf{k},t)}{\omega_e^+(\mathbf{k})}e^{-i\omega_e^+(\mathbf{k})t} + \frac{B(\mathbf{k},t)}{\omega_e^-(\mathbf{k})}e^{-i\omega_e^-(\mathbf{k})t}\right]. \tag{10.123}$$

Using (10.121), LHS of (10.117) is obtained after neglecting the second derivative, $\partial^2 A/\partial t^2$,

$$LHS\ of\ (10.117) = -2i\omega_e^+(\mathbf{k})\,\frac{\partial A}{\partial t}e^{-i\omega_e^+(\mathbf{k})t} - 2i\omega_e^-(\mathbf{k})\,\frac{\partial B}{\partial t}e^{-i\omega_e^-(\mathbf{k})t}.$$

We shall consider the Langmuir and ion sound waves to propagate toward + direction only:

$$\phi^s(\mathbf{k},t) = S^+(\mathbf{k},t)e^{-i\omega_s^+(\mathbf{k})t} \equiv S(\mathbf{k},t)e^{-i\omega_s(\mathbf{k})t}, \tag{10.124}$$

$$\phi^l(\mathbf{k},t) = A(\mathbf{k},t)e^{-i\omega_e^+(\mathbf{k})t} \equiv A(\mathbf{k},t)e^{-i\omega_e(\mathbf{k})t}. \tag{10.125}$$

This simplification makes it easier to see the conspicuous feature of the mode coupling equations. Then (10.117) takes the form

$$i\frac{\partial A(\mathbf{k},t)}{\partial t} = \frac{e\,\omega_{pe}^2}{2T_e}\frac{1}{\omega_e(k)}\int \frac{d^3k'}{(2\pi)^3}\frac{\mathbf{k}\cdot\mathbf{k}'}{k^2} \times \int d^3k'' A(\mathbf{k}',t)S(\mathbf{k}'',t)e^{i[\omega_e(k)-\omega_e(k')-\omega_s(k'')]t}\delta(\mathbf{k}-\mathbf{k}'-\mathbf{k}''). \tag{10.126}$$

The important point of observation is the emergence of the exponential phasor $e^{i[\omega_e(k)-\omega_e(k')-\omega_s(k'')]t}$ (frequency matching condition) and the δ-function, $\delta(\mathbf{k}-\mathbf{k}'-\mathbf{k}'')$ which is the wave vector matching condition.

Next we work on the ion acoustic wave equation. Using (10.123)–(10.125), Eq. (10.119) takes the form in the parallel way

$$i\frac{\partial S(\mathbf{k},t)}{\partial t} = \frac{eT_e}{2\omega_s(k)m_i m_e}\int \frac{d^3k'}{(2\pi)^3}\int d^3k''\,(\mathbf{k}'\cdot\mathbf{k}'')(\mathbf{k}\cdot\mathbf{k}'')\frac{A(\mathbf{k}',t)A(\mathbf{k}'',t)}{\omega_e(k')\omega_e(k'')} \times e^{i[\omega_s(k)-\omega_e(k')-\omega_e(k'')]t}\delta(\mathbf{k}-\mathbf{k}'-\mathbf{k}''). \tag{10.127}$$

Here we again see the frequency and wave vector matching conditions. Equations (10.126) and (10.127) are the desired mode coupling equations in a *turbulent plasma* populated with Langmuir waves and ion acostic waves. The coupling is solely due to the wave vector-matching resonances as represented by the δ functions. In contrast, (9.222)–(9.224) in Chapter 9, the coupled equations due to the wave-particle interaction, have the coefficients that are related with Landau damping rate.

Remark. Let us Fourier transform in time (10.117) by performing $\times \int(\cdots)e^{i\omega t}dt$ to get

$$\begin{aligned}
(-\omega^2+\omega_e^2)\phi^l(\mathbf{k},\omega) &= -\frac{e\,\omega_{pe}^2}{T_e}\int\frac{d^3k'}{(2\pi)^3}\int d^3k''\frac{\mathbf{k}\cdot\mathbf{k}'}{k^2}\delta(\mathbf{k}-\mathbf{k}'-\mathbf{k}'') \\
&\quad\times\int_{-\infty}^{\infty}dte^{i\omega t}\phi^l(\mathbf{k}',t)\phi^s(\mathbf{k}'',t) \\
&= -\frac{e\,\omega_{pe}^2}{T_e}\int\frac{d^3k'}{(2\pi)^3}\int d^3k''\frac{\mathbf{k}\cdot\mathbf{k}'}{k^2}\delta(\mathbf{k}-\mathbf{k}'-\mathbf{k}'') \\
&\quad\times\int_{-\infty}^{\infty}dte^{i\omega t}\int_{-\infty}^{\infty}\frac{d\omega'}{2\pi}e^{-i\omega' t}\phi^l(\mathbf{k}',\omega')\int_{-\infty}^{\infty}\frac{d\omega''}{2\pi}e^{-i\omega'' t}\phi^s(\mathbf{k}'',\omega'').
\end{aligned}$$

Integrating $\int dt$ gives $\delta(\omega - \omega' - \omega'')$. So the above equation can be written in the form

$$(\omega^2 - \omega_e^2)\phi^l(\mathbf{k}, \omega) = \frac{e\,\omega_{pe}^2}{T_e}\int \frac{d^3k'}{(2\pi)^3}\frac{\mathbf{k}\cdot\mathbf{k}'}{k^2} \times \int \frac{d\omega'}{2\pi}\,\phi^l(\mathbf{k}', \omega')\phi^s(\mathbf{k}-\mathbf{k}', \omega-\omega') \tag{10.128}$$

where we have $\omega^2 - \omega_e^2 \simeq 2\omega_e(\omega - \omega_e)$, because $\omega \simeq \omega_e$ per assuming that the nonlinear frequency shift is small. If we include damping of the wave ϕ^l, we put $\omega \to \omega - i\gamma_k$, and (10.128) can be put into the form

$$\omega - \omega_e(k) - i\gamma_k = \frac{e\,\omega_{pe}^2}{2\omega_e(k)T_e}\int \frac{d^3k'}{(2\pi)^3}\frac{\mathbf{k}\cdot\mathbf{k}'}{k^2}\int \frac{d\omega'}{2\pi}\,\phi^l(\mathbf{k}', \omega')\phi^s(\mathbf{k}-\mathbf{k}', \omega-\omega'). \tag{10.129}$$

Exercise. Explain that (10.129) is equivalent to (10.126).

Solution: (10.126) involves implicitly two different time scales. To be clear, it is written explicitly by introducing the slow time ϵt:

$$i\frac{\partial A(\mathbf{k}, \epsilon t)}{\partial t} = \frac{e\,\omega_{pe}^2}{2T_e}\frac{1}{\omega_e(k)}\int \frac{d^3k'}{(2\pi)^3}\frac{\mathbf{k}\cdot\mathbf{k}'}{k^2} \times \int d^3k'' A(\mathbf{k}', \epsilon t)S(\mathbf{k}'', \epsilon t)e^{i[\omega_e(k)-\omega_e(k')-\omega_s(k'')]t}\delta(\mathbf{k}-\mathbf{k}'-\mathbf{k}'')$$

where we put in the notations used in (10.129) the following relations

$$A(\mathbf{k}', \epsilon t)e^{-i\omega_e(k')t} = A(\mathbf{k}', t),$$
$$S(\mathbf{k}'', \epsilon t)e^{-i\omega_e(k'')t} = S(\mathbf{k}'', t).$$

Then we obtain

$$i\frac{\partial A(\mathbf{k}, \epsilon t)}{\partial t}e^{-i\omega_e(k)t} = \frac{e\,\omega_{pe}^2}{2T_e}\frac{1}{\omega_e(k)}\int \frac{d^3k'}{(2\pi)^3}\frac{\mathbf{k}\cdot\mathbf{k}'}{k^2} \times \int d^3k''\delta(\mathbf{k}-\mathbf{k}'-\mathbf{k}'')A(\mathbf{k}', t)S(\mathbf{k}'', t).$$

Multiply both sides by $e^{i\omega t}$ and perform $\int dt$ integral. Then LHS becomes by derivative expansion $[\omega - \omega_e(k)]A(\mathbf{k}, \omega)$. RHS is a convolution integral, giving the form of (10.129).

References:

- B. B. Kadomtsev *Plasma turbulence* (1965) Academic Press London, Chap. II
- Al'tshul L. M. and Karpman V. I. Sov Phys. JETP **20** P. 1043 (1965); Galeev, A. A., Karpman, V. I. and Sagdeev R. Z. Nucl. Fusion **5** P. 20 (1965), for original derivation of Eq. (10.68).
- Camac, M., Kantrowitz, A. R., Litvak, M. M., Patrick, R. M., and Petschek, H. E. *Shock waves in collision-free plasmas*, Nuc. Fusion Supplement **Pt. 2** (1962) p. 423–444, where the authors derived Boltzmann-like kinetic equation for the wave action of a wave under consideration in the distribution of other waves with the collision term provided by the wave-wave merging and wave splitting. The wave-wave interaction term in (10.68) can be further developed to a kinetic equation for the wave action along this line. See also the comment at the end of § 11.6.

10.3. Plasma echo

In this Section, we review the plasma echo theory for an infinite plasma. Plasma echoes have long been known theoretically as well as experimentally. The plasma echoes are also due to the quadratic term in the Vlasov equation. The distribution function of plasma particles $f(k, \omega, v)$ in Fourier space has a singularity at $\omega = kv$ (in addition to other singularities). In its inversion to (x, t) variables, this singularity modulates the distribution function with the exponential phase $e^{ik(x-vt)}$ or $e^{-i\omega(t-x/v)}$. This modulation occurs because of the *free streaming term*, which is called so since $x = vt$ is the characteristic line of a free particle. This term makes the modulation of the distribution function more and more oscillatory as t or x increases, and consequently, $\int f dv$ will become vanishingly small due to almost complete cancellations (phase-mixing). Therefore, the free streaming term yields no appreciable effect on macroscopic variable such as density perturbation. However, if two free streaming terms are multiplied as in nonlinear determination of the second order electric field, it is evident from the expression $e^{ik_1(x_1-vt_1)}e^{ik_2(x_2-vt_2)}$ that a constructive interference can result in at a certain time (temporal echo) (or at a certain spot; spatial echo) such that $k_1x_1 + k_2x_2 = v(k_1t_1 + k_2t_2)$. In this case, the phase of the exponential function of the second order distribution vanishes, and thus the corresponding velocity integral does not phase-mix, resulting in an echo.

For simplicity, we consider echoes in a plasma with ions stationary and forming the uniform background. We also specialize on one-dimensional electrostatic perturbations. Suppose that two pulses are externally applied at time $t = 0$ and $t = \tau$ to perturb the plasma:

$$\phi^{ext}(x,t) = \Phi_1 \ cos(k_1 x)\delta(\omega_p t) + \Phi_2 cos(k_2 x)\delta(\omega_p(t-\tau)). \tag{10.130}$$

In the electrostatic approximation the Vlasov equation reads

$$\frac{\partial}{\partial t} f(x,v,t) + v\frac{\partial f}{\partial x} + \frac{e}{m}\frac{\partial \phi}{\partial x}\frac{\partial f}{\partial v} = 0 \tag{10.131}$$

where ϕ is the total electric potential, the sum of the external and plasma fields: $\phi = \phi^{ext} + \phi^{pla}$. We have the Poisson equation which connects the electron distribution and the plasma electric field:

$$\frac{\partial^2}{\partial x^2}(\phi - \phi^{ext}) = 4\pi e \int f \ d^3v. \tag{10.132}$$

We attempt to solve the simultaneous equations (10.131) and (10.132) for a given ϕ^{ext} as prescribed by (10.130). In mathematical terms, this problem does not constitute a system of homogeneous equations; we have an inhomogeneous system, driven by a source term (10.130); the response f and ϕ should be determined by ϕ^{ext}. We solve the foregoing equations by Fourier-Laplace transform and its inversion which are defined by

$$f(k,v,\omega) = \int_{-\infty}^{\infty} dx \int_0^{\infty} dt e^{-ikx+i\omega t} f(x,v,t)$$

$$f(x,v,t) = \frac{1}{4\pi^2}\int_{-\infty}^{\infty} dk \int_{-\infty+ic}^{\infty+ic} d\omega \ e^{ikx-i\omega t} f(k,v,\omega)$$

where c is large enough so that it runs above all the singularities of f in the complex ω−plane, and the contour is completed by winding the lower-half ω-plane. Then (10.132) is transformed to

$$k^2\phi(k,\omega) = -4\pi e \int dv f(k,v,\omega) + k^2\phi^{ext}(k,\omega). \tag{10.133}$$

The Fourier–Laplace transform of (10.131) goes as follows. First, multiply (10.131) by e^{-ikx} and integrate with respect to x from $-\infty$ to ∞. The Fourier transform of a quadratic term is a convolution integral as shown below.

$$
\begin{aligned}
&FT[A(x)B(x)] \\
&= \int e^{-ikx}A(x)B(x)dx = \frac{1}{2\pi}\int dxe^{-ikx}A(x)\int e^{ik'x}B(k')dk' \\
&= \frac{1}{2\pi}\int dk'B(k')\int e^{-i(k-k')x}A(x)dx = \frac{1}{2\pi}\int dk'B(k')A(k-k') \\
&= \frac{1}{2\pi}\int dk'B(k')\int dk''A(k'')\delta(k''-k+k'). \qquad (10.134)
\end{aligned}
$$

In the above equations, the letters A and B can be interchanged. We obtain

$$
\begin{aligned}
\int_{-\infty}^{\infty} e^{-ikx}f(x)\frac{\partial\phi}{\partial x}dx &= \frac{1}{2\pi}\int dk' f(k')\ i\ (k-k')\phi(k-k') \\
&= \frac{1}{2\pi}\int dk' f(k-k')\ i\ k'\phi(k'). \qquad (10.135)
\end{aligned}
$$

Likewise, we have for Laplace transform:

$$
LT[A(t)B(t)] = \int_L \frac{d\omega'}{2\pi}B(\omega')A(\omega-\omega')
$$

where the subscript L means the Laplace contour.

In (10.131), we put f as the sum of the equilibrium distribution $f_0(v)$ and the perturbation f' (the prime on the f' will be omitted in the following). Then the Fourier–Laplace transform of (10.131) reads

$$
\begin{aligned}
&-i(\omega-kv)f(k,v,\omega) + \frac{e}{m}\ ik\ \phi(k,\omega)\ \frac{\partial f_0}{\partial v} \\
&\quad + \frac{i}{4\pi^2}\frac{e}{m}\int_{-\infty}^{\infty}dk'(k-k')\int_{ic-\infty}^{ic+\infty}d\omega'\ \frac{\partial}{\partial v}f(k',\omega',v)\phi(k-k',\omega-\omega') = 0 \\
&\qquad (10.136)
\end{aligned}
$$

where we put $f(k,v,t=0) = 0$ and c is above all the singularities of $f(k',\omega',v)$ on the complex ω'–plane. The simultaneous equations (13.133) and (10.136) will be solved by successive approximations. We treat the external potential in (10.133) as a first order quantity. The first order solutions will be used to iterate the second order quadratic term. We have from (10.133)

$$
k^2\phi^{(1)}(k,\omega) = -4\pi e\int dv f^{(1)}(k,v,\omega) + k^2\phi^{ext}(k,\omega) \qquad (10.137)
$$

where $\phi^{ext}(k,\omega)$

$$= \frac{\pi\Phi_1}{\omega_p}[\delta(k-k_1)+\delta(k+k_1)] + \frac{\pi\Phi_2}{\omega_p}[\delta(k-k_2)+\delta(k+k_2)]e^{i\omega\tau}. \tag{10.138}$$

The first order equation of (10.136) is

$$-i(\omega-kv)f^{(1)}(k,v,\omega) + \frac{e}{m}\, ik\, \phi^{(1)}(k,\omega)\,\frac{df_0}{dv} = 0 \tag{10.139}$$

$$\text{or} \quad f^{(1)}(k,v,\omega) = \frac{\frac{e}{m}k\phi^{(1)}(k,\omega)f_0'(v)}{\omega-kv} \quad \left(f_0' = \frac{df_o}{dv}\right). \tag{10.140}$$

(10.137) gives

$$\phi^{(1)}(k,\omega) = \frac{\phi^{ext}(k,\omega)}{\varepsilon(k,\omega)}, \quad \varepsilon(k,\omega) = 1 + \frac{\omega_{pe}^2}{k}\int dv \frac{f_0'(v)}{\omega-kv}. \tag{10.141}$$

(10.140) and (10.142) yield

$$f^{(1)}(k,v,\omega) = \frac{e}{m}k\,\frac{\phi^{ext}(k,\omega)}{\varepsilon(k,\omega)}\,\frac{f_0'(v)}{\omega-kv}. \tag{10.142}$$

The second order equations read

$$k^2\phi^{(2)}(k,\omega) = -4\pi e\int dv f^{(2)}(k,v,\omega) \tag{10.143}$$

and

$$-i(\omega-kv)f^{(2)}(k,v,\omega) + \frac{e}{m}\, ik\, \phi^{(2)}(k,\omega)\,\frac{df_0}{dv}$$
$$+ \frac{i}{4\pi^2}\frac{e}{m}\int_{-\infty}^{\infty} dk'(k-k')\int_{ic-\infty}^{ic+\infty} d\omega' \frac{\partial}{\partial v} f^{(1)}(k',\omega',v)\phi^{(1)}(k-k',\omega-\omega') = 0. \tag{10.144}$$

The above two equations yield

$$\phi^{(2)}(k,\omega) = \frac{4\pi ie}{k^2\varepsilon(k,\omega)}\int dv\, \frac{\textit{Last term of above eqn.}}{\omega-kv}. \tag{10.145}$$

The second order solution $\phi^{(2)}$ gives the responses at various k values through the combinations of k_1 and k_2 because of the δ functions in (10.138). We have

$$\frac{\omega_p^2}{\pi^2}\,\phi^{ext}(k',\omega')\,\phi^{ext}(k-k',\omega-\omega')$$
$$=\Big(\Phi_1[\delta(k'-k_1)+\delta(k'+k_1)]+\Phi_2[\delta(k'-k_2)+\delta(k'+k_2)]e^{i\omega'\tau}\Big)$$
$$\times\Big(\Phi_1[\delta(k-k'-k_1)+\delta(k-k'+k_1)]$$
$$+\Phi_2[\delta(k-k'-k_2)+\delta(k-k'+k_2)]e^{i(\omega-\omega')\tau}\Big).$$

Among these, we select the response at $k=k_2-k_1$, which comes from the products $\delta(k'+k_1)\delta(k-k'-k_2)$ and $\delta(k'-k_2)\delta(k-k'+k_1)$. So picking up the relevant terms, we have

$$\phi^{ext}(k',\omega')\,\phi^{ext}(k-k',\omega-\omega')$$
$$=\frac{\pi^2}{\omega_p^2}\,\Phi_1\Phi_2\delta(k+k_1-k_2)\left[\delta(k'+k_1)e^{i(\omega-\omega')\tau}+\delta(k'-k_2)e^{i\omega'\tau}\right].$$

Therefore we obtain from (10.145)

$$\phi^{(2)}(k,\omega)=\frac{e}{4m}\,\Phi_1\Phi_2\frac{k_1k_2}{k^2}\frac{\delta(k+k_1-k_2)}{\varepsilon(k,\omega)}\int\frac{dv}{\omega-kv}\frac{\partial}{\partial v}\int_{ic-\infty}^{ic+\infty}d\omega'\,f_0'(v)$$
$$\times\left[\frac{e^{i(\omega-\omega')\tau}}{(\omega'+k_1v)\varepsilon(-k_1,\omega')\varepsilon(k_2,\omega-\omega')}+\frac{e^{i\omega'\tau}}{(\omega'-k_2v)\varepsilon(k_2,\omega')\varepsilon(-k_1,\omega-\omega')}\right].$$

Integrating by parts with respect to v yields

$$\phi^{(2)}(k,\omega)=\frac{-1}{4}\frac{e}{m}\Phi_1\Phi_2\frac{k_1k_2}{k}\frac{\delta(k+k_1-k_2)}{\varepsilon(k,\omega)}\int dv\,\frac{f_0'(v)}{(\omega-kv)^2}\int_{ic-\infty}^{ic+\infty}d\omega'$$
$$\times\left[\frac{e^{i(\omega-\omega')\tau}}{(\omega'+k_1v)\varepsilon(-k_1,\omega')\varepsilon(k_2,\omega-\omega')}+\frac{e^{i\omega'\tau}}{(\omega'-k_2v)\varepsilon(k_2,\omega')\varepsilon(-k_1,\omega-\omega')}\right].\tag{10.146}$$

Fourier inversion gives

$$\phi^{(2)}(x,\omega) = -\frac{e^{ik_3x}}{8\pi}\,\frac{e}{m}\Phi_1\Phi_2\,\frac{k_1k_2}{k_3}\,\frac{1}{\varepsilon(k_3,\omega)}\int dv\,\frac{f_0'(v)}{(\omega-k_3v)^2}\times\int_{ic-\infty}^{ic+\infty}d\omega' \left[\frac{e^{i(\omega-\omega')\tau}}{(\omega'+k_1v)\varepsilon(-k_1,\omega')\varepsilon(k_2,\omega-\omega')}+\frac{e^{i\omega'\tau}}{(\omega'-k_2v)\varepsilon(k_2,\omega')\varepsilon(-k_1,\omega-\omega')}\right] \tag{10.147}$$

where $k_3 = k_2 - k_1$.

Laplace inversion yields

$$\phi^{(2)}(x,t) = -\frac{e^{ik_3x}}{8\pi}\,\frac{e}{m}\Phi_1\Phi_2\,\frac{k_1k_2}{k_3}\,\frac{1}{2\pi}\int dv\ f_0'(v)\int_{ib-\infty}^{ib+\infty}\frac{d\omega}{\varepsilon(k_3,\omega)}\,\frac{e^{-i\omega t}}{(\omega-k_3v)^2} \times\int_{ic-\infty}^{ic+\infty}d\omega'\left[\frac{e^{i(\omega-\omega')\tau}}{(\omega'+k_1v)\varepsilon(-k_1,\omega')\varepsilon(k_2,\omega-\omega')}+\frac{e^{i\omega'\tau}}{(\omega'-k_2v)\varepsilon(k_2,\omega')\varepsilon(-k_1,\omega-\omega')}\right] \tag{10.148}$$

where b is above all the singularities of the integrand ($0 < c < b$). If τ is long compared with Landau damping time (inverse of Landau damping rate), poles at the roots of the dielectric functions contribute negligibly to the integral, and we neglect them. Picking up the poles at $\omega' = -k_1v$ and $\omega' = k_2v$, $\int d\omega'-$ integral can be done:

$$\phi^{(2)}(x,t) = -i\frac{e^{ik_3x}}{8\pi}\,\frac{e}{m}\Phi_1\Phi_2\,\frac{k_1k_2}{k_3}\int dv f_0'(v)\int_{ib-\infty}^{ib+\infty}\frac{d\omega}{\varepsilon(k_3,\omega)}\,\frac{e^{-i\omega t}}{(\omega-k_3v)^2} \times\left[\frac{e^{i(\omega+k_1v)\tau}}{\varepsilon(-k_1,-k_1v)\varepsilon(k_2,\omega+k_1v)}+\frac{e^{ik_2v\tau}}{\varepsilon(k_2,k_2v)\varepsilon(-k_1,\omega-k_2v)}\right]. \tag{10.149}$$

In $\int d\omega-$integral, $\omega = k_3v$ is a double pole, and the residue is calculated by the formula

$$Res = \frac{d}{d\omega}\left(\frac{e^{-i\omega t}}{\varepsilon(k_3,\omega)}[\cdots]\right)_{\omega=k_3v} \approx -i\ t\left(\frac{e^{-i\omega t}}{\varepsilon(k_3,\omega)}[\cdots]\right)_{\omega=k_3v}$$

where we retained only the $\sim t$ term since we look for the time asymptotic response. Then, we obtain

$$\phi^{(2)}(x,t) = -\frac{i}{2}\frac{e}{m}\Phi_1\Phi_2\,\frac{t\,k_1k_2}{k_3} \times \int dv\; f_0'(v)\,\frac{e^{ik_3x}\;e^{-ik_3vt+ik_2v\tau}}{\varepsilon(k_3,k_3v)\varepsilon(-k_1,-k_1v)\epsilon_2(k_2,k_2v)}. \tag{10.150}$$

The above velocity integral does not phase-mix to zero, but results in echo, when

$$k_3vt \simeq k_2v\tau \quad or \quad t \simeq \frac{k_2}{k_3}\tau. \tag{10.151}$$

This is the echo time that we are looking for.

Remark: Different order of transforms. Let us first Fourier transform (10.131) with respect to x to get

$$\frac{\partial}{\partial t}f(k,v,t) + ikv\;f + \frac{e}{m}ik\phi(k.t)\frac{\partial f_0}{\partial v} + \frac{i\,e}{2\pi m}\frac{\partial}{\partial v}\int_{-\infty}^{\infty}dk'(k-k')f(k',v,t)\phi(k-k',t) = 0.$$

Laplace transforming the above equation gives ($p = -i\omega$)

$$(p+ikv)f(k,v,p) + \frac{e}{m}ik\phi(k,p)\frac{\partial f_0}{\partial v} + \frac{i\,e}{2\pi m}\frac{\partial}{\partial v}\int_{-\infty}^{\infty}dk'(k-k')\int_0^{\infty}dte^{-pt}f(k',v,t)\phi(k-k',t) = 0.$$

In the $dt-$integral, use the inverse Laplace transform to write

$$\int_0^{\infty}(\cdots) = \int_0^{\infty}dte^{-pt}\frac{1}{2\pi i}\int_{a-i\infty}^{a+i\infty}dp_1e^{p_1t}f(k',v,p_1)\frac{1}{2\pi i} \times \int_{b-i\infty}^{b+i\infty}dp_2e^{p_2t}\phi(k-k',p_2).$$

The $dt-$integral can be carried out to give

$$\left(\frac{1}{2\pi i}\right)^2\int_{a-i\infty}^{a+i\infty}dp_1f(k',v,p_1) \times \int_{b-i\infty}^{b+i\infty}\frac{dp_2}{p-p_1-p_2}\phi(k-k',p_2). \quad (Rep > a+b)$$

So the transformed Vlasov equation reads

$$(p + ikv)f(k,v,p) + \frac{e}{m}ik\phi(k,p)\frac{df_0}{dv}$$

$$-\frac{e}{m}\left(\frac{1}{2\pi i}\right)^3 \int_{-\infty}^{\infty} dk'(k-k') \int_{a-i\infty}^{a+i\infty} dp_1 \frac{\partial}{\partial v} f(k',v,p_1)$$

$$\times \int_{b-i\infty}^{b+i\infty} \frac{dp_2}{p-p_1-p_2}\phi(k-k',p_2) = 0.$$

After $\int dp_2$-integral, this becomes (10.144).

Reference: R. W. Gould, T. M. O'Neil, and J. H. Malmberg (1967) *Plasma wave echo*, *Phys. Rev. Lett.*, **19**, p. 219–222.

Chapter 11

Discrete charge effects

Viewing a plasma as the Vlasov fluid is the lowest order description of an actual plasma. In the Vlasov fluid picture, the plasma electric field is the one that is ensemble-averaged, which averages out the fluctuating electric fields associated with individual particles. In the next higher order, the individual discrete particle effect should be included to describe a plasma more realistically, particularly in the events such as radiation and scattering. The latter is not accounted for by the Vlasov fluid in which discreteness of the composing particles is entirely obliterated. The simplest way to include the discrete particle effect is to treat the plasma particles as *dressed particles.* In this Chapter, we show that the calculation of electric field fluctuation is facilitated by the dressed particle approach. The result agrees with the *fluctuation-dissipation theorem.* It is shown that the dressed particle method can be successfully employed to derive the electromagnetic fluctuation as well as the fluctuation in a magnetized plasma.

The fluctuations of the electric field and other plasma properties such as number density and current are all connected with the imaginary part of the plasma dielectric function. This connection is manifestation of the fluctuation-dissipation theorem. In this Chapter, we present a classical derivation of the fluctuation-dissipation theorem from linearized Liouville equation. The latter is solved in a very simple form by the characteristic method, analogously to the *integration along the unperturbed orbit* of the Vlasov equation. The fluctuation-dissipation theorem can be derived most elegantly by quantum mechanical consideration, but the elegance obscures the hidden facet of the theorem; and the classical derivation may be wanted by some readers.

The Cerenkov radiation is emitted by moving charges of constant velocities when the Landau wave-particle resonance conditions are met. It is the inverse process of Landau damping. It is a collective effect, and the emission of the radiation requires the whole plasma medium. The Cerenkov emission from a plasma can be fully accounted for by treating the plasma as aggregate of dressed particles. In an earlier chapter, we derived the Landau damping from the Kramers–Kronig relations and also directly from the causal requirement. It is shown that the Cerenkov emission power is the fluctuation of the electric field generated by dressed particles, and the fundamental reason for the Cerenkov emission can be traced to the causality. According to the fluctuation-dissipation theorem, the nonzeroness of fluctuation is due to the nonzeroness of the imaginary part of the linear dielectric function, which is thus fundamentally attributed to the causality.

The material in this Chapter is drawn from the author's paper: H. J. Lee *Fluctuation of the electric field in a plasma* (2015) J. Korean Phys. Soc. **66** p. 1167–1185.
and other references are:
N. Rostoker, Nucl. Fusion **1**, 101 (1960).
N. Rostoker and M. N. Rosenbluth, Phys. Fluids **3**, 1 (1960).
D. C. Montgomery and D. A. Tidman, *Plasma kinetic theory* (McGraw-Hill, New York, 1964).
K. Nishikawa and M. Wakatani, *Plasma Physics* Springer-Verlag, Heidelberg (1990).
A. Sitenko and V. Malnev *Plasma physics theory* Chapman and Hall, London (1995).

11.1. Potential due to a point charge placed in an isotropic plasma

This is a well-known problem. The answer is Debye potential. Here we consider this problem from two different points of view. First, we begin with an external charge q located at $\mathbf{r}_0$, which is represented by charge density

$$\rho_0(\mathbf{r}) = q\delta(\mathbf{r} - \mathbf{r}_0). \tag{11.1}$$

The potential at $\mathbf{r}$ made by the particles located at $\mathbf{r}'$ is

$$\sum_{\alpha=e,i} \frac{e_\alpha}{|\mathbf{r} - \mathbf{r}'|} \int d^3v f_\alpha(\mathbf{r}', \mathbf{v}, t) + \frac{q\,\delta(\mathbf{r}' - \mathbf{r}_0)}{|\mathbf{r} - \mathbf{r}'|}. \quad (cf.\ \S\ 5.1.2)$$

Integrating this over the whole space $\int d^3r'$ gives the potential at $\mathbf{r}$ made by the plasma and the single charge:

$$\phi(\mathbf{r}, t) = \int d^3r' \sum \frac{e_\alpha}{|\mathbf{r} - \mathbf{r}'|} \int d^3v f_\alpha(\mathbf{r}', \mathbf{v}, t) + \frac{q}{|\mathbf{r} - \mathbf{r}_0|}. \tag{11.2}$$

$$Hence, \quad \nabla\cdot\mathbf{E} = -\nabla^2\phi = 4\pi\sum e_\alpha\int d^3v f_\alpha(\mathbf{r},\mathbf{v},t) + 4\pi q\delta(\mathbf{r}-\mathbf{r}_0). \tag{11.3}$$

The first term on the RHS is the charge density of the plasma particles: $4\pi\rho_c = -4\pi\nabla\cdot\mathbf{P}$ ($\mathbf{P}$, polarization). Fourier transforming (11.3) and using (5.85) give

$$\phi(\mathbf{k}) = \frac{4\pi q}{k^2\varepsilon_L}e^{-i\mathbf{k}\cdot\mathbf{r}_0}.$$

Alternatively, in view of the definition of $\mathbf{D}$, we have

$$\nabla\cdot\mathbf{D} = 4\pi q\ \delta(\mathbf{r}-\mathbf{r}_0). \tag{11.4}$$

Fourier transforming gives

$$ik_iD_i = 4\pi q\ e^{-i\mathbf{k}\cdot\mathbf{r}_0}.$$

Since we have $D_i = \varepsilon_{ij}E_j,\ \ E_j = -ik_j\phi$, it follows that

$$ik_iD_i = k_ik_j\varepsilon_{ij}\phi = k^2\ \varepsilon_L\ \phi = 4\pi q\ e^{-i\mathbf{k}\cdot\mathbf{r}_0}$$

which agrees with the earlier result.

This is a static problem. Using (5.93) with $\omega = 0$ (with $F_{\alpha 0}$ Maxwellian) gives

$$\varepsilon_L = 1 + \sum_\alpha\frac{1}{k^2\lambda_{D\alpha}^2} \equiv 1 + \frac{1}{k^2\lambda_D^2}, \tag{11.5}$$

$$\phi(\mathbf{k}) = \frac{4\pi q\lambda_D^2}{1+k^2\lambda_D^2}e^{-i\mathbf{k}\cdot\mathbf{r}_0}. \tag{11.6}$$

Fourier inversion gives

$$\phi(\mathbf{r}) = \left(\frac{1}{2\pi}\right)^3 4\pi q\int_{-\infty}^{\infty}d^3k\frac{e^{i\mathbf{k}\cdot(\mathbf{r}-\mathbf{r}_0)}}{k^2+\lambda_D^{-2}}.$$

Taking $\mathbf{k}$ along z-direction, using spherical coordinate, and putting $\mathbf{r}_0 = 0$,

$$\phi(\mathbf{r}) = \left(\frac{q}{2\pi^2}\right)\int_0^\infty k^2dk\int sin\theta\ d\theta\frac{e^{ikrcos\theta}}{k^2+\lambda_D^{-2}}\int d\varphi = etc.$$

$$= \frac{2q}{\pi r}\int_0^\infty dk\frac{ksinkr}{k^2+\lambda_D^{-2}}.$$

Picking up the residue at the pole $k = i/\lambda_D$, we finally obtain

$$\phi(\mathbf{r}) = q\ \frac{e^{-r/\lambda_D}}{r}. \tag{11.7}$$

The result is due to the shielding of the point charge field by plasma particles. Comparison with the field in vacuum ($\sim 1/r^2$) indicates that the point charge field in plasma is *less* far reaching (*Debye shielding or screening*).

Secondly, a more direct way of treating this problem goes as follows. In a neutral plasma, an external charge q is introduced; after a sufficient time a thermal equilibrium state is reached in which the electrons and ions are Boltzmann distributed. Then we have the Poisson equation,

$$\nabla^2\phi = -4\pi e n_0(e^{-e\phi/T_i} - e^{e\phi/T_e}) - 4\pi q\delta(\mathbf{r}). \tag{11.8}$$

Expanding in terms of $e\phi/T$,

$$\nabla^2\phi = \frac{\phi}{\lambda_D^2} - 4\pi q\delta(\mathbf{r}). \tag{11.9}$$

Fourier transforming gives

$$\phi(k) = \frac{4\pi q}{k^2 + \lambda_D^{-2}} \tag{11.10}$$

which is identical with (11.6).

Exercise. Write (11.9) in spherical coordinates and solve the differential equation to derive the Debye potential.

11.2. Potential due to an oscillating charge

When an oscillating charge of high frequency is introduced in a Maxwellian plasma, what is the response? Assume an external charge density in the form

$$\rho_0(\mathbf{r}, t) = q\delta(\mathbf{r} - \mathbf{r}_0)cos(\omega_0 t). \tag{11.11}$$

$$Therefore, \quad \nabla \cdot \mathbf{D}(\mathbf{r}, t) = 4\pi q\delta(\mathbf{r} - \mathbf{r}_0)cos(\omega_0 t). \tag{11.12}$$

Fourier transforming gives

$$ik_i D_i(\mathbf{k}, \omega) = 4\pi q e^{-ik\cdot\mathbf{r}_0}\pi\left(\delta(\omega - \omega_0) + \delta(\omega + \omega_0)\right). \tag{11.13}$$

Using the relation $D_i = \varepsilon_{ij}E_j = \varepsilon_{ij}(-ik_j)\phi(\mathbf{k}, \omega)$ in the above gives

$$\phi(\mathbf{k}, \omega) = (2\pi)^2\ q\ \frac{e^{-ik\cdot\mathbf{r}_0}[\delta(\omega - \omega_0) + \delta(\omega + \omega_0)]}{k^2\varepsilon_L(k, \omega)}. \tag{11.14}$$

Fourier inversion in t gives

$$\phi(\mathbf{k}, t) = 4\pi\ q\ cos(\omega_0 t)\ \frac{e^{-ik\cdot\mathbf{r}_0}}{k^2\varepsilon_L(k, \omega_0)} \tag{11.15}$$

where we used $\varepsilon_L(\mathbf{k}, \omega_0) = \varepsilon_L(\mathbf{k}, -\omega_0)$. We assume ω_0 is a high frequency; $\omega_0/k \gg v_{th\alpha}$. Then the ionic response is negligible, and we have $\varepsilon_L(k, \omega_0) = 1 - \omega_0^2/\omega_{pe}^2$. Fourier inverting (11.15) gives, using spherical coordinates in k-space,

$$\phi(\mathbf{r}, t) = \frac{q}{\mid \mathbf{r} - \mathbf{r}_0 \mid} \frac{cos(\omega_0 t)}{(1 - \frac{\omega_0^2}{\omega_{pe}^2})} \tag{11.16}$$

where the integral $\int_{-\infty}^{\infty} dx\ sinx/x = \pi$ was used. Two points are worthy of note: first, in a high frequency field electrons do not form Debye screening; secondly, if the impressing frequency ω_0 is $\sim \omega_{pe}$, the charge resonantly invokes a large amplitude response. The low frequency response is the Debye shielding.

Example: Electric field due to a charge gyrating around a static magnetic field $\hat{\mathbf{z}}B_0$.

The charge density due to the single charge can be represented by $\rho_s(\mathbf{r}, t) = e\ \delta\ (\mathbf{r} - \mathbf{r}_s(t))$ where $\mathbf{r}_s(t)$ is the orbit of the charge. We have

$$\nabla \cdot \mathbf{D}(\mathbf{r}, t) = 4\pi e\ \delta\ (\mathbf{r} - \mathbf{r}_s(t)).$$

Fourier transforming,

$$ik_i D_i(\mathbf{k}, \omega) = 4\pi e \int e^{-i\mathbf{k}\cdot\ \mathbf{r} + i\omega t} \delta\ (\mathbf{r} - \mathbf{r}_s(t)) d^3r\ dt = ik_i \varepsilon_{ij} E_j$$

$$= k_i k_j\ \varepsilon_{ij}\ \phi(\mathbf{k}, \omega) = k^2 \varepsilon_L\ \phi(\mathbf{k}, \omega),$$

$$\phi(\mathbf{k}, \omega) = \frac{4\pi e}{k^2 \varepsilon_L} \int e^{-i\mathbf{k}\cdot\mathbf{r}_s(t) + i\omega t} dt. \tag{11.17}$$

To find the orbit, we solve

$$\frac{d\mathbf{r}}{dt} = \omega_c\ \mathbf{r} \times \hat{\mathbf{z}}$$

to get

$$(x(t), y(t), z(t)) = (r_L\ sin(\omega_c t),\ r_L\ cos(\omega_c t),\ v_\parallel t). \quad (r_L,\ v_\parallel,\ \ constants)$$

In the integral in (11.17), we put $\mathbf{k} = \hat{x}k_\perp + \hat{z}k_\parallel$ without loss of generality. Then

$$\int e^{-i\mathbf{k}\cdot\ \mathbf{r}_s(t) + i\omega t} dt = \Sigma_n J_n(k_\perp r_L) \int_{-\infty}^{\infty} e^{i\omega t - ik_\parallel v_\parallel t - in\omega_c t} dt$$

$$= 2\pi \sum_{n=-\infty}^{\infty} J_n(k_\perp r_L)\ \delta(\omega - k_\parallel v_\parallel - n\omega_c),$$

$$\mathbf{E}(\mathbf{k}, \omega) = -ie8\pi^2 \frac{\mathbf{k}}{k^2 \varepsilon_L} \sum_{n=-\infty}^{\infty} J_n(k_\perp r_L)\ \delta(\omega - k_\parallel v_\parallel - n\omega_c). \tag{11.18}$$

11.3. A moving external charge introduced into plasma

For simplicity, we assume that the external charge moves with a constant velocity $\mathbf{v}_0$. It would be instructive to solve this problem in two ways: i) with the aid of the displacement vector $\mathbf{D}$ and ii) from the Poisson equation. If the initial position of the charge (q) is $\mathbf{r}_0$, the charge density associated with the moving charge is represented by

$$\rho_{ext}(\mathbf{r}, t) = q\ \delta(\mathbf{r} - \mathbf{r}_0 - \mathbf{v}_0 t). \tag{11.19}$$

Fourier transforming (11.19) gives

$$\rho_{ext}(\mathbf{k}, \omega) = q \int e^{-i\mathbf{k}\cdot\mathbf{r}+i\omega t}\delta(\mathbf{r} - \mathbf{r}_0 - \mathbf{v}_0 t)d^3r\ dt$$
$$= qe^{-i\mathbf{k}\cdot\mathbf{r}_0} \int e^{-i\mathbf{k}\cdot\mathbf{v}_0 t+i\omega t}dt = 2\pi q e^{-i\mathbf{k}\cdot\mathbf{r}_0}\delta(\omega - \mathbf{k}\cdot\mathbf{v}_0). \tag{11.20}$$

Now, we use the equation

$$\nabla\cdot\mathbf{D} = 4\pi\rho_{ext} = 4\pi q\ \delta(\mathbf{r} - \mathbf{r}_0 - \mathbf{v}_0 t). \tag{11.21}$$

Fourier transforming the above equation gives

$$\mathbf{k}\cdot\mathbf{D}(\mathbf{k}, \omega) = 4\pi q\ e^{-i\mathbf{k}\cdot\mathbf{r}_0}2\pi\delta(\omega - \mathbf{k}\cdot\mathbf{v}_0), \tag{11.22}$$

$$D_i = \varepsilon_{ij}E_j, \quad E_j = -ik_j\phi(\mathbf{k}, \omega), \quad k_i k_j \varepsilon_{ij} = k^2\varepsilon_L.$$

Therefore we obtain

$$\phi(\mathbf{k}, \omega) = \frac{8\pi^2 q\ e^{-i\mathbf{k}\cdot\mathbf{r}_0}\delta(\omega - \mathbf{k}\cdot\mathbf{v}_0)}{k^2\varepsilon_L(\omega, \mathbf{k})}. \tag{11.23}$$

It is to be noted that, in this derivation, the dielectric function $\varepsilon_L(\mathbf{k}, \omega)$ can take the form obtained either from kinetic equation or fluid equation.

Alternatively, let us use the Poisson equation,

$$\nabla\cdot\mathbf{E}(\mathbf{r}, t) = 4\pi\sum_\alpha e_\alpha \int f_\alpha(\mathbf{r}, \mathbf{v}, t)\ d^3v + 4\pi\ \rho_{ext}(\mathbf{r}, t). \tag{11.24}$$

We assume that the influence of ρ_{ext} is weak enough that the response of the plasma is described by the linear equation,

$$(-i\omega + i\mathbf{k}\cdot\mathbf{v})f_\alpha(\mathbf{k}, \mathbf{v}, \omega) + \frac{e_\alpha}{m_\alpha}\mathbf{E}(\mathbf{k}, \omega)\cdot\frac{\partial f_{\alpha 0}}{\partial\mathbf{v}} = 0$$

$$\textit{or}\quad f_\alpha(\mathbf{k}, \omega) = -\phi(\mathbf{k}, \omega)\ \frac{e_\alpha}{m_\alpha}\ \frac{\mathbf{k}\cdot\frac{\partial f_{0\alpha}}{\partial\mathbf{v}}}{\omega - \mathbf{k}\cdot\mathbf{v}}. \tag{11.25}$$

Fourier transforming the Poisson equation (11.24) gives

$$\phi(\mathbf{k}, \omega) = \frac{8\pi^2 q\ e^{-i\mathbf{k}\cdot\mathbf{r}_0}\delta(\omega - \mathbf{k}\cdot\mathbf{v}_0)}{k^2\varepsilon_L(\omega, \mathbf{k})} \tag{11.26}$$

where ε_L is given by (4.23):

$$\varepsilon_L(\mathbf{k}, \omega) = 1 + \sum \frac{\omega_{p\alpha}^2}{k^2} \int d^3v\ \frac{\mathbf{k}\cdot\frac{\partial f_{\alpha 0}}{\partial \mathbf{v}}}{\omega - \mathbf{k}\cdot\mathbf{v}}. \tag{11.27}$$

To properly deal with the singularity in ε_L, a convenient way is to insert a small positive imaginary part in ω, in accordance with the causal requirement. Then, Fourier inversion of the above equation gives

$$\phi(\mathbf{r}, t) = \frac{q}{2\pi^2} \int d^3k \frac{e^{i\mathbf{k}\cdot(\mathbf{r}-\mathbf{r}_0-\mathbf{v}_0 t)}}{k^2\varepsilon_L(\mathbf{k}\cdot\mathbf{v}_0 + i\epsilon,\ \mathbf{k})}. \tag{11.28}$$

Dressed particle. If $\mathbf{v}_0 = 0$, we recover the static solution. The Debye-shielded potential in the forgoing discussion as well as Eq. (11.28) is the potential made by the 'external' charge plus its accompanying polarization cloud. The entity, the charge *and* the surrounding polarization cloud, was named the *dressed test particle* by Rostoker and Rosenbluth (1960). Now the individual particles in the Vlasov fluid are regarded as 'independent dressed particles'. Treating plasma as a collection of mutually non-interacting dressed particles facilitates explanation of some problems such as scattering and radiation. These problems require inclusion of interaction between plasma particles, and cannot be dealt with in the framework of the *collisionless* Vlasov equation. In fact, the Vlasov equation is the lowest order (in the small parameter $\epsilon = 1/\lambda_D^3 n$) equation for the distribution function, and the crudest. In the next higher order approximation, the dressed particle description is thought to be useful. Equation (11.26) will be used in calculating the fluctuating energy of the electric field in plasma.

11.4. Initial value problem formulation

• **Dynamic screening of plasma against a moving test charge; dressed particle.** In the last Section, we obtained the potential due to a moving test charge by considering the displacement vector **D**— (11.28). Here, we approach the same problem as an initial value problem via Fourier–Laplace transform. Assume that an external charge q_T is suddenly introduced into the plasma at position $\mathbf{r}_0$ with velocity $\mathbf{v}_0$. The subsequent position is given by $\mathbf{r} = \mathbf{r}_0 + \mathbf{v}_0 t$, and the charge density associated with the test particle is represented by $q_T\,\delta(\mathbf{r}-\mathbf{r}_0-\mathbf{v}_0 t)$. The potential made by plasma particles and the test charge is

$$\phi(\mathbf{r},t)=\int d^3r'\sum\frac{e_\alpha}{|\,\mathbf{r}-\mathbf{r}'\,|}\int d^3v f_\alpha(\mathbf{r}',\mathbf{v},t)+\frac{q_T}{|\,\mathbf{r}-\mathbf{r}_0-\mathbf{v}_0 t\,|},$$

$$\nabla\cdot\mathbf{E}=-\nabla^2\phi=4\pi\sum e_\alpha\int d^3v f_\alpha(\mathbf{r},\mathbf{v},t)+4\pi q_T\,\delta(\mathbf{r}-\mathbf{r}_0-\mathbf{v}_0 t). \tag{11.29}$$

In the above f_α is the perturbed distribution function (assuming that the perturbation excited by the introduction of the test charge is small) which satisfies the linearized equation

$$\frac{\partial f_\alpha}{\partial t}+\mathbf{v}\cdot\frac{\partial f_\alpha}{\partial \mathbf{r}}+\frac{e_\alpha}{m_\alpha}\mathbf{E}\cdot\frac{\partial f_{\alpha 0}}{\partial \mathbf{v}}=0 \tag{11.30}$$

with the initial condition

$$f_\alpha(\mathbf{r},\mathbf{v},0)=0. \tag{11.31}$$

(11.31) means that the test charge is suddenly introduced into the plasma at $t=0$. Fourier–Laplace transform of (11.29) gives

$$k^2\phi(\mathbf{k},\omega)=4\pi\sum e_\alpha\int d^3v f_\alpha(\mathbf{k},\mathbf{v},\omega)+4\pi i q_T\,\frac{e^{-i\mathbf{k}\cdot\mathbf{r}_0}}{\omega-\mathbf{k}\cdot\mathbf{v}_0}. \tag{11.32}$$

Fourier–Laplace transform of (11.30) with the initial condition (11.31) gives

$$f_\alpha(\mathbf{k},\mathbf{v},\omega)=-\frac{e_\alpha}{m_\alpha}\,\phi(\mathbf{k},\omega)\frac{\mathbf{k}\cdot\frac{\partial f_{\alpha 0}}{\partial \mathbf{v}}}{\omega-\mathbf{k}\cdot\mathbf{v}}. \tag{11.33}$$

Using (11.33) in (11.32) yields

$$k^2\phi(\mathbf{k},\omega)=4\pi i\;q_T\,\frac{e^{-i\mathbf{k}\cdot\mathbf{r}_0}}{\varepsilon(\mathbf{k},\omega)(\omega-\mathbf{k}\cdot\mathbf{v}_0)} \tag{11.34}$$

$$\textit{where}\quad \varepsilon(\mathbf{k},\omega)=1+\sum_\alpha\frac{\omega_{p\alpha}^2}{k^2}\int d^3v\frac{\mathbf{k}\cdot\frac{\partial f_{\alpha 0}}{\partial \mathbf{v}}}{\omega-\mathbf{k}\cdot\mathbf{v}} \tag{11.35}$$

is the plasma dielectric constant. (11.34) should be compared with (11.23). We see that $\frac{1}{\omega-\mathbf{k}\cdot\mathbf{v}_0}$ in the Laplace formulation in (11.34) corresponds to $\delta(\omega-\mathbf{k}\cdot\mathbf{v}_0)$ in the Fourier formulation. After the inversion in t, the two agrees. The correspondence between the two is readily recognized by invoking the Plemelj formula in (11.34) with neglect of the singularity in ε. Laplace-inversion of (11.34) is written as

$$\phi(\mathbf{k},t)=\int_L\frac{d\omega}{2\pi}e^{-i\omega t}\,\frac{4\pi i q_T\;e^{-i\mathbf{k}\cdot\mathbf{r}_0}}{k^2\varepsilon(\mathbf{k},\omega)(\omega-\mathbf{k}\cdot\mathbf{v}_0)}. \tag{11.36}$$

When $q_T = 0$, in order for (11.36) to be valid we must have $\varepsilon(\mathbf{k},\omega) = 0$ which gives the normal modes of the perturbed plasma. Here in the inversion integral of (11.36), note that we have two kinds of singularities: one kind corresponding to ω's which are the roots of $\varepsilon(\mathbf{k},\omega) = 0$, and the other corresponding to $\omega = \mathbf{k}\cdot\mathbf{v}_0$. We take advantage of the fact that the normal modes are Landau-damped [the associated singularities are in the lower-half ω-plane]. Thus the singularity on the real ω-axis, $\omega = \mathbf{k}\cdot\mathbf{v}_0$, which is uppermost, determines the long time behavior of $\phi(\mathbf{k},t)$. So we asymptotically write

$$\phi(\mathbf{k},t) = 4\pi q_T \frac{e^{-i\mathbf{k}\cdot(\mathbf{r}_0+\mathbf{v}_0 t)}}{k^2\varepsilon(\mathbf{k},\omega)\,|_{\omega=\mathbf{k}\cdot\mathbf{v}_0}}. \tag{11.37}$$

This equation, the main result, gives the potential screened by the moving point charge. (11.37) gives the electric field

$$\mathbf{E}(\mathbf{k},t) = -i\mathbf{k}\phi(\mathbf{k},t) = -4\pi i q_T \frac{\mathbf{k}\ e^{-i\mathbf{k}\cdot(\mathbf{r}_0+\mathbf{v}_0 t)}}{k^2\varepsilon(\mathbf{k},\omega)\,|_{\omega=\mathbf{k}\cdot\mathbf{v}_0}}. \tag{11.38}$$

Inverting (11.38) yields

$$\mathbf{E}(\mathbf{r},t) = \int \frac{d^3k}{8\pi^3} e^{i\mathbf{k}\cdot\mathbf{r}}\mathbf{E}(\mathbf{k},t) = -\frac{iq_T}{2\pi^2}\int d^3k \frac{\mathbf{k}\ e^{i\mathbf{k}\cdot\mathbf{r}}e^{-i\mathbf{k}\cdot(\mathbf{r}_0+\mathbf{v}_0 t)}}{k^2\varepsilon(\mathbf{k},\omega)\,|_{\omega=\mathbf{k}\cdot\mathbf{v}_0}}. \tag{11.39}$$

(11.39) gives the electric field at point $\mathbf{r}$ due to the plasma *and* the test particle whose orbit is $\mathbf{r}_0 + \mathbf{v}_0 t$. If we add up the fields generated at the position $\mathbf{r}$ due to all the *test particles* which are distributed according to a distribution $f_0(\mathbf{r}_0, \mathbf{v}_0)$, we have the ensemble-averaged electric field

$$\langle \mathbf{E}(\mathbf{r},t)\rangle = \int d^3v_0 \int d^3r_0\ \mathbf{E}(\mathbf{r},t;\mathbf{r}_0) f_0(\mathbf{r}_0,\mathbf{v}_0). \tag{11.40}$$

In a uniform and isotropic plasma, there is no favored direction and $f(\mathbf{r}_0,\mathbf{v}_0)$ takes the form of isotropic $f_0(\mathbf{v}_0)$, so (11.40) is expected to vanish. This can be proven by substituting (11.39) into (11.40) and using a Maxwellian distribution for $f_0(\mathbf{v}_0)$:

$$\langle \mathbf{E}(\mathbf{r},t)\rangle = \int d^3v_0 f_0(v_0) \int d^3r_0 \left(-\frac{iq_T}{2\pi^2}\right) \int d^3k \frac{\mathbf{k}\ e^{i\mathbf{k}\cdot\mathbf{r}}e^{-i\mathbf{k}\cdot(\mathbf{r}_0+\mathbf{v}_0 t)}}{k^2\varepsilon(\mathbf{k},\omega)\,|_{\omega=\mathbf{k}\cdot\mathbf{v}_0}}. \tag{11.41}$$

In the above, integral $\int d^3r_0$ yields $\delta(\mathbf{k})$, and $\varepsilon(\mathbf{k},\omega)$ can be evaluated by its static form, $\varepsilon(\mathbf{k},\omega=0) = 1 + 1/k^2\lambda_D^2$. Thus it is immediately seen that (11.41) vanishes upon $\int d^3k$ integration.

• **Fluctuating electric field energy.** Next, we consider the average of E^2, the fluctuating electric field energy of the plasma. It can be calculated in the same fashion as done in (11.40) by taking the ensemble average:

$$W = \frac{1}{8\pi}\langle \mathbf{E}(\mathbf{r},t)\cdot\mathbf{E}(\mathbf{r},t)\rangle = \frac{1}{8\pi}\langle \mathbf{E}(\mathbf{r},t)\cdot\mathbf{E}^*(\mathbf{r},t)\rangle$$

$$= \frac{1}{8\pi}\int d^3v_0 f_0(\mathbf{v}_0)\int d^3r_0 (\frac{q_T}{2\pi^2})^2 \int d^3k \, \frac{\mathbf{k}e^{i\mathbf{k}\cdot\mathbf{r}}e^{-i\mathbf{k}\cdot\mathbf{r}_0}e^{-i\mathbf{k}\cdot\mathbf{v}_0 t}}{k^2\varepsilon(\mathbf{k},\omega)\mid_{\omega=\mathbf{k}\cdot\mathbf{v}_0}}$$

$$\cdot\int d^3k' \frac{\mathbf{k}'\, e^{-i\mathbf{k}'\cdot\mathbf{r}}e^{i\mathbf{k}'\cdot\mathbf{r}_0}e^{i\mathbf{k}'\cdot\mathbf{v}_0 t}}{k'^2\varepsilon^*(\mathbf{k}',\omega)\mid_{\omega=\mathbf{k}'\cdot\mathbf{v}_0}} \tag{11.42}$$

where we used $\mathbf{E}^* = \mathbf{E}$ because $\mathbf{E}(\mathbf{r},t)$ is a real quantity. The integral $\int d^3r_0$ yields $(2\pi)^3\delta(\mathbf{k}-\mathbf{k}')$, and (11.42) becomes upon $\int d^3k'$ integration

$$W = \frac{q_T^2}{4\pi^2}\int d^3v_0 f_0(\mathbf{v}_0)\int \frac{d^3k}{k^2\mid\varepsilon(\mathbf{k},\omega)\mid^2_{\omega=\mathbf{k}\cdot\mathbf{v}_0}}. \tag{11.43}$$

For every value of $\mathbf{k}$, ε is only function of the velocity component $u = \hat{\mathbf{k}}\cdot\mathbf{v}_0 = \frac{\omega}{k}$. (11.43) can be integrated immediately with respect to two velocity components perpendicular to $\mathbf{k}$, giving

$$W = \frac{q_T^2}{4\pi^2}\int_{-\infty}^{\infty} du\; g_0(u)\int \frac{d^3k}{k^2\mid\varepsilon(k,\omega)\mid^2_{\omega=ku}} \tag{11.44}$$

where g_0 is the one-dimensional Maxwellian. Upon change of variable $u \to \omega/k$, (11.44) becomes

$$W = \frac{Nq_T^2}{2\pi}\int_{-\infty}^{\infty}\frac{d\omega}{2\pi}\int d^3k\, \frac{g_0(\frac{\omega}{k})}{k^3\mid\varepsilon(k,\omega)\mid^2}. \tag{11.45}$$

Now every plasma particle is considered as a dressed test particle in the above equation. So the plasma under consideration is regarded as an aggregate of the dressed test particles. In this way, we obtain a more detailed description of a plasma which is absent from the fluid picture of Vlasov description. The functional value $g_0(\frac{\omega}{k})$ can be connected to the imaginary part of ε, i.e. $Im\ \varepsilon$. For a Maxwellian plasma, $Im\ \varepsilon$ is determined by the slope dg_0/du at $u = \omega/k$. we have $g_0(u) = \sqrt{\frac{m}{2\pi T}}e^{-\frac{mu^2}{2T}}$, $dg_0/du = -mug_0(u)/T$ and thus

$$Im\ \varepsilon(k,\omega) = -\pi\,\frac{\omega_{pe}^2}{k^2}\left[\frac{dg_0}{du}\right]_{u=\frac{\omega}{k}} = \pi\frac{\omega_{pe}^2}{k^2}\frac{m}{T}\frac{\omega}{k}\, g_0\left(\frac{\omega}{k}\right) \tag{11.46}$$

where $\omega_{pe}^2 = 4\pi Nq_T^2/m$.

$$Therefore,\quad W = \frac{1}{8\pi}\langle E^2(\mathbf{r},t)\rangle = T\int\frac{d^3k}{(2\pi)^3}\int_{-\infty}^{\infty}\frac{d\omega}{2\pi}\,\frac{Im\ \varepsilon}{\omega\mid\varepsilon(k,\omega)\mid^2}. \tag{11.47}$$

It is noteworthy that the electric field energy is expressed in terms of $Im\ \varepsilon$, the dissipation part of the dielectric function.

(11.47) can be put into the form

$$\langle E^2(\mathbf{r},t)\rangle = \int \frac{d^3k}{(2\pi)^3}\int_{-\infty}^{\infty}\frac{d\omega}{2\pi}\langle |\ E(\mathbf{k},\omega)\ |^2\rangle \tag{11.48}$$

$$with \quad \langle |\ E(\mathbf{k},\omega)\ |^2\rangle = \frac{8\pi\ T\ Im\ \varepsilon}{\omega\ |\ \varepsilon(k,\omega)\ |^2} = -\frac{8\pi T}{\omega}\ Im\Big(\frac{1}{\varepsilon(\mathbf{k},\omega)}\Big). \tag{11.49}$$

(11.48) and (11.49) express the fluctuation-dissipation theorem for the electric field fluctuation. (11.49) is the *spectral density* of the electric field fluctuation. The fluctuation-dissipation theorem will be derived from the Liouville equation in the sequel.

Summary of this section: The fluctuations of the plasma electric field can be calculated by regarding the plasma particles to be the non-interacting dressed particles. The distribution function f_α and the electromagnetic field quantities all can be determined in terms of ρ_{ext}, an ensemble of non-interacting dressed particles. Use (11.23) in (11.25) to get

$$f_\alpha(\mathbf{k},\omega) = -4\pi\ \frac{e_\alpha}{m_\alpha}\ \frac{1}{k^2\varepsilon_L(\mathbf{k},\omega)}\ \frac{\mathbf{k}\cdot\frac{df_{\alpha 0}}{d\mathbf{v}}}{\omega-\mathbf{k}\cdot\mathbf{v}}\rho_{ext}(\mathbf{k},\omega).$$

This equation agrees with Eq. (6.76) in Reference: A. G. Sitenko, *Fluctuations and non-linear wave interactions in plasmas* Pergamon Oxford (1982), upon identifying Sitenko's $\rho^0_{k,\omega} = \rho_{ext}(\mathbf{k},\omega)$ here. See also Reference [N. A. Krall and A. W. Trivelpiece, *Principles of plasma physics* McGraw-Hill (1973) Eq. (11.1.11)]. We briefly mention that Sitenko's zero order ensemble is the extension of (11.19), extended over all the plasma particles. His solution of the Klimontovich kinetic equation (Chapter 11, later) perfectly agrees with the picture that plasma is aggregate of non-interacting dressed particles. Also we have the relation, $\mathbf{E}(\mathbf{k},\omega) = -4\pi i\mathbf{k}\rho_{ext}/k^2\varepsilon_L$, and it is easy to obtain

$$\rho(\mathbf{k},\omega) = \frac{i}{4\pi}\mathbf{k}\cdot\mathbf{E} = \frac{\rho_{ext}(\mathbf{k},\omega)}{\varepsilon_L(\mathbf{k},\omega)}.$$

In order to calculate the spectral density $\langle |\ \rho(\mathbf{k},\omega)\ |^2\rangle$, we can proceed along the line that we followed to obtain (11.49), that is, inverting $\rho(\mathbf{k},\omega)$ for $\rho(\mathbf{r},t)$ and ensemble averaging $\rho^2(\mathbf{r},t)$. Instead, we here show another way of computing the spectral density. Using (11.20) in the above equation, we can write

$$\begin{aligned}
&\langle\rho(\mathbf{k},\omega)\rho^*(\mathbf{k}',\omega')\rangle \\
&= \frac{4\pi^2 q^2}{\varepsilon_L(\mathbf{k},\omega)\varepsilon_L^*(\mathbf{k}',\omega')}\int d^3r_0\int d^3v_0 f_0(v_0)e^{-i(\mathbf{k}-\mathbf{k}')\cdot\mathbf{r}_0} \\
&\quad\times\delta(\omega-\mathbf{k}\cdot\mathbf{v}_0)\delta(\omega'-\mathbf{k}'\cdot\mathbf{v}_0) \\
&= \frac{(2\pi)^3 4\pi^2 q^2}{\varepsilon_L(\mathbf{k},\omega)\varepsilon_L^*(\mathbf{k}',\omega')}\delta(\mathbf{k}-\mathbf{k}')\int d^3v_0 f_0(v_0)\delta(\omega-\mathbf{k}\cdot\mathbf{v}_0)\delta(\omega'-\mathbf{k}'\cdot\mathbf{v}_0) \\
&= \frac{(2\pi)^3 4\pi^2 q^2}{\varepsilon_L(\mathbf{k},\omega)\varepsilon_L^*(\mathbf{k}',\omega')}\delta(\mathbf{k}-\mathbf{k}')\delta(\omega-\omega')\int d^3v_0 f_0(v_0)\delta(\omega-\mathbf{k}\cdot\mathbf{v}_0).
\end{aligned}$$

Now we put $\mathbf{k}=\mathbf{k}'$ and $\omega=\omega'$ in the above equation. In doing so, note that $2\pi\delta(\omega-\omega')\to 1$ and $(2\pi)^3\delta(\mathbf{k}-\mathbf{k}')\to 1$ [see the comment at the end of § 11.8]. Then we obtain

$$\langle|\,\rho(\mathbf{k},\omega)\,|^2\rangle = \frac{2\pi q^2}{|\,\varepsilon_L(\mathbf{k},\omega)\,|^2}\int du g_0(u)\delta(\omega-ku) = \frac{Tk^2}{2\pi\omega}\frac{Im\ \varepsilon_L(\mathbf{k},\omega)}{|\,\varepsilon_L(\mathbf{k},\omega)\,|^2}.$$

This result can be obtained more rapidly, putting the relation, $\rho = i\mathbf{k}\cdot\mathbf{E}/4\pi$ in (11.49). Using the relation $i\omega\mathbf{E}=4\pi\mathbf{J}$ in (11.49) gives the spectral density of current fluctuation,

$$\langle|\,\mathbf{J}(\mathbf{k},\omega)\,|^2\rangle = \frac{T\omega}{2\pi}\ Im\left(\frac{1}{\varepsilon_L(\mathbf{k},\omega)}\right).$$

Now it is clear that we can write

$$\langle|\,\mathbf{E}(\mathbf{k},\omega)\,|^2\rangle = \frac{16\pi^2}{k^2}\frac{\langle|\,\rho_{ext}(\mathbf{k},\omega)\,|^2\rangle}{|\,\varepsilon_L\,|^2},$$

$$\langle|\,\rho(\mathbf{k},\omega)\,|^2\rangle = \frac{\langle|\,\rho_{ext}(\mathbf{k},\omega)\,|^2\rangle}{|\,\varepsilon_L\,|^2}.$$

Repeating the algebra performed in the last page with (11.20), we obtain

$$\langle|\,\rho_{ext}(\mathbf{k},\omega)\,|^2\rangle = 2\pi q^2\int du g_0(u)\delta(\omega-ku) = \frac{Tk^2}{2\pi\omega}\ Im\ \varepsilon_L(\mathbf{k},\omega).$$

The preceding development indicates that the ensemble average in calculating various fluctuations is performed over the initial distribution of the *non-interacting* dressed particles.

Exercise. Prove by direct calculation

$$I \equiv \int dv_x \int dv_z g_0(v_x) g_0(v_z) \delta(\omega - k_x v_x - k_z v_z) = \int du g_0(u) \delta(\omega - ku)$$

where g_0 is one-dimensional Maxwellian and $k = \sqrt{k_x^2 + k_z^2}$.

Hint: $\delta(\omega - k_x v_x - k_z v_z) = \delta(v_x - \frac{\omega - k_z v_z}{k_x}) / \mid k_x \mid$. And carry out the integral

$$I = \frac{m}{2\pi T} exp \left[\frac{m}{2T}\frac{\omega^2}{k^2}\right] \int_{-\infty}^{\infty} dv_z exp \left[-\frac{m}{2T}\frac{k^2}{k_x^2}\left(v_z - \frac{\omega k_z}{k^2}\right)^2\right].$$

This is a proof of the equality for three-dimensional Maxwellian $f_0(\mathbf{v}_0)$

$$\int\int\int f_0(\mathbf{v}_0)\delta(\omega - \mathbf{k}\cdot\mathbf{v}_0)d^3v_0 = \int g_0(u)\delta(\omega - ku)du,$$

$$k = \sqrt{k_x^2 + k_y^2 + k_z^2}.$$

(11.43)–(11.45) can be dealt with the above equality.

11.5. Statistical evaluation of fluctuating variables

• **Random variable.** A random variable $V(t)$, although considered to be varying according to the sequence of time t, is not a definite function of t but is assigned a value only in probabilistic sense: the value $V(t)$ can be ascertained only by definite probability. For example, the '*fine grained*' electric field in a plasma is a random variable which undergoes irregular fluctuations. The fluctuating fields are of statistical nature whose exact details are not of great concern to us; but we are interested in statistical average of certain relevant quantities such as energy density or energy flux. These quantities are quadratic functions of the fluctuating electric and magnetic fields. The statistical average is obtained by taking average over the ensemble. Ensemble can be constructed from the long time-sequence data by cutting the sequence into equal time segments and collecting them; each portion is a member of the ensemble. This way of visualizing ensemble is, in hindsight, no more and no less than rephrasing of ergodicity which states that the time average is equal to the ensemble average.

• **Correlation.** It is often desired to know the degree of the dependence of one random variable $x(t)$ upon another random variable $y(t)$. One way to estimate a possible dependence between these two random variables is to plot

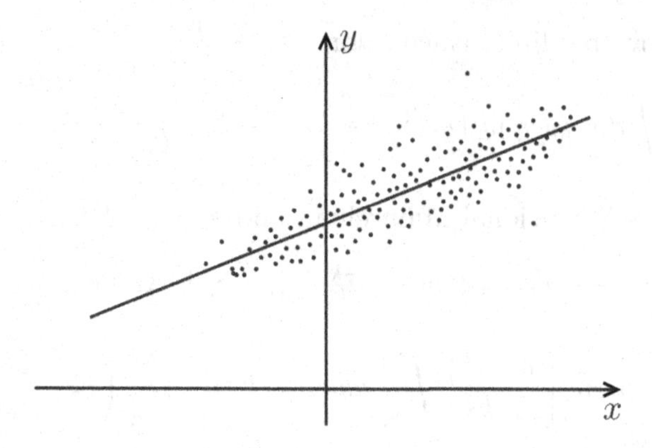

Fig. 11.1. A scatter diagram. Two random variables x and y are correlated around the straight line.

the measurements of these random variables as points in the (x, y) plane and look at the resultant dots(data). Such a plot is known as *scatter diagram* which might have the representation shown in Fig. 11.1. If the two random variables were not dependent upon each other, the dots would be more or less scattered uniformly over the (x, y) plane.

• **Correlation function.** Let $x(t)$ be a random variable. The *auto-correlation* of the random process $x(t)$ is defined by

$$C_x = \langle x(t)x(t')\rangle.$$

The bracket $\langle\,\rangle$ signifies time average or ensemble average. The two averages yield the same result owing to the ergodicity. The average would be a measure how much x at one time is correlated with x at another time. The homogeneity of time means that the average depends only on the difference $t - t' = \tau$ (stationary random process). So the average can be written as, in a stationary random process,

$$C_x = \langle x(t)x(t')\rangle = C_x(t-t') = C_x(\tau) = \lim_{T\to\infty}\frac{1}{T}\int_{-\frac{T}{2}}^{\frac{T}{2}} x(t)x(t-\tau)dt. \tag{11.50}$$

• **Wiener–Khintchine theorem.** We express a real random variable $V(t)$ by a Fourier integral:

$$V(t) = \frac{1}{2\pi}\int_{-\infty}^{\infty} V(\omega)e^{-i\omega t}\,d\omega. \tag{11.51}$$

Since $V(t)$ is real, we have

$$V(t) = V^*(t) = \frac{1}{2\pi}\int_{-\infty}^{\infty} V^*(\omega)e^{i\omega t}\, d\omega$$
$$= \frac{1}{2\pi}\int_{-\infty}^{\infty} V^*(-\omega)e^{-i\omega t}\, d\omega \quad (per\ \omega \to -\omega). \tag{11.52}$$

Comparing (11.51) and (11.52), $V(\omega) = V^*(-\omega)$. Therefore $V(\omega)V^*(\omega) = V^*(-\omega)V(-\omega)$ and $\mid V(\omega) \mid^2$ is an even function of ω. We define the time average as

$$\langle V^2(t)\rangle = \lim_{T\to\infty} \frac{1}{T}\int_{-\frac{T}{2}}^{\frac{T}{2}} dt\; V^2(t). \tag{11.53}$$

Using (11.51) and (11.52) in (11.53) gives

$$\langle V^2(t)\rangle = \lim_{T\to\infty} \frac{1}{T}\int_{-\frac{T}{2}}^{\frac{T}{2}} dt\; \left(\frac{1}{2\pi}\int V(\omega)e^{-i\omega t}d\omega\right)\left(\frac{1}{2\pi}\int V^*(\omega')e^{i\omega' t}d\omega'\right).$$

The time-integral gives $2\pi\delta(\omega - \omega')$ and we have

$$\langle V^2(t)\rangle = \lim_{T\to\infty} \frac{1}{2\pi T}\int_{-\infty}^{\infty} d\omega V(\omega)\int_{-\infty}^{\infty} d\omega' V^*(\omega')\delta(\omega - \omega')$$
$$= \lim_{T\to\infty} \frac{1}{2\pi T}\int_{-\infty}^{\infty} d\omega \mid V(\omega) \mid^2 . \tag{11.54}$$

We define *spectral power density*,

$$G(\omega) = \lim_{T\to\infty} \frac{\mid V(\omega) \mid^2}{2\pi T}. \tag{11.55}$$

Then (11.54) is written as

$$\langle V^2(t)\rangle = \int_{-\infty}^{\infty} G(\omega)d\omega. \tag{11.56}$$

When the quantity $\mid V(\omega) \mid^2 /T$ does not approach a definite limit as $T \to \infty$ but fluctuates with increasing T, we take the ensemble average of both sides of (11.52) without changing the result with the understanding that $G(\omega)$ now is the ensemble-averaged quantity (denoted by over-bar)

$$G(\omega) = \lim_{T\to\infty} \frac{\overline{|V(\omega)|^2}}{2\pi T}. \tag{11.57}$$

We simply state that the time average is equivalent to the ensemble average (ergodicity), and the over-bar is not necessary on the left side of (11.56).

The *auto-correlation function* for the random process $V(t)$ is defined by

$$C(\tau) = \lim_{T\to\infty} \frac{1}{T} \int_{-\frac{T}{2}}^{\frac{T}{2}} dt\, V(t)V(t-\tau). \tag{11.58}$$

The correlation function provides available information about a random process. Using (11.51) in (11.58) gives

$$C(\tau) = \lim_{T\to\infty} \frac{1}{T} \int_{-\frac{T}{2}}^{\frac{T}{2}} dt \left(\frac{1}{2\pi}\int V(\omega)e^{-i\omega t}d\omega\right)\left(\frac{1}{2\pi}\int V(\omega')e^{-i\omega'(t-\tau)}d\omega'\right).$$

$\int dt$-integral gives $2\pi\delta(\omega+\omega')$, and so we obtain

$$C(\tau) = \lim_{T\to\infty} \frac{1}{2\pi T} \int_{-\infty}^{\infty} |V(\omega)|^2 \; e^{-i\omega\tau} d\omega$$

where we used the relation $V(\omega) = V^*(-\omega)$. We may take the ensemble average of this equation, obtaining with (11.57)

$$C(\tau) = \int_{-\infty}^{\infty} G(\omega)e^{-i\omega\tau}d\omega, \tag{11.59}$$

$$G(\omega) = \frac{1}{2\pi}\int_{-\infty}^{\infty} C(\tau)e^{i\omega\tau}d\tau. \tag{11.60}$$

The above two equations are Wiener–Khintchine theorem: the auto-correlation function and the spectral density are a Fourier transform pair. By integrating (11.60) we recover the relation $\int G(\omega)d\omega = \langle V^2(t)\rangle$.

Remark. In our Fourier transform convention, it would be more friendly to write (11.53) and (11.54) as

$$G(\omega) = \lim_{T\to\infty} \frac{|V(\omega)|^2}{T}, \quad \langle V^2(t)\rangle = \int_{-\infty}^{\infty} G(\omega)\frac{d\omega}{2\pi}.$$

This convention will be followed in the sequel.

• **Auto-correlation of fluctuating electric field** is defined by

$$C_E(\tau) = \langle E(t)E(t-\tau)\rangle = \lim_{T\to\infty}\frac{1}{T}\int_{-\frac{T}{2}}^{\frac{T}{2}} dt\ E(t)E(t-\tau)$$

$$= \lim_{T\to\infty}\frac{1}{T}\int_{-\infty}^{\infty} \mid E(\omega)\mid^2\ e^{-i\omega\tau}\frac{d\omega}{2\pi}. \tag{11.61}$$

The *spectral power density* of the fluctuating electric field is defined by

$$C_E(\omega) = \lim_{T\to\infty}\frac{1}{T}\mid E(\omega)\mid^2\ \textit{so that}\ \ C_E(\tau) = \int_{-\infty}^{\infty} C_E(\omega)e^{-i\omega\tau}\frac{d\omega}{2\pi}. \tag{11.62}$$

Inverting (11.62), we obtain the following relation

$$C_E(\omega) = \int_{-\infty}^{\infty} C_E(\tau)e^{i\omega\tau}d\tau.$$

Exercise. Carry out the integral below to verify the first relation of (11.62).

$$C_E(\omega) = \int_{-\infty}^{\infty} d\tau e^{i\omega\tau}\langle E(t)E(t-\tau)\rangle.$$

$$Note:\quad C_E(\tau=0) = \langle E^2(t)\rangle = \int_{-\infty}^{\infty} C_E(\omega)\frac{d\omega}{2\pi}.$$

Integrating the spectral power density over the frequency gives the electric field energy within numerical factor.

11.6. Cerenkov radiation

Equation (11.39) gives the electric field generated in the plasma by an electron whose initial velocity is $\mathbf{v}_0$, in corporation with the rest of the particles in the plasma. The electric field is specifically identified by the parameter $\mathbf{v}_0$ (initial velocity of the dressed particle). Cerenkov radiation is emitted when the Landau resonance condition $\omega = \mathbf{k}\cdot\mathbf{v}$ is met, and is known as the inverse process of the Landau damping. Here we show that the rate of the kinetic energy increase of all the (dressed) particles balances the Landau damping rate of the medium energy T (temperature):

$$\int d^3v_0 f(\mathbf{v}_0)\frac{d}{dt}\frac{1}{2}mv^2 = -e\int d^3v_0 f(\mathbf{v}_0)\mathbf{E}\cdot\mathbf{v}_0 = \gamma T. \tag{11.63}$$

Here we consider only the electron emission assuming that the ions are immobile. In (11.63), γ is the Landau damping rate, $f(\mathbf{v}_0)$ is the initial electron distribution. We have, by using (11.39),

$$-e\mathbf{E}\cdot\mathbf{v}_0 = -\frac{ie^2}{2\pi^2}\int d^3k \frac{\mathbf{k}\cdot\mathbf{v}_0\ e^{i\mathbf{k}\cdot\mathbf{r}}e^{-i\mathbf{k}\cdot\mathbf{r}_0}e^{-i\mathbf{k}\cdot\mathbf{v}_0 t}}{k^2\varepsilon(\mathbf{k},\omega)\mid_{\omega=\mathbf{k}\cdot\mathbf{v}_0}}. \tag{11.64}$$

The integrand in (11.64) without the exponential phasor is the Fourier amplitude in the **k**-space of the quantity $(-e\mathbf{E}\cdot\mathbf{v}_0)$ which is the time rate of the kinetic energy increase. So the ensemble average (or time average) of this quantity is the sum of all the Fourier amplitudes over the **k**-space. Thus we have to prove the following equation:

$$\langle -e\mathbf{E}\cdot\mathbf{v}_0\rangle = \frac{e^2}{2\pi^2}\int d^3v_0 f(\mathbf{v}_0)\int d^3k \frac{-i\ \mathbf{k}\cdot\mathbf{v}_0}{k^2\varepsilon(\mathbf{k},\omega)\mid_{\omega=\mathbf{k}\cdot\mathbf{v}_0}} = \int \frac{d^3k}{(2\pi)^3}\gamma T. \tag{11.65}$$

Clearly, the $\int d^3k$-integral above should be real, and we can show that

$$I = \int d^3k \frac{-i\ \mathbf{k}\cdot\mathbf{v}_0}{k^2\varepsilon(\mathbf{k},\mathbf{k}\cdot\mathbf{v}_0)} = \int d^3k\ \frac{\mathbf{k}\cdot\mathbf{v}_0}{k^2} Im\ \frac{1}{\varepsilon(\mathbf{k},\mathbf{k}\cdot\mathbf{v}_0)}. \tag{11.66}$$

To prove this, let us consider

$$I = \int_{-\infty}^{\infty} dk \frac{-ikv}{\varepsilon(k,kv)} = \int_{-\infty}^{\infty} dk \frac{ikv}{\varepsilon(-k,-kv)} = \int_{-\infty}^{\infty} dk \frac{ikv}{\varepsilon^*(k,kv)}. \tag{11.67}$$

The second expression is obtained via change of variable, $k \to -k$. The third expression is due to the well-known property: $\varepsilon(-k,-\omega) = \varepsilon^*(k,\omega)$. This property ensures that the Fourier transform $\varepsilon(x,t)$ is real. Furthermore it follows that $\varepsilon_r(k)$ (real part) is an even function and $\varepsilon_i(k)$ (imaginary part) is an odd function of k. It follows from (11.67) that $I = I^*$, and therefore I is a real quantity, and (11.66) is obtained. We have

$$Im\ \frac{1}{\varepsilon(k,\omega)} = \frac{-\varepsilon_i(k,\omega)}{\mid\varepsilon(k,\omega)\mid^2} = \frac{-\varepsilon_i(\mathbf{k},\omega)}{\varepsilon_r^2(\mathbf{k},\omega)+\varepsilon_i^2(\mathbf{k},\omega)}. \tag{11.68}$$

Let us expand $\varepsilon_r(\mathbf{k},\omega)$ in the vicinity of $\omega = \omega_r$, the root of $\varepsilon_r(\mathbf{k},\omega) = 0$:

$$\varepsilon_r(\mathbf{k},\omega) \doteq \frac{\partial\varepsilon_r(\mathbf{k},\omega_r)}{\partial\omega_r}(\omega-\omega_r).$$

The Cerenkov emission condition is $\omega_r = \mathbf{k}\cdot\mathbf{v}_0$ which is also the Landau resonance condition. Also introducing the Landau damping rate $\gamma = \varepsilon_i/\frac{\partial\varepsilon_r}{\partial\omega_r}$,

(11.68) becomes, for $\gamma \ll \omega_r$,

$$Im \frac{1}{\varepsilon(k,\omega)} = -\left(\frac{\partial \varepsilon_r(\mathbf{k},\omega_r)}{\partial \omega_r}\right)^{-1} \frac{\gamma}{(\omega-\omega_r)^2+\gamma^2}$$
$$= -\left(\frac{\partial \varepsilon_r(\mathbf{k},\omega_r)}{\partial \omega_r}\right)^{-1} \pi\delta(\omega-\omega_r). \tag{11.69}$$

Therefore (11.65) is written as

$$\langle -e\mathbf{E}\cdot\mathbf{v}_0\rangle = \frac{e^2}{2\pi^2}\int d^3v_0 f(\mathbf{v}_0)\int d^3k \frac{\mathbf{k}\cdot\mathbf{v}_0}{k^2}\left(\frac{\partial \varepsilon_r(\mathbf{k},\omega_r)}{\partial \omega_r}\right)^{-1} \pi\delta(\mathbf{k}\cdot\mathbf{v}_0-\omega_r)$$
$$= \frac{e^2}{2\pi}\int d^3k \, \frac{\omega_r}{k^2}\frac{\varepsilon_i(\mathbf{k},\omega_r)}{\frac{\partial\varepsilon_r(\mathbf{k},\omega_r)}{\partial\omega_r}} \frac{\int d^3v_0 f(\mathbf{v}_0)\delta(\mathbf{k}\cdot\mathbf{v}_0-\omega_r)}{\varepsilon_i(\mathbf{k},\omega_r)}$$
$$= \frac{e^2}{2\pi}\int d^3k \, \gamma \, \frac{\omega_r}{k^2} \, \frac{\frac{1}{k}g_0(\frac{\omega_r}{k})}{\pi \, \frac{\omega_{pe}^2}{k^2}\left[\frac{dg_0(v)}{dv}\right]_{\frac{\omega_r}{k}}} = \int \frac{d^3k}{(2\pi)^3}\gamma T \tag{11.70}$$

where we used (11.46).

For this subsection, the Reference is: K. Nishikawa and M. Wakatani, *Plasma Physics* Springer-Verlag, Heidelberg (1990).

• **Usefulness of Equation (11.34).** (11.34) gives the fluctuating electric potential due to a test charge moving in a plasma. The plasma is regarded as an aggregate of the test particles, distributed over the following distribution (*Klimontovich distribution*; see Chapter 12)

$$F(\mathbf{r},\mathbf{v},t=0) = \sum_{i=1}^{N}\delta(\mathbf{r}-\mathbf{r}_{i0})\delta(\mathbf{v}-\mathbf{v}_{i0})$$

with the subscript 0 denoting the initial vales. Therefore (11.34) can be used to obtain the fluctuating electric field of the plasma by summing over the initial distribution, assuming that the (test) particles are non-interacting. We obtain

$$\delta E(\mathbf{k},\omega) = \frac{\mathbf{k}}{k^2}\frac{4\pi e}{\varepsilon(k,\omega)}\sum_{i=1}^{N}\frac{e^{-i\mathbf{k}\cdot\mathbf{r}_{i0}}}{\omega-\mathbf{k}\cdot\mathbf{v}_{i0}} \tag{11.71}$$

which can be used to calculate the fluctuating electric energy of a plasma. See Chapter 12.

• **A useful formula.** As an application of (11.69), let us write $\varepsilon = \varepsilon' + i\varepsilon''$ $(\varepsilon'' \ll \varepsilon')$. Then, with the aid of (11.69), we have

$$\frac{1}{\varepsilon(\omega,k)} = \lim_{\varepsilon''\to 0} \frac{1}{\varepsilon' + i\varepsilon''} = P\,\frac{1}{\varepsilon'(k,\omega_r)} - i\pi\,\frac{\delta(\omega-\omega_r)}{\partial\varepsilon'/\partial\omega_r} \qquad (A)$$

where $\varepsilon(\omega,k) = 0$ at $\omega = \omega_r$. This formula is useful for further development of Eq. (10.68), Chapter 10. Resonant wave-wave interaction can be developed to an analytic expression by picking up the contribution from those terms on the RHS of (10.68) that satisfy $\varepsilon^{(1)*}(\mathbf{k}' + \mathbf{k}'', \omega_{k'} + \omega_{k''}) = 0$ and $\varepsilon^{(1)}(\mathbf{k} - \mathbf{k}', \omega_k - \omega_{k'}) = 0$. Treating those terms in the manner of (A), authors set up wave kinetic equation which governs the evolution of the wave action. Because of the δ-function in (A), the resonant terms become explicit and play the role of source in the wave action kinetic equation. The wave kinetic equation developed from (10.68) can be seen in Soviet literature. Reference for wave kinetic equation: Camac, M., Kantrowitz, A. R., Litvak, M. M., Patrick, R. M., and Petschek, H. E. (1962). *Shock waves in collision-free plasmas, Nuc. Fusion Supplement*, **Pt. 2**, p. 423–444.

11.7. Fluctuation-dissipation theorem

References: G. Bekefi, *Radiation processes in plasmas*, Wiley New York (1966); R. Kubo, *Statistical-mechanical theory of irreversible processes I. General theory and simple applications to magnetic and conduction problems* J. Phys. Soc. Japan **12**, p. 570-586 (1957); P. P. J. M. Schram, *Kinetic theory of gases and plasmas* Kluwer Academic Dordrecht (1991).

Beginning remark. The important equation in the first half of this subsection is (11.97) below, which is derived by characteristic method. In the literature, the perturbed Liouville distribution is commonly expressed by the exponential operator; to change the operator to ordinary integration a considerable effort must be expended. (11.97) is convenient because it is already in the integral form, and physically transparent.

Let us perturb an isolated system in thermal equilibrium whose Hamiltonian is H_0. The perturbation is caused by an external '*force*' $F(t)$ applied to the system. This will give rise to a small change in the energy

$$H_1 = -A(\Gamma, 0)F(t). \qquad (11.72)$$

Recall that a Hamiltonian is the energy expressed in terms of the phase space coordinates. For example, $F(t)$ may be an external force perturbing the system and $A(\Gamma)$ is the response of the system resulted in by the action of $F(t)$ on the system. $A(\Gamma)$ is the physical quantity conjugate to the force F.

The symbol Γ in $A(\Gamma)$ represents the coordinate in the Γ–space which consists of the position and momentum, $\mathbf{x}_i$ and $\mathbf{p}_i$, $i = 1, 2, \cdots N$ (N is the number of particles). F may be considered to be an electric field acting on the medium that gives rise to a current and A can be identified as some quantity connected to the current. Note that both A and F are real. Because A refers to a response function, its value at t depends on the values of F at $t' < t$. This is causality and is expressed by writing

$$A(t) = \int_{-\infty}^{t} \chi(t-\tau)F(\tau)d\tau = \int_{0}^{\infty} \chi(t')F(t-t')dt' \tag{11.73}$$

where $\chi(t)$ is the susceptibility which is a function of the medium properties. Time-dependence of $A(\Gamma, t)$ is developed by the action of $F(t)$, in response of the medium. The rate of change of energy of the system is

$$\frac{dW}{dt} = \frac{\partial H_1}{\partial t} = -A\,\frac{dF}{dt}. \tag{11.74}$$

We require two kinds of average of dW/dt: the time average and the ensemble average. In evaluating the time average, $A(\Gamma)$ should be read as an explicit function of t to write

$$\left\langle \frac{dW}{dt} \right\rangle_{T_0} = -\lim_{T_0 \to \infty} \frac{1}{T_0} \int_{-\frac{T_0}{2}}^{\frac{T_0}{2}} A(t)\frac{dF}{dt}dt \tag{11.75}$$

where the time interval T_0 is long enough so that the system at each time can be reckoned as a member of the ensemble of the system. The integral is usually evaluated from $-\infty$ to $+\infty$. Fourier transforming (11.73) gives

$$A(\omega) = \chi(\omega)F(\omega), \tag{11.76}$$

$$\chi(\omega) = \int_{0}^{\infty} \chi(t)e^{i\omega t}dt. \tag{11.77}$$

Since $\chi(t)$ is real, $\chi(-\omega) = \chi^*(\omega)$. Using the Fourier inversion expressions for $A(t)$ and dF/dt and taking the real part of the integral, we obtain

$$\left\langle \frac{dW}{dt} \right\rangle_{T_0} = \frac{1}{2\pi T_0} \int_{-\infty}^{\infty} d\omega\omega \mid F(\omega) \mid^2 \; Im\ \chi(\omega). \tag{11.78}$$

On the other hand, we calculate the ensemble average of dW/dt. Then the desired fluctuation-dissipation theorem follows by equating the two results (by ergodicity). To obtain the ensemble average, we need the Liouville distribution function $D(\Gamma, t)$ whose behavior is described by the Liouville equation,

$$\frac{\partial D(\Gamma,t)}{\partial t} + [D,H] = 0, \tag{11.79}$$

$$where \quad [D,H] = \frac{\partial D}{\partial x_i}\frac{\partial H}{\partial p_i} - \frac{\partial H}{\partial x_i}\frac{\partial D}{\partial p_i} \tag{11.80}$$

is the Poisson bracket, and we used the summation convention, and H is the total Hamiltonian. It is convenient to introduce the Liouville operator

$$L = \frac{\partial H}{\partial p_i}\frac{\partial}{\partial x_i} - \frac{\partial H}{\partial x_i}\frac{\partial}{\partial p_i}. \tag{11.81}$$

Then, $[D,H] = LD$, and (11.79) can be written as

$$\frac{\partial D(\Gamma,t)}{\partial t} + LD = 0. \tag{11.82}$$

In equilibrium, the Hamiltonian is independent of t, and represented by $H_0(\Gamma)$. The corresponding thermodynamic equilibrium state with temperature T has the Liouville density

$$D_0 = \frac{1}{Z}\, e^{-H_0/T} \quad \left(Z = \int d\Gamma e^{-H_0/T}\right). \tag{11.83}$$

When external agencies are applied to the system, the Hamiltonian is perturbed and we write

$$H(\Gamma,t) = H_0(\Gamma) + H_1(\Gamma,t). \tag{11.84}$$

Correspondingly, we have

$$D(\Gamma,t) = D_0(\Gamma) + D_1(\Gamma,t), \tag{11.85}$$

$$L(\Gamma,t) = L_0(\Gamma) + L_1(\Gamma,t), \tag{11.86}$$

where

$$L_0 = \frac{\partial H_0}{\partial p_i}\frac{\partial}{\partial x_i} - \frac{\partial H_0}{\partial x_i}\frac{\partial}{\partial p_i}, \tag{11.87}$$

$$L_1 = \frac{\partial H_1}{\partial p_i}\frac{\partial}{\partial x_i} - \frac{\partial H_1}{\partial x_i}\frac{\partial}{\partial p_i}. \tag{11.88}$$

Then (11.82) is linearized in the form

$$\frac{\partial D_1}{\partial t} + L_0 D_1 = -L_1 D_0. \tag{11.89}$$

Historically, (11.89) was solved by treating the operator L_0 as an algebraic quantity. Then, a formal solution of (11.89) is obtained in terms of the exponential operator $e^{iL_0 t}$. See Reference: R. Kubo, J. Phys. Soc. Japan **12**, 570 (1957), where a quantum mechanical derivation of the fluctuation-dissipation theorem is also treated. Even earlier, the operator method was employed in the kinetic theory literature of chemical physics. Instead, we use the characteristic method to derive a simple integral which fully determines D_1 in terms of the source term $L_1 D_0$. Using Hamilton's equations,

$$\frac{\partial H_0}{\partial p_i} = \dot{x}_i, \quad \frac{\partial H_0}{\partial x_i} = -\dot{p}_i, \tag{11.90}$$

the operator L_0 becomes

$$L_0 = \dot{x}_i \frac{\partial}{\partial x_i} + \dot{p}_i \frac{\partial}{\partial p_i}. \tag{11.91}$$

(11.72) and (11.88) give

$$L_1 = F(t)\left(-\frac{\partial A}{\partial p_i}\frac{\partial}{\partial x_i} + \frac{\partial A}{\partial x_i}\frac{\partial}{\partial p_i}\right). \tag{11.92}$$

Using (11.83) in the above equation yields

$$L_1 D_0 = -\frac{1}{T} D_0(\Gamma) F(t) \left(\dot{x}_i \frac{\partial A}{\partial x_i} + \dot{p}_i \frac{\partial A}{\partial p_i}\right). \tag{11.93}$$

So the perturbed Liouville equation (11.89) takes the form

$$\frac{\partial D_1}{\partial t} + \left(\dot{x}_i \frac{\partial}{\partial x_i} + \dot{p}_i \frac{\partial}{\partial p_i}\right) D_1 = \frac{1}{T} D_0(\Gamma) F(t) \left(\dot{x}_i \frac{\partial}{\partial x_i} + \dot{p}_i \frac{\partial}{\partial p_i}\right) A(\Gamma). \tag{11.94}$$

This equation can be solved by characteristic method. The characteristic equations are:

$$\frac{dt}{1} = \frac{dx_i}{\dot{x}_i} = \frac{dp_i}{\dot{p}_i} = \frac{dD_1}{\frac{1}{T} D_0 F(t) \left(\dot{x}_i \frac{\partial}{\partial x_i} + \dot{p}_i \frac{\partial}{\partial p_i}\right) A(\Gamma)}. \tag{11.95}$$

$$\textit{Therefore,} \quad \frac{dD_1}{dt} = \frac{1}{T} D_0 F(t) \left(\dot{x}_i \frac{\partial}{\partial x_i} + \dot{p}_i \frac{\partial}{\partial p_i}\right) A(\Gamma) = \frac{1}{T} D_0 F(t) \, \frac{dA}{dt}. \tag{11.96}$$

(11.96) is the Lagrangian representation (as opposed to the Eulerian representation) of the time rate of the change of D_1. The development presented here is analogous to the integration along the unperturbed orbit of Vlasov equation. The characteristic equations (11.95) ascertain that $\dot{x}_i = dx_i/dt$ and $\dot{p}_i = dp_i/dt$. Integrating (11.96) yields

$$D_1(t) = \frac{1}{T} D_0(\Gamma) \int_{-\infty}^{t} F(t') \frac{dA}{dt'} dt'. \tag{11.97}$$

This is a self-standing solution of (11.89) (as contrasted with the operator solution). Now, the ensemble average of dW/dt is evaluated by

$$\begin{aligned} \left\langle \frac{dW}{dt} \right\rangle_{D_1} &= -\int d\Gamma \; D_1 \; A(\Gamma, 0) \; \frac{dF}{dt} \\ &= -\frac{1}{T} \int d\Gamma D_0 A(\Gamma, 0) \; \frac{dF(t)}{dt} \int_{-\infty}^{t} F(t') \frac{dA(\Gamma, t')}{dt'} dt'. \end{aligned} \tag{11.98}$$

Let us introduce the *correlation function*

$$C(t) = \int d\Gamma D_0(\Gamma) A(\Gamma, 0) A(\Gamma, t). \tag{11.99}$$

Then (11.98) becomes

$$\left\langle \frac{dW}{dt} \right\rangle_{D_1} = -\frac{1}{T} \frac{dF}{dt} \int_{-\infty}^{t} F(t') \frac{dC(t')}{dt'} dt'. \tag{11.100}$$

Take time average of the above equation:

$$\left\langle \frac{dW}{dt} \right\rangle_{D_1} = -\frac{1}{TT_0} \int_{-\infty}^{\infty} dt \; \frac{dF}{dt} \int_{-\infty}^{t} F(t') \frac{dC(t')}{dt'} dt' \tag{11.101}$$

where $T_0 \to \infty$ is implied. Change variable, $t' \to t - \tau$:

$$\left\langle \frac{dW}{dt} \right\rangle_{D_1} = \frac{1}{TT_0} \int_{-\infty}^{\infty} dt \; \frac{dF}{dt} \; I(t) \tag{11.102}$$

$$\text{where} \quad I(t) = \int_{0}^{\infty} d\tau F(t - \tau) \frac{dC(t - \tau)}{d\tau}. \tag{11.103}$$

In the above, we use $C(t - \tau) = C(-\tau)$ (the correlation function is independent of the origin of t in a homogeneous plasma) and put the inverse

Fourier relations for $F(t)$ and $C(\tau)$. Then, we obtain

$$I(t) = \int_0^\infty d\tau \int_{-\infty}^\infty \frac{d\omega_2}{2\pi} F(\omega_2) e^{-i\omega_2(t-\tau)} \int_{-\infty}^\infty i\,\omega_1 \frac{d\omega_1}{2\pi} C(\omega_1) e^{i\omega_1\tau}$$

$$= -\int_{-\infty}^\infty \frac{d\omega_1}{2\pi} \omega_1 C(\omega_1) \int_{-\infty}^\infty \frac{d\omega_2}{2\pi} \frac{e^{-i\omega_2 t}}{\omega_1 + \omega_2} F(\omega_2). \tag{11.104}$$

Using (11.104) in (11.102) yields

$$\left\langle \frac{dW}{dt} \right\rangle_{D_1} = \frac{i}{TT_0} \int_{-\infty}^\infty \frac{d\omega}{2\pi}\, \omega \mid F(\omega) \mid^2 \xi(\omega) \tag{11.105}$$

$$where \quad \xi(\omega) = \int_{-\infty}^\infty \frac{d\omega_1}{2\pi}\, \omega_1\, \frac{C(\omega_1)}{\omega_1 - \omega}$$

$$= \frac{i}{2}\, \omega\, C(\omega) + \frac{1}{2\pi}\, P \int \frac{\omega_1\, C(\omega_1)}{\omega_1 - \omega} d\omega_1. \tag{11.106}$$

Note that $C(t)$ is real and has the symmetry, $C(t) = C(-t)$. Then, $C(\omega)$ is real. Equating the real part of (11.105) to (11.78) gives (justified by the ergodic hypothesis)

$$C_A(\omega) = \frac{2T}{\omega}\, Im\, \chi(\omega). \tag{11.107}$$

This equation is the fluctuation-dissipation theorem. Recall that we have introduced $C(\omega)$ as the Fourier transform,

$$C_A(\omega) = \int_{-\infty}^\infty C(\tau) e^{i\omega\tau} d\tau. \tag{11.108}$$

Integrating over $d\omega$ gives

$$C_A(\tau = 0) = \int_{-\infty}^\infty C_A(\omega) \frac{d\omega}{2\pi}. \tag{11.109}$$

From the definition of the correlation function,

$$C_A(\tau) = \int d\Gamma D_0(\Gamma) A(\Gamma, 0) A(\Gamma, \tau) = \langle A(t) A(t+\tau) \rangle \tag{11.110}$$

where the average is taken over the ensemble or over the time ($\frac{1}{T_0} \int_{-\infty}^\infty dt(\cdots)$), we can express the fluctuation-dissipation theorem in the form

$$\langle A^2(t) \rangle = C_A(\tau = 0) = \int_{-\infty}^\infty C_A(\omega)\, \frac{d\omega}{2\pi} = \frac{T}{\pi} \int_{-\infty}^\infty d\omega\, \frac{Im\, \chi(\omega)}{\omega}. \tag{11.111}$$

In literature, $C(\omega)$, which is the Fourier transform of the correlation function $C(\tau)$, is called the *spectral power density* (Wiener–Khintchine theorem). It is also useful to express the theorem in (11.111) in terms of the Fourier amplitude $A(\omega)$:

$$\begin{aligned}\langle A^2(t)\rangle &= \langle A(t)A^*(t)\rangle \\ &= \frac{1}{4\pi^2}\int_{-\infty}^{\infty} d\omega \int_{-\infty}^{\infty} d\omega' \langle A(\omega)A^*(\omega')\rangle\, e^{-i(\omega-\omega')t}\end{aligned} \tag{11.112}$$

where the bracket means the ensemble average. Comparison of (11.111) and (11.112) leads to the relation

$$\langle A(\omega)A^*(\omega')\rangle = 2\pi\delta(\omega-\omega')C_A(\omega). \tag{11.113}$$

In particular, we have by putting $\omega = \omega'$ in (11.113)

$$\langle \mid A(\omega) \mid^2\rangle = C_A(\omega). \tag{11.114}$$

This equation gives the physical meaning of the spectral density in terms of the ensemble average of $\mid A(\omega) \mid^2$. Note that $2\pi\delta(\omega-\omega') \to 1$ when we put $\omega = \omega'$. This matter will be touched upon again in the sequel. Integrating (11.114) over $d\omega$ and using (11.109) and (11.111) yield

$$\langle A^2(t)\rangle = \int_{-\infty}^{\infty} \frac{d\omega}{2\pi}\langle \mid A(\omega) \mid^2\rangle.$$

The Wiener–Khintchine theorem relates to essentially the same physical content in terms of the time average. In either case of time average or ensemble average, the important point is that the correlation function and the spectral density are a Fourier transform pair.

Now it remains to identify the functions A, F in the Hamiltonian with the electromagnetic fields. This identification will be automatically taken care of if we derive the fluctuation-dissipation theorem in the alternative way that we do in the next section. Here we write the following correspondences

$$F \to E_0, \quad i\omega A(\omega) \to J(\omega), \quad \chi(\omega) \to \frac{\sigma_{ext}}{\omega}$$

where E_0 is the external electric field, σ_{ext} is given by later equation (11.132). Then (11.107) and (11.114) give the current spectral density in the form

$$\langle \mid J(\omega) \mid^2\rangle = \frac{T\omega}{2\pi}\, Im\left(\frac{1}{\varepsilon}\right).$$

11.8. Alternative derivation

In the preceding section, the fluctuation-dissipation theorem was derived by using the ergodic hypothesis that the ensemble average is equivalent to the time average. In this Section, the theorem will be derived more directly, without the assistance of the ergodic hypothesis, by ensemble averaging the electrodynamic equations. The following development is owed to Schram [P. P. J. M. Schram, *Kinetic theory of gases and plasmas* (1991) Kluwer Academic Dordrecht]. Let us consider the ensemble average over D_1 of the time derivative of a phase function which we denote by $\dot{A} = \frac{dA}{dt}$:

$$\langle \dot{A} \rangle_1 = \int d\Gamma \dot{A}(\Gamma, 0) D_1(\Gamma, t). \tag{11.115}$$

Schram used Kubo's operator solution but the simple solution, (11.97), is used here to write the above equation in the form

$$\langle \dot{A} \rangle_1 = \frac{1}{T} \int_0^\infty d\tau F(t-\tau) \int d\Gamma D_0(\Gamma) \dot{A}(\Gamma, 0) \dot{A}(\Gamma, t-\tau). \tag{11.116}$$

The integral $\int d\Gamma(\cdots)$ is the correlation function of the quantity $\dot{A}(\Gamma)$ which we express as

$$C_{\dot{A}}(t-\tau) = \int d\Gamma D_0(\Gamma) \dot{A}(\Gamma, 0) \dot{A}(\Gamma, t-\tau) = \langle \dot{A}(\Gamma, 0) \dot{A}(\Gamma, t-\tau) \rangle_0. \tag{11.117}$$

Since we have $C(t-\tau) = C(-\tau) = C(\tau)$, (11.116) can be written as

$$\langle \dot{A} \rangle_1 = \frac{1}{T} \int_0^\infty d\tau F(t-\tau) \langle \dot{A}(\Gamma, 0) \dot{A}(\Gamma, \tau) \rangle_0 \tag{11.118}$$

where $\langle \ \rangle_1$ and $\langle \ \rangle_0$ respectively denote the ensemble average over the perturbed and equilibrium Liouville distribution.

In order to obtain the current correlation function, we consider the charge conservation equation

$$\frac{\partial \rho}{\partial t} + \nabla \cdot \mathbf{J} = 0 \tag{11.119}$$

from which Schram identified as

$$A(\Gamma) = -\rho(\Gamma, \mathbf{r}), \quad \dot{A} = \nabla \cdot \mathbf{J}(\Gamma, \mathbf{r}),$$
$$F(t) = \phi_0(t) \ (\textit{external potential}) \tag{11.120}$$

$$\langle \nabla \cdot \mathbf{J}(\mathbf{r}, \Gamma) \rangle_1 = \frac{1}{T} \int_0^\infty d\tau \int d^3 r' \phi_0(\mathbf{r}', t-\tau) \langle \nabla' \cdot \mathbf{J}(\mathbf{r}', 0, \Gamma) \nabla \cdot \mathbf{J}(\mathbf{r}, \tau, \Gamma) \rangle_0. \tag{11.121}$$

Integrate by parts with respect to r' to obtain

$$\langle \nabla \cdot \mathbf{J}(\mathbf{r}) \rangle_1 = \frac{-1}{T} \int_0^\infty d\tau \int d^3 r' \frac{\partial \phi_0(\mathbf{r}')}{\partial \mathbf{r}'} \cdot \langle \mathbf{J}(\mathbf{r}') \nabla \cdot \mathbf{J}(\mathbf{r}, \tau) \rangle_0. \tag{11.122}$$

$$\textit{Therefore,} \quad \langle \mathbf{J}(\mathbf{r}) \rangle_1 = \frac{1}{T} \int_0^\infty d\tau \int d^3 r' \mathbf{E}_0(\mathbf{r}') \cdot \langle \mathbf{J}(\mathbf{r}') \mathbf{J}(\mathbf{r}, \tau) \rangle_0 \tag{11.123}$$

where the dot product is performed between the r'-variables, and $\mathbf{E}_0 = -\nabla \phi_0$. We introduce the current correlation function:

$$C_{ij}(\mathbf{r}', \mathbf{r}, \tau) = \langle J_i(\mathbf{r}', 0, \Gamma) J_j(\mathbf{r}, \tau, \Gamma) \rangle_0. \tag{11.124}$$

Then, (11.123) can be written as

$$\langle J_j(\mathbf{r}) \rangle_1 = \frac{1}{T} \int_0^\infty d\tau \int d^3 r' E_{0i}(\mathbf{r}', t - \tau)\, C_{ij}(\mathbf{r}', \mathbf{r}, \tau). \tag{11.125}$$

Next, we Fourier transform the above equation:

$$\langle J_j(\mathbf{k}, \omega) \rangle_1 = \frac{1}{T} \int_\infty^\infty d^3 r \int_{-\infty}^\infty dt\, e^{-i\mathbf{k}\cdot\mathbf{r} + i\omega t}$$
$$\times \int d^3 r' \int_0^\infty d\tau E_{0i}(\mathbf{r}', t - \tau)\, C_{ij}(\mathbf{r}', \mathbf{r}, \tau).$$

Substitute in the above equation the corresponding Fourier expressions:

$$E_0(\mathbf{r}', t - \tau) = \int \frac{d^3 k_2}{(2\pi)^3} \int \frac{d\omega_2}{2\pi} E_0(\mathbf{k}_2, \omega_2) e^{i\mathbf{k}_2 \cdot \mathbf{r}'} e^{-i\omega_2 (t - \tau)},$$

$$C(\mathbf{r}, \mathbf{r}', \tau) = C(\mathbf{r} - \mathbf{r}', \tau) = \int \frac{d^3 k_1}{(2\pi)^3} \int \frac{d\omega_1}{2\pi} C(\mathbf{k}_1, \omega_1) e^{i\mathbf{k}_1 \cdot (\mathbf{r} - \mathbf{r}')} e^{-i\omega_1 \tau}.$$

Then, we obtain

$$\langle J_j(\mathbf{k}, \omega) \rangle_1 = \frac{1}{T} \int_0^\infty d\tau \int_{-\infty}^\infty \frac{d\omega_1}{2\pi} e^{i(\omega - \omega_1)\tau} E_{0i}(\mathbf{k}, \omega)\, C_{ij}(\mathbf{k}, \omega_1).$$

Carrying out $\int d\tau$-integral yields

$$\langle J_j(\mathbf{k}, \omega) \rangle_1 = \frac{1}{T} \frac{1}{2\pi i} \int_{-\infty}^\infty d\omega_1 \frac{C_{ij}(\mathbf{k}, \omega_1)}{\omega_1 - \omega} E_{0i}(\mathbf{k}, \omega) \equiv \sigma_{ij}(\mathbf{k}, \omega) E_{0i}(\mathbf{k}, \omega) \tag{11.126}$$

$$\textit{where} \quad \sigma_{ij} = \frac{1}{T} \frac{1}{2\pi i} \int_{-\infty}^\infty d\omega_1 \frac{C_{ij}(\mathbf{k}, \omega_1)}{\omega_1 - \omega}. \tag{11.127}$$

In (11.126), we connected the plasma current with the external electric field E_0. So the conductivity σ_{ij} in (11.127) is appropriately designated as *external conductivity.* In electrostatic interaction, the correlation function and the conductivity become scalar quantities and we write

$$\langle J_L(k,\omega)\rangle_1 = \sigma_{ext}(k,\omega)E_0(k,\omega) \tag{11.128}$$

$$\sigma_{ext}(k,\omega) = \frac{1}{T}\frac{1}{2\pi i}\int_{-\infty}^{\infty} d\omega' \frac{C(k,\omega')}{\omega'-\omega}. \tag{11.129}$$

The subscript L means *longitudinal.* The plasma current J_L can be represented in two ways: (11.128) and

$$\langle J_L\rangle = \sigma E \tag{11.130}$$

where σ is related with the dielectric function ε by $\varepsilon = 1 + \frac{4\pi i}{\omega}\sigma$. The electrostatic Maxwell equation takes the form

$$-i\omega(E - E_0) + 4\pi\langle J\rangle = 0. \tag{11.131}$$

Note that $(E - E_0)$ is *total minus external* electric field which is produced by plasma current J_L. The foregoing equations, (11.128)–(11.131), give the relation

$$\sigma_{ext} = \frac{i\omega}{4\pi}\left(\frac{1}{\varepsilon} - 1\right). \tag{11.132}$$

Then the real part of (11.129) is obtained

$$Re\,[\,\sigma_{ext}(k,\omega)] = -\frac{\omega}{4\pi}\,Im\left[\frac{1}{\varepsilon(k,\omega)}\right]. \tag{11.133}$$

The real part of RHS of (11.129) is contributed from the δ-function part of the integral, the half-residue at $\omega' = \omega$ (while the imaginary part is given by the principal part of the integral). Noting that $C(k,\omega')$ itself is real, we have

$$Re\,[\,RHS\ of\ (11.129)\,] = \frac{1}{2T}C(k,\omega). \tag{11.134}$$

Equating (11.133) and (11.134) gives

$$C(k,\omega) = -\frac{T\omega}{2\pi}\,Im\left(\frac{1}{\varepsilon(k,\omega)}\right) \equiv C_J(k,\omega). \tag{11.135}$$

This is the *fluctuation-dissipation theorem* for the current spectral density (Fourier transform of $C(r,\tau)$).

Next we consider the current correlation function in Fourier space. With the aid of the Fourier expression,

$$J_L(k,\omega) = \int dx' \int dt' J_L(x',t')e^{-ikx'+i\omega t'},$$

and its complex conjugate (note that $J_L(x,t)$ itself is real)

$$J_L^*(k',\omega') = \int dx \int dt J_L(x,t)e^{ik'x-i\omega' t},$$

let us consider the correlation function

$$\langle J_L(k,\omega)J_L^*(k',\omega')\rangle_0 = \int dx \int dt \int dx' \times \int dt' e^{-ikx'+i\omega t'} e^{ik'x-i\omega' t}\langle J_L(x,t)J_L(x',t')\rangle_0.$$

Introduce new variables $\tau = t' - t, \quad s = x' - x$ to change $t' \to \tau \quad x' \to s$ to write

$$\langle J_L(k,\omega)J_L^*(k',\omega')\rangle_0 = \int dx \int dt \int ds \int d\tau e^{-i(k-k')x+i(\omega-\omega')t} \times \langle J_L(x,t)J_L(x+s,t+\tau)\rangle_0 e^{-iks+i\omega\tau}. \tag{11.136}$$

The correlation function in the real space, $\langle J_L(x,t)J_L(x+s,t+\tau)\rangle_0$ is independent of the origin (x,t), that is, $\langle J_L(x,t)J_L(x+s,t+\tau)\rangle_0 = \langle J_L(0,0)J_L(s,\tau)\rangle_0 = C(s,\tau)$, because of the uniformity of the equilibrium. Thus, we arrive at the expression

$$\langle J_L(k,\omega)J_L^*(k',\omega')\rangle_0 = (2\pi)^2\delta(k-k')\delta(\omega-\omega')C_J(k,\omega). \tag{11.137}$$

Putting $k = k'$ and $\omega = \omega'$ in (11.137) gives

$$\langle | J_L(k,\omega) |^2\rangle_0 = C_J(k,\omega). \tag{11.138}$$

From the charge conservation equation,

$$\frac{\partial\rho}{\partial t} + \nabla\cdot\mathbf{J} = 0 \quad or \quad \rho(k,\omega) = \frac{k}{\omega}J(k,\omega)$$

we obtain the charge density correlation function

$$C_\rho(k,\omega) = \frac{k^2}{\omega^2}C_J(k,\omega) = -\frac{Tk^2}{2\pi\omega}\, Im\left(\frac{1}{\varepsilon(k,\omega)}\right). \tag{11.139}$$

The Poisson equation, $\nabla^2\phi = -4\pi\rho$ or $\phi(k,\omega) = \frac{4\pi}{k\omega}J(k,\omega)$ gives

$$C_\phi(k,\omega) = \frac{(4\pi)^2}{k^2\omega^2}C_J(k,\omega) = -\frac{8\pi T}{k^2\omega}\, Im\left(\frac{1}{\varepsilon(k,\omega)}\right) \tag{11.140}$$

which is the electric potential correlation function. The Maxwell equation, $i\omega E = 4\pi J$ gives

$$C_E(k,\omega) = -\frac{16\pi^2}{\omega^2} C_J(k,\omega) = 8\pi\ \frac{T}{\omega}\ Im\left(\frac{1}{\varepsilon(k,\omega)}\right). \tag{11.141}$$

The correlation functions in the Fourier space (spectral density) is often denoted as, for example, $C_E(k,\omega) = \langle | E(k,\omega) |^2 \rangle$. It is instructive to derive (11.138) directly; not via (11.136) and (11.137). We have

$$\begin{aligned}
&\langle J_L(k,\omega) J_L^*(k,\omega)\rangle \\
&\quad = \int dx \int dt \int dx' \int dt' e^{-ikx'+i\omega t'} e^{ikx-i\omega t} \langle J_L(x,t) J_L(x',t')\rangle \\
&\quad = \int ds \int d\tau \int dx \int dt \langle J_L(x,t) J_L(s+x,\tau+t)\rangle e^{-iks+i\omega\tau}.
\end{aligned}$$

The integral $\int dx \int dt \langle J_L(x,t) J_L(s+x,\tau+t)\rangle$ divided by length (or volume in three-dimension) and time interval T_0 is the correlation function $C_J(s,\tau)$. Therefore, (11.138) follows from the above equation. As mentioned earlier, $2\pi\delta(\omega-\omega')$ should be put to unity when $\omega \to \omega'$ in (11.137).

11.9. Electromagnetic fluctuation

The current associated with a dressed particle can be represented as

$$\mathbf{J}_{ext}(\mathbf{r},t) = q\mathbf{v}_0 \delta(\mathbf{r} - \mathbf{r}_0 - \mathbf{v}_0 t) \quad cf.\ (11.19). \tag{11.142}$$

The full electromagnetic Maxwell equations should be used with the above *external* current as the source term. Then, the electromagnetic fields determined in terms of $\mathbf{J}_{ext}$ will be ensemble-averaged over the distribution of dressed particles which is the equilibrium plasma distribution itself. We have

$$\nabla \times \mathbf{E} = -\frac{1}{c}\frac{\partial \mathbf{B}}{\partial t}, \tag{11.143}$$

$$\nabla \times \mathbf{B} = \frac{1}{c}\frac{\partial \mathbf{E}}{\partial t} + \frac{4\pi}{c}\sum_\alpha e_\alpha \int \mathbf{v} f_\alpha d^3v + \frac{4\pi}{c}\mathbf{J}_{ext}\,. \tag{11.144}$$

The plasma distribution function f_α obeys the Vlasov equation

$$\frac{\partial}{\partial t} f_\alpha(\mathbf{r},\mathbf{v},t) + \mathbf{v}\cdot\frac{\partial f_\alpha}{\partial \mathbf{r}} + \frac{e_\alpha}{m_\alpha}\left[\mathbf{E}(\mathbf{r},t) + \frac{\mathbf{v}}{c}\times\mathbf{B}(\mathbf{r},t)\right]\cdot\frac{\partial f_{\alpha 0}}{\partial \mathbf{v}} = 0. \tag{11.145}$$

After Fourier transforming the above equations and eliminating **B** we obtain

$$\mathbf{k}\times(\mathbf{k}\times\mathbf{E})+\frac{\omega^2}{c^2}\mathbf{E}=-\frac{4\pi i\omega}{c^2}(\mathbf{J}+\mathbf{J}_{ext}) \tag{11.146}$$

where

$$\mathbf{J}=-i\,E_j\sum_\alpha\frac{e_\alpha^2}{m_\alpha}\int d^3v\,\frac{\mathbf{v}\,\frac{\partial f_{\alpha 0}}{\partial v_j}}{\omega-\mathbf{k}\cdot\mathbf{v}} \tag{11.147}$$

where index j is summed over.

$$J_{ext}(\mathbf{k},\omega)=2\pi q\mathbf{v}_0e^{-i\mathbf{k}\cdot\mathbf{r}_0}\delta(\omega-\mathbf{k}\cdot\mathbf{v}_0). \tag{11.148}$$

(11.146) can be put into the form,

$$\left[n^2\left(\frac{k_ik_j}{k^2}-\delta_{ij}\right)+\varepsilon_{ij}\right]E_j=-\frac{4\pi i}{\omega}J_i^{ext} \tag{11.149}$$

where $n^2=\frac{c^2k^2}{\omega^2}$ is the refractive index, and ε_{ij} is the dielectric tensor:

$$\varepsilon_{ij}=\delta_{ij}+\sum\frac{\omega_{p\alpha}^2}{\omega}\int d^3v\,\frac{v_i\,\frac{\partial f_{\alpha 0}}{\partial v_j}}{\omega-\mathbf{k}\cdot\mathbf{v}}. \tag{11.150}$$

We have the following important relation for ε_{ij} which is valid if $f_{\alpha 0}$ is isotropic (see § 4.3):

$$\varepsilon_{ij}=\delta_{ij}\varepsilon_T+(\varepsilon_L-\varepsilon_T)\frac{k_ik_j}{k^2}, \tag{11.151}$$

$$\varepsilon_L=1+\Sigma\frac{\omega_{p\alpha}^2}{k^2}\int\frac{\mathbf{k}\cdot\frac{\partial f_{\alpha 0}}{\partial \mathbf{v}}}{\omega-\mathbf{k}\cdot\mathbf{v}}d^3v:\ \textit{longitudinal dielectric constant}, \tag{11.152}$$

$$\varepsilon_T=1-\Sigma\frac{\omega_{p\alpha}^2}{\omega}\int\frac{f_{\alpha 0}}{\omega-\mathbf{k}\cdot\mathbf{v}}d^3v:\ \textit{transverse dielectric constant}. \tag{11.153}$$

(11.149) can be inverted in the form

$$E_i(\mathbf{k},\omega)=-\frac{4\pi i}{\omega}\,G_{ij}(\mathbf{k},\omega)\,J_j^{ext}(\mathbf{k},\omega) \tag{11.154}$$

$$\textit{with}\quad G_{ij}=\frac{k_ik_j}{k^2}\frac{1}{\varepsilon_L}+\left(\delta_{ij}-\frac{k_ik_j}{k^2}\right)\frac{1}{\varepsilon_H},\quad(\varepsilon_H=\varepsilon_T-n^2). \tag{11.155}$$

We form the correlation spectral density

$$\langle E_i(\mathbf{k},\omega)E_j^*(\mathbf{k}',\omega')\rangle = \frac{16\pi^2}{\omega\omega'}\int d^3r_0 \int d^3v_0 f_0(v_0) \times G_{il}(\mathbf{k},\omega)J_l^{ext}(\mathbf{k},\omega)G_{jm}^*(\mathbf{k}',\omega')J_m^{ext*}(\mathbf{k}',\omega'). \tag{11.156}$$

Using (11.148) and carrying out the $\int d^3r_0$–integral, the RHS of the above equation becomes

$$\frac{16\pi^2}{\omega^2}(2\pi)^5 q^2 \delta(\mathbf{k}-\mathbf{k}')\delta(\omega-\omega')\frac{T\omega}{m\pi\omega_p^2}G_{il}(\mathbf{k},\omega)G_{jm}^*(\mathbf{k}',\omega')Im\ \varepsilon_{lm} \tag{11.157}$$

where we used

$$\int d^3v f_0(v)v_l v_m \delta(\omega - \mathbf{k}\cdot\mathbf{v}) = \frac{T\omega n_0}{m\pi\omega_p^2}\ Im\ \varepsilon_{lm} \tag{11.158}$$

which comes from the definition of ε_{ij} [see (11.148)]. (11.156) becomes, upon putting $\mathbf{k} = \mathbf{k}'$ and $\omega = \omega'$,

$$\begin{aligned}\langle E_i(\mathbf{k},\omega)E_j^*(\mathbf{k},\omega)\rangle \\ = \frac{8\pi T}{\omega}\left[\frac{k_ik_l}{k^2}\frac{1}{\varepsilon_L} + \left(\delta_{il} - \frac{k_ik_l}{k^2}\right)\frac{1}{\varepsilon_H}\right] \\ \times \left[\frac{k_jk_m}{k^2}\frac{1}{\varepsilon_L^*} + \left(\delta_{jm} - \frac{k_jk_m}{k^2}\right)\frac{1}{\varepsilon_H^*}\right] \\ \times \left[\delta_{lm} Im\ \varepsilon_T + \frac{k_lk_m}{k^2}(Im\ \varepsilon_L - Im\ \varepsilon_T)\right]\end{aligned} \tag{11.159}$$

where we used (11.151) for $Im\ \varepsilon_{lm}$. It is straightforward to carry out the above multiplication. At the expense of some algebra, we obtain the final result

$$\begin{aligned}\langle E_i(\mathbf{k},\omega)E_j^*(\mathbf{k},\omega)\rangle \\ = \frac{8\pi T}{\omega}\left[\frac{k_ik_j}{k^2}\frac{Im\ \varepsilon_L(\mathbf{k},\omega)}{|\ \varepsilon_L(\mathbf{k},\omega)\ |^2} + \left(\delta_{ij} - \frac{k_ik_j}{k^2}\right)\frac{Im\ \varepsilon_T(\mathbf{k},\omega)}{|\ \varepsilon_H(\mathbf{k},\omega)\ |^2}\right].\end{aligned} \tag{11.160}$$

[Cf. Eq. (10.3.2), A. F. Alexandrov, L. S. Bogdankevich, A. A. Rukhadze, *Principles of plasma electrodynamics*, Springer Berlin (1984)]

- **Electron density fluctuations in Maxwellian plasma.**

In § 11.4, we obtained the spectral density of the electron density fluctuation $\langle | \rho(\mathbf{k},\omega) |^2 \rangle$:

$$\langle | \rho(\mathbf{k},\omega) |^2 \rangle = \frac{Tk^2}{2\pi\omega} \frac{Im\ \varepsilon_L(\mathbf{k},\omega)}{| \varepsilon_L(\mathbf{k},\omega) |^2}. \tag{11.161}$$

This quantity determines the scattering cross section of the incoherent scattering of electromagnetic waves by plasma. Let us further calculate (11.161) for a Maxwellian plasma. Let us recall [(5.93) and (5.94)]

$$\varepsilon_L(\mathbf{k},\omega) = 1 + \frac{1}{k^2\lambda_e^2}\left[1 - \varphi(\zeta) + i\sqrt{\pi}\zeta e^{-\zeta^2} - \varphi(\mu\zeta) + i\sqrt{\pi}\ \mu\ \zeta e^{-\mu^2\zeta^2}\right], \tag{11.162}$$

$$\varepsilon_T(\mathbf{k},\omega) = 1 - \frac{\omega_{pe}^2}{\omega^2}[\ \varphi(\zeta) - i\sqrt{\pi}\zeta e^{-\zeta^2}], \tag{11.163}$$

where

$$\varphi(\zeta) = 2\zeta e^{-\zeta^2}\int_0^\zeta dx\ e^{x^2} = -\zeta Z(\zeta) + i\sqrt{\pi}\zeta e^{-\zeta^2} \tag{11.164}$$

with $Z(\zeta)$ the plasma dispersion function. In (11.162), $\mu = m_e/m_i$, and $\mu\zeta$ terms are the ion contributions which should be included in the low frequency region. Using (11.162) and (11.163) in (11.161) gives

$$\langle | \rho(\mathbf{k},\omega) |^2 \rangle = \sqrt{2\pi} n_0 \frac{\lambda_e^3 k^3}{\omega_{pe}} \frac{e^{-\zeta^2} + \mu e^{-\mu^2\zeta^2}}{[2 + \lambda_e^2 k^2 - \varphi(\zeta) - \varphi(\mu\zeta)]^2 + \pi\zeta^2[e^{-\zeta^2} + \mu e^{-\mu^2\zeta^2}]^2}. \tag{11.165}$$

The spectral density in (11.165) is plotted in Fig. 11.2: after Reference [A. Sitenko and V. Malnev, *Plasma physics theory*, Chapman and Hall London (1995) p. 270]; our λ_e and ζ correspond, respectively, to a and z in the reference.

Example. Derive the spectral density of the electric current fluctuation $\langle J_i(\mathbf{k},\omega) J_j^*(\mathbf{k},\omega)\rangle$. Solution First, (11.147) and (11.150) can verify (A) below:

$$J_i = \frac{-i\omega}{4\pi}(\varepsilon_{ij} - \delta_{ij})E_j = \frac{-i\omega}{4\pi}\left[(\varepsilon_T - 1)\delta_{ij} + (\varepsilon_L - \varepsilon_T)\frac{k_i k_j}{k^2}\right]E_j \quad (A)$$

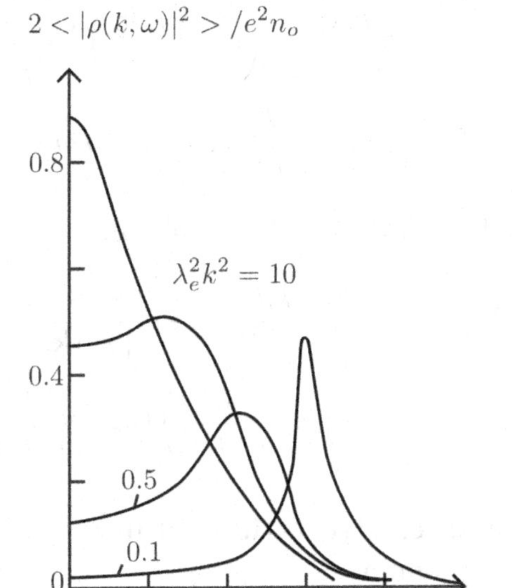

Fig. 11.2. The spectral distribution of the charge density. After A. Sitenko and V. Malnev, *Plasma physics theory* Chapman and Hall, London (1995) p. 270.

where we used (11.151). Using (11.154) in (A) gives

$$J_i = \left[(1-\varepsilon_T)\delta_{ij} + (\varepsilon_T - \varepsilon_L)\frac{k_i k_j}{k^2}\right] \; G_{jl} J_l^{ext} \equiv H_{jl} J_l^{ext}$$

where

$$H_{ij} = \frac{1-\varepsilon_L}{\varepsilon_L}\frac{k_i k_j}{k^2} + \left(\delta_{ij} - \frac{k_i k_j}{k^2}\right)\frac{1-\varepsilon_T}{\varepsilon_H}, \quad (\varepsilon_H = \varepsilon_T - n^2).$$

Analogously with (11.156) we calculate

$$\langle J_i(\mathbf{k},\omega) J_j^*(\mathbf{k}',\omega')\rangle$$

$$= \int d^3 r_0 \int d^3 v_0 f_0(v_0) H_{il}(\mathbf{k},\omega) J_l^{ext}(\mathbf{k},\omega) H_{jm}^*(\mathbf{k}',\omega') J_m^{ext*}(\mathbf{k}',\omega'). \quad (B)$$

The succeeding algebra goes parallel with (11.157) and (11.158), and we write the intermediate result

$$\langle J_i(\mathbf{k},\omega) J_j^*(\mathbf{k},\omega)\rangle$$

$$= \frac{\omega T}{2\pi}\left[\frac{1-\varepsilon_L}{\varepsilon_L}\frac{k_i k_j}{k^2} + \left(\delta_{ij} - \frac{k_i k_j}{k^2}\right)\frac{1-\varepsilon_T}{\varepsilon_H}\right]$$

$$\times \left[\frac{1 - \varepsilon_L^*}{\varepsilon_L^*} \frac{k_i k_j}{k^2} + \left(\delta_{ij} - \frac{k_i k_j}{k^2} \right) \frac{1 - \varepsilon_T^*}{\varepsilon_H^*} \right]$$

$$\times \left[\delta_{lm} Im \ \varepsilon_T + \frac{k_l k_m}{k^2} (Im \ \varepsilon_L - Im \ \varepsilon_T) \right]. \quad (C)$$

After straightforward algebra with (C) we obtain

$$\langle J_i(\mathbf{k}, \omega) J_j^*(\mathbf{k}, \omega) \rangle = \frac{\omega T}{2\pi} \left[\frac{k_i k_j}{k^2} Im \ \varepsilon_L \frac{| 1 - \varepsilon_L |^2}{| \varepsilon_L(\mathbf{k}, \omega) |^2} \right.$$
$$\left. + \left(\delta_{ij} - \frac{k_i k_j}{k^2} \right) Im \ \varepsilon_T \frac{| 1 - \varepsilon_T |^2}{| \varepsilon_H(\mathbf{k}, \omega) |^2} \right].$$

11.10. Fluctuation of electrostatic field in a magnetized plasma

The particle orbit of species α in a magnetized plasma with the static magnetic field $\mathbf{B}_0 = \hat{\mathbf{z}} B_0$ is described by

$$\mathbf{r}(t) = \mathbf{r}_0 + \mathbf{R}_\alpha(t, \mathbf{v}_0), \tag{11.166}$$

$$R_x = \frac{1}{\omega_{c\alpha}} \left[v_{0x} sin \ \omega_{c\alpha} t - v_{0y} (cos \ \omega_{c\alpha} t - 1) \right], \tag{11.167}$$

$$R_y = \frac{1}{\omega_{c\alpha}} \left[v_{0x} (cos \ \omega_{c\alpha} t - 1) + v_{0y} \ sin \ \omega_{c\alpha} t \right], \tag{11.168}$$

$$R_z = v_{0z} t, \tag{11.169}$$

where the subscript 0 quantities are the initial values at $t = 0$. The charge density associated with a dressed particle can be written as

$$\rho_{ext}(\mathbf{r}, t) = q\delta(\mathbf{r} - \mathbf{r}(t)) = q\delta(\mathbf{r} - \mathbf{r}_0 - \mathbf{R}_\alpha(t, \mathbf{v}_0)). \tag{11.170}$$

Its Fourier transform is

$$\rho_{ext}(\mathbf{k}, \omega) = q \ e^{-i\mathbf{k} \cdot \mathbf{r}_0} \int_{-\infty}^{\infty} dt \ exp \left[i\omega t - i\mathbf{k} \cdot \mathbf{R}_\alpha(t, \mathbf{v}_0) \right]. \tag{11.171}$$

We have the Poisson equation for the single dressed particle

$$\nabla \cdot \mathbf{E}(\mathbf{r}, t) = 4\pi \sum_\alpha \int f_\alpha(\mathbf{r}, \mathbf{v}, t) d^3 v + 4\pi \rho_{ext}(\mathbf{r}, t) \quad (\alpha = e, \ i) \tag{11.172}$$

where f_α is governed by the linearized Vlasov equation

$$\frac{\partial}{\partial t} f_\alpha(\mathbf{r}, \mathbf{v}, t) + \mathbf{v} \cdot \frac{\partial f_\alpha}{\partial \mathbf{r}} + \frac{e_\alpha}{m_\alpha} \frac{\mathbf{v} \times \mathbf{B_0}}{c} \cdot \frac{\partial f_\alpha}{\partial \mathbf{v}}$$

$$= -\frac{e_\alpha}{m_\alpha} \mathbf{E}(\mathbf{r}, t) \cdot \frac{\partial f_{\alpha 0}}{\partial \mathbf{v}}. \tag{11.173}$$

(11.173) can be solved by characteristic method with the result expressed by Bessel function series. Consulting Chapter 6, we obtain the following solution:

$$f_\alpha(\mathbf{k}, \omega, \mathbf{v}) = -\frac{e_\alpha}{m_\alpha} \phi(\mathbf{k}, \omega) \sum_{n=-\infty}^{\infty} \left[k_\| \frac{\partial f_{\alpha 0}}{\partial v_\|} + \frac{n\omega_{c\alpha}}{v_\perp} \frac{\partial f_{\alpha 0}}{\partial v_\perp} \right] \frac{[J_n(\frac{k_\perp v_\perp}{\omega_{c\alpha}})]^2}{\omega - k_\| v_\| - n\omega_{c\alpha}}. \tag{11.174}$$

Putting the above equation into the Fourier-transformed version of (11.172) yields for the electric potential contributed by a single dressed particle

$$\phi(\mathbf{k}, \omega) = \frac{4\pi}{k^2} \frac{\rho_{ext}(\mathbf{k}, \omega)}{\varepsilon_L(\mathbf{k}, \omega)} \tag{11.175}$$

where the dielectric function ε_L is

$$\varepsilon_L = 1 + \sum_\alpha \frac{\omega_{p\alpha}^2}{k^2} \sum_{n=-\infty}^{\infty} \int d^3v \left[k_\| \frac{\partial f_{\alpha 0}}{\partial v_\|} + \frac{n\omega_{c\alpha}}{v_\perp} \frac{\partial f_{\alpha 0}}{\partial v_\perp} \right] \frac{[J_n(\frac{k_\perp v_\perp}{\omega_{c\alpha}})]^2}{\omega - k_\| v_\| - n\omega_{c\alpha}} \tag{11.176}$$

where $d^3v = 2\pi v_\perp dv_\perp dv_\|$.

Next, we calculate the correlation spectral density $\langle \phi(\mathbf{k}, \omega) \phi^*(\mathbf{k}', \omega') \rangle$ where the ensemble average is taken over the entire plasma particles, both species of $\alpha = e, i$. Using (11.171) and (11.175), we can write after $\int d^3 r_0$–integral

$$\langle \phi(\mathbf{k}, \omega) \phi^*(\mathbf{k}', \omega') \rangle$$

$$= \frac{16\pi^2}{k^4} (2\pi)^3 \delta(\mathbf{k} - \mathbf{k}') \frac{1}{\varepsilon_L(\mathbf{k}, \omega)} \frac{1}{\varepsilon_L^*(\mathbf{k}, \omega')} \sum_\alpha \int d^3 v_0 f_{0\alpha}(\mathbf{v}_0)$$

$$\times \int_{-\infty}^{\infty} dt \; exp \left[i\omega t - i\mathbf{k} \cdot \mathbf{R}_\alpha(\mathbf{v}_0, t) \right]$$

$$\times \int_{-\infty}^{\infty} dt' \; exp \left[-i\omega' t' + i\mathbf{k} \cdot \mathbf{R}_\alpha(\mathbf{v}_0, t') \right]. \tag{11.177}$$

To carry out the time-integral in (11.177), we calculate

$$\mathbf{k}\cdot\mathbf{R}_\alpha(t,\mathbf{v}_0) = k_z v_{0z} t$$
$$+\frac{1}{\omega_{c\alpha}}\Big[(k_x v_{0x} + k_y v_{0y})\sin\,\omega_{c\alpha}t + (\cos\,\omega_{c\alpha}t - 1)(k_y v_{0x} - k_x v_{0y})\Big].$$

Introduce cylindrical coordinates:

$$v_{0x} = v_\perp \cos\,\varphi_1,\quad v_{0y} = v_\perp \sin\,\varphi_1,\quad k_x = k_\perp\,\cos\,\varphi_2,$$
$$k_y = k_\perp\,\sin\,\varphi_2. \tag{11.178}$$

Then, we have

$$\mathbf{k}\cdot\mathbf{R}(t,\mathbf{v}_0) = k_z v_{0z} t + a_\alpha[\sin\,(\omega_{c\alpha}t+\varphi) - \sin\,\varphi] \tag{11.179}$$

$$where\ \ \varphi = \varphi_2 - \varphi_1,\quad a_\alpha = \frac{k_\perp v_\perp}{\omega_{c\alpha}}. \tag{11.180}$$

Then the product of two time–integrals in (11.177) can be written as

$$\int_{-\infty}^{\infty} dt e^{i(\omega - k_z v_{0z})t} \sum_{n=-\infty}^{\infty} J_n(a_\alpha) e^{in(\omega_{c\alpha}t+\varphi)} \int_{-\infty}^{\infty} dt' e^{-i(\omega' - k_z v_{0z})t'}$$
$$\times \sum_{m=-\infty}^{\infty} J_m(a_\alpha) e^{-im(\omega_{c\alpha}t'+\varphi)}.$$

Note that only the $n = m$ terms survive in the double series because of the $\int d\varphi$–integral [in the velocity integral in (11.177)]. Thus, the above expression becomes

$$\sum_{n=-\infty}^{\infty} J_n^2(a_\alpha)(2\pi)^2\;\delta(\omega - k_z v_{0z} + n\omega_{c\alpha})\delta(\omega' - k_z v_{0z} + n\omega_{c\alpha}).$$

Therefore, (11.177) takes the form

$$\langle\phi(\mathbf{k},\omega)\phi^*(\mathbf{k}',\omega')\rangle = \frac{16\pi^2}{k^4}(2\pi)^3\delta(\mathbf{k}-\mathbf{k}')\delta(\omega-\omega')\frac{1}{|\;\varepsilon_L(\mathbf{k},\omega)\;|^2}$$
$$\times\sum_\alpha q_\alpha^2\int d^3v_0 f_{0\alpha}(v_{\perp 0}, v_{\parallel 0})\sum_{n=-\infty}^{\infty} J_n^2(a_\alpha)(2\pi)^2\;\delta(\omega - k_z v_{0z} + n\omega_{c\alpha}). \tag{11.181}$$

Putting $\omega = \omega'$ and $\mathbf{k} = \mathbf{k}'$ in the above equation, we obtain

$$\langle | \phi(\mathbf{k}, \omega) |^2 \rangle = \frac{16\pi^2 2\pi}{k^4} \frac{1}{| \varepsilon_L(\mathbf{k}, \omega) |^2}$$
$$\times \sum_\alpha q_\alpha^2 \int d^3 v_0 f_{0\alpha}(v_{\perp 0}, v_{\parallel 0}) \sum_{n=-\infty}^{\infty} J_n^2(a_\alpha)\, \delta(\omega - k_z v_{0\parallel} + n\omega_{c\alpha}). \tag{11.182}$$

This is the spectral density of the electric field fluctuation in a magnetized plasma.

Chapter 12

Kinetic theory: Liouville, Klimontovich, Lenard–Balescu

The fundamental concept of statistical description of plasma begins with the Liouville equation which governs the dynamical behavior of the *ensemble* of the system in the phase space. The ensemble is represented by the Liouville distribution function $F^{(N)} = F^{(N)}(\mathbf{r}_1, \mathbf{r}_2, \ldots, \mathbf{r}_N, \mathbf{p}_1, \mathbf{p}_2, \ldots, \mathbf{p}_N, t)$ (N is the number of particles in the system), which is (6N+1)-variable function. The Liouville distribution function, which is governed by the Liouville equation or the Klimontovich equation, involves a chain of multi-particle distribution functions which is called the BBGKY hierarchy (§12.3). A higher hierarchy distribution function is *fine grained* while a lower hierarchy distribution function is *coarse grained.* The hierarchy stems from the potential energy of the Hamiltonian of a system. The potential energy of a particle in the system is influenced by all the rest of the particles in the community. The hierarchy cannot be unchained without introducing approximation, and the task of the kinetic theory is to invent a proper approximation. The approximate treatment of this hierarchy is facilitated by expanding in a small parameter as referred to *plasma parameter*, the inverse of the number of particles in the Debye sphere. The development makes it possible to invent one unchained equation for the one-particle distribution function by expressing the two-particle distribution functions in terms of one-particle distribution function. We have famous kinetic equations for one particle distribution function: Lenard–Balescu equation or Boltzmann equation with Landau collision term. References are supplied at the end of the Chapter.

12.1. Klimontovich equation

We start with the microscopic distribution for a single system of N particles (say, composed of electrons only), called Klimontovich distribution function or density

$$K(\mathbf{r}, \mathbf{v}, t) = \sum_{i=1}^{N} \delta(\mathbf{r} - \mathbf{r}_i(t))\delta(\mathbf{v} - \mathbf{v}_i(t)) \tag{12.1}$$

where $\mathbf{r}_i(t), \mathbf{v}_i(t)$ represent the exact orbit of the i-th particle as determined by the forces $\mathbf{F}_i$ acting on it. We need the function K for each species, and should write in (12.1) K_α ($\alpha = e,\ i$). For notational simplicity, the species index α will be omitted if convenient, but will be reinstated if necessary. Note that if K is integrated over the whole six-dimensional phase space $(\mathbf{r}, \mathbf{v})$ of the system the result is N, the total number of particles. From the equation of motion for individual particle

$$\dot{\mathbf{r}_i} = \mathbf{v}_i, \quad \dot{\mathbf{v}_i} = \frac{\mathbf{F}_i}{m_i}, \tag{12.2}$$

it follows that the Klimontovich density satisfies the equation below

$$\frac{\partial}{\partial t} K_\alpha(\mathbf{r}, \mathbf{v}, t) + \mathbf{v} \cdot \frac{\partial K_\alpha}{\partial \mathbf{r}} + \frac{\mathbf{F}_\alpha}{m_\alpha} \cdot \frac{\partial K_\alpha}{\partial \mathbf{v}} = 0 \tag{12.3}$$

where $\mathbf{F}_\alpha$ is the self-consistent force that a particle of species α at $(\mathbf{r},\ \mathbf{v})$ would experience. (12.3) is elaborated in the following. Differentiating (12.1) with respect to t gives

$$\begin{aligned}\frac{\partial K}{\partial t} &= \sum_{i=1}^{N} \left[\frac{d\mathbf{r}_i}{dt} \cdot \frac{\partial}{\partial \mathbf{r}_i} \delta(\mathbf{r} - \mathbf{r}_i(t))\, \delta(\mathbf{v} - \mathbf{v}_i(t)) + \frac{d\mathbf{v}_i}{dt} \cdot \frac{\partial}{\partial \mathbf{v}_i} \delta(\mathbf{r} - \mathbf{r}_i(t))\, \delta(\mathbf{v} - \mathbf{v}_i(t))\right] \\ &= -\sum_{i=1}^{N} \left[\mathbf{v}_i \cdot \frac{\partial}{\partial \mathbf{r}} \delta(\mathbf{r} - \mathbf{r}_i(t))\delta(\mathbf{v} - \mathbf{v}_i(t)) + \frac{\mathbf{F}_i}{m_i} \cdot \frac{\partial}{\partial \mathbf{v}} \delta(\mathbf{r} - \mathbf{r}_i(t))\delta(\mathbf{v} - \mathbf{v}_i(t))\right].\end{aligned}$$

In the first term inside Σ, $\mathbf{v}_i$ can be replaced by $\mathbf{v}$ owing to the delta function. Also

$$\mathbf{F}_i = \mathbf{F}_i(\mathbf{r}_i, \mathbf{v}_i) = e_\alpha(\mathbf{E}(\mathbf{r}_i) + \frac{1}{c}\mathbf{v}_i \times \mathbf{B}(\mathbf{r}_i)) = e_\alpha(\mathbf{E}(\mathbf{r}, t) + \frac{1}{c}\mathbf{v} \times \mathbf{B}(\mathbf{r}, t)) \equiv \mathbf{F}_\alpha \tag{12.4}$$

due to the δ functions. In (12.4), $\mathbf{r}_i$ and $\mathbf{v}_i$ are Lagrangian variables; in the last expression of (12.4) we have put them in terms of Eulerian coordinates $\mathbf{r}$ and t $[\mathbf{r}_i = \mathbf{r}_i(\mathbf{r}, t)]$. Thus we obtain

$$\frac{\partial K_\alpha(\mathbf{r}, \mathbf{v}, t)}{\partial t} + \mathbf{v} \cdot \frac{\partial K_\alpha}{\partial \mathbf{r}} + \frac{\mathbf{F}_\alpha(\mathbf{r}, t)}{m_\alpha} \cdot \frac{\partial K_\alpha}{\partial \mathbf{v}} = 0. \tag{12.5}$$

The self-consistent fields, **E** and **B**, obey Maxwell's equations

$$\nabla \cdot \mathbf{E} = 4\pi\rho \tag{12.6}$$

$$\nabla \times \mathbf{B} = \frac{4\pi}{c}\mathbf{J} + \frac{1}{c}\frac{\partial \mathbf{E}}{\partial t} \tag{12.7}$$

where the source terms are generated by all the particles in the system:

$$\rho(\mathbf{x}, t) = \sum_{\alpha} e_{\alpha} \int d^3v\ K_{\alpha}(\mathbf{r}, \mathbf{v}, t) = \sum_{\alpha} e_{\alpha} \sum_{i=1}^{N} \delta(\mathbf{r} - \mathbf{r}_i^{\alpha}(t)), \tag{12.8}$$

$$\mathbf{J}(\mathbf{r}, t) = \sum_{\alpha} e_{\alpha} \int d^3v\ \mathbf{v}K_{\alpha}(\mathbf{r}, \mathbf{v}, t) = \sum_{\alpha} e_{\alpha} \sum_{i=1}^{N} \mathbf{v}_i^{\alpha}(t)\delta(\mathbf{r} - \mathbf{r}_i^{\alpha}(t)). \tag{12.9}$$

Equation (12.3) is apparently same as the Vlasov equation but its content is very different from the latter. (12.3) requires knowledge of exact orbit which is impossible to obtain. (12.3) is necessary to be reduced to workable equation on the basis of statistical consideration, which is the task in the sequel.

12.2. Liouville equation for a many-particle distribution function

Liouville distribution function of a system of N particles

$$F^{(N)} = F^{(N)}(\mathbf{r}_1, \mathbf{r}_2, \ldots, \mathbf{r}_N, \mathbf{p}_1, \mathbf{p}_2, \ldots, \mathbf{p}_N, t)$$

specifies the distribution of systems over a statistical *ensemble.* $F^{(N)}$ may be thought of in terms of the trajectory of a phase point in 6N-dimensional phase space (3N coordinate space plus 3N momentum space; called Γ-space), as is shown later. In the course of time this point would move along a path that would be determined by the equation of motion. Instead of considering a trajectory belonging to a single phase point [the complete knowledge of the orbit of the phase point in the Γ-space is impossible to obtain], the system is represented by a collection, or an ensemble, of phase points, each of which is a possible state of the system with the same macroscopic characteristics. With compatible initial conditions $F^{(N)}$ satisfies the Liouville equation

$$\frac{\partial F^{(N)}}{\partial t} + \sum_{i=1}^{N} \left(\frac{\partial H}{\partial \mathbf{p}_i} \cdot \frac{\partial F^{(N)}}{\partial \mathbf{r}_i} - \frac{\partial F^{(N)}}{\partial \mathbf{p}_i} \cdot \frac{\partial H}{\partial \mathbf{r}_i} \right) = 0 \tag{12.10}$$

where the Hamiltonian H (considering only electrostatic interaction) is given by

$$H = \frac{1}{2m}\sum_{i=1}^{N}\left(\mathbf{p}_i - \frac{e}{c}\mathbf{A}(\mathbf{r}_i, t)\right)^2 + e\sum_{i=1}^{N}\phi(\mathbf{r}_i, t) + \frac{e^2}{2}\sum_{i=1}^{N}\sum_{j(\neq i)=1}^{N}\frac{1}{\mid \mathbf{r}_i - \mathbf{r}_j \mid}, \tag{12.11}$$

where $\mathbf{A}(\mathbf{r}_i, t)$ and $\phi(\mathbf{r}_i, t)$ are vector and scalar potentials from which the *external* magnetic and electric fields acting on the i-th particle are derived. Here 'external' means that its origin has nothing to do with the plasma particles composing the system. The last term represents the electrostatic inter-particle interaction. We have

$$\mathbf{B} = \nabla \times \mathbf{A},$$

$$\mathbf{E} = -\nabla\phi - \frac{1}{c}\frac{\partial \mathbf{A}}{\partial t}.$$

The quantity

$$F^{(N)}(\mathbf{r}_1, \mathbf{p}_1, \mathbf{r}_2, \mathbf{p}_2, \ldots, \mathbf{r}_N, \mathbf{p}_N, t)d^3r_1 d^3r_2 \cdots d^3r_N d^3p_1 d^3p_2 \cdots d^3p_N$$

is the probability at time t that particle 1 is at the state $(\mathbf{r}_1 + d\mathbf{r}_1, \mathbf{p}_1 + d\mathbf{p}_1)$ and that particle 2 is at the state $(\mathbf{r}_2 + d\mathbf{r}_2, \mathbf{p}_2 + d\mathbf{p}_2) \cdots$, and that particle N is at the state $(\mathbf{r}_N + d\mathbf{r}_N, \mathbf{p}_N + d\mathbf{p}_N)$.

(12.10) is a lengthy first order partial differential equation involving (6N+1) partial derivatives. The characteristic equation of (12.10) reads

$$\begin{aligned} dt &= \frac{dx_1}{\frac{\partial H}{\partial p_{1x}}} = \frac{dy_1}{\frac{\partial H}{\partial p_{1y}}} = \frac{dz_1}{\frac{\partial H}{\partial p_{1z}}} = \frac{-dp_{1x}}{\frac{\partial H}{\partial x_1}} = \frac{-dp_{1y}}{\frac{\partial H}{\partial y_1}} = \frac{-dp_{1z}}{\frac{\partial H}{\partial z_1}} \\ &= \frac{dx_2}{\frac{\partial H}{\partial p_{2x}}} = \frac{dy_2}{\frac{\partial H}{\partial p_{2y}}} = \frac{dz_2}{\frac{\partial H}{\partial p_{2z}}} = \frac{-dp_{2x}}{\frac{\partial H}{\partial x_2}} = \frac{-dp_{2y}}{\frac{\partial H}{\partial y_2}} = \frac{-dp_{2z}}{\frac{\partial H}{\partial z_2}} = \cdots \\ &= \frac{dx_N}{\frac{\partial H}{\partial p_{Nx}}} = \frac{dy_N}{\frac{\partial H}{\partial p_{Ny}}} = \frac{dz_N}{\frac{\partial H}{\partial p_{Nz}}} = \frac{-dp_{Nx}}{\frac{\partial H}{\partial x_N}} = \frac{-dp_{Ny}}{\frac{\partial H}{\partial y_N}} = \frac{-dp_{Nz}}{\frac{\partial H}{\partial z_N}}. \end{aligned}$$

The above expression can be grouped into a compact and physically transparent form

$$\frac{d\mathbf{r}_i}{dt} = \frac{\partial H}{\partial \mathbf{p}_i}, \quad \frac{d\mathbf{p}_i}{dt} = -\frac{\partial H}{\partial \mathbf{r}_i}, \quad (i = 1, 2, \ldots, N) \tag{12.12}$$

which are Hamilton's equations of motion.

The trajectory of the phase point in the 6N-dimensional Γ-space is determined by the vector sum of the individual path of each particle composing the N-particle system. So we conclude that the Liouville function $F^{(N)}(\mathbf{r}_1, \mathbf{r}_2, \ldots, \mathbf{r}_N, \mathbf{p}_1, \mathbf{p}_2, \ldots, \mathbf{p}_N, t)$ is constant along the phase point trajectory which is the vector sum of the individual trajectories which are determined by equations of motion (12.12).

It would be instructive to show that Hamilton's equations (12.12) take the familiar form of the equation of motion, which is shown below. Using (12.11) in the first equation of (12.12) gives

$$\frac{\partial H}{\partial \mathbf{p}_i} = \frac{d\mathbf{r}_i}{dt} = \mathbf{v}_i = \frac{1}{m}\left(\mathbf{p}_i - \frac{e}{c}\mathbf{A}(\mathbf{r}_i, t)\right). \tag{12.13}$$

We calculate from (12.11) [using the identity $\frac{1}{2}\nabla F^2 = \mathbf{F} \times \nabla \times \mathbf{F} + \mathbf{F} \cdot \nabla \mathbf{F}$]

$$\frac{\partial H}{\partial \mathbf{r}_i} = \frac{1}{m}\left(\mathbf{p}_i - \frac{e}{c}\mathbf{A}(\mathbf{r}_i, t)\right) \times \frac{\partial}{\partial \mathbf{r}_i} \times \left(\mathbf{p}_i - \frac{e}{c}\mathbf{A}(\mathbf{r}_i, t)\right)$$

$$+ \frac{1}{m}\left(\mathbf{p}_i - \frac{e}{c}\mathbf{A}(\mathbf{r}_i, t)\right) \cdot \frac{\partial}{\partial \mathbf{r}_i}\left(\mathbf{p}_i - \frac{e}{c}\mathbf{A}(\mathbf{r}_i, t)\right) + e\frac{\partial}{\partial \mathbf{r}_i}\left[\phi(\mathbf{r}_i, t) + \frac{e}{2}\sum_{j(\neq i)=1}^{N} \frac{1}{|\mathbf{r}_i - \mathbf{r}_j|}\right].$$

Noting that $\mathbf{p}$ operated by $\frac{\partial}{\partial \mathbf{r}_i}$ vanishes and using (12.13), the vector and scalar potential terms on the RHS simplify, and the above equation yields

$$\frac{\partial H}{\partial \mathbf{r}_i} = -\frac{e}{c}\mathbf{v}_i \times \mathbf{B}(\mathbf{r}_i, t) - \frac{e}{c}\mathbf{v}_i \cdot \nabla \mathbf{A}(\mathbf{r}_i, t) + e\frac{\partial}{\partial \mathbf{r}_i}\left[\phi(\mathbf{r}_i, t) + \frac{e}{2}\sum_{j(\neq i)=1}^{N} \frac{1}{|\mathbf{r}_i - \mathbf{r}_j|}\right]. \tag{12.14}$$

On the other hand,

$$\frac{d\mathbf{p}_i}{dt} = m\frac{d\mathbf{v}_i}{dt} + \frac{e}{c}\frac{d\mathbf{A}}{dt} = m\frac{d\mathbf{v}_i}{dt} + \frac{e}{c}\left(\frac{\partial \mathbf{A}}{\partial t} + \mathbf{v}_i \cdot \nabla \mathbf{A}\right). \tag{12.15}$$

Using (12.14) and (12.15) in (12.12) yields

$$m\frac{d\mathbf{v}_i}{dt} = e\left[\mathbf{E}(\mathbf{r}_i, t) + \frac{1}{c}\mathbf{v}_i \times \mathbf{B}(\mathbf{r}_i, t)\right] \tag{12.16}$$

and

$$\mathbf{E}(\mathbf{r}_i, t) = -\frac{1}{c}\frac{\partial}{\partial t}\mathbf{A}(\mathbf{r}_i, t) - \frac{\partial}{\partial \mathbf{r}_i}\left[\phi(\mathbf{r}_i, t) + \frac{e}{2}\sum_{j(\neq i)=1}^{N} \frac{1}{|\mathbf{r}_i - \mathbf{r}_j|}\right] \equiv \mathbf{E}_i(\mathbf{r}_i, t) \tag{12.17}$$

which is the total (internal plus external) electric field acting on the particle i. Note that $\mathbf{B}$ in (12.16) is the external magnetic field because we excluded the magnetic interaction between plasma particles in the Hamiltonian (12.11).

The characteristic equation of (12.16) from which $\mathbf{p}_i$ is absent reads

$$dt = \frac{d\mathbf{r}_i}{\mathbf{v}_i} = \frac{m d\mathbf{v}_i}{e(\mathbf{E}_i + \frac{1}{c}\mathbf{v}_i \times \mathbf{B}_i)}. \quad (i = 1, 2, \ldots, N) \tag{12.18}$$

This equation would result from the familiar form of the Liouville equation

$$\frac{\partial F^{(N)}}{\partial t} + \sum_{i=1}^{N} \left[\mathbf{v}_i \cdot \frac{\partial F^{(N)}}{\partial \mathbf{r}_i} + \mathbf{a}_i \cdot \frac{\partial F^{(N)}}{\partial \mathbf{v}_i} \right] = 0 \tag{12.19}$$

$$\text{where} \quad \mathbf{a}_i = \frac{e}{m} \left[\mathbf{E}_i(\mathbf{r}_i, t) + \frac{1}{c} \mathbf{v}_i \times \mathbf{B}_i(\mathbf{r}_i, t) \right].$$

Exercise. Derive (12.19) from (12.10) rigorously.

Solution. Using (12.13) and (12.14) in (12.10), the typical term (omitting the particle index) is written as

$$\frac{\partial F^{(N)}}{\partial t} + \cdots + \mathbf{v}.\frac{\partial F^{(N)}}{\partial \mathbf{r}} - \left(-\frac{e}{c}\mathbf{v} \times \mathbf{B} - \frac{e}{c}\mathbf{v} \cdot \nabla \mathbf{A} + e\nabla\phi_T \right) \cdot \frac{\partial F^{(N)}}{\partial \mathbf{p}} + \cdots = 0 \tag{12.20}$$

where ϕ_T is the bracketed quantity in (12.17), the total potential. We carry out the change of variables $(\mathbf{r}, \mathbf{p}, t) \to (\mathbf{r}', \mathbf{v}, t')$ in which

$$t = t', \quad \mathbf{r} = \mathbf{r}', \quad \mathbf{p} = m\mathbf{v} + \frac{e}{c}\mathbf{A}(\mathbf{r}', t').$$

So the differential operators on $F^{(N)}$ in (12.20) changes according to

$$\frac{\partial}{\partial t} = \frac{\partial}{\partial t'} + \frac{\partial}{\partial \mathbf{v}} \cdot \frac{\partial \mathbf{v}}{\partial t'} = \frac{\partial}{\partial t'} - \frac{e}{mc}\frac{\partial \mathbf{A}}{\partial t'} \cdot \frac{\partial}{\partial \mathbf{v}},$$

$$\frac{\partial}{\partial \mathbf{p}} = \frac{1}{m}\frac{\partial}{\partial \mathbf{v}},$$

$$\frac{\partial}{\partial \mathbf{r}} = \frac{\partial}{\partial \mathbf{r}'} + \frac{\partial \mathbf{v}}{\partial \mathbf{r}'} \cdot \frac{\partial}{\partial \mathbf{v}} = \frac{\partial}{\partial \mathbf{r}'} - \frac{e}{mc}\frac{\partial A_j}{\partial x_i'}\frac{\partial}{\partial v_j},$$

from which it follows that

$$\mathbf{v} \cdot \frac{\partial}{\partial \mathbf{r}} = \mathbf{v} \cdot \frac{\partial}{\partial \mathbf{r}'} - \frac{e}{mc}(\mathbf{v} \cdot \nabla') \, \mathbf{A} \cdot \frac{\partial}{\partial \mathbf{v}}.$$

Substituting the above relations into (12.20) yields for a typical term

$$\frac{\partial F^{(N)}}{\partial t'} + \cdots + \mathbf{v} \cdot \frac{\partial F^{(N)}}{\partial \mathbf{r}'} + \mathbf{a} \cdot \frac{\partial F^{(N)}}{\partial \mathbf{v}} + \cdots = 0. \tag{12.21}$$

In the above equation, primes can be removed; similar transformations can be carried out for all the terms to obtain (12.19). When $\frac{\partial}{\partial \mathbf{v}} \cdot \mathbf{a} = 0$, (12.19) can be also written as

$$\frac{\partial F^{(N)}}{\partial t} + \sum_{i=1}^{N} \left[\frac{\partial}{\partial \mathbf{r}_i} \cdot (\mathbf{v}_i F^{(N)}) + \frac{\partial}{\partial \mathbf{v}_i} \cdot (\mathbf{a}_i F^{(N)}) \right] = 0. \tag{12.22}$$

In the above we saw that the Liouville equation conserves the Liouville function (or Liouville probability density) along the phase point trajectory in the Γ-space, which is characteristic of the Liouville equation. The Liouville function is the probability of the N particles being in the state specified in the argument of $F^{(N)}$. Conversely, we might start from the *Liouville theorem* which states that the probability density of the ensemble is conserved as the phase points move about in the Γ-space. For notational brevity, let the representative point of a single phase point state be denoted by $\mathbf{X}(t)$. The dynamical evolution of an individual system is determined by the individual orbit equations (12.18). Let $\mathbf{X}_0$ be the initial state of the representative point: $\mathbf{X}_0 = \mathbf{X}(t = 0)$. Then the initial probability density is the distribution of the aggregate of the representative points, which is $F^{(N)}(\mathbf{X}_0)$. As time increases, the initial phase point $\mathbf{X}_0$ traces an orbit $\mathbf{X}(t) = \mathbf{X}(\mathbf{X}_0, t)$. The probability density at time t which is the distribution of the re-distributed representative points is $F^{(N)}(\mathbf{X}(t))$. The Liouville theorem simply states that

$$F^{(N)}(\mathbf{X}_0) = F^{(N)}(\mathbf{X}(t)). \tag{12.23}$$

The Liouville equation is deduced from the Liouville theorem in (12.23) by taking the Lagrangian derivative of $F^{(N)}(\mathbf{X}(t))$ (or differentiating along the trajectory), thus obtaining (12.19).

Remark. A device to write $F^{(N)}$ in terms of the Eulerian coordinate is to introduce the Klimontovich distribution function:

$$F^{(N)}(\mathbf{X}, t) = \int d\tau_0 F^{(N)}(\mathbf{X}_0) \prod_{i=1}^{N} \delta(\mathbf{r}_i - \mathbf{r}_i(\mathbf{r}_{i0}, t)\ \delta(\mathbf{v}_i - \mathbf{v}_i(\mathbf{v}_{i0}, t)). \tag{12.24}$$

Here $\mathbf{X}$ with no argument denotes the Eulerian coordinate while Lagrangian coordinate is denoted by explicitly attaching or containing '(t)' as in (12.23). Subscript 0 denotes the initial values. $d\tau_0$ is the volume element surrounding the point $\mathbf{X}_0$ in the Γ-space. Differentiating (12.24) with $(\frac{\partial}{\partial t})$, we obtain the differential form of the Liouville theorem (12.19) with the aid of orbit equations (12.18). Carry out the algebra as an exercise. [$\mathbf{r}_i, \mathbf{v}_i$ are Eulerian variables.]

12.3. BBGKY hierarchy equation

For the sake of simplicity, we shall consider only electrostatic forces in the Hamiltonian. Then putting $\mathbf{A} = 0$ in the preceding discussion, the Liouville equation is written, in place of (12.10),

$$\frac{\partial F^{(N)}}{\partial t} + \frac{1}{m}\sum_{i=1}^{N}\left(\frac{\partial H}{\partial \mathbf{v}_i}\cdot\frac{\partial F^{(N)}}{\partial \mathbf{r}_i} - \frac{\partial F^{(N)}}{\partial \mathbf{v}_i}\cdot\frac{\partial H}{\partial \mathbf{r}_i}\right) = 0$$

where the Hamiltonian H is given by

$$H(\mathbf{v}_i, \mathbf{r}_i, t) = \frac{m}{2}\sum_{i=1}^{N} v_i^2 + e\sum_{i=1}^{N}\phi_T(\mathbf{r}_i, t),$$

$$\phi_T(\mathbf{r}_i, t) = \phi(\mathbf{r}_i, t) + \frac{e}{2}\sum_{j(\neq i)=1}^{N}\frac{1}{\mid \mathbf{r}_i - \mathbf{r}_j \mid}.$$

Using $\frac{\partial H}{\partial \mathbf{v}_i} = m\mathbf{v}_i$ yields

$$\frac{\partial F^{(N)}}{\partial t} + \sum_{i=1}^{N}\left(\mathbf{v}_i \cdot \frac{\partial F^{(N)}}{\partial \mathbf{r}_i} - \frac{e}{m}\frac{\partial \phi_T}{\partial \mathbf{r}_i}\cdot\frac{\partial F^{(N)}}{\partial \mathbf{v}_i}\right) = 0 \tag{12.25}$$

where $F^{(N)} = F^{(N)}(\mathbf{r}_1, \mathbf{v}_1, \mathbf{r}_2, \mathbf{v}_2, \ldots, \mathbf{r}_N, \mathbf{v}_N, t)$.

- **1-particle distribution function:**

$$F^{(1)}(\mathbf{r}_1, \mathbf{v}_1, t) = \int F^{(N)}(\mathbf{r}_1, \mathbf{v}_1, \mathbf{r}_2, \mathbf{v}_2, \ldots, \mathbf{r}_N, \mathbf{v}_N, t) d^3r_2 d^3v_2 d^3r_3 d^3v_3 \cdots d^3r_N d^3v_N \tag{12.26}$$

is the probability density at time t to find the particle 1 at the state $(\mathbf{r}_1 + d\mathbf{r}_1,\ \mathbf{v}_1 + d\mathbf{v}_1)$, regardless of the whereabouts of the rest of the particles. Likewise

$$F^{(1)}(\mathbf{r}_2, \mathbf{v}_2, t) = \int F^{(N)}(\mathbf{r}_1, \mathbf{v}_1, \mathbf{r}_2, \mathbf{v}_2, \ldots, \mathbf{r}_N, \mathbf{v}_N, t) d^3r_1 d^3v_1 d^3r_3 d^3v_3 \cdots d^3r_N d^3v_N$$

is the probability density at time t to find particle 2 at the state $(\mathbf{r}_2 + d\mathbf{r}_2,\ \mathbf{v}_2 + d\mathbf{v}_2)$, regardless of whereabouts of the rest of the particles. Since the particles are indistinguishable, the above two probability densities are the same, and we can state that $F^{(1)}(\mathbf{r}, \mathbf{v}, t)$ which is obtained by integrating $F^{(N)}$ over all the particles except any one particle is the probability density at time t to find a particle at the state $(\mathbf{r} + d\mathbf{r},\ \mathbf{v} + d\mathbf{v})$. So $F^{(1)}(\mathbf{r}, \mathbf{v}, t)$ has the same meaning as the $f(\mathbf{r}, \mathbf{v}, t)$ in the familiar Vlasov equation.

- **2-particle distribution function:**

$$F^{(2)}(\mathbf{r}_1, \mathbf{v}_1, \mathbf{r}_2, \mathbf{v}_2, t) = \int F^{(N)}(\mathbf{r}_1, \mathbf{v}_1, \mathbf{r}_2, \mathbf{v}_2, \ldots, \mathbf{r}_N, \mathbf{v}_N, t) d^3r_3 d^3v_3 d^3r_4 d^3v_4 \cdots d^3r_N d^3v_N \tag{12.27}$$

is the joint probability density at time t to find a particle at the state $(\mathbf{r}_1 + d\mathbf{r}_1,\ \mathbf{v}_1 + d\mathbf{v}_1)$ and to find simultaneously another particle at the state $(\mathbf{r}_2 + d\mathbf{r}_2,\ \mathbf{v}_2 + d\mathbf{v}_2)$, regardless of the whereabouts of the rest of the particles. $F^{(2)}$ is obtained by integrating $F^{(N)}$ over all the particles except *any* two particles.

- **s-particle distribution function $F^{(s)}$:**

$$F^{(s)}(\mathbf{r}_1, \mathbf{v}_1, \mathbf{r}_2, \mathbf{v}_2, \ldots, \mathbf{r}_s, \mathbf{v}_s, t) = \int F^{(N)}(\mathbf{r}_1, \mathbf{v}_1, \mathbf{r}_2, \mathbf{v}_2, \ldots, \mathbf{r}_N, \mathbf{v}_N, t) d^3r_{s+1} d^3v_{s+1} \cdots d^3r_N d^3v_N \tag{12.28}$$

which is the joint probability density at time t to simultaneously find s particles at each of s phase space locations in the Γ-space.

Now we take Liouville's equation (12.25) and integrate over coordinates and velocities of all the particles except those for particle 1 (say). Then the immediately integrated terms (corresponding to $i = 2, 3, \ldots, N$) vanish (no particles on the infinite surface; at the boundary) to give

$$\frac{\partial}{\partial t} F^{(1)}(\mathbf{r}_1, \mathbf{v}_1, t) + \mathbf{v}_1 \cdot \frac{\partial F^{(1)}}{\partial \mathbf{r}_1} - \frac{e}{m} \frac{\partial \phi(\mathbf{r}_1, t)}{\partial \mathbf{r}_1} \cdot \frac{\partial F^{(1)}}{\partial \mathbf{v}_1} = \frac{e^2}{m} \int d^3r_2 d^3v_2 \cdots d^3r_N d^3v_N \left(\frac{\partial}{\partial \mathbf{r}_1} \sum_{j=2}^{N} \frac{1}{\mid \mathbf{r}_1 - \mathbf{r}_j \mid} \right) \cdot \frac{\partial F^{(N)}}{\partial \mathbf{v}_1}. \tag{12.29}$$

Since the particles are indistinguishable ($F^{(N)}$ is symmetric with respect to interchange of particle index), the summation on j yields $(N-1)$ identical terms (multiplied by $F^{(2)}$), and the right hand side becomes

$$\frac{(N-1)e^2}{m}\int d^3r_2 d^3v_2 \left(\frac{\partial}{\partial \mathbf{r}_1}\frac{1}{\mid \mathbf{r}_1-\mathbf{r}_2\mid}\right)\cdot\frac{\partial}{\partial \mathbf{v}_1}F^{(2)}(\mathbf{r}_1,\mathbf{v}_1,\mathbf{r}_2,\mathbf{v}_2,t).$$

To improve visuality, the above equation is rewritten, suppressing the subscript 1,

$$\frac{\partial}{\partial t}F^{(1)}(\mathbf{r},\mathbf{v},t)+\mathbf{v}\cdot\frac{\partial F^{(1)}}{\partial \mathbf{r}}-\frac{e}{m}\frac{\partial\phi(\mathbf{r},t)}{\partial \mathbf{r}}\cdot\frac{\partial F^{(1)}}{\partial \mathbf{v}}$$
$$=\frac{(N-1)e^2}{m}\int d^3r_2 d^3v_2\left(\frac{\partial}{\partial \mathbf{r}}\frac{1}{\mid \mathbf{r}-\mathbf{r}_2\mid}\right)\cdot\frac{\partial}{\partial \mathbf{v}}F^{(2)}(\mathbf{r},\mathbf{v},\mathbf{r}_2,\mathbf{v}_2,t). \tag{12.30}$$

(12.30) is not closed: the equation for $F^{(1)}$ contains $F^{(2)}$. If we set up the equation for $F^{(2)}$ by integrating the Liouville equation by analogous procedure, the $F^{(2)}$-equation will contain $F^{(3)}$ function. Thus this chain of connection should be terminated at some stage by invoking some physical argument. One way to terminate the chain of the hierarchial equations would be to find some approximation for $F^{(2)}$ in terms of $F^{(1)}$. Then the equation for $F^{(1)}$ will be closed to give a description of the time evolution of $F^{(1)}$. As a first approximation, let us put

$$F^{(2)}(\mathbf{r},\mathbf{v},\mathbf{r}_2,\mathbf{v}_2,t)=F^{(1)}(\mathbf{r},\mathbf{v},t)\ F^{(1)}(\mathbf{r}_2,\mathbf{v}_2,t). \tag{12.31}$$

(12.31) means: the probability at time t that a particle is to be found at the phase space location $(\mathbf{r},\mathbf{v})$ is independent from (uncorrelated with) the probability that another particle is simultaneously found at other phase space location $(\mathbf{r}',\mathbf{v}')$. In other words, the particles in the plasma behave as if they were not influenced by the other particles located at other phase points in the system. More refined kinetic theory requires an extra small term of joint probability on the RHS of (12.31). Using (12.31) in (12.30) yields

$$\frac{\partial}{\partial t}F^{(1)}(\mathbf{r},\mathbf{v},t)+\mathbf{v}\cdot\frac{\partial F^{(1)}}{\partial \mathbf{r}}-\frac{e}{m}\frac{\partial\phi(\mathbf{r},t)}{\partial \mathbf{r}}\cdot\frac{\partial F^{(1)}}{\partial \mathbf{v}}$$
$$-\frac{(N-1)e}{m}\left[\int d^3r_2 d^3v_2\left(\frac{\partial}{\partial \mathbf{r}}\frac{e}{\mid \mathbf{r}-\mathbf{r}_2\mid}\right)F^{(1)}(\mathbf{r}_2,\mathbf{v}_2,t)\right]\cdot\frac{\partial}{\partial \mathbf{v}}F^{(1)}(\mathbf{r},\mathbf{v},t)=0. \tag{12.32}$$

Here we recall that ϕ in the above equation is external. The quantity in the large square bracket is the electric field produced at $\mathbf{r}$ by the rest of the electrons in the plasma (see (5.19)). This is the ensemble-averaged force acting on the particle at $\mathbf{r}$ exerted by all the rest of the particles in the plasma. (12.32) is (5.21), which is the Vlasov equation and is obtained as a crude approximation that the 2-particle distribution function is a product of two 1-particle distribution functions.

Remark. The interaction potential which signifies that the force acting on the i-th particle is contributed by the j-th ($i \neq j$) particle necessitates the introduction of the two-particle distribution function $F^{(2)}$ in the reduction of the Liouville equation. The kinetic equation (12.30) can be expressed by

$$\frac{\partial f}{\partial t} + \mathbf{v} \cdot \frac{\partial f}{\partial \mathbf{r}} + \mathbf{a}_{ext} \cdot \frac{\partial f}{\partial \mathbf{v}} = I(f)$$

provided $F^{(2)}$ is expressed in terms of $F^{(1)}$. The major problem of kinetic theory is to find an approximate form for $I(f)$. To close (12.30), the interaction part contributing to the change of f should be found in terms of f itself. The Boltzmann collision term is a coarse grained equation for f which applies to diffuse gases in which strong but impulsive localized interaction occurs. Recall Eq. (1.30) in Chapter 1:

$$I_{Boltzmann}(f) = \int d\Omega \int d^3v_2 g\sigma(g,\theta)\Big[f(v_1')f(v_2') - f(v_1)f(v_2)\Big].$$

For an ionized gas where the inter-particle interactions are weak but long ranged, $I(f)$ is a Fokker–Planck type which takes the form of a Landau collision integral.

Exercise. In contrast to the Klimontovich distribution function in 6-dimension phase space

$$K(\mathbf{r},\mathbf{v},t) = \sum_{i=1}^{N} \delta(\mathbf{r} - \mathbf{r}_i(t))\delta(\mathbf{v} - \mathbf{v}_i(t)),$$

the Liouville distribution function in 6N-dimension phase space can be represented by δ functions as

$$F^{(N)}(\mathbf{r}_1,\mathbf{v}_1,\mathbf{r}_2,\mathbf{v}_2,\ldots,\mathbf{r}_N,\mathbf{v}_N,t) = \prod_{i=1}^{N} \delta(\mathbf{r}_i - \mathbf{r}_i(t))\delta(\mathbf{v}_i - \mathbf{v}_i(t)). \tag{12.33}$$

Show that (12.33) satisfies the Liouville equation (12.19). Note that $\mathbf{r}_i(t)$, $\mathbf{v}_i(t)$ are Lagrangian coordinates.

Next we derive the equation for $F^{(2)}$. We take Liouville's equation (12.25) and integrate over coordinates and velocities of all the particles except those for particles 1 and 2. Then the immediately integrated terms (corresponding to $i = 3, 4, \ldots, N$) vanish, and the resulting intermediate equation can be written as

$$\frac{\partial}{\partial t}F^{(2)}(\mathbf{r}_1, \mathbf{v}_1, \mathbf{r}_2, \mathbf{v}_2, t) + \mathbf{v}_1 \cdot \frac{\partial F^{(2)}}{\partial \mathbf{r}_1} + \mathbf{v}_2 \cdot \frac{\partial F^{(2)}}{\partial \mathbf{r}_2}$$
$$- \frac{e}{m}\left[\frac{\partial \phi(\mathbf{r}_1, t)}{\partial \mathbf{r}_1} \cdot \frac{\partial F^{(2)}}{\partial \mathbf{v}_1} + \frac{\partial \phi(\mathbf{r}_2, t)}{\partial \mathbf{r}_2} \cdot \frac{\partial F^{(2)}}{\partial \mathbf{v}_2}\right]$$
$$+ \text{interaction potential term } (IPT) = 0, \tag{12.34}$$

$$IPT = -\frac{e^2}{m}\int \frac{\partial}{\partial \mathbf{r}_1}\left(\frac{1}{|\,\mathbf{r}_1 - \mathbf{r}_2\,|} + \frac{1}{|\,\mathbf{r}_1 - \mathbf{r}_3\,|} + \frac{1}{|\,\mathbf{r}_1 - \mathbf{r}_4\,|} + \cdots\right) \cdot \frac{\partial F^{(N)}}{\partial \mathbf{v}_1} d^3r_3 d^3v_3 \cdots d^3r_N d^3v_N$$
$$- \frac{e^2}{m}\int \frac{\partial}{\partial \mathbf{r}_2}\left(\frac{1}{|\,\mathbf{r}_2 - \mathbf{r}_1\,|} + \frac{1}{|\,\mathbf{r}_2 - \mathbf{r}_3\,|} + \frac{1}{|\,\mathbf{r}_2 - \mathbf{r}_4\,|} + \cdots\right) \cdot \frac{\partial F^{(N)}}{\partial \mathbf{v}_2} d^3r_3 d^3v_3 \cdots d^3r_N d^3v_N. \tag{12.35}$$

Here we make use of the symmetry of $F^{(N)}$ with respect to the interchange of particle index. In the integrands, the particle index 4 can be replaced by 3 (at the same time, with the interchange of 3 and 4 in $F^{(N)}$). In this way, we see that there are $(N-2)$ identical terms. Therefore we have

$$IPT = -\frac{e^2}{m}\int \frac{\partial}{\partial \mathbf{r}_1}\left(\frac{1}{|\,\mathbf{r}_1 - \mathbf{r}_2\,|} + \frac{N-2}{|\,\mathbf{r}_1 - \mathbf{r}_3\,|}\right) \cdot \frac{\partial F^{(N)}}{\partial \mathbf{v}_1} d^3r_3 d^3v_3 \cdots d^3r_N d^3v_N$$
$$- \frac{e^2}{m}\int \frac{\partial}{\partial \mathbf{r}_2}\left(\frac{1}{|\,\mathbf{r}_2 - \mathbf{r}_1\,|} + \frac{N-2}{|\,\mathbf{r}_2 - \mathbf{r}_3\,|}\right) \cdot \frac{\partial F^{(N)}}{\partial \mathbf{v}_2} d^3r_3 d^3v_3 \cdots d^3r_N d^3v_N$$
$$= -\frac{e^2}{m}\left(\frac{\partial}{\partial \mathbf{r}_1}\frac{1}{|\,\mathbf{r}_1 - \mathbf{r}_2\,|} \cdot \frac{\partial}{\partial \mathbf{v}_1} + \frac{\partial}{\partial \mathbf{r}_2}\frac{1}{|\,\mathbf{r}_2 - \mathbf{r}_1\,|} \cdot \frac{\partial}{\partial \mathbf{v}_2}\right) F^{(2)}$$
$$- \frac{(N-2)e^2}{m}\int\left(\frac{\partial}{\partial \mathbf{r}_1}\frac{1}{|\,\mathbf{r}_1 - \mathbf{r}_3\,|} \cdot \frac{\partial}{\partial \mathbf{v}_1} + \frac{\partial}{\partial \mathbf{r}_2}\frac{1}{|\,\mathbf{r}_2 - \mathbf{r}_3\,|} \cdot \frac{\partial}{\partial \mathbf{v}_2}\right) F^{(3)} d^3r_3 d^3v_3. \tag{12.36}$$

We rewrite the above results as a single equation (omitting the external potential term):

$$\frac{\partial F^{(2)}}{\partial t} + \mathbf{v}_1 \cdot \frac{\partial F^{(2)}}{\partial \mathbf{r}_1} + \mathbf{v}_2 \cdot \frac{\partial F^{(2)}}{\partial \mathbf{r}_2} - \frac{e^2}{m}\left(\frac{\partial}{\partial \mathbf{r}_1}\frac{1}{|\,\mathbf{r}_1 - \mathbf{r}_2\,|} \cdot \frac{\partial}{\partial \mathbf{v}_1} + \frac{\partial}{\partial \mathbf{r}_2}\frac{1}{|\,\mathbf{r}_2 - \mathbf{r}_1\,|} \cdot \frac{\partial}{\partial \mathbf{v}_2}\right) F^{(2)}$$
$$- \frac{(N-2)e^2}{m}\int\left(\frac{\partial}{\partial \mathbf{r}_1}\frac{1}{|\,\mathbf{r}_1 - \mathbf{r}_3\,|} \cdot \frac{\partial}{\partial \mathbf{v}_1} + \frac{\partial}{\partial \mathbf{r}_2}\frac{1}{|\,\mathbf{r}_2 - \mathbf{r}_3\,|} \cdot \frac{\partial}{\partial \mathbf{v}_2}\right) F^{(3)} d^3r_3 d^3v_3 = 0. \tag{12.37}$$

Exercise. The equation governing general $F^{(s)}$ is obtained from the Liouville equation (12.25) by integrating over the variables associated with (N-s) particles. Derive the following equation called BBGKY hierarchy equation (Bogoliubov, Born, Green, Kirkwood, and Yvon).

$$\frac{\partial F^{(s)}}{\partial t} + \sum_{i=1}^{s} \mathbf{v}_i \cdot \frac{\partial F^{(s)}}{\partial \mathbf{r}_i} + \sum_{i=1}^{s}\sum_{j=1}^{s} \mathbf{a}_{ij} \cdot \frac{\partial F^{(s)}}{\partial \mathbf{v}_i}$$

$$+ (N-s)\sum_{i=1}^{s} \int d^3r_{s+1} d^3v_{s+1} \mathbf{a}_{i,s+1} \cdot \frac{\partial F^{(s+1)}}{\partial \mathbf{v}_i} = 0 \qquad (12.38)$$

where

$$F^{(s)}(\mathbf{r}_1, \mathbf{v}_1, \mathbf{r}_2, \mathbf{v}_2, , \ldots, \mathbf{r}_s, \mathbf{v}_s, t)$$

$$= \int F^{(N)}(\mathbf{r}_1, \mathbf{v}_1, \mathbf{r}_2, \mathbf{v}_2, \ldots, \mathbf{r}_s, \mathbf{v}_s, \ldots, \mathbf{r}_N, \mathbf{v}_N, t) d^3r_{s+1} d^3v_{s+1} \cdots d^3r_N d^3v_N,$$

$$\mathbf{a}_{ij} = \frac{e^2}{m} \frac{\mathbf{r}_i - \mathbf{r}_j}{\mid \mathbf{r}_i - \mathbf{r}_j \mid^3} = -\frac{e^2}{m} \frac{\partial}{\partial \mathbf{r}_i} \frac{1}{\mid \mathbf{r}_i - \mathbf{r}_j \mid}$$

is the interacting acceleration. We use $\mathbf{a}_{ij} = 0$ when $i = j$.

(12.38) can be written in terms of the dimensionless variables by expressing length in the unit of the Debye length and time in the unit of the electron plasma frequency. Then the dimensionless equation of (12.38) takes the form

$$\frac{\partial F^{(s)}}{\partial t} + \sum_{i=1}^{s} \mathbf{v}_i \cdot \frac{\partial F^{(s)}}{\partial \mathbf{r}_i} + \frac{1}{g} \sum_{i=1}^{s}\sum_{j=1}^{s} \mathbf{a}_{ij} \cdot \frac{\partial F^{(s)}}{\partial \mathbf{v}_i}$$

$$+ (N-s)\sum_{i=1}^{s} \int d^3r_{s+1} d^3v_{s+1} \mathbf{a}_{i,s+1} \cdot \frac{\partial F^{(s+1)}}{\partial \mathbf{v}_i} = 0.$$

Notice the emergence of the factor g, which is very large ($\gg 1$) in plasma: g turns out to be equal to particle number density (n_0) times the force range. Usually the latter is taken as the Debye length (λ_D) and $g = 4\pi n_0 \lambda_D^3$. Plasma is characterized by the property that the number of particles in the Debye sphere is very large. So the small parameter $1/g$ in the BBGKY equation allows one to approach the equation by expanding $F^{(s)}$ in perturbation series expanded in the powers of $1/g$. [See p. 405 P. C. Clemmow and J. P. Daugherty, *Electrodynamics of particles and Plasmas* (1969) Addison-Wesley London].

12.4. Klimontovich hierarchy

Here we consider the ensemble average of products of Klimontovich distribution functions or the moments of Klimontovich distribution function, and show the latter can be expressed in terms of the Liouville distribution functions $F^{(s)}$. Here we introduce short hand notation: $(\mathbf{r}, \mathbf{v}) \equiv \mathbf{x}$. Consider the ensemble average

$$\langle K(\mathbf{x},t)\rangle = \int F^{(N)}(\mathbf{x}_1,\mathbf{x}_2,\ldots,\mathbf{x}_N,t)K(\mathbf{x},t)d^6x_1d^6x_2\cdots d^6x_N$$
$$= \int F^{(N)}(\mathbf{x}_1,\mathbf{x}_2,\ldots,\mathbf{x}_N,t)\Big[\delta(\mathbf{x}-\mathbf{x}_1)+\delta(\mathbf{x}-\mathbf{x}_2)+\cdots+\delta(\mathbf{x}-\mathbf{x}_N)\Big]d^6x_1d^6x_2\cdots d^6x_N. \tag{12.39}$$

Term $\delta(\mathbf{x}-\mathbf{x}_1)$ gives 1-particle distribution $V^{-1}F^{(1)}$ ($\mathbf{x}$,t) (V is the volume of the box containing the system). Term $\delta(\mathbf{x}-\mathbf{x}_2)$ gives the same. This is the consequence of indistinguishability of particles. Thus, we have

$$\langle K(\mathbf{x},t)\rangle = \frac{N}{V}F^{(1)}(\mathbf{x},t). \tag{12.40}$$

Next, consider the second moment

$$\langle K(\mathbf{x},t)K(\mathbf{x}',t)\rangle = \int \sum_{i=1}^{N}\sum_{j=1}^{N}\delta(\mathbf{x}-\mathbf{x}_i)\delta(\mathbf{x}'-\mathbf{x}_j)F^{(N)}d^6x_1d^6x_2\cdots d^6x_N$$
$$= \frac{N}{V}\int \delta(\mathbf{x}-\mathbf{x}_1(t))\delta(\mathbf{x}'-\mathbf{x}_1(t))F^{(1)}(\mathbf{x}_1,t)d^6x_1$$
$$+\frac{N(N-1)}{V^2}\int \delta(\mathbf{x}-\mathbf{x}_1(t))\delta(\mathbf{x}'-\mathbf{x}_2(t))F^{(2)}(\mathbf{x}_1,\mathbf{x}_2,t)d^6x_1d^6x_2$$
$$= \frac{N}{V}F^{(1)}(\mathbf{x},t)\delta(\mathbf{x}-\mathbf{x}') + \frac{N(N-1)}{V^2}F^{(2)}(\mathbf{x},\mathbf{x}',t) \tag{12.41}$$

where the first term is due to the terms $i = j$ and the second term stems from the cross terms. Similarly, we obtain for the third moment

$$\langle K(\mathbf{x},t)K(\mathbf{x}',t)K(\mathbf{x}'',t)\rangle$$
$$= \int \sum_{i=1}^{N}\sum_{j=1}^{N}\sum_{k=1}^{N}\delta(\mathbf{x}-\mathbf{x}_i)\delta(\mathbf{x}'-\mathbf{x}_j)\delta(\mathbf{x}''-\mathbf{x}_k)F^{(N)}d^6x_1d^6x_2\cdots d^6x_N$$
$$= \frac{N}{V}F^{(1)}(\mathbf{x},t)\delta(\mathbf{x}-\mathbf{x}')\delta(\mathbf{x}-\mathbf{x}'') + \frac{N(N-1)}{V^2}[F^{(2)}(\mathbf{x},\mathbf{x}',t)\delta(\mathbf{x}-\mathbf{x}'')$$
$$+F^{(2)}(\mathbf{x},\mathbf{x}'',t)\delta(\mathbf{x}-\mathbf{x}') + F^{(2)}(\mathbf{x},\mathbf{x}'',t)\delta(\mathbf{x}''-\mathbf{x}')]$$
$$+\frac{1}{V^3}N(N-1)(N-2)F^{(3)}(\mathbf{x},\mathbf{x}',\mathbf{x}'',t). \tag{12.42}$$

The ensemble averages of products of Klimontovich distribution functions or the moments of Klimontovich distribution function can be expressed in terms of the Liouville functions $F^{(s)}$.

12.4.1. *Klimontovich moment equations*

We consider Klimontovich equation (12.3) for particles of one species (electrons) interacting through electrostatic forces:

$$\frac{\partial}{\partial t}K(\mathbf{r},\mathbf{v},t)+\mathbf{v}\cdot\frac{\partial K}{\partial \mathbf{r}}-\frac{e}{m}\left[\int\frac{\partial}{\partial \mathbf{r}}\frac{e}{\mid \mathbf{r}-\mathbf{r}'\mid}K(\mathbf{r}',\mathbf{v}',t)d^3r'd^3v'\right]\cdot\frac{\partial K(\mathbf{r},\mathbf{v},t)}{\partial \mathbf{v}}=0. \tag{12.43}$$

(12.43) is exact without any approximation, and requires to know all the details of orbits of the N particles, as expressed by the δ-functions. In order to reduce it to a usable equation, the statistical average is taken by multiplying it with $F^{(N)}$ and by integrating over all the $\mathbf{r}_i$ and $\mathbf{v}_i$. Noting that the differential operators in (12.43) are independent of the integration variables, the averaged equation takes the form

$$\frac{\partial\langle K\rangle}{\partial t}+\mathbf{v}\cdot\frac{\partial\langle K\rangle}{\partial \mathbf{r}}-\frac{e}{m}\int\frac{\partial}{\partial \mathbf{r}}\frac{e}{\mid \mathbf{r}-\mathbf{r}'\mid}\cdot\frac{\partial\langle KK'\rangle}{\partial \mathbf{v}}d^3r'd^3v'=0 \tag{12.44}$$

where K' has the obvious meaning. When we substitute (12.40) and (12.41) into (12.44), we discard the $\delta(\mathbf{x}-\mathbf{x}')$ term in (12.41) because $\mathbf{r}\neq\mathbf{r}'$ in (12.43). This amounts to rejecting the self-force by putting

$$\int\delta(\mathbf{r}-\mathbf{r}')\frac{\partial}{\partial \mathbf{r}}\frac{1}{\mid \mathbf{r}-\mathbf{r}'\mid}d^3r'=0. \tag{12.45}$$

The singular function in (12.45) repeatedly appears in the subsequent Klimontovich development and we shall set it to zero without further mention. We obtain

$$\frac{\partial}{\partial t}F^{(1)}(\mathbf{r},\mathbf{v},t)+\mathbf{v}\cdot\frac{\partial F^{(1)}}{\partial \mathbf{r}}-\frac{eN}{m}\int d^3r'd^3v'\frac{\partial}{\partial \mathbf{r}}\frac{e}{\mid \mathbf{r}-\mathbf{r}'\mid}\cdot\frac{\partial}{\partial \mathbf{v}}F^{(2)}(\mathbf{r},\mathbf{v},\mathbf{r}',\mathbf{v}',t)=0 \tag{12.46}$$

which is the same as (12.30) with neglect of the external potential in the latter.

Next, to obtain equation for $\langle K(\mathbf{r},\mathbf{v},t)K'(\mathbf{r}',\mathbf{v}',t)\rangle$, note that

$$\frac{\partial}{\partial t}\langle KK'\rangle=\left\langle\frac{\partial}{\partial t}(KK')\right\rangle=\left\langle K'\frac{\partial K}{\partial t}\right\rangle+\left\langle K\frac{\partial K'}{\partial t}\right\rangle.$$

Using (12.43), we construct two equations:

$$K'\left(\frac{\partial}{\partial t}+\mathbf{v}\cdot\frac{\partial}{\partial \mathbf{r}}\right)K=\frac{e}{m}K'\left[\int\frac{\partial}{\partial \mathbf{r}}\frac{e}{\mid \mathbf{r}-\mathbf{r}''\mid}K''d^3r''d^3v''\right]\cdot\frac{\partial K}{\partial \mathbf{v}}=0,$$

$$K\left(\frac{\partial}{\partial t}+\mathbf{v}'\cdot\frac{\partial}{\partial \mathbf{r}'}\right)K'=\frac{e}{m}K\left[\int\frac{\partial}{\partial \mathbf{r}'}\frac{e}{\mid \mathbf{r}'-\mathbf{r}''\mid}K''d^3r''d^3v''\right]\cdot\frac{\partial K'}{\partial \mathbf{v}'}=0.$$

Adding the above two equations and taking the average give

$$\frac{\partial}{\partial t}\langle KK'\rangle+\left(\mathbf{v}\cdot\frac{\partial}{\partial \mathbf{r}}+\mathbf{v}'\cdot\frac{\partial}{\partial \mathbf{r}'}\right)\langle KK'\rangle$$
$$=\int d^3r''d^3v''\left[\frac{\partial}{\partial \mathbf{r}}\frac{e}{\mid \mathbf{r}-\mathbf{r}''\mid}\cdot\frac{\partial}{\partial \mathbf{v}}+\frac{\partial}{\partial \mathbf{r}'}\frac{e}{\mid \mathbf{r}'-\mathbf{r}''\mid}\cdot\frac{\partial}{\partial \mathbf{v}'}\right]\langle KK'K''\rangle. \tag{12.47}$$

Substituting (12.41) and (12.42) into (12.47) yields

$$\left(\frac{\partial}{\partial t}+\mathbf{v}\cdot\frac{\partial}{\partial \mathbf{r}}+\mathbf{v}'\cdot\frac{\partial}{\partial \mathbf{r}'}\right)F^{(2)}(\mathbf{x},\mathbf{x}',t)$$
$$=\frac{N}{m}\left[\frac{\partial}{\partial \mathbf{r}}\frac{e^2}{\mid \mathbf{r}-\mathbf{r}'\mid}\cdot\frac{\partial}{\partial \mathbf{v}}+\frac{\partial}{\partial \mathbf{r}'}\frac{e^2}{\mid \mathbf{r}'-\mathbf{r}\mid}\cdot\frac{\partial}{\partial \mathbf{v}'}\right]F^{(2)}(\mathbf{x},\mathbf{x}',t)$$
$$+\frac{N}{m}\int d^3r''d^3v''\left[\frac{\partial}{\partial \mathbf{r}}\frac{e^2}{\mid \mathbf{r}-\mathbf{r}''\mid}\cdot\frac{\partial}{\partial \mathbf{v}}+\frac{\partial}{\partial \mathbf{r}'}\frac{e^2}{\mid \mathbf{r}'-\mathbf{r}''\mid}\cdot\frac{\partial}{\partial \mathbf{v}'}\right]F^{(3)}(\mathbf{x},\mathbf{x}',\mathbf{x}'',t) \tag{12.48}$$

where $\mathbf{x}=(\mathbf{r},\mathbf{v})$. (12.48) is the same as (12.37). (12.46), (12.48), and the higher Klimontovich moment equations are in complete agreement with the BBGKY hierarchy equations (12.38).

Exercise. Carry out the algebra leading to (12.48).

Solution: The following steps will be useful. Write

$$\left(\frac{\partial}{\partial t}+\mathbf{v}\cdot\frac{\partial}{\partial \mathbf{r}}+\mathbf{v}'\cdot\frac{\partial}{\partial \mathbf{r}'}\right)\left[\frac{N}{V}F^{(1)}(\mathbf{x},t)\delta(\mathbf{x}-\mathbf{x}')+\frac{N^2}{V^2}F^{(2)}(\mathbf{x},\mathbf{x}',t)\right]$$
$$=\frac{1}{m}\int d^3r''d^3v''\left[\frac{\partial}{\partial \mathbf{r}}\frac{e^2}{\mid \mathbf{r}-\mathbf{r}''\mid}\cdot\frac{\partial}{\partial \mathbf{v}}^{a}+\frac{\partial}{\partial \mathbf{r}'}\frac{e^2}{\mid \mathbf{r}'-\mathbf{r}''\mid}\cdot\frac{\partial}{\partial \mathbf{v}'}^{b}\right]\times$$
$$\left(\frac{N}{V}F^{(1)}(\mathbf{x},t)\delta(\mathbf{x}-\mathbf{x}')\delta(\mathbf{x}-\mathbf{x}'')^{\copyright}+\frac{N^2}{V^2}F^{(2)}(\mathbf{x},\mathbf{x}',t)\delta(\mathbf{x}-\mathbf{x}'')\right.$$
$$\left.+\frac{N^2}{V^2}F^{(2)}(\mathbf{x},\mathbf{x}'',t)\delta(\mathbf{x}-\mathbf{x}')+\frac{N^2}{V^2}F^{(2)}(\mathbf{x},\mathbf{x}'',t)\delta(\mathbf{x}''-\mathbf{x}')+\frac{N^3}{V^3}F^{(3)}(\mathbf{x},\mathbf{x}',\mathbf{x}'',t)\right).$$

Note $\delta(\mathbf{x}-\mathbf{x}') = \delta(\mathbf{r}-\mathbf{r}')\delta(\mathbf{v}-\mathbf{v}')$ to see that product terms $a© = b© = 0$.

Also, $\left(\mathbf{v}\cdot\frac{\partial}{\partial \mathbf{r}} + \mathbf{v}'\cdot\frac{\partial}{\partial \mathbf{r}'}\right)[F^{(1)}(\mathbf{x},t)\delta(\mathbf{x}-\mathbf{x}')] = \delta(\mathbf{x}-\mathbf{x}')\mathbf{v}\cdot\frac{\partial}{\partial \mathbf{r}}F^{(1)}(\mathbf{x},t)$.

Omitting the $F^{(3)}$-term, we have, using (12.46),

$$\left(\frac{\partial}{\partial t} + \mathbf{v}\cdot\frac{\partial}{\partial \mathbf{r}} + \mathbf{v}'\cdot\frac{\partial}{\partial \mathbf{r}'}\right)\frac{N^2}{V^2}F^{(2)}(\mathbf{x},\mathbf{x}',t)$$
$$+\frac{N^2}{V^2}\delta(\mathbf{x}-\mathbf{x}')\frac{1}{m}\int d^3r''d^3v''\frac{\partial}{\partial \mathbf{r}}\frac{e^2}{|\,\mathbf{r}-\mathbf{r}''\,|}\cdot\frac{\partial}{\partial \mathbf{v}}F^{(2)}(\mathbf{x},\mathbf{x}'',t)^d$$
$$=\frac{1}{m}\int d^3r''d^3v''\left[\frac{\partial}{\partial \mathbf{r}}\frac{e^2}{|\,\mathbf{r}-\mathbf{r}''\,|}\cdot\frac{\partial}{\partial \mathbf{v}}^{a} + \frac{\partial}{\partial \mathbf{r}'}\frac{e^2}{|\,\mathbf{r}'-\mathbf{r}''\,|}\cdot\frac{\partial}{\partial \mathbf{v}'}^{b}\right]$$
$$\times\frac{N^2}{V^2}(F^{(2)}(\mathbf{x},\mathbf{x}',t)\delta(\mathbf{x}-\mathbf{x}'') + F^{(2)}(\mathbf{x},\mathbf{x}'',t)\delta(\mathbf{x}-\mathbf{x}')^{©}$$
$$+F^{(2)}(\mathbf{x},\mathbf{x}'',t)\delta(\mathbf{x}''-\mathbf{x}')).$$

Product term $a©$ produces two terms one of which cancels term d, the other term cancels $b©$. Also we have vanishing terms due to (12.45), and the remaining terms give (12.48).

12.4.2. *Correlation function*

So far we obtained two lowest order equations, (12.30) and (12.48). Two-particle correlation function g is defined by

$$F^{(2)}(\mathbf{x},\mathbf{x}',t) = F^{(1)}(\mathbf{x},t)F^{(1)}(\mathbf{x}',t) + g(\mathbf{x},\mathbf{x}',t). \tag{12.49}$$

Three-particle correlation function (h) is defined by the *Mayor cluster expansion*:

$$F^{(3)}(\mathbf{x},\mathbf{x}',\mathbf{x}'',t) = F^{(1)}(\mathbf{x},t)F^{(1)}(\mathbf{x}',t)F^{(1)}(\mathbf{x}'',t) + F^{(1)}(\mathbf{x},t)g(\mathbf{x}',\mathbf{x}'',t)$$
$$+F^{(1)}(\mathbf{x}',t)g(\mathbf{x},\mathbf{x}'',t) + F^{(1)}(\mathbf{x}'',t)g(\mathbf{x},\mathbf{x}',t) + h(\mathbf{x},\mathbf{x}',\mathbf{x}'',t). \tag{12.50}$$

(12.30), (12.48), (12.49), and (12.50) constitute a closed set of equations for $F^{(1)}$ and g if the three-particle correlation h is neglected. We rewrite (12.30) by using (12.49) as

$$\left(\frac{\partial}{\partial t} + \mathbf{v}\cdot\frac{\partial}{\partial \mathbf{r}}\right)F^{(1)}(\mathbf{x})$$
$$=\frac{eN}{m}\int d^3r''d^3v''\frac{\partial}{\partial \mathbf{r}}\frac{e}{|\,\mathbf{r}-\mathbf{r}''\,|}\cdot\frac{\partial}{\partial \mathbf{v}}\left[g(\mathbf{x},\mathbf{x}'') + F^{(1)}(\mathbf{x})F^{(1)}(\mathbf{x}'')\right] \tag{12.51}$$

where we suppressed t for notational simplicity. Interchanging the unprimed symbols to single primed symbols, we have

$$\left(\frac{\partial}{\partial t} + \mathbf{v}' \cdot \frac{\partial}{\partial \mathbf{r}'}\right) F^{(1)}(\mathbf{x}')$$

$$= \frac{eN}{m} \int d^3r'' d^3v'' \frac{\partial}{\partial \mathbf{r}'} \frac{e}{\mid \mathbf{r}' - \mathbf{r}'' \mid} \cdot \frac{\partial}{\partial \mathbf{v}'} \left[g(\mathbf{x}', \mathbf{x}'') + F^{(1)}(\mathbf{x}')F^{(1)}(\mathbf{x}'') \right]. \quad (12.52)$$

Putting (12.49) and (12.50) into (12.48) and carrying out the differentiation, we have

$$\left(\frac{\partial F}{\partial t} + \mathbf{v} \cdot \frac{\partial F}{\partial \mathbf{r}}\right) F' \,© + \left(\frac{\partial F'}{\partial t} + \mathbf{v}' \cdot \frac{\partial F'}{\partial \mathbf{r}'}\right) F^{\otimes} + \left(\frac{\partial}{\partial t} + \mathbf{v} \cdot \frac{\partial}{\partial \mathbf{r}} + \mathbf{v}' \cdot \frac{\partial}{\partial \mathbf{r}'}\right) g(\mathbf{x}, \mathbf{x}')$$

$$= \frac{\partial}{\partial \mathbf{r}} \frac{e^2/m}{\mid \mathbf{r} - \mathbf{r}' \mid} \cdot \frac{\partial}{\partial \mathbf{v}} (FF' + g(\mathbf{x}, \mathbf{x}')) + \frac{\partial}{\partial \mathbf{r}'} \frac{e^2/m}{\mid \mathbf{r} - \mathbf{r}' \mid} \cdot \frac{\partial}{\partial \mathbf{v}'} (FF' + g(\mathbf{x}, \mathbf{x}'))$$

$$+ \int d^3r'' d^3v'' \frac{\partial}{\partial \mathbf{r}} \frac{e}{\mid \mathbf{r} - \mathbf{r}'' \mid} \cdot \frac{\partial}{\partial \mathbf{v}} [FF''F' \,© + F'g(\mathbf{x}, \mathbf{x}'')\,© + Fg(\mathbf{x}', \mathbf{x}'') + F''g(\mathbf{x}, \mathbf{x}')]$$

$$+ \int d^3r'' d^3v'' \frac{\partial}{\partial \mathbf{r}'} \frac{e}{\mid \mathbf{r}' - \mathbf{r}'' \mid} \cdot \frac{\partial}{\partial \mathbf{v}'} [FF'F''^{\otimes} + Fg(\mathbf{x}', \mathbf{x}'')^{\otimes} + F'g(\mathbf{x}, \mathbf{x}'') + F''g(\mathbf{x}, \mathbf{x}')] = 0$$

where F, F', F'' have the obvious meaning. The three terms indicated by © add to zero due to (12.51). The three terms indicated by $\otimes$ add to zero due to (12.52). Thus the final equation for g becomes

$$\left(\frac{\partial}{\partial t} + \mathbf{v} \cdot \frac{\partial}{\partial \mathbf{r}} + \mathbf{v}' \cdot \frac{\partial}{\partial \mathbf{r}'}\right) g(\mathbf{x}, \mathbf{x}', t) = \frac{\partial}{\partial \mathbf{r}} \frac{e^2/m}{\mid \mathbf{r} - \mathbf{r}' \mid} \cdot \frac{\partial}{\partial \mathbf{v}} \left[F^{(1)}(\mathbf{x}, t) F^{(1)}(\mathbf{x}', t)^a \right.$$

$$\left. + g(\mathbf{x}, \mathbf{x}', t)^e \right] + \frac{\partial}{\partial \mathbf{r}'} \frac{e^2/m}{\mid \mathbf{r} - \mathbf{r}' \mid} \cdot \frac{\partial}{\partial \mathbf{v}'} \left[F^{(1)}(\mathbf{x}, t) F^{(1)}(\mathbf{x}', t)^b + g(\mathbf{x}, \mathbf{x}', t)^f \right]$$

$$+ \int d^3r'' d^3v'' \frac{\partial}{\partial \mathbf{r}} \frac{e^2/m}{\mid \mathbf{r} - \mathbf{r}'' \mid} \cdot \frac{\partial}{\partial \mathbf{v}} \left[F^{(1)}(\mathbf{x}, t) g(\mathbf{x}', \mathbf{x}'', t)^c + F^{(1)}(\mathbf{x}'', t) g(\mathbf{x}, \mathbf{x}', t)^g \right]$$

$$+ \int d^3r'' d^3v'' \frac{\partial}{\partial \mathbf{r}'} \frac{e^2/m}{\mid \mathbf{r}' - \mathbf{r}'' \mid} \cdot \frac{\partial}{\partial \mathbf{v}'} \left[F^{(1)}(\mathbf{x}', t) g(\mathbf{x}, \mathbf{x}'', t)^d + F^{(1)}(\mathbf{x}'', t) g(\mathbf{x}, \mathbf{x}', t)^h \right]. \quad (12.53)$$

Note that half of the right hand side of (12.53) is obtained by interchanging "unprime" and "single prime". (12.51) and (12.53) are two coupled equations which will be further simplified, and then solved. The crucial simplification is to put $\frac{\partial}{\partial \mathbf{v}} g(\mathbf{v}) = 0$. Reference: D. R. Nicholson, *Introduction to Plasma theory*, Wiley (1983). Then we discard the second terms in each line of the RHS of (12.53). We can argue that term $a \gg$ term e, term $b \gg$ term f. Terms h, g are the so called electric field term, which vanish in a homogeneous plasma. Reference: D. C. Montgomery and D. A, Tidman, *Plasma kinetic theory*, McGraw-Hill (1964).

12.5. Lenard–Balescu equation

Reference: D. C. Montgomery and D. A, Tidman, *Plasma kinetic theory*, McGraw-Hill (1964)

In the RHS of (12.53), we retain only the terms, a, b, c, d. Also, we neglect the nonlinear term in (12.51). Then we have the following set of coupled integro-differential equations:

$$\begin{aligned}&\left(\frac{\partial}{\partial t}+\mathbf{v}\cdot\frac{\partial}{\partial \mathbf{r}}+\mathbf{v}'\cdot\frac{\partial}{\partial \mathbf{r}'}\right)g(\mathbf{x},\mathbf{x}',t)\\&=\left[\frac{\partial}{\partial \mathbf{r}}\frac{e^2/m}{\mid \mathbf{r}-\mathbf{r}'\mid}\cdot\frac{\partial}{\partial \mathbf{v}}+\frac{\partial}{\partial \mathbf{r}'}\frac{e^2/m}{\mid \mathbf{r}-\mathbf{r}'\mid}\cdot\frac{\partial}{\partial \mathbf{v}'}\right]F(\mathbf{v})F(\mathbf{v}')\\&\quad+\frac{\partial F(\mathbf{v})}{\partial \mathbf{v}}\cdot\int d^3v''\int d^3r''\left(\frac{\partial}{\partial \mathbf{r}}\frac{e^2/m}{\mid \mathbf{r}-\mathbf{r}''\mid}\right)g(\mathbf{x}',\mathbf{x}'',t)\\&\quad+\frac{\partial F(\mathbf{v}')}{\partial \mathbf{v}'}\cdot\int d^3v''\int d^3r''\left(\frac{\partial}{\partial \mathbf{r}'}\frac{e^2/m}{\mid \mathbf{r}'-\mathbf{r}''\mid}\right)g(\mathbf{x},\mathbf{x}'',t),\end{aligned} \tag{12.54}$$

and

$$\frac{\partial}{\partial t}F(\mathbf{v},t)=\frac{eN}{m}\int d^3r''d^3v''\frac{\partial}{\partial \mathbf{r}}\frac{e}{\mid \mathbf{r}-\mathbf{r}''\mid}\cdot\frac{\partial}{\partial \mathbf{v}}g(\mathbf{r},\mathbf{r}'',\mathbf{v},\mathbf{v}'',t). \tag{12.55}$$

In (12.54), we assumed that $F(\mathbf{v})\equiv F^{(1)}(\mathbf{x},t)$ is homogeneous in space, and its time-dependence is weak. Consult § 5.7 for the following analysis.

• Development of (12.54); the steps leading to (12.79) below are explained in the following.

The homogeneous part of (12.54) consists of two parts which resemble the Vlasov equations:

$$\left(\frac{\partial}{\partial t}+\mathbf{v}\cdot\frac{\partial}{\partial \mathbf{r}}\right)g(\mathbf{x},\mathbf{x}',t)-\frac{\partial F(\mathbf{v})}{\partial \mathbf{v}}\cdot\int d^3v''\int d^3r''\left(\frac{\partial}{\partial \mathbf{r}}\frac{e^2/m}{\mid \mathbf{r}-\mathbf{r}''\mid}\right)g(\mathbf{x}',\mathbf{x}'',t)=0, \tag{12.56}$$

$$\left(\frac{\partial}{\partial t}+\mathbf{v}'\cdot\frac{\partial}{\partial \mathbf{r}'}\right)g(\mathbf{x},\mathbf{x}',t)-\frac{\partial F(\mathbf{v}')}{\partial \mathbf{v}'}\cdot\int d^3v''\int d^3r''\left(\frac{\partial}{\partial \mathbf{r}'}\frac{e^2/m}{\mid \mathbf{r}'-\mathbf{r}''\mid}\right)g(\mathbf{x},\mathbf{x}'',t)=0. \tag{12.57}$$

The above equations will be *opratorially* combined in the form (12.58) below. In reading (12.56), the variable $\mathbf{x}'$ is regarded as a parameter or simply ignored if preferred. [In reading (12.57), the variable $\mathbf{x}$ is regarded as a parameter or simply ignored if preferred.] Then one notices that it is the Vlasov equation in which $g(\mathbf{x}') = g(\mathbf{r}', \mathbf{v}', t)$ [$g(\mathbf{x}) = g(\mathbf{r}, \mathbf{v}, t)$] plays the role of the perturbed distribution function. It is useful to write (12.54) in the following form (to symmetrize the equations, we change notations from primes to numeric subscripts):

$$\frac{\partial}{\partial t} g(\mathbf{x}_1, \mathbf{x}_2, t) + Hg = S, \tag{12.58}$$

$$S = \left[\frac{\partial}{\partial \mathbf{r}_1} \frac{e^2/m}{\mid \mathbf{r}_1 - \mathbf{r}_2 \mid} \cdot \frac{\partial}{\partial \mathbf{v}_1} + \frac{\partial}{\partial \mathbf{r}_2} \frac{e^2/m}{\mid \mathbf{r}_1 - \mathbf{r}_2 \mid} \cdot \frac{\partial}{\partial \mathbf{v}_2}\right] F(\mathbf{v}_1) F(\mathbf{v}_2), \tag{12.59}$$

$$H = H_1 + H_2.$$

H_1 and H_2 are linear operators which act on arbitrary spatial function $h(\mathbf{r}_1, \mathbf{r}_2)$ in the following manner:

$$H_1 h(\mathbf{r}_1, \mathbf{r}_2) = \mathbf{v}_1 \cdot \frac{\partial}{\partial \mathbf{r}_1} h(\mathbf{r}_1, \mathbf{r}_2) - \frac{\partial F(\mathbf{v}_1)}{\partial \mathbf{v}_1} \cdot \int d^3 v_3 \int d^3 r_3 \frac{\partial}{\partial \mathbf{r}_1} \frac{e^2/m}{\mid \mathbf{r}_1 - \mathbf{r}_3 \mid} h(\mathbf{r}_2, \mathbf{r}_3), \tag{12.60}$$

$$H_2 h(\mathbf{r}_1, \mathbf{r}_2) = \mathbf{v}_2 \cdot \frac{\partial}{\partial \mathbf{r}_2} h(\mathbf{r}_1, \mathbf{r}_2) - \frac{\partial F(\mathbf{v}_2)}{\partial \mathbf{v}_2} \cdot \int d^3 v_3 \int d^3 r_3 \frac{\partial}{\partial \mathbf{r}_2} \frac{e^2/m}{\mid \mathbf{r}_2 - \mathbf{r}_3 \mid} h(\mathbf{r}_1, \mathbf{r}_3). \tag{12.61}$$

To solve (12.58), we introduce an operator $p(t)$ which is a solution of the homogeneous operator equation

$$\frac{\partial p(t)}{\partial t} + Hp = 0; \quad p(t) = e^{-Ht} \tag{12.62}$$

subject to the initial condition $p(0) = 1$, with an inverse defined by

$$p^{-1} p = p\, p^{-1} = 1.$$

If such an operator is available, it follows that the solution of (12.58) can be written as

$$g(t) = p(t) \int_0^t p^{-1}(\tau) S(\tau) d\tau + p(t) g(t = 0). \tag{12.63}$$

(12.63) can be proved by direct substitution into (12.58). Since the operators H_1 and H_2 commute, the solution to (12.62) can also be written as

$$p(t) = e^{-Ht} = e^{-(H_1 + H_2)t} = e^{-H_1 t} e^{-H_2 t} \tag{12.64}$$

and we define operators

$$p_1(t) = e^{-H_1 t}, \quad p_2(t) = e^{-H_2 t} \tag{12.65}$$

which satisfy, respectively,

$$\frac{\partial p_1}{\partial t} + H_1 p_1 = 0, \quad \frac{\partial p_2}{\partial t} + H_2 p_2 = 0. \tag{12.66}$$

Now we substitute (12.64) into (12.63) with putting $g(t = 0) = 0$:

$$\begin{aligned} g(t) &= e^{-Ht} \int_0^t e^{H\tau}\, S(\tau) d\tau = \int_0^t e^{-H(t-\tau)} S(\tau) d\tau \\ &= \int_0^t p(t-\tau) S(\tau) d\tau = \int_0^t p(\tau) S(t-\tau) d\tau \\ &= \int_0^t p_1(\tau) p_2(\tau) S(t-\tau) d\tau. \end{aligned}$$

Furthermore, we shall assume S to be time-independent, as given by (12.59). Then we have

$$g(t) = \int_0^t p_1(\tau) p_2(\tau) S\, d\tau. \tag{12.67}$$

A word of caution: S, although it is time independent, it is an object of the operator p, so it must follow p. Now we need an explicit expression for the integrand of (12.67). So we consider the effect of the operator $p_1(t)$ when it operates on an arbitrary function $h(\mathbf{r}_1, \mathbf{v}_1)$. We state: *if we apply the operator $p_1(t)$ to such a function $h(\mathbf{r}_1, \mathbf{v}_1)$, the result is a generation of a time-dependent function $\tilde{h}(\mathbf{r}_1, \mathbf{v}_1, t)$ which satisfies the equation*

$$\frac{\partial}{\partial t} \tilde{h}(\mathbf{r}_1, \mathbf{v}_1, t) + H_1 \tilde{h}(\mathbf{r}_1, \mathbf{v}_1, t) = 0. \tag{12.68}$$

The above remark is mathematically expressed as

$$p_1(t) h(\mathbf{r}_1, \mathbf{v}_1) = \tilde{h}(\mathbf{r}_1, \mathbf{v}_1, t) \tag{12.69}$$

where $\tilde{h}(\mathbf{r}_1, \mathbf{v}_1, t)$ satisfies (12.68). The proof goes as follows. Differentiating both sides of (12.69) with respect to t gives

$$\frac{\partial p_1}{\partial t}\, h(\mathbf{r}_1, \mathbf{v}_1) = \frac{\partial}{\partial t} \tilde{h}(\mathbf{r}_1, \mathbf{v}_1, t),$$

$$LHS = -H_1 p_1(t) h(\mathbf{r}_1, \mathbf{v}_1). \quad (per\ (12.66))$$

By the assertion (12.69), LHS in the above $= -H_1 \tilde{h}(\mathbf{r}_1, \mathbf{v}_1, t)$, from which (12.68) follows.

Returning to (12.67), we first consider the action of $p_1(t)$ on S (p_1 and p_2 commute).

$$p_1(t) S(\mathbf{r}_1, \mathbf{v}_1; \mathbf{x}_2) = \tilde{S}(\mathbf{r}_1, \mathbf{v}_1, t; \mathbf{x}_2), \quad (\mathbf{x}_2 = (\mathbf{r}_2, \mathbf{v}_2) \textit{ is a parameter})$$

where $\tilde{S}(\mathbf{r}_1, \mathbf{v}_1, t; \mathbf{x}_2)$ satisfies the equation below

$$\begin{aligned} &\frac{\partial}{\partial t}\tilde{S}(\mathbf{r}_1, \mathbf{v}_1, t; \mathbf{x}_2) + \mathbf{v}_1 \cdot \frac{\partial}{\partial \mathbf{r}_1}\tilde{S}(\mathbf{r}_1, \mathbf{v}_1, t; \mathbf{x}_2) \\ &\quad = \frac{\partial F(\mathbf{v}_1)}{\partial \mathbf{v}_1} \cdot \int d^3 v_3 d^3 r_3 \frac{\partial}{\partial \mathbf{r}_1} \frac{e^2/m}{\mid \mathbf{r}_1 - \mathbf{r}_3 \mid} \tilde{S}(\mathbf{r}_3, \mathbf{v}_3, t; \mathbf{x}_2). \end{aligned} \tag{12.70}$$

(12.70) is nothing but the linearized Vlasov equation which can be inverted in terms of the initial value of $\tilde{S}(t = 0)$. Here the initial value of $\tilde{S}(t = 0)$ is equal to $S(\mathbf{r}_1, \mathbf{v}_1; \mathbf{x}_2)$. Thus the solution of (12.70) in $(\mathbf{k}_1, \omega_1)$-space is written as

$$\begin{aligned} &\tilde{S}(\omega_1, \mathbf{k}_1, \mathbf{v}_1; \mathbf{x}_2) \\ &= \frac{i}{\omega_1 - \mathbf{k}_1 \cdot \mathbf{v}_1}\left[1 - \frac{4\pi e^2}{m k_1^2 \varepsilon(\mathbf{k}_1, \omega_1)} \mathbf{k}_1 \cdot \frac{\partial F(\mathbf{v}_1)}{\partial \mathbf{v}_1} \int \frac{d^3 v_1}{\omega_1 - \mathbf{k}_1 \cdot \mathbf{v}_1}\right] S(\mathbf{k}_1, \mathbf{v}_1; \mathbf{x}_2) \\ &\equiv L(1) S(\mathbf{k}_1, \mathbf{v}_1; \mathbf{x}_2) \end{aligned} \tag{12.71}$$

where $L(1)$ is an operator defined by

$$L(1)(\cdots) = \frac{i}{\omega_1 - \mathbf{k}_1 \cdot \mathbf{v}_1}\left[(\cdots) - \frac{4\pi e^2}{m k_1^2 \varepsilon(\mathbf{k}_1, \omega_1)} \mathbf{k}_1 \cdot \frac{\partial F(\mathbf{v}_1)}{\partial \mathbf{v}_1} \int \frac{d^3 v_1 \,(\cdots)}{\omega_1 - \mathbf{k}_1 \cdot \mathbf{v}_1}\right].$$

Note that v_1 inside the integral above is dummy. And

$$\varepsilon(\mathbf{k}_1, \omega_1) = 1 + \frac{4\pi e^2}{m k_1^2} \int d^3 v_1 \frac{\mathbf{k}_1 \cdot \frac{\partial F}{\partial \mathbf{v}_1}}{\omega_1 - \mathbf{k}_1 \cdot \mathbf{v}_1} \tag{12.72}$$

and $S(\mathbf{k}_1, \mathbf{v}_1; \mathbf{x}_2)$ is the Fourier transform of $\tilde{S}(t = 0) = S(\mathbf{r}_1, \mathbf{v}_1; \mathbf{x}_2)$. Inverting (12.71) gives

$$\tilde{S}(\mathbf{r}_1, \mathbf{v}_1, t; \mathbf{x}_2) = \int_{-\infty + ic_1}^{\infty + ic_1} \frac{d\omega_1}{2\pi} \int \frac{d^3 k_1}{(2\pi)^3} e^{i\mathbf{k}_1 \cdot \mathbf{r}_1 - i\omega_1 t} L(1) S(\mathbf{k}_1, \mathbf{v}_1; \mathbf{x}_2). \tag{12.73}$$

Next, we apply the operator $p_2(t)$ on (12.73):

$$p_2(t)\tilde{S}(\mathbf{r}_1, \mathbf{v}_1, t; \mathbf{x}_2) = \int_{-\infty+ic_1}^{\infty+ic_1} \frac{d\omega_1}{2\pi} \int \frac{d^3k_1}{(2\pi)^3} e^{i\mathbf{k}_1\cdot\mathbf{r}_1 - i\omega_1 t} L(1)p_2(t)S(\mathbf{k}_1, \mathbf{v}_1; \mathbf{x}_2) \tag{12.74}$$

where $p_2(t)S(\mathbf{k}_1, \mathbf{v}_1; \mathbf{x}_2) \equiv S'(\mathbf{k}_1, \mathbf{v}_1, \mathbf{x}_2, t)$ satisfies

$$\frac{\partial}{\partial t} S'(\mathbf{k}_1, \mathbf{v}_1, \mathbf{x}_2, t) + \mathbf{v}_2 \cdot \frac{\partial}{\partial \mathbf{r}_2} S'(\mathbf{k}_1, \mathbf{v}_1, \mathbf{x}_2, t) = \frac{\partial F(\mathbf{v}_2)}{\partial \mathbf{v}_2} \cdot \int d^6x_3 \frac{\partial}{\partial \mathbf{r}_2} \frac{e^2/m}{\mid \mathbf{r}_2 - \mathbf{r}_3 \mid} S'(\mathbf{k}_1, \mathbf{v}_1, \mathbf{x}_3, t). \tag{12.75}$$

The Fourier–Laplace solution of (12.75) in terms of the Fourier–Laplace variables, $(\mathbf{k}_2, \omega_2)$, is

$$S'(\mathbf{k}_1, \mathbf{v}_1, \mathbf{k}_2, \mathbf{v}_2, \omega_2) = L(2)S(\mathbf{k}_1, \mathbf{v}_1, \mathbf{k}_2, \mathbf{v}_2) \tag{12.76}$$

where $S(\mathbf{k}_1, \mathbf{v}_1, \mathbf{k}_2, \mathbf{v}_2)$ is the spatial Fourier transform of the initial value as given by (12.59), and $L(2)$ is defined by (12.71) with replacement of $1 \to 2$. Inverting (12.76) yields

$$p_2(t)S(\mathbf{k}_1, \mathbf{v}_1; \mathbf{x}_2) = \int_{-\infty+ic_2}^{\infty+ic_2} \frac{d\omega_2}{2\pi} \int \frac{d^3k_2}{(2\pi)^3} e^{i\mathbf{k}_2\cdot\mathbf{r}_2 - i\omega_2 t} L(2)S(\mathbf{k}_1, \mathbf{v}_1, \mathbf{k}_2, \mathbf{v}_2). \tag{12.77}$$

Substituting (12.77) into (12.74) gives

$$\begin{aligned} p_2(t)\tilde{S}(\mathbf{r}_1, \mathbf{v}_1, t; \mathbf{x}_2) &= p_1(t)p_2(t)S(\mathbf{r}_1, \mathbf{v}_1, \mathbf{r}_2, \mathbf{v}_2) \\ &= \int_{-\infty+ic_2}^{\infty+ic_2} \frac{d\omega_2}{2\pi} \int \frac{d^3k_2}{(2\pi)^3} \int_{-\infty+ic_1}^{\infty+ic_1} \frac{d\omega_1}{2\pi} \int \frac{d^3k_1}{(2\pi)^3} \\ &\quad \times e^{i(\mathbf{k}_1\cdot\mathbf{r}_1 + \mathbf{k}_2\cdot\mathbf{r}_2)} e^{-i(\omega_1+\omega_2)t} L(1)L(2)S(\mathbf{k}_1, \mathbf{v}_1, \mathbf{k}_2, \mathbf{v}_2). \end{aligned} \tag{12.78}$$

Integrating (12.78) with respect to t via (12.67), we obtain g the solution of (12.58), in the form

$$\begin{aligned} g(\mathbf{r}_1, \mathbf{v}_1, \mathbf{r}_2, \mathbf{v}_2, t) &= \int_{-\infty+ic_2}^{\infty+ic_2} \frac{d\omega_2}{2\pi} \int \frac{d^3k_2}{(2\pi)^3} \int_{-\infty+ic_1}^{\infty+ic_1} \frac{d\omega_1}{2\pi} \int \frac{d^3k_1}{(2\pi)^3} \frac{e^{-i(\omega_1+\omega_2)t} - 1}{-i(\omega_1 + \omega_2)} \\ &\quad \times e^{i(\mathbf{k}_1\cdot\mathbf{r}_1 + \mathbf{k}_2\cdot\mathbf{r}_2)} L(1)L(2)S(\mathbf{k}_1, \mathbf{v}_1, \mathbf{k}_2, \mathbf{v}_2). \end{aligned} \tag{12.79}$$

This equation should be used in (12.55) which we write in the form

$$\frac{\partial}{\partial t}F(\mathbf{v}_1,t) = N\int d^3r_2 d^3v_2\, \frac{\partial}{\partial \mathbf{r}_1}\frac{e^2/m}{|\,\mathbf{r}_1-\mathbf{r}_2\,|}\cdot\frac{\partial}{\partial \mathbf{v}_1}g(\mathbf{r}_1,\mathbf{v}_1,\mathbf{r}_2,\mathbf{v}_2,t). \tag{12.80}$$

• (12.79) and (12.80) are the objects of our further development. Recall that $F(\mathbf{v},t)$ is a slowly varying function of t. To match the slow time scale of the LHS of (12.80) the RHS is evaluated in the time asymptotic limit of $g(\mathbf{r}_1,\mathbf{v}_1,\mathbf{r}_2,\mathbf{v}_2,t\to\infty)$. Our next development points to (12.94) below.

Let us consider the spatial Fourier transform of (12.79):

$$\begin{aligned} &g(\mathbf{k}_1,\mathbf{v}_1,\mathbf{k}_2,\mathbf{v}_2,t)\\ &\quad= i\int_{-\infty+ic_2}^{\infty+ic_2}\frac{d\omega_2}{2\pi}\int_{-\infty+ic_1}^{\infty+ic_1}\frac{d\omega_1}{2\pi}\frac{e^{-i(\omega_1+\omega_2)t}-1}{\omega_1+\omega_2}\,L(1)L(2)S(\mathbf{k}_1,\mathbf{v}_1,\mathbf{k}_2,\mathbf{v}_2). \end{aligned} \tag{12.81}$$

Exercise. Show that

$$\int L(2)S(\mathbf{k}_1,\mathbf{v}_1,\mathbf{k}_2,\mathbf{v}_2)d^3v_2 = \frac{i}{\varepsilon(\mathbf{k}_2,\omega_2)}\int\frac{d^3v_2}{\omega_2-\mathbf{k}_2\cdot\mathbf{v}_2}\,S(\mathbf{k}_1,\mathbf{v}_1,\mathbf{k}_2,\mathbf{v}_2). \tag{12.82}$$

This exercise is useful to calculate:

$$\begin{aligned} &\int g(\mathbf{k}_1,\mathbf{v}_1,\mathbf{k}_2,\mathbf{v}_2,t)d^3v_2\\ &\quad= -i\int_{-\infty+ic_2}^{\infty+ic_2}\frac{d\omega_2}{2\pi}\int_{-\infty+ic_1}^{\infty+ic_1}\frac{d\omega_1}{2\pi}\,\frac{[e^{-i(\omega_1+\omega_2)t}-1]}{(\omega_1+\omega_2)\varepsilon(\mathbf{k}_2,\omega_2)(\omega_1-\mathbf{k}_1\cdot\mathbf{v}_1)}\\ &\qquad\times\left[\int\frac{d^3v_2'S(\mathbf{k}_1,\mathbf{v}_1,\mathbf{k}_2,\mathbf{v}_2')}{\omega_2-\mathbf{k}_2\cdot\mathbf{v}_2'}-\frac{\omega_p^2}{k_1^2\varepsilon(\mathbf{k}_1,\omega_1)}\mathbf{k}_1\cdot\frac{\partial F(\mathbf{v}_1)}{\partial\mathbf{v}_1}\right.\\ &\qquad\left.\times\int\frac{d^3v_1'}{\omega_1-\mathbf{k}_1\cdot\mathbf{v}_1'}\int\frac{d^3v_2'}{\omega_2-\mathbf{k}_2\cdot\mathbf{v}_2'}S(\mathbf{k}_1,\mathbf{v}_1',\mathbf{k}_2,\mathbf{v}_2')\right]. \end{aligned} \tag{12.83}$$

It is reminded that in (12.83) c_1 and c_2 are real and positive, and are large enough so that all the singularities of the integrand lie below the horizontal paths in their respective complex frequency planes. We can omit the term (-1) in the factor $[e^{-i(\omega_1+\omega_2)t}-1]$ in (12.83) because that term vanishes upon integration. Here is the reason: the exponential term requires the contour to be completed by the infinite semi-circle surrounding the lower-half $\omega_{1,2}$-planes. However, for the term (-1), we can surround the upper-half planes, which contain no singularities, and the integral vanishes. In this case, the contribution from the infinite semi-circle is $O(\frac{1}{|\omega_1|^2})$ or $O(\frac{1}{|\omega_2|^2})$, which goes to zero on the infinite circle.

Exercise. Obtain Fourier transform of (12.59). Solution: for the first term,

$$S(\mathbf{k}_1, \mathbf{v}_1, \mathbf{k}_2, \mathbf{v}_2) = \iint \frac{\partial}{\partial \mathbf{r}_1} \frac{e^2/m}{|\mathbf{r}_1 - \mathbf{r}_2|} e^{-i\mathbf{k}_1\cdot\mathbf{r}_1} e^{-i\mathbf{k}_2\cdot\mathbf{r}_2} d^3r_1 d^3r_2 \ (put\ \mathbf{r}_1 - \mathbf{r}_2 = \mathbf{r})$$

$$= \iint \frac{\partial}{\partial \mathbf{r}} \frac{e^2/m}{r} e^{-i\mathbf{k}_1\cdot\mathbf{r}} e^{-i\mathbf{k}_1\cdot\mathbf{r}_2} e^{-i\mathbf{k}_2\cdot\mathbf{r}_2} d^3r d^3r_2 = (2\pi)^3 \delta(\mathbf{k}_1 + \mathbf{k}_2)\, \frac{i\omega_p^2}{k_1^2}\, \mathbf{k}_1.$$

For the second term of (12.59), interchange subscripts 1 and 2 to obtain the answer in the form

$$S(\mathbf{k}_1, \mathbf{v}_1, \mathbf{k}_2, \mathbf{v}_2) = (2\pi)^3 \delta(\mathbf{k}_1 + \mathbf{k}_2) \frac{i\omega_p^2}{k_1^2} \left(\mathbf{k}_1 \cdot \frac{\partial}{\partial \mathbf{v}_1} + \mathbf{k}_2 \cdot \frac{\partial}{\partial \mathbf{v}_2} \right) F(\mathbf{v}_1) F(\mathbf{v}_2). \tag{12.84}$$

Now we substitute (12.84) into (12.83). The lengthy expression can be made somewhat compact by defining the following functions.

$$U(\mathbf{k}, \omega) = \int d^3v\, \frac{F(\mathbf{v})}{\omega - \mathbf{k}\cdot\mathbf{v}} = \int_{-\infty}^{\infty} du\, \frac{G(u)}{\omega - ku} \tag{12.85}$$

where the one-dimensional distribution $G(u)$ is derived from the three-dimensional distribution $F(\mathbf{v})$ by $G(v_x) = \int F(v_x, v_y, v_z) dv_y dv_z$ or $G(u) = \int d^3v F(\mathbf{v}) \delta(u - \mathbf{k}\cdot\mathbf{v}/k)$.

$$\varepsilon(\mathbf{k}, \omega) = 1 + \frac{\omega_p^2}{k^2} \int d^3v\, \frac{\mathbf{k}\cdot\frac{\partial F}{\partial \mathbf{v}}}{\omega - \mathbf{k}\cdot\mathbf{v}} = 1 + \frac{\omega_p^2}{k^2} \int du\, \frac{k\frac{\partial G}{\partial u}}{\omega - ku}. \tag{12.86}$$

In terms of U and ε, we have

$$\int \frac{d^3v_2' S(\mathbf{k}_1, \mathbf{v}_1, \mathbf{k}_2, \mathbf{v}_2')}{\omega_2 - \mathbf{k}_2\cdot\mathbf{v}_2'}$$

$$= i(2\pi)^3 \delta(\mathbf{k}_1 + \mathbf{k}_2) \left(\frac{\omega_p^2}{k_1^2}\, \mathbf{k}_1 \cdot \frac{\partial F(\mathbf{v}_1)}{\partial \mathbf{v}_1} U(\mathbf{k}_2, \omega_2) + F(\mathbf{v}_1)[\varepsilon(\mathbf{k}_2, \omega_2) - 1] \right).$$

$$\int \frac{d^3v_1'}{\omega_1 - \mathbf{k}_1\cdot\mathbf{v}_1'} \int \frac{d^3v_2'}{\omega_2 - \mathbf{k}_2\cdot\mathbf{v}_2'} S(\mathbf{k}_1, \mathbf{v}_1', \mathbf{k}_2, \mathbf{v}_2')$$

$$= i(2\pi)^3 \delta(\mathbf{k}_1 + \mathbf{k}_2) \Big(U(\mathbf{k}_2, \omega_2)[\varepsilon(\mathbf{k}_1, \omega_1) - 1] + U(\mathbf{k}_1, \omega_1)[\varepsilon(\mathbf{k}_2, \omega_2) - 1] \Big).$$

Substituting the above two equations into (12.83) gives

$$\int g(\mathbf{k}_1, \mathbf{v}_1, \mathbf{k}_2, \mathbf{v}_2, t) d^3 v_2$$

$$= \int_{-\infty+ic_1}^{\infty+ic_1} \frac{d\omega_1}{2\pi} \int_{-\infty+ic_2}^{\infty+ic_2} \frac{d\omega_2}{2\pi} \frac{e^{-i(\omega_1+\omega_2)t}}{(\omega_1+\omega_2)(\omega_1 - \mathbf{k}_1 \cdot \mathbf{v}_1)} (2\pi)^3 \delta(\mathbf{k}_1 + \mathbf{k}_2)$$

$$\times \left[F(\mathbf{v}_1) \left(1 - \frac{1}{\varepsilon(\mathbf{k}_2, \omega_2)} \right) - \frac{\omega_p^2}{k_1^2} \mathbf{k}_1 \cdot \frac{\partial F(\mathbf{v}_1)}{\partial \mathbf{v}_1} \frac{1}{\varepsilon(\mathbf{k}_1, \omega_1)} \right.$$

$$\left. \times \left(U(\mathbf{k}_1, \omega_1) - \frac{U(\mathbf{k}_1, \omega_1) + U(\mathbf{k}_2, \omega_2)}{\varepsilon(\mathbf{k}_2, \omega_2)} \right) \right]. \tag{12.87}$$

Now we consider the time-asymptotic limit ($t \to \infty$) of (12.87) and perform the $\int d\omega_2$-integral first. Here the contour in $\int d\omega_2$-integral is taken to be the horizontal line running above the singularity $\omega_2 = -\omega_1$, completed with the infinite semi-circle winding the lower-half ω_2-plane. The dielectric function $\varepsilon(\mathbf{k}, \omega)$ has only stable roots [$\varepsilon(\mathbf{k}, \omega) = 0$ only for $Im\ \omega \equiv \omega_i < 0$], and therefore all the singularities of $1/\varepsilon(\mathbf{k}_2, \omega_2)$ contribute to exponentially damped Landau-responses as $t \to \infty$. So the only surviving contribution to the asymptotic limit comes from the pole $\omega_2 = -\omega_1$. Therefore we have

$$\int g(\mathbf{k}_1, \mathbf{v}_1, \mathbf{k}_2, \mathbf{v}_2, t \to \infty) d^3 v_2$$

$$= -i(2\pi)^3 \delta(\mathbf{k}_1 + \mathbf{k}_2) \int_{-\infty+ic_1}^{\infty+ic_1} \frac{d\omega_1}{2\pi} \frac{1}{(\omega_1 - \mathbf{k}_1 \cdot \mathbf{v}_1)}$$

$$\times \left[F(\mathbf{v}_1) \left(1 - \frac{1}{\varepsilon(\mathbf{k}_2, -\omega_1)} \right)^R - \frac{\omega_p^2}{k_1^2} \mathbf{k}_1 \cdot \frac{\partial F(\mathbf{v}_1)}{\partial \mathbf{v}_1} \frac{1}{\varepsilon(\mathbf{k}_1, \omega_1)} \right.$$

$$\left. \times \left(U(\mathbf{k}_1, \omega_1)^{\copyright} - \frac{U(\mathbf{k}_1, \omega_1) + U(\mathbf{k}_2, -\omega_1)}{\varepsilon(\mathbf{k}_2, -\omega_1)} \right) \right] \tag{12.88}$$

where the $(-)$ sign following the equality sign is due to traversing the contour (encircling the lower-half plane) in the clockwise sense. Now let us take a look at the integrand in (12.88) on the complex ω_1-plane. We take the horizontal contour $c_1 > 0$ to be close to the real ω_1-axis. We have $\varepsilon(\mathbf{k}_2, -\omega_1) = \varepsilon(-\mathbf{k}_1, -\omega_1) = \varepsilon^*(\mathbf{k}_1, \omega_1)$ due to the delta function $\delta(\mathbf{k}_1 + \mathbf{k}_2)$. So the singularity $\varepsilon(\mathbf{k}_1, \omega_1) = 0$ accompanies the mirror image which is the singularity of $\varepsilon(\mathbf{k}_2, -\omega_1) = 0$, in $\int d\omega_2$-integral, on the upper plane, separated by the contour c_1. First, in order to calculate term ©, let us wind around the upper-half ω_1-plane (this is a matter of convenience). Then the region inside the contour contains no singularities, and the integral vanishes. The term R is conveniently calculated by completing the contour with encircling the lower-half ω_1-plane (thus excluding the singularity associated with $\varepsilon(\mathbf{k}_2, -\omega_1) = 0$ but containing the pole at $\omega_1 = \mathbf{k}_1 \cdot \mathbf{v}_1$). Thus (12.88) is left with:

$$\int g(\mathbf{k}_1, \mathbf{v}_1, \mathbf{k}_2, \mathbf{v}_2, t \to \infty) d^3 v_2 = -(2\pi)^3 F(\mathbf{v}_1) \left[1 - \frac{1}{\varepsilon(-\mathbf{k}_1, -\mathbf{k}_1 \cdot \mathbf{v}_1)}\right] + \\ - i(2\pi)^3 \delta(\mathbf{k}_1 + \mathbf{k}_2) \frac{\omega_p^2}{k_1^2} \mathbf{k}_1 \cdot \frac{\partial F(\mathbf{v}_1)}{\partial \mathbf{v}_1} \\ \times \int_{-\infty + ic_1}^{\infty + ic_1} \frac{d\omega_1}{2\pi} \frac{1}{(\omega_1 - \mathbf{k}_1 \cdot \mathbf{v}_1)} \frac{U(\mathbf{k}_1, \omega_1) + U(\mathbf{k}_2, -\omega_1)}{\varepsilon(\mathbf{k}_1, \omega_1) \varepsilon(\mathbf{k}_2, -\omega_1)}. \tag{12.89}$$

The integral in (12.89) can be written as

$$I \equiv \int_{-\infty + ic_1}^{\infty + ic_1} \frac{d\omega_1}{2\pi} \frac{1}{\mid \varepsilon(\mathbf{k}_1, \omega_1) \mid^2} \frac{1}{(\omega_1 - \mathbf{k}_1 \cdot \mathbf{v}_1)} \\ \times \int d^3 v_2 F(\mathbf{v}_2) \left[\frac{1}{\omega_1 - \mathbf{k}_1 \cdot \mathbf{v}_2}^{\oplus} + \frac{1}{-\omega_1 + \mathbf{k}_1 \cdot \mathbf{v}_2}^{\ominus}\right]. \tag{12.90}$$

It appears that the integral I vanishes because of the apparent cancellation of $\oplus$ and $\ominus$ terms. However, care should be exercised. ω_1 in the $\oplus$ function is below the horizontal contour c_1, and the $\ominus$ function should be analytically continued into the region above the line c_1. We take the line c_1 coincide with the real ω_1 axis ($c_1 \to 0$). [This arrangement is also a matter of convenience; by doing so (12.90) reduces to legitimate compact expression.] In the limit of $c_1 \to 0$, the singularity of the $\oplus$ function pushes up the contour into a convexed-up indentation while the singularity of $\ominus$ function pushes down the contour into a concaved-down indentation. All these indentations of contour point to the fact that the singularities cannot cross over the line c_1. All these deformations of contour are tantamount to analytic continuations of the relevant functions into the expanded regions. Equivalently, Plemelj formulas can be used:

$$\frac{1}{\omega_1 - \mathbf{k}_1 \cdot \mathbf{v}_1} = P\, \frac{1}{\omega_1 - \mathbf{k}_1 \cdot \mathbf{v}_1} - i\,\pi\delta(\omega_1 - \mathbf{k}_1 \cdot \mathbf{v}_1),$$

$$\oplus = \frac{1}{\omega_1 - \mathbf{k}_1 \cdot \mathbf{v}_2} = P\, \frac{1}{\omega_1 - \mathbf{k}_1 \cdot \mathbf{v}_2} - i\,\pi\delta(\omega_1 - \mathbf{k}_1 \cdot \mathbf{v}_2),$$

$$\ominus = \frac{1}{-\omega_1 + \mathbf{k}_1 \cdot \mathbf{v}_2} = \frac{-1}{\omega_1 - \mathbf{k}_1 \cdot \mathbf{v}_2}$$

$$= -\left[P\, \tfrac{1}{\omega_1 - \mathbf{k}_1 \cdot \mathbf{v}_2} + i\,\pi\delta(\omega_1 - \mathbf{k}_1 \cdot \mathbf{v}_2)\right],$$

$$\oplus + \ominus = -2i\pi\; \delta(\omega_1 - \mathbf{k}_1 \cdot \mathbf{v}_2).$$

What we need is the imaginary part of the RHS of (12.89). So we calculate:

$$Re\; I = \int \frac{d\omega_1}{2\pi} \frac{1}{\mid \varepsilon(\mathbf{k}_1, \omega_1) \mid^2}(-i\pi)\delta(\omega_1 - \mathbf{k}_1 \cdot \mathbf{v}_1)(-2i\pi) \int d^3v_2 F(\mathbf{v}_2)\delta(\omega_1 - \mathbf{k}_1 \cdot \mathbf{v}_2)$$

$$= \frac{-\pi}{\mid \varepsilon(\mathbf{k}_1, \mathbf{k}_1 \cdot \mathbf{v}_1) \mid^2} \int d^3v_2 F(\mathbf{v}_2)\delta(\mathbf{k}_1 \cdot \mathbf{v}_1 - \mathbf{k}_1 \cdot \mathbf{v}_2).$$

Im 2nd line of (12.89)

$$= (2\pi)^3\, \frac{\omega_p^2}{k_1^2}\, \mathbf{k}_1 \cdot \frac{\partial F(\mathbf{v}_1)}{\partial \mathbf{v}_1} \frac{i\pi}{\mid \varepsilon(\mathbf{k}_1, \mathbf{k}_1 \cdot \mathbf{v}_1) \mid^2} \int d^3v_2 F(\mathbf{v}_2)\delta[\mathbf{k}_1 \cdot (\mathbf{v}_1 - \mathbf{v}_2)]. \tag{12.91}$$

To calculate the imaginary part of the first line of the RHS of (12.89), we use

$$Im\left[\frac{1}{\varepsilon(\mathbf{k}, \omega)}\right] = -i\, \frac{Im[\varepsilon(\mathbf{k}, \omega)]}{\mid \varepsilon(\mathbf{k}, \omega) \mid^2}$$

$$Im[\varepsilon(\mathbf{k}, \omega)] = -i\pi\, \frac{\omega_p^2}{k^2} \int d^3v_2\; \mathbf{k} \cdot \frac{\partial F(\mathbf{v}_2)}{\partial \mathbf{v}_2} \delta(\omega - \mathbf{k} \cdot \mathbf{v}_2)$$

$$Im\left[\frac{1}{\varepsilon(-\mathbf{k}_1, -\mathbf{k}_1 \cdot \mathbf{v}_1)}\right] = Im\left[\frac{1}{\varepsilon^*(\mathbf{k}_1, \mathbf{k}_1 \cdot \mathbf{v}_1)}\right] = i\, \frac{Im[\varepsilon(\mathbf{k}_1, \mathbf{k}_1 \cdot \mathbf{v}_1)]}{\mid \varepsilon(\mathbf{k}_1, \mathbf{k}_1 \cdot \mathbf{v}_1) \mid^2}$$

Im 1st line of (12.89)

$$= (2\pi)^3\, \frac{\omega_p^2}{k_1^2} F(\mathbf{v}_1)\, \frac{i\pi}{\mid \varepsilon(\mathbf{k}_1, \mathbf{k}_1 \cdot \mathbf{v}_1) \mid^2} \int d^3v_2 \mathbf{k}_1 \cdot \frac{\partial F(\mathbf{v}_2)}{\partial \mathbf{v}_2} \delta[\mathbf{k}_1 \cdot (\mathbf{v}_1 - \mathbf{v}_2)]. \tag{12.92}$$

Adding (12.91) and (12.92) gives the RHS of (12.89). Thus we have

$$\int g(\mathbf{k}_1, \mathbf{v}_1, \mathbf{k}_2, \mathbf{v}_2, t \to \infty) d^3 v_2 = \frac{\omega_p^2}{k_1^2} \frac{-i\pi(2\pi)^3}{|\, \varepsilon(\mathbf{k}_1, \mathbf{k}_1 \cdot \mathbf{v}_1)\,|^2} \int d^3 v_2 \delta[\mathbf{k}_1 \cdot (\mathbf{v}_1 - \mathbf{v}_2)]$$
$$\times \mathbf{k}_1 \cdot \left(F(\mathbf{v}_1) \frac{\partial F(\mathbf{v}_2)}{\partial \mathbf{v}_2} - F(\mathbf{v}_2) \frac{\partial F(\mathbf{v}_1)}{\partial \mathbf{v}_1} \right) \delta(\mathbf{k}_1 + \mathbf{k}_2). \tag{12.93}$$

Inverse Fourier transform of (12.93) with respect to $\mathbf{k}_2$ is obtained by performing $\times \int d^3 k_2 e^{i\mathbf{k}_2 \cdot \mathbf{r}_2}$:

$$\int g(\mathbf{k}_1, \mathbf{v}_1, \mathbf{r}_2, \mathbf{v}_2, t \to \infty) d^3 v_2 = \frac{\omega_p^2}{k_1^2} \frac{-i\pi(2\pi)^3}{|\, \varepsilon(\mathbf{k}_1, \mathbf{k}_1 \cdot \mathbf{v}_1)\,|^2} e^{-i\mathbf{k}_1 \cdot \mathbf{r}_2}$$
$$\times \int d^3 v_2 \delta[\mathbf{k}_1 \cdot (\mathbf{v}_1 - \mathbf{v}_2)]\ \mathbf{k}_1 \cdot \left(F(\mathbf{v}_1) \frac{\partial F(\mathbf{v}_2)}{\partial \mathbf{v}_2} - F(\mathbf{v}_2) \frac{\partial F(\mathbf{v}_1)}{\partial \mathbf{v}_1} \right). \tag{12.94}$$

The spatial Fourier transform of (12.80) can be written as

$$\frac{\partial}{\partial t} F(\mathbf{v}_1, t)$$
$$= n \int d^3 k_1 \int d^3 r_2 \left(\frac{\partial}{\partial \mathbf{r}_1} \frac{e^2/m}{|\, \mathbf{r}_1 - \mathbf{r}_2 \,|} \cdot \frac{\partial}{\partial \mathbf{v}_1} \right) e^{i\mathbf{k}_1 \cdot \mathbf{r}_1} \int d^3 v_2 g(\mathbf{k}_1, \mathbf{v}_1, \mathbf{r}_2, \mathbf{v}_2, t \to \infty). \tag{12.95}$$

Now we substitute (12.94) into (12.95). We first calculate $\int d^3 r_2 (\cdots)$:

$$\int d^3 r_2 \frac{\partial}{\partial \mathbf{r}_1} \frac{e^2/m}{|\, \mathbf{r}_1 - \mathbf{r}_2 \,|} e^{i\mathbf{k}_1 \cdot (\mathbf{r}_1 - \mathbf{r}_2)}$$
$$= \int d^3 R \frac{\partial}{\partial \mathbf{R}} \frac{e^2/m}{R} e^{i\mathbf{k}_1 \cdot \mathbf{R}} = -i\mathbf{k}_1 \int d^3 R \frac{e^2/m}{R} e^{i\mathbf{k}_1 \cdot \mathbf{R}} = -i\mathbf{k}_1 \frac{4\pi e^2/m}{k_1^2}.$$

So (12.95) takes the final form

$$\frac{\partial}{\partial t} F(\mathbf{v}_1, t) = -\frac{\partial}{\partial \mathbf{v}_1} \cdot \int d^3 k_1\ \mathbf{k}_1\ \frac{\omega_p^2}{k_1^2} \frac{\omega_p^2}{k_1^2} \frac{\pi(2\pi)^3}{|\, \varepsilon(\mathbf{k}_1, \mathbf{k}_1 \cdot \mathbf{v}_1)\,|^2}$$
$$\times \int d^3 v_2 \delta[\mathbf{k}_1 \cdot (\mathbf{v}_1 - \mathbf{v}_2)]\ \mathbf{k}_1 \cdot \left(F(\mathbf{v}_1) \frac{\partial F(\mathbf{v}_2)}{\partial \mathbf{v}_2} - F(\mathbf{v}_2) \frac{\partial F(\mathbf{v}_1)}{\partial \mathbf{v}_1} \right) \tag{12.96}$$

which is the **Lenard–Balescu equation**.

The $\int d^3v_2$-integral can be formally carried out as follows. Take $\mathbf{k}_1 = \hat{\mathbf{x}}k_1$ without loss of generality. Then the velocity integral in (12.96) becomes

$$\int d^3v_2\ \delta(k_1v_{1x} - k_1v_{2x})k_1 \left(F(\mathbf{v}_1)\frac{\partial F(\mathbf{v}_2)}{\partial v_{2x}} - F(\mathbf{v}_2)\frac{\partial F(\mathbf{v}_1)}{\partial v_{1x}}\right).$$

In terms of $G(v_{2x}) \equiv \int dv_{2y}dv_{2z}F(\mathbf{v}_2)$, the above expression becomes

$$\int dv_{2x}\ \delta(k_1v_{1x} - k_1v_{2x})k_1 \left(F(\mathbf{v}_1)\frac{\partial G(v_{2x})}{\partial v_{2x}} - G(v_{2x})\frac{\partial F(\mathbf{v}_1)}{\partial v_{1x}}\right)$$
$$= F(\mathbf{v}_1)\frac{\partial G(v_{1x})}{\partial v_{1x}} - G(v_{1x})\frac{\partial F(\mathbf{v}_1)}{\partial v_{1x}}.$$

Put $v_{1x} \to u = \mathbf{v}_1 \cdot \mathbf{k}_1/k_1$ to obtain the Lenard–Balescu equation in the form

$$\frac{\partial}{\partial t}F(\mathbf{v}_1, t) = -N\frac{\partial}{\partial \mathbf{v}_1} \cdot \int d^3k_1\ \mathbf{k}_1\ \frac{\omega_p^2}{k_1^2}\frac{\omega_p^2}{k_1^2}\frac{\pi(2\pi)^3}{|\ \varepsilon(\mathbf{k}_1, \mathbf{k}_1\cdot\mathbf{v}_1)\ |^2}$$
$$\times \left[F(\mathbf{v}_1)\frac{\partial G(u)}{\partial u} - G(u)\ \hat{\mathbf{k}_1} \cdot \frac{\partial F}{\partial \mathbf{v}_1}\right] \qquad (12.97)$$
$$\text{where} \quad G(u) = \int d^3vF(\mathbf{v})\delta\left(u - \frac{\mathbf{k}\cdot\mathbf{v}}{k}\right).$$

12.6. Alternative derivation of Lenard–Balescu equation

In the last section, we considered the lowest two Liouville distributions $F^{(1)}$ and $F^{(2)}$ or equivalently $F^{(1)}$ and the correlation function g, in the BBGKY hierarchy. After a lengthy algebra with the simultaneous equations for $F^{(1)}$ and g, employing the operational method for solving differential equation, we ended up with the Lenard–Balescu equation. In this Section, we begin with the Klimontovich hierarchy of the Liouville equation. The key element of analysis is to separate the Klimontovich distribution function into two parts: the slowly varying (time-wise) part and the fluctuating part. The formal procedure of the analysis is very similar to that for the quasilinear theory in Chapter 10.

References: D. C. Montgomery *Theory of unmagnetized plasma*, Gordon and Beach, New York (1971); E. M. Lifshitz and L. P. Pitaevskii, *Physical kinetics*, Landau and Lifshitz Course of theoretical physics Volume 10 Butterworth-Heinemann (1981).

We begin with the Klimontovich equation (12.43) and the Poisson equation:

$$\frac{\partial K(\mathbf{r},\mathbf{v},t)}{\partial t}+\mathbf{v}\cdot\frac{\partial K}{\partial \mathbf{r}}-\frac{e}{m}\frac{\partial \phi}{\partial \mathbf{r}}\cdot\frac{\partial K}{\partial \mathbf{v}}=0, \tag{12.98}$$

$$\phi(\mathbf{r},t)=\int\frac{e}{\mid\mathbf{r}-\mathbf{r}'\mid}K(\mathbf{r}',\mathbf{v}',t)d^3r'd^3v', \tag{12.99}$$

$$\nabla^2\phi=4\pi e\int d^3v\ K. \tag{12.100}$$

Separate the functions in the manner

$$K=\overline{K}+\delta K,\quad \phi=\overline{\phi}+\delta\phi$$

where the overbar and δ denote, respectively, ensemble averaged (by means of $F^{(N)}$) and fluctuating parts. If a δ-quantity is averaged, the result is zero. $\overline{K}=\langle K\rangle$ is given by (12.40). We have

$$\begin{aligned}&\frac{\partial}{\partial t}(\overline{K}+\delta K)+\mathbf{v}\cdot\frac{\partial}{\partial \mathbf{r}}(\overline{K}+\delta K)\\&\quad=\frac{e}{m}\frac{\partial\overline{\phi}}{\partial \mathbf{r}}\cdot\frac{\partial}{\partial \mathbf{v}}\overline{K}+\frac{e}{m}\frac{\partial\overline{\phi}}{\partial \mathbf{r}}\cdot\frac{\partial}{\partial \mathbf{v}}\delta K+\frac{e}{m}\frac{\partial\delta\phi}{\partial \mathbf{r}}\cdot\frac{\partial}{\partial \mathbf{v}}\overline{K}+\frac{e}{m}\frac{\partial\delta\phi}{\partial \mathbf{r}}\cdot\frac{\partial}{\partial \mathbf{v}}\delta K.\end{aligned} \tag{12.101}$$

The first two terms on the RHS vanish because the averaged electric field in a homogeneous plasma is zero. The homogeneous plasma is characterized by $\frac{\partial}{\partial \mathbf{r}}\overline{K}=\frac{\partial}{\partial \mathbf{r}}F^{(1)}(\mathbf{r},\mathbf{v},t)=0$. If Eq. (12.101) is averaged with the remaining terms, we are left with

$$\frac{\partial}{\partial t}\overline{K}=\frac{e}{m}\overline{\frac{\partial}{\partial \mathbf{v}}\ \delta K\cdot\frac{\partial}{\partial \mathbf{r}}\ \delta\phi}. \tag{12.102}$$

So we should collect from (12.101) to write

$$\left(\frac{\partial}{\partial t}+\mathbf{v}\cdot\frac{\partial}{\partial \mathbf{r}}\right)\delta K=\frac{e}{m}\ \frac{\partial}{\partial \mathbf{v}}\overline{K}\cdot\frac{\partial}{\partial \mathbf{r}}\ \delta\phi. \tag{12.103}$$

Exercise. Show that the averaged electric field in homogeneous plasma is zero. This is physically almost obvious, and can be proven as follows.

$$\begin{aligned}\frac{\partial}{\partial \mathbf{r}}\overline{\phi}(\mathbf{r},t)&=\int d^3r'd^3v'\overline{K'}\frac{\partial}{\partial \mathbf{r}}\frac{e}{\mid\mathbf{r}-\mathbf{r}'\mid}\\&=-\int d^3r'd^3v'\overline{K'}\frac{\partial}{\partial \mathbf{r}'}\frac{e}{\mid\mathbf{r}-\mathbf{r}'\mid}=\int d^3r'd^3v'\frac{e}{\mid\mathbf{r}-\mathbf{r}'\mid}\frac{\partial}{\partial \mathbf{r}'}\overline{K'}=0.\end{aligned}$$

We shall seek the solution of the two coupled equations, (12.102) and (12.103). (12.103) is identical in form to the linearized Vlasov equation. If $\overline{K}$ is time-independent, the mathematical machinery that we used for solving the Vlasov equation can be employed here. We make a compromise in this respect: while the time dependence of $\overline{K}$ is retained in (12.102), we hold $\overline{K}$ constant in (12.103). This approximate scheme turns out to be acceptable when the plasma parameter is small ($\ll 1$), as justified by the multiple time perturbation analysis of (12.102) and (12.103). We first solve (12.103) which we rewrite by changing into more comfortable notations ($\delta K \to \delta f$, $\overline{K} \to F$)

$$\left(\frac{\partial}{\partial t} + \mathbf{v}\cdot\frac{\partial}{\partial \mathbf{r}}\right)\delta f(\mathbf{r},\mathbf{v},t) = \frac{1}{m}\frac{\partial F(\mathbf{v})}{\partial \mathbf{v}}\cdot\int d^3r' d^3v' \delta f(\mathbf{r}',\mathbf{v}',t)\frac{\partial\phi(\mathbf{r}-\mathbf{r}')}{\partial \mathbf{r}} \tag{12.104}$$

where $\phi(\mathbf{r}-\mathbf{r}') = \frac{e^2}{|\mathbf{r}-\mathbf{r}'|}$. The Laplace–Fourier transform is defined by

$$\delta f(\omega,\mathbf{k},\mathbf{v}) = \int_0^\infty dt e^{i\omega t}\int d^3r\ e^{-i\mathbf{k}\cdot\mathbf{r}}\delta f(\mathbf{r},\mathbf{v},t),$$

$$\delta f(\mathbf{r},\mathbf{v},t) = \int_{-\infty+ic}^{\infty+ic}\frac{d\omega}{2\pi}e^{-i\omega t}\int d^3k\ e^{i\mathbf{k}\cdot\mathbf{r}}\delta f(\omega,\mathbf{k},\mathbf{v}).$$

Recall that in $\int d\omega$ integral above, the integration path runs above all the singularities of $\delta f(\omega,\mathbf{k},\mathbf{v})$. Laplace transforming (12.104) gives

$$\begin{aligned}&-i\omega\ \delta f(\omega,\mathbf{r},\mathbf{v}) - \delta f(0,\mathbf{r},\mathbf{v}) + \mathbf{v}\cdot\frac{\partial}{\partial \mathbf{r}}\delta f(\omega,\mathbf{r},\mathbf{v})\\ &\quad = \frac{1}{m}\frac{\partial F}{\partial \mathbf{v}}\cdot\int d^3r' d^3v'\ \delta f(\mathbf{r}',\mathbf{v}',\omega)\frac{\partial\phi(\mathbf{r}-\mathbf{r}')}{\partial \mathbf{r}}\end{aligned} \tag{12.105}$$

where $\delta f(0,\mathbf{r},\mathbf{v})$ is the initial value. The Fourier transform of the integral part is a convolution integral:

$$\begin{aligned}&\int d^3r\ e^{-i\mathbf{k}\cdot\mathbf{r}}\int d^3r' d^3v'\ \delta f(\mathbf{r}',\mathbf{v}',\omega)\frac{\partial\phi(\mathbf{r}-\mathbf{r}')}{\partial \mathbf{r}}\\ &\quad = i\mathbf{k}\int d^3r\ e^{-i\mathbf{k}\cdot\mathbf{r}}\int d^3r' d^3v' \delta f(\mathbf{r}',\mathbf{v}',\omega)\phi(\mathbf{r}-\mathbf{r}')\\ &\quad = i\mathbf{k}\ \phi(k)\int \delta f(\mathbf{k},\mathbf{v}',\omega)d^3v'.\end{aligned}$$

Using this in (12.105) gives

$$\delta f(\omega, \mathbf{k}, \mathbf{v}) = i\ \frac{\delta f(0, \mathbf{k}, \mathbf{v})}{\omega - \mathbf{k}\cdot\mathbf{v}} - \frac{\phi(k)}{m}\frac{\mathbf{k}\cdot\frac{\partial F}{\partial \mathbf{v}}}{\omega - \mathbf{k}\cdot\mathbf{v}}\int \delta f(\mathbf{k}, \mathbf{v}', \omega) d^3v'.$$

Integrating both sides with $\int d^3v(\cdots)$ and substituting the result yield

$$\delta f(\omega, \mathbf{k}, \mathbf{v}) = \frac{i}{\omega - \mathbf{k}\cdot\mathbf{v}}\left[\delta f(0, \mathbf{k}, \mathbf{v}) - \frac{\omega_{pe}^2}{k^2\varepsilon(\mathbf{k}, \omega)}\ \mathbf{k}\cdot\frac{\partial F}{\partial \mathbf{v}}\int\frac{\delta f(0, \mathbf{k}, \mathbf{v})}{\omega - \mathbf{k}\cdot\mathbf{v}}d^3v\right] \tag{12.106}$$

$$\varepsilon(\mathbf{k}, \omega) = 1 + \frac{\omega_{pe}^2}{k^2}\int d^3v\ \frac{\mathbf{k}\cdot\frac{\partial F}{\partial \mathbf{v}}}{\omega - \mathbf{k}\cdot\mathbf{v}} \tag{12.107}$$

where we used

$$\phi(k) = \int\frac{e^2}{r}e^{-i\mathbf{k}\cdot\mathbf{r}}d^3r = \frac{4\pi e^2}{k^2}$$

$$\textit{Note:}\quad \int\left(\frac{\partial}{\partial \mathbf{r}}\ \frac{1}{r}\right)e^{-i\mathbf{k}\cdot\mathbf{r}}d^3r = i\mathbf{k}\int\frac{1}{r}\ e^{-i\mathbf{k}\cdot\mathbf{r}}d^3r$$

$$LHS = -\int\frac{\mathbf{r}}{r^3}e^{-i\mathbf{k}\cdot\mathbf{r}}d^3r = \frac{4\pi i}{k^2}\ \mathbf{k}.\ \ (\textit{see}\ (5.22))$$

Exercise. Calculate directly the Fourier transform of $\frac{1}{r}$. The answer can be written down by taking the limit $\lambda_D \to \infty$ in (11.6).

So far we obtained $\delta f(\omega, \mathbf{k}, \mathbf{v})$ in terms of $\delta f(0, \mathbf{k}, \mathbf{v})$ and $F(\mathbf{v})$ by performing the analogous analysis as we did for the Vlasov equation. The functions $\delta f(\omega, \mathbf{k}, \mathbf{v})$ and $\varepsilon(\mathbf{k}, \omega)$ are defined in the regions $\omega_i > 0$, but for un-pathological functions F and $\delta f(0)$, the functions can be analytically continued into the regions $\omega_i < 0$.

Next, we calculate the electric field $\delta\mathbf{E}(\mathbf{k}, \omega) = -i\mathbf{k}\delta\phi(\mathbf{k}, \omega)$ associated with the fluctuating distribution function $\delta f(\omega, \mathbf{k}, \mathbf{v})$. We have the Poisson equation

$$k^2\delta\phi(\mathbf{k}, \omega) = 4\pi e\int d^3v\ \ \delta f(\omega, \mathbf{k}, \mathbf{v}).$$

Integrating (12.105), we obtain

$$\int d^3v\ \ \delta f(\omega, \mathbf{k}, \mathbf{v}) = \frac{i}{\varepsilon(\mathbf{k}, \omega)}\int d^3v\ \frac{\delta f(0, \mathbf{k}, \mathbf{v})}{\omega - \mathbf{k}\cdot\mathbf{v}}.$$

$$\text{Therefore,}\quad \delta\mathbf{E}(\mathbf{k}, \omega) = \frac{\mathbf{k}}{k^2}\ \frac{4\pi e}{\varepsilon(\mathbf{k}, \omega)}\int d^3v\ \frac{\delta f(0, \mathbf{k}, \mathbf{v})}{\omega - \mathbf{k}\cdot\mathbf{v}}. \tag{12.108}$$

So far the development is parallel to the analysis of the initial value problem of the Vlasov equation. Now we pay attention to the initial value of δf : $\delta f(0, \mathbf{k}, \mathbf{v})$. At $t = 0$, the initial Klimontovich distribution function is

$$K(0, \mathbf{r}, \mathbf{v}) = \sum_{i=1}^{N} \delta(\mathbf{r} - \mathbf{r}_{i0})\delta(\mathbf{v} - \mathbf{v}_{i0}). \tag{12.109}$$

Therefore, $$\delta f(0, \mathbf{r}, \mathbf{v}) = \sum_{i=1}^{N} \delta(\mathbf{r} - \mathbf{r}_{i0})\delta(\mathbf{v} - \mathbf{v}_{i0}) - F(\mathbf{v}) \tag{12.110}$$

where $F(\mathbf{v})$ is the ensemble average of the function in (12.109) over the Liouville function $F^{(N)}(\mathbf{x}_{10}, \mathbf{x}_{20}, \cdots, \mathbf{x}_{N0})$. Fourier transform of (12.110) is

$$\delta f(0, \mathbf{k}, \mathbf{v}) = \sum_{i=1}^{N} e^{-i\mathbf{k}\cdot\mathbf{r}_{i0}}\delta(\mathbf{v} - \mathbf{v}_{i0}) - (2\pi)^3\delta(\mathbf{k})F(\mathbf{v}). \tag{12.111}$$

Using (12.111) in (12.106) and (12.108) gives

$$\delta\mathbf{E}(\mathbf{k}, \omega) = \frac{\mathbf{k}}{k^2}\frac{4\pi e}{\varepsilon(\mathbf{k}, \omega)}\sum_{i=1}^{N}\frac{e^{-i\mathbf{k}\cdot\mathbf{r}_{io}}}{\omega - \mathbf{k}\cdot\mathbf{v}_{i0}}, \tag{12.112}$$

$$\begin{aligned}\delta f(\omega, \mathbf{k}, \mathbf{v}) = {} & \frac{i}{\omega - \mathbf{k}\cdot\mathbf{v}}\left[\sum_{i=1}^{N} e^{-i\mathbf{k}\cdot\mathbf{r}_{i0}}\delta(\mathbf{v} - \mathbf{v}_{i0}) - (2\pi)^3\delta(\mathbf{k})F(\mathbf{v})\right] \\ & - \frac{i}{\omega - \mathbf{k}\cdot\mathbf{v}}\left[\frac{4\pi e^2}{mk^2\varepsilon(\mathbf{k}, \omega)}\mathbf{k}\cdot\frac{\partial F}{\partial \mathbf{v}}\sum_{i=1}^{N}\frac{e^{-i\mathbf{k}\cdot\mathbf{r}_{io}}}{\omega - \mathbf{k}\cdot\mathbf{v}_{i0}}\right].\end{aligned} \tag{12.113}$$

If we Laplace-invert the above equations, we obtain $\delta\mathbf{E}(\mathbf{k}, t)$ and $\delta f(\mathbf{k}, \mathbf{v}, t)$. We consider the time-asymptotic ($t \to \infty$) forms of $\delta\mathbf{E}$ and δf. It is clear that the time-asymptotic forms of these functions are determined by the *upper most* singularities of the integrands. There are two kinds of singularities: the first kind associated with $\varepsilon(\mathbf{k}, \omega) = 0$ and the second, those on the real ω-axis. We assume that the plasma is stable (only Landau-damped), then the first kind makes the perturbation damp away since the singular points have negative imaginary parts. The asymptotic behavior is determined by the singularities on the real ω-axis: $\omega = \mathbf{k}\cdot\mathbf{v}$ and $\omega = \mathbf{k}\cdot\mathbf{v}_{i0}$.

Thus we have as $t \to \infty$:

$$\delta \mathbf{E}(\mathbf{k}, t) \longrightarrow -i\, \frac{4\pi e \mathbf{k}}{k^2} \sum_{i=1}^{N} \frac{e^{-i\mathbf{k}\cdot(\mathbf{r}_{i0}+\mathbf{v}_{i0}t)}}{\varepsilon(\mathbf{k},\ \mathbf{k}\cdot\mathbf{v}_{i0})}, \tag{12.114}$$

$$\begin{aligned}\delta f(\mathbf{k}, \mathbf{v}, t) \longrightarrow & \sum_{i=1}^{N} e^{-i\mathbf{k}\cdot(\mathbf{r}_{i0}+\mathbf{v}_{i0}t)}\delta(\mathbf{v}-\mathbf{v}_{i0}) - (2\pi)^3 F(\mathbf{v})\delta(\mathbf{k}) \\ & -\frac{4\pi e^2}{mk^2}\,\mathbf{k}\cdot\frac{\partial F}{\partial \mathbf{v}} \sum_{i=1}^{N} \frac{e^{-i\mathbf{k}\cdot\mathbf{r}_{i0}}}{\mathbf{k}\cdot(\mathbf{v}_{io}-\mathbf{v})}\left[\frac{e^{-i\mathbf{k}\cdot\mathbf{v}_{i0}}}{\varepsilon(\mathbf{k},\ \mathbf{k}\cdot\mathbf{v}_{i0}t)} - \frac{e^{-i\mathbf{k}\cdot\mathbf{v}t}}{\varepsilon(\mathbf{k},\ \mathbf{k}\cdot\mathbf{v})}\right].\end{aligned} \tag{12.115}$$

12.6.1. *Ensemble average $\langle \delta f \delta \mathbf{E} \rangle$ and Lenard–Balescu equation*

The Lenard–Balescu equation is derived from (12.102) which is written in the changed notations

$$\frac{\partial}{\partial t} F(\mathbf{v}, t) = \frac{e}{m}\frac{\partial}{\partial \mathbf{v}} \cdot \langle \delta f \delta \mathbf{E} \rangle \tag{12.116}$$

where the ensemble average of the RHS is to be calculated by using (12.114) and (12.115). We write

$$\begin{aligned}&\delta f(\mathbf{k}, \mathbf{v}, t)\delta \mathbf{E}(\mathbf{k}', t) \\ &= -i\,\frac{4\pi e\mathbf{k}'}{k'^2} \sum_{j=1}^{N} \frac{e^{-i\mathbf{k}'\cdot(\mathbf{r}_{j0}+\mathbf{v}_{j0}t)}\,^{©}}{\varepsilon(\mathbf{k}',\ \mathbf{k}'\cdot\mathbf{v}_{j0})} \\ &\quad\times\left[\sum_{i=1}^{N} e^{-i\mathbf{k}\cdot(\mathbf{r}_{i0}+\mathbf{v}_{i0}t)}\delta(\mathbf{v}-\mathbf{v}_{i0})^{©'} - \frac{4\pi e^2}{mk^2}\,\mathbf{k}\cdot\frac{\partial F}{\partial \mathbf{v}}\sum_{i=1}^{N}\frac{e^{-i\mathbf{k}\cdot\mathbf{r}_{i0}}}{\mathbf{k}\cdot(\mathbf{v}_{io}-\mathbf{v})}\right. \\ &\quad\times\left.\left(\frac{e^{-i\mathbf{k}\cdot\mathbf{v}_{i0}t}\ ^{\oplus}}{\varepsilon(\mathbf{k},\ \mathbf{k}\cdot\mathbf{v}_{i0})} - \frac{e^{-i\mathbf{k}\cdot\mathbf{v}t}\ ^{\otimes}}{\varepsilon(\mathbf{k},\ \mathbf{k}\cdot\mathbf{v})}\right)\right].\end{aligned} \tag{12.117}$$

Taking the ensemble average of the above quantity operationally means multiplying

$$F^{(N)}(\mathbf{x}_{10}, \mathbf{x}_{20}, \ldots, \mathbf{x}_{N0}) = F^{(1)}(\mathbf{v}_{10})F^{(1)}(\mathbf{v}_{20})\cdots F^{(1)}(\mathbf{v}_{N0})$$

and performing integration with respect to $d\mathbf{r}_{i0}d\mathbf{v}_{i0}$ and $d\mathbf{r}_{j0}d\mathbf{v}_{j0}$. Let us consider the average of the product $\langle ©©' \rangle$. Only the $i = j$ terms survive the spatial integration, and the sum $\sum_1^N$ gives N identical terms. We have two δ functions; $\delta(\mathbf{k}+\mathbf{k}')$ and $\delta(\mathbf{v}-\mathbf{v}_{i0})$. Thus we obtain

$$\langle ©©' \rangle = -4\pi i e n_0 \frac{\mathbf{k}'}{k^2}\frac{\delta(\mathbf{k}+\mathbf{k}')}{\varepsilon(\mathbf{k}',\mathbf{k}'\cdot\mathbf{v})}F(\mathbf{v}).$$

Next we consider the product $\langle © \bigoplus \rangle$. Here again only the terms $i = j$ survive the spatial integration and the summation gives N identical terms, and $\delta(\mathbf{k}+\mathbf{k}')$ is produced; the velocity integral cannot be completed.

$$\begin{aligned}\langle © \bigoplus \rangle = \frac{-4\pi i e n_0}{k^2}\frac{4\pi e^2}{mk^2}\left(\mathbf{k}\cdot\frac{\partial F}{\partial \mathbf{v}}\right)\delta(\mathbf{k}+\mathbf{k}')\,\mathbf{k} \\ \times\int d^3v'\frac{1}{\mathbf{k}\cdot(\mathbf{v}'-\mathbf{v})}\frac{F(\mathbf{v}')}{|\,\varepsilon(\mathbf{k},\mathbf{k}\cdot\mathbf{v}')\,|^2}.\end{aligned}$$

Finally the product $© \bigotimes$ is calculated likewise:

$$\begin{aligned}\langle © \bigotimes \rangle = \frac{-4\pi i e n_0}{k^2}\frac{4\pi e^2}{mk^2}\left(\mathbf{k}\cdot\frac{\partial F}{\partial \mathbf{v}}\right)\delta(\mathbf{k}+\mathbf{k}')\,\mathbf{k} \\ \times\int d^3v'\frac{-e^{-i\mathbf{k}\cdot(\mathbf{v}-\mathbf{v}')t}}{\mathbf{k}\cdot(\mathbf{v}'-\mathbf{v})}\frac{F(\mathbf{v}')}{|\,\varepsilon(\mathbf{k},\mathbf{k}\cdot\mathbf{v}')\,|^2}\end{aligned}$$

where we put $\varepsilon(-\mathbf{k},-\mathbf{k}\cdot\mathbf{v}') = \varepsilon(-\mathbf{k},-\mathbf{k}\cdot\mathbf{v})$ because the velocity integral is predominantly contributed from the region around $\mathbf{v}'=\mathbf{v}$. Collecting the above results, we write

$$\begin{aligned}\langle \delta f(\mathbf{k},\mathbf{v},t)\delta\mathbf{E}(\mathbf{k}',t)\rangle = 4\pi i e n_0\delta(\mathbf{k}+\mathbf{k}')\frac{\mathbf{k}}{k^2} \\ \times\left[\frac{F(\mathbf{v})}{\varepsilon(-\mathbf{k},-\mathbf{k}\cdot\mathbf{v})}+\frac{4\pi e^2}{mk^2}\mathbf{k}\cdot\frac{\partial F}{\partial \mathbf{v}}\int d^3v'\frac{1-e^{-i\mathbf{k}\cdot(\mathbf{v}-\mathbf{v}')t}}{\mathbf{k}\cdot(\mathbf{v}-\mathbf{v}')}\frac{F(\mathbf{v}')}{|\,\varepsilon(\mathbf{k},\mathbf{k}\cdot\mathbf{v})\,|^2}\right]\end{aligned} \tag{12.118}$$

where time t is understood to be approaching ∞. Using the formula

$$\lim_{t\to\infty}\frac{1-e^{-i\mathbf{k}\cdot(\mathbf{v}-\mathbf{v}')t}}{\mathbf{k}\cdot(\mathbf{v}-\mathbf{v}')} = i\pi\delta(\mathbf{k}\cdot(\mathbf{v}-\mathbf{v}'))$$

(12.118) becomes

$$\lim_{t\to\infty}\langle\delta f(\mathbf{k},\mathbf{v},t)\delta\mathbf{E}(\mathbf{k}',t)\rangle = 4\pi i e n_0 \delta(\mathbf{k}+\mathbf{k}')\tfrac{\mathbf{k}}{k^2}$$

$$\times\left[\frac{F(\mathbf{v})}{\varepsilon(-\mathbf{k},-\mathbf{k}\cdot\mathbf{v})}+\frac{\omega_p^2}{k^2}\,\mathbf{k}\cdot\frac{\partial F}{\partial\mathbf{v}}i\pi\int d^3v'\;\delta(\mathbf{k}\cdot\mathbf{v}-\mathbf{k}\cdot\mathbf{v}')\frac{F(\mathbf{v}')}{\mid\varepsilon(\mathbf{k},\mathbf{k}\cdot\mathbf{v})\mid^2}\right] \tag{12.119}$$

where the first term inside the square bracket is

$$\frac{\varepsilon(\mathbf{k},\mathbf{k}\cdot\mathbf{v})}{\mid\varepsilon(\mathbf{k},\mathbf{k}\cdot\mathbf{v})\mid^2}F(\mathbf{v}) \tag{12.120}$$

whose imaginary part only is needed. Using the formula

$$Im\;\varepsilon(\mathbf{k},\omega) = -i\;\pi\frac{\omega_p^2}{k^2}\int d^3v'\mathbf{k}\cdot\frac{\partial F(\mathbf{v}')}{\partial\mathbf{v}'}\delta(\omega-\mathbf{k}\cdot\mathbf{v}')$$

the imaginary part of (12.120) is

$$-\frac{\omega_p^2}{k^2}\;i\pi\frac{F(\mathbf{v})}{\mid\varepsilon(\mathbf{k},\mathbf{k}\cdot\mathbf{v})\mid^2}\int d^3v'\;\delta(\mathbf{k}\cdot\mathbf{v}-\mathbf{k}\cdot\mathbf{v}')\;\mathbf{k}\cdot\frac{\partial F(\mathbf{v}')}{\partial\mathbf{v}'}. \tag{12.121}$$

Using (12.121) in (12.118) gives

$$\lim_{t\to\infty}\langle\delta f(\mathbf{k},\mathbf{v},t)\delta\mathbf{E}(\mathbf{k}',t)\rangle = 4\pi i e n_0\delta(\mathbf{k}+\mathbf{k}')\frac{\mathbf{k}}{k^2}$$

$$\times\,i\pi\;\frac{\omega_p^2}{k^2}\;\frac{\mathbf{k}\cdot}{\mid\varepsilon(\mathbf{k},\mathbf{k}\cdot\mathbf{v})\mid^2}\int d^3v'\;\delta(\mathbf{k}\cdot\mathbf{v}-\mathbf{k}\cdot\mathbf{v}')\left(F(\mathbf{v}')\frac{\partial F(\mathbf{v})}{\partial\mathbf{v}}-F(\mathbf{v})\frac{\partial f(\mathbf{v}')}{\partial\mathbf{v}'}\right). \tag{12.122}$$

After Fourier inversion of (12.122) by performing $\int d^3k d^3k'(\cdots)e^{i\mathbf{k}\cdot\mathbf{r}}e^{i\mathbf{k}'\cdot\mathbf{r}}$ (12.116) gives the Lenard–Balescu equation in the form

$$\frac{\partial}{\partial t}F(\mathbf{v},t) = -N\;\frac{\partial}{\partial\mathbf{v}}\cdot\int d^3k\;\mathbf{k}\;\frac{\omega_p^2}{k^2}\frac{\omega_p^2}{k^2}\frac{\pi(2\pi)^3}{\mid\varepsilon(\mathbf{k},\mathbf{k}\cdot\mathbf{v})\mid^2}$$

$$\times\int d^3v'\delta[\mathbf{k}\cdot(\mathbf{v}-\mathbf{v}')]\;\mathbf{k}\cdot\left(F(\mathbf{v})\frac{\partial F(\mathbf{v}')}{\partial\mathbf{v}'}-F(\mathbf{v}')\frac{\partial F(\mathbf{v})}{\partial\mathbf{v}}\right). \tag{12.123}$$

12.6.2. *Similarity with the quasilinear theory*

Consider a stable plasma which consists of N_α ($\alpha = e,\ i$) particles and has average field $\mathbf{E} = 0$ (the plasma is homogeneous). The distribution of particles of each species is described by the Klimontovich distribution function

$$F_\alpha(\mathbf{r}, \mathbf{v}, t) = \sum_{i=1}^{N_\alpha} \delta(\mathbf{r} - \mathbf{r}_i(t))\delta(\mathbf{v} - \mathbf{v}_i(t)) \quad (\alpha = e,\ i) \tag{12.124}$$

with the initial distribution function

$$F_\alpha(\mathbf{r}, \mathbf{v}, t = 0) = \sum_{i=1}^{N_\alpha} \delta(\mathbf{r} - \mathbf{r}_{i0})\delta(\mathbf{v} - \mathbf{v}_{i0}). \tag{12.125}$$

The distribution function satisfies the Liouville equation

$$\frac{\partial F_\alpha}{\partial t} + \mathbf{v} \cdot \frac{\partial F_\alpha}{\partial \mathbf{r}} + \frac{e_\alpha}{m_\alpha}\mathbf{E} \cdot \frac{\partial F_\alpha}{\partial \mathbf{v}} = 0 \tag{12.126}$$

and the Poisson equation

$$\nabla \cdot \mathbf{E} = 4\pi \sum_\alpha n_{0\alpha} e_\alpha \int F_\alpha(\mathbf{r}, \mathbf{v}, t) d^3v \quad (n_{0\alpha} = N_\alpha/V,\ \ V:\ volume). \tag{12.127}$$

Fourier transforming the above equations with respect to $\mathbf{r}$ gives

$$F_\alpha(\mathbf{k}, \mathbf{v}, t) = \sum_{i=1}^{N_\alpha} \delta(\mathbf{v} - \mathbf{v}_i(t)) e^{-i\mathbf{k}\cdot\mathbf{r}_i(t)} \tag{12.128}$$

$$F_\alpha(\mathbf{k}, \mathbf{v}, t = 0) = \sum_{i=1}^{N_\alpha} \delta(\mathbf{v} - \mathbf{v}_{i0}) e^{-i\mathbf{k}\cdot\mathbf{r}_{i0}} \tag{12.129}$$

$$i\mathbf{k} \cdot \mathbf{E}(\mathbf{k}, t) = 4\pi \sum_\alpha n_{0\alpha} e_\alpha \int F_\alpha(\mathbf{k}, \mathbf{v}, t) d^3v.$$

To write Fourier transform of (12.126), let us use the convolution formula:

$$\int d^3r e^{-i\mathbf{k}\cdot\mathbf{r}} \mathbf{E}(\mathbf{r}) F(\mathbf{r}) = \left(\frac{1}{2\pi}\right)^3 \int d^3k' \mathbf{E}(\mathbf{k}') F(\mathbf{k} - \mathbf{k}').$$

If one wishes to express the above in Fourier series, put $\left(\frac{1}{2\pi}\right)^3 \int d^3k' \to \sum_{\mathbf{k}'}$. Then (12.126) yields

$$\frac{\partial}{\partial t} F_\alpha(\mathbf{k}, \mathbf{v}, t) + i\mathbf{k} \cdot \mathbf{v} F_\alpha + \frac{e_\alpha}{m_\alpha} \mathbf{E}(\mathbf{k}, t) \cdot \frac{\partial F_\alpha(\mathbf{k} = 0, \mathbf{v}, t)}{\partial \mathbf{v}}$$

$$= -\frac{e_\alpha}{m_\alpha} \sum_{\mathbf{k}'}{}' \mathbf{E}(\mathbf{k}', t) \cdot \frac{\partial}{\partial \mathbf{v}} F_\alpha(\mathbf{k} - \mathbf{k}', \mathbf{v}, t) \tag{12.130}$$

where we separated the $k = 0$ term from the product term and the prime on $\sum$ denotes that the $\mathbf{k}' = \mathbf{k}$ term is excluded in the summation. Let us assume in (12.130) that $F_\alpha(\mathbf{k} = 0, \mathbf{v}, t) \gg F_\alpha(\mathbf{k}, \mathbf{v}, t)$ and neglect the RHS:

$$\frac{\partial}{\partial t} F_\alpha(\mathbf{k}, \mathbf{v}, t) + i\mathbf{k} \cdot \mathbf{v} F_\alpha + \frac{e_\alpha}{m_\alpha} \mathbf{E}(\mathbf{k}, t) \cdot \frac{\partial F_\alpha(\mathbf{k} = 0, \mathbf{v}, t)}{\partial \mathbf{v}} = 0. \quad (12.131)$$

The crucial assumption here is that the spatially uniform part of the distribution is much greater than the spatially fluctuating part ($F_\alpha(\mathbf{k} = 0, \mathbf{v}, t) \gg F_\alpha(\mathbf{k}, \mathbf{v}, t)$). Neglect of the RHS of (12.130), the 'collision term', appears to be acceptable in view of the Vlasov equation which was also made 'collisionless' by neglecting the higher order correlation term. More refinement over the Vlasov-like equation, (12.131), is to take account of the time dependance of $F_\alpha(\mathbf{k} = 0, \mathbf{v}, t)$. The equation for $F_\alpha(\mathbf{k} = 0, \mathbf{v}, t)$ can be set up by extracting the terms corresponding to $\mathbf{k} = 0$ in (12.130):

$$\frac{\partial}{\partial t} F_\alpha(\mathbf{k} = 0, \mathbf{v}, t) = -\frac{e_\alpha}{m_\alpha} \sum_{\mathbf{k}} \mathbf{E}(-\mathbf{k}, t) \cdot \frac{\partial}{\partial \mathbf{v}} F_\alpha(\mathbf{k}, \mathbf{v}, t). \quad (12.132)$$

Together with (12.131) and (12.132), we have two more equations:

$$i\mathbf{k} \cdot \mathbf{E}(\mathbf{k}, t) = 4\pi \sum_\alpha n_{0\alpha} e_\alpha \int F_\alpha(\mathbf{k}, \mathbf{v}, t) d^3v, \quad (12.133)$$

$$F_\alpha(\mathbf{k}, \mathbf{v}, t = 0) = \sum_{i=1}^{N_\alpha} \delta(\mathbf{v} - \mathbf{v}_{i0}) e^{-i\mathbf{k} \cdot \mathbf{r}_{i0}}, \quad (12.134)$$

where $\mathbf{r}_{i0}$ and $\mathbf{v}_{i0}$ are the initial position and velocity of the i-th particle of species α (the α-index omitted). Note the similarity between the above equations and the quasilinear equations.

Consulting the discussion in Section 12.6.1, we have the following notational correspondences: $F(\mathbf{k}, \mathbf{v}, t)$ *in* (12.131) $\leftrightarrow \delta f$ *in* (12.104); $F(\mathbf{k} = 0, \mathbf{v}, t)$ *in* (12.132) $\leftrightarrow F(\mathbf{v}, t)$ *in* (12.104), (12.116); $\mathbf{E}(\mathbf{k}, t)$ *in* (12.133) $\leftrightarrow$ $\delta\mathbf{E}$ *in* (12.108). The remaining discussion is to show explicitly that the present formulation expressed by (12.131)–(12.134) is equivalent to the analysis in Section 12.6.1.

We solve (12.131)–(12.133) by Laplace transform with initial condition (12.134). The slow time variation of $F_\alpha(\mathbf{k} = 0, \mathbf{v}, t)$ in (12.131) will be neglected by holding that quantity constant [that quantity will be expressed as $F_\alpha(\mathbf{k} = 0, \mathbf{v}, \epsilon t)$ with the understanding that 'ϵt' represents the slow time].

Laplace transform of (12.131) yields

$$F_\alpha(\mathbf{k},\mathbf{v},\omega) = \frac{-i}{\omega - \mathbf{k}\cdot\mathbf{v}}\left[\frac{e_\alpha}{m_\alpha}\mathbf{E}(\mathbf{k},\omega)\cdot\frac{\partial}{\partial\mathbf{v}}F_\alpha(\mathbf{k}=0,\mathbf{v},\epsilon t) - F_\alpha(\mathbf{k},\mathbf{v},t=0)\right]. \tag{12.135}$$

Laplace transform of (12.133) and use of (12.135) give

$$\begin{aligned}\mathbf{k}\cdot\mathbf{E}(\mathbf{k},\omega) &= \frac{4\pi}{\varepsilon(\mathbf{k},\omega)}\sum_\alpha n_{0\alpha}e_\alpha\int d^3v\frac{F_\alpha(\mathbf{k},\mathbf{v},t=0)}{\omega-\mathbf{k}\cdot\mathbf{v}}\\ &= \frac{4\pi}{\varepsilon(\mathbf{k},\omega)}\sum_\alpha n_{0\alpha}e_\alpha\sum_{i=1}^{N_\alpha}\frac{e^{-i\mathbf{k}\cdot\mathbf{r}_{i0}}}{\omega-\mathbf{k}\cdot\mathbf{v}_{i0}} \qquad [cf.\ (11.79),\ \ (12.112)]\end{aligned} \tag{12.136}$$

$$\text{where} \quad \varepsilon(\mathbf{k},\omega) = 1 + \sum_\alpha\frac{\omega_{p\alpha}^2}{k^2}\int d^3v\frac{\mathbf{k}\cdot\frac{\partial}{\partial\mathbf{v}}F_\alpha(\mathbf{k}=0,\mathbf{v},\ \epsilon t)}{\omega-\mathbf{k}\cdot\mathbf{v}}. \tag{12.137}$$

Using (12.134) and (12.136) in (12.135) gives

$$\begin{aligned}F_\alpha(\mathbf{k},\mathbf{v},\omega) = &\frac{-i}{\omega-\mathbf{k}\cdot\mathbf{v}}\frac{1}{k^2\varepsilon(\mathbf{k},\omega)}\sum_\alpha\omega_{p\alpha}^2\sum_{i=1}^{N_\alpha}\frac{e^{-i\mathbf{k}\cdot\mathbf{r}_{i0}}}{\omega-\mathbf{k}\cdot\mathbf{v}_{i0}}\mathbf{k}\cdot\frac{\partial F_\alpha(\mathbf{k}=0,\mathbf{v},\ \epsilon t)}{\partial\mathbf{v}}\\ &+\frac{i}{\omega-\mathbf{k}\cdot\mathbf{v}}\sum_{i=1}^{N_\alpha}\delta(\mathbf{v}-\mathbf{v}_{i0})e^{-i\mathbf{k}\cdot\mathbf{r}_{i0}}.\end{aligned} \tag{12.138}$$

Inverse Laplace transform of (12.136) and (12.138) yields

$$\mathbf{k}\cdot\mathbf{E}(\mathbf{k},t) = 4\pi i\sum_\alpha n_{0\alpha}e_\alpha\sum_{i=1}^{N_\alpha}\frac{e^{-i\mathbf{k}\cdot(\mathbf{r}_{i0}+\mathbf{v}_{i0}t)}}{\varepsilon(\mathbf{k},\mathbf{k}\cdot\mathbf{v}_{i0})}, \tag{12.139}$$

$$\begin{aligned}F_\alpha(\mathbf{k},\mathbf{v},t) = &\sum_{i=1}^{N_\alpha}\delta(\mathbf{v}-\mathbf{v}_{i0})e^{-i\mathbf{k}\cdot(\mathbf{r}_{i0}+\mathbf{v}t)}+\\ &-\frac{1}{k^2}\sum_\alpha\omega_{p\alpha}^2\sum_{i=1}^{N_\alpha}\frac{\mathbf{k}\cdot\frac{\partial}{\partial\mathbf{v}}F_\alpha(\mathbf{k}=0,\mathbf{v},\ \epsilon t)}{\mathbf{k}\cdot\mathbf{v}-\mathbf{k}\cdot\mathbf{v}_{i0}}\left(\frac{e^{-i\mathbf{k}\cdot(\mathbf{r}_{i0}+\mathbf{v}t)}}{\varepsilon(\mathbf{k},\mathbf{k}\cdot\mathbf{v})}-\frac{e^{-i\mathbf{k}\cdot(\mathbf{r}_{i0}+\mathbf{v}_{i0}t)}}{\varepsilon(\mathbf{k},\mathbf{k}\cdot\mathbf{v}_{i0})}\right),\end{aligned} \tag{12.140}$$

which is identical with (12.115) [$\delta(\mathbf{k})$-term in the latter disappears eventually due to $\mathbf{k}\delta(\mathbf{k}) = 0$]. (12.139) gives upon using the electrostatic nature of $\mathbf{E}$

$$\mathbf{E}(-\mathbf{k}, t) = -\frac{4\pi i\mathbf{k}}{k^2}\sum_{\alpha} n_{0\alpha}e_{\alpha}\sum_{i=1}^{N_{\alpha}} \frac{e^{i\mathbf{k}\cdot(\mathbf{r}_{i0}+\mathbf{v}_{i0}t)}}{\varepsilon(-\mathbf{k}, -\mathbf{k}\cdot\mathbf{v}_{i0})}. \tag{12.141}$$

Putting $\mathbf{k} \to -\mathbf{k}$, (12.141) is identical with (12.114) for one-species plasma (electrons). Substitute (12.140) and (12.141) into (12.132):

$$\frac{\partial}{\partial t}F_{\alpha}(\mathbf{k}=0, \mathbf{v}, \epsilon t) = -\frac{\partial}{\partial \mathbf{v}}\cdot\sum_{\mathbf{k}}\frac{e_{\alpha}}{m_{\alpha}}\langle \mathbf{E}(-\mathbf{k}, t)F_{\alpha}(\mathbf{k}, \mathbf{v}, t)\rangle \tag{12.142}$$

where the ensemble average should be taken as is indicated by the angular bracket. (12.140)–(12.142) are entirely identical with (12.114)–(12.116), and the Lenard–Balescu equation follows by repeating the same algebra as in § 12.6.1.

12.7. Electric field fluctuation

In this Section, we calculate the electric fluctuation by employing the quasilinear theory of the Liouville equation developed in the previous section. The result agrees with that obtained in Chapter 11. We assume a spatially uniform ensemble: $\frac{\partial\langle K\rangle}{\partial \mathbf{r}} = 0$, $\langle \mathbf{E}\rangle = 0$. Also $\mathbf{E}(t)$ is assumed to be a stationary random process, so that the fluctuation $\langle \mathbf{E}(\mathbf{r}, t)\mathbf{E}(\mathbf{r}', t')\rangle$ depends only upon $\tau = t - t'$ and $\mathbf{R} = \mathbf{r} - \mathbf{r}'$. Therefore we have

$$\langle \mathbf{E}(\mathbf{r}, t)\ \mathbf{E}(\mathbf{r}', t')\rangle = \langle \mathbf{E}(\mathbf{r}, t)\ \mathbf{E}(\mathbf{r} - \mathbf{R}, t - \tau)\rangle$$

which is independent of $\mathbf{r}$ and t but is dependent only on $\mathbf{R}$ and τ. Since $\langle \mathbf{E}\rangle = 0$, $\mathbf{E} = \langle \mathbf{E}\rangle + \delta\mathbf{E} = \delta\mathbf{E}$, we have

$$\langle \mathbf{E}(\mathbf{r}, t)\ \mathbf{E}(\mathbf{r} - \mathbf{R}, t - \tau)\rangle = \langle \delta\mathbf{E}(\mathbf{r}, t)\ \delta\mathbf{E}(\mathbf{r} - \mathbf{R}, t - \tau)\rangle. \tag{12.143}$$

Let us evaluate the above quantity in the limit of $t \to \infty$, with the aid of (12.114). The ensemble average in the above expression is taken over the distribution

$$F^{(N)}(\mathbf{x}_{10}, \mathbf{x}_{20}, \ldots, \mathbf{x}_{N0}) = F^{(1)}(\mathbf{v}_{10})F^{(1)}(\mathbf{v}_{20})\cdots F^{(1)}(\mathbf{v}_{N0}).$$

[The ensemble is spatially uniform and is free of correlation.] Using (12.114), we write

$$\langle \delta\mathbf{E}(\mathbf{k}, t)\ \delta\mathbf{E}(\mathbf{k}', t')\rangle = \left\langle -i\frac{4\pi e\mathbf{k}}{k^2}\sum_{i=1}^{N}\frac{e^{-i\mathbf{k}\cdot(\mathbf{r}_{i0}+\mathbf{v}_{i0}t)}}{\varepsilon(\mathbf{k},\ \mathbf{k}\cdot\mathbf{v}_{i0})}(-i)\frac{4\pi e\mathbf{k}'}{k'^2}\sum_{j=1}^{N}\frac{e^{-i\mathbf{k}'\cdot(\mathbf{r}_{j0}+\mathbf{v}_{j0}t')}}{\varepsilon(\mathbf{k}',\ \mathbf{k}'\cdot\mathbf{v}_{j0})}\right\rangle.$$

The operation $\langle\cdots\rangle$ includes spatial integration from $-\infty$ to $+\infty$ (with density $n_0 = N/V$ constant while both of N and V go to ∞). So we have the relations

$$\langle \mathbf{k}e^{-i\mathbf{k}\cdot\mathbf{r}_{i0}}\rangle \sim \mathbf{k}\,\delta(\mathbf{k}) = 0, \quad \langle \mathbf{k}e^{-i\mathbf{k}\cdot\mathbf{r}_{i0}}e^{-i\mathbf{k}'\cdot\mathbf{r}_{j0}}\rangle = 0 \text{ when } i \neq j,$$
$$\langle \mathbf{k}e^{-i\mathbf{k}\cdot\mathbf{r}_{i0}}e^{-i\mathbf{k}'\cdot\mathbf{r}_{j0}}\rangle \sim \mathbf{k}\,\delta(\mathbf{k}+\mathbf{k}') \text{ when } i = j.$$

Therefore, only products of terms $i = j$ survive the averaging operation to give

$$\begin{aligned}
&\langle \delta\mathbf{E}(\mathbf{k},t)\ \delta\mathbf{E}(\mathbf{k}',t')\rangle \\
&= -\frac{4\pi e\mathbf{k}}{k^2}\frac{4\pi e\mathbf{k}'}{k'^2}\delta(\mathbf{k}+\mathbf{k}')\sum_{i=1}^{N}\frac{e^{-i\mathbf{k}\cdot\mathbf{v}_{i0}t}}{\varepsilon(\mathbf{k},\ \mathbf{k}\cdot\mathbf{v}_{i0})}\frac{e^{-i\mathbf{k}'\cdot\mathbf{v}_{i0}t'}}{\varepsilon(\mathbf{k}',\ \mathbf{k}'\cdot\mathbf{v}_{i0})} \\
&\quad\times\int F^{(1)}(\mathbf{v}_{10})F^{(1)}(\mathbf{v}_{20})\cdots F^{(1)}(\mathbf{v}_{N0})d^3v_{10}d^3v_{20}\cdots d^3v_{N0}.
\end{aligned}$$

The summation yields N identical terms. Note that $\mathbf{v}_{i0}$ is a dummy integration variable which we can put as $\mathbf{v}$. With normalization constant $(volume)^{-1}$ which we omitted so far, the integration yields the factor n_0, the density, and we can write

$$\begin{aligned}
&\langle \delta\mathbf{E}(\mathbf{k},t)\ \delta\mathbf{E}(\mathbf{k}',t-\tau)\rangle \\
&= -\delta(\mathbf{k}+\mathbf{k}')\left(\frac{4\pi e}{k^2}\right)^2 n_0\ \mathbf{k}\mathbf{k}'\int d^3vF(\mathbf{v})\frac{e^{-i\mathbf{k}\cdot\mathbf{v}\tau}}{|\ \varepsilon(\mathbf{k},\ \mathbf{k}\cdot\mathbf{v})\ |^2}. \qquad (12.144)
\end{aligned}$$

The integral in (12.144) can be written as

$$\int du\int d^3v\frac{e^{-iku\tau}}{|\ \varepsilon(\mathbf{k},ku)\ |^2}\delta\left(u-\frac{\mathbf{k}\cdot\mathbf{v}}{k}\right)F(\mathbf{v}).$$

Thus (12.144) reads

$$\langle \delta\mathbf{E}(\mathbf{k},t)\ \delta\mathbf{E}(\mathbf{k}',t')\rangle = \delta(\mathbf{k}+\mathbf{k}')\left(\frac{4\pi e}{k^2}\right)^2 n_0\ \mathbf{k}\mathbf{k}'\int_{-\infty}^{\infty} duG(u)\ \frac{e^{-iku\tau}}{|\ \varepsilon(\mathbf{k},ku)\ |^2} \quad (\tau = t-t'),$$
$$G(u) = \int F(\mathbf{v})\delta\left(u-\frac{\mathbf{k}\cdot\mathbf{v}}{k}\right)d^3v.$$

Make a change of variable in the above $u \to \omega/k$ to write

$$\langle \delta\mathbf{E}(\mathbf{k},t)\ \delta\mathbf{E}(\mathbf{k}',t-\tau)\rangle = -\delta(\mathbf{k}+\mathbf{k}')\left(\frac{4\pi e}{k^2}\right)^2 n_0\ \frac{\mathbf{k}\mathbf{k}'}{k}\int_{-\infty}^{\infty} d\omega\ G(\frac{\omega}{k})\ \frac{e^{-i\omega\tau}}{|\ \varepsilon(\mathbf{k},\omega)\ |^2}. \qquad (12.145)$$

If the ensemble is homogeneous both in space and time, the correlation tensor is calculated by

$$C_{ij}(\mathbf{R},\tau) \equiv \langle \delta\mathbf{E}(\mathbf{r},t)\ \delta\mathbf{E}(\mathbf{r}',t')\rangle = \langle \delta\mathbf{E}(\mathbf{r},t)\delta\mathbf{E}(\mathbf{r}-\mathbf{R},t-\tau\rangle. \quad (12.146)$$

The Fourier transform of $C_{ij}(\mathbf{R},\tau)$ is the *spectral density*, $C_{ij}(\mathbf{k},\omega)$:

$$C_{ij}(\mathbf{R},\tau) = \int \frac{d^3k}{(2\pi)^3} \int \frac{d\omega}{2\pi} C_{ij}(\mathbf{k},\omega) e^{i\mathbf{k}\cdot\mathbf{R}-i\omega\tau}.$$

Note that (12.145) represents $C_{ij}(\mathbf{k},\tau)$, and thus we can write

$$C_{ij}(\mathbf{R},\tau) = F.T.\ of\ (12.145) = \int \frac{d^3k}{(2\pi)^3} e^{i\mathbf{k}\cdot\mathbf{R}} \left(\frac{4\pi e}{k^2}\right)^2 n_0 \frac{\mathbf{kk}}{k} \times \int_{-\infty}^{\infty} d\omega\ G\left(\frac{\omega}{k}\right) \frac{e^{-i\omega\tau}}{|\ \varepsilon(\mathbf{k},\omega)\ |^2}.$$

Comparing the above two equations enables one to identify

$$C_{ij}(\mathbf{k},\omega) = n_0 \left(\frac{4\pi e}{k^2}\right)^2 \frac{\mathbf{kk}}{k} 2\pi \frac{G(\frac{\omega}{k})}{|\ \varepsilon(\mathbf{k},\omega)\ |^2}. \quad (cf.\ (11.69)) \qquad (12.147)$$

The autocorrelation function defined in (12.146) enables one to calculate the average energy of the particle executing random motion as a consequence of being acted on by random fluctuating electric field. The trace of the spectral density, $C_{ii}(\mathbf{k},\omega)$, is proportional to the electrostatic energy per unit volume per unit frequency per unit wave number; that is

$$\int \frac{d^3k}{(2\pi)^3} \int \frac{d\omega}{2\pi} \langle |\ E(k,\omega)\ |^2\rangle = C_{ii}(R=0,\tau=0) = \int \frac{d^3k}{(2\pi)^3} \int \frac{d\omega}{2\pi} \times (12.147).$$

Therefore, we obtain

$$\langle |\ E(k,\omega)\ |^2\rangle = n_0\ (2\pi)^2 \frac{(4\pi e)^2}{k^3} \frac{G(\frac{\omega}{k})}{|\ \varepsilon(\mathbf{k},\omega)\ |^2}. \quad (cf.\ (11.69)) \qquad (12.148)$$

This result agrees with (11.69) in H. J. Lee (2015), which was obtained by the dressed particle approach.

References to Chapter 12

Clemmow P. C. and Daugherty J. P. *Electrodynamics of Particles and Plasmas*, Addison-Wesley London, 1969.

Lifshitz E. M. and Pitaevskii L. P. *Physical Kinetics*, Landau and Lifshitz Course of theoretical physics Volume 10 Butterworth-Heinemann, 1981.
Lee H. J. Fluctuation of the electric field in a plasma, *J. Korean Phys. Soc.* **66**(8) (2015) pp. 1167–1185.
Montgomery D. C. and Tidman D. A. *Plasma Kinetic Theory*, McGraw-Hill, 1964.
Montgomery, D. C. *Theory of Unmagnetized Plasma*, Gordon and Beach New York, 1971.
Nicholson D. R. *Introduction to Plasma theory*, Wiley, 1983.

Chapter 13

Surface wave

Surface waves propagate along the interface of two different media; the amplitude of a surface wave is rapidly decreasing in the perpendicular direction to the interface. Therefore the associated wave energy propagates along the interface, narrowly confined about the interfacial layer. Laboratory and natural plasmas are bounded in some form or another. The localized character of the surface wave energy is relevant to heating of the solar corona. Surface waves have application in low temperature surface physics.

If we deal with a bounded plasma, the governing equations should be solved, respectively, for the plasma region and the non-plasma region, and the solutions thus obtained should be matched through appropriate matching conditions. Applying correct boundary conditions at the interface is important to identify legitimate surface wave modes. The boundary conditions are of two kinds; electromagnetic and kinematic. Additional dynamic boundary condition must be considered if we use warm fluid equation since the pressure becomes another dynamical variable. If the Vlasov equation is used, the kinematic boundary condition will be taken as the *specular reflection condition*, which means $f(v_x, v_y, v_z, x, y, z = 0) = f(v_x, v_y, -v_z, x, y, z = 0)$ where $z = 0$ is the interface. The electromagnetic field components also should satisfy certain jump conditions across the interface — electromagnetic boundary conditions. As a rule, we can obtain meaningful jump conditions from the governing equations by integrating the latter across the interface over the infinitesimally thin layer. This is a useful mathematical scheme to find the necessary matching conditions. This method is applicable when the interfacial layer is sharp. Assuming a sharp *fixed* boundary is an idealization of a real interfacial boundary because the surface is undulating simultaneously with the wave motion. Assuming a fixed interface amounts to linearizing the nonlinear kinematic boundary condition. A refined theory of water surface wave requires consideration of nonlinear kinematic boundary condition at the interface. In this Chapter we shall consider a semi-infinite plasma and a plasma slab bounded by a vacuum.

If we use a *warm* fluid equation, the wave equation is of a higher order differential equation as compared with the *cold* fluid wave equation. Thus the number of boundary conditions (kinematic plus dynamic) for a warm plasma increases by one as compared with a cold plasma. This mathematical distinction can be conveniently translated into physical terms which states that *surface charges* are formed on a cold plasma boundary while such entity is nonexistent on a warm boundary. The extra necessary boundary condition for a warm plasma is (13.34) below which is the fluid version of the specular reflection condition.

It is convenient to write the Maxwell equations by decomposing the electromagnetic fields into two directions: perpendicular and parallel components with respect to the interface. The Maxwell equations then decouple as TM (*transverse magnetic*) and TE (*transverse electric*) modes. We can deal with the reduced sets of equations independently. However, the actual propagation of the modes requires satisfaction of the interfacial boundary conditions for both cases.

13.1. Surface waves derived from the two-fluid equations

The characteristic feature of a surface wave is that it is localized in the layer in the vicinity of the interface, with the amplitude of the wave decaying exponentially along the coordinate departing away from the surface. The surface waves are called *pure* if the fields in both plasma and vacuum are evanescent. The surface wave can be of a mixed type, which means that the field in one medium is evanescent while spatially oscillating in the other. We shall consider only pure surface wave boundary condition. The important concept is concerned with the surface charge and the surface current. Through a rigorous mathematical analysis of fluid equations, we show that the surface charges and the surface currents disappear in a plasma of nonzero temperature but exist in a cold (zero temperature) plasma. We draw this conclusion from the solvability of the matching conditions obtained from the governing equations of the two media.

13.1.1. *Transverse magnetic mode (TM mode)*

Here we study the TM mode (transverse magnetic; the magnetic field is *transverse* to the direction of propagation, so the magnetic field is perpendicular to the interface) surface wave propagating on a warm electron plasma occupying the region $x > 0$ (ions are assumed to be immobile) bounded by a vacuum ($x < 0$). The governing equations for the plasma are:

$$\frac{\partial \mathbf{v}}{\partial t} + \mathbf{v} \cdot \nabla \mathbf{v} = -\frac{e}{m}\left(\mathbf{E} + \frac{1}{c}\mathbf{v} \times \mathbf{B}\right) - \frac{\beta^2}{N}\nabla n \tag{13.1}$$

where N (n) is the equilibrium (perturbation) electron number density, and β is the electron thermal velocity ($\beta^2 = \gamma_e T_e/m_e$, with γ_e the ratio of specific heats).

$$\frac{\partial n}{\partial t} + \nabla \cdot (n + N)\mathbf{v} = 0, \tag{13.2}$$

$$\nabla \cdot \mathbf{E} = -4\pi e n, \tag{13.3}$$

$$\nabla \times \mathbf{E} = -\frac{1}{c}\frac{\partial \mathbf{B}}{\partial t}, \tag{13.4}$$

$$\nabla \times \mathbf{B} = \frac{1}{c}\frac{\partial \mathbf{E}}{\partial t} - \frac{4\pi e}{c}(N + n)\mathbf{v}. \tag{13.5}$$

In the above equations, the nonlinear terms are kept for future use.

• **Plasma solution.** We suppose that the surface wave propagates along the z-direction; the typical wave quantity varies in the form $f(x)e^{ikz - i\omega t}$. Then the y-direction has the translational invariance and can be ignored ($\partial/\partial y = 0$). In this coordinate system the TM mode surface wave set consists of (E_x, E_z, B_y) [Note that B_y represents the transverse magnetic field component]. All the other field components can be set equal to zero because the latter components are not coupled to the TM-mode set. Eliminating **B** between (13.4) and (13.5) yields

$$\left(1 - \frac{\beta^2}{c^2}\right)\frac{\partial^2 E_z}{\partial x \partial z} + L_x E_x = -Q_x + \frac{4\pi e}{c^2}\frac{\partial}{\partial t}(n v_x), \tag{13.6}$$

$$\left(1 - \frac{\beta^2}{c^2}\right)\frac{\partial^2 E_x}{\partial x \partial z} + L_z E_z = -Q_z + \frac{4\pi e}{c^2}\frac{\partial}{\partial t}(n v_z), \tag{13.7}$$

$$L_x = \frac{1}{c^2}\left(\frac{\partial^2}{\partial t^2} + \omega_p^2\right) - \frac{\partial^2}{\partial z^2} - \frac{\beta^2}{c^2}\frac{\partial^2}{\partial x^2}, \tag{13.8}$$

$$L_z = \frac{1}{c^2}\left(\frac{\partial^2}{\partial t^2} + \omega_p^2\right) - \frac{\partial^2}{\partial x^2} - \frac{\beta^2}{c^2}\frac{\partial^2}{\partial z^2}, \tag{13.9}$$

$$\mathbf{Q} = -\frac{4\pi e N}{c^2}\left(\mathbf{v} \cdot \nabla \mathbf{v} + \frac{e}{mc}\mathbf{v} \times \mathbf{B}\right). \tag{13.10}$$

Combining (13.6) and (13.7) with neglect of the nonlinear terms gives

$$\left[L_x L_z - \left(1 - \frac{\beta^2}{c^2}\right)^2 \frac{\partial^4}{\partial z^2 \partial x^2}\right] E_{x,z} = 0. \tag{13.11}$$

After Fourier transforming with respect to t and z, (13.11) is a linear differential equation in the variable x, whose solutions are $\sim e^{-\gamma x}$. So putting $E_{x,z} \sim e^{ikz - i\omega t - \gamma x}$ (satisfying the boundary condition at $x = \infty$) in (13.11) gives the following two values for γ, the *attenuation constants* of the surface wave,

$$\gamma_{\pm}^2 = \frac{c^2}{2\beta^2}\left[\frac{2k^2\beta^2}{c^2} + \frac{\omega_p^2 - \omega^2}{c^2}\left(1 + \frac{\beta^2}{c^2} \pm \left(1 - \frac{\beta^2}{c^2}\right)\right)\right]. \tag{13.12}$$

We rename the symbols:

$$\gamma_+^2 \equiv \gamma^2 = k^2 + \frac{\omega_p^2 - \omega^2}{\beta^2}, \tag{13.13}$$

$$\gamma_-^2 \equiv \alpha^2 = k^2 + \frac{\omega_p^2 - \omega^2}{c^2}. \tag{13.14}$$

Then the linear solutions read

$$E_x = Ae^{-\gamma x} + Ge^{-\alpha x}, \tag{13.15}$$

$$E_z = -\frac{ik}{\gamma} Ae^{-\gamma x} - \frac{i\alpha}{k} Ge^{-\alpha x}, \tag{13.16}$$

$$B_y = \frac{\omega^2 - \omega_p^2}{ck\omega} Ge^{-\alpha x}, \tag{13.17}$$

where A and G are constants of integration.

- **Vacuum solution** The wave equation in the vacuum reads

$$\left(\nabla^2 + \frac{\omega^2}{c^2}\right)\mathbf{E} = 0. \quad (\nabla \cdot \mathbf{E} = 0) \tag{13.18}$$

Its solution satisfying the boundary condition at $x = -\infty$ is

$$E_x = Fe^{\lambda x}, \tag{13.19}$$

$$E_z = \frac{i\lambda}{k} Fe^{\lambda x}, \tag{13.20}$$

$$B_y = \frac{\omega}{ck} Fe^{\lambda x}, \quad (\lambda = \sqrt{k^2 - \omega^2/c^2}). \tag{13.21}$$

The three undetermined constants of integration in the above plasma and vacuum solutions, A, G, and F, should be determined by the matching conditions at the interface $x = 0$.

• **Boundary conditions.** In the linear approximation, we shall consider the interface to be a fixed plane $x = 0$. In general, the plasma-vacuum boundary is a rippling surface subject to the perturbation associated with the wave. The kinematics of the rippling boundary requires nonlinear consideration which will be taken up later. The boundary conditions connecting the plasma fields to the vacuum fields are obtained by integrating the governing equations across the infinitesimal segment $x = -\Delta$ to $x = +\Delta$, where Δ is implied to be approaching zero. We consider (13.3) as an example. The infinitesimal integration gives

$$\int_{-\Delta}^{\Delta} dx \left(\frac{\partial E_x}{\partial x} + \frac{\partial E_y}{\partial y} + \frac{\partial E_z}{\partial z} \right) = -4\pi e \int_{-\Delta}^{\Delta} n \, dx.$$

On the LHS, only the first term (integrable term, $\partial E_x/\partial x$) survives in the limit of $\Delta \to 0$. When n is finite (volume-distributed), the RHS vanishes in the limit of $\Delta \to 0$, and we have that E_x is continuous when it crosses the interface $x = 0$. Discontinuity of E_x occurs when there are surface charges on $x = 0$, i.e. when $n = n^*\delta(x)$, i.e. particles are surface-distributed around the interface. We define the surface charge and surface current as

$$\sigma^* = -e \int_{-\Delta}^{\Delta} n dx \;\; = -en(0)2\Delta \equiv -en^*, \quad \mathbf{J}^* = -e \int_{-\Delta}^{\Delta} N\mathbf{v} dx.$$

Clearly, in order for $n(0)2\Delta$ to be finite in the limit $\Delta \to 0$, $n(x)$ should be a delta-function $\sim \delta(x)$. We show later that surface charges and surface currents are present in cold plasmas but those quantities are non-existent in warm plasmas. Then operating $\lim_{\Delta \to 0} \int_{-\Delta}^{\Delta} (...)dx$ with (13.4), (13.3), and (13.5), respectively, yields

$$[E_z] \equiv E_z(x = 0^+) - E_z(x = 0^-) = 0, \tag{13.22}$$

$$[E_x] = 4\pi \, \sigma^*, \tag{13.23}$$

$$[B_y] = \frac{4\pi}{c} \, J_z^*. \tag{13.24}$$

Infinitesimal integration of (13.2), gives

$$\frac{\partial \sigma^*}{\partial t} = e \int_{-\Delta}^{\Delta} N \frac{\partial v_x}{\partial x} dx + e \int_{-\Delta}^{\Delta} N \frac{\partial v_z}{\partial z} dx.$$

The first integral on the RHS can be integrated out to give $-J_x = -eNv_x$ at $x = 0$, since N = 0 in the vacuum side. The second integral is by definition $\partial J_z^*/\partial z$. Thus we have

$$\frac{\partial \sigma^*}{\partial t} + J_x(0) + \frac{\partial J_z^*}{\partial z} = 0. \tag{13.25}$$

The surface current J_z^* is evaluated as follows. Equation (13.1) gives

$$-i\omega v_z = -\frac{e}{m}E_z - \frac{ik\beta^2}{N}n$$

which yields upon the infinitesimal integration

$$J_z^* = \frac{k\beta^2}{\omega}\sigma^*. \tag{13.26}$$

(13.23), (13.25), and (13.26) give

$$J_x(0) = \frac{i\omega}{4\pi}\left(1 - \frac{k^2\beta^2}{\omega^2}\right)[E_x]. \tag{13.27}$$

(13.23), (13.24), and (13.26) give

$$[B_y] = \frac{k\beta^2}{c\omega}[E_x]. \tag{13.28}$$

(13.22), (13.27), and (13.28) are the three matching conditions which connect the plasma fields to the vacuum fields. Equation (13.22) is the well-known formula which states that *the tangential component of the electric field is continuous across the boundary.* We also have $[B_x] = 0$ per $\nabla \cdot \mathbf{B} = 0$. Using the plasma and the vacuum solutions in the above three matching conditions, one obtains

$$[E_z] = 0: \quad \frac{kA}{\gamma} + \frac{\alpha G}{k} = -\frac{\lambda}{k}F, \tag{13.29}$$

$$(13.27): \quad \omega_p^2 G + \omega^2 A = (\omega^2 - k^2\beta^2)(A + G - F), \tag{13.30}$$

$$(13.28): \quad (\omega^2 - \omega_p^2)G - \omega^2 F = k^2\beta^2(A + G - F). \tag{13.31}$$

(13.30) and (13.31) are redundant. This demonstrates algebraically that (13.23) and (13.28) are redundant. Thus one more boundary condition should be worked out.

At this juncture, we digress to consider a cold plasma, which is a degenerate case from the more general warm plasma. Putting $\beta = 0$ gives

$\gamma \to \infty$ in (13.13), and then the plasma solution is obtained by putting $A = 0$ in (13.15) and (13.16). Hence, the unknown coefficients in (13.29)–(13.30) are only G and F. No more boundary condition is necessary. Hence, one obtains the dispersion relation in the form

$$\alpha + \left(1 - \frac{\omega_{pe}^2}{\omega^2}\right)\lambda = 0 : \quad \textit{cold plasma surface wave.}$$

• **Specular reflection condition.** Instead of using (13.1) to find one more boundary condition for warm plasma, we invoke kinetic consideration. In his famous paper, Reference [L. Landau *On the vibrations of the electronic plasma*, J. Physics (USSR) **10** (1946) p. 25.], Landau proposed a boundary condition on the interface for the distribution function f such that

$$f(x = 0, y, z, v_x, v_y, v_z, t) = f(x = 0, y, z, -v_x, v_y, v_z, t). \tag{13.32}$$

This condition, the *specular reflection condition*, is visualized as that the particles are mirror-reflected upon impinging upon the interface. This is an extra kinematic boundary condition to be satisfied on the boundary, in addition to the electromagnetic boundary conditions that are obtained from the Maxwell equations. We calculate the mean velocity (fluid velocity) on the boundary from the above distribution to ascertain that

$$\int_{-\infty}^{\infty} v_x f(v_x) dv_x = -\int_{-\infty}^{\infty} v_x f(v_x) dv_x \quad (by \;\; v_x \to -v_x). \tag{13.33}$$

So the specular reflection condition amounts in fluid description to that the normal component of the fluid velocity is zero on the interface:

$$v_x(x = 0) = 0. \tag{13.34}$$

[With a slight danger of confusion, we used the same notation v_x both for the fluid velocity and the kinetic velocity variable. v_x in (13.34) is the fluid velocity, the velocity moment.] The specular reflection condition (13.34) serves the independent boundary condition in place of (13.30) or (13.31). But (13.34) makes $\sigma^* = J_z^* = 0$ (as is evident from (13.25) and (13.26)): *there is no surface charge and surface current in a warm plasma.* [We should be careful to generalize this to plasmas: J_z^* here is strictly the surface current in the context of our governing equations (13.1) to (13.5). When there is a beam in a cold plasma, surface currents as well as the surface charges exist even in warm plasma.] In summary, we have the matching conditions:

$$[E_z] = 0,$$

$$v_x(x = 0) = 0,$$

$$[E_x] = 0, \tag{13.35}$$

$$[B_y] = 0. \tag{13.36}$$

Now (13.34) gives

$$\omega_p^2 G + \omega^2 A = 0. \tag{13.37}$$

To prove this, use (13.5) or the plasma equation in the example below. (13.35) and (13.36) give, respectively,

$$A + G = F, \tag{13.38}$$

$$(\omega^2 - \omega_p^2)G - \omega^2 F = 0. \tag{13.39}$$

We can use any two of the three equations (13.37), (13.38), and (13.39). Any one of the three is reproduced by combination of the other two. The dispersion relation is obtained from (13.29), (13.37), (13.38):

$$\omega^2(\alpha + \lambda) - \lambda\, \omega_P^2 - \frac{k^2}{\gamma}\, \omega_p^2 = 0. \tag{13.40}$$

Letting $\gamma \to \infty$ in the above recovers the cold plasma result, earlier.

In summary, the necessary boundary conditions are: $[E_z] = 0$ and any two of the three: $v_x|_0 = 0$, $[E_x] = 0$, $[B_y] = 0$. [For TE mode quantities, use $[E_y] = 0$, $[B_x] = 0$, and $[B_z] = 0$.]

Exercise. Prove that the following two equations are redundant:

$$[B_y] = \frac{4\pi}{c} J_z^* \quad and \quad [E_x] = 4\pi\sigma^*.$$

Solution: Take the x-component of (13.5) at $x = 0^{\pm}$ to write

$$B_y = \frac{\omega}{ck} E_x + \frac{4\pi i}{ck} J_x, \quad or$$

$$B_y(0^+) = \frac{\omega}{ck} E_x(0^+) + \frac{4\pi i}{ck} J_x(0^+) \quad (plasma)$$

$$B_y(0^-) = \frac{\omega}{ck} E_x(0^-) \quad (vacuum).$$

$$Therefore\ B_y(0^+) - B_y(0^-) = [B_y] = \frac{\omega}{ck}[E_x] + \frac{4\pi i}{ck} J_x(0^+).$$

Clearly $[B_y] = \frac{4\pi}{c} J_z^*$ and $\frac{\omega}{ck}[E_x] + \frac{4\pi i}{ck} J_x(0^+) = \frac{4\pi}{c} J_z^*$ are redundant. Using (13.26) and (13.27) in the latter equation above yields $[E_x] = 4\pi\sigma^*$. Q.E.D.

Exercise. Show that using $v_x|_{x=0} = 0$ in (13.1) yields (13.37). Sol. Use $n = -\nabla \cdot \mathbf{E}/4\pi e$ and (13.15) and (13.16). This exercise is a consistency check. All the equations are consistent with the boundary conditions that we obtained.

13.1.2. *Cold plasma surface wave and surface charges*

More discussion about cold plasma. The cold plasma equation corresponds to the case $\beta = 0$ in (13.7). The attenuation constant γ in (13.13) becomes ∞ while α is intact. Thus we have from (13.15) to (13.17)

$$E_x = Ge^{-\alpha x}, \tag{13.41}$$

$$E_z = -\frac{i\alpha}{k} Ge^{-\alpha x}, \tag{13.42}$$

$$B_y = \frac{\omega^2 - \omega_p^2}{ck\omega} Ge^{-\alpha x}. \tag{13.43}$$

The vacuum solutions are still given by (13.19) to (13.21). Note that the number of undetermined constants is reduced by one as compared to the case of warm plasma, and we only need two matching conditions. In a cold plasma, we have

$$-i\omega v_z = -\frac{e}{m} E_z. \tag{13.44}$$

Integrating this over the infinitesimal layer about the interface gives

$$J_z^* = 0 \tag{13.45}$$

$$therefore \quad \frac{\partial \sigma^*}{\partial t} = -J_x(0) = eNv_x(0). \tag{13.46}$$

The boundary conditions applicable for cold plasmas are

$$[E_z] = 0, \tag{13.47}$$

$$[E_x] = 4\pi\sigma^*, \tag{13.48}$$

$$[B_y] = 0. \tag{13.49}$$

These equations are obtained from (13.27) and (13.28) by putting $\beta = 0$. There is no need or room to impose the extra kinematic boundary condition in a cold plasma, because $f(v_x) \sim \delta(v_x)$, which is already an even function in v_x everywhere. Equation (13.46) indicates that *cold plasmas form surface charges*. Again the nullity of the surface currents (Eq. (13.45)) is valid strictly only in the context of present governing equation (13.1). In a drifting cold plasma, the surface current is existent, and the tangential component of the magnetic field B_y is discontinuous across the interface.

Next, let us use in (13.46)

$$-i\omega v_x = -\frac{e}{m}E_x. \tag{13.50}$$

Then, the surface charge is obtained in the form

$$\sigma^* = \frac{Ne^2}{m\omega^2}\,E_x(0^+). \tag{13.51}$$

Now (13.48) becomes

$$[E_x] = \frac{\omega_p^2}{\omega^2}E_x(0^+) \quad or \quad E_x(0^+)\left(1 - \frac{\omega_p^2}{\omega^2}\right) = E_x(0^-). \tag{13.52}$$

The electric displacement $\mathbf{D}$ is defined by, per (13.5),

$$\frac{1}{c}\frac{\partial \mathbf{D}}{\partial t} = \frac{1}{c}\frac{\partial \mathbf{E}}{\partial t} - \frac{4\pi e N}{c}\,\mathbf{v}. \tag{13.53}$$

The x-component of this gives, with the aid of (13.50),

$$D_x = E_x\left(1 - \frac{\omega_p^2}{\omega^2}\right). \tag{13.54}$$

So (13.52) is written equivalently as

$$[D_x] = 0 \tag{13.55}$$

i.e. *the normal component of the electric displacement is continuous across the interface*. Applying the derived matching conditions, one obtains from (13.47) and (13.49) (or (13.55)), respectively,

$$\alpha G + \lambda F = 0, \quad (\omega^2 - \omega_p^2)G = \omega^2 F. \tag{13.56}$$

It turns out that (13.49) and (13.52) (or 13.55) are redundant. Equation (13.56) gives the dispersion relation

$$\alpha + \lambda\left(1 - \frac{\omega_{pe}^2}{\omega^2}\right) = 0 \quad or \quad \sqrt{k^2 + \frac{\omega_{pe}^2 - \omega^2}{c^2}} + \left(1 - \frac{\omega_{pe}^2}{\omega^2}\right)\sqrt{k^2 - \frac{\omega^2}{c^2}} = 0, \tag{13.57}$$

which can be recovered from (13.40) by letting $\gamma \to \infty$. In the electrostatic limit, $c \to \infty$, (13.57) becomes

$$2 - \frac{\omega_{pe}^2}{\omega^2} = 0 \tag{13.58}$$

which is known as Ritchie dispersion relation. In summary, the electromagnetic boundary conditions in a cold plasma are $[E_z] = 0$ and $[B_y] = 0$. If electrostatic waves are analyzed, $[E_z] = 0$ and $[D_x] = 0$ should be used.

Exercise. (13.57) represents the wave known as *surface polariton.* This dispersion relation is very similar to that of a two-stream instability, Eq. (8.16). Squaring and solving by quadrature give

$$\omega^2 = c^2k^2 + \frac{\omega_{pe}^2}{2} - \frac{1}{2}\sqrt{\omega_{pe}^4 + 4c^4k^4}. \tag{13.59}$$

13.1.3. *Surface ion acoustic wave*

This is an electrostatic wave in which the ions provide the inertia while the electrons provide the pressure, restoring force. Introducing potential, $\mathbf{E} = -\nabla\phi$, we write

$$\frac{\partial \mathbf{v}_\alpha}{\partial t} = -\frac{e_\alpha}{m_\alpha}\nabla\phi - \frac{\beta_\alpha^2}{N}\nabla n_\alpha, \quad (\alpha = e,\ i) \tag{13.60}$$

$$\frac{\partial n_\alpha}{\partial t} + N\nabla\cdot\mathbf{v}_\alpha = 0, \tag{13.61}$$

$$\nabla^2\phi = -4\pi\sum_\alpha e_\alpha n_\alpha. \tag{13.62}$$

We shall assume that ions are cold ($\beta_i = 0$). The above equations yield the relations,

$$n_e = \left(1 - \frac{\omega^2}{\omega_{pi}^2}\right) n_i = \frac{1 - \frac{\omega_{pi}^2}{\omega^2}}{4\pi e}\,\nabla^2\phi \tag{13.63}$$

$$therefore, \quad \beta^2\nabla^2\nabla^2\phi + \omega^2\,\frac{\omega^2 - \omega_{pe}^2 - \omega_{pi}^2}{\omega^2 - \omega_{pi}^2}\,\nabla^2\phi = 0. \tag{13.64}$$

The vacuum and plasma solutions are obtained

$$\phi = Ae^{kx+ikz-i\omega t}, \quad (x < 0) \tag{13.65}$$

$$\phi = (Ce^{-\gamma x} + De^{-kx})e^{ikz-i\omega t}, \quad (x > 0) \tag{13.66}$$

$$\gamma = \left[k^2 + \frac{\omega_{pe}^2 + \omega_{pi}^2 - \omega^2}{\beta^2(1 - \omega_{pi}^2/\omega^2)}\right]^{\frac{1}{2}}. \tag{13.67}$$

The immediate boundary condition is per $[E_z] = 0$

$$[\phi] = 0. \tag{13.68}$$

Operating the infinitesimal integration with (13.62) gives

$$\left[\frac{\partial \phi}{\partial x}\right] = -4\pi\ (\sigma_i^* + \sigma_e^*), \quad where \tag{13.69}$$

$$\sigma_i^* = \frac{-iNe}{\omega}\ v_{ix}(0), \quad \sigma_e^* = \frac{iNe/\omega}{1 - \beta^2 k^2/\omega^2}\ v_{ex}(0). \tag{13.70}$$

We still need one more condition since we have three undetermined constants, A, C, and D. For the warm species electron, we invoke the specular reflection condition

$$v_{ex}(0) = 0. \tag{13.71}$$

And again we have $\sigma_e^* = 0$ for the warm electrons. Using (13.68) to (13.71), one obtains dispersion relation

$$\frac{k}{\gamma} = \left(1 - \frac{\omega_{pi}^2}{\omega^2}\right) \frac{2\omega^2 - \omega_{pe}^2 - \omega_{pi}^2}{\omega_{pe}^2}. \tag{13.72}$$

In the zero electron mass limit ($m_e \to 0$, or $\omega_{pe} \to \infty$) (this limit amounts to Boltzmann distributed electrons), the dispersion relation becomes

$$\frac{k}{\gamma} = \frac{\omega_{pi}^2}{\omega^2} - 1 = \left[1 + \frac{1}{k^2\lambda_e^2(1 - \omega_{pi}^2/\omega^2)}\right]^{-\frac{1}{2}}, \quad where \tag{13.73}$$

$$\lambda_e = \sqrt{\frac{\gamma_e T_e}{4\pi N e^2}}$$

is the electron Debye length.

Exercise. Squaring and solving by quadrature (13.73), derive

$$\omega^2 = \lambda_e^2 k^2 \omega_{pi}^2 + \frac{\omega_{pi}^2}{2} - \frac{1}{2}(\omega_{pi}^4 + 4k^4\lambda_e^4\omega_{pi}^4)^{\frac{1}{2}}. \tag{13.74}$$

Note the similarity with the TM mode wave (13.59) ($c \longleftrightarrow \lambda_e \omega_{pi}$). Also compare this with the two-stream wave dispersion relation in (8.16).

Remark. In a cold plasma, the normal component of the electric field E_x is discontinuous by the amount of the surface charge formed on the interfacial layer. We can put $E_x(x) \sim S(x)$ where $S(x)$ is a step function. Then $\partial E(x)/\partial x \sim \delta(x) \sim n(x)$, i.e. the density must be surface distributed. The inverse is also true.

13.1.4. *Landau's analysis of TM mode surface wave*

In the coordinate system where the interface normal is x-direction and the surface wave propagates in the z-direction, the y- coordinate is ignorable. When the electric (magnetic) field of the electromagnetic surface wave is parallel to the y-direction the polarization is called TE(TM) wave. In the book by L. D. Landau and E. M. Lifshitz, *Electrodynamics of continuous media*, 1960 Pergamon p. 288, TE mode is E-wave and TM mode is H-wave [In plasma, the magnetic permeability $\mu = 1$, and H=B.] The Maxwell equations can be written as

$$\nabla \times \mathbf{E} = \frac{i\omega}{c}\mathbf{B},$$

$$\nabla \times \mathbf{B} = -\frac{i\omega}{c}\mathbf{D} = -\frac{i\omega}{c}\varepsilon\mathbf{E},$$

where $D_i = \varepsilon_{ij}E_j = \varepsilon\delta_{ij}E_j = \varepsilon E_i$ with ε dielectric constant, in isotropic medium. The above two equations yield

$$\nabla^2\mathbf{E} + \frac{\omega^2}{c^2}\varepsilon\mathbf{E} - \nabla\nabla\cdot\mathbf{E} = 0, \tag{13.75}$$

$$\nabla^2\mathbf{B} + \frac{\omega^2}{c^2}\varepsilon\mathbf{B} = 0. \tag{13.76}$$

Here we assume that the interface $x = 0$ separates two media of different dielectric constant, $\varepsilon = \varepsilon_1$ $(x > 0)$, and $\varepsilon = \varepsilon_2$ $(x < 0)$ and focus on the TM-wave. Putting $\mathbf{B} = \hat{\mathbf{y}}B_y(x)e^{ikz-i\omega t}$ in (13.76) gives

$$\left(\frac{\partial^2}{\partial x^2} - k^2\right)B_y + \frac{\omega^2}{c^2}\varepsilon_1 B_y = 0$$

which is solved by (dropping the phasor)

$$B_y = A_1 e^{-\alpha_1 x},$$

$$E_z = \frac{ic}{\omega\varepsilon_1}\frac{\partial B_y}{\partial x} = -A_1\frac{ic\alpha_1}{\omega\varepsilon_1}e^{-\alpha_1 x}. \quad (x > 0)$$

In the medium 2, $(x < 0)$, one obtains

$$B_y = A_2 e^{\alpha_2 x},$$

$$E_z = A_2\frac{ic\alpha_2}{\omega\varepsilon_2}e^{\alpha_2 x},$$

$$\alpha_{1,2} = \sqrt{k^2 - \frac{\omega^2}{c^2}\varepsilon_{1,2}}. \tag{13.77}$$

The matching condition, $[B_y] = 0$ yields $A_1 = A_2$. $[E_z] = 0$ gives

$$-\frac{\alpha_1}{\varepsilon_1} = \frac{\alpha_2}{\varepsilon_2}. \tag{13.78}$$

Note that a plasma of stationary ions has $\varepsilon = 1 - \frac{\omega_{pe}^2}{\omega^2}$. If medium 2 (or 1) is vacuum, (13.78) agrees with (13.57). In order to satisfy (13.78), ε_1 and ε_2 should have opposite signs: $\varepsilon_1\varepsilon_2 < 0$. Then (13.78) can be satisfied if

$$\varepsilon_1 < |\varepsilon_2|.$$

The dispersion relation (13.78) can be written as

$$c^2k^2 = \omega^2\varepsilon_1|\varepsilon_2|(|\varepsilon_2| - \varepsilon_1). \tag{13.79}$$

- **Impossibility of TE mode.** The y-component of (13.75) gives

$$\left(\frac{\partial^2}{\partial x^2} - k^2\right) E_y + \frac{\omega^2}{c^2}\,\varepsilon E_y = 0$$

which is solved by

$$E_y = G_1 e^{-\alpha_1 x}, \quad (x > 0)$$

$$E_y = G_2 e^{\alpha_2 x} \quad (x < 0)$$

where $\alpha_{1,2}$ are given by (13.77). The continuity of the tangential component of the electric field yields $G_1 = G_2 \equiv G$. $\nabla \times \mathbf{E} = \frac{i\omega}{c}\mathbf{B}$ gives

$$B_x = -\frac{ck}{\omega} G e^{-\alpha_1 x},$$

$$B_z = \frac{ic\alpha_1}{\omega} G e^{-\alpha_1 x}, \quad (x > 0)$$

$$B_x = -\frac{ck}{\omega} G e^{\alpha_2 x},$$

$$B_z = -\frac{ic\alpha_2}{\omega} G e^{\alpha_2 x}. \quad (x < 0)$$

The boundary condition $[B_x] = 0$ gives an identity, and $[B_z] = 0$ yield $\alpha_1 = -\alpha_2$ which is impossible to be satisfied. So we conclude that TE mode surface wave does not exist in isotropic plasmas.

References

Gradov, O. M. and Stenflo, L. Linear theory of a cold bounded plasma, *Phys. Reports*, **94** (1983) p. 111–137.

Landau, L. D. and Lifshitz, E. M. *Electrodynamics of Continuous Media*, 1960 Pergamon p. 288.
Lee, H. J. and Cho, S. H. Boundary conditions for surface waves propagating along the interface of plasma flow and free space, *J. Plasma Phys.* **58** pt.3 (1997) p. 409–419.
Lee, H. J. Surface waves in magnetized thermal plasma, *J. Korean Phys. Soc.* **28** (1995) p. 51–59.
Kaw, P. K. and McBride, J. B. Surface waves on a plasma, *Phys. Fluids* **13** (1970) p. 1784–1793.
McBride, J. B. and Kaw, P. K. Low frequency surface waves on a warm plasma, *Phys. Lett.* **33A** (1970) p. 72–73.

13.2. Surface waves in a moving plasma: $\mathbf{v}_0||\mathbf{k}$

We consider a moving plasma relative to a half-space stationary vacuum. In this case, surface currents as well as surface charges are present on the plasma-vacuum interface when the plasma is cold. The important boundary conditions are (13.94) and (13.96) which contrast $[B_y] = [D_x] = 0$ in the stationary plasma. See (13.49) and (13.55). The difference stems from the nonzero surface current in a moving plasma, while the surface current in a stationary plasma is zero (see (13.45)). The drift velocities are denoted by $\mathbf{v}_{0\alpha}(\alpha = e, i)$ and are assumed to be directed in the direction of wave propagation, z-direction (i.e. $\mathbf{v}_0||\mathbf{k}$). We have the following governing equations (linearized),

$$\frac{\partial \mathbf{v}_\alpha}{\partial t} + \mathbf{v}_{0\alpha} \cdot \nabla \mathbf{v}_\alpha = \frac{e_\alpha}{m_\alpha}\left(\mathbf{E} + \frac{\mathbf{v}_{0\alpha}}{\mathbf{c}} \times \mathbf{B}\right) - \frac{\beta_\alpha^2}{N}\nabla n_\alpha, \quad (\beta_i = 0) \tag{13.80}$$

$$\frac{\partial n_\alpha}{\partial t} + \mathbf{v}_{0\alpha} \cdot \nabla n_\alpha + N\nabla \cdot \mathbf{v}_\alpha = 0, \tag{13.81}$$

$$\nabla \cdot \mathbf{E} = 4\pi \sum_\alpha e_\alpha n_\alpha, \tag{13.82}$$

$$\nabla \times \mathbf{E} = -\frac{1}{c}\frac{\partial \mathbf{B}}{\partial t}, \tag{13.83}$$

$$\nabla \times \mathbf{B} = \frac{1}{c}\frac{\partial \mathbf{E}}{\partial t} + \frac{4\pi}{c}\sum_\alpha e_\alpha(N\mathbf{v}_\alpha + n\mathbf{v}_{0\alpha}). \tag{13.84}$$

13.2.1. *Electrostatic wave*

We derive the electrostatic dispersion relation for a moving plasma, following the steps indicated below. The readers are expected to complete the algebra that has been omitted.

i) Assume the wave quantities vary $\propto e^{ikz-i\omega t}$ and show that

$$n_i = -\frac{\omega_{pi}^2}{\omega_{i*}^2}\frac{\nabla^2\phi}{4\pi e}, \quad n_e = \left(1-\frac{\omega_{pi}^2}{\omega_{i*}^2}\right)\frac{\nabla^2\phi}{4\pi e}, \quad (\mathbf{E} = -\nabla\phi) \tag{13.85}$$

where $\omega_{\alpha*} = \omega - kv_{0\alpha}$, the Doppler-shifted frequency.

ii) The vacuum and plasma solutions are obtained by (13.65) and (13.66) with γ given by

$$\gamma = \left[k^2 + \frac{\omega_{pe}^2\omega_{i*}^2 - \omega_{e*}^2(\omega_{i*}^2 - \omega_{pi}^2)}{\beta^2(\omega_{i*}^2 - \omega_{pi}^2)}\right]^{\frac{1}{2}}. \tag{13.86}$$

iii) The immediate boundary condition at $x = 0$ is $[\phi] = 0$. Explain why. Performing the infinitesimal integration with the Poisson equation, derive

$$\left(1-\frac{\omega_{pi}^2}{\omega_{i*}^2}\right)\frac{\partial\phi}{\partial x}(0^+) = \frac{\partial\phi}{\partial x}(0^-). \tag{13.87}$$

iv) We need the specular reflection condition $v_{ex}(0) = 0$ for warm electrons. This makes the electron surface charge vanish, hence (13.87). Show that the final dispersion relation takes the form

$$\frac{k}{\gamma} = \varepsilon_i\left[\frac{\omega_{e*}^2}{\omega_{pe}^2}(1+\varepsilon_i) - 1\right] \quad \textit{where} \quad \varepsilon_i = 1 - \frac{\omega_{pi}^2}{\omega_{i*}^2}. \tag{13.88}$$

13.2.2. *Electromagnetic boundary conditions for a moving plasma*

Using the governing equations (13.80)–(13.84), electromagnetic boundary conditions will be derived for a plasma moving relative to a vacuum. For simplicity, ions will be assumed immobile. Integrating (13.84) over the infinitesimal layer, we obtain

$$[B_y] = \frac{4\pi}{c}J_z^*, \quad J_z^* = -eN\int_{-\delta}^{\delta} v_z dx - ev_0\int_{-\delta}^{\delta} n_e dx. \tag{13.89}$$

Using (13.80) in the above gives

$$J_z^* = \left(v_0 + \frac{k\beta^2}{\omega_{e*}}\right)\sigma_e^*. \tag{13.90}$$

Integrating (13.81) gives

$$\omega_{e*}\sigma_e^*\left(1-\frac{k^2\beta^2}{\omega_{e*}^2}\right) - ieNv_{ex}(0) = 0. \tag{13.91}$$

Integrating Poisson equation gives

$$[E_x] = 4\pi\sigma_e^*. \tag{13.92}$$

If electrons are warm, the specular reflection condition should be invoked: $v_{ex}(0) = 0$, which gives in turn $\sigma_e^* = J_z^* = 0$, and

$$[E_x] = [B_y] = 0. \tag{13.93}$$

We also have the continuity of E_z: $[E_z] = 0$.

For cold electrons, the surface current is not vanishing:

$$J_z^* = v_0\, \sigma_e^*, \quad [B_y] = \frac{4\pi}{c} v_0 \sigma_e^* = \frac{v_0}{c}[E_x], \quad \textit{where} \tag{13.94}$$

$$\sigma_e^* = \frac{ieN}{\omega_{e*}} v_{ex}(0). \tag{13.95}$$

The x-component of (13.84) gives the relation, upon infinitesimal integration,

$$[B_y] = \frac{\omega}{ck}[D_x]. \tag{13.96}$$

The normal component of the displacement vector as well as the tangential component of the magnetic field is not continuous in a cold drifting plasma.

The y-component of (13.84) gives, upon infinitesimal integration,

$$[B_z] = -\frac{4\pi}{c} J_y^* = \frac{4\pi e}{c} \int_{-\delta}^{\delta} N v_y dx.$$

It can be seen that $J_y^* = 0$ from (13.80) and (13.81), and we conclude that $[B_z] = 0$ in a drifting plasma when $\mathbf{v}_0$ is parallel to the surface wave vector $\mathbf{k}$. Further development of the surface waves in moving plasma is continued to § 13.2.4. In the mean time we discuss the kinematics of rippling interface.

13.2.3. *Kinematics of mobile interface*

References: L. I. Sedov, *A course in continuum mechanics* Vol. 2, Wolters-Noordhoff Publishing Groningen Chapter 7 (1972); H. J. Lee and Y. K. Lim, *Modulation of a TM mode surface wave on a free boundary between plasma and vacuum*, Plasma Phys. and Control. Fusion, **40** p. 1619–1633 (1998).

So far we assumed that the plasma-vacuum interface is fixed, ignoring the rippling of the boundary surface associated with plasma motion. If the normal component (here the word 'normal' is referred to the interface, i.e. normal with respect to the moving interface) of the velocity of the moving interface is much less than the other characteristic velocity of plasma, the fixed boundary approximation would be valid. Here, we consider a fluid-fluid interface. The interface between two fluids is defined by the property that the fluids do not cross it. The normal component of velocities of either fluid must be equal to the normal component of the interface velocity. Let the interface be described by $F(x, y, z, t) = 0$. The normal component of the interface velocity u is

$$u = -\frac{\partial F/\partial t}{|\nabla F|} \quad (\textit{see Exercise below}). \tag{13.97}$$

The normal component of the fluid velocity is

$$\frac{\mathbf{v} \cdot \nabla F}{|\nabla F|}. \tag{13.98}$$

The condition that these two be equal is

$$\frac{dF}{dt} = \frac{\partial F}{\partial t} + \mathbf{v} \cdot \nabla F = 0 \quad (\textit{equation of moving interface}). \tag{13.99}$$

This is the way that a moving interface is interwoven with fluid motion. To put it into an alternative form, let the interface be described by $x = \xi(y, z, t)$, and write $F(x, y, z, t) = \xi(y, z, t) - x$ in (13.99). This gives the equation of the moving interface in the form

$$\frac{d\xi}{dt} = v_x. \tag{13.100}$$

A consequence of (13.99) is that the normal components of the fluid velocities are continuous across the interface. The proof of this goes as follows. Both fluids on either side of the interface satisfy (13.99):

$$\frac{\partial F}{\partial t} + \mathbf{v}_1 \cdot \nabla F = 0, \quad \frac{\partial F}{\partial t} + \mathbf{v}_2 \cdot \nabla F = 0. \tag{13.101}$$

$$\textit{Therefore} \quad (\mathbf{v}_1 - \mathbf{v}_2) \cdot \nabla F = 0 \quad \textit{or} \quad \hat{\mathbf{n}} \cdot (\mathbf{v}_1 - \mathbf{v}_2) = 0 \tag{13.102}$$

where $\hat{\mathbf{n}}$ is the unit vector normal to the interface. However, the tangential components need not be continuous. Kelvin–Helmholtz instability occurs when the tangential components of the velocities are different.

Exercise. Prove (13.97).

Solution: The analogy with the wave velocity (phase velocity) of a plane wave $e^{i\phi(x,t)}$ would be helpful. The wave front of a plane wave is given by $\phi(x(t),t) = \text{const}$. The wave velocity with which the wave front moves is obtained from $\frac{\partial \phi}{\partial t} + \frac{\partial \phi}{\partial x}\frac{dx}{dt} = 0$, giving

$$\mathbf{v}_{wave} = \hat{\mathbf{x}}\frac{dx}{dt} = -\hat{\mathbf{x}}\,\frac{\partial \phi/\partial t}{\partial \phi/\partial x}.$$

The three-dimensional extension is straightforward. The velocity $\mathbf{v}_s$ with which the surface $F(x,y,z,t) = 0$ moves is obtained from

$$\frac{\partial F}{\partial t} + \nabla F \cdot \mathbf{v}_s = 0. \tag{13.103}$$

The unit vector normal to the surface is $\hat{\mathbf{n}} = \nabla F/|\nabla F|$. Substituting this expression into the above gives

$$\frac{\partial F}{\partial t} + |\nabla F|\ \hat{\mathbf{n}} \cdot \mathbf{v}_s = 0$$

from which (13.97) follows. In fact, the kinematic condition (13.99) for the moving surface is obtained by putting $\mathbf{v}_s = \mathbf{v}$ in (13.103). This amounts to saying that the interface coincides with the stream lines.

- **Alternative derivation of (13.97).**

Consider a mobile surface S described by $F(x,y,z,t) = 0$. At time $t + \Delta t$, S moves and deforms into S' (see Fig. 13.1).

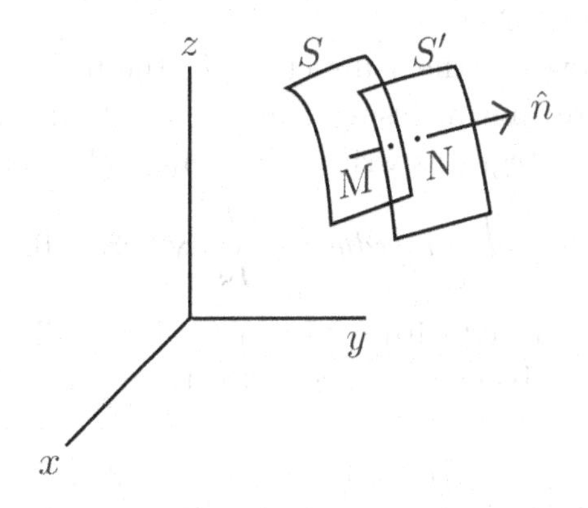

Fig. 13.1. A moving surface S deforms to S'.

Select on S a point M and let $\hat{n}$ be the normal unit vector to S. Let this direction intersect the displaced surface S' at N. The velocity of the displacement of S in the direction normal to S at M is the vector $\mathbf{v}_s$ which is defined by

$$\mathbf{v}_s = \hat{\mathbf{n}} \lim_{\Delta t \to 0} \frac{\overline{MN}}{\Delta t}.$$

The displaced surface S' is described by $F(x+\Delta x, y+\Delta y, z+\Delta z, t+\Delta t) = 0$, with $\Delta x = \overline{MN}\ \hat{\mathbf{n}} \cdot \hat{\mathbf{x}}$, $\Delta y = \overline{MN}\ \hat{\mathbf{n}} \cdot \hat{\mathbf{y}}$, $\Delta z = \overline{MN}\ \hat{\mathbf{n}} \cdot \hat{\mathbf{z}}$. Expanding gives

$$F(x,y,z,t) + \frac{\partial F}{\partial x}\Delta x + \frac{\partial F}{\partial y}\Delta y + \frac{\partial F}{\partial z}\Delta z + \frac{\partial F}{\partial t}\Delta t = 0.$$

Solving for $\overline{MN}$ (note that $F(x,y,z,t) = 0$),

$$\overline{MN} = \frac{-\Delta t\ \partial F/\partial t}{n_x\ \partial F/\partial x + n_y\ \partial F/\partial y + n_z\ \partial F/\partial z}$$
$$= \frac{-\Delta t\ \partial F/\partial t}{|\nabla F|(n_x^2 + n_y^2 + n_z^2)} \quad (n_x = \hat{\mathbf{n}} \cdot \hat{\mathbf{x}},\ etc.).$$

Note that $n_x^2 + n_y^2 + n_z^2 = 1$. Thus we have

$$\mathbf{v}_s = \hat{\mathbf{n}}\ \frac{-\partial F/\partial t}{|\nabla F|} \quad Q.E.D.$$

• **Boundary conditions at a mobile interface** are again obtained by integrating the governing equations across the infinitesimal thickness in the normal direction.

• **Continuity equation.** We integrate (13.2) over a *fixed* volume of an infinitesimal cubicle of dimension $dxdydz$ in which $x = 0$ is the interface at time $t = 0$. See Fig. 13.2.

This surface moves in the x-direction. In the figure, two surfaces inside the fixed volume are shown, respectively, at $t = 0$ and infinitesimally later time t. The volume integral is written by Gauss' theorem as

$$\frac{\partial}{\partial t}\int_{vol} n\ dxdydz + \int_S n\mathbf{v} \cdot d\mathbf{S} = 0. \tag{13.104}$$

('n' should not be confused with $\hat{\mathbf{n}}$.) Later the limit will be taken of reducing the thickness of the cubicle, dx, to zero. In the above, the volume integral is written

$$\frac{\partial}{\partial t}\int_{vol} n\ dxdydz = \frac{\partial}{\partial t}\left[\left(\frac{dx}{2} - ut\right)n_1 + \left(\frac{dx}{2} + ut\right)n_2\right] dydz$$

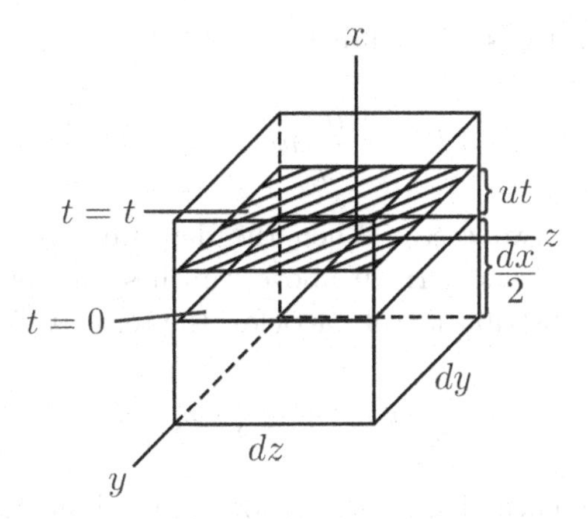

Fig. 13.2. Integration of the continuity equation across a moving interface.

where u is the normal velocity of the interface; $u = \hat{\mathbf{n}} \cdot \mathbf{v}_s$. We consider two cases.

i) The densities are volume-distributed. In this case $n \propto dx$ and $dx/2$-terms vanish as dx reduces to an infinitesimal thickness, and we are left with

$$\frac{\partial}{\partial t}\int_{vol} n\ dxdydz = (-un_1 + un_2)dS \quad (dS = dydz). \tag{13.105}$$

ii) The densities are surface-distributed. In this case $n_1 = n_2 \propto dx^{-1}$ or $\propto \delta(x)n^*$, and we have

$$\frac{\partial}{\partial t}\int_{vol} n\ dxdydz = \frac{\partial n^*}{\partial t}dS \quad (n^*,\ \textit{surface density}). \tag{13.106}$$

Next we turn to the surface integral in (13.104). In this integral, the flux through two surfaces perpendicular to the x-axis becomes, in the limit of $dx \to 0$,

$$[(nv_x)|_{0^+} - (nv_x)|_{0^-}]\, dS. \tag{13.107}$$

Fluxes through the other faces of the cubicle are obtained by Taylor expansion,

$$\left[(nv_y)_0 + \frac{\partial}{\partial y}(nv_y)\frac{dy}{2}\right] dxdz - \left[(nv_y)_0 - \frac{\partial}{\partial y}(nv_y)\frac{dy}{2}\right] dxdz$$
$$+ \left[(nv_z)_0 + \frac{\partial}{\partial z}(nv_z)\frac{dz}{2}\right] dxdy - \left[(nv_z)_0 - \frac{\partial}{\partial y}(nv_z)\frac{dz}{2}\right] dxdy. \tag{13.108}$$

After cancellation, the surface integral in (13.104) is obtained as

$$[(nv_x)|_{0^+} - (nv_x)|_{0^-}]\, dS + \left[\frac{\partial}{\partial y}(nv_y) + \frac{\partial}{\partial z}(nv_z)\right]^{@} dS\, dx. \tag{13.109}$$

Again we consider two cases. When the densities are volume-distributed ($n_1, n_2 \propto dx$), the second term @ above vanishes rapidly as $dx \to 0$. When we have a surface distribution of particles, the term @ yields, as before,

$$\left[\frac{\partial}{\partial y}(n^* v_y) + \frac{\partial}{\partial z}(n^* v_z)\right] dS \tag{13.110}$$

which is the divergence of the surface flow on the boundary. The boundary conditions obtained from the foregoing analysis of the continuity equation is summarized as follows.

i) When there is no surface distribution of particles,

$$-u[n] + [nv_n] = 0 \quad (v_n : \ normal\ fluid\ velocity). \tag{13.111}$$

ii) When there is surface density,

$$\frac{\partial n^*}{\partial t} + \nabla_{\|} \cdot (n^* \mathbf{v}_{\|}) + n_1 v_{1n} - n_2 v_{2n} = 0. \tag{13.112}$$

Equation (13.111) is an identity $0=0$, since $u = v_{n1} = v_{n2}$.

Remark. The preceding analysis shows that a term $\frac{\partial q}{\partial t}$ yields, upon the infinitesimal integration, $-u[q]$ (if the quantity q has no corresponding surface distribution). Also the infinitesimal integration of $\nabla \cdot \mathbf{Q}$ across the interface yields $\hat{\mathbf{n}} \cdot [\mathbf{Q}]$.

Exercise. Using the Gauss theorem,

$$\int_V \nabla\, \phi\, d\tau = \int_S \phi\, d\mathbf{S},$$

show that the infinitesimal integration of $\nabla\phi$ yields $\hat{\mathbf{n}}[\phi]$.

Exercise. Using the Gauss theorem,

$$\int_V \nabla \times \mathbf{Q}\, d\tau = \int_S d\mathbf{S} \times \mathbf{Q},$$

show that the infinitesimal integration of $\nabla \times \mathbf{Q}$ yields $\hat{\mathbf{n}} \times [\mathbf{Q}]$.

These Exercises indicate that the ∇-operator corresponds to $\hat{\mathbf{n}}\frac{\partial}{\partial s}$ with s being the coordinate normal to the surface. Therefore the boundary conditions stemming from the infinitesimal integrations can be performed more quickly by writing, as an example, for a term $\nabla\phi$,

$$\lim_{\delta\to 0}\int_{-\delta}^{\delta}\nabla\phi ds = \lim_{\delta\to 0}\int_{-\delta}^{\delta}\hat{n}\frac{d\phi}{ds}\,ds = \hat{\mathbf{n}}[\phi] \tag{13.113}$$

similarly for other ∇-related terms. Let us consider the infinitesimal integration of the total derivative term $\frac{d\rho}{dt}$,

$$\frac{d\rho}{dt} = \frac{\partial\rho}{\partial t} + \mathbf{v}\cdot\nabla\rho \to -u[\rho] + \lim_{\delta\to 0}\int_{-\delta}^{\delta} v_n\,\frac{\partial\rho}{\partial s}ds = -u[\rho] + v_n[\rho] = 0$$

since $v_{n1} = v_{n2} = u$. Physically, the total derivative is the time rate of change of ρ seen by an observer moving with the fluid. In the frame of reference of the boundary layer, this rate remains finite and contributes nothing when integrated over the infinitesimal thickness.

Let us consider the electron equation of motion in its nonlinear form,

$$\rho\frac{d\mathbf{v}}{dt} = \sigma\mathbf{E} + \frac{1}{c}\mathbf{J}\times\mathbf{B} - \nabla p, \tag{13.114}$$

$$\rho = m_e n_e,\ \sigma = -en_e,\quad \mathbf{J} = -en_e\mathbf{v},\quad p = \gamma_e m_e T_e n_e.$$

Using the continuity equation, we have

$$\rho\frac{dv_i}{dt} = \frac{\partial}{\partial t}(\rho v_i) + \frac{\partial}{\partial x_j}(\rho v_i v_j).$$

Carrying out the infinitesimal volume integration using the cubicle in Fig. 13.2, this total derivative term in the left hand side of (13.114) yields

$$-u[\rho v_i] + [\rho v_i v_n],\quad (i = x, y, z;\ n = normal) \tag{13.115}$$

which is 0 because $v_{n1} = v_{n2} = u$. For the other terms, we calculate

$$\lim_{\Delta\to 0}\int_{-\Delta}^{\Delta}\sigma\mathbf{E}\,ds. \tag{13.116}$$

Here we allow for surface charge and surface current by putting $\sigma = \sigma^*\ \delta(s)$ and $\mathbf{J} = \mathbf{J}^*\delta(s)$ with $\delta(s)$ Dirac delta function. Also putting $\delta(s) = \frac{1}{2\Delta}$, we have for the expression (13.116),

$$\sigma^*\lim_{\Delta\to 0}\frac{1}{2\Delta}\int_{-\Delta}^{\Delta}\mathbf{E}\,ds = \sigma^*\frac{1}{2\Delta}\frac{2\Delta}{2}(\mathbf{E}(0^-) + \mathbf{E}(0^+)) = \sigma^*\langle\mathbf{E}\rangle \tag{13.117}$$

where we used the trapezoidal rule to calculate the area and the symbol $\langle \cdots \rangle$ signifies the arithmetic mean of the boundary values at 0^+ and 0^-. In this way, (13.109) yields the boundary condition

$$\sigma^* \langle \mathbf{E} \rangle + \frac{1}{c}\, \mathbf{J}^* \times \langle \mathbf{B} \rangle - \hat{\mathbf{n}}\, [p] = 0. \tag{13.118}$$

Equations (13.3)–(13.5) yields the boundary conditions

$$\hat{\mathbf{n}} \cdot [\mathbf{E}] = 4\pi\sigma^*, \tag{13.119}$$

$$\hat{\mathbf{n}} \times [\mathbf{E}] = \frac{u}{c}[\mathbf{B}], \tag{13.120}$$

$$\hat{\mathbf{n}} \times [\mathbf{B}] = -\frac{u}{c}[\mathbf{E}] + \frac{4\pi}{c}\mathbf{J}^*. \tag{13.121}$$

Also (13.119) and (13.121) give $u\sigma^* = \hat{\mathbf{n}} \cdot \mathbf{J}^*$. One can prove

$$\frac{\partial \sigma^*}{\partial t} + \hat{\mathbf{n}} \cdot [\mathbf{J}^*] = -\nabla_{\parallel} \cdot \mathbf{J}^*_{\parallel} \quad (cf. Eq.\ (13.25)) \tag{13.122}$$

where the symbol $\parallel$ means the component parallel to the interface. Putting $u = 0$ recovers the boundary conditions for immobile and fixed interface.

• **More kinematics of deformable boundary.** In the following, we ignore y-coordinate, and write the equation of interface in the form

$$F(x, z, t) = x - \xi(z, t) = 0, \quad \nabla F = \hat{\mathbf{x}} - \hat{\mathbf{z}}\frac{\partial \xi}{\partial z},$$

$$|\nabla F| = \sqrt{1 + \left(\frac{\partial \xi}{\partial z}\right)^2}.$$

$$Therefore \quad \hat{\mathbf{n}} = \frac{\nabla F}{|\nabla F|} = \left[1 + \left(\frac{\partial \xi}{\partial z}\right)^2\right]^{-1/2} \left(\hat{\mathbf{x}} - \hat{\mathbf{z}}\frac{\partial \xi}{\partial z}\right), \tag{13.123}$$

$$u = \hat{\mathbf{n}} \cdot \mathbf{v} = \left[1 + \left(\frac{\partial \xi}{\partial z}\right)^2\right]^{-1/2} \frac{\partial \xi}{\partial t} \quad (cf.\ \ Eq.\ (13.97)), \tag{13.124}$$

$$v_x = u\, \hat{\mathbf{n}} \cdot \hat{\mathbf{x}} = u \left[1 + \left(\frac{\partial \xi}{\partial z}\right)^2\right]^{-1/2},$$

$$v_z = u\, \hat{\mathbf{n}} \cdot \hat{\mathbf{z}} = -u\, \frac{\partial \xi}{\partial z}\left[1 + \left(\frac{\partial \xi}{\partial z}\right)^2\right]^{-1/2}. \tag{13.125}$$

Also the relation $u = \hat{\mathbf{n}} \cdot \mathbf{v}$ (use Eq. (13.123) here) gives, as expected,

$$v_x = \frac{\partial \xi}{\partial t} + v_z \frac{\partial \xi}{\partial z} = \frac{d\xi}{dt}. \tag{13.126}$$

Exercise. Prove that the unit normal vector $\hat{\mathbf{n}}$ satisfies

$$\frac{d\hat{\mathbf{n}}}{dt} = \hat{\mathbf{n}} \times (\hat{\mathbf{n}} \times \nabla u). \tag{13.127}$$

Solution: By direct differentiation of $u = -\frac{\partial F}{\partial t}/|\nabla F|$, we get

$$\nabla u = \frac{\hat{\mathbf{z}}}{|\nabla F|^2} \left(|\nabla F| \frac{\partial}{\partial t} \frac{\partial \xi}{\partial z} - \frac{\partial \xi}{\partial t} \frac{\partial}{\partial z} |\nabla F| \right).$$

Using (13.123) and the above equation in (13.127) gives

$$\left(\frac{d\hat{\mathbf{n}}}{dt} \right)_x = \frac{-\partial \xi/\partial z}{|\nabla F|^3} \left(\frac{\partial}{\partial t} \frac{\partial \xi}{\partial z} - \frac{\frac{\partial \xi}{\partial t} \frac{\partial}{\partial z} |\nabla F|}{|\nabla F|} \right), \tag{13.128}$$

$$\left(\frac{d\hat{\mathbf{n}}}{dt} \right)_z = \frac{-1}{|\nabla F|^3} \left(\frac{\partial}{\partial t} \frac{\partial \xi}{\partial z} - \frac{\frac{\partial \xi}{\partial t} \frac{\partial}{\partial z} |\nabla F|}{|\nabla F|} \right). \tag{13.129}$$

On the other hand, direct differentiation of (13.123) yields

$$\left(\frac{d\hat{\mathbf{n}}}{dt} \right)_x = \frac{-1}{|\nabla F|^3} \frac{\partial \xi}{\partial z} \frac{d}{dt} \frac{\partial \xi}{\partial z}, \tag{13.130}$$

$$\left(\frac{d\hat{\mathbf{n}}}{dt} \right)_z = \frac{-1}{|\nabla F|^3} \frac{d}{dt} \frac{\partial \xi}{\partial z}. \tag{13.131}$$

Equality of $\frac{d}{dt} \frac{\partial \xi}{\partial z}$ and the term in the parentheses in (13.128) can be proven by means of (13.125). Equation (13.127) can be alternatively proven by drawing a figure as in Reference: T. H. Stix, *The theory of plasma waves*, McGraw-Hill p. 71 (1962).

• **Linearization of boundary conditions.** (13.118)–(13.123) (one can use (13.127) instead of (13.123)) are the exact nonlinear boundary conditions. In order to linearize these conditions, we start with (13.123), which gives

$$\mathbf{n} = \mathbf{n}^{(0)} + \epsilon\, \mathbf{n}^{(1)} + \ldots \quad \textit{where} \quad \mathbf{n}^{(0)} = \hat{\mathbf{x}}, \quad \mathbf{n}^{(1)} = -\hat{\mathbf{z}} \frac{\partial \xi}{\partial z} \tag{13.132}$$

where ξ is a first order quantity (this will be assumed in the sequel).

$$u = \epsilon \frac{\partial \xi}{\partial t} + \cdots, \tag{13.133}$$

$$v_x|_{x=\xi} = \epsilon\, v_x|_{x=0} + \cdots = \epsilon\, \frac{\partial \xi}{\partial t} + \ldots \quad \longrightarrow \quad \xi = \frac{i}{\omega} v_x|_{x=0}. \tag{13.134}$$

Exercise. Obtain (13.132) by linearizing (13.127).

Solution: In breaking down (13.118)–(13.121) order by order, note that σ^*, u, and $\mathbf{J}^*$ are first order quantities. We have in the lowest order,

$$\hat{\mathbf{x}} \cdot [\mathbf{E}] = 4\pi\sigma^*, \quad (cf \quad Eq.\ (13.23)) \tag{13.135}$$

$$\hat{\mathbf{x}} \times [\mathbf{E}] = 0, \quad (cf \quad Eq.\ (13.22)) \tag{13.136}$$

$$\hat{\mathbf{x}} \times [\mathbf{B}] = \frac{4\pi}{c}\mathbf{J}^*, \quad (cf \quad Eq.\ (13.24)) \tag{13.137}$$

which are the same as the boundary conditions at a immobile interface. In breaking down (13.118), we should read it as a second order balance equation in view of (13.114). Since $p \propto \epsilon\ n$, we rewrite it

$$\sigma^* \langle \mathbf{E} \rangle + \frac{1}{c}\ \mathbf{J}^* \times \langle \mathbf{B} \rangle - \mathbf{n}^{(1)}\ [p\ ] = 0. \tag{13.138}$$

We assume that there are no parallel surface currents, and put the right hand side of (13.122) equal to zero. Then we have

$$\sigma^* = eN\ \xi = \frac{i}{\omega}\ eNv_x|_{x=0}. \tag{13.139}$$

For warm plasmas, $\sigma^* = \mathbf{J}^* = 0$, and (13.138) gives $[\mathrm{p}] = 0$. Also we have $v_x|_{x=0} = 0$, the specular reflection condition. If a plasma is cold, $\sigma^* \langle \mathbf{E} \rangle + \frac{1}{c}\ \mathbf{J}^* \times \langle \mathbf{B} \rangle = 0$ as well as (13.139) is valid.

Exercise. Take the z-component of (13.138) and see that it is satisfied by $v_x|_{x=0} = 0$.

- **Alternative derivation of Eq. (13.118).**

Eliminate σ and $\mathbf{J}$ in (13.114) by means of $\nabla \cdot \mathbf{E} = 4\pi\sigma$ and $\nabla \times \mathbf{B} = \frac{4\pi}{c}\mathbf{J} + \frac{1}{c}\frac{\partial \mathbf{E}}{\partial t}$ to get

$$\rho\frac{d\mathbf{v}}{dt} = \frac{1}{4\pi}(\mathbf{E}\nabla \cdot \mathbf{E} + \mathbf{E} \cdot \nabla\mathbf{E} + \mathbf{B}\nabla \cdot \mathbf{B} + \mathbf{B} \cdot \nabla\mathbf{B}) - \nabla\left(p + \frac{E^2}{8\pi} + \frac{B^2}{8\pi}\right)$$
$$-\frac{1}{4\pi c}\frac{\partial}{\partial t}(\mathbf{E} \times \mathbf{B}) + \rho\mathbf{g}. \tag{13.140}$$

Carry out the infinitesimal integration after putting $\nabla = \mathbf{n}\ \frac{\partial}{\partial s}$ to get

$$\frac{1}{4\pi}[E_n\mathbf{E} + B_n\mathbf{B}] - \mathbf{n}\ \left[p + \frac{E^2}{8\pi} + \frac{B^2}{8\pi}\right] + \frac{u}{4\pi c}[\mathbf{E} \times \mathbf{B}] = 0. \tag{13.141}$$

Here write the bracketed terms in terms of the quantities 1 and 2, and use the identity $a_2b_2 - a_1b_1 = \frac{1}{2}(a_2 - a_1)(b_2 + b_1) + \frac{1}{2}(a_2 + a_1)(b_2 - b_1)$, and reinstate σ^* and $\mathbf{J}^*$ by means of (13.119)–(13.121). Also use $B_{1n} = B_{2n}$ $(\nabla \cdot \mathbf{B} = 0)$. The indicated manipulations lead to (13.118).

13.2.4. *Separation of TM and TE mode in a drifting plasma*

It is advised to read this section in continuation to § 13.2.2. The governing equations of drifting plasma are (13.80)–(13.84). Here we assume that the ions are stationary with infinite mass ($\omega_{pi} \to 0, v_{0i} = 0$), and $\mathbf{v}_{0e} = \hat{\mathbf{z}}v_0$ directed in the direction of the wave propagation ($\sim e^{ikz - i\omega t}$). The x-direction is the direction of inhomogeneity, and the y-direction is the infinite direction which has the translational invariance ($\partial/\partial y = 0$). (13.83) gives in components

$$\frac{i\omega}{c}B_x = -ikE_y, \tag{13.142}$$

$$\frac{i\omega}{c}B_y = ikE_x - \frac{\partial E_z}{\partial x}, \tag{13.143}$$

$$\frac{i\omega}{c}B_z = \frac{\partial E_y}{\partial x}. \tag{13.144}$$

$\nabla \times \mathbf{B} = -\frac{i\omega}{c}\mathbf{E} + \frac{4\pi}{c}\mathbf{J}$ yields in components

$$-ikB_y = -\frac{i\omega}{c}E_x + \frac{4\pi}{c}J_x, \tag{13.145}$$

$$ikB_x - \frac{\partial B_z}{\partial x} = -\frac{i\omega}{c}E_y + \frac{4\pi}{c}J_y, \tag{13.146}$$

$$\frac{\partial B_y}{\partial x} = -\frac{i\omega}{c}E_z + \frac{4\pi}{c}J_z. \tag{13.147}$$

$\mathbf{J} = -eN\mathbf{v} - en\mathbf{v}_0$ gives in components

$$J_x = -eNv_x = \frac{iNe^2}{m\omega'}\left(E_x - \frac{v_0}{c}B_y\right), \tag{13.148}$$

$$J_y = -eNv_y = \frac{iNe^2}{m\omega'}\left(E_y + \frac{v_0}{c}B_x\right), \tag{13.149}$$

$$J_z = -eNv_z - env_0 = \frac{iNe^2\omega}{m\omega'^2}E_z + \frac{e^2Nv_0}{m\omega'^2}\left(\frac{\partial E_x}{\partial x} - \frac{v_0}{c}\frac{\partial B_y}{\partial x}\right), \tag{13.150}$$

where $\omega' = \omega - kv_0$, the Doppler-shifted frequency, and we used (13.80) (with $\beta = 0$) and (13.81). Inspection of (13.142)–(13.150) indicates that the equations are separated into two groups: the TM mode set (B_y, E_x, E_z) in (13.143), (13.145), (13.147), (13.148), (13.150) and the TE mode set in (E_y, B_x, B_z) (13.142), (13.144), (13.146), (13.149).

13.2.5. *TM mode surface wave in a drifting cold plasma:* $\mathbf{v}_0 \| \mathbf{k}$

Reference: H. J. Lee and S. H. Cho, *Boundary conditions for surface waves propagating along the interface of plasma flow and free space*, J. Plasma Phys. **58** pt.3 p. 409–419 (1997) for §13.2.5–§13.2.7.

In § 13.2.2, we have established the boundary conditions that must be satisfied on the interface of a moving plasma. In this section, we derive the dispersion relation of TM mode electromagnetic surface wave in a moving cold plasma. Surface currents as well as the surface charges exist in a cold drifting plasma, as indicated by the key equation (13.94). Use of (13.145) and (13.148) gives the relation

$$\kappa\, B_y = \frac{\omega}{c}\left(1 - \frac{\omega_p^2}{\omega\omega'}\right) E_x. \tag{13.151}$$

Use of (13.143) and (13.151) yields

$$i\,\alpha^2 E_x - \kappa\,\frac{\partial E_z}{\partial x} = 0. \tag{13.152}$$

Substituting (13.150) into (13.147) and using (13.151) therein give

$$\frac{\partial E_x}{\partial x} + i\kappa\, E_z = 0. \tag{13.153}$$

In obtaining (13.153), we neglected $v_0^2/c^2 \ll 1$ and

$$\kappa = k - \frac{v_0\omega_p^2}{c^2\omega'}, \quad \alpha = \sqrt{k^2 + \frac{\omega_p^2 - \omega^2}{c^2}}. \tag{13.154}$$

Use of (13.152) and (13.153) yields

$$\frac{\partial^2 E_{x,z}}{\partial x^2} - \alpha^2 E_{x,z} = 0.$$

$$Therefore \quad E_x = Ae^{-\alpha x}. \quad (x > 0, \quad A: const.) \tag{13.155}$$

Using (13.155) in (13.151) and (13.153) give

$$B_y = \frac{\omega}{c\kappa}\left(1 - \frac{\omega_p^2}{\omega\omega'}\right) A\, e^{-\alpha x}, \tag{13.156}$$

$$E_z = -i\, \frac{\alpha}{\kappa}\, A\, e^{-\alpha x}. \tag{13.157}$$

(13.155)–(13.157) are the plasma solutions. The vacuum solutions are

$$B_y = F\, e^{\lambda x} \quad (x < 0, \quad F:\ const.), \tag{13.158}$$

$$E_x = \frac{ck}{\omega} F\, e^{\lambda x}, \tag{13.159}$$

$$E_z = i\frac{c\lambda}{\omega} F\, e^{\lambda x}, \tag{13.160}$$

where $\lambda = \sqrt{k^2 - \omega^2/c^2}$. The boundary condition $[E_z] = 0$ gives

$$A = -\frac{c\lambda\kappa}{\omega\alpha}\, F. \tag{13.161}$$

Next we use the boundary condition (13.94): $[B_y] = \frac{v_0}{c}\,[E_x]$

$$\frac{\omega}{c\kappa}\left(1 - \frac{\omega_p^2}{\omega\omega'}\right) A - F = \frac{v_0}{c}\left(A - \frac{ck}{\omega} F\right). \tag{13.162}$$

Using (13.161) in (13.162) gives

$$\frac{\lambda}{\alpha}\left(\frac{\kappa v_0}{\omega} - 1 + \frac{\omega_p^2}{\omega\omega'}\right) = \frac{\omega'}{\omega}.$$

Using (13.154) for κ in the above equation yields the TM mode dispersion relation

$$\alpha + \lambda\left(1 - \frac{\omega_p^2}{\omega'^2}\right) = 0. \tag{13.163}$$

In obtaining (13.163), we neglected $v_0^2/c^2 \ll 1$. If the plasma is stationary ($v_0 = 0$), (13.163) reduces to (13.57).

13.2.6. *Nonexistence of TE mode wave in a drifting cold plasma:* $\mathbf{v}_0 \| \mathbf{k}$

The nonexistence of TE mode in isotropic plasma was discussed earlier. In a drifting plasma, surface current relevant for the TE mode still does not exist and the TE mode surface wave is nonexistent even in drifting plasma.

This matter will be discussed in this subsection. The relevant equations are (13.142), (13.144), (13.146), and (13.149). Use of (13.149) in (13.146) results in

$$i\kappa B_x - \frac{\partial B_z}{\partial x} = -\frac{i\,\omega}{c}\left(1 - \frac{\omega_p^2}{\omega\omega'}\right)E_y.$$

Using (13.142) and (13.144) in the above equation gives

$$\frac{\partial^2 E_y}{\partial x^2} - \alpha^2 E_y = 0 \tag{13.164}$$

which is solved by

$$E_y = A_1\, e^{-\alpha x} \quad (x < 0). \tag{13.165}$$

The magnetic field components are obtained

$$B_x = -\frac{ck}{\omega}A_1\, e^{-\alpha x}, \tag{13.166}$$

$$B_z = \frac{ic\alpha}{\omega}A_1 e^{-\alpha x}. \tag{13.167}$$

The vacuum equation $(\nabla^2 + \frac{\omega^2}{c^2})\mathbf{E} = 0$ or $\left(\frac{\partial^2}{\partial x^2} - k^2 + \frac{\omega^2}{c^2}\right)E_y = 0$ is solved by

$$E_y = A_2 e^{\lambda x} \quad \left(x < 0,\ \ \lambda = \sqrt{k^2 - \frac{\omega^2}{c^2}}\right), \tag{13.168}$$

$$B_x = -\frac{ck}{\omega}A_2 e^{\lambda x}, \tag{13.169}$$

$$B_z = -\frac{ic\lambda}{\omega}A_2 e^{\lambda x}. \tag{13.170}$$

The immediate boundary condition is $[E_y] = 0$ which gives $A_1 = A_2$. The boundary condition $[B_x] = 0$ is an identity, and we should consider the boundary condition involving B_z; B_z is again continuous owing to zero surface current, $J_y^* = 0$. Thus we have $\alpha = -\lambda$ which is impossible to be satisfied. Thus we conclude that TE mode surface wave is nonexistent in a drifting plasma with $\mathbf{v}_0 \| \mathbf{k}$.

13.2.7. *Surface wave in a drifting cold plasma:* $\mathbf{v}_0 \perp \mathbf{k}$

When $\mathbf{v}_0 \| \mathbf{k}$, we saw that the field variables can be separated into two sets, TM and TE variables. However, when $\mathbf{v}_0 \perp \mathbf{k}$, the two sets are coupled, as shown below, and the analysis of surface wave requires more laborious work. In the following, we take the drift velocity as $\mathbf{v}_0 = \hat{\mathbf{y}} v_0$ while the wave propagates in the z-direction ($\sim e^{ikz-i\omega t}$). Note that the y-coordinate is still ignorable. We have (13.142)–(13.147) as governing equations. The equation motion reads in components

$$v_x = -\frac{ie}{\omega m}\Big(E_x + \frac{v_0}{c}B_z\Big), \tag{13.171}$$

$$v_y = -\frac{ie}{\omega m}E_y, \tag{13.172}$$

$$v_z = -\frac{ie}{\omega m}\Big(E_z - \frac{v_0}{c}B_x\Big). \tag{13.173}$$

We also have the Poisson equation

$$\nabla \cdot \mathbf{E} = -4\pi e n. \tag{13.174}$$

The components of the current $\mathbf{J} = -eN\mathbf{v} - en\mathbf{v}_0$ take the expressions

$$J_x = \frac{iNe^2}{\omega m}\Big(E_x + \frac{v_0}{c}B_y\Big), \tag{13.175}$$

$$J_y = \frac{iNe^2}{\omega m}E_y + \frac{Ne^2 v_0}{m\omega^2}\left(\nabla \cdot \mathbf{E} + \frac{v_0}{c}\left(\frac{\partial B_z}{\partial x} - ikB_x\right)\right), \tag{13.176}$$

$$J_z = \frac{iNe^2}{\omega m}\Big(E_z - \frac{v_0}{c}B_x\Big). \tag{13.177}$$

Using (13.175)–(13.177) in (13.142)–(13.147) gives

$$\frac{i\omega}{c}B_x = -ikE_y, \tag{13.178}$$

$$\frac{i\omega}{c}B_y = ikE_x - \frac{\partial E_z}{\partial x}, \tag{13.179}$$

$$\frac{i\omega}{c}B_z = \frac{\partial E_y}{\partial x}, \tag{13.180}$$

$$-ikB_y = -\frac{i\omega}{c}E_x + \frac{i\omega_p^2}{\omega c}\Big(E_x + \frac{v_0}{c}B_y\Big), \tag{13.181}$$

$$ikB_x - \frac{\partial B_z}{\partial x} = -\frac{i\omega}{c}E_y + \frac{i\omega_p^2}{\omega c}E_y + \frac{v_0}{c}\frac{\omega_p^2}{\omega^2}\left(\nabla \cdot \mathbf{E} + \frac{v_0}{c}\left(\frac{\partial B_z}{\partial x} - ikB_x\right)\right). \tag{13.182}$$

$$\frac{\partial B_y}{\partial x} = -\frac{i\omega}{c}E_z + \frac{i\omega_p^2}{c\omega}\Big(E_z - \frac{v_0}{c}B_x\Big). \tag{13.183}$$

All the field variables are interwoven, and separation of TM and TE variables are impossible. (13.178) and (13.180) imply $\nabla\cdot\mathbf{B}=0$. Equating $n=-i\frac{N}{\omega}\nabla\cdot\mathbf{v}=-\frac{1}{4\pi e}\nabla\cdot\mathbf{E}$, one can derive

$$\nabla\cdot\mathbf{E}=\frac{iv_0\,\omega_p^2}{\varepsilon\omega^2\,ck}\left(\frac{\partial^2}{\partial x^2}-k^2\right)B_x \tag{13.184}$$

where $\varepsilon=1-\omega_p^2/\omega^2$ and $\nabla\cdot\mathbf{B}=0$ was used. (13.181) can be reduced to

$$\varepsilon E_x=\frac{c\kappa}{\omega}B_y\quad\left(\kappa=k+\frac{v_0\omega_p^2}{\omega c^2}\right). \tag{13.185}$$

Using (13.184) in (13.183) gives

$$\left(\frac{\partial^2}{\partial x^2}-k^2\right)B_y=-\frac{i\omega\varepsilon}{c}\frac{\partial E_z}{\partial x}-\frac{\omega k\varepsilon}{c}E_x. \tag{13.186}$$

(13.179), (13.185), and (13.186) are the three equations for three TM mode variables (B_y,E_x,E_z). Eliminating B_y in (13.179) and (13.185) yields

$$i\left(\varepsilon-\frac{c^2}{\omega^2}k\kappa\right)E_x=-\frac{c^2\kappa}{\omega^2}\frac{\partial E_z}{\partial x}. \tag{13.187}$$

Eliminating B_y in (13.179) and (13.186) gives

$$\left(\frac{\partial^2}{\partial x^2}-k^2+\frac{\omega^2\varepsilon}{c^2}\right)\left(\frac{\partial E_z}{\partial x}-ikE_x\right)=0. \tag{13.188}$$

Using (13.187) in (13.188) yields

$$\left(\frac{\partial^2}{\partial x^2}-k^2+\frac{\omega^2\varepsilon}{c^2}\right)E_x=0. \tag{13.189}$$

Using $\nabla\cdot\mathbf{B}=0$ in (13.182) gives

$$\left(1+\frac{v_0^2}{c^2}\frac{\omega_p^2}{\omega^2}\right)\left(\frac{\partial^2}{\partial x^2}-k^2\right)B_x=\frac{k\omega\varepsilon}{c}E_y+\frac{ikv_0}{c}\frac{\omega_p^2}{\omega^2}\nabla\cdot\mathbf{E}$$

which becomes upon using (13.178) and (13.184)

$$\left(\frac{\partial^2}{\partial x^2}-k^2+\frac{\omega^2\varepsilon}{c^2}\left(1+\frac{v_0^2}{c^2}\frac{\omega_p^2}{\varepsilon\omega^2}\right)^{-1}\right)B_x=0. \tag{13.190}$$

In the above equation, we shall neglect $\frac{v_0^2}{c^2}\frac{\omega_p^2}{\varepsilon\omega^2}\ll 1$. Therefore we obtain

$$E_x=Ae^{-\alpha x}, \tag{13.191}$$

$$B_x=Ge^{-\alpha x} \tag{13.192}$$

where $\alpha = \sqrt{k^2 - \frac{\omega^2\varepsilon}{c^2}}$, and A and G are constants. Now all the field components can be determined:

$$E_y = -\frac{\omega}{ck} G e^{-\alpha x}, \tag{13.193}$$

$$B_z = -i\,\frac{\alpha}{k}\,G\,e^{-\alpha x}, \tag{13.194}$$

$$\frac{\partial E_x}{\partial x} + ikE_z = -i\frac{\omega_p^2 v_0}{c^3 k} G e^{-\alpha x}, \tag{13.195}$$

$$E_z = \left(-i\frac{\alpha A}{k} - \frac{\omega_p^2 v_0}{c^3 k^2} G\right) e^{-\alpha x}, \tag{13.196}$$

$$B_y = \frac{e^{-\alpha x}}{\omega ck} \left(\omega^2 \varepsilon A + \frac{i\alpha v_0 \omega_p^2}{ck}\,G\right). \tag{13.197}$$

Vacuum solutions:

$$E_x = Fe^{\lambda x}, \quad E_y = He^{\lambda x}, \quad E_z = \frac{i\lambda}{k}\,Fe^{\lambda x},$$

$$B_x = -\frac{ck}{\omega} He^{\lambda x}, \quad B_y = \frac{\omega}{ck} Fe^{\lambda x}, \quad B_z = -\frac{ic\lambda}{\omega} He^{\lambda x}. \tag{13.198}$$

One can check that $\nabla \cdot \mathbf{E} = 0$ and $\nabla \cdot \mathbf{B} = 0$ in the above.

Boundary conditions. The immediate ones are $[E_y] = 0$ and $[E_z] = 0$. Infinitesimal integration of (13.183) yields $[B_y] = 0$. Finally let us use (13.182) to show the terms that survive the infinitesimal integration:

$$\int_{-\delta}^{\delta} \left[\left(1 + \frac{v_0^2}{c^2}\frac{\omega_p^2}{\omega^2}\right)\frac{\partial B_z}{\partial x} + \frac{v_0}{c}\frac{\omega_p^2}{\omega^2}\frac{\partial E_x}{\partial x}\right] dx = 0.$$

In the above integral, ω_p is regarded as a step function: $\omega_p = \omega_p$ for $x > 0$ and $\omega_p = 0$ for $x < 0$. After integration by parts, we obtain

$$\left(1 + \frac{v_0^2}{c^2}\frac{\omega_p^2}{\omega^2}\right) B_z(0^+) - B_z(0^-) + \frac{v_0}{c}\frac{\omega_p^2}{\omega^2} E_x(0^+) = 0. \tag{13.199}$$

Example. To derive (13.199) in an alternative way, use $\nabla \times \mathbf{B} = \frac{1}{c}\frac{\partial \mathbf{E}}{\partial t} - \frac{4\pi e}{c}(N\mathbf{v} + n\mathbf{v}_0)$ whose y-component yields upon infinitesimal integration

$$[B_z] = -\frac{4\pi}{c} J_y^*$$

where the surface current

$$J_y^* = -eN \int v_y dx + v_0 \sigma^*$$

with the surface charge $\sigma^* = -e \int n dx$. Infinitesimal integration of (13.172) gives $\int v_y dx = 0$. Thus we have

$$[B_z] = -\frac{4\pi}{c} \, v_0 \sigma^*.$$

The continuity equation $-i\omega n + N\nabla \cdot \mathbf{v} = 0$ yields upon infinitesimal integration

$$-i \, \omega \int n dx + N v_x(0^+) + iNk \int v_z dx = 0.$$

(13.173) indicates that the last integral in the above vanishes.

$$Therefore \quad \sigma^* = \frac{ieN}{\omega} v_x(0^+), \quad [B_z] = -\frac{4\pi v_0}{c} \frac{ieN}{\omega} v_x(0^+).$$

Here we use (13.171) to write

$$B_z(0^+) - B_z(0^-) = -\frac{v_0}{c} \frac{\omega_p^2}{\omega^2} \left(E_x(0^+) + \frac{v_0}{c} B_z(0^+) \right)$$

which is (13.199).

Enforcing the boundary conditions, we have

$$[E_y] = 0: \quad -\frac{\omega}{ck} G = H, \tag{13.200}$$

$$[E_z] = 0: \quad -i \, \frac{\alpha}{k} \, A - \frac{v_0 \, \omega_p^2}{c^3 k^2} \, G = \frac{i\lambda}{k} F, \tag{13.201}$$

$$[B_y] = 0: \quad \frac{1}{\omega ck} \left(\omega^2 \varepsilon A + i \frac{\alpha v_0 \omega_p^2}{ck} \, G \right) = \frac{\omega}{ck} \, F, \tag{13.202}$$

$$(13.199): \quad -i \, \frac{\alpha}{k} \left(1 + \frac{v_0^2}{c^2} \frac{\omega_p^2}{\omega^2} \right) G + i \, \frac{c\lambda}{\omega} \, H = -\frac{v_0}{c} \frac{\omega_p^2}{\omega^2} A. \tag{13.203}$$

(13.200) and (13.203) give

$$\frac{v_0}{c} \frac{\omega_p^2}{\omega^2} \, A - \frac{i}{k} \left(\lambda + \left(1 + \frac{v_0^2}{c^2} \frac{\omega_p^2}{\omega^2} \right) \alpha \right) \, G = 0. \tag{13.204}$$

(13.201) and (13.202) yield

$$i\,(\alpha+\varepsilon\lambda)\,A+\frac{v_0}{c}\,\frac{\omega_p^2}{k}\left(\frac{1}{c^2}-\frac{\alpha\lambda}{\omega^2}\right)\,G=0. \tag{13.205}$$

The solvability condition of (13.204) and (13.205) give, after some algebra,

$$\frac{v_0^2}{c^2}\,\frac{\omega_p^2}{\omega^2}\left(\frac{\omega_p^2}{c^2}-\alpha^2-\alpha\lambda\right)=(\alpha+\lambda)(\alpha+\varepsilon\lambda)$$

which reduces to

$$\alpha+\lambda\left(1-\frac{\omega_p^2}{\omega^2}\left(1-\frac{v_0^2}{c^2}\right)\right)=0.$$

In this non-relativistic theory, $v_0^2/c^2 \ll 1$ can be neglected, giving

$$\alpha+\varepsilon\lambda=0 \tag{13.206}$$

which is the same dispersion relation as that in the isotropic plasma. We conclude that when $v_0 \ll c$ the streaming effect of the plasma is nil for $\mathbf{v}_0 \perp \mathbf{k}$.

Remark. It is emphasized that boundary conditions are contained in the governing equations (dynamic and electromagnetic equations) that are used in the problem under investigation. This was shown in (13.199). As another example, note that (13.143) gives upon infinitesimal integration, $[E_z]=0$. (13.147) is rewritten, using (13.150), as

$$\frac{\partial B_y}{\partial x}\left(1+\frac{v_0^2}{c^2}\,\frac{\omega_p^2}{\omega'^2}\right)=-\frac{i\omega}{c}\left(1-\frac{\omega_p^2}{\omega'^2}\right)E_z+\frac{\omega_p^2 v_0}{c\,\omega'^2}\,\frac{\partial E_x}{\partial x}$$

which yields upon infinitesimal integration (ω_p should be regarded as a step function in x)

$$\left(1+\frac{v_0^2}{c^2}\,\frac{\omega_p^2}{\omega'^2}\right)B_y(0^+)-B_y(0^-)=\frac{v_0}{c}\,\frac{\omega_p^2}{\omega'^2}E_x(0^+).$$

One can show that this matching condition and $[E_z]=0$ yield the dispersion relation

$$\frac{\alpha}{\lambda}+1-\frac{\omega_p^2}{\omega'^2}\left(1-\frac{v_0^2}{c^2}\right)=0$$

which is (13.163).

13.3. Surface waves in single fluid

In the previous Sections, we considered surface waves in two-fluid plasma consisting of electron and ion fluids. In this Section, surface waves in single neutral fluid and magnetohydrodynamic fluid will be considered. The boundary conditions derived earlier will be applied. In particular, Rayleigh–Taylor instability and Kelvin–Helmholtz instability will be discussed. Also the shallow water equation which leads to the KdV equation will be derived.

13.3.1. *Gravity wave*

Gravity wave is a surface wave propagating on the surface of an incompressible inviscid fluid of depth H. It is the wave observed propagating on a lake surface when the water is disturbed. The perpendicular coordinate is y (gravity $\mathbf{g} \parallel \hat{\mathbf{y}}$); the z-coordinate is the infinite direction on the horizontal plane and can be ignored ($\partial/\partial z = 0$); the surface wave is assumed to propagate in the x-direction (phasor: $e^{ikx-i\omega t}$). See Fig. 13.3.

The disturbed motion of the fluid particles are parallel to the plane $z = 0$, and thus we can take the perturbed fluid velocity as $\mathbf{v} = (v_x, v_y, 0)$. The motion has a vertical component. Neglecting the second order term $O(v^2)$, the momentum equation reads

$$\rho \frac{\partial v_x}{\partial t} = -\frac{\partial p}{\partial x}, \tag{13.207}$$

$$\rho \frac{\partial v_y}{\partial t} = -\frac{\partial p}{\partial y} - \rho g. \tag{13.208}$$

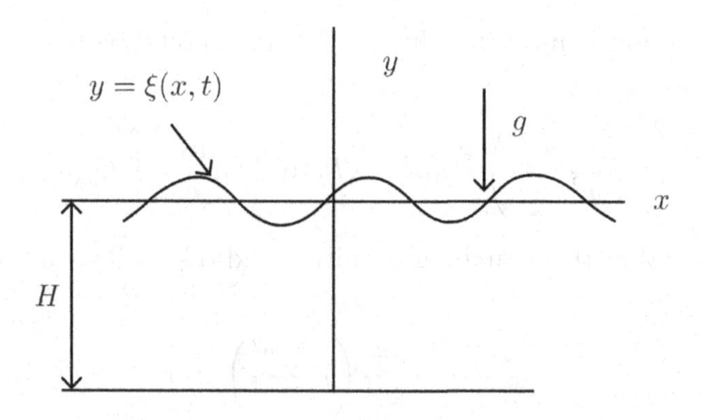

Fig. 13.3. Surface water wave.

The following continuity equation is due to incompressibility

$$\nabla \cdot \mathbf{v} = \frac{\partial v_x}{\partial x} + \frac{\partial v_y}{\partial y} = 0 \tag{13.209}$$

with $\rho = const.$. We have the above three equations for three quantities, v_x, v_y, and p. It appears that we have an inhomogeneous system, but differentiating (13.208) with respect to x produces a homogeneous set. The first two equations yield

$$\frac{\partial v_x}{\partial y} - \frac{\partial v_y}{\partial x} = 0 \tag{13.210}$$

which is nothing but the relation $\nabla \times \mathbf{v} = 0$. (13.209) and (13.210) are combined to give

$$\left(\frac{\partial^2}{\partial x^2} + \frac{\partial^2}{\partial y^2}\right)(v_x, v_y) = 0. \tag{13.211}$$

So both v_x and v_y are solutions of the Laplace equation. We seek the solution in the form

$$(v_x, v_y, p) = (V_x(y), V_y(y), P(y))e^{ikx - i\omega t}. \tag{13.212}$$

The boundary conditions would be relevant to the y coordinate only. Substituting this into (13.211), we have

$$\frac{d^2 V_y}{dy^2} - k^2 V_y = 0 \tag{13.213}$$

whose solution satisfying the boundary condition $V_y\big|_{y=-H} = 0$ is

$$V_y(y) = V_0 \; sinh[k(y + H)] \quad (V_0 : \; const.). \tag{13.214}$$

Using (13.212) in (13.208) yields

$$i\omega \; \rho V_y = \frac{\partial P}{\partial y} + \rho \; g \; e^{-ikx + i\omega t} \tag{13.215}$$

which is integrated in the form

$$i\rho V_0 \frac{\omega}{k} \; cosh[k(y + H)] = P + \rho \; g \; y \; e^{-ikx + i\omega t} \tag{13.216}$$

where we set the integration constant zero. Let $\xi(x, t)$ be the elevation of fluid particles undulating about the equilibrium free surface $y = 0$. Then in accordance with (13.212), the undulating surface can be written as

$$y = \xi(x, t) = \xi_0 e^{ikx - i\omega t} \tag{13.217}$$

where ξ_0 is the amplitude of the elevation. Note that we have $p = 0$ on the surface (free surface). Note that the RHS of (13.216) is the total pressure, perturbed pressure P plus the zero order pressure $\rho g \xi_0$. Assuming $H \gg \xi_0$, (13.216) evaluated on $y = \xi$ yields

$$iV_0 \frac{\omega}{k}\ cosh(kH) = g\xi_0 \quad (dynamic\ boundary\ condition). \tag{13.218}$$

Also,

$$v_y\big|_{y=\xi} = \frac{d\xi}{dt} = \frac{\partial \xi}{\partial t} + (\mathbf{v}\cdot\nabla)\xi.$$

Since both $\mathbf{v}$ and ξ are perturbed quantities, the convective term can be neglected. Thus (13.214) and (13.217) give

$$V_0\ sinh(kH) = -i\omega\ \xi_0 \quad (kinematic\ boundary\ condition). \tag{13.219}$$

We have (13.218) and (13.219) for two integration constants V_0 and ξ_0. The solvability condition yields the dispersion relation

$$\omega^2 = g\ k\ tanh(kH). \tag{13.220}$$

The energy associated with a wave packet propagates with the group velocity. It can be shown from (13.220) that

$$\frac{\partial \omega}{\partial k} = \frac{1}{2}\sqrt{\frac{g}{k}}, \quad kH \gg 1, \tag{13.221}$$

$$\frac{\partial \omega}{\partial k} = \sqrt{gH}, \quad kH \ll 1. \tag{13.222}$$

Problem. Find the phase velocity of gravity wave when the depth of water is infinite. Find the particle orbit oscillating on the wave motion.

13.3.2. *Rayleigh–Taylor instability*

The Rayleigh–Taylor instability occurs when a heavy incompressible fluid is supported against gravity by another fluid of smaller density. If the plane interface between the two fluids is disturbed, the instability takes place. More generally the Rayleigh–Taylor instability can occur in an inhomogeneous continuous fluid if the equilibrium density has an inverted density gradient.

Consider an incompressible fluid with an equilibrium density with a gradient along the vertical direction (y-direction)

$$\rho = \rho(y)$$

acted upon by gravity force

$$\mathbf{g} = -\hat{\mathbf{y}}\, g.$$

The governing fluid equations are

$$\rho\left(\frac{\partial \mathbf{v}}{\partial t} + \mathbf{v}\cdot\nabla\mathbf{v}\right) = -\nabla p - \rho g\hat{\mathbf{y}}, \tag{13.223}$$

$$\frac{\partial \rho}{\partial t} + \mathbf{v}\cdot\nabla\rho = 0, \tag{13.224}$$

$$\nabla\cdot\mathbf{v} = 0. \tag{13.225}$$

In the equilibrium state ($\mathbf{v} = 0$, $\rho = \rho_0$, $p = p_0$), the pressure force is balanced by the gravity:

$$\frac{\partial p_0}{\partial y} = -\rho_0 g \quad or \quad p_0(y) = -\rho_0 g y + const. \tag{13.226}$$

Now we assume arbitrary small perturbations (denoted by primes) propagating in the x-direction with amplitude varying as functions of y. Then, the z-direction is ignorable. We have

$$\mathbf{v}(x,y) = \mathbf{v}'(y)e^{ikx-i\omega t}, \tag{13.227}$$

$$p(x,y) = p_0(y) + p'(y)e^{ikx-i\omega t}, \tag{13.228}$$

$$\rho(x,y) = \rho_0(y) + \rho'(y)e^{ikx-i\omega t}. \tag{13.229}$$

Substituting into (13.227)–(13.229) and linearizing with respect to the primed quantities, we obtain

$$\omega\rho_0 v_x'(y) = kp'(y), \tag{13.230}$$

$$i\omega\rho_0 v_y'(y) = \frac{\partial p'}{\partial y} + \rho'(y)g, \tag{13.231}$$

$$-i\omega\rho' + v_y'\frac{\partial \rho_0}{\partial y} = 0, \tag{13.232}$$

$$ikv_x' + \frac{\partial v_y'}{\partial y} = 0. \tag{13.233}$$

Now we consider the case of two uniform fluids of constant density separated by a horizontal boundary $y = 0$. Let the densities of two fluids be ρ_1 and ρ_2. That is, ρ_0 is a step function: $\rho_0 = \rho_2$ for $y > 0$ and $\rho_0 = \rho_1$ for $y < 0$.

In this case, (13.232) gives $\rho' = 0$ on either side of $y = 0$. Then for both fluids, the above equations can be reduced to

$$\frac{\partial^2 v'_y}{\partial y^2} = k^2 v'_y \tag{13.234}$$

whose general solution is

$$v'_y = Ae^{ky} + Be^{-ky} \quad (A,\ B: \quad const.). \tag{13.235}$$

Since v'_y must vanish when $|y| \to \infty$, we have

$$v'_{y1} = Ae^{ky} \quad (y < 0), \quad v'_{y2} = Be^{-ky} \quad (y > 0). \tag{13.236}$$

Now suppose that the interface $y = 0$ is slightly perturbed in the form

$$y = \xi(x,t) = \xi_0 e^{ikx - i\omega t}. \tag{13.237}$$

The boundary conditions for the two fluids to satisfy at the interface can be obtained from $d\xi/dt = v_y$. So,

$$v_{y1}\Big|_{y=\xi} = \frac{d\xi}{dt} \simeq \frac{\partial \xi}{\partial t} = -i\omega\xi_0 e^{ikx-i\omega t}. \tag{13.238}$$

The same expression holds for v_{y2}. That is,

$$v_{y1}\Big|_{y=\xi} = v_{y2}\Big|_{y=\xi} = -i\omega\xi_0 e^{ikx-i\omega t} \tag{13.239}$$

$$v'_{y1}(\xi) = v'_{y2}(\xi) = -i\omega\xi_0. \tag{13.240}$$

Assuming $k\xi_0 \ll 1$, the above relation requires A = B in (13.236). [This conclusion amounts to applying the matching condition at $y = 0$.] The balance of total pressure at the rippled surface demands that

$$-\rho_1 g\xi + p'_1(\xi)e^{ikx-i\omega t} = -\rho_2 g\xi + p'_2(\xi)e^{ikx-i\omega t} \tag{13.241}$$

$$-\rho_1 g\xi_0 + p'_1(\xi) = -\rho_2 g\xi_0 + p'_2(\xi) \tag{13.242}$$

where we used (13.226). One finds from (13.230)–(13.233) that p' is related to v'_y

$$p'(y) = \frac{i\omega\rho_0}{k^2}\frac{\partial v'_y}{\partial y}. \tag{13.243}$$

Thus we have

$$p'_1(\xi) = \frac{\omega^2}{k}\rho_1\xi_0, \quad p'_2(\xi) = -\frac{\omega^2}{k}\rho_2\xi_0. \tag{13.244}$$

Using (13.244) in (13.242) yields

$$\omega^2 = gk \, \frac{\rho_1 - \rho_2}{\rho_1 + \rho_2}. \tag{13.245}$$

If $\rho_1 < \rho_2$, ω is imaginary and the perturbation grows in time (instability). The dispersion relation of the surface wave (13.245) finds its application in the propagation of tsunamis. Since the sea water density (ρ_1) is much greater than that of air (ρ_2), the phase velocity is approximately $\frac{\omega}{k} = \sqrt{\frac{g}{k}}$ ant its group velocity is one-half of it. It is immediately seen that the longer wave moves faster. The physical reason for the Rayleigh–Taylor instability is that more potential energy is released by displacing some of the heavy fluid downward than is absorbed by displacing a corresponding amount of light fluid upward. If the surface tension acts at the interface, it has a stabilizing effect, particularly for small ripples.

An electrically conducting fluid can also be supported against gravity by magnetic field. In this case the supporting force is due to currents at the boundary interacting with the magnetic field ($\mathbf{J} \times \mathbf{B}$ force). The boundary is again unstable; the instability is referred to as *Kruskal–Schwarzschild instability* and it is analogous to the Rayleigh–Taylor instability.

Next we shall consider the more general case where the density variation $\rho_0(y)$ is continuous, without a sharp interface. (13.230)–(13.233) can be arranged in the form

$$\left(\frac{\omega}{k}\right)^2 \frac{\partial}{\partial y}\left(\rho_0 \, \frac{\partial v'_y}{\partial y}\right) = \left(\omega^2 \rho_0 + g \, \frac{\partial \rho_0}{\partial y}\right) \, v'_y \tag{13.246}$$

which should be supplemented by two boundary conditions at two boundary surfaces $y = a$ and $y = b$:

$$v'_y(a) = v'_y(b) = 0 \quad or \quad \left.\frac{dv'_y}{dy}\right|_{y=a} = \left.\frac{dv'_y}{dy}\right|_{y=b} = 0 \tag{13.247}$$

or some mixed type.

In general, $\rho_0(y)$ is specified as a function of y, and (13.246) with the boundary condition (13.247) can be solved numerically. Before discussing general features of this eigen value problem, we consider as an illustration, a simple case allowing for an analytic solution. The density is taken to be

$$\rho_0 = S \; exp \, \left(\frac{y}{\lambda}\right),$$

λ being a constant giving the density scale length. Using this in (13.246) gives

$$\frac{d^2 v_y}{dy^2} + \frac{1}{\lambda}\frac{dv_y}{dy} - k^2 v_y - \frac{gk^2}{\lambda\omega^2} = 0$$

where we suppressed the prime on v_y for notational simplicity. Taking the boundary condition to be $v_y = 0$ at $y = 0$ and $y = h$, this equation is solved by

$$v_y = e^{-y/2\lambda}\, sin\left(\frac{n\pi y}{h}\right)$$

where n is an integer. The eigen frequency satisfies the dispersion relation

$$\omega^2 = -\frac{g}{d}, \tag{13.248}$$

$$d = \lambda\left[1 + \left(\frac{1}{2k\lambda}\right)^2 + \left(\frac{n\pi}{kh}\right)^2\right].$$

It is seen that the growth rate in this case has a very different dependence on k from that of two separated fluids. As $k \to \infty$ (13.248) gives a limiting finite growth rate, while the sharp boundary case (13.245) gives $\omega \to \infty$.

We recognize that (13.246) and (13.247) constitute the classical Sturm–Liouville eigen value problem. We state, without proof, the standard theorem of the subject: the eigen values $(k/\omega)^2$ are all positive if $d\rho_0/dy$ is everywhere negative; they are all negative if $d\rho_0/dy$ is everywhere positive; if $d\rho_0/dy$ changes sign somewhere in the interval (a, b), then both positive and negative eigen values of ω^2 exist. Thus the inhomogeneous stratified fluid is unstable if and only if $d\rho_0/dy$ is positive somewhere (i.e. not everywhere negative).

We multiply (13.246) by v_y^* and integrate over y, enforcing the boundary condition, to derive

$$\omega^2 = -g\int_a^b |v_y|^2\frac{\partial\rho_0}{\partial y}dy \div \int_a^b \rho_0\left(\frac{1}{k^2}\left|\frac{\partial v_y}{\partial y}\right|^2 + |v_y|^2\right)dy. \tag{13.249}$$

(13.249) is the variational form for the eigen value problem under consideration. That is, ω takes an extremum value for $v_y(y)$ which is actual solution of the Sturm–Liouville boundary value problem (13.246) and (13.247) [see example below]. Its usefulness is due to the easy estimate for ω^2 by using a trial solution which satisfies the boundary conditions, but which is of any form. The more closely the trial function approximates an eigen function, the better the estimate for eigen value ω^2. However, since (13.249) depends only on the integral properties of v_y, even a crude guess gives good approximation as far as it is compatible with the boundary conditions. The approximate eigen value may be improved by allowing the trial solution to contain a number of arbitrary constants A_j $(j = 1, 2, \cdots N)$ and minimizing the trial eigen value with respect to A_j. Then we have the conditions

$$\frac{\partial \omega^2}{\partial A_j} = 0 \quad (j = 1, 2, \cdots N).$$

So $A_j's$ can be determined by inversion of a matrix. This process of approximation is known as *Rayleigh–Ritz method.*

We point out that the dispersion relation (13.245) can be derived from (13.246) by specializing to the case of two uniform fluids of constant density separated by a horizontal interface $y = 0$. Integrate (13.246) over an infinitesimal interval $(-\epsilon,\ \epsilon)$ across the boundary $y = 0$. In this operation, only the perfect differential terms survive. It is important to have terms integrated by parts whenever it is possible, not to lose any integrable terms. [Or put $\partial\rho_0/\partial y = (\rho_2 - \rho_1)\delta(y)$ and integrate.] By this procedure, we get from (13.246)

$$\left(\frac{\omega}{k}\right)^2 \left[\rho_0 \frac{\partial v_y}{\partial y}\right] = g\left[\rho_0 v_y\right] \tag{13.250}$$

where $[Q] = Q_2 - Q_1$. Using (13.236) with $A = B$ in (13.250) gives (13.245).

Example. Variational form of Sturm–Liouville eigen value problem. Consider the standard form of Sturm–Liouville differential equation

$$\frac{d}{dx}\left(f(x)\frac{dy}{dx}\right) + [\lambda - g(x)]y = 0 \tag{13.251}$$

subject to the boundary conditions $y(0) = y(1) = 0$. Multiply (13.251) by y and integrate from $x = 0$ to $x = 1$ to obtain

$$\lambda = \int_0^1 \left(f\left(\frac{dy}{dx}\right)^2 + gy^2\right) dx \div \int_0^1 y^2 dx. \tag{13.252}$$

(13.252) is equivalent to (13.251). Let λ_0 and y_0 be respectively the eigen value and the eigen function of (13.251). Let $y(x)$ be the variation of the eigen function $y_0(x)$ by an amount $\delta y(x)$: $y(x) = y_0(x) + \delta y(x)$, which also satisfies the boundary conditions. That is, $\delta y(0) = \delta y(1) = 0$. We are interested in finding the shift of the eigen value $\delta\lambda = \lambda(y) - \lambda_0(y_0)$. From (13.252), we can write

$$d\lambda = \frac{\int\ [f(y_0' + (\delta y)')^2 + g(y_0 + \delta y)^2]\ dx}{\int (y_0 + \delta y)^2\ dx} - \frac{\int\ [f(y_0')^2 + gy_0^2]\ dx}{\int y_0^2\ dx}$$

where prime denotes derivative with respect to x. The above expression becomes, after neglecting terms of order $(\delta y)^2$,

$$d\lambda = 2\frac{\int [fy_0'(\delta y)' + (g - \lambda_0)y_0\delta y]\, dx}{\int (y_0 + \delta y)^2\, dx}$$

where we used

$$\lambda_0 = \frac{\int [f(y_0')^2 + gy_0^2]\, dx}{\int y_0^2\, dx}.$$

After integrating by parts, we obtain

$$d\lambda = -2\,\frac{\int [(fy_0')' + (\lambda_0 - g)y_0]\delta y\, dx}{\int (y_0 + \delta y)^2\, dx}.$$

$\delta\lambda$ vanishes when $y_0(x)$ satisfies (13.251). The conclusion is that minimizing the expression (13.252) is equivalent to solving the Sturm–Liouville problem (13.252) with the boundary conditions.

13.3.3. *Kelvin–Helmholtz instability*

Surface waves propagating on an interface between two fluids flowing relative to each other can be subject to instability which is called the Kelvin–Helmholtz instability. More generally, in a stratified inhomogeneous fluid with different layers in relative motion, the Kelvin–Helmholtz instability can arise. The existence of velocity shear in the flow is essential for the excitation of the instability. If a jet of fluid is injected into a stationary fluid, the kinetic energy of the jet is converted into turbulent energy. Good examples are those spectacular jets observed in some quasars, active galactic nuclei, and some radio galaxies. The Kelvin–Helmholtz instability can also occur in a sheared magnetized fluid.

We consider an inviscid incompressible fluid with stratified density, each layer streaming with different velocity $\hat{\mathbf{x}}u_0(y)$. The equilibrium state is still represented by (13.234). In the perturbed state, we can write

$$\mathbf{v} = (u_0(y) + v_x, v_y, 0) \quad \rho = \rho_0(y) + \rho'.$$

The z-coordinate can be ignored. Expressing the variables as in (13.227)–(13.229), the linearized equations of (13.223)–(13.225) now take the form

$$\rho_0\left[(-\omega + ku_0)v_x' - iv_y'\frac{\partial u_0}{\partial y}\right] = -kp', \tag{13.253}$$

$$i\rho_0(-\omega + ku_0)v_y' = -\frac{\partial p'}{\partial y} - g\rho', \tag{13.254}$$

$$i(-\omega + ku_0)\ \rho' + v'_y\ \frac{\partial \rho_0}{\partial y} = 0, \tag{13.255}$$

$$ikv'_x + \frac{\partial v'_y}{\partial y} = 0. \tag{13.256}$$

Notice that, if $u_0 = 0$, the above equations reduce to (13.230)–(13.233). Let the two fluids of densities ρ_1 and ρ_2 be separated by a horizontal interface $y = 0$. The streaming velocities of two fluids are respectively u_1 and u_2. In each of the two regions, (13.253)–(13.256) reduce to

$$\frac{\partial^2 v'_y}{\partial y^2} - k^2 v'_y = 0$$

whose solutions are given by

$$v'_{y1} = A\ e^{ky} \quad (y < 0), \quad v'_{y2} = B\ e^{-ky} \quad (y > 0). \tag{13.257}$$

We assume that the interface is perturbed in the form $\xi(x,t) = \xi_0\ e^{ikx - i\omega t}$. Then

$$v'_y(\xi)e^{ikx-i\omega t} = \frac{d\xi}{dt} = \frac{\partial \xi}{\partial t} + u_0 \frac{\partial \xi}{\partial x} = (-i\omega + iku_0)\xi_0 e^{ikx-i\omega t},$$

$$\frac{v'_{y1}(\xi)}{\omega - ku_1} = \frac{v'_{y2}(\xi)}{\omega - ku_2} = -i\xi_0. \tag{13.258}$$

(13.257) and (13.258) give

$$A = -i\xi_0(\omega - ku_1), \quad B = -i\xi_0(\omega - ku_2). \tag{13.259}$$

Using the relation

$$p'(y) = i\ \frac{\rho_0}{k^2}(\omega - ku_0)\frac{\partial v'_y}{\partial y} \tag{13.260}$$

in the pressure balance equation (13.242), one obtains the dispersion relation

$$\rho_1(\omega - ku_1)^2 + \rho_2(\omega - ku_2)^2 = gk(\rho_1 - \rho_2). \tag{13.261}$$

If $u_1 = u_2 = 0$, the above equation reduces to Rayleigh–Taylor dispersion relation (13.245). (13.261) can be solved by quadrature, giving

$$\omega = k\ \frac{\rho_1 u_1 + \rho_2 u_2}{\rho_1 + \rho_2} \pm \left[gk\frac{\rho_1 - \rho_2}{\rho_1 + \rho_2} - k^2\ \frac{\rho_1\rho_2(u_1 - u_2)^2}{(\rho_1 + \rho_2)^2}\right]^{\frac{1}{2}}. \tag{13.262}$$

It can be seen that no instability arises if $u_1 = u_2$ when $g = 0$.

Next we shall derive (13.262) by an alternative method. Assume that the gradient of the streaming velocity is continuous: $u_0 = u_0(y)$ is not a step function. (13.253)–(13.256) can be arranged in the form

$$\frac{\partial}{\partial y}\left[\rho_0(-\omega + ku_0(y))\frac{\partial v_y'}{\partial y} - k\rho_0(y)\frac{\partial u_0}{\partial y}v_y'\right]$$

$$+k^2\rho_0(\omega - ku_0)v_y' + gk^2\frac{v_y'\frac{\partial \rho_0}{\partial y}}{\omega - ku_0} = 0. \tag{13.263}$$

Two boundary conditions must be given for v_y'. Then this is a general eigen value problem. We can specialize (13.263) to the situation where two fluids with different streaming velocities are separated by interface $y = 0$. In this case by integrating (13.263) over the infinitesimal interval $(-\epsilon, \epsilon)$ across $y = 0$, one obtains

$$\left[\rho_0(-\omega + ku_0)\,\frac{\partial v_y'}{\partial y}\right]_{1,2} = gk^2\left[\frac{\rho_0 v_y'}{-\omega + ku_0}\right]_{1,2}. \tag{13.264}$$

The symbol $[\,]_{1,2}$ means the difference of the bracketed quantity. The normal component of velocity on each side is given by

$$v_{y1}' = -i\xi_0(\omega - ku_1)e^{ky}, \quad v_{y2}' = -i\xi_0(\omega - ku_2)e^{-ky}. \tag{13.265}$$

Substituting (13.265) into (13.264) gives the dispersion relation (13.261).

Problem. For given u_1 and u_2, determine the critical wave length beyond which no instability can occur.

13.4. Surface waves in MHD fluid

13.4.1. *Magnetohydrodynamic Rayleigh–Taylor instability*

We now consider the MHD Rayleigh–Taylor instability with a uniform static magnetic field $\mathbf{B}_0$ permeated in inhomogeneous fluid, stratified in y-direction. The governing equations are

$$\rho\left(\frac{\partial \mathbf{v}}{\partial t} + \mathbf{v}\cdot\nabla\mathbf{v}\right) = -\nabla p + \frac{1}{c}\mathbf{J}\times\mathbf{B} - \rho g\hat{\mathbf{y}}, \tag{13.266}$$

$$\frac{\partial \mathbf{B}}{\partial t} = \nabla\times(\mathbf{v}\times\mathbf{B}), \tag{13.267}$$

$$\mathbf{J} = \frac{c}{4\pi}\,\nabla\times\mathbf{B}, \tag{13.268}$$

$$\frac{\partial \rho}{\partial t} + \mathbf{v} \cdot \nabla \rho = 0, \tag{13.269}$$

$$\nabla \cdot \mathbf{v} = 0. \tag{13.270}$$

In the equilibrium ($\mathbf{v} = 0$), the pressure is balanced by the gravity: $\frac{\partial p_0}{\partial y} = -\rho_0 g$ [see (13.226)]. Now we assume small perturbations (denoted by primes) propagating in x-direction with amplitudes varying as functions of y. The z-direction is ignorable ($\partial/\partial z = 0$). We have

$$\mathbf{v}(x, y) = \mathbf{v}'(y) e^{ikx - i\omega t}, \tag{13.271}$$

$$p(x, y) = p_0(y) + p'(y) e^{ikx - i\omega t}, \tag{13.272}$$

$$\rho(x, y) = \rho_0(y) + \rho'(y) e^{ikx - i\omega t}, \tag{13.273}$$

$$\mathbf{B} = \mathbf{B}_0 + \mathbf{B}'(y) e^{ikx - i\omega t}, \tag{13.274}$$

$$\mathbf{J}' = \frac{c}{4\pi} \nabla \times \mathbf{B}'. \tag{13.275}$$

We first consider the case where $\mathbf{B}_0 \parallel y = 0$, the interface. Let us assume $\mathbf{B}_0 = \hat{\mathbf{x}} B_0$. Then (13.226)–(13.270) give

$$\omega \rho_0 v_x' = k p', \tag{13.276}$$

$$-i\omega \rho_0 v_y' = -\frac{\partial p'}{\partial y} + g\rho' + \frac{B_0}{c} J_z', \tag{13.277}$$

$$i\omega B_x' = B_0 \frac{\partial v_y'}{\partial y}, \tag{13.278}$$

$$-i\omega B_y' = k B_0 v_y', \tag{13.279}$$

$$J_x' = \frac{c}{4\pi} \frac{\partial B_z'}{\partial y}, \tag{13.280}$$

$$J_z' = \frac{c}{4\pi} \left(ikB_y' - \frac{\partial B_x'}{\partial y} \right), \tag{13.281}$$

$$-i\omega \rho + v_y' \frac{\partial \rho_0}{\partial y} = 0, \tag{13.282}$$

$$ikv_x' + \frac{\partial v_y'}{\partial y} = 0. \tag{13.283}$$

The perturbed current density is found to be

$$J_z' = \frac{i}{4\pi\omega} \frac{B_0 c}{} \left(\frac{\partial^2}{\partial y^2} - k^2 \right) v_y'. \tag{13.284}$$

Using this result, (13.277)–(13.283) can be reduced to

$$\frac{\partial}{\partial y}\left(\rho_0 \frac{\partial v_y'}{\partial y}\right) - k^2\rho_0 v_y' - \frac{k^2 g}{\omega^2}\frac{d\rho_0}{dy}v_y' - \frac{B_0^2 k^2}{4\pi\omega^2}\left(\frac{\partial^2}{\partial y^2} - k^2\right)v_y' = 0. \tag{13.285}$$

Compare (13.285) with (13.246). The last term of (13.285) is new and represents the effect of the magnetic field. (13.285) can be used to obtain the dispersion relation for a surface wave undulating the interface $y = 0$ which separates a fluid of density ρ_2 from a fluid of density ρ_1 underneath. Proceeding as described in the Section on Rayleigh–Taylor instability, one obtains

$$\omega^2 = kg\,\frac{\rho_1 - \rho_2}{\rho_1 + \rho_2} + \frac{B_0^2 k^2}{2\pi(\rho_1 + \rho_2)}. \tag{13.286}$$

The first term is the instability driving term obtained previously, and the second term containing the magnetic field is seen to act to stabilizing effect. Physically, this stabilizing effect results from the energy required to bend magnetic field lines, as the frozen field lines should move with the fluid. It is seem from (13.286) that the stabilizing effect of the magnetic field increases as k increases. Modes are stable if their wave lengths are sufficiently short that

$$k > \frac{2\pi g(\rho_1 - \rho_2)}{B_0^2}.$$

In deriving (13.286), it was assumed that the wave vector $\mathbf{k} \parallel \mathbf{B}_0$. For more general direction of propagation (with both $\mathbf{k}$ and $\mathbf{B}_0$ parallel with the plane $y = 0$), it is left as an exercise to show that the only modification is to replace kB_0 by $\mathbf{k}\cdot\mathbf{B}_0$ in (13.285) and (13.286). Thus, the case $\mathbf{k} \parallel \mathbf{B}_0$ is most stable. When $\mathbf{k} \perp \mathbf{B}_0$ and $\rho_2 \gg \rho_1$, the dispersion relation is obtained in the form

$$\omega^2 = -kg$$

which results in instability (*Kruskal–Schwarzschild instability*). In this case the stabilizing effect is lost completely because the fluid is able to move without causing any bending of the field lines.

Problem. Derive the dispersion relation (13.286).

13.4.2. *Kruskal–Schwarzschild instability*

The Kruskal–Schwarzschild instability is due to $\mathbf{g}\times\mathbf{B}$ drift. In this subsection we present an alternative view which connects the Kruskal–Schwarzschild instability with the $\mathbf{g} \times \mathbf{B}$ drift in orbit theory. We envisage the situation depicted where the quiescent interface $z = 0$ separates plasma ($z > 0$) and a vacuum ($z < 0$). A static magnetic field $\mathbf{B}_0 = \hat{\mathbf{y}} B_0$ is permeated in the whole space while the gravity acts downward $\mathbf{g} = -\hat{\mathbf{z}} g$. Assume that the interface $z = 0$ is perturbed in the form

$$F(x, z, t) = z - \eta(x, t) = 0, \quad \eta(x, t) = \xi(t) coskx. \tag{13.287}$$

In this geometry, the y-coordinate is ignorable ($\partial/\partial y = 0$). The gravity and the magnetic field give rise to the drift

$$\mathbf{V}_\alpha = \frac{cm_\alpha}{e_\alpha B_0^2}\, \mathbf{g} \times \mathbf{B}_0 = \hat{\mathbf{x}}\, \frac{gcm_\alpha}{e_\alpha B_0} \quad (\alpha = e,\ i). \tag{13.288}$$

The current associated with this drift velocity which we designate by subscript 0 is regarded as an external current flowing through the plasma:

$$\mathbf{J}_0 = \sum_\alpha e_\alpha n_\alpha \mathbf{V}_\alpha = \hat{\mathbf{x}} \frac{cg\rho_m}{B_0} \quad (\rho_m = m_i n_i + m_e n_e,\ \ mass\ density). \tag{13.289}$$

We have charge (ϱ_0) conservation relation for this 'external current',

$$\nabla \cdot \mathbf{J}_0 + \frac{\partial \varrho_0}{\partial t} = 0. \tag{13.290}$$

The Maxwell equation $\nabla \times \mathbf{B} = \frac{1}{c}\frac{\partial \mathbf{E}}{\partial t} + \frac{4\pi}{c}(\mathbf{J} + \mathbf{J}_0) = \frac{1}{c}\frac{\partial \mathbf{D}}{\partial t} + \frac{4\pi}{c}\mathbf{J}_0$ gives the relation (by taking divergence),

$$\nabla \cdot (\varepsilon \mathbf{E}) = 4\pi \varrho_0. \quad (\varepsilon:\ dielectric\ const.) \tag{13.291}$$

In a cold plasma a surface charge is formed in a thin layer about the interface: $\sigma^* = \int_{-\epsilon}^{\epsilon} \varrho_0 dz$. Performing infinitesimal integration with (13.290) across the rippled interface given by (13.287) yields

$$\frac{\partial \sigma^*}{\partial t} = -\hat{\mathbf{n}} \cdot \mathbf{J}_0 \tag{13.292}$$

where $\hat{\mathbf{n}}$ is the unit normal to the rippled interface

$$\hat{\mathbf{n}} = \frac{-\hat{\mathbf{x}}\, \frac{\partial \eta}{\partial x} + \hat{\mathbf{z}}}{\sqrt{1 + (\frac{\partial \eta}{\partial x})^2}} \simeq -\hat{\mathbf{x}}\, \frac{\partial \eta}{\partial x} + \hat{\mathbf{z}}. \tag{13.293}$$

Using (13.289) and (13.293) in (13.292) gives

$$\sigma^* = -k\, \frac{cg\rho_m}{B_0} sinkx \int_{-\infty}^{t} \xi dt. \tag{13.294}$$

Performing infinitesimal integration with (13.291) across the interface (here we put $\hat{\mathbf{n}} \simeq \hat{\mathbf{z}}$ as the first approximation), we obtain

$$\varepsilon E_z^p - E_z^v = 4\pi\sigma^* \quad at \quad z = 0 \tag{13.295}$$

where the small vacuum field E_z^v can be neglected, and we have

$$E_z|_{z=0} = \frac{4\pi\sigma^*}{\varepsilon}. \tag{13.296}$$

Since ε is constant, the plasma field satisfies $\nabla \cdot \mathbf{E} = 0$ and $\nabla \times \mathbf{E} = 0$, and therefore the solution for E_z for $z > 0$ which satisfies $\nabla^2 E_z = 0$ and the boundary value (13.296) is obtained as

$$E_z = \frac{4\pi\sigma^*}{\varepsilon} e^{-kz} = -\frac{4\pi}{\varepsilon}\frac{cg\rho_m}{B_0} k\ sinkx\ e^{-kz} \int \xi(t)dt. \tag{13.297}$$

From $\nabla \cdot \mathbf{E} = 0$, one obtains

$$E_x = \frac{4\pi}{\varepsilon}\frac{cg\rho_m}{B_0} k\ coskx\ e^{-kz} \int \xi(t)dt. \tag{13.298}$$

The above electric field again gives rise to $\mathbf{E} \times \mathbf{B}$-drift, $\mathbf{V} = c\mathbf{E} \times \mathbf{B}_0/B_0^2$,

$$V_x = \frac{4\pi}{\varepsilon}\frac{c^2 g\rho_m}{B_0^2} k\ sinkx\ e^{-kz} \int \xi(t)dt, \tag{13.299}$$

$$V_z = \frac{4\pi}{\varepsilon}\frac{c^2 g\rho_m}{B_0^2} k\ coskx\ e^{-kz} \int \xi(t)dt. \tag{13.300}$$

Since we have $\frac{\partial z}{\partial t}\Big|_{z=0} = V_z\Big|_{z=0}$, it follows that

$$\frac{d^2\xi}{dt^2} = \frac{4\pi}{\varepsilon}\frac{\rho_m c^2}{B_0^2} gk\xi \simeq gk\xi \tag{13.301}$$

where we use $\varepsilon = 1 + \frac{4\pi\rho_m c^2}{B_0^2}$, and neglect 1. (13.301) gives the growth rate $\xi(t) \sim e^{\sqrt{gk}\ t}$ for the instability.

13.4.3. *Magnetohydrodynamic Kelvin–Helmholtz instability*

We shall now consider the Kelvin–Helmholtz instability in a magnetized fluid. It is advantageous to analyze the governing equations in term of the displacement vector $\vec{\xi}$ which is defined by $\mathbf{v} = \frac{d\vec{\xi}}{dt}$. We first linearize (13.266)–(13.269) and denote the equilibrium quantities and the perturbations by the subscripts 0 and 1, respectively.

$$\frac{\partial \rho_1}{\partial t} + \nabla \cdot (\rho_0 \mathbf{v}_1) = 0, \tag{13.302}$$

$$\rho_0 \frac{\partial \mathbf{v}_1}{\partial t} + \nabla p_1 + \frac{1}{4\pi}[\mathbf{B}_1 \times (\nabla \times \mathbf{B}_0) + \mathbf{B}_0 \times (\nabla \times \mathbf{B}_1)] = 0, \tag{13.303}$$

$$\frac{\partial \mathbf{B}_1}{\partial t} = \nabla \times (\mathbf{v}_1 \times \mathbf{B}_0). \tag{13.304}$$

Writing $\mathbf{v}_1 = \frac{\partial \vec{\xi}}{\partial t}$, we have

$$\rho_1 = -\nabla \cdot (\rho_0 \vec{\xi}), \quad \mathbf{B}_1 = \nabla \times (\vec{\xi} \times \mathbf{B}_0). \tag{13.305}$$

Using the above relations, (13.271)–(13.275) becomes

$$\rho_0 \frac{\partial^2 \vec{\xi}}{\partial t^2} = -\nabla p_1 + \frac{1}{4\pi}[(\nabla \times \mathbf{B}_0) \times \mathbf{B}_1 + (\nabla \times \mathbf{B}_1 \times \mathbf{B}_0]. \tag{13.306}$$

Using the vector identity

$$\nabla(\mathbf{A} \cdot \mathbf{B}) = (\mathbf{A} \cdot \nabla)\mathbf{B} + (\mathbf{B} \cdot \nabla)\mathbf{A} + \mathbf{A} \times (\nabla \times \mathbf{B}) + \mathbf{B} \times (\nabla \times \mathbf{A}).$$

(13.285) can be put into the form

$$\rho_0 \frac{\partial^2 \vec{\xi}}{\partial t^2} = -\nabla P + \frac{1}{4\pi}[-\mathbf{B}_1 \cdot \nabla \mathbf{B}_0 + \mathbf{B}_0 \cdot \nabla \mathbf{B}_1] \tag{13.307}$$

where $P = p_1 + \frac{1}{4\pi}\mathbf{B}_0 \cdot \mathbf{B}_1$ is the 'total pressure'. Assuming that ρ_0 and $\mathbf{B}_0$ are uniform and the fluid is incompressible so that $\nabla \cdot \mathbf{v}_1 = \nabla \cdot \vec{\xi} = 0$, the divergence of (13.286) yields

$$\nabla^2 P = 0. \tag{13.308}$$

Fluid in medium 2 is at rest, and fluid in medium 1 is moving in x-direction with velocity u_1, and y-coordinate is perpendicular to the interface. The z-direction is ignorable. Then, the solution of Laplace equation (13.308) appropriate to the surface waves in each region is

$$P = P_0 \, e^{-k|y|} \, e^{ikx - i\omega t}. \tag{13.309}$$

The displacement vector corresponding to this pressure perturbation is obtained from (13.305) and (13.306):

$$\rho_0 \, \frac{\partial^2 \vec{\xi}}{\partial t^2} = -\nabla P + \frac{1}{4\pi}(\mathbf{B}_0 \cdot \nabla)^2 \vec{\xi} \tag{13.310}$$

where $\mathbf{B}_0$ was assumed constant and $\nabla \cdot \vec{\xi} = 0$ was used. In accordance with (13.294), the displacement can be assumed to be of the form $\vec{\xi} = \vec{\xi}_0 e^{ikx-i\omega t}$. Then we have in region 1, after using (13.310),

$$\vec{\xi} = \frac{\nabla P}{\rho_1[(\omega - ku_1)^2 - k^2 v_{A1}^2)]} \tag{13.311}$$

where $v_A = \frac{B_{0x}}{\sqrt{4\pi\rho}}$ is the Alfven velocity. The corresponding vector in region 2 is obtained by replacing $\rho_1 \to \rho_2$ and $u_1 \to 0$. The boundary conditions at the interface $y = 0$ are the continuity of P and continuity of the normal component of $\vec{\xi}$, i.e. ξ_y. Applying these conditions to (13.309) and (13.311) gives the dispersion relation

$$\frac{1}{\rho_1[(\omega - ku_1)^2 - k^2 v_{A1}^2]} + \frac{1}{\rho_2[(\omega - ku_2)^2 - k^2 v_{A2}^2]} = 0. \tag{13.312}$$

Solving this equation for ω, we get

$$\omega = \frac{k}{\rho_1 + \rho_2}[\rho_1 u_1 \pm [(\rho_1 + \rho_2)(\rho_1 v_{A1}^2 + \rho_2 v_{A2}^2) - \rho_1 \rho_2 u_1^2]^{1/2}]. \tag{13.313}$$

Compare this equation with (13.262). Note that no instability occurs when $u_1 - 0$. The instability is the consequence of the relative drift u_1 of the two fluids along the interface. The condition of instability is obtained from (13.313) as

$$u_1^2 > \frac{(\rho_1 + \rho_2)}{\rho_1 \rho_2}(\rho_1 v_{A1}^2 + \rho_2 v_{A2}^2). \tag{13.314}$$

When there is no magnetic field, the instability always occur as long as there is a relative drift. The magnetic field is seen to act to stabilizing effect when it is directed along the direction of wave propagation. When the magnetic field is perpendicular to $\mathbf{k}$, it has no influence on the stability.

References for §13.4: M. Kruskal and M. Schwarzschild, *Some instabilities of a completely ionized plasma,* Proc. Royal Soc. London **2** (1954) p. 348; S. Chandrasekhar, *Hydrodynamic and hydromagnetic stability,* Dover New York (1961) Chapter X, XI.

13.5. Theory of shallow water wave

The waves on shallow water surface are a good example of dispersive wave observed in nature. The necessary analysis of the shallow water wave requires careful application of the boundary condition on the undulating surface

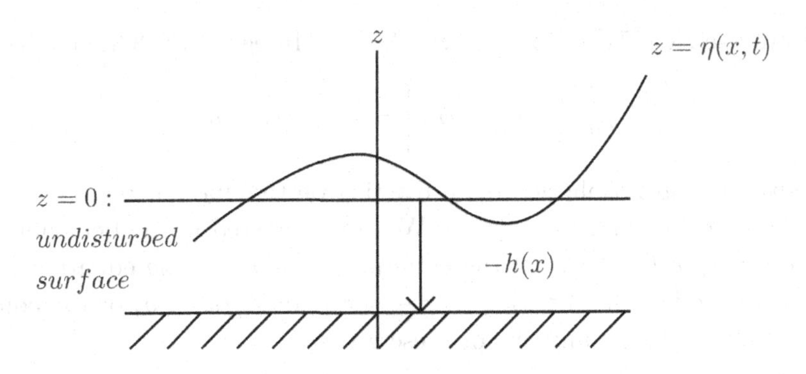

Fig. 13.4. Shallow water wave.

between the water and air (see § 13.2.3). The famous Korteweg–de Vries equation can be derived from the equations of the shallow water wave. The soliton solution that is obtained from the KdV equation was earlier observed by J. S. Russell in a Scottland open channel when he was riding a horse along the channel (1834).

Let us consider an inviscid, incompressible fluid of density ρ whose surface $z = \eta(x,t)$ is interfaced with air. The bottom of the channel is expressed by $z = -h(x)$. See Fig. 13.4. Here we assumed that the y-coordinate is infinite and has the translational invariance ($\partial/\partial y = 0$), and was ignored.

Because of the incompressibility we have the continuity equation in the form

$$\nabla \cdot \mathbf{v} = \frac{\partial v_x}{\partial x} + \frac{\partial v_z}{\partial z} = 0. \tag{13.315}$$

The equation of motion is

$$\rho\left(\frac{\partial}{\partial t} + \mathbf{v}\cdot\nabla\right)\mathbf{v} = -\nabla p - g\rho\hat{\mathbf{z}}. \tag{13.316}$$

We have only these two equations for the moment but later the boundary conditions will enter, assuming the status of the governing equation. It is convenient to introduce the velocity potential $\mathbf{v} = \nabla\phi$ (per $\nabla \times \mathbf{v} = 0$). Then the above equations take the forms

$$\nabla^2\phi = 0, \tag{13.317}$$

$$\rho\left[\frac{\partial}{\partial t}\nabla\phi + \nabla\frac{1}{2}(\nabla\phi)^2\right] = -\nabla p - g\rho\hat{\mathbf{z}}, \tag{13.318}$$

where we used $\mathbf{v}\cdot\nabla\mathbf{v}+\mathbf{v}\times\nabla\times\mathbf{v}=\nabla v^2/2$. Integrating (3.318) gives

$$\rho\left[\frac{\partial\phi}{\partial t}+\frac{1}{2}(\nabla\phi)^2\right]=-p+p_0-g\rho z \tag{13.319}$$

where p_0 is the atmospheric pressure acting on the undisturbed surface ($z=0$) of the fluid in equilibrium state. We are interested in the behavior of the surface $z=\eta(x,t)$. We need the equation for $\eta(x,t)$. The equation for the wavy interface is (13.99) with $F=z-\eta(x,t)$ or $d\eta/dt=v_z$ or the equation below (13.218). Using any one of these gives

$$v_z=\frac{\partial\eta}{\partial t}+v_x\frac{\partial\eta}{\partial x},\quad at\ z=\eta(x,t). \tag{13.320}$$

Likewise at the bottom of the channel, we have the boundary condition

$$v_z=-v_x\frac{\partial h}{\partial x},\quad at\ z=-h(x). \tag{13.321}$$

• **Hydrostatic approximation.** Instead of using the exact equation of motion (13.318), we assume that the vertical acceleration of the fluid is negligible. Then the pressure satisfies the hydrostatic relation

$$p=\rho g(\eta-z). \tag{13.322}$$

Then the equation of motion (13.316) gives for the x-component

$$\frac{\partial v_x}{\partial t}+v_x\frac{\partial v_x}{\partial x}=-g\frac{\partial\eta}{\partial x}. \tag{13.323}$$

Equations (13.315), (13.320) and (13.321) can be combined into a simple expression under the hydrostatic approximation. Let us integrate (13.315):

$$\int_{-h(x)}^{\eta}\frac{\partial v_x}{\partial x}dz+v_z\Big|_{z=\eta}-v_z\Big|_{z=-h}$$

$$=\int_{-h(x)}^{\eta}\frac{\partial v_x}{\partial x}dz+\frac{\partial\eta}{\partial t}+v_x\Big|_{\eta}\frac{\partial\eta}{\partial x}+v_x\Big|_{-h}\frac{\partial h}{\partial x}=0 \tag{13.324}$$

where we used (13.320) and (13.321). (13.324) can be put into the integral

$$\frac{\partial}{\partial x}\int_{-h}^{\eta}v_x dz=-\frac{\partial\eta}{\partial t}. \tag{13.325}$$

We again assume that v_x changes negligibly along z, and integrate to obtain

$$\frac{\partial}{\partial x}\left[v_x(\eta+h)\right]+\frac{\partial\eta}{\partial t}=0. \tag{13.326}$$

(13.323) and (13.326) are the two shallow water wave equations obtained under the hydrostatic approximation. When $h(x) = const.$, we can simply write the water wave equations as one-dimensional equations

$$\frac{\partial v}{\partial t} + v\,\frac{\partial v}{\partial x} = -g\,\frac{\partial \eta}{\partial x} \tag{13.327}$$

$$\frac{\partial \eta}{\partial t} + \frac{\partial}{\partial x}(\eta v) = 0. \tag{13.328}$$

We obtained the last equation after going through many steps. But in hindsight there is a very quick way to write it. Starting with the continuity equation for a compressible fluid, $\frac{\partial \rho}{\partial t} + \frac{\partial}{\partial x}(\rho u) = 0$, one replaces $\rho \to \eta\rho_0$ with ρ_0 const., and then obtains (13.328). We can see that the temporal rate of change $\partial\eta/\partial t$ is produced by the difference of fluxes ηu through the sections at x and $x + dx$. Let us compare the above two equations and the one-dimensional isentropic flow in gas dynamics: $\frac{\partial v}{\partial t} + v\,\frac{\partial v}{\partial x} = -\frac{1}{\rho}\,\frac{\partial p}{\partial x}$, $\frac{\partial \rho}{\partial t} + \frac{\partial}{\partial x}(\rho v) = 0$, $p = A\rho^{\gamma}$, where γ is the adiabatic constant. The water wave equation corresponds to the gas dynamic equation with $\gamma = 2$. The solutions of the gas dynamic isentropic flow equations are well known. Reference: R. Courant and K. O. Friedrichs, *Supersonic flow and shock waves*, Springer New York (1948); H. J. Lee, *Nonlinear analysis of the cold fluid-Poisson plasmas by using the characteristic method*, J. Korean Phys. Soc. **69** (2016) p. 1191–1211. The nonlinear equations keep on steepening the wave profile and eventually the wave overturns. This is because the equations have only nonlinear terms but are without a dispersive term. If the surface tension is included in the momentum equation, we have the dispersive effect which counter balances the nonlinear steepening effect, thus resulting in a stationary form of wave profile. In the sequel, we abandon the hydrostatic approximation and consider the non-approximated equations (13.317)–(13.321).

Assuming that $p = p_0$ on $z = \eta$, (13.319) takes the form

$$\frac{\partial \phi}{\partial t} + \frac{1}{2}(\nabla\phi)^2 = -g\eta \quad on \quad z = \eta. \tag{13.329}$$

Also we have (13.320):

$$\frac{\partial \phi}{\partial z} = \frac{\partial \eta}{\partial t} + \frac{\partial \phi}{\partial x}\,\frac{\partial \eta}{\partial x}, \quad on \quad z = \eta. \tag{13.330}$$

Assuming zero slope of the bottom,

$$\frac{\partial \phi}{\partial z} = 0 \quad on\ z = -h. \tag{13.331}$$

In the fluid ($-h < z < \eta$) we have

$$\frac{\partial^2 \phi}{\partial x^2} + \frac{\partial^2 \phi}{\partial z^2} = 0 \quad -h < z < \eta. \tag{13.332}$$

The above four equations are the complete set of water wave equations.

• **Linear solution.** In linearizing the above equations, we consider that the elevation η of the boundary surface is a quantity of $O(\epsilon)$. Then (13.329) and (13.330) are regarded as holding on $z = 0$. The two equations give after linearizing

$$\frac{\partial^2 \phi}{\partial t^2} + g\,\frac{\partial \phi}{\partial z} = 0 \quad on \;\; z = 0. \tag{13.333}$$

The above three equations are the linearized equations for ϕ. After solving them for ϕ, the surface is found as

$$\eta = -\frac{1}{g}\,\frac{\partial \phi}{\partial t}\bigg|_{z=0}. \tag{13.334}$$

We look for the solution in the form

$$\phi(x, z) = \tilde{\phi}(z) e^{ikx - i\omega t}. \tag{13.335}$$

Using this form in (13.332) gives the solution

$$\tilde{\phi}(z) = C cosh[k(z + h)] \tag{13.336}$$

which satisfies the boundary condition (13.331). (13.334) yields

$$\eta(x, t) = \frac{i\omega}{g}\tilde{\phi}(0) e^{ikx - i\omega t}. \tag{13.337}$$

Using (13.335) and (13.336) in (13.333) gives the dispersion relation of the surface water wave in the form

$$\omega^2 = gk\; tanh[kh]. \tag{13.338}$$

The linear dispersion relation in (13.338) is suggestive of the existence of progressive solitary wave, or soliton. To look into such a possibility, let us expand the dispersion relation for $kh \ll 1$, i.e. the wave length of the water wave is $\gg$ than the depth of the water. We have for such shallow water

$$\omega = \sqrt{gh}\,\left(k - \frac{1}{6}h^2 k^3 + \cdots\right). \tag{13.339}$$

The $\sim k^3$–term is the dispersive term. With the $\sim k$–term only the wave form is simply translated. This is easily seen by making correspondence,

$\omega \to i\ \partial/\partial t$, $\quad k \to -i\ \partial/\partial x$. Then we can write a wave equation from (13.339)

$$\frac{\partial \phi}{\partial t} + \sqrt{gh}\ \frac{\partial \phi}{\partial x} + \frac{h^2}{6}\ \sqrt{gh}\ \frac{\partial^3 \phi}{\partial x^3} = 0. \tag{13.340}$$

If we include a nonlinear term which was omitted in the course of the linearization, we can write the KdV equation in the form

$$\frac{\partial \phi}{\partial t} + \beta\ \phi\ \frac{\partial \phi}{\partial x} + \frac{h^2}{6}\ \sqrt{gh}\ \frac{\partial^3 \phi}{\partial x^3} = 0 \tag{13.341}$$

which is the KdV equation. Note that the first two terms in (13.340) can be put into simply $\partial\phi/\partial t$ in a different Galilean frame. In the following, we shall address the nonlinear equations (13.329)–(13.332) and discuss a preliminary arrangement for perturbation analysis.

Remark. In order to perform the perturbation analysis of (13.329)–(13.332), it is more convenient to put the boundary surface elevation (η) into the main equations, and forget about the boundary conditions. This is because the boundary is an unknown function just like other field quantities. Also we take the bottom of the channel to be $z = 0$ and the quiescent surface $z = 0 \to z = h$. Let us denote the free surface by

$$z^*(x,t) = h(1 + \epsilon\eta(x,t)). \tag{13.342}$$

Now η is dimensionless, and the old η should be replaced by $h\eta$. Any function of z in (13.329) and (13.330) should be replaced by z^* and we expand

$$f(x, z^*, t) = f(x,h,t) + f_z(x,h,t)\epsilon h\eta + \cdots = f_0 + [f_z]_0\ \epsilon h\eta + \cdots \tag{13.343}$$

where the subscript z means differentiation and the subscript 0 means that the z coordinate is evaluated at $z = h$ (at $\eta = 0$). $[f_z]_0 = \frac{\partial f}{\partial z}|_{z=h}$. Also we put $\phi = \epsilon\ \psi$. Then (13.329) and (13.330) are written as

$$[\psi_t]_0 + gh\eta = -\frac{\epsilon}{2}\left[[\psi_x]_0^2 + [\psi_z]_0^2 + 2h\eta[\psi_{zt}]_0\right] + O(\epsilon^2), \tag{13.344}$$

$$h\eta_t - [\psi_z]_0 = \epsilon h\left[[\psi_{zz}]_0\eta - [\psi_x]_0\eta_x\right] + O(\epsilon^2). \tag{13.345}$$

(13.332) and (13.331) take the forms

$$\psi_{xx} + \psi_{zz} = 0, \tag{13.346}$$

$$\psi_z(x, 0, t) = 0. \tag{13.347}$$

The above four equations are the water wave equations in which the boundary conditions are incorporated. These equations need to be further expanded by introducing stretched length and time coordinates ξ and τ:

$$\xi = \sqrt{\epsilon}(x - c_0 t), \quad \tau = \epsilon^{3/2} t. \tag{13.348}$$

The Laplace equation (13.346) should be expanded in the powers of z. As compared to the similar perturbation analysis of nonlinear ion acoustic wave (see Appendix C) the water wave equations are much more involved. One can obtain the KdV equation only after lengthy algebra.

References: S. Leibovich and A. R. Seebass, *Nonlinear waves*, Cornell University Press (1974) *Examples of dissipative and dispersive systems leading to the Burgers and the Korteweg-de Vries equations*; G. B. Whitham, *Linear and nonlinear waves*, Wiley (1974) Chapter 13.

13.6. Nonlinear surface waves and the operator method

In § 13.3.1, we investigated a linear water wave resulting from satisfaction of two boundary conditions, at the bottom of the water and on the free surface, with the free surface equation linearized. In § 13.5, two nonlinear terms were included in the water wave analysis: $\mathbf{v}\cdot\nabla\mathbf{v}$ term and the nonlinear motion of the free surface (13.320). We can extract the KdV equation from the nonlinear water wave equations by a suitable perturbational analysis. On a plasma-vacuum interface, a KdV soliton can propagate or modulation of surface wave can occur if the nonlinear terms are included. The usual mathematical procedure is to derive a KdV equation or a nonlinear Schrödinger equation from the nonlinear surface wave equations. In the nonlinear analysis of plasma surface waves, the operator method is useful, as will be shown below.

Usually a wave function $\phi(x, z, t)$ which obeys the wave equation and in terms of which the boundary conditions are prescribed is sought in the form

$$\phi(x, z, t) = F(x)G(z, t) \tag{13.349}$$

where we note that $x = 0$ is the quiescent interface. The y coordinate is ignorable. The function G takes the form $G(z, t) = G(z - \lambda t)$, a progressive wave form both in KdV equation and nonlinear Schrödinger equation. If we have $\nabla^2\phi = 0$ (say, in vacuum), the solution is usually sought in the form involving a power series expansion

$$\phi(x, z, t) = \sum_{n=0}^{\infty} x^n G_n(z, t). \tag{13.350}$$

The power n implies the hierarchy of the perturbation series. Since this is a perturbational series, the function G in (13.349) cannot be determined independently from the power x^n. This is an algebraic cumbersomeness underlying in the above expression. If (13.350) is substituted into $\nabla^2\phi = 0$, we have the recurrence relation in the form

$$\frac{\partial^2 G_n}{\partial z^n} + (n+2)(n+1)G_{n+2} = 0. \tag{13.351}$$

The operator method exempts one from the labor of working with the type of recurrence relation in the analysis. Particularly, the operator method is advantageous in carrying out the multiple scale perturbational series expansion; it saves labor in dealing with term by term algebraic manipulations. The essential ingredient of the method is to treat the operators $\frac{\partial}{\partial t}$ and $\frac{\partial}{\partial z}$ as algebraic quantities and to expand the operators by using the derivative expansion. For example, we can assume $\phi(x,z,t) = X(x)Z(z,t)$ in the Laplace equation, and we have

$$\frac{1}{X}\frac{d^2X}{dx^2} + \frac{1}{Z}\frac{d^2Z}{dz^2} = 0. \tag{13.352}$$

Let us put

$$\frac{1}{X}\frac{d^2X}{dx^2} = p^2 = -\frac{1}{Z}\frac{d^2Z}{dz^2}. \tag{13.353}$$

Hence, $p = -i\,\partial/\partial z$ automatically solves the second equation, and $X = e^{px}$. Therefore the Laplace equation is solved in operatorial form

$$\phi(x,z,t) = e^{-ix\frac{\partial}{\partial z}} Z(z,t) = \sum_{n=0}^{\infty} \frac{1}{n!}\left(-ix\,\frac{\partial}{\partial z}\right)^n Z(z,t). \tag{13.354}$$

In a multi-time scale analysis, we expand as

$$\frac{\partial}{\partial t} = \frac{\partial}{\partial t_0} + \epsilon\,\frac{\partial}{\partial t_1} + \cdots \tag{13.355}$$

where the numeric subscripts denote the order. The inverse operator $(\partial/\partial t)^{-1}$ is calculated as

$$\left(\frac{\partial}{\partial t}\right)^{-1} = \left(\frac{\partial}{\partial t_0}\right)^{-1}\left(1+\epsilon\,\frac{\partial/\partial t_1}{\partial/\partial t_0}+\cdots\right)^{-1} = \frac{i}{\omega}\left(1-\epsilon\,\frac{i}{\omega}\,\frac{\partial}{\partial t_1}+\cdots\right) \tag{13.356}$$

where we put $\partial/\partial t_0 = -i\,\omega$. Therefore, the inverse operator makes perfect sense.

As an example, the nonlinear TM mode wave in a cold plasma can be investigated by using (13.6)–(13.10) with $\beta = 0$. Omitting the details, we need to engage the operator

$$A\left(\frac{\partial}{\partial t}, \frac{\partial}{\partial z}\right) = \left[\frac{1}{c^2}\left(\frac{\partial^2}{\partial t^2} + \omega_p^2\right) - \frac{\partial^2}{\partial z^2}\right]^{1/2} \tag{13.357}$$

which yields the electric field component by the operator equation

$$E_z = e^{-xA}\, F(z,t). \tag{13.358}$$

In the perturbation series expansion, the exponential operatorial function must also be expanded, which goes as follows:

$$e^{-xA} = e^{-x(A_0+\epsilon A_1+\cdots)} = e^{-xA_0}(1 - \epsilon\; x\; A_1 + \cdots) \tag{13.359}$$

where

$$A_0 = \left[\frac{1}{c^2}\left(\partial_{t_0}^2 + \omega_p^2\right) - \partial_{z_0}^2\right]^{1/2}, \tag{13.360}$$

$$A_1 = \frac{\partial A_0}{\partial \partial_{t_0}}\frac{\partial}{\partial t_1} + \frac{\partial A_0}{\partial \partial_{z_0}}\frac{\partial}{\partial z_1}, \tag{13.361}$$

which are the two terms of the Taylor expansion of A, and $\partial_{t_0} = \partial/\partial t_0 = -i\omega$, $\partial_{z_0} = \partial/\partial z_0 = i\; k$. The power function, fractional function, and the exponential function of A are perfectly well defined by regarding the $\partial/\partial t$, $\partial/\partial z$ as algebraic quantities.

This operator method has been used to derive the nonlinear Schrödinger equation of the TM mode surface wave in the Reference [H. J. Lee and Y. K. Lim, *Modulation of a TM mode surface wave*, Plasma Phys. Control. Fusion (1998) p. 1619–1633] and Kadomtsev–Petviashvili equation of ion acoustic surface wave by stretched (strained) coordinate perturbation analysis in Reference [H. J. Lee, *Kadomtsev–Petviashvili equation of an ion acoustic surface wave*, J. of Korean Phys. Soc. **38** (2001) p. 794–797]. Also, KdV equation of ion acoustic surface wave was derived applying the operator method and stretched coordinate perturbation analysis. See Reference: H. S. Hong and H. J. Lee, *Korteweg-de Vries equation of ion acoustic surface waves*, Phys. Plasmas, **6** (1999) pp. 3422–3424.

General References for Chapter 13

Gradov, O. M. and Stenflo, L. *J. Plasma Phys.* **65** (2001) p. 73.
Kaw, P. K. and McBride, J. B. Surface waves on a plasma, *Phys. Fluids* **13** (1970) pp. 1784–1793.

Landau, L. D. and Lifshitz, E. M. *Fluid Mechanics.* Pergamon Oxford, (1959).
McBride, J. B. and Kaw, P. K. Low frequency surface waves on a warm plasma, *Phys. Lett.* **33A** (1970) pp. 72–73.
Stenflo, L. and Gradov, O. M. *Phys. Rev. A* **44** (1991) p. 5320.
Stenflo, L. and Gradov, O. M. *Phys. Plasmas* **3** (1996) p. 2467.
Stenflo, L. *Phys. Scr. T* **63** (1996) p. 59.
Shivarova, A. and Zhelyazkov, I. *Plasma Physics* **20** (1978) p. 1049.

Chapter 14

Kinetic theory of surface wave

In plasma physics, the results of wave analysis obtained from the fluid equation and the kinetic equation overlap in sufficient proximity if proper approximation is taken in the latter. This is also the case for the surface waves. The kinetic analysis of surface wave requires more ingenuity to solve the Vlasov equation and would be interesting in its own right and also reveals details that are screened by somewhat crude description by fluid equations. An important element for the kinetic wave analysis for a bounded plasma is the boundary condition that should be applied to the distribution function f on the boundary. There are two types of boundary conditions; the diffusive and the specular reflective boundary conditions. If we use the *specular reflection boundary condition* for f on the interface, we have close correspondence with the fluid equation result. The specular reflection assumes perfect reflection of the particles off the boundary. The more general boundary condition includes diffuse scattering in which the particles suffer complete loss of memory about their previous velocities. It is a Markovian process. In this book, we only use the specular reflection condition.

For the purpose of enlightenment, we write the general form of the kinetic boundary condition on the interface $x = 0$ [see B. Buti, *Nonlinear surface phenomena in plasmas*, Advances in space plasma physics, World Scientific Publishing Singapore (1985)]

$$v_x f(\mathbf{r}, v_x, v_y, v_z, t)\big|_{x=0} = \int_{v_x<0} d^3v' P(\mathbf{v}, \mathbf{v}') v'_x f(\mathbf{r}, v'_x, v_y, v_z, t)\big|_{x=0}$$

where

$$P(\mathbf{v}, \mathbf{v}') = \chi\delta(v'_x + v_x)\delta(v'_y - v_y)\delta(v'_z - v_z) + (1 - \chi)P_d(\mathbf{v})$$

is the probability of a particle with velocity $\mathbf{v}'$ having velocity $\mathbf{v}$ after scattering. The first (second) term represents perfect reflection (diffuse scattering). The parameter χ gives the relative fraction of perfect and diffuse scattering. In this book, we take $\chi = 1$.

14.1. Semi-infinite plasma

We begin with an isotropic plasma (no magnetic field imbedded, no streaming of plasma) to derive the surface wave dispersion relation kinetically. The kinetic dispersion relation will be shown to yield the fluid dispersion relation obtained earlier upon taking appropriate cold plasma approximation. We consider a semi-bounded plasma which occupies the region ($x > 0$) bounded by a vacuum ($x < 0$). We use the linearized Vlasov equation

$$\frac{\partial}{\partial t} f_\alpha(\mathbf{r}, \mathbf{v}, t) + \mathbf{v} \cdot \frac{\partial f_\alpha}{\partial \mathbf{r}} + \frac{e_\alpha}{m_\alpha}\left(\mathbf{E} + \frac{1}{c}\mathbf{v} \times \mathbf{B}\right) \cdot \frac{\partial f_{\alpha 0}}{\partial \mathbf{v}} = 0 \tag{14.1}$$

where the zero order distribution function $f_{\alpha 0}(v)$ is isotropic. Note that the last term vanishes for isotropic f_0 since $\frac{\partial f_{\alpha 0}}{\partial \mathbf{v}} \parallel \mathbf{v}$. The plasma current

$$\mathbf{J} = \sum e_\alpha \int \mathbf{v} f_\alpha d^3 v \tag{14.2}$$

gives rise to electric and magnetic fields

$$\nabla \times \mathbf{E} = -\frac{1}{c}\frac{\partial \mathbf{B}}{\partial t}, \tag{14.3}$$

$$\nabla \times \mathbf{B} = \frac{1}{c}\frac{\partial \mathbf{E}}{\partial t} + \frac{4\pi}{c}\mathbf{J}, \tag{14.4}$$

$$\nabla \times \nabla \times \mathbf{E} + \frac{1}{c^2}\frac{\partial^2 \mathbf{E}}{\partial t} + \frac{4\pi}{c^2}\frac{\partial \mathbf{J}}{\partial t} = 0. \tag{14.5}$$

Solving (14.1) in a finite plasma in general (semi-bounded here) requires a boundary condition on f at the boundary (at the interface $x = 0$ here) which we take to be a sharp boundary. In reality the zero order distribution f_0 will have a finite spatial gradient in the vicinity of $x = 0$, which renders obtaining analytic solutions of surface wave difficult. But we assume a sharp boundary, which may be justified if the characteristic lengths of the plasma are much greater than the scale length of the inhomogeneity around the interface. We assume the *specular reflection condition* in which the particles undergo a mirror reflection such that

$$f_\alpha(v_x, v_y, v_z, t, x = 0, y, z) = f_\alpha(-v_x, v_y, v_z, t, x = 0, y, z). \tag{14.6}$$

On a diffuse boundary, the incident particles suffer from complete loss of past memory and scatter. In specular reflections, the particles elastically bounce off the interface with a velocity determined by the colliding velocity. Since we have, under the condition of (14.6),

$$\int_{-\infty}^{\infty} v_x f_\alpha(v_x) dv_x = 0 \quad at\ x = 0$$

the normal component of the *fluid velocity* at the interface is zero. The distribution function f, which satisfies the following reflectional symmetry

$$f(v_x, v_y, v_z, t, x, y, z) = f(-v_x, v_y, v_z, t, -x, y, z) \tag{14.7}$$

satisfies (14.6). Then both (14.1) and (14.7) can be satisfied by making (14.1) invariant under the reflection $(x \to -x, v_x \to -v_x)$. This is achieved by extending **E** and **B** (defined in the region $x > 0$ in (14.1)) appropriately into the region $x < 0$. How? To facilitate inspection, we fully write (14.1), exhibiting all the terms in components:

$$\begin{aligned}
&\frac{\partial f}{\partial t} + v_x \frac{\partial f}{\partial x} + v_y \frac{\partial f}{\partial y} + v_z \frac{\partial f}{\partial z} + \frac{e}{m}\left[E_x \frac{\partial f_0}{\partial v_x} + E_y \frac{\partial f_0}{\partial v_y} + E_z \frac{\partial f_0}{\partial v_z}\right] \\
&\quad + \frac{e}{mc}\left[(v_y B_z - B_y v_z)\frac{\partial f_0}{\partial v_x} + (v_z B_x - v_x B_z)\frac{\partial f_0}{\partial v_y}\right. \\
&\quad \left. + (v_x B_y - v_y B_x)\frac{\partial f_0}{\partial v_z}\right] = 0. \qquad (14.1)
\end{aligned}$$

Now we see that the following prescription makes (14.1) invariant under the reflection $(x \to -x,\ v_x \to -v_x)$:

$$E_x(-x) = -E_x(x), \quad E_y(-x) = E_y(x), \quad E_z(-x) = E_z(x), \tag{14.8}$$

$$B_x(-x) = B_x(x), \quad B_y(-x) = -B_y(x), \quad B_z(-x) = -B_z(x). \tag{14.9}$$

The parallel components of **E** (with respect to the interface) should be continued evenly; the perpendicular component E_x oddly; the parallel components of **B** should be continued oddly; the perpendicular component B_x evenly. We now solve (14.1) by Fourier transform with the extended fields. See Fig. 14.1.

When we Fourier-transform $\mathbf{E}(\mathbf{r}, t)$ and $\mathbf{B}(\mathbf{r}, t)$, we must be careful in dealing with the discontinuities of the field components or their derivatives at $x = 0$. These discontinuities arise due to the artificial extensions as

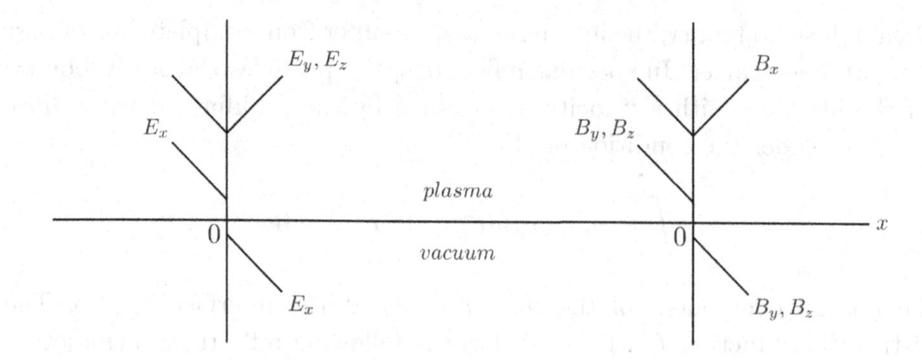

Fig. 14.1. Extension of electric and magnetic fields in a semi-infinite plasma.

prescribed in (14.8) and (14.9). Fourier transform of (14.1) gives

$$f_\alpha(\mathbf{k}, \mathbf{v}, \omega) = -i\frac{e_\alpha}{m_\alpha}\frac{\mathbf{E}(\mathbf{k}, \omega)}{\omega - \mathbf{k}\cdot\mathbf{v}} \cdot \frac{\partial f_{\alpha 0}}{\partial \mathbf{v}}. \tag{14.10}$$

Fourier transform of (14.5) yields

$$\mathbf{k} \times (\mathbf{k} \times \mathbf{E}) + \frac{\omega^2}{c^2}\mathbf{E} + \frac{4\pi i\omega}{c^2}\mathbf{J} = \hat{\mathbf{z}}M(k_z, \omega) \tag{14.11}$$

where

$$M = 2ik_z E_x(x = 0^-, k_z, \omega) - 2\left[\frac{\partial}{\partial x}E_z(x, k_z, \omega)\right]_{x=0^-} \tag{14.12}$$

is the boundary term which is absent in an infinite plasma.

Exercise. Derive (14.11). Suppress the y-coordinate, which is ignorable when the wave propagates along the z-direction, $\sim e^{ik_z z - i\omega t}$. Also put $E_y = 0$ which decouples from the TM mode set (E_x, E_z, B_y).

Now the plasma fields will be determined in terms of $\hat{\mathbf{z}}M(k_z, \omega)$ in (14.11), which is the relevant boundary quantity. Introducing the conductivity tensor per $J_i = \sigma_{ij}E_j$, the dielectric tensor $\varepsilon_{ij} = \delta_{ij} + \frac{4\pi i}{\omega}\sigma_{ij}$ is obtained as (see Chapter 4)

$$\varepsilon_{ij} = \delta_{ij} + \Sigma\frac{\omega_{p\alpha}^2}{\omega}\int \frac{v_i}{\omega - \mathbf{k}\cdot\mathbf{v}}\frac{\partial f_{\alpha 0}}{\partial v_j}\, d^3v. \tag{14.13}$$

Further manipulation of the integral in (14.13), including integration by parts, allows us to write it in the form

$$\varepsilon_{ij} = \varepsilon_L\frac{k_i k_j}{k^2} + \varepsilon_T\left(\delta_{ij} - \frac{k_i k_j}{k^2}\right) \tag{14.14}$$

where

$$\varepsilon_L = 1 + \Sigma \frac{\omega_{p\alpha}^2}{k^2} \int \frac{k_j}{\omega - \mathbf{k}\cdot\mathbf{v}} \frac{\partial f_{\alpha 0}}{\partial v_j} d^3v = 1 + \Sigma \frac{\omega_p^2}{\omega^2} 2\zeta^2(1 + \zeta Z(\zeta)),$$

$$\varepsilon_T = 1 - \Sigma \frac{\omega_{p\alpha}^2}{\omega} \int \frac{f_{\alpha 0}}{\omega - \mathbf{k}\cdot\mathbf{v}} d^3v = 1 + \Sigma \frac{\omega_p^2}{\omega^2} \zeta Z(\zeta),$$

$$Z(\zeta) = \frac{1}{\sqrt{\pi}} \int_{-\infty}^{\infty} \frac{e^{-q^2}}{q - \zeta} dq \quad \left(\zeta = \frac{\omega/k}{\sqrt{2T/m}}\right).$$

The dielectric tensor ε_{ij} in an isotropic medium can always be written as a linear combination of two basic tensors δ_{ij} and $k_i k_j/k^2$. Then (14.11) is easily inverted to give

$$E_x(\mathbf{k}, \omega) = -\frac{M}{\Delta}\left(\varepsilon_{xz} + \frac{c^2}{\omega^2} k_x k_z\right), \tag{14.15}$$

$$E_z(\mathbf{k}, \omega) = \frac{M}{\Delta}\left(\varepsilon_{xx} - \frac{c^2}{\omega^2} k_z^2\right), \tag{14.16}$$

$$\Delta = \varepsilon_L \left(\varepsilon_T - \frac{c^2 k^2}{\omega^2}\right); \quad k^2 = k_x^2 + k_z^2. \tag{14.17}$$

Using the above equations in (14.3) yields

$$B_y(\mathbf{k}, \omega) = -\frac{M}{\Delta} \frac{c}{\omega} k_x \varepsilon_L. \tag{14.18}$$

To reinstate the x-dependance, the Fourier inversion integral $\int dk_x$ must be performed on (14.15)–(14.18). This can be done by using the contour integral and evaluating the residues at the poles. To ensure, say, $B_y \to 0$ as $x \to +\infty$, the inversion integral should be carried out by encircling the upper-half k_x-plane.

The vacuum solutions satisfy $\nabla \cdot \mathbf{E} = 0$ and $\nabla^2 \mathbf{E} + (\omega^2/c^2)\mathbf{E} = 0$, and we obtain

$$E_z(x, k_z, \omega) = i\,\lambda F e^{\lambda x} \qquad F:\ const. \quad \lambda = \sqrt{k_z^2 - \omega^2/c^2}, \tag{14.19}$$

$$E_x = k_z F e^{\lambda x}, \qquad B_y(x, k_z, \omega) = \frac{\omega}{c} F e^{\lambda x}, \tag{14.20}$$

where we omitted the phasor $exp(ik_z z - i\omega t)$. The plasma solutions and the vacuum solutions should be matched at the interface $x = 0$ according to the *electromagnetic boundary conditions*: the continuity of E_z and B_y. We have two undetermined constants, F and M (k_x-independent). Enforcing the two boundary conditions, we have

$$\int_{-\infty}^{\infty} \frac{dk_x}{\Delta} \left[\varepsilon_{xx} - \frac{c^2}{\omega^2} k_z^2 + i\lambda \frac{c^2}{\omega^2} \varepsilon_L k_x \right] = 0. \tag{14.21}$$

Using (14.17) and the relation $\varepsilon_{xx} - c^2 k_z^2/\omega^2 = [k_x^2 \varepsilon_L + k_z^2(\varepsilon_T - c^2k^2/\omega^2)]/k^2$ in the above equation gives the dispersion relation in the form

$$\int_{-\infty}^{\infty} \frac{dk_x}{k^2} \left(\frac{k_x^2}{\varepsilon_T - c^2k^2/\omega^2} + \frac{k_z^2}{\varepsilon_L} \right) + i\lambda \frac{c^2}{\omega^2} \int_{-\infty}^{\infty} \frac{dk_x \, k_x}{\varepsilon_T - c^2k^2/\omega^2} = 0. \tag{14.22}$$

This is the general dispersion relation of electromagnetic surface wave propagating in the semi-bounded plasma.

- **Alternative derivation of (14.22).** The TM mode wave equations are

$$\frac{\partial E_x}{\partial z} - \frac{\partial E_z}{\partial x} = \frac{i\omega}{c} B_y, \tag{14.23}$$

$$-i\frac{\omega}{c} E_x + \frac{4\pi}{c} J_x = -\frac{\partial B_y}{\partial z}, \tag{14.24}$$

$$-\frac{i\omega}{c} E_z + \frac{4\pi}{c} J_z = \frac{\partial B_y}{\partial x}, \tag{14.25}$$

whose Fourier-transformed versions are

$$k_x E_z - k_z E_x + \frac{\omega}{c} B_y = 0, \tag{14.26}$$

$$ik_z B_y - i\frac{\omega}{c} E_x + \frac{4\pi}{c} J_x = 0, \tag{14.27}$$

$$ik_x B_y + i\frac{\omega}{c} E_z - \frac{4\pi}{c} J_z = B_y(0^+) - B_y(0^-) \equiv A. \tag{14.28}$$

The last term in (14.28) comes from the Fourier transform of $\partial B_y/\partial x$ as $B_y(x)$ is oddly continued at $x = 0$. The relation $J_i = \sigma_{ij} E_j = \frac{\omega}{4\pi i}(\varepsilon_{ij} - \delta_{ij}) E_j$ gives

$$J_x = \frac{\omega}{4\pi i}(\varepsilon_{xx} E_x + \varepsilon_{xz} E_z - E_x),$$

$$J_z = \frac{\omega}{4\pi i}(\varepsilon_{zx} E_x + \varepsilon_{zz} E_z - E_z).$$

Using these equations, (14.27) and (14.28) become

$$-\frac{c}{\omega} k_z B_y + \varepsilon_{xx} E_x + \varepsilon_{xz} E_z = 0, \tag{14.29}$$

$$\frac{c}{\omega} k_x B_y + \varepsilon_{zx} E_x + \varepsilon_{zz} E_z = -i\frac{c}{\omega} A. \tag{14.30}$$

From (14.26), (14.29), and (14.30), we can determine:

$$E_z = -i\frac{c}{\omega}\frac{A}{\Delta}\left(\varepsilon_{xx} - \frac{c^2}{\omega^2}k_z^2\right), \tag{14.31}$$

$$B_y = i\frac{c^2}{\omega^2}\frac{A}{\Delta}(k_x\varepsilon_{xx} + k_z\varepsilon_{xz}), \tag{14.32}$$

$$\Delta = \varepsilon_{xx}\varepsilon_{zz} - \varepsilon_{xz}\varepsilon_{zx} - \frac{c^2}{\omega^2}\left[\varepsilon_{xx}k_x^2 + 2k_xk_z\varepsilon_{zx} + k_z^2\varepsilon_{zz}\right]. \tag{14.33}$$

Exercise. Show that Δ in (14.33) is the same as the Δ in (14.17) (use (14.14)). Show $k_z\varepsilon_{xz} + k_x\varepsilon_{xx} = \varepsilon_L k_x$.

According to the Exercise above, (14.31)–(14.33) are essentially identical with (14.16)–(14.18), respectively. So we end up with the dispersion relation (14.22). In this derivation, the A–term in (14.28) is derived from the even-odd continuation of the relevant field components across the interface in order to make the Vlasov equation reflectionally invariant.

• **Electrostatic dispersion relation.** The electrostatic surface wave dispersion relation in a semi-infinite plasma can be derived by taking the limit $c \to \infty$ in (14.22) which is the electromagnetic surface wave dispersion relation in a semi-infinite plasma. Note λ in (14.22) becomes k_z in the $c \to \infty$ limit. Thus (14.22) yields

$$\int dk_x \frac{k_z^2}{k^2\varepsilon_L} - ik_z\int dk_x\frac{k_x}{k^2} = 0.$$

The second term can be contour-integrated, and the above equation reduces to

$$\int_{-\infty}^{\infty} dk_x \frac{k_z}{k^2\varepsilon_L} + \pi = 0. \tag{14.34}$$

Example. Independent derivation of the electrostatic dispersion relation; not via $c \to \infty$. It would be instructive to derive the above result independently by using the electrostatic equations from the beginning. In the following, we develop the electrostatic theory. We start with

$$\frac{\partial f}{\partial t} + v_x\frac{\partial f}{\partial x} + v_z\frac{\partial f}{\partial z} + \frac{e}{m}\left(E_x\frac{\partial f_0}{\partial v_x} + E_z\frac{\partial f_0}{\partial v_z}\right) = 0. \quad (A)$$

The specular reflection condition $f(v_x, x = 0) = f(-v_x, x = 0)$ can be satisfied with the extension $E_x(-x) = -E(x)$, $E_z(-x) = E_z(x)$. Then, solving (A) in the whole domain $-\infty < x < \infty$ with the extended field gives

$$f_\alpha(\mathbf{k}, \mathbf{v}, \omega) = -i\frac{e_\alpha}{m_\alpha}\frac{E_j\frac{\partial f_{\alpha 0}}{\partial v_j}}{\omega - \mathbf{k}\cdot\mathbf{v}}. \quad (B)$$

Fourier-transforming the Poisson equation yields

$$ik_x E_x + ik_z E_z = 4\pi\Sigma\ e_\alpha \int f_\alpha d^3v + N; \quad N = 2E_x(k_z, \omega, x = 0^+). \quad (C)$$

Note: In (C), $N = 2E_x(k_z, \omega, x = 0^+) = -2E_x(k_z, \omega, x = 0^-)$. The latter has nothing to do with the vacuum field value. Using (C) and the equation

$$k_z E_x(k_x, k_z) = k_x E_z(k_x, k_z); \quad \nabla\times\mathbf{E} = 0, \quad (D)$$

we obtain

$$E_x = -i\ \frac{k_x N}{\varepsilon_L k^2}, \quad E_z = -i\ \frac{k_z N}{\varepsilon_L k^2} \quad (E)$$

$$where \quad \varepsilon_L = 1 + \Sigma\frac{\omega_{p\alpha}^2}{k^2}\int d^3v\ \frac{\mathbf{k}\cdot\frac{\partial f_{\alpha 0}}{\partial \mathbf{v}}}{\omega - \mathbf{k}\cdot\mathbf{v}}.$$

We need the electric displacement $D_x(k_x, k_z, \omega) = E_x(k_x.k_z, \omega) + \frac{4\pi i}{\omega}J_x$ where

$$\frac{4\pi i}{\omega}J_x = \frac{4\pi i}{\omega}\Sigma e_\alpha \int v_x f_\alpha d^3v = \Sigma\frac{\omega_{p\alpha}^2}{\omega}\int v_x\ \frac{E_j\frac{\partial f_{\alpha 0}}{\partial v_j}}{\omega - \mathbf{k}\cdot\mathbf{v}}.$$

The RHS of the above equation turns out to be $(\varepsilon_L - 1)E_x$, so we have $D_x = \varepsilon_L E_x$. This statement can be most easily proven by assuming f_0 a Maxwellian. Use $\frac{\partial f_{\alpha 0}}{\partial v_j} = -\frac{m_\alpha}{T_\alpha}v_j f_{\alpha 0}$ and $\mathbf{E}\cdot\mathbf{v} = \frac{E_x}{k_x}\mathbf{k}\cdot\mathbf{v}$ to write

$$RHS = -E_x\Sigma\frac{\omega_{p\alpha}^2}{\omega}\frac{T_\alpha}{m_\alpha}\int\frac{v_x}{k_x}\ \frac{\mathbf{k}\cdot\mathbf{v}}{\omega - \mathbf{k}\cdot\mathbf{v}}f_{\alpha 0}\ d^3v.$$

Put $\frac{\mathbf{k}\cdot\mathbf{v}}{\omega - \mathbf{k}\cdot\mathbf{v}} = -1 + \frac{\omega}{\omega - \mathbf{k}\cdot\mathbf{v}}$. (-1)–term vanishes upon integration, and we have

$$RHS = -E_x\Sigma\frac{\omega_{p\alpha}^2}{k_x}\frac{T_\alpha}{m_\alpha}\int\frac{v_x f_{\alpha 0}}{\omega - \mathbf{k}\cdot\mathbf{v}}d^3v = E_x\Sigma\frac{\omega_{p\alpha}^2}{k_x}\int\frac{\frac{\partial f_{\alpha 0}}{\partial v_x}}{\omega - \mathbf{k}\cdot\mathbf{v}}d^3v$$

$$= -E_x\Sigma\ \omega_{p\alpha}^2\int\frac{f_{\alpha 0}\ d^3v}{(\omega - \mathbf{k}\cdot\mathbf{v})^2} = E_x\Sigma\frac{\omega_{p\alpha}^2}{k^2}\int d^3v\ \frac{\mathbf{k}\cdot\frac{\partial f_{\alpha 0}}{\partial \mathbf{v}}}{\omega - \mathbf{k}\cdot\mathbf{v}}.$$

Therefore, we have from (E)

$$D_x(k_x, k_z, \omega) = -i\frac{k_x}{k^2}N.$$

The vacuum solutions are:

$$E_x(k_z,\omega,x) = -i\ G(k_z,\omega)e^{k_x x}, \quad E_z(k_z,\omega,x) = G(k_z,\omega)e^{k_x x}$$

where G should satisfy $\nabla\cdot\mathbf{E}=0$ and $\nabla\times\mathbf{E}=0$, and we find $G\sim e^{ik_x z-i\omega t}$. The two boundary conditions on the interface $x=0$ are the continuity of E_z and D_x. So let us calculate:

$$E_z(k_z,x=0^+) = \lim_{x\to 0}\int_{-\infty}^{\infty}\frac{dk_x}{2\pi}E_z(k_x,k_z)e^{ik_x x} = -\frac{iN}{2\pi}\int_{-\infty}^{\infty}\frac{k_z dk_x}{\varepsilon_L k^2} = G,$$

$$D_x(k_z,x=0^+) = -\frac{iN}{2\pi}\int_{-\infty}^{\infty}\frac{k_x dk_x}{k^2} = \frac{N}{2} = -iG.$$

The above two equations give the dispersion relation in the form (14.34).

Exercise. In the vacuum, the potential ϕ ($\mathbf{E}=-\nabla\phi$) satisfies $\partial^2\phi/\partial x^2 + \partial^2\phi/\partial z^2 = 0$, which is solved by either $\phi = \tilde{\phi}exp(ik_x z + k_x x)$ or $\phi = \tilde{\phi}exp(ik_z z + k_z x)$. The surface wave dispersion relation is obtained by using any of these two.

One-dimensional electric field. In the preceding analysis, we obtained a two-dimensional electric field in a semi-bounded plasma that satisfies the legitimate interfacial boundary conditions. One-dimensional non-propagating electric field that satisfies the boundary condition can be deduced from the above result by taking $k_z\to 0$. Suppose $\mathbf{E}=\hat{\mathbf{x}}E_x$ in the above semi-bounded plasma. Then equation (E) gives

$$E_x = -i\frac{N}{\varepsilon_L k_x}, \quad E_z = 0.$$

The vacuum solution can be written as $E_x = E_0 exp(k_x x - i\omega t)$. We need only one boundary condition $[D_x]=0$ at $x=0$. Thus,

$$D_x(x=0^+) = -\frac{iN}{2\pi}\int_{-\infty}^{\infty}\frac{dk_x}{k_x}e^{ik_x 0^+} = \frac{N}{2}. \qquad (cf.\ Appendix)$$

Therefore $E_0 e^{-i\omega t} = N/2$, $N = 2E_0 e^{-i\omega t}$. The plasma solution is obtained in the form

$$E_x(x,\omega) = -\frac{iE_0}{\pi}\ e^{-i\omega t}\int_{-\infty}^{\infty}\frac{dk_x}{\varepsilon_L k_x}e^{ik_x x}. \qquad (cf.\ (14.284),\ later)$$

Reference for kinetic treatment of plasma surface wave; H. C. Barr and T. J. M. Boyd, *Surface waves in hot plasmas*, J. Phys. A: Gen. Phys. **5**, p. 1108–1118 (1972).

14.2. Plasma slab

A slab plasma requires more complex analysis as compared to a semi-bounded plasma because a slab has two interfaces on which the boundary conditions must be enforced. The dispersion relation of a surface wave propagating in a slab has a two-mode structure: *symmetric* and *anti-symmetric* mode. Fluid treatment of slab surface waves is thoroughly reviewed in Reference: O. M. Gradov and L. Stenflo, *Linear theory of a cold bounded plasma*, Phys. Reports **94**, p. 111–137 (1983). Kinetic dispersion relations for the surface waves in a slab have been derived in Reference: H. J. Lee and Y. K. Lim, *Kinetic theory of surface waves in a plasma slab*, J. Korean Phys. Soc. **50**, p. 1056–1061 (2007). In the following we closely follow Lee and Lim (2007). In a kinetic treatment of a slab plasma, the specular reflection conditions must be satisfied on two interfaces *simultaneously*, and the extended field components exhibit infinite repetitions. See Fig. 14.2. The field components are obtained by summing the infinite series but the agreement between the fluid and kinetic results can be demonstrated.

Let us assume that a plasma occupies the region ($0 < x < L$, $-\infty < y$, $z < \infty$) bounded by a vacuum ($x < 0$, $x > L$). The slab has two boundaries, $x = 0$ and $x = L$, on which the kinematic boundary condition, i.e., the specular reflection condition, should be satisfied:

$$f_\alpha(v_x, v_y, v_z, t, x = 0, y, z) = f_\alpha(-v_x, v_y, v_z, t, x = 0, y, z), \qquad (14.35)$$

$$f_\alpha(v_x, v_y, v_z, t, x = L, y, z) = f_\alpha(-v_x, v_y, v_z, t, x = L, y, z). \qquad (14.36)$$

The fields in the Vlasov equation (14.1) are defined in the region $0 < x < L$. We extend these fields into the regions $x < 0$ and $x > L$ in such a way that (14.1) is invariant under the simultaneous reflections ($x \to -x$, $v_x \to -v_x$) and ($x \to 2L - x$, $v_x \to -v_x$). Then the solution of (14.1), $f(x, v_x)$, solved in the infinite region $-\infty < x < \infty$ is formally identical to $f(-x, -v_x)$ and $f(2L - x, -v_x)$, and the specular reflection condition (14.35) and (14.36) are automatically satisfied. The recipes that make (14.1) invariant under such reflections are:

$$E_x(-x) = -E_x(x), \quad E_x(2L - x) = -E_x(x), \qquad (14.37)$$

$$E_z(-x) = E_z(x), \quad E_z(2L - x) = E_z(x), \qquad (14.38)$$

$$B_y(-x) = -B_y(x), \quad B_y(2L - x) = -B_y(x). \qquad (14.39)$$

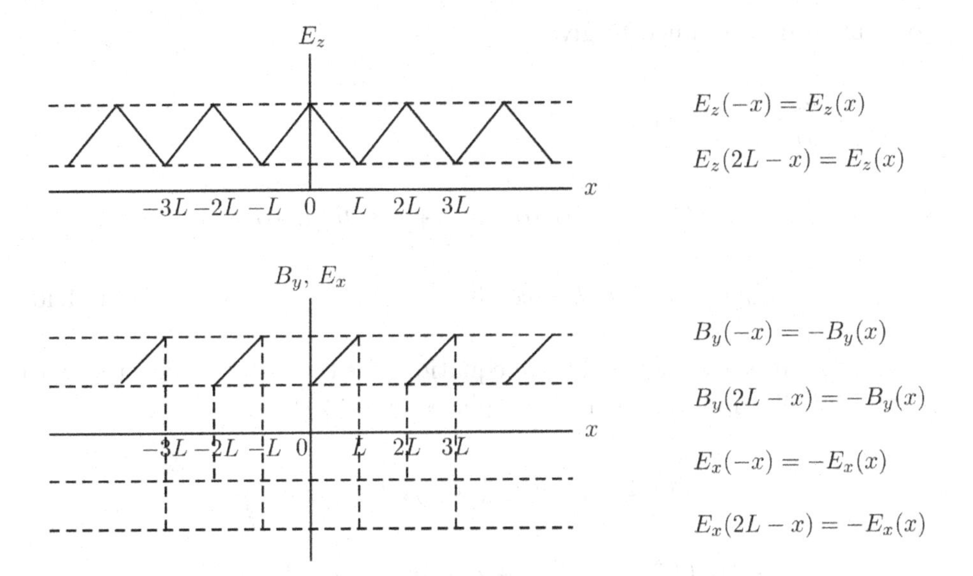

Fig. 14.2. Extension of electric and magnetic fields in a plasma slab ($0 < x < L$).

The profiles of $E_z(x)$ and $B_y(x)$ near $x = 0$ and $x = L$ as defined by the above extensions are plotted in Fig. 14.2.

In Fourier transforming (14.23)–(14.25), note that $\frac{\partial}{\partial z} = ik_z$ simply, but special attention should be paid to $\frac{\partial}{\partial x}$-terms. For the even function $E_z(x)$, the piecewise integration yields

$$\int_{-\infty}^{\infty} \frac{\partial E_z(x)}{\partial x} e^{-ik_x x} dx = ik_x E_z(k_x, k_z, \omega).$$

However, algebraic labor is required for the integration of the odd function $B_y(x)$ because it should be done by piecewise:

$$\begin{aligned}\int_{-\infty}^{\infty} \frac{\partial B_y(x)}{\partial x} e^{-ik_x x} dx &= ik_x B_y(k_x, k_z, \omega) \\ &+ \left(B_y(x)e^{-ik_x x}\right) \left(\Big|_{0^+}^{L^-} + \Big|_{L^+}^{2L^-} + \Big|_{2L^+}^{3L^-} + \cdots \right) \\ &- \left(B_y(x)e^{-ik_x x}\right) \left(\Big|_{0^-}^{-L^+} + \Big|_{-L^-}^{-2L^+} + \Big|_{-2L^-}^{-3L^+} + \cdots \right)\end{aligned}$$

where the superscript $\pm = \pm\epsilon$, with ϵ a positive infinitesimal, e.g. $L^{\pm} = L \pm \epsilon$, $-2L^{\pm} = -2L \pm \epsilon$. Since we have $B_y(0^+) = B_y(\pm 2L^+) = B_y(\pm 4L^+) = \cdots$, and $B_y(0^-) = B_y(\pm 2L^-) = B_y(\pm 4L^-) = \cdots = -B_y(0^+)$, the series in the above integral is summed to give

$$\int_{-\infty}^{\infty} \frac{\partial B_y(x)}{\partial x} e^{-ik_x x} dx$$

$$= ik_x B_y(k_x, k_z, \omega) + 4B_y(0^-) \left(\frac{1}{2} + \cos\, 2k_x L + \cos\, 4k_x L + \cdots \right)$$

$$+ 4B_y(L^-) \left(\cos\, k_x L + \cos\, 3k_x L + \cos\, 5k_x L + \cdots \right). \quad (14.40)$$

Thus, the Fourier transformed wave equations for the slab are identical with (14.26), (14.27), and (14.28) if one replaces in (14.28)

$$A \to 4B_y(0^-) \left(\frac{1}{2} + \cos\, 2k_x L + \cos\, 4k_x L + \cdots \right)$$

$$+ 4B_y(L^-) \left(\cos\, k_x L + \cos\, 3k_x L + \cos\, 5k_x L + \cdots \right).$$

If all the cosine-terms are thrown away, we recover the wave equations for the semi-bounded plasma.

To reinstate the x-dependance, the Fourier integral $\int dk_x e^{ixk_x}(\cdots)$ must be performed on (14.31) and (14.32) where A is replaced by the series as given by (14.40). We calculate:

$$E_z(k_z, \omega, x) = \int_{-\infty}^{\infty} \frac{dk_x}{\Delta} \left(\varepsilon_{xx} - \frac{c^2 k_z^2}{\omega^2} \right) e^{ixk_x}$$

$$\times \left[A \left(\frac{1}{2} + \cos\, 2k_x L + \cos\, 4k_x L + \cdots \right) \right.$$

$$\left. + B \left(\cos\, k_x L + \cos\, 3k_x L + \cos\, 5k_x L + \cdots \right) \right]$$

where A and B are new '*constants*'. Here let us consider the integral

$$I \equiv \int_{-\infty}^{\infty} \frac{dk_x}{\Delta} \left(\varepsilon_{xx} - \frac{c^2 k_z^2}{\omega^2} \right) e^{ixk_x} \cos\, nk_x L$$

$$= \frac{1}{2} \int_{-\infty}^{\infty} \frac{dk_x}{\Delta} \left(\varepsilon_{xx} - \frac{c^2 k_z^2}{\omega^2} \right) [e^{ik_x(x+nL)} + e^{ik_x(x-nL)}]$$

where $n = 1, 2, 3, \ldots$. In the second integral we replace $k_x \to -k_x$ and note that $(\varepsilon_{xx} - \frac{c^2k_z^2}{\omega^2})/\Delta$ is an even function of k_x to obtain

$$I = \int_{-\infty}^{\infty} \frac{dk_x}{\Delta}\left(\varepsilon_{xx} - \frac{c^2k_z^2}{\omega^2}\right) e^{ik_x nL} \cos k_x x.$$

Therefore we obtain

$$E_z(k_z, \omega, x = 0) = \int_{-\infty}^{\infty} \frac{dk_x}{\Delta}\left(\varepsilon_{xx} - \frac{c^2k_z^2}{\omega^2}\right)$$
$$\times\left[A\left(\frac{1}{2} + e^{2ik_xL} + e^{4ik_xL} + \cdots\right) + B(e^{ik_xL} + e^{3ik_xL} + e^{5ik_xL} + \cdots)\right], \tag{14.41}$$

$$\begin{aligned}
&E_z(k_z, \omega, x = L) \\
&= \int_{-\infty}^{\infty} \frac{dk_x}{\Delta}\left(\varepsilon_{xx} - \frac{c^2k_z^2}{\omega^2}\right)\left[A\left(\frac{1}{2}e^{ik_xL} + \cos k_xL\ (e^{2ik_xL} + e^{4ik_xL} + \cdots)\right.\right. \\
&\quad \left. + B\ \cos k_xL(e^{ik_xL} + e^{3ik_xL} + e^{5ik_xL} + \cdots)\right] \\
&= \int_{-\infty}^{\infty} \frac{dk_x}{\Delta}\left(\varepsilon_{xx} - \frac{c^2k_z^2}{\omega^2}\right)\left[A\ (e^{ik_xL} + e^{3ik_xL} + e^{5ik_xL} + \cdots)\right. \\
&\quad \left. + B\left(\frac{1}{2} + e^{2ik_xL} + e^{4ik_xL} + \cdots\right)\right].
\end{aligned} \tag{14.42}$$

Next we consider

$$\begin{aligned}
B_y(k_z, \omega, x) = -\frac{c}{\omega}\int_{-\infty}^{\infty} \frac{dk_x}{\Delta}\varepsilon_L k_x e^{ixk_x}&\left[A\left(\frac{1}{2} + \cos 2k_xL + \cos 4k_xL + \cdots\right)\right. \\
&\left. + B\ (\cos k_xL + \cos 3k_xL + \cos 5k_xL + \cdots)\right].
\end{aligned}$$

The typical term in the above,

$$I \equiv \int_{-\infty}^{\infty} \frac{dk_x}{\Delta}\varepsilon_L k_x e^{ixk_x}\cos nk_xL = \int_{-\infty}^{\infty} \frac{dk_x}{2\Delta}\varepsilon_L k_x(e^{ik_x(x+nL)} + e^{ik_x(x-nL)})$$

with $n = 1, 2, 3, \ldots$. In the second integral, replace $k_x \to -k_x$ and note that ε_L is an even function of k_x to obtain

$$I = i\int_{-\infty}^{\infty} \frac{dk_x}{\Delta}\varepsilon_L k_x e^{ik_x nL}\sin k_x x.$$

Therefore,

$$B_y(k_z, \omega, x = 0) = -\frac{A}{2}\,\frac{c}{\omega}\int_{-\infty}^{\infty}\frac{dk_x}{\Delta}\varepsilon_L k_x \tag{14.43}$$

$$\begin{aligned}&B_y(k_z, \omega, x = L)\\ &\quad = -\frac{c}{\omega}\int_{-\infty}^{\infty}\frac{dk_x}{\Delta}\varepsilon_L k_x\Bigg[A\left(\frac{1}{2}e^{ik_xL} + isin\ k_xL\ (e^{2ik_xL} + e^{4ik_xL} + \cdots)\right)\\ &\qquad + B\ i\ sin\ k_xL(e^{ik_xL} + e^{3ik_xL} + \cdots)\Bigg].\end{aligned}$$

The coefficient of A is summed to zero and the coefficient of B is added to $-\frac{1}{2}$.

$$\textit{Therefore}\ B_y(k_z, \omega, x = L) = \frac{B}{2}\frac{c}{\omega}\int_{-\infty}^{\infty}\frac{dk_x}{\Delta}\varepsilon_L k_x. \tag{14.44}$$

In the vacuum region, the solutions are:

$$\begin{aligned}E_z(x, k_z, \omega) &= -i\lambda F_1 e^{-\lambda x} \quad (x > L),\\ E_z(x, k_z, \omega) &= i\lambda F_2 e^{\lambda x} \quad (x < 0),\end{aligned} \tag{14.45}$$

$$\begin{aligned}B_y(x, k_z, \omega) &= \frac{\omega}{c}F_1 e^{-\lambda x} \quad (x > L),\\ B_y(x, k_z, \omega) &= \frac{\omega}{c}F_2 e^{\lambda x} \quad (x < 0),\end{aligned} \tag{14.46}$$

where $\lambda = \sqrt{k_z^2 - \omega^2/c^2}$. The continuity of E_z and B_y at $x = 0$ and $x = L$ gives four algebraic equations for four constants, A, B, F_1, and F_2 and the solvability condition yields the dispersion relation in the form

$$\begin{aligned}&\int_{-\infty}^{\infty}\frac{dk_x}{\Delta}\left(\varepsilon_{xx} - \frac{c^2k_z^2}{\omega^2}\right)\left[\frac{1}{2} \pm e^{iLk_x} + e^{2ik_xL} \pm e^{3iLk_x} + e^{4ik_xL} \pm \cdots\right]\\ &\qquad\qquad + i\,\frac{\lambda}{2}\,\frac{c^2}{\omega^2}\int_{-\infty}^{\infty}\frac{dk_x}{\Delta}\varepsilon_L k_x = 0.\end{aligned} \tag{14.47}$$

The surface waves in a plasma slab always have two modes as represented by the $\pm$ signs, which correspond to symmetric and anti-symmetric modes, respectively. The series consisting of the exponential terms, although they are not convergent as they are, clearly converge by picking up the poles in the upper-half k_x plane from the denominator Δ. When the two series (the $\pm$ series) are summed, the compact form of the dispersion relation reads

$$\int \frac{dk_x}{k^2}\left(\frac{k_x^2}{\varepsilon_T - c^2k^2/\omega^2} + \frac{k_z^2}{\varepsilon_L}\right)\left(\frac{1 \mp e^{ik_xL}}{1 \pm e^{ik_xL}}\right) + i\lambda\frac{c^2}{\omega^2}\int \frac{dk_x\, k_x}{\varepsilon_T - c^2k^2/\omega^2} = 0 \tag{14.48}$$

where we used the relation $\varepsilon_{xx} - c^2k_z^2/\omega^2 = [k_x^2\varepsilon_L + k_z^2(\varepsilon_T - c^2k^2/\omega^2)]/k^2$. In (14.48), the singularities associated with $1 \pm e^{ik_xL} = 0$ should simply be ignored in the contour integration. When $L \to \infty$, the exponential factor vanishes, and both modes reduce to the dispersion relation of the surface waves in a semi-infinite plasma.

Remark. If we choose the coordinate system in such a way that $z = 0$ is the interface of a semi-infinite plasma, expressions for some of the surface wave-related quantities are changed. The geometry of $z = 0$ interface may be necessary if we consider a magnetized plasma in which the direction of the static magnetic field is taken to be the z-direction (in the literature, the magnetic field direction is invariably the z-direction). The following changes should be made.

Specular reflection condition (14.6):

$$f(v_x, v_y, v_z, t, x, y, z = 0) = f(v_x, v_y, -v_z, t, x, y, z = 0).$$

Extension of the field components:

$$E_x(-z) = E_x(z), \quad E_y(-z) = E_y(z), \quad E_z(-z) = -E_z(z),$$
$$B_x(-z) = -B_x(z), \quad B_y(-z) = -B_y(z), \quad B_z(-z) = B_z(z).$$

Electric field equation:

$$\mathbf{k} \times (\mathbf{k} \times \mathbf{E}) + \frac{\omega^2}{c^2}\mathbf{E} + \frac{4\pi i\omega}{c^2}\mathbf{J} = \mathbf{M}(k_x, k_y, \omega) \tag{14.49}$$

$$\begin{aligned}\mathbf{M}(k_x, k_y, \omega) = &-2i(\hat{\mathbf{x}}k_x + \hat{\mathbf{y}}k_y)E_z(k_x, k_y, \omega, z = 0^+)\\ &+ 2\left[\frac{\partial}{\partial z}\mathbf{E}_{\|}(k_x, k_y, \omega, z)\right]_{z=0^+}\end{aligned} \tag{14.50}$$

where $E_{\|} = \hat{\mathbf{x}}E_x + \hat{\mathbf{y}}E_y$. If the surface wave is assumed to propagate along x-direction ($\sim e^{ik_xx - i\omega t}$), the y-direction is ignorable, and we have $k_y = 0$. Also E_y can be set to zero since it decouples from the TM mode set

$(E_x,\ E_z,\ B_y)$. Then we have

$$\mathbf{M} = \hat{\mathbf{x}}M(k_x,\omega) = -2i\hat{\mathbf{x}}k_x E_z(k_x,\omega,z=0^+) + 2\hat{\mathbf{x}}\left[\frac{\partial}{\partial z}\mathbf{E}_x(k_x,\omega,z)\right]_{z=0^+}.$$

Surface wave fields are:

$$E_x = \frac{M}{\Delta}\left(\varepsilon_{zz} - \frac{c^2}{\omega^2}k_x^2\right), \tag{14.51}$$

$$E_z = -\frac{M}{\Delta}\left(\varepsilon_{xz} + \frac{c^2}{\omega^2}k_x k_z\right), \tag{14.52}$$

$$B_y = \frac{M}{\Delta}\frac{c}{\omega}(k_x\varepsilon_{xz} + k_z\varepsilon_{zz}), \tag{14.53}$$

$$k_x\varepsilon_{xz} + k_z\varepsilon_{zz} = \varepsilon_L k_z, \quad \Delta = \varepsilon_L\left(\varepsilon_T - \frac{c^2k^2}{\omega^2}\right). \tag{14.54}$$

Note that the above field components are obtained from (14.15)–(14.18) by interchanging the subscripts $x \leftrightarrow z$. Therefore the surface wave dispersion relation is also obtained from (14.22) by interchanging the subscripts $x \leftrightarrow z$. This is also true for the slab dispersion relation (14.48).

• **Cold plasma surface wave in a slab.** Let us consider waves in a cold plasma slab, with ions immobile. Using the relations $\varepsilon_L = \varepsilon_T = 1-\omega_{pe}^2/\omega^2 \equiv \varepsilon(\omega)$ in (14.48), we can write it as

$$\int\frac{dk_x}{k^2}\left(\frac{-\omega^2k_x^2/c^2}{k_x^2+\gamma^2} + \frac{k_z^2}{\varepsilon(\omega)}\right)\left(\frac{1\mp e^{ik_xL}}{1\pm e^{ik_xL}}\right) - i\lambda\int\frac{k_x dk_x}{k_x^2+\gamma^2} = 0$$

$$\textit{where}\quad \gamma = \left[k_z^2 + \frac{\omega_{pe}^2-\omega^2)}{c^2}\right]^{\frac{1}{2}}, \qquad \lambda = \sqrt{k_z^2 - \omega^2/c^2}.$$

The first integral yields two terms due to the simple poles at $k_x = ik_z$ and $k_x = i\gamma$. It can be seen that the contribution from the pole at ik_z cancels the second integral, giving the results

$$1 + \frac{\lambda}{\gamma}\,\varepsilon(\omega)\ tanh\left(\frac{\gamma L}{2}\right) = 0, \quad \textit{symmetric mode},$$

$$1 + \frac{\lambda}{\gamma}\,\varepsilon(\omega)\ coth\left(\frac{\gamma L}{2}\right) = 0, \quad \textit{anti-symmetric mode}.$$

If $L \to \infty$, both modes coalesce into the surface polariton mode as given by (13.57) and (13.59).

The above dispersion relations are the *TM mode surface waves in a slab.*

Exercise. Derive the above results from the fluid equations.

Solution: The governing equations are (13.1)–(13.5) with $\beta = 0$, which we write below after linearization:

$$\frac{\partial n}{\partial t} + N\nabla \cdot \mathbf{v} = 0,$$

$$\frac{\partial \mathbf{v}}{\partial t} = -\frac{e}{m}\mathbf{E},$$

$$\nabla \cdot \mathbf{E} = -4\pi e n,$$

$$\nabla \times \mathbf{E} = -\frac{1}{c}\frac{\partial \mathbf{B}}{\partial t},$$

$$\nabla \times \mathbf{B} = \frac{1}{c}\frac{\partial \mathbf{E}}{\partial t} - \frac{4\pi e N}{c}\mathbf{v}.$$

The TM mode variables are: v_x, v_z, E_x, E_z, B_y. The y-coordinate is ignorable. Assume the wave $\sim e^{ikz-i\omega t}$. One can derive

$$\frac{\partial^2 E_z}{\partial x^2} = \gamma^2 E_z, \quad E_x = -\frac{ik}{\gamma^2}\frac{\partial E_z}{\partial x}, \quad \gamma = \sqrt{k^2 + \frac{\omega_{pe}^2 - \omega^2}{c^2}}.$$

Write down the solutions for the three regions: plasma region $(-a < x < a)$, upper vacuum $(x > a)$, and lower vacuum $(x < -a)$. Apply the boundary conditions on $x = 0$ and $x = L$: continuity of E_z and B_y. We obtain for the plasma solutions

$$E_z = G_1 e^{\gamma x} + G_2 e^{-\gamma x},$$

$$E_x = \frac{-ik}{\gamma}(G_1 e^{\gamma x} - G_2 e^{-\gamma x}),$$

$$B_y = \frac{ic}{\gamma\omega}(\gamma^2 - k^2)(G_1 e^{\gamma x} - G_2 e^{-\gamma x}).$$

The vacuum equation, $(\nabla^2 + \omega^2/c^2)\mathbf{E} = 0$ is solved to yield

$$E_x = H_1 e^{\lambda x}, \quad E_z = \frac{i\lambda}{k}H_1 e^{\lambda x}, \quad B_y = \frac{\omega}{ck}H_1 e^{\lambda x}, \quad (x < -a),$$

$$E_x = H_2 e^{-\lambda x}, \quad E_z = -\frac{i\lambda}{k}H_2 e^{-\lambda x}, \quad B_y = \frac{\omega}{ck}H_2 e^{-\lambda x}, \quad (x > a),$$

where G's and H's are constants to be determined. Applying the boundary conditions gives

$$-\frac{i\lambda}{k}H_2e^{-a\lambda} = G_1e^{a\gamma} + G_2e^{-a\gamma},$$

$$\frac{i\lambda}{k}H_1e^{-a\lambda} = G_1e^{-a\gamma} + G_2e^{a\gamma},$$

$$\frac{\omega}{ck}H_2e^{-a\lambda} = \frac{ic}{\omega\gamma}(\gamma^2 - k^2)(G_1e^{a\gamma} - G_2e^{-a\gamma}),$$

$$\frac{\omega}{ck}H_1e^{-a\lambda} = \frac{ic}{\omega\gamma}(\gamma^2 - k^2)(G_1e^{-a\gamma} - G_2e^{a\gamma}).$$

Eliminating H_2 gives

$$G_1e^{a\gamma}(1 - R) + G_2e^{-a\gamma}(1 + R) = 0.$$

Eliminating H_1 gives

$$G_1e^{-a\gamma}(1 + R) + G_2e^{a\gamma}(1 - R) = 0,$$

$$R = \frac{\lambda c^2}{\gamma\omega^2}(\gamma^2 - k^2) = -\frac{\lambda}{\gamma}\,\varepsilon(\omega).$$

The solvability condition for G_1 and G_2 yields the dispersion relation in the form

$$\Phi(\gamma) = \pm\Phi(-\gamma) \quad with \quad \Phi(\gamma) = e^{a\gamma}[1 - R(\gamma)].$$

This agrees with the results obtained from the cold plasma limit of the kinetic dispersion relation. The $\pm$ sign corresponds to the symmetric and anti–symmetric modes obtained above. We have

$$\frac{E_z(-a)}{E_z(a)} = -\frac{H_1}{H_2} = \frac{G_1e^{-a\gamma} + G_2e^{a\gamma}}{G_1e^{a\gamma} + G_2e^{-a\gamma}},$$

$$\frac{G_2}{G_1} = -\frac{\Phi(\gamma)}{\Phi(-\gamma)} = -\frac{\Phi(-\gamma)}{\Phi(\gamma)}.$$

Using the above relations, we can show that the symmetric and anti-symmetric modes respectively correspond to the boundary relation $E_z(a) = \pm E_z(-a)$. The two-mode structure is characteristic of slab dispersion relation in general.

• **Electrostatic slab dispersion relation.** Taking the limit $c \to \infty$ in (14.48), the dispersion relation for electrostatic wave in a slab reads

$$\pi + \int_{-\infty}^{\infty} \frac{dk_x}{k^2} \frac{k_z}{\varepsilon_L} \left(\frac{1 \mp e^{ik_x L}}{1 \pm e^{ik_x L}} \right) = 0.$$

In the following, this result will be derived from the electrostatic equations (not via $c \to \infty$ limit).

The plasma occupies the region $0 < x < L$ bounded by vacuum ($x < 0,\ x > L$). The specular reflection conditions should be satisfied both at $x = 0$ and $x = L$:

$$f_\alpha(v_x, v_y, v_z, x = 0, y, z, t) = f_\alpha(-v_x, v_y, v_z, x = 0, y, z, t),$$
$$f_\alpha(v_x, v_y, v_z, x = L, y, z, t) = f_\alpha(-v_x, v_y, v_z, x = L, y, z, t).$$

These conditions can be satisfied if

$$f_\alpha(x, v_x) = f_\alpha(-x, -v_x) \quad and \quad f_\alpha(x, v_x) = f_\alpha(2L - x, -v_x).$$

Therefore, the electric field should be extended as

$$E_x(-x) = -E_x(x), \quad E_x(2L - x) = -E_x(x),$$
$$E_z(-x) = E_z(x), \quad E_z(2L - x) = E_z(x),$$

so that the Vlasov equation becomes invariant under such reflections. The electric field components extended in the above fashion are plotted in Fig. 14.3 in which one can ascertain $E_x(2L-0^+) = -E_x(0^+)$, $E_x(2L-0^-) = -E_x(0^-)$, etc. In the Fourier-transform of

$$\nabla \cdot \mathbf{E}(\mathbf{r}, t) = 4\pi e_\alpha \int f_\alpha d^3v$$

the jumps of E_x occurring at the discontinuities should be carefully accounted for. We write the Fourier transform

$$i\mathbf{k} \cdot \mathbf{E}(k_x, k_z, \omega) + 4E_x(k_z, x = 0^-, \omega) \left(\frac{1}{2} + cos\ 2k_x L + cos\ 4k_x L + \cdots \right)$$
$$+ 4E_x(k_z, x = L^-, \omega)(cos\ k_x L + cos\ 3k_x L + \cdots)$$
$$= 4\pi \sum_\alpha e_\alpha \int f_\alpha(\mathbf{k}, \mathbf{v}, \omega) d^3v.$$

Using (14.10) for f gives

$$E_i(k_x, k_z, \omega) = \frac{4ik_i}{k^2\varepsilon_L}\left[A\left(\frac{1}{2} + cos2k_xL + cos4k_xL + \cdots\right)\right.$$
$$\left. + B(cosk_xL + cos3k_xL + \cdots)\right]$$

where the subscript $i = x, z$, $A = E_x(k_z, x = 0^-, \omega)$, $B = E_x(k_z, x = L^-, \omega)$, and ε_L is given by (14.14). Let us invert the above to obtain $E_i(x)$. It is important to do this term by term. After algebra we get

$$E_x(x, k_z, \omega) = A'\int_{-\infty}^{\infty}\frac{k_x dk_x}{k^2\varepsilon_L}\left[e^{ik_x x} + 2i\ sink_x x(e^{2ik_xL} + e^{4ik_xL} + \cdots)\right]$$
$$+ B'\int_{-\infty}^{\infty}\frac{k_x dk_x}{k^2\varepsilon_L} 2i\ sink_x x(e^{ik_xL} + e^{3ik_xL} + \cdots),$$

$$E_z(x, k_z, \omega) = A'\int_{-\infty}^{\infty}\frac{k_z dk_x}{k^2\varepsilon_L}\left[e^{ik_x x} + 2\ cosk_x x(e^{2ik_xL} + e^{4ik_xL} + \cdots)\right]$$
$$+ B'\int_{-\infty}^{\infty}\frac{k_z dk_x}{k^2\varepsilon_L} 2\ cosk_x x(e^{ik_xL} + e^{3ik_xL} + \cdots),$$

where $A' = A/2$ and $B' = B/2$, and we used $\varepsilon_L(-k_x) = \varepsilon_L(k_x)$.

The normal component of the electric displacement, $D_x = \varepsilon_L E_x$ becomes

$$D_x(k_x, k_z, \omega) = \frac{4ik_x}{k^2}\left[A\left(\frac{1}{2} + cos2k_xL + cos4k_xL + \cdots\right)\right.$$
$$\left. + B(cosk_xL + cos3k_xL + \cdots)\right].$$

Inverting this equation gives

$$D_x(x, k_z, \omega) = A'\int_{-\infty}^{\infty}\frac{k_x dk_x}{k^2}\left[e^{ik_x x} + 2i\ sink_x x(e^{2ik_xL} + e^{4ik_xL} + \cdots)\right]$$
$$+ B'\int_{-\infty}^{\infty}\frac{k_x dk_x}{k^2} 2i\ sink_x x(e^{ik_xL} + e^{3ik_xL} + \cdots).$$

Clearly, we have

$$\bullet \quad D_x(0, k_z, \omega) = A' \int_{-\infty}^{\infty} \frac{k_x dk_x}{k^2} = i\pi A'.$$

$$D_x(L, k_z, \omega) = A' \int_{-\infty}^{\infty} \frac{k_x dk_x}{k^2} \left[e^{ik_x L} + 2i\ sink_x L(e^{2ik_x L} + e^{4ik_x L} + \cdots) \right]$$

$$+ B' \int_{-\infty}^{\infty} \frac{k_x dk_x}{k^2} 2i\ sink_x L(e^{ik_x L} + e^{3ik_x L} + \cdots).$$

Picking up the pole, the above series converges. Summing the series, we find the coefficient of A' vanishes and the coefficient of B' gives after successive cancellations

$$\bullet \quad D_x(L, k_z, \omega) = -i\pi B'.$$

The vacuum solutions for the regions $x < 0$ and $x > L$ can be written as

$$E_x(x, k_z, \omega) = F_1 e^{-k_z x}, \quad E_z(x, k_z, \omega) = -iF_1 e^{-k_z x} \quad (x > L),$$

$$E_x(k_z, x, \omega) = F_2 e^{k_z x}, \quad E_z(x, k_z, \omega) = iF_2 e^{k_z x} \quad (x < 0),$$

where F_1 and F_2 are: $F_i = \tilde{F}_i exp(ik_z z - i\omega t)$, $i = 1, 2$. The plasma solutions and the vacuum solutions should be matched at the interfaces $x = 0$ and $x = L$ by the electromagnetic boundary conditions which are the continuity of E_z and D_x.

• Boundary conditions.

$[E_z] = 0$ at $x = 0$ gives

$$A' \int_{-\infty}^{\infty} \frac{k_z dk_x}{k^2 \varepsilon_L} \left[1 + 2\ (e^{2ik_x L} + e^{4ik_x L} + \cdots) \right]$$

$$+ B' \int_{-\infty}^{\infty} \frac{k_z dk_x}{k^2 \varepsilon_L} 2\ (e^{ik_x L} + e^{3ik_x L} + \cdots) = iF_2.$$

$[D_x] = 0$ at $x = 0$ gives

$$i\pi A' = F_2.$$

$[E_z] = 0$ at $x = L$ gives

$$A' \int_{-\infty}^{\infty} \frac{k_z dk_x}{k^2 \varepsilon_L} \left[e^{ik_x L} + 2 \cos k_x L (e^{2ik_x L} + e^{4ik_x L} + \cdots) \right]$$

$$+ B' \int_{-\infty}^{\infty} \frac{k_z dk_x}{k^2 \varepsilon_L} 2 \cos k_x L (e^{ik_x L} + e^{3ik_x L} + \cdots) = -iF_1 e^{-k_z L}.$$

$[D_x] = 0$ at $x = L$ gives

$$-i\pi B' = F_1 \, e^{-k_z L}.$$

In the above four equations, F_1 and F_2 can be easily eliminated, and the resulting two equations for A' and B' can be put into the form

$$c_1 A' + c_2 B' = 0, \quad c_2 A' + c_1 B' = 0$$

with

$$c_1 = \pi + \int_{-\infty}^{\infty} \frac{k_z dk_x}{k^2 \varepsilon_L} [1 + 2 \, (e^{2ik_x L} + e^{4ik_x L} + \cdots)],$$

$$c_2 = \int_{-\infty}^{\infty} \frac{k_z dk_x}{k^2 \varepsilon_L} 2 \, (e^{ik_x L} + e^{3ik_x L} + \cdots).$$

The surface wave dispersion relation is obtained in the form $c_1 = \pm c_2$, or

$$\pi + \int_{-\infty}^{\infty} dk_x \, \frac{k_z}{k^2 \varepsilon_L} \left(\frac{1 \mp e^{ik_x L}}{1 \pm e^{ik_x L}} \right) = 0 \quad (for \ slab).$$

The surface waves in a plasma slab always have two modes as represented by the $\pm$ signs, which correspond to symmetric (upper sign) and anti-symmetric (lower sign) mode, respectively. The apparent singularities located at $1 \pm e^{ik_z L} = 0$ in the second integral should be simply ignored. When $L \to \infty$, the exponential factor vanishes (by picking up the poles in the upper-half k_z-plane), and both modes coalesce to the dispersion relation of the surface wave in the semi-infinite plasma, Eq. (14.34).

14.3. Evaluation of the semi-infinite plasma electrostatic surface wave dispersion relation

Reference: M. J. Lee and H. J. Lee (2011), *Landau damping of surface plasma waves* The Open Plasma Physics Journal 2011 Vol 4 pp 55-62.

In order to obtain the plot of the normal modes of surface waves in kinetic description, we should first carry out the Fourier inversion integral in the dispersion relation like (13.34); this particular dispersion relation is valid for the electrostatic mode in a semi-infinite plasma. The integrals can be done by contour integration. The issue of causality has been already settled at the stage of determining ε_L, so no new causality issue is arisen in connection with the surface waves. The imaginary part of ε_L and its real part are simply transported to damping of the surface waves.

In this Section, the electrostatic dispersion relation (14.34) for the semi-bounded plasma is evaluated for high and low frequency ranges.

$$(14.34): \quad 1 + \frac{k_x}{\pi} \lim_{z \to 0} \int dk_z \frac{e^{ik_z z}}{k^2 \varepsilon_L} = 1 + \frac{k_x}{\pi} \int_{-\infty}^{\infty} \frac{dk_z}{k^2 \varepsilon_L} = 0. \tag{14.55}$$

The above equation is written by interchanging the subscripts z and x in (14.34); $z = 0$ is now taken to be the interface (this is only a matter of convenience). Here we discover two normal modes of surface wave: high frequency electron plasma surface wave and low frequency ion acoustic surface wave. The high frequency surface wave (as well as its corresponding bulk wave) is such that $\omega \gg \mathbf{k} \cdot \mathbf{v}_e$ while the low frequency ion acoustic surface wave as well as its corresponding bulk ion acoustic wave has frequencies such that $\mathbf{k} \cdot \mathbf{v}_i \ll \omega \ll \mathbf{k} \cdot \mathbf{v}_e$ ($\mathbf{v}_\alpha$ is the thermal velocity of species α). An interesting result is that the surface ion acoustic wave and the bulk ion acoustic wave are identical or indistinguishable in as much as the real frequency and the damping rate (Landau damping) of both waves are identical.

14.3.1. *Surface ion acoustic wave and Landau damping*

In the low frequency range, we have $\varepsilon_L = \varepsilon_r + i\varepsilon_i$ with

$$\varepsilon_r = 1 + \frac{1}{k^2 \lambda_e^2} \left(1 - \frac{T_e}{m_i} \frac{k^2}{\omega^2}\right) = 1 - \frac{\omega_{pi}^2}{\omega^2} + \frac{1}{k^2 \lambda_e^2}, \tag{14.56}$$

$$\varepsilon_i = \sum_{\alpha=e,i} \frac{\sqrt{\pi}}{k^2 \lambda_\alpha^2} \frac{\omega}{k} \sqrt{\frac{m_\alpha}{2T_\alpha}}\, exp \left[\frac{-m_\alpha \omega^2}{2T_\alpha k^2}\right]. \tag{14.57}$$

Our general procedure toward obtaining the surface wave dispersion relation and the Landau damping is as follows. Assuming $\varepsilon_r \gg \varepsilon_i$, we adopt a kind of perturbation approach: the zero order dispersion relation would be obtained by neglecting ε_i in (14.55). The real frequency determined by ε_r would be used, if necessary, to simplify the integral. We put $\varepsilon_L = \varepsilon_r(1 + i\varepsilon_i/\varepsilon_r)$ where the second term is $\ll 1$. Then the dispersion relation integral in (14.55) becomes

$$1 + \frac{k_x}{\pi} \int \frac{dk_z}{k^2 \varepsilon_r} \left(1 - i\frac{\varepsilon_i}{\varepsilon_r}\right) = 0. \tag{14.58}$$

Approximate eigenfrequency is obtained from the real part,

$$D_r \equiv 1 + \frac{k_x}{\pi} \int \frac{dk_z}{k^2 \varepsilon_r} = 0 \tag{14.59}$$

which takes the form upon using (14.56)

$$1 + \frac{k_x}{\pi q(\omega)} \int \frac{dk_z}{k_z^2 + \gamma^2} = 0 \tag{14.60}$$

$$q(\omega) = 1 - \frac{\omega_{pi}^2}{\omega^2}, \; \gamma = \sqrt{k_x^2 + \frac{1}{q\lambda_e^2}}$$

with ion plasma frequency $\omega_{pi} = \sqrt{\frac{4\pi Ne^2}{m_i}}$, and electron Debye length $\lambda_e = \sqrt{\frac{T_e}{4\pi Ne^2}}$. Now the integral in (14.60) can be carried out by picking up the residue at the pole $k_z = i\gamma$ to yield

$$D_r = 1 + \frac{1}{q(\omega)} \sqrt{\frac{q\kappa}{1 + q\kappa}} = 0. \quad (\kappa = k_x^2 \lambda_e^2) \tag{14.61}$$

The quantity γ is the attenuation constant, giving the z-dependence, $\sim e^{-\gamma z}$. Clearly, as is seen from (14.61), $q(\omega)$ should be negative to have the ion acoustic surface wave supported. Also we need $1 + q\kappa < 0$:

$$q < 0, \quad q + \frac{1}{\kappa} < 0. \tag{14.62}$$

Squaring (14.61) gives $\kappa q^2 + q - \kappa = 0$ which is solved by

$$q = \frac{1}{2\kappa}(-1 \pm \sqrt{1 + 4\kappa^2}). \tag{14.63}$$

Here we consider two limiting cases of κ.

i) $\kappa \ll 1$: In this case, we have $q \simeq \frac{1}{2\kappa}[-1 \pm (1 + 2\kappa^2)]$ in which the positive sign should be rejected in view of (14.62), while the $(-)$ sign is in accord with (14.62). Neglecting the κ^2-term, we obtain

$$\omega^2 = \frac{k_x^2 \lambda_e^2 \omega_{pi}^2}{1 + k_x^2 \lambda_e^2}. \left(q \simeq -\frac{1}{\kappa}, \quad \kappa \ll 1\right) \tag{14.64}$$

Equation (14.64) is the well-known dispersion relation. The surface ion acoustic wave dispersion relation is of the same form as the bulk ion acoustic wave.

ii) $\kappa \gg 1$: In this case, we obtain $q \simeq \frac{1}{2\kappa}(-1 \pm 2\kappa)$. Only the $(-)$ sign is acceptable and we have the solution

$$\omega^2 = \frac{\omega_{pi}^2}{2}. \quad (q \simeq -1, \quad \kappa \gg 1) \tag{14.65}$$

It should be noted that the surface ion acoustic wave does not asymptote to the line $\omega^2 = \omega_{pi}^2$, but approaches the line $\omega^2 = \frac{\omega_{pi}^2}{2}$.

We write for future reference

$$\frac{\gamma}{k_x} = \sqrt{\frac{\kappa^2}{1+\kappa^2}} \ll 1 \quad when\ \kappa \ll 1, \tag{14.66}$$

$$\frac{\gamma}{k_x} = \sqrt{\frac{2\kappa - 1}{2\kappa + 1}} \simeq 1 \quad when\ \kappa \gg 1. \tag{14.67}$$

Next we calculate the imaginary part of (14.58):

$$D_i \equiv -i\frac{k_x}{\pi}\int dk_z\, \frac{\varepsilon_i}{k^2\varepsilon_r^2}.$$

Using the relation $k^2\varepsilon_r = q(k_z^2 + \gamma^2)$ and (14.57), the above expression can be put into the form

$$D_i = -i\frac{k_x}{\sqrt{\pi}}\frac{\omega}{q^2}\int_{-\infty}^{\infty} \frac{dk_z}{(k_z^2+\gamma^2)^2\sqrt{k_z^2+k_x^2}} \sum \frac{1}{\lambda_\alpha^2}\sqrt{\frac{m_\alpha}{2T_\alpha}}exp\left[-\frac{m_\alpha\omega^2}{2T_\alpha(k_z^2+k_x^2)}\right]. \tag{14.68}$$

We first deal with the ion term, and consider the integral

$$I_i = \int_{-\infty}^{\infty} \frac{dk_z\ exp[-\frac{\xi_i}{k_z^2+k_x^2}]}{(k_z^2+\gamma^2)^2\sqrt{k_z^2+k_x^2}}. \quad \left(\xi_i = \frac{m_i\omega^2}{2T_i}\right) \tag{14.69}$$

We integrate (14.69) along the real k_z-axis without resorting to contour integration. The following change of variable is very convenient to carry out the asymptotic evaluation of the integral in (14.69):

$$y = \frac{k_z}{\sqrt{k_z^2+\gamma^2}}, \quad or \quad k_z = \frac{\gamma y}{\sqrt{1-y^2}}. \tag{14.70}$$

Then, integral I_i becomes

$$I_i = \frac{2}{\gamma^3}\int_0^1 dy \frac{(1-y^2)exp[-\frac{\xi_i(1-y^2)}{y^2(\gamma^2-k_x^2)+k_x^2}]}{\sqrt{y^2(\gamma^2-k_x^2)+k_x^2}}. \qquad (14.71)$$

Equation (14.71) is appropriate for asymptotic analysis for the limiting cases.

i) Case of $\kappa \ll 1$

In this case, we have $\gamma \ll k_x$. Neglecting γ^2 as compared to k_x^2, we obtain

$$I_i = \frac{2}{k_x\gamma^3}e^{-\frac{\xi_i}{k_x^2}} I_0, \quad I_0 = \int_0^1 \sqrt{(1-y^2)}\, dy = \frac{\pi}{4}.$$

Here we have $\frac{\xi_i}{k_x^2} = \frac{m_i}{2T_i}\lambda_e^2\omega_{pi}^2 = \frac{T_e}{2T_i}$. Thus,

$$I_i = \frac{\pi}{2k_x\gamma^3}\, e^{-\frac{T_e}{2T_i}}. \qquad (14.72)$$

The corresponding electron integral is calculated similarly by replacing $\xi_i \to \xi_e = \frac{m_e\omega^2}{2T_e}$ in (14.71). Here $\frac{\xi_e}{k_x^2} = \frac{m_e}{2m_i} \ll 1$ and the exponential term can be put to unity. We obtain

$$I_e = \frac{\pi}{2k_x\gamma^3}. \qquad (14.73)$$

Then (14.68), (18.72), (14.73), and the relations $q = -1/\kappa$, $\omega = \omega_{pi}k_x\lambda_e$, $\gamma = \kappa k_x$ give

$$D_i = -i\sqrt{\frac{\pi}{8}}\,\frac{1}{k_x^4\lambda_e^4}\left[\left(\frac{T_e}{T_i}\right)^{\frac{3}{2}} e^{-\frac{T_e}{2T_i}} + \sqrt{\frac{m_e}{m_i}}\right]. \qquad (14.74)$$

The Landau damping rate is obtained from the formula

$$\omega_i = -\frac{D_i(\omega)}{\partial D_r(\omega)/\partial\omega}. \qquad (14.75)$$

We calculate from (14.61):

$$\frac{\partial D_r}{\partial\omega} = \frac{\partial D_r}{\partial q}\frac{\partial q}{\partial\omega}$$

$$\frac{\partial D_r}{\partial q} = \frac{1}{q} - \frac{1}{2\kappa} \simeq -\frac{1}{2\kappa}, \quad \frac{\partial q}{\partial\omega} \simeq \frac{2}{\omega_{pi}k_x^3\lambda_e^3}.$$

Using the above relations in (14.75) yield the Landau damping rate for $\kappa \ll 1$

$$\omega_i = -\omega_{pi}k_x\lambda_e\sqrt{\frac{\pi}{8}}\left[\left(\frac{T_e}{T_i}\right)^{\frac{3}{2}} e^{-\frac{T_e}{2T_i}} + \sqrt{\frac{m_e}{m_i}}\right]. \qquad (14.76)$$

In (14.76), the mass ratio term comes from the electron term in ε_i. It is interesting to see that surface ion acoustic wave Landau damping in (14.76) is equal to the corresponding Landau damping of the bulk ion acoustic wave.

ii) Case of $\kappa \gg 1$
In this case, we have $\gamma \simeq k_x$, so I_i in (14.71) reduces to

$$I_i = \frac{2}{k_x \gamma^3} e^{-\frac{\xi_i}{k_x^2}} J_i, \tag{14.77}$$

$$J_i = \int_o^1 dy(1-y^2) e^{\frac{\xi_i y^2}{k_x^2}}. \tag{14.78}$$

Integral J_i can be obtained by integration by parts and after some algebra:

$$J_i = -\frac{1}{2x^2} e^{x^2} + \frac{F(x)}{x}\left(1 + \frac{1}{2x^2}\right) \tag{14.79}$$

where $F(x)$ is the error function:

$$F(x) = \int_0^x e^{t^2} dt; \quad x = \frac{\sqrt{\xi_i}}{k_x} = \frac{1}{2}\sqrt{\frac{T_e}{T_i}} \frac{1}{k_x \lambda_e}. \tag{14.80}$$

The quantity x in (14.80), which is the ratio of the surface wave velocity to the ion thermal velocity, is assumed to be $\gg 1$ from the beginning when we wrote (14.57). The limiting conditions $x \gg 1$ and $k_x \lambda_e \gg 1$ restrict the validity of the present analysis ($\kappa \gg 1$) to the regions

$$\sqrt{\frac{T_e}{T_i}} \gg k_x \lambda_e \gg 1. \tag{14.81}$$

Using (14.79) in (14.77) gives

$$I_i = \frac{2}{k_x \gamma^3}\left[-\frac{1}{2x^2} + \frac{e^{-x^2}}{x}\left(1 + \frac{1}{2x^2}\right) F(x)\right].$$

The error function $F(x)$ has asymptotic expansion for $x \gg 1$:

$$F(x) = \int_0^x e^{t^2} dt \simeq \frac{e^{x^2}}{2x}\left(1 + \frac{1}{2x^2} + \frac{3}{4x^4} + \cdots\right). \quad (x \gg 1)$$

The above two equations give, using $\gamma \simeq k_x$, $\omega = \omega_{pi}/\sqrt{2}$,

$$I_i \simeq \frac{1}{k_x \gamma^3} \frac{1}{x^4}\left(1 + \frac{1}{x^2}\right) = 16 \left(\frac{T_i}{T_e}\right)^2 \lambda_e^4. \tag{14.82}$$

Neglecting the electron contribution in (14.57) as compared to the above ion contribution, we obtain (14.89) later.

Exercise. Show that the electron term in ε_i in (14.57) contributes negligibly to ion acoustic wave Landau damping when $k_x\lambda_e \gg 1$. This is due to the large disparity of electron and ion mass.

Solution: To calculate the electron integral for the case of $\kappa \gg 1$, we redefine the symbol x:

$$x = \frac{\sqrt{\xi_e}}{k_x} = \frac{1}{2}\sqrt{\frac{m_e}{m_i}}\frac{1}{k_x\lambda_e} \ll 1 \tag{14.83}$$

which is the ratio of the surface wave velocity to the electron thermal velocity. The electron integral corresponding to (14.69) is

$$I_e = \frac{2}{k_x\gamma^3}e^{-x^2}J_e, \tag{14.84}$$

$$J_e = \int_0^1 dy(1-y^2)e^{\frac{\xi_e y^2}{k_x^2}} = -\frac{1}{2x^2}e^{x^2} + \frac{F(x)}{x}\left(1+\frac{1}{2x^2}\right),$$

which is of the same form as (14.79), but x is now $\ll 1$ and $F(x)$ is represented by different asymptotic series:

$$F(x) = xe^{x^2}\left(1 - \frac{2}{3}x^2 + \frac{4}{15}x^4 - \cdots\right). \quad (x \ll 1)$$

Using the above two equations in (14.84), we obtain a simple result with neglect of higher order terms,

$$I_e = \frac{2}{k_x\gamma^3}. \tag{14.85}$$

Using (14.82) and (14.85) in (14.68) gives

$$D_i = \frac{-8i}{\sqrt{\pi}}\sqrt{\frac{T_i}{T_e}}k_x\lambda_e - \frac{i}{\sqrt{\pi}}\sqrt{\frac{m_e}{m_i}}\frac{1}{k_x^3\lambda_e^3}. \tag{14.86}$$

To apply formula (14.75), we calculate

$$\frac{\partial D_r}{\partial q} \simeq -1, \quad \frac{\partial q}{\partial \omega} \simeq \frac{4\sqrt{2}}{\omega_{pi}}. \tag{14.87}$$

Using the above values, we obtain for Landau damping rate when $\kappa \gg 1$

$$\omega_i = -\sqrt{\frac{2}{\pi}}\sqrt{\frac{T_i}{T_e}}\,\omega_{pi}k_x\lambda_e - \frac{1}{8}\sqrt{\frac{2}{\pi}}\sqrt{\frac{m_e}{m_i}}\frac{\omega_{pi}}{k_x^3\lambda_e^3} \equiv \omega_i^i + \omega_i^e \tag{14.88}$$

where the superscripts denote ion and electron contribution from ε_i. The ratio of electron contribution to ion's is

$$\frac{\omega_i^i}{\omega_i^e} = 8\sqrt{\frac{m_i}{m_e}}\sqrt{\frac{T_i}{T_e}}k_x^4\lambda_e^4 \gg 1.$$

So the electron term in ε_i in (14.57) contributes negligibly to ion acoustic wave Landau damping when $k_x\lambda_e \gg 1$. Thus, we conclude

$$\omega_i = -\sqrt{\frac{2}{\pi}}\sqrt{\frac{T_i}{T_e}}\,\omega_{pi}k_x\lambda_e. \quad (k_x\lambda_e \gg 1) \tag{14.89}$$

14.3.2. *Surface electron plasma wave and Landau damping*

For the high frequency wave ($\omega \gg \mathbf{k}\cdot\mathbf{v}_e$), $\varepsilon_L = \varepsilon_r + i\varepsilon_i$ has the expression

$$\varepsilon_r = 1 - \frac{\omega_{pe}^2}{\omega^2} - 3\frac{\omega_{pe}^4}{\omega^4}k^2\lambda_e^2, \tag{14.90}$$

$$\varepsilon_i = \sqrt{\frac{\pi}{2}}\left(\frac{m_e}{T_e}\right)^{\frac{3}{2}}\frac{\omega\omega_{pe}^2}{k^3}exp\left[-\frac{m_e}{2T_e}\frac{\omega^2}{k^2}\right]. \tag{14.91}$$

The real part of dispersion integral (14.58) can be evaluated as follows.

$$\int\frac{dk_z}{k^2\varepsilon_r} = \int\frac{dk_z}{k^2(1-\frac{\omega_{pe}^2}{\omega^2}-3\frac{\omega_{pe}^4}{\omega^4}k^2\lambda_e^2)} = \frac{-\omega^4}{3\omega_{pe}^4\lambda_e^2}\int\frac{dk_z}{(k_z^2+k_x^2)(k_z^2+\gamma^2)}$$

$$\gamma = \sqrt{k_x^2 - \frac{\omega^4(1-\frac{\omega_{pe}^2}{\omega^2})}{3\omega_{pe}^4\lambda_e^2}}. \tag{14.92}$$

Picking up the residues at the poles $k_z = ik_x$ and $k_z = i\gamma$, the integral has the value

$$\pi\cdot\frac{1/k_x - 1/\gamma}{1-\omega_{pe}^2/\omega^2}.$$

Thus, the real part of the dispersion relation reads

$$D_r = 1 + \frac{1}{q(\omega)}\left(1-\frac{k_x}{\gamma}\right), \quad q(\omega) = 1 - \frac{\omega_{pe}^2}{\omega^2}. \tag{14.93}$$

The real frequency is obtained from $D_r = 0$. We seek the solution with the condition $q < 0$, which insures that the quantity in the square root of (14.92) is positive. Accordingly, we have $\gamma > k_x$. The relation $D_r = 0$ gives, after squaring,

$$-1 + (2-x)^2 - \frac{(1-x)(2-x)^2}{3k_x^2\lambda_e^2\, x^2} = 0. \qquad \left(x = \frac{\omega_{pe}^2}{\omega^2}\right)$$

It is seen that $x = 2$ is an approximate solution when $k_x^2\lambda_e^2 \ll 1$. Note that $x = 1$ or $q = 0$ (in this case $\gamma = k_x$) is not a root of $D_r = 0$. Let us look for a solution of the above equation in the vicinity of $x = 2$. So put $x = 2 + \delta$ ($\delta \ll 1$), and obtain δ in the region of $k_x\lambda_e \ll 1$. One obtains

$$\omega = \frac{\omega_{pe}}{\sqrt{2}}\left(1 + \frac{\sqrt{3}}{2}\, k_x\lambda_e\right). \tag{14.94}$$

This frequency is called *Ritchie mode.* Using (14.94) in (14.92) gives

$$\frac{k_x}{\gamma} = 2\sqrt{3}\, k_x\lambda_e \ll 1. \tag{14.95}$$

Next we calculate the Landau damping corresponding to the mode (14.94). The imaginary part of the dispersion integral (14.58) upon using (14.90) and (14.91) takes the form

$$D_i = -iAI, \quad A = \frac{1}{9\sqrt{2\pi}}\sqrt{\frac{m_e}{T_e}}\,\frac{k_x}{\lambda_e^6}\,\frac{\omega^9}{\omega_{pe}^8}, \tag{14.96}$$

$$I = \int_{-\infty}^{\infty} dk_z\, \frac{exp\left[-\frac{\alpha}{k_z^2+k_x^2}\right]}{(k_z^2+k_x^2)^{\frac{5}{2}}(k_z^2+\gamma^2)^2} \qquad \left(\alpha = \frac{m_e\omega^2}{2T_e}\right) \tag{14.97}$$

It is useful to make a change of variable via (cf. Eq. (14.70))

$$y = \frac{k_z}{\sqrt{k_z^2+k_x^2}} \quad \text{or} \quad k_z = \frac{k_x y}{\sqrt{1-y^2}}.$$

Then the integral I in (14.97) becomes

$$I = \frac{2}{k_x^4}e^{-x^2}\int_0^1 dy(1-y^2)^3\frac{e^{x^2y^2}}{[k_x^2y^2+\gamma^2(1-y^2)]^2}. \qquad \left(x^2 = \frac{\alpha}{k_x^2}\right) \tag{14.98}$$

In the region where $k_x \ll \gamma$, integral I takes simple form

$$I = \frac{2}{k_x^4\gamma^4}e^{-x^2}J, \quad J = \int_0^1 dy(1-y^2)\, e^{x^2y^2}, \tag{14.99}$$

$$x = \frac{\sqrt{\alpha}}{k_x} = \frac{1}{2}\frac{1}{k_x\lambda_e} \gg 1. \tag{14.100}$$

After integration by parts in (14.99), one obtains

$$J = \frac{1}{x}F(x)\left(1 + \frac{1}{2x^2}\right) - \frac{1}{2x^2}e^{x^2}.$$

Using the asymptotic series for the error function $F(x)$ ($x \gg 1$), we obtain

$$J - \frac{e^{x^2}}{2x^4}\left(1 + \frac{1}{x^2}\right). \tag{14.101}$$

Using (14.99) and (14.101) in (14.99) yields

$$I = \frac{1}{k_x^4\gamma^4x^4}\left(1 + \frac{1}{x^2}\right) = 9 \cdot 2^8 \cdot \lambda_e^8(1 + 4k_x^2\lambda_e^2)$$

where the last term will be neglected. Then (14.96) gives

$$D_i = -i\frac{8}{\sqrt{\pi}}k_x\lambda_e. \tag{14.102}$$

Landau damping rate will be obtained by the formula (14.75). Equation (14.92) gives

$$\frac{\partial D_r}{\partial \omega} = -\frac{1}{q^2}\frac{\partial q}{\partial \omega}\left(1 - \frac{k_x}{\gamma}\right) + \frac{k_x}{q}\frac{1}{\gamma^2}\frac{\partial \gamma}{\partial \omega}. \tag{14.103}$$

We have $\frac{\partial q}{\partial \omega} \simeq \frac{4\sqrt{2}}{\omega_{pe}}$, $q \simeq -1$, $\gamma \gg k_x$. The first term of (14.103) is $\simeq -\frac{4\sqrt{2}}{\omega_{pe}}$, and the second term is $\simeq \frac{12\sqrt{2}}{\omega_{pe}}k_x^2\lambda_e^2$ which will be neglected. So we finally obtain for the Landau damping rate

$$\omega_i = -\sqrt{\frac{2}{\pi}}k_x\lambda_e\omega_{pe}. \tag{14.104}$$

14.4. Surface waves in magnetized Vlasov plasma

References: H. J. Lee and C.-G. Kim, *Kinetic theory of electrostatic surface waves in magnetized plasmas*, J. Korean Phys. Soc. **54** (2009) p. 85-93; H. J. Lee, *Electrostatic surface waves in a magnetized two-fluid plasma*, Plasma Phys. Control. Fusion **37** (1995) p. 755-762.

In this Section, we investigate surface waves in magnetized semi-infinite and slab plasmas by means of the Vlasov equation. We restrict consideration to the electrostatic waves. First, we note that there are three orientations involved in the problem: i) the direction of $\mathbf{B}_0$, the static magnetic field which we will take as $\mathbf{B}_0 = \hat{\mathbf{z}} B_0$, ii) the interface normal $\hat{\mathbf{n}}$, iii) the wave vector $\mathbf{k} \perp \hat{\mathbf{n}}$. So we will investigate the following three different cases:

$$1)\ \mathbf{B}_0 = \hat{\mathbf{z}} B_0; \quad \hat{\mathbf{n}} = \hat{\mathbf{z}}; \quad \mathbf{k} \parallel \hat{\mathbf{x}} \ \ i.e\ \ wave \ \sim e^{ik_x x - i\omega t}; \ \ x-wave; \ \ \frac{\partial}{\partial y} = 0.$$

$$2)\ \mathbf{B}_0 = \hat{\mathbf{z}} B_0; \quad \hat{\mathbf{n}} = \hat{\mathbf{x}}; \quad \mathbf{k} \parallel \hat{\mathbf{z}} \ \ i.e\ \ wave \ \sim e^{ik_z z - i\omega t}; \ \ z-wave; \ \ \frac{\partial}{\partial y} = 0.$$

$$3)\ \mathbf{B}_0 = \hat{\mathbf{z}} B_0; \quad \hat{\mathbf{n}} = \hat{\mathbf{x}}; \quad \mathbf{k} \parallel \hat{\mathbf{y}} \ \ i.e\ \ wave \ \sim e^{ik_y y - i\omega t}; \ \ y-wave; \ \ \frac{\partial}{\partial y} \neq 0.$$

For the cases 1) and 2), the y-coordinate is ignorable. The names, $x-$, $y-$, $z-$ waves refer to earlier work by Lee (Plasma Phys. Control. Fusion 1995). The recipe to make the Vlasov equation invariant, with the static magnetic field term included, under the specular reflection, is investigated by inspecting all the terms of the Vlasov equation, for each case of $x-$, $y-$, $z-$ waves. Confining to electrostatic waves, the recipe is: *the perpendicular (parallel) component of the electric field to the interface is oddly (evenly) continued* for all the three $x-$, $y-$, $z-$ waves. Now we solve each case above. The governing equations are:

$$\frac{\partial}{\partial t} f_\alpha(\mathbf{r}, \mathbf{v}, t) + \mathbf{v} \cdot \frac{\partial f_\alpha}{\partial \mathbf{r}} + \frac{e_\alpha}{m_\alpha c} \mathbf{v} \mathbf{B}_0 \cdot \frac{\partial}{\partial \mathbf{v}} f_\alpha + \frac{e_\alpha}{m_\alpha} \mathbf{E} \cdot \frac{\partial f_{\alpha 0}}{\partial \mathbf{v}} = 0. \quad (\alpha = e,\ i), \tag{14.105}$$

$$\nabla \cdot \mathbf{E}(\mathbf{r}, t) = 4\pi \sum_\alpha e_\alpha \int f_\alpha d^3 v, \tag{14.106}$$

$$f_{\alpha 0}(\mathbf{v}) = \left(\frac{m_\alpha}{2\pi T_{\alpha \parallel}} \right)^{1/2} \frac{m_\alpha}{2\pi T_{\alpha \perp}} e^{-\frac{m_\alpha v_z^2}{2T_{\alpha\parallel}}} e^{-\frac{m_\alpha}{2T_{\alpha\perp}}(v_x^2 + v_y^2)}, \tag{14.107}$$

where the usual symbols $\parallel$ and $\perp$ refer to the direction of the static magnetic field $\mathbf{B}_0$.

14.4.1. *Semi-infinite plasma:* $\mathbf{B}_0 = \hat{\mathbf{z}} B_0$; $\hat{\mathbf{n}} = \hat{\mathbf{z}}$; $\mathbf{k} \parallel \hat{\mathbf{x}}$*: x–wave*

The plasma occupying the region ($z > 0$) is bounded by vacuum ($z < 0$). The surface wave propagates along the x- direction ($\propto e^{ik_x x - i\omega t}$). The kinematic boundary condition for f is

$$f_\alpha(v_x, v_y, v_z, x, y, z = 0, t) = f_\alpha(v_x, v_y, -v_z, x, y, z = 0, t). \qquad (14.108)$$

(14.105) is rewritten in the form to exhibit all the components

$$\frac{\partial}{\partial t} f_\alpha(v_x, v_y, v_z, x, z, t) + v_x \frac{\partial f_\alpha}{\partial x} + v_z \frac{\partial f_\alpha}{\partial z} + \omega_{c\alpha} v_y \frac{\partial f_\alpha}{\partial v_x} - \omega_{c\alpha} v_x \frac{\partial f_\alpha}{\partial v_y}$$
$$+ \frac{e_\alpha}{m_\alpha}\left(E_x(x,z)\frac{\partial f_{0\alpha}}{\partial v_x} + E_z(x,z)\frac{\partial f_{0\alpha}}{\partial v_z}\right) = 0 \quad \left(\omega_{c\alpha} = \frac{e_\alpha B_0}{m_\alpha c}\right) \qquad (14.109)$$

where we put $E_y = -\partial\phi/\partial y = 0$.

The condition (14.108) is automatically satisfied if $f_\alpha(v_z, z) = f_\alpha(-v_z, -z)$ i.e. if (14.109) is invariant under the reflections $(z \to -z, v_z \to -v_z)$. By inspection, it is immediately seen that the electric field components should be extended in the following way to furnish (14.109) with the invariance:

$$E_x(-z) = E_x(z), \quad E_z(-z) = -E_z(z) \quad for\ x - wave. \qquad (14.110)$$

(14.109) can be solved by introducing cylindrical coordinates in the velocity space: $v_x = v_\perp\ cos\varphi,\ \ v_y = v_\perp\ sin\varphi$. Also Eq. (14.109) is Fourier-transformed with respect to the variables, x, z, and t in the infinite region $-\infty < z < \infty$ with the extended field (14.110). Consulting Chapter 5, one obtains

$$f_\alpha(\mathbf{k}, \mathbf{v}, \omega) = \frac{ie_\alpha}{m_\alpha \omega_{c\alpha}} \sum_{n=-\infty}^{\infty} \sum_{l=-\infty}^{\infty} e^{i(n-l)\varphi} \frac{J_n(a_\alpha) J_l(a_\alpha)}{b_\alpha + l}$$
$$\times \left[E_x \frac{l}{a_\alpha}\frac{\partial f_{\alpha 0}}{\partial v_\perp} + E_z \frac{\partial f_{\alpha 0}}{\partial v_z}\right], \qquad (14.111)$$

$$a_\alpha = \frac{k_x v_\perp}{\omega_{c\alpha}}, \quad b_\alpha = \frac{k_z v_z - \omega}{\omega_{c\alpha}}.$$

Fourier-transforming Eq. (14.106) gives

$$ik_x E_x(k_x, k_z, \omega) + ik_z E_z(k_x, k_z, \omega) = 4\pi \sum_\alpha e_\alpha \int f_\alpha(\mathbf{k}, \mathbf{v}, \omega) d^3 v + M(k_x, \omega) \qquad (14.112)$$

where $M(k_x, \omega) = 2E_z(k_x, \omega, z = 0^+)$ comes from the discontinuity of $E_z(z)$ at $z = 0$. Equations (14.111) and (14.112) yield

$$E_j(k_x, k_z, \omega) = -i\frac{k_j}{k^2\varepsilon_L} M(k_x, \omega) \quad (j = x,\ z) \tag{14.113}$$

where $k_x E_z = k_z E_x$ was used and

$$\varepsilon_L = 1 + \sum_\alpha \frac{\omega_{p\alpha}^2}{k^2} \sum_{n=-\infty}^{\infty} \int d^3v \frac{J_n^2(a_\alpha)}{\omega - k_z v_z - n\omega_{c\alpha}} \left[k_z \frac{\partial f_{\alpha 0}}{\partial v_z} + \frac{n\omega_{c\alpha}}{v_\perp}\frac{\partial f_{\alpha 0}}{\partial v_\perp}\right] \tag{14.114}$$

where $\omega_{p\alpha}$ is the plasma frequency. The quantity ε_L is the longitudinal dielectric constant. In infinite plasma, the electrostatic dispersion relation is written $\varepsilon_L = 0$. We need the normal component of the electric displacement,

$$D_z(k_x, k_z, \omega) = E_z(k_x, k_z, \omega) + \frac{4\pi i}{\omega} \sum_\alpha e_\alpha \int v_z f_\alpha(\mathbf{k}, \mathbf{v}, \omega) d^3v. \tag{14.115}$$

Using Eq. (14.111) in the above equation gives

$$D_z(k_x, k_z, \omega) = \varepsilon_z(k_x, k_z, \omega) E_z(k_x, k_z, \omega) = -i\frac{k_z\varepsilon_z}{k^2\varepsilon_L} M(k_x, \omega) \tag{14.116}$$

where

$$\varepsilon_z = 1 + \sum_\alpha \frac{\omega_{p\alpha}^2}{\omega k_z} \sum_{n=-\infty}^{\infty} \int d^3v \frac{v_z\ J_n^2(a_\alpha)}{\omega - k_z v_z - n\omega_{c\alpha}} \left[k_z \frac{\partial f_{\alpha 0}}{\partial v_z} + \frac{n\omega_{c\alpha}}{v_\perp}\frac{\partial f_{\alpha 0}}{\partial v_\perp}\right]. \tag{14.117}$$

The vacuum equations, $\nabla \cdot \mathbf{E} = 0$ and $\nabla \times \mathbf{E} = 0$, are solved by

$$E_x(k_x, k_z, z) = F(k_x, \omega)e^{k_x z}, \tag{14.118}$$

$$E_z(k_x, k_z, z) = -iF(k_x, \omega)e^{k_x z}, \tag{14.119}$$

where F is an arbitrary function of (k_x, ω). The two boundary conditions at the interface $z = 0$ are the continuity of E_x and D_z. To reinstate the z-dependence, the Fourier inversion integral should be performed on

Eqs. (14.113) and (14.116). Then the boundary conditions at $z = 0$ give

$$M\,\frac{(-i)}{2\pi}\int_{-\infty}^{\infty} dk_z\,\frac{k_x}{k^2\varepsilon_L} = F, \tag{14.120}$$

$$M\,\frac{(-i)}{2\pi}\int_{-\infty}^{\infty} dk_z\,\frac{k_z\varepsilon_z}{k^2\varepsilon_L} = -iF, \tag{14.121}$$

from which we obtain the kinetic surface wave dispersion relation,

$$\int_{-\infty}^{\infty} \frac{ik_x + k_z\varepsilon_z}{k^2\varepsilon_L}\,dk_z = 0. \tag{14.122}$$

• **Cold plasma limit.** To check the correctness of (14.122), its cold plasma limit will be considered. We put

$$f_{\alpha 0}(\mathbf{v}) = \frac{\delta(v_\perp)}{v_\perp}\delta(v_z)\delta(\varphi)$$

into (14.114) for ε_L. In the limit of $a \to 0$, $J_n(a) \to \delta_{n,\,0}$, and the first term integral involving $\partial f_{\alpha 0}/\partial v_z$ yields, after integration by parts,

$$-\sum_\alpha \frac{\omega_{p\alpha}^2}{\omega^2}\,\frac{k_z^2}{k^2}.$$

For the second term involving $\partial f_{\alpha 0}/\partial v_\perp$, we integrate by parts with respect to $v_\perp$, and use $\frac{2n}{a}J_n(a)J_n'(a) \to \frac{1}{2}(\delta_{n,1} - \delta_{n,-1})$ to obtain

$$-\sum_\alpha \frac{\omega_{p\alpha}^2}{\omega^2 - \omega_{c\alpha}^2}\,\frac{k_x^2}{k^2}.$$

Adding the above two terms give

$$\varepsilon_L^{cold} = 1 - \sum_\alpha \frac{\omega_{p\alpha}^2}{k^2}\left(\frac{k_z^2}{\omega^2} + \frac{k_x^2}{\omega^2 - \omega_{c\alpha}^2}\right). \tag{14.123}$$

A similar calculation for (14.117) yields

$$\varepsilon_z^{cold} = 1 - \sum_\alpha \frac{\omega_{p\alpha}^2}{\omega^2}. \tag{14.124}$$

Putting (14.123) and (14.124) into (14.122), the cold plasma dispersion relation is obtained

$$\int_{-\infty}^{\infty} dk_z \frac{k_z + \frac{ik_x}{1-\sum \omega_{p\alpha}^2/\omega^2}}{k_z^2 + \lambda_z^2} = 0 \tag{14.125}$$

where

$$\lambda_z = k_x \left[\frac{1 - \sum \frac{\omega_{p\alpha}^2}{\omega^2 - \omega_{c\alpha}^2}}{1 - \sum \frac{\omega_{p\alpha}^2}{\omega^2}} \right]^{1/2} . \tag{14.126}$$

(14.126) is the attenuation constant of the surface wave, as will be seen if one inverts Eq. (14.113), and agrees with that obtained in earlier work [H. J. Lee, Plasma Phys. Control. Fusion (1995), Eq. (9)]. Carrying out the contour integral in Eq. (14.125), the final cold plasma dispersion relation is written

$$\frac{k_x}{\lambda_z} + 1 - \sum_\alpha \frac{\omega_{p\alpha}^2}{\omega^2} = 0 \tag{14.127}$$

which agrees with the dispersion relation obtained earlier from the cold fluid equations [H. J. Lee, Plasma Phys. Control. Fusion (1995), Eq. (16)].

• **Plasma slab.** Next, we consider electrostatic surface waves in a magnetized slab which occupies the region $0 < z < L$ bounded by vacuum ($z < 0,\ z > L$). The specular reflection conditions should be satisfied both at $z = 0$ and $z = L$:

$$f_\alpha(v_x, v_y, v_z, x, y, z = 0, t) = f_\alpha(v_x, v_y, -v_z, x, y, z = 0, t), \tag{14.128}$$

$$f_\alpha(v_x, v_y, v_z, x, y, z = L, t) = f_\alpha(v_x, v_y, -v_z, x, y, z = L, t). \tag{14.129}$$

In order to satisfy (14.128), the electric field should be extended as in (14.110):

$$E_x(-z) = E_x(z), \quad E_z(-z) = -E_z(z). \tag{14.130}$$

To enforce (14.129), we note that Eq. (14.109) is invariant under the reflections ($z \to 2L - z, v_z \to -v_z$) provided the electric field is extended as

$$E_x(2L - z) = E_x(z), \quad E_z(2L - z) = -E_z(z). \tag{14.131}$$

Then the solution of (14.109), $f(z, v_z)$, solved in the infinite region $-\infty < z < \infty$ with the extended fields given by (14.130) and (14.131), formally satisfy $f(z, v_z) = f(-z, -v_z)$ *and* $f(z, v_z) = f(2L - z, -v_z)$, and the

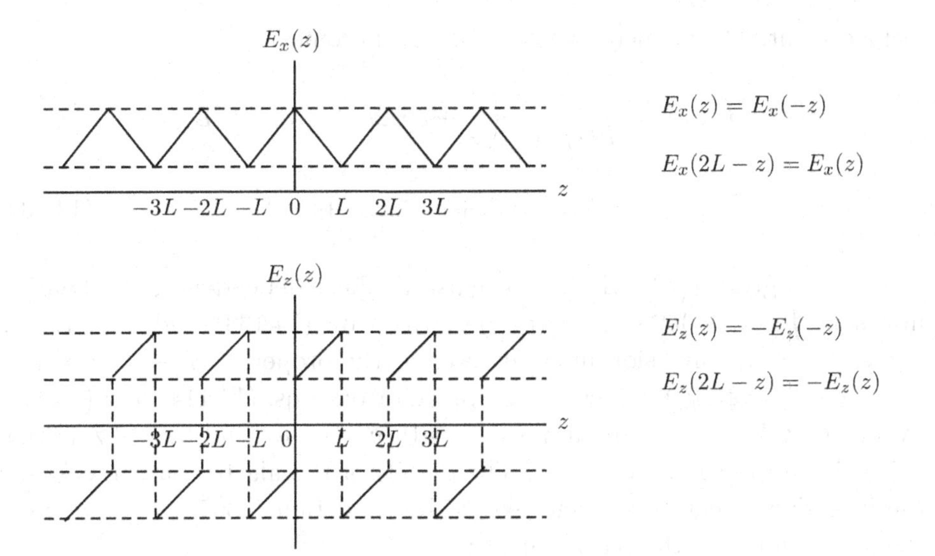

Fig. 14.3. Extension of electric and magnetic fields in a plasma slab ($0 < z < L$).

conditions (14.128) and (14.129) are automatically satisfied. The electric field components extended in the fashion of (14.130) and (14.131) are plotted in Fig. 14.3 in which one can ascertain $E_z(2L-0^+) = -E_z(0^+)$, $E_z(2L-0^-) = -E_z(0^-)$, etc. In the Fourier transform of (14.109), the jumps of E_z occurring at the discontinuities should be carefully accounted for.

We write the Fourier transform of (14.106):

$$
\begin{aligned}
i\mathbf{k}\cdot\mathbf{E}(k_x,k_z,\omega) &+ 4E_z(k_x, z=0^-,\omega)\left(\frac{1}{2} + cos\ 2k_zL + cos\ 4k_zL + \cdots\right)\\
&+ 4E_z(k_x, z=L^-,\omega)(cos\ k_zL + cos\ 3k_zL + \cdots)\\
&= 4\pi\sum_\alpha e_\alpha \int f_\alpha(\mathbf{k},\mathbf{v},\omega)d^3v. \qquad (14.132)
\end{aligned}
$$

If all the *cosine* functions are simply discarded, (14.132) reduces to (14.112). Using (14.111) (with $n = l$) in (14.132) gives

$$
\begin{aligned}
E_i(k_x,k_z,\omega) = \frac{4ik_i}{k^2\varepsilon_L}\bigg[&A\left(\frac{1}{2} + cos2k_z + cos4k_zL + \cdots\right)\\
&+ B(cosk_zL + cos3k_zL + \cdots)\bigg] \qquad (14.133)
\end{aligned}
$$

where the subscript $i = x, z$, $A = E_z(k_x, z = 0^-,\omega), B = E_z(k_x, z = L^-,\omega)$, and ε_L is given by (14.114). The normal component of the electric displacement, D_z, defined by (14.115), now becomes

$$D_z(k_x, k_z, \omega) = \frac{4ik_z\varepsilon_z}{k^2\varepsilon_L}\left[A\left(\frac{1}{2} + cos2k_zL + cos4k_zL + \cdots\right)\right.$$
$$\left. + B(cosk_zL + cos3k_zL + \cdots)\right] \tag{14.134}$$

where ε_z is given by (14.117). To reinstate the z-dependence, the Fourier-inversion $\int dk_z(\cdots)e^{ik_z z}$ must be performed on (14.133) and (14.134). In carrying out the inversion integrals, we use the property, $\varepsilon_L(-k_z) = \varepsilon_L(k_z)$ and $\varepsilon_z(-k_z) = \varepsilon_z(k_z)$, which can be proven from Eqs. (14.114) and (14.117) by changing $k_z \to -k_z$ simultaneously with $v_z \to -v_z$, and by utilizing that $f_0(v_z)$ is an even function of v_z. The details of the algebra in carrying out the inversion integrals are referred to [Lee and Lim (2007)], and we write the final results for the plasma fields:

$$E_x(k_x, \omega, z = 0) = \frac{2ik_x}{\pi}\int_{-\infty}^{\infty}\frac{dk_z}{k^2\varepsilon_L}\left[A\left(\frac{1}{2} + e^{2ik_zL} + e^{4ik_zL} + \cdots\right)\right.$$
$$\left. + B(e^{ik_zL} + e^{3ik_zL} + \cdots)\right], \tag{14.135}$$

$$E_x(k_x, \omega, z = L) = \frac{2ik_x}{\pi}\int_{-\infty}^{\infty}\frac{dk_z}{k^2\varepsilon_L}\left[B\left(\frac{1}{2} + e^{2ik_zL} + e^{4ik_zL} + \cdots\right)\right.$$
$$\left. + A(e^{ik_zL} + e^{3ik_zL} + \cdots)\right], \tag{14.136}$$

$$D_z(k_x, \omega, z = 0) = \frac{i}{\pi}A\int_{-\infty}^{\infty} dk_z\,\frac{k_z\varepsilon_z}{k^2\varepsilon_L}, \tag{14.137}$$

$$D_z(k_x, \omega, z = L) = \frac{-i}{\pi}B\int_{-\infty}^{\infty} dk_z\,\frac{k_z\varepsilon_z}{k^2\varepsilon_L}. \tag{14.138}$$

The vacuum solutions for the regions $z < 0$ and $z > L$ are

$$E_x(k_x, z, \omega) = F_1e^{-k_x z}, \quad E_z(k_x, z, \omega) = iF_1e^{-k_x z}, \quad (z > L) \tag{14.139}$$

$$E_x(k_x, z, \omega) = F_2e^{k_x z}, \quad E_z(k_x, z, \omega) = -iF_2e^{k_x z}, \quad (z < 0) \tag{14.140}$$

where F_1 and F_2 are arbitrary functions of (k_x, ω). The plasma solutions and the vacuum solutions should be matched at the interfaces $z = 0$ and $z = L$ by the electromagnetic boundary conditions which are the continuity of E_x and D_z. These conditions give four algebraic equations for four constants,

F_1, F_2, A, and B. After some algebra we can write the final result

$$\int_{-\infty}^{\infty} dk_z \, \frac{k_z \varepsilon_z}{k^2 \varepsilon_L} + \int_{-\infty}^{\infty} dk_z \, \frac{ik_x}{k^2 \varepsilon_L} \left(\frac{1 \mp e^{ik_z L}}{1 \pm e^{ik_z L}} \right) = 0 \quad (for\ x\text{−}wave,\ slab). \tag{14.141}$$

The surface waves in a plasma slab always have two modes as represented by the $\pm$ signs, which correspond to symmetric (upper sign) and anti-symmetric (lower sign) mode, respectively. The apparent singularities located at $1 \pm e^{ik_z L} = 0$ in the second integral should be simply ignored. When $L \to \infty$, the exponential factor vanishes (by picking up the poles in the upper-half k_z-plane), and both modes in (14.141) coalesce to the dispersion relation of x–wave in the semi-infinite plasma, Eq. (14.122).

• **Cold plasma limit.** The cold plasma limit of the slab dispersion relation of x-wave is obtained by substituting (14.123) and (14.124) into (14.141):

$$\int_{-\infty}^{\infty} \frac{dk_z k_z}{k_z^2 + \lambda_z^2} + \frac{ik_x}{1 - \sum \frac{\omega_{p\alpha}^2}{\omega^2}} \int_{-\infty}^{\infty} \frac{dk_z}{k_z^2 + \lambda_z^2} \left(\frac{1 \mp e^{ik_z L}}{1 \pm e^{ik_z L}} \right) = 0 \tag{14.142}$$

where λ_z is defined by (14.126). Carrying out the above contour integral, the cold slab dispersion relation for 1-wave is obtained

$$1 + \frac{k_x/\lambda_z}{1 - \sum \frac{\omega_{p\alpha}^2}{\omega^2}} \, tanh \frac{\lambda_z L}{2} = 0 \quad (symmetric\ mode), \tag{14.143}$$

$$1 + \frac{k_x/\lambda_z}{1 - \sum \frac{\omega_{p\alpha}^2}{\omega^2}} \, coth \frac{\lambda_z L}{2} = 0 \quad (anti\text{-}symmetric\ mode). \tag{14.144}$$

If $L \to \infty$, (14.143) and (14.144) both reduce to (14.127).

14.4.2. *Semi-infinite plasma:* $\mathbf{B}_0 = \hat{\mathbf{z}} B_0$; $\hat{\mathbf{n}} = \hat{\mathbf{x}}$; $\mathbf{k} \parallel \hat{\mathbf{z}}$: *z–wave*

The plasma occupying the region $(x > 0)$ is bounded by vacuum $(x < 0)$. The surface wave propagates along the z- direction $(\propto e^{ik_z z - i\omega t})$. The kinematic boundary condition for f is

$$f_\alpha(v_x, v_y, v_z, x = 0, y, z, t) = f_\alpha(-v_x, v_y, v_z, x = 0, y, z, t). \tag{14.145}$$

At first sight, the boundary condition (14.145) cannot be satisfied by making (14.109) invariant under the reflections $(x \to -x, v_x \to -v_x)$ by simple extension of E_x and E_z, because of the $\omega_{c\alpha}$-dependent two terms. However, we should notice that f_α is an even function with respect to simultaneous reflections of v_x and v_y (as is evident from (14.111)), independently of the manner that the electric field is extended, i.e.

$$f_\alpha(v_x, v_y) = f_\alpha(-v_x, -v_y); \quad \textit{inherent symmetry}. \tag{14.146}$$

Therefore, the specular reflection condition (14.145) can be satisfied provided the invariant below is established by extension of the field components:

$$f_\alpha(v_x, v_y, v_z, x, y, z, t) = f_\alpha(-v_x, -v_y, v_z, -x, y, z, t). \tag{14.147}$$

It can be immediately seen that Eq. (14.109) is invariant under the reflections $(x \to -x, v_x \to -v_x, v_y \to -v_y)$ by extending the electric field as

$$E_x(-x) = -E_x(x), \quad E_z(-x) = E_z(x) \quad for\ z-wave. \tag{14.148}$$

(14.111), the solution of (14.109) for the infinite region $-\infty < x, z < \infty$ with the extended fields can be also used for the analysis of this subsection, but the Fourier transform of the Poisson equation, (14.106) becomes different due to different extension, (14.148). Note in (14.111) that $\varphi = tan^{-1}\frac{v_y}{v_x}$ and f_α is an even function of (v_x, v_y) under the simultaneous reflections $v_x \to -v_x$ and $v_y \to -v_y$, thus satisfying (14.146). We also expect that the wave variable k_x and the Fourier-inversion variable k_z in x–wave interchange the roles for z–wave. Fourier transform of (14.106) gives

$$ik_xE_x(k_x, k_z, \omega) + ik_zE_z(k_x, k_z, \omega) = 4\pi \sum_\alpha e_\alpha \int f_\alpha(\mathbf{k}, \mathbf{v}, \omega)d^3v + N(k_z, \omega) \tag{14.149}$$

where $N(k_z, \omega) = 2E_x(k_z, \omega, x = 0^+)$ comes from the discontinuity of $E_x(x)$ at $x = 0$. (14.111) and (14.149) give

$$E_j(k_x, k_z, \omega) = -i\frac{k_j}{k^2\varepsilon_L}\, N(k_z, \omega) \quad (j = x,\ z) \tag{14.150}$$

where ε_L is given by Eq. (14.114). We need the normal component of the electric displacement,

$$D_x(k_x, k_z, \omega) = E_x(k_x, k_z, \omega) + \frac{4\pi i}{\omega}\sum_\alpha e_\alpha \int v_x f_\alpha(\mathbf{k}, \mathbf{v}, \omega)d^3v. \tag{14.151}$$

Using Eq. (14.111) in the above equation gives

$$D_x(k_x, k_z, \omega) = \varepsilon_x(k_x, k_z, \omega) E_x(k_x, k_z, \omega) = -i\, \frac{k_x \varepsilon_x}{k^2 \varepsilon_L} N(k_z, \omega), \quad (14.152)$$

$$\varepsilon_x = 1 + \sum_\alpha \frac{\omega_{p\alpha}^2}{\omega k_x} \sum_{n=-\infty}^{\infty} \sum_{l=-\infty}^{\infty} \int d^3v\, \frac{e^{i(n-l)\varphi} v_\perp\, cos\varphi J_n(a_\alpha) J_l(a_\alpha)}{\omega - k_z v_z - l\omega_{c\alpha}} \times \left[k_z\, \frac{\partial f_{\alpha 0}}{\partial v_z} + \frac{l\omega_{c\alpha}}{v_\perp} \frac{\partial f_{\alpha 0}}{\partial v_\perp} \right].$$

The integral is non-vanishing only when $n = l \pm 1$. Using the formula $J_{l-1}(a) + J_{l+1}(a) = \frac{2l}{a} J_l(a)$, one obtains

$$\varepsilon_x = 1 + \sum_\alpha \frac{\omega_{p\alpha}^2}{\omega k_x} \int d^3v \sum_{n=-\infty}^{\infty} \frac{\frac{n}{a_\alpha} J_n^2(a_\alpha)}{\omega - k_z v_z - n\omega_{c\alpha}} \left[n\omega_{c\alpha} \frac{\partial f_{\alpha 0}}{\partial v_\perp} + k_z v_\perp\, \frac{\partial f_{\alpha 0}}{\partial v_z} \right]. \quad (14.153)$$

The vacuum solutions are slightly different from (14.118) and (14.119):

$$E_x(k_z, \omega, x) = -iG(k_z, \omega) e^{k_z x}, \quad (14.154)$$

$$E_z(k_z, \omega, x) = G(k_z, \omega) e^{k_z x}, \quad (14.155)$$

where G is an arbitrary function of (k_z, ω). The two boundary conditions on the interface $x = 0$ are the continuity of E_z and D_x. Omitting the intermediate steps, we write the dispersion relation,

$$\int_{-\infty}^{\infty} \frac{ik_z + k_x \varepsilon_x}{k^2 \varepsilon_L}\, dk_x = 0. \quad (14.156)$$

(14.156) should be compared with (14.122). They have an identical structure, but the difference between ε_x and ε_z should be noted; compare (14.117) and (14.153).

• **Cold plasma limit.** We need to calculate the cold plasma limit of ε_x, (14.153), as was done earlier. Owing to the relation, $\lim_{a \to 0} \frac{n}{a} J_n^2(a) = 0$, the second integral involving $\partial f_{\alpha 0}/\partial v_z$ vanishes. The first integral is integrated by parts and use is made of the relation, $\frac{2n}{a} J_n(a) J_n'(a) \to \frac{1}{2}(\delta_{n,1} - \delta_{n,-1})$

to obtain

$$\varepsilon_x^{cold} = 1 - \sum_\alpha \frac{\omega_{p\alpha}^2}{\omega^2 - \omega_{c\alpha}^2}. \tag{14.157}$$

Using (14.123) and (14.157) in (14.156) yields the cold plasma dispersion relation,

$$\int_{-\infty}^{\infty} dk_x \, \frac{k_x + \frac{ik_z}{1-\sum \frac{\omega_{p\alpha}^2}{\omega^2-\omega_{c\alpha}^2}}}{k_x^2 + \lambda_x^2} = 0 \tag{14.158}$$

where

$$\lambda_x = k_z \left[\frac{1 - \sum \frac{\omega_{p\alpha}^2}{\omega^2}}{1 - \sum \frac{\omega_{p\alpha}^2}{\omega^2 - \omega_{c\alpha}^2}} \right]^{1/2} \tag{14.159}$$

agrees with the attenuation constant obtained in [H. J. Lee, Plasma Phys. Control. Fusion (1995) Eq. (30)]. Carrying out the contour integral in (14.158) gives the dispersion relation

$$\frac{k_z}{\lambda_x} + 1 - \sum_\alpha \frac{\omega_{p\alpha}^2}{\omega^2 - \omega_{c\alpha}^2} = 0 \tag{14.160}$$

which agrees with the dispersion relation obtained in [H. J. Lee, Plasma Phys. Control. Fusion (1995) Eq. (33)]. Finally we notice from Eqs. (14.126) and (14.159) that the ratio of the attenuation constant to the wave number (λ/k) of x– and z–waves is reciprocal to each other. This reciprocity is discussed in [H. J. Lee Plasma, Phys. Control. Fusion (1995)].

• **Plasma slab.** The slab dispersion relation for z–wave can be written down from the x–wave slab dispersion relation without going through the detail of steps by interchanging the subscripts x and z:

$$\int_{-\infty}^{\infty} dk_x \, \frac{k_x \varepsilon_x}{k^2 \varepsilon_L} + \int_{-\infty}^{\infty} dk_x \, \frac{ik_z}{k^2 \varepsilon_L} \left(\frac{1 \mp e^{ik_x L}}{1 \pm e^{ik_x L}} \right) = 0 \quad (for\ z\text{–}wave\ slab). \tag{14.161}$$

Here again the upper (lower) sign corresponds to the symmetric (anti-symmetric) mode. If $L \to \infty$, (14.161) reduces to (14.156).

• **Cold plasma limit of the slab dispersion relation.** The cold plasma limit of of the slab dispersion relation for z–wave is obtained by substituting

(14.123) and (14.157) into (14.161) and by carrying out the contour integral:

$$1+\frac{k_z/\lambda_x}{1-\sum\frac{\omega_{p\alpha}^2}{\omega^2-\omega_{c\alpha}^2}}\ tanh\frac{\lambda_x L}{2}=0 \quad (symmetric\ mode) \tag{14.162}$$

$$1+\frac{k_z/\lambda_x}{1-\sum\frac{\omega_{p\alpha}^2}{\omega^2-\omega_{c\alpha}^2}}\ coth\frac{\lambda_x L}{2}=0 \quad (anti\text{-}symmetric\ mode) \tag{14.163}$$

where λ_x is defined by (14.159). If $L \to \infty$, (14.162) and (14.163) both reduce to (14.160).

Exercise. Check the correctness of (14.162) and (14.163) by deriving them from fluid equations.

Solution: the electric potential for z–wave $\phi(x,z,t)=\phi(x)e^{ik_z z-i\omega t}$ satisfies, in the plasma region,

$$\frac{d^2\phi}{dx^2}-\lambda_x^2\phi=0 \quad (0<x<L) \tag{14.164}$$

where λ_x is given by Eq. (14.159). Solving (14.164) yields

$$\phi(x)=A_1e^{-\lambda_x x}+A_2e^{\lambda_x x} \quad (0<x<L) \tag{14.165}$$

with constants A_1 and A_2. In the vacuum region, we have $\nabla^2\phi=0$, which is solved by

$$\phi(x)=F_1e^{-k_z x}, \quad (x>L) \tag{14.166}$$

$$\phi(x)=F_2e^{k_z x}, \quad (x<0) \tag{14.167}$$

with constants F_1 and F_2. The boundary conditions, the continuity of ϕ at $x=0$ and $x=L$ gives

$$F_2=A_1+A_2, \tag{14.168}$$

$$F_1e^{-k_z L}=A_1e^{-\lambda_x L}+A_2e^{\lambda_x L}. \tag{14.169}$$

We have for the normal component of the electric displacement,

$$D_x=\varepsilon_x^{cold}\frac{\partial\phi}{\partial x}$$

in the plasma region. The continuity of D_x gives

$$F_2k_z=\lambda_x\ \varepsilon_x^{cold}(A_2-A_1), \tag{14.170}$$

$$F_1k_ze^{-k_z L}=\lambda_x\ \varepsilon_x^{cold}(A_1e^{-\lambda_x L}-A_2e^{\lambda_x L}). \tag{14.171}$$

The solvability condition for the above four equations yields

$$1 + \frac{\lambda_x}{k_z}\varepsilon_x^{cold} = \pm e^{-\lambda_x L}\left(1 - \frac{\lambda_x}{k_z}\varepsilon_x^{cold}\right) \tag{14.172}$$

which can be arranged into the form (14.162) and (14.163).

Remark. We have three important electromagnetic coefficients involved in the surface wave dispersion relations: ε_x, ε_z, and ε_L. In our specification of the coordinate system ($k_y = 0$) and electric field ($E_y = 0$), the dielectric tensor ε_{ij} ($i, j = x, y, z$) has nonzero elements only at the four corners; a tensor element attached anywhere by the subscript 'y' is zero. The dielectric tensors in the magnetized plasma are obtained in Chapter 6, and we here use the expressions derived therein.

$$\varepsilon_{xx} = 1 + \sum_\alpha \frac{\omega_{p\alpha}^2}{\omega^2}\sum_n \int d^3v \frac{v_\perp \frac{n^2 J_n^2(a_\alpha)}{a_\alpha^2}\left[(\omega - k_z v_z)\frac{\partial f_{\alpha 0}}{\partial v_\perp} + k_z v_\perp \frac{\partial f_{\alpha 0}}{\partial v_z}\right]}{\omega - k_z v_z - n\omega_{c\alpha}},$$

$$\varepsilon_{xz} = \sum_\alpha \frac{\omega_{p\alpha}^2}{\omega^2}\sum_n \int d^3v \frac{v_\perp \frac{n J_n^2(a_\alpha)}{a_\alpha}\left[(\omega - n\omega_{c\alpha})\frac{\partial f_{\alpha 0}}{\partial v_z} + n\omega_{c\alpha}\frac{v_z}{v_\perp}\frac{\partial f_{\alpha 0}}{\partial v_\perp}\right]}{\omega - k_z v_z - n\omega_{c\alpha}},$$

$$\varepsilon_{zz} = 1 + \sum_\alpha \frac{\omega_{p\alpha}^2}{\omega^2}\sum_n \int d^3v \frac{v_z J_n^2(a_\alpha)\left[(\omega - n\omega_{c\alpha})\frac{\partial f_{\alpha 0}}{\partial v_z} + n\omega_{c\alpha}\frac{v_z}{v_\perp}\frac{\partial f_{\alpha 0}}{\partial v_\perp}\right]}{\omega - k_z v_z - n\omega_{c\alpha}},$$

$$\varepsilon_{zx} = \sum_\alpha \frac{\omega_{p\alpha}^2}{\omega^2}\sum_n \int d^3v \frac{v_z \frac{n J_n^2(a_\alpha)}{a_\alpha}\left[(\omega - k_z v_z)\frac{\partial f_{\alpha 0}}{\partial v_\perp} + k_z v_\perp \frac{\partial f_{\alpha 0}}{\partial v_z}\right]}{\omega - k_z v_z - n\omega_{c\alpha}}.$$

In the above equations, it appears that ε_{xz} is different from ε_{zx}. But it can be proven that they are the same and ε_{ij} satisfies Onsager's relation. According to the relation, $D_i = \varepsilon_{ij}E_j$, we have

$$D_x = \varepsilon_{xx}E_x + \varepsilon_{xz}E_z \equiv \varepsilon_x E_x, \tag{14.173}$$

$$D_z = \varepsilon_{zx}E_x + \varepsilon_{zz}E_z \equiv \varepsilon_z E_z. \tag{14.174}$$

Using the above tensor elements, (14.173) and (14.174) can be shown to agree with (14.153) and (14.117), respectively. Also ε_L given by (14.114) can be legitimately called the longitudinal dielectric permittivity, since it agrees with the general definition $\varepsilon_L = k_i k_j \varepsilon_{ij}/k^2$.

14.4.3. *Slab plasma:* $\mathbf{B}_0 = \hat{\mathbf{z}}B_0$; $\hat{\mathbf{n}} = \hat{\mathbf{x}}$; $\mathbf{k} \parallel \hat{\mathbf{y}}$: *y–wave*

We consider a slab plasma occupying the region ($0 < x < L$) which is bounded by vacuum ($x < 0$, $x > L$). The surface wave is assumed to propagate on the (y, z)-plane with the phasor $\propto e^{ik_y y + ik_z z - i\omega t}$. We deal with a slab plasma, which is a more complicated problem than the semi-infinite problem. By taking the limit the slab thickness $\to \infty$, we will obtain the dispersion relation of the semi-infinite plasma. Solving the dispersion relation and reducing it to the friendly integral form requires ingenuity. Success of this complex case confirms the correctness of our theoretical method. Inclusion of the z-dependence is a slight generalization. The kinematic boundary condition for f is

$$f_\alpha(v_x, v_y, v_z, x = 0, y, z, t) = f_\alpha(-v_x, v_y, v_z, x = 0, y, z, t), \tag{14.175}$$

$$f_\alpha(v_x, v_y, v_z, x = L, y, z, t) = f_\alpha(-v_x, v_y, v_z, x = L, y, z, t). \tag{14.176}$$

We Fourier transform (14.105) and (14.106) with respect to the spatial coordinates y and z and time t:

$$\begin{aligned} &v_x \frac{\partial}{\partial x} f_\alpha(x, k_y, k_z, v_x, v_y, v_z, \omega) - i(\omega - k_y v_y - k_z v_z) f_\alpha \\ &\quad + \omega_{c\alpha}\left(v_y \frac{\partial}{\partial v_x} - v_x \frac{\partial}{\partial v_y}\right) f_\alpha \\ &= \frac{e_\alpha}{m_\alpha}\left(\frac{\partial f_{\alpha 0}}{\partial v_x}\frac{\partial}{\partial x} + ik_y \frac{\partial f_{\alpha 0}}{\partial v_y} + ik_z \frac{\partial f_{\alpha 0}}{\partial v_z}\right)\phi(x, k_y, k_z, \omega), \quad \left(\omega_{c\alpha} = \frac{e_\alpha B_0}{m_\alpha c}\right) \end{aligned} \tag{14.177}$$

$$\frac{\partial^2 \phi}{\partial x^2} - (k_y^2 + k_z^2)\phi(x, k_y, k_z, \omega) = -4\pi \sum_\alpha e_\alpha \int d^3 v f_\alpha(x, k_y, k_z, \mathbf{v}, \omega). \tag{14.178}$$

Now the specular reflection conditions take the following forms in which the spatial coordinates y, z, and time t are replaced by the Fourier variables k_y, k_z, and ω, respectively:

$$f_\alpha(v_x, v_y, v_z, x = 0, k_y, k_z, \omega) = f_\alpha(-v_x, v_y, v_z, x = 0, k_y, k_z, \omega), \tag{14.179}$$

$$f_\alpha(v_x, v_y, v_z, x = L, k_y, k_z, \omega) = f_\alpha(-v_x, v_y, v_z, x = L, k_y, k_z, \omega). \tag{14.180}$$

(14.179) is automatically satisfied if $f_\alpha(x, v_x) = f_\alpha(-x, -v_x)$ (with other variables untouched), i.e., if (14.177) is invariant under the simple reflection $(x \to -x, v_x \to -v_x)$. It is easy to see that this simple reflection does not work. So we reflect as

$$x \to -x, \quad v_x \to -v_x, \quad v_y \to -v_y, \quad k_y \to -k_y.$$

Then the function $f_\alpha(-x, -k_y, k_z, -v_x, -v_y, v_z, \omega)$ satisfies the same equation (14.177) *provided* we require

$$\phi(x, k_y, k_z, \omega) = \phi(-x, -k_y, k_z, \omega). \tag{14.181}$$

Note that the above relation is equivalent to $E_x(x) = -E_x(-x)$. In fact, we satisfy the specular reflection boundary condition on $x = 0$, because we have

$$f_\alpha(v_x, v_y, v_z, x = 0, k_y, k_z, \omega) = f_\alpha(-v_x, -v_y, v_z, x = 0, -k_y, k_z, \omega).$$

We see that f_α has the symmetry with respect to the variables (x, k_y, v_x, v_y):

$$f_\alpha(x, k_y, k_z, v_x, v_y, v_z, \omega) = f_\alpha(-x, -k_y, k_z, -v_x, -v_y, v_z, \omega).$$

The above symmetry is observed from (14.188) which is the solution of (14.177) in terms of the Fourier variables: f_α is an even function with respect to simultaneous reflections of v_x and v_y, and also an even function with respect to simultaneous reflections of k_x and k_y, because f_α is a function of $v_\perp = \sqrt{v_x^2 + v_y^2}$, $\varphi = tan^{-1}\frac{v_y}{v_x}$, and $k_\perp = \sqrt{k_x^2 + k_y^2}$, $\theta = tan^{-1}\frac{k_y}{k_x}$.

To enforce (14.176), we note that (14.177) is invariant under the reflections $(x \to 2L - x, v_x \to -v_x, k_y \to -k_y, v_y \to -v_y)$, provided the potential is extended as

$$\phi(x, k_y, k_z, \omega) = \phi(2L - x, -k_y, k_z, \omega). \tag{14.182}$$

The consequences of the extensions made in (14.181) and (14.182) should be carefully collected in the Fourier transform of (14.148). It is sufficient to consider the x-dependence of $\phi(x, k_y)$ in carrying out the Fourier transform of (14.178) over the entire region $-\infty < x < \infty$. The function $\phi(x)$ which has the properties

$$\phi(x) = \phi(-x) \quad and \quad \phi(x) = \phi(2L - x) \tag{14.183}$$

has discontinuities of $\frac{\partial \phi}{\partial x}$ at $x = 0,\ 2L,\ 4L, \cdots$ of the same value. The derivative $\frac{\partial \phi}{\partial x}$ has another value of jump at $x = L,\ 3L,\ 5L, \cdots$. Thus we have for the Fourier transform (with respect to x) of (14.178),

$$k^2\phi(\mathbf{k},\omega) - 4\pi\sum_\alpha e_\alpha \int f_\alpha(\mathbf{k},\mathbf{v},\omega)d^3v$$
$$= A(k_y,k_z,\omega)\left(\frac{1}{2} + cos2k_xL + cos4k_xL + \cdots\right)$$
$$+ B(k_y,k_z,\omega)(cosk_xL + cos3k_xL + \cdots) \tag{14.184}$$

where $A = 4\,\frac{\partial\phi}{\partial x}\,|_{x=0^-}$ and $B = 4\,\frac{\partial\phi}{\partial x}\,|_{x=L^-}$. We introduce cylindrical coordinates in the velocity space such that $v_x = v_\perp cos\varphi$, $v_y = v_\perp sin\varphi$, $k_x = k_\perp cos\theta$, $k_y = k_\perp sin\theta$. (14.177) is integrated to yield [consult Chapter 6]

$$f_\alpha(\mathbf{k},\mathbf{v},\omega) = -\frac{e_\alpha\phi(\mathbf{k},\omega)}{m_\alpha}\sum_{n=-\infty}^{\infty}\sum_{l=-\infty}^{\infty}\left(\frac{\partial f_{\alpha 0}}{\partial v_\perp}\frac{l\omega_{c\alpha}}{v_\perp} + k_z\frac{\partial f_{\alpha 0}}{\partial v_z}\right)$$
$$\times J_n(a_\alpha)J_l(a_\alpha)\frac{e^{i(l-n)\theta}e^{i(n-l)\varphi}}{\omega - k_zv_z - l\omega_{c\alpha}} \tag{14.185}$$

where $a_\alpha = \frac{k_\perp v_\perp}{\omega_{c\alpha}}$. Substituting (14.185) into (14.184), we can obtain

$$\phi(\mathbf{k},\omega) = \frac{R(k_y,k_z,\omega)}{k^2\varepsilon_L} \tag{14.186}$$

where $R(k_y,k_z,\omega)$ is the right hand side of (14.184) and ε_L, the dielectric constant of the magnetized plasma under consideration, is given by (14.114). In an infinite plasma, $R = 0$ and $\varepsilon_L = 0$ gives the dispersion relation of the electrostatic waves. From (14.186), one obtains the electric field components,

$$E_j(\mathbf{k},\omega) = -i\,\frac{k_jR}{k^2\varepsilon_L}\quad (j = x,y,z). \tag{14.187}$$

We need the normal component of the electric displacement.

$$D_x(\mathbf{k},\omega) = E_x(\mathbf{k},\omega) + \frac{4\pi i}{\omega}\sum_\alpha e_\alpha \int v_xf_\alpha(\mathbf{k},\mathbf{v},\omega)d^3v. \tag{14.188}$$

Using (14.185) in the above equation, with the aid of $\phi = i\frac{E_x}{k_x}$, gives

$$D_x = E_x + \frac{E_x}{\omega k_x}\sum_\alpha \frac{\omega_{p\alpha}^2}{2}\int d^3v\, v_\perp \sum_n\sum_l \left(\frac{\partial f_{\alpha 0}}{\partial v_\perp}\frac{l\omega_{c\alpha}}{v_\perp} + k_z\frac{\partial f_{\alpha 0}}{\partial v_z}\right)$$
$$\times \frac{J_n J_l\, e^{i(l-n)\theta}}{\omega - k_z v_z - l\omega_{c\alpha}}[e^{i(n-l+1)\varphi} + e^{i(n-l-1)\varphi}]. \qquad (14.189)$$

In the above integral, only the terms of $n = l \pm 1$ survive in the $d\varphi$-integral, and we obtain

$$D_x(\mathbf{k},\omega) = \varepsilon_x(\mathbf{k},\omega)E_x(\mathbf{k},\omega) = -i\frac{k_x\varepsilon_x}{k^2\varepsilon_L}R(k_y,k_z,\omega) \qquad (14.190)$$

where

$$\varepsilon_x = 1 + \sum_\alpha \frac{\omega_{p\alpha}^2}{\omega k_\perp}\int d^3v\, v_\perp \sum_{l=-\infty}^{\infty}\left(\frac{\partial f_{\alpha 0}}{\partial v_\perp}\frac{l\omega_{c\alpha}}{v_\perp} + k_z\frac{\partial f_{\alpha 0}}{\partial v_z}\right)$$
$$\times \frac{1}{\omega - k_z v_z - l\omega_{c\alpha}}\left(\frac{lJ_l^2}{a_\alpha} + i\frac{k_y}{k_x}J_l J_l'\right)\quad \left(J_l' = \frac{dJ_l(a_\alpha)}{da_\alpha}\right). \qquad (14.191)$$

• **Vacuum solution.** The vacuum equations $\nabla^2\phi = 0$ is solved for $x < 0$ and $x > L$

$$x < 0: \quad \phi(x,k_y,k_z,\omega) = F_1 e^{\kappa x + ik_y y + ik_z z - i\omega t}, \qquad (14.192)$$
$$x > L: \quad \phi(x,k_y,k_z,\omega) = F_2 e^{-\kappa x + ik_y y + ik_z z - i\omega t}, \qquad (14.193)$$

where F_1 and F_2 are arbitrary constants, and

$$\kappa = \sqrt{k_y^2 + k_z^2}.$$

The electric components are given by

$$x < 0: \quad E_x = -\kappa F_1 e^{\kappa x},\ E_y = -ik_y F_1 e^{\kappa x},\ E_z = -ik_z F_1 e^{\kappa x}, \qquad (14.194)$$
$$x > L: \quad E_x = \kappa F_2 e^{-\kappa x},\ E_y = -ik_y F_2 e^{-\kappa x},\ E_z = -ik_z F_2 e^{-\kappa x}, \qquad (14.195)$$

where we omitted the phasor $e^{ik_y y + ik_z z - i\omega t}$. The boundary conditions at the interface $x = 0$ and $x = L$ are the continuity of E_y (or E_z) and D_x. The conditions on E_z are redundant to the conditions on E_y. To reinstate the x-dependence, the Fourier inversion integral should be performed on (14.187) and (14.190). Let us first consider the integral:

$$E_y(k_y, k_z, \omega, x) = -ik_y \int_{-\infty}^{\infty} \frac{dk_x e^{ik_x x}}{k^2 \varepsilon_L} \left[A \left(\frac{1}{2} + cos2k_x L + cos4k_x L + \cdots \right) \right.$$
$$\left. + B(cosk_x L + cos3k_x L + cos5k_x L + \cdots) \right]. \quad (14.196)$$

It is important to do the integral of the series term by term. Here we have

$$\int_{-\infty}^{\infty} dk_x \, \frac{e^{ixk_x}}{k^2 \varepsilon_L} cos(nk_x L)$$
$$= \frac{1}{2} \int_{-\infty}^{\infty} dk_x \, \frac{e^{ixk_x}}{k^2 \varepsilon_L} (e^{ink_x L} + e^{-ink_x L}). \quad (n = 1, 2, 3, \ldots)$$

In the second integral, we make a change of variable $k_x \to -k_x$ and utilize that ε_L is an even function of k_x. Then we have

$$\int_{-\infty}^{\infty} dk_x \, \frac{e^{ixk_x}}{k^2 \varepsilon_L} cos(nk_x L) = \int_{-\infty}^{\infty} dk_x \, \frac{e^{ik_x nL}}{k^2 \varepsilon_L} cos(k_x x). \quad (14.197)$$

Using (14.197) in (14.196) gives

$$E_y(k_y, k_z, \omega, x = 0) = -ik_y \int_{-\infty}^{\infty} \frac{dk_x}{k^2 \varepsilon_L} \left[A \left(\frac{1}{2} + e^{2ik_x L} + e^{4ik_x L} + \cdots \right) \right.$$
$$\left. + B(e^{ik_x L} + e^{3ik_x L} + e^{5ik_x L} + \cdots) \right], \quad (14.198)$$

$$E_y(k_y, k_z, \omega, x = L)$$
$$= -ik_y \int_{-\infty}^{\infty} dk_x \left[A \left(\frac{1}{2} e^{iLk_x} + cosk_x L(e^{2ik_x L} + e^{4ik_x L} + \cdots) \right. \right.$$
$$\left. + B \; cosk_x L(e^{ik_x L} + e^{3ik_x L} + e^{5ik_x L} + \cdots) \right].$$

Writing the *cosine* functions in terms of *exponential* functions, one can easily obtain

$$E_y(k_y, k_z, \omega, x = L) = -ik_y \int_{-\infty}^{\infty} \frac{dk_x}{k^2 \varepsilon_L}$$
$$\times \left[B \left(\frac{1}{2} + e^{2ik_x L} + e^{4ik_x L} + \cdots \right) + A(e^{ik_x L} + e^{3ik_x L} + e^{5ik_x L} + \cdots) \right].$$
$$(14.199)$$

Next, we calculate the integral

$$D_x(k_y, k_z, \omega, x) = -i \int_{-\infty}^{\infty} dk_x e^{ik_x x} \frac{k_x \varepsilon_x}{k^2 \varepsilon_L}$$
$$\times \left[A \left(\frac{1}{2} + cos2k_x L + cos4k_x L + \cdots \right) + B(cosk_x L + cos3k_x L + cos5k_x L + \cdots) \right]. \tag{14.200}$$

Here we consider a typical term

$$\int_{-\infty}^{\infty} dk_x e^{ik_x x} \frac{k_x \varepsilon_x}{k^2 \varepsilon_L} \, cos(nk_x L) \quad (n = 1, 2, 3, \ldots) \tag{14.201}$$

where ε_x consists of two parts, the even function part and the odd function part, with respect to k_y: $\varepsilon_x = \varepsilon^E + \varepsilon^O$ with

$$\varepsilon^E = 1 + \sum_\alpha \frac{\omega_{p\alpha}^2}{\omega k_\perp} \int d^3v \, v_\perp \sum_{l=-\infty}^{\infty} \left(\frac{\partial f_{\alpha 0}}{\partial v_\perp} \frac{l\omega_{c\alpha}}{v_\perp} + k_z \frac{\partial f_{\alpha 0}}{\partial v_z} \right) \frac{lJ_l^2/a_\alpha}{\omega - k_z v_z - l\omega_{c\alpha}}, \tag{14.202}$$

$$\varepsilon^O = \sum_\alpha \frac{\omega_{p\alpha}^2}{\omega k_\perp} \int d^3v \, v_\perp \sum_{l=-\infty}^{\infty} \left(\frac{\partial f_{\alpha 0}}{\partial v_\perp} \frac{l\omega_{c\alpha}}{v_\perp} + k_z \frac{\partial f_{\alpha 0}}{\partial v_z} \right) \frac{ik_y J_l J_l'/k_x}{\omega - k_z v_z - l\omega_{c\alpha}}. \tag{14.203}$$

Writing in (14.201) the cosine function in terms of exponential functions and making change of variable $k_x \to -k_x$ as before, we have

$$\int_{-\infty}^{\infty} dk_x e^{ik_x x} \frac{k_x \varepsilon_x}{k^2 \varepsilon_L} \, cos(nk_x L) = \int_{-\infty}^{\infty} dk_x e^{ik_x nL} \frac{k_x}{k^2 \varepsilon_L} (\varepsilon^E \, i \, sink_x x + \varepsilon^O cosk_x x). \tag{14.204}$$

Using (14.204) in (14.200) gives after some algebra

$$\begin{aligned} &D_x(k_y, k_z, \omega, x = 0) \\ &\quad = -iA \int_{-\infty}^{\infty} dk_x \frac{k_x}{k^2 \varepsilon_L} \left[\frac{1}{2} \, \varepsilon^E + \varepsilon^O \left(\frac{1}{2} + e^{2ik_x L} + e^{4ik_x L} + \cdots \right) \right] \\ &\qquad - iB \int_{-\infty}^{\infty} dk_x \frac{k_x}{k^2 \varepsilon_L} \varepsilon^O (e^{ik_x L} + e^{3ik_x L} + e^{5ik_x L} + \cdots)], \end{aligned} \tag{14.205}$$

$$D_x(k_y, k_z, \omega, x = L)$$
$$= -iB\int_{-\infty}^{\infty} dk_x \frac{k_x}{k^2\varepsilon_L}\left[-\frac{1}{2}\,\varepsilon^E + \varepsilon^O\left(\frac{1}{2} + e^{2ik_xL} + e^{4ik_xL} + \cdots\right)\right]$$
$$- iA\int_{-\infty}^{\infty} dk_x \frac{k_x}{k^2\varepsilon_L}\varepsilon^O(e^{ik_xL} + e^{3ik_xL} + e^{5ik_xL} + \cdots)]. \quad (14.206)$$

Enforcing the boundary conditions on $x = 0$ and $x = L$ give four algebraic equations for four undetermined constants, A, B, F_1, and F_2. The solvability condition yields the relation

$$\mu^2 - \gamma^2 = \beta^2 - \delta^2 \quad (14.207)$$

where

$$\mu = \int \frac{dk_x}{k^2\varepsilon_L}\left(\frac{1}{2}k_x\varepsilon^E + i\kappa\left[\frac{1}{2} + e^{2ik_xL} + e^{4ik_xL} + \cdots\right]\right), \quad (14.208)$$

$$\beta = \int \frac{dk_x k_x}{k^2\varepsilon_L}\varepsilon^O\left(\frac{1}{2} + e^{2ik_xL} + e^{4ik_xL} + \cdots\right), \quad (14.209)$$

$$\gamma = \int \frac{i\kappa dk_x}{k^2\varepsilon_L}\left(e^{ik_xL} + e^{3ik_xL} + e^{5ik_xL} + \cdots\right), \quad (14.210)$$

$$\delta = \int \frac{dk_x k_x}{k^2\varepsilon_L}\varepsilon^O\left(e^{ik_xL} + e^{3ik_xL} + e^{5ik_xL} + \cdots\right). \quad (14.211)$$

Here obtaining solutions of the dispersion relation, (14.207), is facilitated by the observation that it is intact if we replace $k_y \to -k_y$. This means that if $\xi(k_y)$ is a solution of (14.207), $\xi(-k_y)$ also solves (14.207). To take advantage of this even parity in k_y of (14.207), we write it as

$$(\beta + \mu)(\beta - \mu) = (\delta + \gamma)(\delta - \gamma). \quad (14.212)$$

We seek the solutions of (14.212) by equating as

$$\beta + \mu = \delta + \gamma. \quad (14.213)$$

It is immediately seen that (14.213) indeed solves (14.212) because changing $k_y \to -k_y$ in (14.213) yields the relation $-\beta + \mu = -\delta + \gamma$, which is the remaining relation to be satisfied in (14.212). The second reduction of the dispersion relation, (14.207), is obtained by writing it as

$$(\beta + \mu)(\beta - \mu) = (-1)(\delta + \gamma) \cdot (-1)(\delta - \gamma). \quad (14.214)$$

We can see that the relation

$$\beta - \mu = -(\delta - \gamma) = \gamma - \delta \tag{14.215}$$

contains another set of solutions of (14.207) because the rest of the relation of (45) is automatically satisfied. In summary, we have two independent reductions of the dispersion relation (14.207):

$$\mu \pm \gamma = \delta \pm \beta. \tag{14.216}$$

It turns out that the lower (upper) sign corresponds to the symmetric (anti-symmetric) mode. The solutions obtained from (14.216) by replacing $k_y \to -k_y$ are also legitimate solutions of the slab dispersion relation.

Using the expressions in (14.208)–(14.211), the dispersion relation is written

$$\int_{-\infty}^{\infty} \frac{dk_x}{k^2 \varepsilon_L} \left[\frac{k_x \varepsilon^E}{2} + (i\kappa + k_x \varepsilon^O) \right.$$
$$\left. \times \left(\frac{1}{2} \pm e^{iLk_x} + e^{2iLk_x} \pm e^{3iLk_x} + e^{4iLk_x} \pm \cdots \right) \right] = 0.$$

Clearly the series in the parentheses converge by picking the poles of $k^2 \varepsilon_L = 0$ in the upper-half k_x-plane and the exponential terms vanish when $L \to \infty$. We can sum the series formally to make the expression compact. It can be easily shown that we have the following expressions for the series,

$$e^{iLk_x} + e^{2iLk_x} + e^{3iLk_x} + \cdots = \frac{e^{iLk_x}}{1 - e^{iLk_x}},$$
$$e^{iLk_x} + e^{3iLk_x} + e^{5iLk_x} + \cdots = \frac{e^{iLk_x}}{1 - e^{2iLk_x}},$$
$$e^{2iLk_x} + e^{4iLk_x} + e^{6iLk_x} + \cdots = \frac{e^{2iLk_x}}{1 - e^{2iLk_x}}.$$

Thus we obtain for the dispersion relation in Eq. (14.216)

$$\int_{-\infty}^{\infty} \frac{dk_x}{k^2 \varepsilon_L} \left[k_x \varepsilon^E + (i\kappa + k_x \varepsilon^O) \frac{1 \pm e^{ik_x L}}{1 \mp e^{ik_x L}} \right] = 0, \tag{14.217}$$

where the apparent singularities associated with the denominator $1 \mp e^{ik_x L} = 0$ should be simply disregarded. If $k_y = 0$, we have $\varepsilon^O = 0$ and (14.217) agrees with the result obtained in earlier work in Reference: [H. J. Lee and C.-G. Kim, J. Korean Phys. Soc. **54** p. 85 (2009) Eq. (55)]).

If $L \to \infty$, the factor e^{ik_xL} vanishes (by piking the poles in the upper-half k_x-plane), and (14.217) becomes

$$\int dk_x \, \frac{k_x \varepsilon_x + i\kappa}{k^2 \varepsilon_L} = 0. \tag{14.218}$$

(14.218) recovers the electrostatic dispersion relation for a semi-infinite plasma obtained in Ref. [H. J. Lee and C.-G. Kim, J. Korean Phys. Soc. **54** p. 85 (2009) Eq. (37)].

• **Cold fluid limit.** To further check the correctness of (14.217), its cold plasma limit will be considered. We put

$$f_{\alpha 0}(\mathbf{v}) = \frac{\delta(v_\perp)}{v_\perp}\delta(v_z)\delta(\varphi)$$

into (14.114) for ε_L. We integrate each term by parts, and use

$$J_n(0) = \delta_{n,0}, \quad \frac{2n}{a} J_n(a) J_n'(a) \to \frac{1}{2}(\delta_{n,1} - \delta_{n,-1}) \tag{14.219}$$

in the limit of $a \to 0$, to obtain

$$\varepsilon_L^{cold} = 1 - \sum_\alpha \frac{\omega_{p\alpha}^2}{k^2}\left(\frac{k_z^2}{\omega^2} + \frac{k_\perp^2}{\omega^2 - \omega_{c\alpha}^2}\right). \tag{14.220}$$

The cold plasma limit of ε_x (14.191) can be calculated similarly. It can be easily shown that the terms involving $\frac{\partial f_{\alpha 0}}{\partial v_z}$ vanish in the limit of $a_\alpha \to 0$. The first term of the velocity integral containing J_l^2 in the rest of (14.191) yields

$$-\sum_\alpha \frac{\omega_{p\alpha}^2}{\omega^2 - \omega_{c\alpha}^2}. \tag{14.221}$$

The second term involving $J_l J_l'$ becomes after integration by parts with respect to $v_\perp$

$$-i\sum_\alpha \frac{\omega_{p\alpha}^2}{\omega k_\perp}\frac{k_y}{k_x}\sum_l \frac{l\,\omega_{c\alpha}}{\omega - l\,\omega_{c\alpha}}\int dv_\perp \frac{\delta(v_\perp)}{v_\perp}\frac{d}{dv_\perp}[v_\perp J_l J_l']. \tag{14.222}$$

Denoting the $\int dv_\perp$–integral by I_α, we have

$$I_\alpha = \int da_\alpha \frac{1}{a_\alpha}\delta\left(a_\alpha \frac{\omega_{c\alpha}}{k_\perp}\right)\frac{d}{da_\alpha}[a_\alpha J_l J_l'] = \frac{k_\perp}{\omega_{c\alpha}} \lim_{a_\alpha \to 0}\left[\frac{J_l J_l'}{a_\alpha} + J_l' J_l' + J_l J_l''\right]. \tag{14.223}$$

Upon using the asymptotic values of the Bessel functions, the last term involving J_l'' vanishes. Using (14.211) and $J_l' \to \frac{1}{2}(\delta_{l,1} - \delta_{l,-1})$, we obtain

$$l\, I_\alpha = -\frac{k_\perp}{2\omega_{c\alpha}}(\delta_{l,1} - \delta_{l,-1}).$$

Collecting the preceding results, we finally obtain for the cold plasma limit of ε_x

$$\varepsilon_x^{cold} = 1 - \sum_\alpha \left(1 - \frac{ik_y}{k_x}\frac{\omega_{c\alpha}}{\omega}\right)\frac{\omega_{p\alpha}^2}{\omega^2 - \omega_{c\alpha}^2}, \tag{14.224}$$

$$\varepsilon^E = 1 - \sum_\alpha \frac{\omega_{p\alpha}^2}{\omega^2 - \omega_{c\alpha}^2} \equiv P, \quad \varepsilon^O = \frac{ik_y}{k_x}\sum_\alpha \frac{\omega_{c\alpha}}{\omega}\frac{\omega_{p\alpha}^2}{\omega^2 - \omega_{c\alpha}^2} \equiv \frac{ik_y}{k_x}Q. \tag{14.225}$$

Substituting (14.220) and (14.225) into (14.217) and carrying out the contour integrals, we can obtain the cold fluid limit of the electrostatic slab dispersion relation. The first term of (14.217) is calculated to obtain

$$\int dk_x \frac{k_x \varepsilon^E}{k^2 \varepsilon_L^{cold}} = \int dk_x \frac{Pk_x}{P(k_x^2 + \lambda^2)} = i\pi \tag{14.226}$$

where

$$\lambda = \sqrt{k_y^2 + k_z^2\frac{C}{P}}, \quad C = 1 - \sum_\alpha \frac{\omega_{p\alpha}^2}{\omega^2}. \tag{14.227}$$

Similar calculation for the rest of the terms in (14.217) yields the slab dispersion relations

$$P\lambda + (Qk_y + \kappa)\, coth\left(\frac{\lambda L}{2}\right) = 0, \quad (anti\text{-}symmetric\ mode), \tag{14.228}$$

$$P\lambda + (Qk_y + \kappa)\, tanh\left(\frac{\lambda L}{2}\right) = 0, \quad (symmetric\ mode). \tag{14.229}$$

If $L \to \infty$, the two modes coalesce to

$$P\lambda + Qk_y + \kappa = 0 \tag{14.230}$$

which is the cold plasma electrostatic dispersion relation in the magnetized semi-infinite plasma. Furthermore, if $k_z = 0$, we have $\lambda = \kappa = k_y$, and the dispersion relation takes the form

$$2 + \sum_\alpha \frac{\omega_{p\alpha}^2}{\omega^2 - \omega_{c\alpha}^2}\left(\frac{\omega_{c\alpha}}{\omega} - 1\right) = 0. \tag{14.231}$$

Equation (14.231) agrees with the dispersion relation obtained earlier from the cold fluid equations in Ref. [H. J. Lee, Plasma Phys. Control. Fusion (1995) **37** p. 755 Eq. (40)].

• **Fluid treatment.** We use the following two-fluid equations to obtain the plasma solutions of the electrostatic field for $0 < x < L$:

$$-i\omega \mathbf{v}_\alpha = -\frac{e_\alpha}{m_\alpha}\nabla\phi + \omega_{c\alpha}\mathbf{v}_\alpha \times \hat{\mathbf{z}}, \tag{14.232}$$

$$-i\omega n_\alpha + N\nabla \cdot \mathbf{v}_\alpha = 0, \tag{14.233}$$

$$\nabla^2\phi = 4\pi e(n_e - n_i), \tag{14.234}$$

where $\phi = \phi(\omega, \mathbf{r})$, and N is the equilibrium density of the plasma. One can obtain from the above equations

$$v_{\alpha x} = \frac{ie_\alpha}{m_\alpha}\frac{\omega}{\omega_{c\alpha}^2 - \omega^2}\left(\frac{\partial\phi}{\partial x} + i\,\frac{\omega_{c\alpha}}{\omega}\frac{\partial\phi}{\partial y}\right), \tag{14.235}$$

$$\frac{\partial^2\phi}{\partial x^2} - \lambda^2\phi = 0, \tag{14.236}$$

where λ is defined by (14.227). Solving the above equation gives

$$\phi(x, y, z, \omega) = (A_1 e^{-\lambda x} + A_2 e^{\lambda x})e^{ik_y y + ik_z z - i\omega t} \tag{14.237}$$

which in turn yields (omitting the phasor)

$$E_x = \lambda(A_1 e^{-\lambda x} - A_2 e^{\lambda x}), \tag{14.238}$$

$$E_y = -ik_y(A_1 e^{-\lambda x} + A_2 e^{\lambda x}), \tag{14.239}$$

$$E_z = -ik_z(A_1 e^{-\lambda x} + A_2 e^{\lambda x}), \tag{14.240}$$

where A_1 and A_2 are constants. The vacuum solutions can be written down as

$$x < 0: \quad E_x = -\kappa F_1 e^{\kappa x},\ E_y = -ik_y F_1 e^{\kappa x},\ E_z = -ik_z F_1 e^{\kappa x}, \tag{14.241}$$

$$x > L: \quad E_x = \kappa F_2 e^{-\kappa x},\ E_y = -ik_y F_2 e^{-\kappa x},\ E_z = -ik_z F_2 e^{-\kappa x}, \tag{14.242}$$

where F_1 and F_2 are constants.

We need the normal component of the electric displacement vector, D_x, in the plasma:

$$D_x = E_x + \frac{4\pi Ni}{\omega}\sum_\alpha e_\alpha v_{\alpha x}. \tag{14.243}$$

Using Eqs. (14.235) and (14.237) in the above equation gives

$$D_x = PE_x - iQE_y = A_1 e^{-\lambda x}(\lambda P - Qk_y) - A_2 e^{\lambda x}(\lambda P + Qk_y) \quad (14.244)$$

where P and Q are defined by (14.225).

The continuity of E_y at $x = 0$ and $x = L$ yields

$$A_1 + A_2 = F_1, \quad (14.245)$$

$$A_1 e^{-\lambda L} + A_2 e^{\lambda L} = F_2 e^{-\kappa L}. \quad (14.246)$$

The continuity of D_x at $x = 0$ and $x = L$ yields

$$-\kappa F_1 = A_1(P\lambda - Qk_y) - A_2(P\lambda + Qk_y), \quad (14.247)$$

$$\kappa F_2 e^{-\kappa L} = A_1 e^{-\lambda L}(P\lambda - Qk_y) - A_2 e^{\lambda L}(P\lambda + Qk_y). \quad (14.248)$$

The solvability condition of the above four equations gives

$$e^{\frac{\lambda L}{2}}(P\lambda + \kappa + Qk_y)\, e^{\frac{\lambda L}{2}}(P\lambda + \kappa - Qk_y)$$
$$= e^{-\frac{\lambda L}{2}}(P\lambda - \kappa + Qk_y)\, e^{\frac{-\lambda L}{2}}(P\lambda - \kappa - Qk_y) \quad (14.249)$$

which is the slab dispersion relation obtained from the fluid equations. (14.249) has the even parity in k_y: (14.249) is intact if we replace $k_y \to -k_y$. We can obtain the solutions of (14.249) by equating as

$$e^{\frac{\lambda L}{2}}(P\lambda + \kappa + Qk_y) = e^{-\frac{\lambda L}{2}}(P\lambda - \kappa - Qk_y) \quad (14.250)$$

because the rest of (14.249) is nothing but the relation obtained by replacing $k_y \to -k_y$ in the above equation. Another *independent* relation is obtained by equating as

$$e^{\frac{\lambda L}{2}}(P\lambda + \kappa + Qk_y) = -e^{-\frac{\lambda L}{2}}(P\lambda - \kappa - Qk_y). \quad (14.251)$$

Equations (14.250) and (14.251) can be arranged into the forms of (14.228) and (14.229), showing complete agreement with the cold fluid limit of the kinetic results.

14.5. Echoes in a semi-bounded plasma

In § 10.3, echo in infinite plasma was investigated. Echo is given rise to as a consequence of nonlinear (second order) mixing of two waves when a plasma is impressed by two pulses separated by a time delay (temporal echo) or launched simultaneously at two different spots separated by a distance (spatial echo). In this Section, *spatial echo* theory is developed in a semi-bounded plasma. The presence of the boundary surface gives the extra complication because the distribution function has to satisfy the kinematic boundary condition. In a bounded plasma, echoes can occur in diversity of spots (for the case of spatial echoes) as compared with an infinite plasma. Echoes in a bounded plasma would be of practical interest since laboratory plasmas are bounded.

14.5.1. *Landau's treatment of the semi-infinite plasma under the specular reflection condition*

In his famous paper, Landau [L. Landau (1946) *On the vibrations of the electronic plasma*, J. Physics (USSR) **10**, 25] solved a boundary value problem in a semi-bounded plasma interfaced with a vacuum, under the condition of the specular reflection condition. We first review the development by Landau for the self-consistent response function in the presence of boundary. The approach is useful for considering the surface wave echo. He solves a first order differential equation in x coordinate for the Vlasov equation; not using the recipe of making the Vlasov equation reflectionally invariant. But he also ended up with the necessity of suitable extension of the electric field components. We restrict ourselves to electrostatic waves:

$$\frac{\partial}{\partial t} f_\alpha(\mathbf{r}, \mathbf{v}, t) + \mathbf{v} \cdot \frac{\partial f_\alpha}{\partial \mathbf{r}} + \frac{e_\alpha}{m_\alpha} \mathbf{E} \cdot \frac{\partial f_{\alpha 0}}{\partial \mathbf{v}} = 0, \tag{14.252}$$

$$\nabla \cdot \mathbf{E} = 4\pi \sum_\alpha e_\alpha \int f_\alpha d^3 v, \quad \nabla \times \mathbf{E} = 0. \tag{14.253}$$

The space is divided by the interface $x = 0$: the region $x > 0$ is plasma and the region $x < 0$ is vacuum. The specular reflection boundary condition on f is assumed:

$$f_\alpha(v_x, x = 0) = f_\alpha(-v_x, x = 0). \tag{14.254}$$

The wave has the phasor $e^{ik_z z - i\omega t}$ and the wave amplitude varies along the perpendicular direction x and the y-coordinate is ignorable. From now on, ions will be assumed to form stationary neutralizing background; only electron perturbation will be considered. First we Fourier transform the above equations with respect to t and z ($\sim e^{ik_z z - i\omega t}$) to write (14.252) in the form

$$\frac{\partial}{\partial x} f(\omega, x, k_z, v_x, v_z) - i\frac{\psi}{v_x} f - \frac{e}{m v_x} \mathbf{E}(x, k_z, \omega) \cdot \frac{\partial f_0}{\partial \mathbf{v}} = 0, \quad \psi = \omega - k_z v_z. \tag{14.255}$$

The above first order differential equation for x is solved in the form (see Appendix A.32)

$$f(x,v_x) = e^{i\frac{\psi}{v_x}x}\left[f(x_0,v_x)e^{-i\frac{\psi}{v_x}x_0} + \int_{x_0}^{x} dx' \frac{e}{mv_x}\mathbf{E}(x')\cdot\frac{\partial f_0}{\partial \mathbf{v}}e^{-i\frac{\psi}{v_x}x'}\right] \tag{14.256}$$

where x_0 is the reference point of the boundary condition. It should be remembered that $\omega(=\omega_r + i\epsilon)$ has an infinitesimal positive imaginary part $(\epsilon > 0)$ in order to have legitimate time-Fourier transform (see Chapter 4). Note that the first term inside the bracket [ ] goes to zero as $x_0 \longrightarrow \infty$ provided $v_x < 0$. So we write

$$f^N(x,v_x) = \int_{\infty}^{x} dx' \frac{e}{mv_x}\mathbf{E}(x')\cdot\frac{\partial f_0}{\partial \mathbf{v}}e^{i\frac{\psi}{v_x}(x-x')} \quad for\ v_x < 0. \tag{14.257}$$

Thus, we have succeeded in constructing the solution that satisfies the boundary condition $f(v_x,\infty) = 0$ provided $v_x < 0$. For $v_x > 0$, we take the reference point to be zero $(x_0 = 0)$ and enforce $f(0,v_x) = f(0,-v_x)$, since we now know $f(0,-v_x)$.

$$\begin{aligned} f^P(x,v_x) =& -\int_0^{\infty} dx' \frac{e}{mv_x}\left(E_x(x')\frac{\partial f_0}{\partial v_x} - E_z(x')\frac{\partial f_0}{\partial v_z}\right)e^{i\frac{\psi}{v_x}(x+x')} \\ &+ \int_0^{x} dx' \frac{e}{mv_x}\mathbf{E}(x')\cdot\frac{\partial f_0}{\partial \mathbf{v}}e^{i\frac{\psi}{v_x}(x-x')} \quad for\ v_x > 0. \end{aligned} \tag{14.258}$$

Clearly, $f(x,v_x)$ in (14.258) goes to zero as $x \to \infty$ owing to the exponential factor $e^{i\frac{\psi}{v_x}x}$. The above two equations both satisfy $f(x,v_x) \to 0$ as $x \to \infty$ and the condition $f^N(0,v_x) = f^P(0,-v_x)$ or $f^P(0,v_x) = f^N(0,-v_x)$. The expression for the distribution function valid for the entire range of v_x can be written in the form

$$f(x,v_x) = \frac{1}{2}(1 - H(v_x))f^N(x,v_x) + \frac{1}{2}(1 + H(v_x))f^P(x,v_x) \tag{14.259}$$

where $H(v_x)$ is the step function: $H(v) = 1$ for $v > 0$ and $H(v) = -1$ for $v < 0$. It can be easily shown that (14.259) satisfies $f(0,v_x) = f(0,-v_x)$. It is left as an exercise to check that

$$f^N(-x,-v_x) = f^P(x,v_x), \quad f^P(-x,-v_x) = f^N(x,v_x) \tag{14.260}$$

if $E(-x) = -E(x)$. Therefore, (14.259) satisfies

$$f(x, v_x) = f(-x, -v_x) \quad provided \; E_x(-x) = -E_x(x). \qquad (14.261)$$

So, in the following, we use the extension that $E_x(x)$ is oddly continued into the region $x < 0$.

• Calculation of $f(k_x, v_x) = \int_{-\infty}^{\infty} dx e^{-ik_x x} f(x, v_x)$. We have three terms in (14.257) and (14.258) to invert. First, putting the relation

$$\mathbf{E}(x', k_z) = \int_{-\infty}^{\infty} \mathbf{E}(k_x', k_z) e^{ik_x' x'} \frac{dk_x'}{2\pi}$$

into (14.257), we write

$$f^N(k_x, v_x) = \frac{e}{mv_x} \int_{-\infty}^{\infty} dx e^{-ik_x x} \int_{\infty}^{x} dx' e^{i\frac{\psi}{v_x}(x-x')} \int_{-\infty}^{\infty} \frac{dk_x'}{2\pi} e^{ik_x' x'} \mathbf{E}(k_x', k_z) \cdot \frac{\partial f_0}{\partial \mathbf{v}}.$$

$\int dx'$-integral can be done to yield

$$-i \frac{e^{i(k_x' - \psi/v_x)x}}{k_x' - \psi/v_x}.$$

In obtaining the above result, we used $v_x < 0$ and ω has a small positive imaginary part [so the integrated term vanishes at $x' = +\infty$]. Next, carry out the $\int dx$-integral, which yields $2\pi\delta(k_x - k_x')$. Thus, we finally obtain

$$f^N(k_x, v_x) = \frac{-ie}{mv_x} \frac{1}{k_x - \frac{\psi}{v_x}} \mathbf{E}(k_x, k_z) \cdot \frac{\partial f_0}{\partial \mathbf{v}} \quad (v_x < 0). \qquad (14.262)$$

Next, for the first term of (14.258), the Fourier transform becomes

$$\frac{-ie}{mv_x} \delta(k_x - \frac{\psi}{v_x}) \int_{-\infty}^{\infty} dk_x' \frac{1}{k_x' - \frac{\psi}{v_x}} \mathbf{E}(k_x', k_z) \cdot \frac{\partial f_0}{\partial \mathbf{v}} \qquad (14.263)$$

where we used $v_x > 0$ and the relations $E_x(-k_x) = -E_x(k_x)$ and $E_z(-k_x) = E_z(k_x)$ (odd continuation of $E_x(x)$). The Fourier transform of the second term of (14.258) can be derived in a similar manner. The integral $\int_0^x dx'$-integral produces two terms, corresponding to the lower limit 0 and the upper limit x. The term corresponding to the lower limit cancels the term in (14.263), and the final result for f^P is obtained in the same form as f^N in (14.262):

$$f^P(k_x, v_x) = \frac{-ie}{mv_x} \frac{1}{k_x - \frac{\psi}{v_x}} \mathbf{E}(k_x, k_z) \cdot \frac{\partial f_0}{\partial \mathbf{v}} \quad (v_x > 0) \tag{14.264}$$

where we used $v_x > 0$ and ω having a small positive imaginary part. The detailed steps leading to the above equation are left as exercise. The above equation can be put into the familiar form

$$f^P(\mathbf{k}, \mathbf{v}, \omega) = f^N(\mathbf{k}, \mathbf{v}, \omega) = \frac{ie}{m} \frac{\mathbf{E} \cdot \frac{\partial f_0}{\partial \mathbf{v}}}{\omega - \mathbf{k} \cdot \mathbf{v}} \tag{14.265}$$

which is of the same form as that of infinite plasma. So far we presented Landau's story. The surface wave dispersion relation can be derived in the same fashion as in Section 14.1 by suitably extending the electric field.

Summary and the inverse relations of (14.262) and (14.264). For simplicity, let us consider one-dimensional variation:

$$-i\omega f(x,v) + v\frac{\partial f}{\partial x} - \frac{e}{m}E(x)\frac{df_0}{dv} = 0, \quad f(0,v) = f(0,-v) \quad (x > 0)$$

whose solution satisfying the specular reflection condition can be written as the following two equations:

$$f^N(x,v) = \frac{e}{mv}\frac{df_0}{dv}\int_{\infty}^{x} d\xi\ E(\xi)e^{i\omega(x-\xi)/v} \quad (v<0), \quad (A)$$

$$\begin{aligned} f^P(x,v) &= \frac{e}{mv}\frac{df_0}{dv}\, e^{i\omega x/v}\left[\int_0^x d\xi\ E(\xi)\ e^{-i\omega\xi/v} - \int_0^\infty d\xi\ E(\xi)\ e^{i\omega\xi/v}\right] \\ &= \frac{e}{mv}\frac{df_0}{dv}\int_{-\infty}^{x} d\xi\ E(\xi)\ e^{i\omega(x-\xi)/v} \quad (v>0). \quad (B) \end{aligned}$$

In writing (B), we used $E(-\xi) = -E(\xi)$. Both f^N and f^P are Fourier transformed in the same form

$$f^{N,P}(k,v) = \frac{ie}{m}\frac{df_0}{dv}\frac{E(k)}{\omega - kv}. \quad (C)$$

It is instructive to invert (C) to retrieve (A) and (B). First,

$$f^N(x,v) = \int_{-\infty}^{\infty} \frac{dk}{2\pi}\, e^{ikx} f^N(k,v) = -\frac{1}{2\pi}\,\frac{ie}{mv}\,\frac{df_0}{dv}\int_{-\infty}^{\infty} dk\; e^{ikx}\,\frac{E(k)}{k-\omega/v}.$$

Substitute into the above equation $E(k) = \int_{-\infty}^{\infty} E(\xi)e^{-ik\xi}d\xi$:

$$f^N(x,v) = -\frac{1}{2\pi}\,\frac{ie}{mv}\,\frac{df_0}{dv}\int_{-\infty}^{\infty} E(\xi)d\xi \int_{-\infty}^{\infty} dk\,\frac{e^{ik(x-\xi)}}{k-\frac{\omega}{v}}. \quad (v<0)$$

The $\int dk$-integral above $(v < 0)$ should follow the convex up path going around the singularity $k = \omega/v$. [When $v > 0$, $\int dk$-integral should follow the concave down path to be used for f^P inversion.] So the $\int dk$-integral is the sum of the principal part plus the delta function term:

$$\int_{-\infty}^{\infty} dk\,\frac{e^{ik(x-\xi)}}{k-\frac{\omega}{v}} = P\int_{-\infty}^{\infty} dk\,\frac{e^{ik(x-\xi)}}{k-\frac{\omega}{v}} \mp i\pi\; exp\left[i\,\frac{\omega}{v}(x-\xi)\right]$$

where $-$ $(+)$ sign corresponds to $v < 0$ $(v > 0)$. The principal value integral is

$$P\int_{-\infty}^{\infty} dk\,\frac{e^{ik(x-\xi)}}{k-\frac{\omega}{v}} = i\pi\; exp\left[i\,\frac{\omega}{v}(x-\xi)\right] H(x-\xi)$$

where H is the step function defined as $H(x) = 1$ for $x > 0$ and $H(x) = -1$ for $x < 0$. Thus, we finally obtain

$$f^N(x,v) = -\frac{e}{mv}\,\frac{df_0}{dv}\int_{x}^{\infty} d\xi E(\xi)\; exp\left[i\,\frac{\omega}{v}(x-\xi)\right]$$

in agreement with (A). The expression for f^P in (B) can also be derived in similar fashion after straight manipulation.

14.5.2. *A semi-infinite plasma placed against an oscillating external electric field*

Landau (1946) solved a boundary value problem: against an oscillating field $E_0e^{-i\omega t}$ (at $x = \infty$) is placed a semi-infinite plasma occupying the region $0 < x < \infty$ with $x = 0$ being a specular reflective wall. Landau obtained the field inside the plasma from the Vlasov and Poisson equations. Landau called this problem finding "the law of the penetration of the field inside".

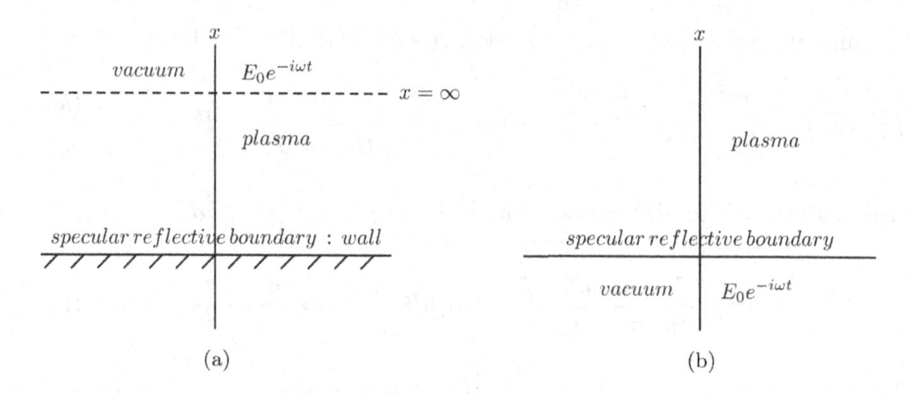

Fig. 14.4. Two equivalent placements of a plasma and a vacuum.

See Fig. 14.4(a). This problem turns out to be equivalent to that posed by Fig. 14.4(b), which is a more familiar arrangement.

We seek the solution for Fig. 14.4(a). At $x = \infty$, the displacement (D) is continuous to the vacuum electric field E_0:

$$\varepsilon E(x = \infty) = E_0, \quad \varepsilon = 1 - \frac{\omega_p^2}{\omega^2} \tag{14.266}$$

where ε is the plasma dielectric constant. Landau determined the plasma electric field $E(x)$ from the Vlasov and Maxwell equations under the above boundary condition. In this subsection, we review Landau's treatment of this inhomogeneous problem as outlined above, obtaining $E(x)$ in terms of E_0. The plasma response to the oscillating vacuum electric field of frequency ω is governed by the one-dimensional Vlasov equation

$$-i\omega f(x, u) + u\frac{\partial f}{\partial x} + \frac{e}{m}E(x)\frac{df_0}{du} = 0, \quad f(0, u) = f(0, -u). \quad (x > 0) \tag{14.267}$$

Note that the transverse coordinates parallel to the wall have translational invariance; E and the perturbed distribution f vary along the perpendicular coordinate x only. We also have one-dimensional Maxwell equation, $\frac{d}{dx}(J - \frac{i\omega}{4\pi}E(x)) = 0$, or

$$-i\omega E(x) + 4\pi J(x) = -i\omega E_0, \quad J(x) = e\int_{-\infty}^{\infty} uf(u, x)du. \tag{14.268}$$

The boundary condition on the electric field expressed by (14.266) should be supplemented. Mathematically, the above two equations constitute an integral equation for $E(x)$, with E_0 being a source term. The integral equation can be solved by Fourier transform. (14.267) was solved in the last subsection, and we can write immediately (with putting $k_z = 0$ there)

$$\begin{aligned} J(x) = & \int_{-\infty}^{0} du \frac{df_0}{du} \int_{x}^{\infty} d\xi \, \frac{e^2}{m} E(\xi) \, e^{i\omega(x-\xi)/u} \\ & + \int_{0}^{\infty} du \frac{df_0}{du} \int_{0}^{\infty} d\xi \, \frac{e^2}{m} E(\xi) \, e^{i\omega(x+\xi)/u} \\ & - \int_{0}^{\infty} du \frac{df_0}{du} \int_{0}^{x} d\xi \, \frac{e^2}{m} E(\xi) \, e^{i\omega(x-\xi)/u}. \end{aligned} \tag{14.269}$$

Now we split $E(x)$:

$$E(x) = E_1(x) + \frac{E_0}{\varepsilon}. \tag{14.270}$$

Let us write $J(x)$ in (14.269) as $J(x, E)$. Then J is also split into two parts:

$$J(x) = J_1(x, E_1(x)) + J_0\left(x, \frac{E_0}{\varepsilon}\right). \tag{14.271}$$

J_0 can be obtained by carrying out the integrals since $E(\xi)$ is replaced by constant E_0/ε which can be taken out of the integral. After some algebra, we obtain

$$\begin{aligned} J_0(x) &= \frac{e^2}{m} \frac{E_0}{\varepsilon} \frac{2i}{\omega} \int_0^\infty duu \frac{df_0}{du} \left(e^{i\omega x/u} - 1\right) \\ &= \frac{e^2}{m} \frac{E_0}{\varepsilon} \frac{2i}{\omega} \left[\int_0^\infty duu \frac{df_0}{du} \, e^{i\omega x/u} + \frac{N}{2}\right] \end{aligned} \tag{14.272}$$

where N is the total number of particles and we assumed that ω has an inherent infinitesimal positive imaginary part. Using (14.271) and (14.272) in (14.268) gives

$$E_1(x) + \frac{4\pi i}{\omega} J_1(x) = 2\frac{E_0}{\varepsilon} \frac{\omega_p^2}{\omega^2} \int_0^\infty duu \frac{df_0}{du} \, e^{i\omega x/u}. \quad (x > 0) \tag{14.273}$$

This is an integral equation as is evident from (14.269). It is evident (from the derivation explained in the last subsection) that the above equation is valid for $x > 0$. We need the equation valid in the region $x < 0$. First, $E_1(x)$ is oddly continued into the region $x < 0$. In the expression (14.269), we can show that $J(-x) = -J(x)$ provided $E(-\xi) = -E(\xi)$. Changing $x \to -x$ in

(14.273) gives

$$E_1(x) + \frac{4\pi i}{\omega} J_1(x) = -2\frac{E_0}{\varepsilon}\frac{\omega_p^2}{\omega^2}\int_0^\infty duu\frac{df_0}{du}\, e^{-i\omega x/u} \quad (x < 0) \qquad (14.274)$$

which is valid in the region $-\infty < x < 0$. The above two equations can be put into single equation as

$$E_1(x) + \frac{4\pi i}{\omega} J_1(x) = 2\frac{E_0}{\varepsilon}\frac{\omega_p^2}{\omega^2} H(x)\int_0^\infty duu\frac{df_0}{du}\, e^{i\omega|x|/u} \quad (-\infty < x < \infty) \qquad (14.275)$$

where $H(x)$ is the step function. Now we can work with the Fourier variables $E_1(k)$ and $J_1(k)$. $J_1(k)$ is nothing but the Fourier transform of $J(x)$ in (14.268) (with $E(\xi)$ replaced by $E_1(\xi)$), with $f(u,k)$ obtained in the last subsection. We have

$$J_1(k) = e\int_{-\infty}^\infty uf(u,k)du = -i\,\frac{e^2}{m}E_1(k)\int_{-\infty}^\infty du\ \frac{u\,\frac{df_0}{du}}{\omega - ku}. \qquad (14.276)$$

The fourier transform of the RHS of (14.275) ($\equiv R$) is the sum of two terms, $\int_{-\infty}^0 dxe^{-ikx}(\cdots) + \int_0^\infty dxe^{-ikx}(\cdots)$. After some algebra, we obtain

$$R(k) = 2\frac{E_0}{\varepsilon}\frac{\omega_p^2}{\omega^2}\left[-\frac{2i}{k}\int_0^\infty duu\ \frac{df_0}{du} + \frac{\omega}{k}\int_0^\infty du\frac{df_0}{du}\int_{-\infty}^\infty dxe^{-ikx}e^{i\omega|x|/u}\right] \qquad (14.277)$$

which can be further integrated to yield

$$R(k) = 2\frac{E_0}{\varepsilon}\frac{\omega_p^2}{\omega^2}\frac{i}{k}\left[1 + \omega\int_{-\infty}^\infty du\ \frac{u\,\frac{df_0}{du}}{\omega - ku}\right]. \qquad (14.278)$$

In Landau's notation, we define

$$K(k) = \frac{\omega_p^2}{\omega}\int_{-\infty}^\infty du\ \frac{u\,\frac{df_0}{du}}{ku - \omega}, \quad K_0 = K(k=0) = \frac{\omega_p^2}{\omega^2} = 1 - \varepsilon. \qquad (14.279)$$

Then the Fourier transform of (14.275) reads

$$E_1(k)(1 - K(k)) = \frac{2iE_0}{\varepsilon}\,\frac{K_0 - K(k)}{k}. \qquad (14.280)$$

Exercise. Derive (14.278). Hint: In $\int dx$-integral, write $e^{-ikx} = \frac{i}{k}\frac{d}{dx}e^{-ikx}$ and integrate by parts.

Remark. J(k) in (14.276) is an odd function of k, $J(-k) = -J(k)$, if E(k) is an odd function of k. Thus it follows that J(x) is also an odd function of x, which we used in writing (14.274).

It is immediately seen that

$$1 - K(k) = \varepsilon_L = 1 + \frac{\omega_p^2}{k}\int_{-\infty}^{\infty} du \, \frac{\frac{df_0}{du}}{\omega - ku} \quad (\textit{dielectric function}).$$

So (14.280) can be put into the form

$$E_1(k) = 2i\,\frac{E_0}{\varepsilon k}\left(1 - \frac{\varepsilon}{\varepsilon_L(k,\omega)}\right). \tag{14.281}$$

Inverting this equation, we obtain

$$E_1(x,\omega) = -\frac{E_0}{\varepsilon} - \frac{iE_0}{\pi}\int_{-\infty}^{\infty} e^{ikx}\,\frac{dk}{k\,\varepsilon_L(k,\omega)} \quad (x > 0) \tag{14.282}$$

where we used

$$\int_{-\infty}^{\infty}\frac{e^{ikx}}{k}dk = i\int_{-\infty}^{\infty} dk\,\frac{sinkx}{k} = i\pi H(x) \quad H(x):\ \textit{step function} \tag{14.283}$$

where $H(x) = 1\ (x > 0),\ -1\ (x < 0)$. Therefore the plasma electric field defined by (14.270) is the second term of the above,

$$E(x,\omega) = -\frac{iE_0}{\pi}\int_{-\infty}^{\infty} e^{ikx}\,\frac{dk}{k\,\varepsilon_L(k,\omega)}. \tag{14.284}$$

In Landau's paper, spatial damping is derived from the above equation [see his Eq. (4.2)]. The contribution to the integral comes from $k = 0$ and zeros of $\varepsilon_L(k,\omega)$. The latter poles give rise to spatial damping accompanying the penetration of the electric field into the plasma, since $\varepsilon_L(k,\omega)$ should be evaluated along the Landau contour. This is the important point of Landau's analysis above. An interesting point of Landau's analysis above is the separation of the electric field (14.270) to satisfy the boundary condition (14.266). In the following, we present an alternative derivation of the above result by using Poisson's equation without the separation of the electric field.

• **Alternative derivation.** According to Landau's arrangement, the plasma electric field is connected to the vacuum electric field E_0 at $x = +\infty$ (Eq. (14.266)) while the plane $x = 0$ serves as the reflecting wall of the plasma particles. We opt for a more familiar separation of the vacuum and the plasma regions. Let the reflecting plane $x = 0$ separate the vacuum ($x < 0$) and the plasma ($x > 0$), and the vacuum field E_0 is matched to the plasma field at $x = 0$. Physically, the two arrangements for space division are equivalent. Now we can solve the problem in an easier way.

We Fourier transform (14.267) with the plasma electric field $E(x)$ ($x > 0$) continued into the $x < 0$ region in the odd function manner. Then (14.267), having made valid over the entire region ($-\infty < x < \infty$), gives the Fourier solution

$$f(k, u) = \frac{ie}{m} \frac{\frac{df_0}{du}}{\omega - ku} E(k) \tag{14.285}$$

where we changed e in (14.267)$\longrightarrow -e$ (for electron). The Poisson equation reads

$$\frac{\partial E}{\partial x} = -4\pi e \int_{-\infty}^{\infty} f(x, u) du. \tag{14.286}$$

Fourier transforming this gives

$$E(k) = \frac{iN}{k\, \varepsilon_L(\omega, k)}, \quad \varepsilon_L(\omega, k) = 1 + \frac{\omega_p^2}{k} \int_{-\infty}^{\infty} du \frac{\frac{df_0}{du}}{\omega - ku} \tag{14.287}$$

where $N = E(0^-) - E(0^+)$ comes from the discontinuity of $E(x)$ at $x = 0$. Next, we calculate the displacement $D(k) = E(k) + \frac{4\pi i}{\omega} J(k)$, $J(k) = -e \int du u f(k, u)$. We obtain

$$D(k) = E(k) \left(1 + \frac{\omega_p^2}{\omega} \int_{-\infty}^{\infty} du \frac{u \frac{df_0}{du}}{\omega - ku} \right) = \varepsilon_L E(k) = \frac{iN}{k},$$

$$D(\omega, x = 0^+) = \lim_{x \to 0+} \int \frac{dk}{2\pi} \frac{iN}{k} e^{ikx} = \frac{iN}{2\pi} i\pi H(x = 0^+) = -\frac{N}{2}.$$

[The principal part is zero since the integrand is an odd function. The δ-function part yields $i\pi$. See Appendix.] Using the boundary condition, $D(x = 0^+) = D(x = 0^-) = E_0$, yields $N = -2E_0$. Then inverting (14.287) gives the plasma electric field in the form (14.282). The symbol N here is not to be confused with the total number of particles.

• **Inversion integral in (14.284).** The dominant contribution to the integral comes from the pole at $k = 0$. So we neglect the contribution from the zeroes of ε_L. At $k = 0$, the real k-axis path should be deformed into the principal part of the integral plus a concave-down shape infinitesimal semi-circle (being pushed downward by the approaching $k = 0$ singularity).

This deformed contour gives the analytic continuation of the integral which is analytic in the upper-half k–plane. The value of the integral is $2\pi i$ times residue at $k = 0$, i.e. $2\pi i/\varepsilon_L(0,\omega)$ [which is the principal value $i\pi$ plus the infinitesimal semi-circle contribution $i\pi$]. Thus we obtain

$$E(x,\omega) = \frac{2E_0}{1-\omega_p^2/\omega^2} \quad for\ x > 0, \quad E(x,\omega) = 0 \quad for\ x < 0 \qquad (14.288)$$

where we used

$$\varepsilon_L = 1 + \frac{\omega_p^2}{k}\int du\, \frac{\frac{df_0}{du}}{\omega - ku} = 1 + \frac{\omega_p^2}{\omega}\int du\, \frac{u\frac{df_0}{du}}{\omega - ku}.$$

If we take into account the singularities associated with $\varepsilon_L(k,\omega)$ in (14.284), the velocity integral in ε_L should be evaluated along the Landau contour which will produce imaginary part to the integral. This will be the collisionless damping in the k-space.

14.5.3. *Spatial echoes in a semi-bounded plasma*

Reference: H. J. Lee and M. J. Lee, (2015), *Echo in a semi-bounded plasma*, The Open Plasma Phys. Journal **8** (Bentham) p. 1–7.

We consider a plasma consisting of electrons and stationary ions, the latter forming the uniform background. The plasma occupies the half-space $x \geq 0$ with the region $x < 0$ being a vacuum. In this Section, we investigate echoes arising by two oscillating charges placed on two vertical locations, vertical to the interface. Then, only the vertical coordinate x will be non-ignorable: the perturbed electron distribution function $f(x,v,t)$ and the electric field $E(x,t)$ will depend on x-coordinate only while y- and z-coordinates have translational invariance. In this specification we are restricting to the echoes that are not associated with surface wave. We look for the plasma response in the second order when the plasma is impressed by two external oscillating charges prescribed by (14.291). Plasma echoes occurring in an infinite plasma was studied in § 10.3. Echoes arise when the product of two linear distribution functions (satisfying the Vlasov and Poisson equations) does not phase-mix. "Phase mix" involves the velocity space integral which is contributed by the singularity at $\omega = kv$ (other singularities, say, those associated with $\varepsilon_L = 0$ contribute negligibly to echo generation). The $\omega = kv$ singularity modulates $f(x,v,t)$ by the exponential phase $e^{ik(x-vt)}$ or $e^{-i\omega(t-x/v)}$. Linearly, this modulation phase-mixes to zero in the velocity integral, yielding zero contribution to the electric field or the density perturbation. But in nonlinear product, it is possible for the second order quantities not to phase-mix, giving rise to echoes. In semi-infinite plasma, the linear solutions are obtained by satisfying the boundary conditions and, due to the fulfilment of that, the solution contains extra terms as compared to infinite plasma. See (14.309) below and its following equation. Thus clearly echoes can occur in a diversity of spots in semi-bounded plasma.

We have the nonlinear Vlasov equation and the Poisson equation to describe the electrostatic perturbation:

$$\frac{\partial}{\partial t} f(v,x,t) + v\frac{\partial f}{\partial x} - \frac{e}{m}E(x,t)\,\frac{\partial f}{\partial v} = 0, \tag{14.289}$$

$$\frac{\partial E}{\partial x} = 4\pi\left(-e\int dv f + \rho_0(x,t)\right), \tag{14.290}$$

where ρ_0 represents the external charges:

$$\rho_0(x,t) = \rho_1 e^{i\omega_1 t}\delta[k_0(x-L_1)] + \rho_2 e^{-i\omega_2 t}\,\delta[k_0(x-L_2)]. \tag{14.291}$$

k_0 is introduced to make the argument of the δ–function dimensionless. We solve the simultaneous equations (14.289) and (14.290) with a given source $\rho_0(x,t)$ as prescribed by (14.291). In mathematical terms, we have an inhomogeneous system, driven by the source term. The responses f and E should be determined by ρ_0. The kinematic boundary condition is the specular reflection condition

$$f(v,0) = f(-v,0). \tag{14.292}$$

We need only one electric boundary condition in the present one-dimensional problem: the electric displacement $D(x)$ is continuous across the interface. This can be seen from

$$\frac{\partial D(x,\omega)}{\partial x} = \frac{\partial}{\partial x}\left(E + \frac{4\pi i}{\omega}J\right) = 4\pi\rho_0(x,\omega).$$

Since ρ_0 does not contribute to the surface current, the infinitesimal integration gives

$$D|_{x=0^+} = D|_{x=0^-} = E_0 : \ \textit{vacuum electric field}.$$

Assuming that the external perturbation is small, we solve (14.289) and (14.290) by successive approximation. Let us Fourier transform the above equations with respect to t to write

$$-i\omega f(v,x,\omega)+v\frac{\partial f}{\partial x}-\frac{e}{m}\int_{-\infty}^{\infty}\frac{d\omega'}{2\pi}E(x,\omega-\omega')\frac{\partial}{\partial v}f(x,v,\omega')=0,\quad (x>0) \tag{14.293}$$

$$\frac{\partial}{\partial x}E(x,\omega)=4\pi\left[-e\int_{-\infty}^{\infty}f(x,v,\omega)dv+\rho_0(x,\omega)\right],\quad (x>0) \tag{14.294}$$

$$\rho_0(x,\omega)=2\pi[\rho_1\delta(\omega+\omega_1)\delta(k_0(x-L_1))+\rho_2\delta(\omega-\omega_2)\,\delta(k_0(x-L_2))]. \tag{14.295}$$

The last three equations constitute a set of nonlinear simultaneous equations, which we solve by successive approximation in terms of perturbation series:

$$f(x,v,\omega)=f_0(v)+f^{(1)}(x,v,\omega)+f^{(2)}(x,v,\omega)+\cdots,$$
$$E(x,\omega)=E^{(1)}(x,\omega)+E^{(2)}(x,\omega)+\cdots.$$

Breaking down (14.293) and (14.294) order by order, we have

$$-i\omega f^{(1)}(x,v,\omega)+v\frac{\partial f^{(1)}}{\partial x}=\frac{e}{m}\;E^{(1)}(x,\omega)\;\frac{df_0}{dv},\quad (x>0) \tag{14.296}$$

$$\frac{\partial}{\partial x}E^{(1)}(x,\omega)=4\pi\left[-e\int_{-\infty}^{\infty}f^{(1)}(x,v,\omega)dv+\rho_0(x,\omega)\right],\quad (x>0) \tag{14.297}$$

$$-i\omega f^{(2)}(x,v,\omega)+v\frac{\partial f^{(2)}}{\partial x}=\frac{e}{m}\;E^{(2)}(x,\omega)\;\frac{df_0}{dv}$$
$$+\frac{e}{m}\int_{-\infty}^{\infty}\frac{d\omega'}{2\pi}E^{(1)}(x,\omega-\omega')\;\frac{\partial}{\partial v}f^{(1)}(x,v,\omega'), \tag{14.298}$$

$$\frac{\partial}{\partial x}E^{(2)}(x,\omega)=-4\pi e\int_{-\infty}^{\infty}f^{(2)}(x,v,\omega)dv.\quad (x>0) \tag{14.299}$$

• **Linear solution.** The simultaneous equations (14.296) and (14.297) can be solved by Fourier transform provided $E(x)$ is continued into region $x<0$ in an odd function manner: $E(-x)=-E(x)$. We obtain

$$f^{(1)}(k,\omega,v)=\frac{ie}{m}\;\frac{1}{\omega-kv}\;\frac{df_0}{dv}E^{(1)}(k,\omega). \tag{14.300}$$

Multiplying (14.297) by e^{-ikx} and integrating piecewisely from $-\infty$ to 0^- and 0^+ to ∞, one obtains

$$ikE^{(1)}(k,\omega) + N = -4\pi e \int_{-\infty}^{\infty} f^{(1)}(k,\omega,v)dv + 4\pi\rho_0(k,\omega) \quad (14.301)$$

$$where \quad N = E^{(1)}(0^-,\omega) - E^{(1)}(0^+,\omega).$$

$$\rho_0(k,\omega) = \frac{2\pi}{k_0}[\,\rho_1 e^{-ikL_1}\delta(\omega+\omega_1) + \rho_2 e^{-ikL_2}\delta(\omega-\omega_2)]. \quad (14.302)$$

Then, the electric field is given by

$$E^{(1)}(k,\omega) = \frac{i}{k\;\varepsilon(k,\omega)}[N - 4\pi\rho_0(k,\omega)], \quad (14.303)$$

$$\varepsilon(k,\omega) = 1 + \frac{\omega_p^2}{k}\int_{-\infty}^{\infty} dv\, \frac{\frac{df_0}{dv}}{\omega - kv}, \quad (14.304)$$

which is the dielectric function. N is determined from the electric boundary condition as shown in the following.

We need the electric displacement $D(x)$ to enforce the electric boundary condition. By definition, $D(k,\omega) = E(k,\omega) + \frac{4\pi i}{\omega}J(k,\omega)$ where J is the current, $J(k,\omega) = -e\int_{-\infty}^{\infty} dv\; v\; f(k,\omega,v)$. Using (14.300), we obtain

$$D^{(1)}(k,\omega) = \varepsilon(k,\omega)\; E^{(1)}(k,\omega) = \frac{i}{k}[N - 4\pi\rho_0(k,\omega)]. \quad (14.305)$$

Inverting (14.305) gives in the region $x > 0$

$$D^{(1)}(x,\omega) = -\frac{1}{2}\,NH(x) + \frac{4\pi^2}{k_0}[\rho_1\delta(\omega+\omega_1)H(x-L_1) + \rho_2\delta(\omega-\omega_2)H(x-L_2)] \quad (14.306)$$

where we used the formula

$$\int_{-\infty}^{\infty} dk\, \frac{e^{ikx}}{k} = i\pi H(x)$$

where $H(x)$ is the step function: $H(x) = 1$ if $x > 0$ or -1 if $x < 0$. Therefore, we obtain

$$D^{(1)}(0^+,\omega) = -\frac{N}{2} - \frac{4\pi^2}{k_0}[\rho_1\delta(\omega+\omega_1) + \rho_2\delta(\omega-\omega_2)]. \quad (14.307)$$

Setting this quantity to E_0, we have

$$N = -2E_0 - \frac{8\pi^2}{k_0}[\rho_1\delta(\omega+\omega_1) + \rho_2\delta(\omega-\omega_2)]. \tag{14.308}$$

Using (14.308) in (14.303) gives

$$E^{(1)}(k,\omega) = \frac{-2i}{k\ \varepsilon(k,\omega)}\left[E_0 + \frac{4\pi^2}{k_0}(\rho_1\delta(\omega+\omega_1)(1+e^{-ikL_1}) \right.$$
$$\left. + \rho_2\delta(\omega-\omega_2)(1+e^{-ikL_2}))\right]. \tag{14.309}$$

In (14.309), we determined $E^{(1)}$ in terms of the vacuum field E_0 and the external charges. E_0 and $'1'$ in the factors $(1+e^{-ikL_1})$ and $(1+e^{-ikL_2})$ come from the boundary condition which will be absent from an infinite plasma. In an infinite plasma without boundary, we would obtain

$$E^{(1)}(k,\omega) = \frac{-4\pi i\rho_0(k,\omega)}{k\ \varepsilon(k,\omega)} = \frac{-i}{k\varepsilon(k,\omega)}$$
$$\times \frac{8\pi^2}{k_0}(\rho_1\delta(\omega+\omega_1)e^{-ikL_1} + \rho_2\delta(\omega-\omega_2)e^{-ikL_2}).$$

• **Second order solution and echo occurrence.** Next, we proceed to obtain the second order solutions from (14.298) and (14.299). After Fourier transforming with respect to x, we obtain

$$f^{(2)}(k,\omega,v) = \frac{i}{\omega-kv}\frac{e}{m}\left(\frac{\partial}{\partial v}Q(k,\omega,v) + \frac{df_0}{dv}E^{(2)}(k,\omega)\right) \tag{14.310}$$

where Q stands for

$$Q(k,\omega,v) = \int_{-\infty}^{\infty}\frac{d\omega'}{2\pi}\int_{-\infty}^{\infty}\frac{dk'}{2\pi}E^{(1)}(\omega-\omega',k-k')f^{(1)}(\omega',k',v). \tag{14.311}$$

Fourier transforming (14.299) and using (14.310) give

$$E^{(2)}(k,\omega) = \frac{1}{k\varepsilon(k,\omega)}\left[-\frac{4\pi e^2}{m}\int_{-\infty}^{\infty}\frac{dv}{\omega-kv}\frac{\partial Q}{\partial v} + i\ N^{(2)}\right]$$

where $N^{(2)}$ is the jump of $E^{(2)}(x)$ at $x=0$. This quantity needs not to be carried with because the field discontinuity requirement has been already satisfied by N. After integration by parts, this equation becomes

$$E^{(2)}(k,\omega) = \frac{4\pi^2 e^2}{m\varepsilon(k,\omega)} \int_{-\infty}^{\infty} dv\, \frac{Q(k,\omega,v)}{(\omega - kv)^2}. \tag{14.312}$$

Let us summarize the equations that will be used to investigate the echo occurrence condition:

$$E^{(2)}(k,\omega) = \frac{ie^3}{m^2}\frac{1}{\varepsilon(k,\omega)} \int_{-\infty}^{\infty} dv \frac{\frac{df_0}{dv}}{(\omega - kv)^2} \int_{-\infty}^{\infty} d\omega' \times \int_{-\infty}^{\infty} dk' \,\frac{E^{(1)}(\omega',k',v)E^{(1)}(\omega-\omega',k-k')}{\omega' - k'v}, \tag{14.313}$$

$$E^{(1)}(k,\omega) = \frac{-2i}{k\varepsilon(k,\omega)}\left[E_0 + \frac{4\pi^2}{k_0}(\rho_1(b + e^{-ikL_1})\delta(\omega+\omega_1) + \rho_2(b + e^{-ikL_2})\delta(\omega-\omega_2))\right], \tag{14.314}$$

where the letter 'b' stands for unity; it will be eventually put to 1 but indicates that it is a boundary term. We write the product terms which survive the phase-mixing if a suitable condition is met:

$$\frac{E^{(1)}(\omega',k')E^{(1)}(\omega-\omega',k-k')}{\omega'-k'v} = \frac{-\alpha}{k'(k-k')(\omega'-k'v)\varepsilon(k',\omega')\varepsilon(k-k',\omega-\omega')}\times$$

$$\rho_1\rho_2\delta(\omega'-\omega_2)\delta(\omega-\omega'+\omega_1)e^{-ik'L_2}e^{-i(k-k')L_1} \tag{14.315}$$

$$\rho_1\rho_2\delta(\omega'+\omega_1)\delta(\omega-\omega'-\omega_2)e^{-ik'L_1}e^{-i(k-k')L_2} \tag{14.316}$$

$$b\ \rho_1^2\delta(\omega'+\omega_1)\delta(\omega-\omega'+\omega_1)e^{-i(k-k')L_1} \tag{14.317}$$

$$b\ \rho_2^2\delta(\omega'-\omega_2)\delta(\omega-\omega'-\omega_2)e^{-i(k-k')L_2} \tag{14.318}$$

$$b\ \rho_1\rho_2\delta(\omega'-\omega_2)\delta(\omega-\omega'+\omega_1)e^{-i(k-k')L_1} \tag{14.319}$$

$$b\ \rho_1\rho_2\delta(\omega'+\omega_1)\delta(\omega-\omega'-\omega_2)e^{-i(k-k')L_2} \tag{14.320}$$

where $\alpha = (8\pi^2/k_0)^2$. Terms (14.315) to (14.320) are to be added. The terms multiplied by b are boundary terms which are irrelevant to an infinite plasma. When these terms are Fourier-inverted to determine the echo spots, two algebraic points need to be mentioned: (i) pick up only the free streaming singularity $\omega = kv$ (ii) in evaluating the double pole differentiate only the exponential function.

We show that two terms (14.315) and (14.316) can yield an echo resonance. Adding the two terms, we have

$$\frac{E^{(1)}(\omega', k')E^{(1)}(\omega - \omega', k - k')}{\omega' - k'v} = -\alpha\rho_1\rho_2\delta(\omega + \omega_1 - \omega_2) \tag{14.321}$$

$$\times \frac{\delta(\omega' + \omega_1)e^{-ik'L_1}\, e^{-i(k-k')L_2} + \delta(\omega' - \omega_2)e^{-ik'L_2}\, e^{-i(k-k')L_1}}{k'(k-k')(\omega' - k'v)\varepsilon(k', \omega')\varepsilon(k-k', \omega - \omega')}. \tag{14.322}$$

Integral $\int d\omega'$ in (14.313) can be done easily owing to the δ–functions:

$$\begin{aligned} E^{(2)}(k, \omega) = -i\, A \int_{-\infty}^{\infty} \frac{dv}{(\omega - kv)^2} \frac{df_0}{dv} \frac{\delta(\omega + \omega_1 - \omega_2)}{\varepsilon(k, \omega)} \\ \times \int_{-\infty}^{\infty} \frac{dk'}{k'(k-k')} \left[\frac{e^{-ik'L_2}\, e^{-i(k-k')L_1}}{(\omega_2 - k'v)\varepsilon(k', \omega_2)\varepsilon(k - k', -\omega_1)} \right. \\ \left. - \frac{e^{-ik'L_1}\, e^{-i(k-k')L_2}}{(\omega_1 + k'v)\varepsilon(k', -\omega_1)\varepsilon(k-k', \omega_2)} \right], \end{aligned} \tag{14.323}$$

$$A = \alpha\, \frac{e}{m}\, \frac{\omega_p^2}{4\pi}\rho_1\rho_2.$$

In $\int dk'$–integral, poles at $k' = 0$ and $k' = k$ do not produce the echo resonance terms; poles at the roots of the dielectric functions contribute negligibly to the integral, long after the Landau damping time; poles at $k' = \omega_2/v$ and $k' = -\omega_1/v$ produce the resonance terms. Therefore, picking up the residues at $k' = \omega_2/v$ and $k' = -\omega_1/v$, $\int dk'$–integral can be done:

$$\begin{aligned} E^{(2)}(k, \omega) = 2\pi A \int_{-\infty}^{\infty} \frac{dv}{(\omega - kv)^2} \frac{df_0}{dv} \frac{\delta(\omega + \omega_1 - \omega_2)}{\varepsilon(k, \omega)} \\ \times \left[-\frac{e^{-iL_2\frac{\omega_2}{v}}\, e^{-i(k - \frac{\omega_2}{v})L_1}}{\omega_2(k - \frac{\omega_2}{v})\varepsilon(\frac{\omega_2}{v}, \omega_2)\varepsilon(k - \frac{\omega_2}{v}, -\omega_1)} + \frac{e^{i\frac{\omega_1}{v}L_1}\, e^{-i(k + \frac{\omega_1}{v})L_2}}{\omega_1(k + \frac{\omega_1}{v})\varepsilon(-\frac{\omega_1}{v}, -\omega_1)\varepsilon(k + \frac{\omega_1}{v}, \omega_2)} \right]. \end{aligned} \tag{14.324}$$

$E^{(2)}(x,t)$ can be obtained by inverting the above equation. $\int d\omega$-integral can be done immediately, and we write

$$E^{(2)}(x,t) = \frac{A}{2\pi}\, e^{-i\omega_3 t} \int_{-\infty}^{\infty} dv\, \frac{df_0}{dv} \int_{-\infty}^{\infty} dk\, \frac{e^{ikx}}{\varepsilon(k,\omega_3)(\omega_3 - kv)^2}$$
$$\times \left[-\frac{e^{-iL_2 \frac{\omega_2}{v}}\, e^{-i(k-\frac{\omega_2}{v})L_1}}{\omega_2 (k - \frac{\omega_2}{v})\varepsilon(\frac{\omega_2}{v},\omega_2)\varepsilon(k-\frac{\omega_2}{v},-\omega_1)} + \frac{e^{i\frac{\omega_1}{v}L_1}\, e^{-i(k+\frac{\omega_1}{v})L_2}}{\omega_1(k+\frac{\omega_1}{v})\varepsilon(-\frac{\omega_1}{v},-\omega_1)\varepsilon(k+\frac{\omega_1}{v},\omega_2)} \right] \tag{14.325}$$

where $\omega_3 = \omega_2 - \omega_1$. In the above $\int dk$–integral, the non-phase-mixing contribution comes from the double pole at $k = \omega_3/v$. The residue is obtained by differentiating the integrand with respect to k and putting $k = \omega_3/v$. Here we differentiate only the exponential factor since it gives the asymptotically dominant term. The two terms in the large bracket are nicely combined upon putting $k = \omega_3/v$ to yield

$$E^{(2)}(x,t) = -Ae^{-i\omega_3 t}\, \frac{2x - L_1 - L_2}{\omega_1\omega_2} \int_{-\infty}^{\infty} \frac{dv}{v}\, \frac{df_0}{dv}\, \frac{exp\left[i\frac{\omega_1}{v}L_1 - i\frac{\omega_2}{v}L_2 + i\frac{\omega_3}{v}x\right]}{\varepsilon(\frac{\omega_3}{v},\omega_3)\varepsilon(\frac{\omega_2}{v},\omega_2)\varepsilon(\frac{-\omega_1}{v},-\omega_1)}. \tag{14.326}$$

Echo is given rise to where the exponential argument becomes zero, thus the velocity integral is not phase-mixed:

$$x_{echo} = \frac{\omega_2 L_2 - \omega_1 L_1}{\omega_3}, \tag{14.327}$$

$$E^{(2)}(x_{echo},t) = Ae^{-i\omega_3 t}\, \frac{(\omega_1+\omega_2)(L_1 - L_2)}{\omega_1\omega_2\omega_3}$$
$$\times \int_{-\infty}^{\infty} \frac{dv}{v}\, \frac{\frac{df_0}{dv}}{\varepsilon(\frac{\omega_3}{v},\omega_3)\varepsilon(\frac{\omega_2}{v},\omega_2)\varepsilon(\frac{-\omega_1}{v},-\omega_1)}. \tag{14.328}$$

The above echo given by (14.327) and (14.328) can arise in an infinite plasma as well.

Exercise. Repeating similar algebra performed in the foregoing analysis, obtain the following echo locations and the corresponding electric fields. These echoes do not occur in an infinite plasma.

At $x_{echo} = \frac{L_1}{2}$:

$$E^{(2)}(x_{echo}, t) = -\alpha\rho_1^2 \frac{eL_1}{8\pi m} \frac{\omega_p^2}{\omega_1^2} e^{2i\omega_1 t}$$

$$\times \int_{-\infty}^{\infty} dv \frac{\frac{1}{v}\frac{df_0}{dv}}{\varepsilon(-\frac{2\omega_1}{v}, -2\omega_1)\varepsilon(-\frac{\omega_1}{v}, -\omega_1), \varepsilon(-\frac{\omega_1}{v}, -\omega_1)}.$$

At $x_{echo} = \frac{L_2}{2}$:

$$E^{(2)}(x_{echo}, t) = -\alpha\rho_2^2 \frac{eL_2}{8\pi m} \frac{\omega_p^2}{\omega_2^2} e^{-2i\omega_2 t}$$

$$\times \int_{-\infty}^{\infty} dv \frac{\frac{1}{v}\frac{df_0}{dv}}{\varepsilon(\frac{2\omega_2}{v}, 2\omega_2)\varepsilon(\frac{\omega_2}{v}, \omega_2), \varepsilon(\frac{\omega_2}{v}, \omega_2)}.$$

At $x_{echo} = -\frac{L_1\omega_1}{\omega_3}$ $(\omega_3 = \omega_2 - \omega_1)$:

$$E^{(2)}(x_{echo}, t) = \alpha\rho_1\rho_2 \frac{eL_1}{4\pi m} \frac{\omega_p^2}{\omega_1\omega_3} e^{-i\omega_3 t}$$

$$\times \int_{-\infty}^{\infty} dv \frac{\frac{1}{v}\frac{df_0}{dv}}{\varepsilon(\frac{-\omega_1}{v}, -\omega_1)\varepsilon(\frac{\omega_2}{v}, \omega_2)\varepsilon(\frac{\omega_3}{v}, \omega_3)}.$$

At $x_{echo} = \frac{L_2\omega_2}{\omega_3}$:

$$E^{(2)}(x_{echo}, t) = -\alpha\rho_1\rho_2 \frac{eL_2}{4\pi m} \frac{\omega_p^2}{\omega_2\omega_3} e^{-i\omega_3 t}$$

$$\times \int_{-\infty}^{\infty} dv \frac{\frac{1}{v}\frac{df_0}{dv}}{\varepsilon(\frac{-\omega_1}{v}, -\omega_1)\varepsilon(\frac{\omega_2}{v}, \omega_2)\varepsilon(\frac{\omega_3}{v}, \omega_3)}.$$

The above equations predict possible echo spots and the corresponding electric fields. The bulk of this section is taken from the author's paper [H. J. Lee and M. J. Lee, The Open Plasma Phys. Journal **8** (2015)]. If the external charges are prescribed in the form

$$\rho_0(x, z, t) = \rho_1 e^{i\omega_1 t}\delta\left[k_0(x - L_1)\right]\delta\left[k_0(z - S_1)\right] + \rho_2 e^{i\omega_2 t}\delta\left[k_0(x - L_2)\right]\delta\left[k_0(z - S_2)\right]$$

the z-coordinate is not ignorable in the governing equations, and we may talk about *surface wave echo.* A Reference is: H. J. Lee and M. J. Lee, *Surface wave echo in a semi-bounded plasma,* J. of Modern Phys. **7** (2016) p. 1400–1412.

14.6. Penetration of electric field into a plasma slab

In § 14.5.2, we considered penetration of electric field into a semi-bounded plasma placed against an oscillating electric field. This problem was first considered by L. Landau (1946), and we reviewed his theory in § 14.5.2. In this Section, we address a similar problem of finding the electric field penetrated into a *plasma slab* that is inserted in a vacuum with a constant electric field (or we look for the response from the vacuum field $\sim E_0 e^{i\omega t}$ for a particular frequency, when a plasma slab is inserted). This problem will be of practical interest. We can approach this problem as a boundary value problem as formulated in § 14.2. We shall consider the one-dimensional response by neglecting the horizontal coordinates y and z. The machinery to solve this problem is all prepared in § 14.2. We have one-dimensional Vlasov and Poisson equations

$$\frac{\partial}{\partial t} f(x,t,v) + v\frac{\partial f}{\partial x} - \frac{e}{m}E(x,t)\,\frac{\partial f_0}{\partial v} = 0,$$

$$\frac{\partial E}{\partial x} = -4\pi e \int dv f.$$

The plasma electric fields at $x = 0$ and $x = L$ should be connected to the vacuum field by the continuity conditions $D(x)|_0 = E_0, D(x)|_L = E_0$ for a particular frequency ω. In accord with the specular reflection condition $f(v,x)|_{x=0} = f(-v,x)|_{x=0}$, $f(v,x)|_{x=L} = f(-v,x)|_{x=L}$, the electric field is extended in the fashion $E(-x) = -E(x)$, $E(2L-x) = -E(x)$. Then the Poisson equation is Fourier-transformed in the form

$$ikE(k,\omega) + N(k,\omega) = -4\pi e \int f(k,v,\omega)dv \qquad (14.329)$$

$$\text{where} \quad N(k,\omega) = 2A(\omega)\left(\tfrac{1}{2} + cos2kL + cos4kL + \cdots\right) + 2B(\omega)(coskL + cos3kL + \cdots).$$

[See Section § 14.2 *Electrostatic slab dispersion relation*]. (14.329) gives

$$E(k,\omega) = \frac{iN(k,\omega)}{k\varepsilon(k,\omega)} \qquad (14.330)$$

with

$$\varepsilon(k,\omega) = 1 + \frac{\omega_p^2}{k}\int dv\,\frac{\frac{df_0}{dv}}{\omega - kv}.$$

The constants A and B in N will be determined from the electric boundary conditions as shown in the following. Then, E is determined entirely in terms of the vacuum field E_0. The electric boundary condition is the continuity of the electric displacement, $D(x)$. The latter in the plasma side can be obtained from $D(k,\omega) = \varepsilon(k,\omega)E(k,\omega)$. Thus we obtain

$$D(k,\omega) = \frac{2i}{k}\left[A\left(\frac{1}{2} + cos2kL + cos4kL + \cdots\right)\right.$$
$$\left. + B(coskL + cos3kL + \cdots)\right]. \qquad (14.331)$$

To invert (14.331), we write

$$D(x,\omega) = \int_{-\infty}^{\infty} \frac{dk}{2\pi}\, e^{ikx} \frac{2i}{k}\left[A\left(\frac{1}{2} + cos2kL + cos4kL + \cdots\right)\right.$$
$$\left. + B(coskL + cos3kL + \cdots)\right]. \qquad (14.332)$$

It is important to integrate term by term, and sum over the integrated terms. Let us consider a term

$$\int_{-\infty}^{\infty} \frac{dk}{k} e^{ikx}\, cos\; nkL = \frac{1}{2}\int_{-\infty}^{\infty} \frac{dk}{k} e^{ikx}\,(e^{inkL} + e^{-inkL})$$
$$= i\int_{-\infty}^{\infty} \frac{dk}{k}\, e^{inkL} sin\; kx. \quad (n = 1,2,3,\cdots) \qquad (14.333)$$

To obtain the last equality, we changed variable $k \to -k$. Therefore, (14.332) takes the form

$$D(x,\omega) = A'\int_{-\infty}^{\infty} \frac{dk}{k}\; [e^{ikx} + 2isin\; kx\; (e^{2ikL} + e^{4ikL} + \cdots)$$
$$+ B'\int_{-\infty}^{\infty} \frac{dk}{k}\; 2isin\; kx(e^{ikL} + e^{3ikL} + \cdots). \qquad (14.334)$$

Likewise we can write

$$E(x,\omega) = A'\int_{-\infty}^{\infty} \frac{dk}{k\varepsilon}\; [e^{ikx} + 2isin\; kx\; (e^{2ikL} + e^{4ikL} + \cdots)]$$
$$+ B'\int_{-\infty}^{\infty} \frac{dk}{k\varepsilon}\; 2isin\; kx(e^{ikL} + e^{3ikL} + \cdots) \qquad (14.335)$$

where A' and B' are different 'constants'. (14.334) and (14.335) are the plasma solutions in which the constants A$'$ and B$'$ should be determined from the boundary conditions. In particular, the boundary values are as follows.

$$D(0,\omega) = i\pi A' \tag{14.336}$$

where we used

$$\int_{-\infty}^{\infty} dk\, \frac{e^{ikx}}{k} = i\pi, \quad for\ x > 0. \tag{14.337}$$

At $x = L$, (14.334) gives quite a simple result.

$$D(L,\omega) = A' \int_{-\infty}^{\infty} \frac{dk}{k}\, [e^{ikL} + 2isin\ kL\ (e^{2ikL} + e^{4ikL} + \cdots)]$$
$$+ B' \int_{-\infty}^{\infty} \frac{dk}{k}\, 2isinkL\ (e^{ikL} + e^{3ikL} + \cdots). \tag{14.338}$$

Writing $sinkL$ in terms of exponential functions, we have successive cancelations, and the coefficient of A$'$ becomes zero. The coefficient of B$'$ gives

$$D(L,\omega) = -i\pi B'. \tag{14.339}$$

The continuity of $D(x = 0)$ across the interface gives

$$A' = -\frac{i}{\pi}\, E_0 = -B'. \tag{14.340}$$

Therefore, the electric field in the slab is obtained as

$$E(x,\omega) = -\frac{i}{\pi} E_0 \int_{-\infty}^{\infty} \frac{dk}{k\varepsilon}\, [e^{ikx} + 2isin\ kx\ (e^{2ikL} + e^{4ikL} + \cdots)]$$
$$+ \frac{i}{\pi} E_0 \int_{-\infty}^{\infty} \frac{dk}{k\varepsilon}\, 2isin\ kx(e^{ikL} + e^{3ikL} + \cdots). \tag{14.341}$$

It is easy to check the boundary values $E(0,\omega)$ and $E(L,\omega)$. Clearly,

$$E(0,\omega) = -\frac{i}{\pi} E_0 \int_{-\infty}^{\infty} \frac{dk}{k\varepsilon}\, e^{ik0^+}$$

which gives $\varepsilon E(0,\omega) = E_0$ upon using (14.337). To check $E(L,\omega)$, note
$2i\ sinkL(e^{2ikL} + e^{4ikL} + \cdots) = (e^{ikL} - e^{-ikL})(e^{2ikL} + e^{4ikL} + \cdots) = -e^{ikL}$ after successive cancellations, and the first integral in (14.341) vanishes. Also we have
$2i\ sinkL(e^{ikL} + e^{3ikL} + \cdots) = (e^{ikL} - e^{-ikL})(e^{ikL} + e^{3ikL} + \cdots) = -1$, and the second integral reduces to

$$E(L,\omega) = -\frac{i}{\pi}E_0 \int_{-\infty}^{\infty} \frac{dk}{k\varepsilon}$$

which yields $\varepsilon E(L,\omega) = E_0$.

Equation (14.341) can be put into a compact form by changing $k \to -k$ in the $sin\ kx$ function. We shall consider the second integral in (14.341):

$$\frac{i}{\pi}E_0 \int_{-\infty}^{\infty} \frac{dk}{k\varepsilon}(e^{ikx} - e^{-ikx})(e^{ikL} + e^{3ikL} + \cdots)$$
$$= \frac{i}{\pi}E_0 \int_{-\infty}^{\infty} \frac{dk}{k\varepsilon}e^{ikx}(e^{ikL} + e^{3ikL} + \cdots + e^{-ikL} + e^{-3ikL} + \cdots)$$

where we used $\varepsilon(-k) = \varepsilon(k)$. In this way, (14.341) can be written in the form

$$E(x,\omega) = -\frac{i}{\pi}E_0 \int_{-\infty}^{\infty} \frac{dk}{k\varepsilon}e^{ikx}\left[\frac{1}{2} - e^{ikL} + e^{2ikL} - e^{3ikL} + e^{4ikL} - \cdots\right] +$$
$$-\frac{i}{\pi}E_0 \int_{-\infty}^{\infty} \frac{dk}{k\varepsilon}e^{ikx}\left[\frac{1}{2} - e^{-ikL} + e^{-2ikL} - e^{-3ikL} + e^{-4ikL} - \cdots\right].$$

Summing the series formally, we obtain the final result:

$$E(x,\omega) = -\frac{i}{\pi}E_0 \int_{-\infty}^{\infty} \frac{dk}{k\varepsilon}e^{ikx}\left[\frac{1 - e^{ikL}}{2(1 + e^{ikL})} + \frac{1 - e^{-ikL}}{2(1 + e^{-ikL})}\right]. \qquad (14.342)$$

The series in (14.342) converges by picking up the poles. $k = 0$ is not a singularity; it is only apparent. We only need the singularities at $\varepsilon(k) = 0$ in the contour integration. The zeros at $(1 + e^{ikL})$ should be simply ignored. The remaining job is to carry out the integral for a given $\varepsilon(k,\omega)$.

Exercise. Prove the formal sum

$$\frac{1}{2} + e^{ikL} + e^{2ikL} + e^{3ikL} + e^{4ikL} + \cdots = \frac{1 + e^{ikL}}{2(1 - e^{ikL})} = -\frac{1}{2}\ coth\left(\frac{ikL}{2}\right).$$

Remark. The author's choice of interesting subjects which are not dealt with in this Chapter include parametric instabilities in a bounded plasma, and fluctuation of electric field in a bounded plasma, and spatial echoes in a plasma slab. For these topics, References are:

L. Stenflo, *Nonlinear interaction of electrostatic waves in a plasma*, Ann. Physik. **7**. Folge, Bd. 27, Heft 3, S. p. 289-295 (1971)–this is a kinetic theory.
A. G. Sitenko, *Fluctuations and Non-linear Wave Interactions in Plasmas*, Pergamon, Oxford, (1982) Chapter 11.
H. J. Lee and Y. K. Lim, *Surface wave echo in a plasma slab*, Physical Science International Journal **12(2)** (2016) p. 1–16 PSIJ.28647

Chapter 15

Dusty plasmas

Dusty plasmas are such plasmas that contain dust particles (in the size range of 10 nm–100 μm) which are usually negatively charged (10^3–10^4 electronic charge). Dusty plasmas are present in the space environment, e.g., planetary rings, cometary tails, astroid zones, and lower ionosphere of the earth as well as in laboratory plasmas and plasma processing devices. In a plasma, the mere presence of dust particulates (grains) can influence the dispersion of plasma normal modes through the modification of the charge neutrality condition, even though the grains are assumed to be dynamically inactive. Since the grains are very massive (the mass of 1 μm grain is of the order of 10^9–10^{12} proton mass), the assumption that the grains are stationary species forming a portion of the negative charge background is a good assumption for wave motions of high frequency greater than the ion plasma frequency. However, a very low frequency electrostatic mode, called a dust acoustic wave, has been identified in a plasma of thermalized electrons and ions when the collective motion of the dust fluid is taken into account.

15.1. Charging process in dusty plasmas

A grain immersed in a plasma can be charged by collisions with electrons and ions, photoionizations, and secondary emissions, etc. The charges on the dust grains fluctuate since the charging currents depend on the variable plasma quantities. Therefore, the electric charge on the dust particles is time-dependent and should be considered as an additional dynamical variable in dusty plasma dynamics. The grain charge as a self-consistent dynamical variable gives rise to a new mode characteristic of a dusty plasma. In the following, we consider the charging process on the grain by collisions with electrons and ions in the Coulomb field.

15.1.1. *Charging cross section*

When a negatively charged dust grain is approached by an electron, the free path of the latter is deflected off the grain; an ion would curl inwards. If the grain is not charged, the collision cross section would be πR^2 both for electron and ion if the grain is assumed to be a sphere of radius R. Here, we calculate the collision cross sections of electron and ion when the grain is charged. The charge on the grain can be expressed by the potential $\psi = -eZ/R$, where Z is the number of charges on the grain. The cross sections are obtained from two conservation relations of angular momentum and energy.

$$\textit{electron}: \quad m_e v_g R = m_e v b \tag{15.1}$$

where b is the impact distance, v_g is the grazing velocity (the velocity at the grain surface of the tangential orbit).

$$\textit{electron}: \quad \frac{1}{2}m_e v^2 = \frac{1}{2}m_e v_g^2 + \frac{Ze^2}{R}. \tag{15.2}$$

The above two equations give

$$b^2 = R^2\left(1 - \frac{2e^2 Z}{m_e R v^2}\right). \tag{15.3}$$

Therefore, the collision cross section for an electron is given by $\sigma(v) = \pi b^2 = \pi R^2(1 - \frac{2e^2 Z}{m_e R v^2})$. However, we have an additional provision for electron cross section. In order for an electron to graze the grain surface, the electron velocity should be greater than the minimum velocity which can be obtained by putting $v_g = 0$ in (15.2): $v_{min} = \sqrt{2e^2 Z/R m_e}$. Thus, the electron cross section can be written as

$$\sigma_e(v, Z) = \pi R^2\left(1 - \frac{2e^2 Z}{m_e R v^2}\right) \quad \textit{for} \quad v^2 > \frac{2e^2 Z}{R m_e}, \tag{15.4}$$

$$\sigma_e(v, Z) = 0 \quad \textit{for} \quad v^2 < \frac{2e^2 Z}{R m_e}. \tag{15.5}$$

The ion conservation relations are

$$m_i v_g R = m_i v b, \quad \frac{1}{2}m_i v^2 = \frac{1}{2}m_i v_g^2 - \frac{Ze^2}{R}. \tag{15.6}$$

There is no minimum velocity restriction for ions, and the ion cross section is written

$$\sigma_i(v, Z) = \pi R^2\left(1 + \frac{2e^2 Z}{m_i R v^2}\right), \tag{15.7}$$

• Current on the grain surface produced by collisions with electrons and ions

The collision cross sections obtained above can be used to calculate the currents on the grain surface. The electron current is

$$I_e(Z) = -e \int_{v_{min}}^{\infty} d^3v\ v\ \sigma_e(v, Z) f_{e0}(v), \quad d^3v = 4\pi v^2 dv, \qquad (15.8)$$

$$f_{e0}(v) = N_e \left(\frac{m_e}{2\pi T_e}\right)^{3/2} exp\left(-\frac{m_e v^2}{2T_e}\right) \qquad (15.9)$$

where N_e is the equilibrium electron number density of the plasma. It is straightforward to calculate the integral by using (15.4), and we obtain

$$I_e(Z) = -e\sqrt{8\pi T_e/m_e}\ R^2 N_e\ exp\left(\frac{e\psi}{T_e}\right). \quad (\psi = -eZ/R) \qquad (15.10)$$

Likewise, the ion current is obtained by

$$I_i(Z) = e \int_0^{\infty} d^3v\ v\ \sigma_i(v, Z) f_{i0}(v), \qquad (15.11)$$

$$f_{i0}(v) = N_i \left(\frac{m_i}{2\pi T_i}\right)^{3/2} exp\left(-\frac{m_i v^2}{2T_i}\right), \qquad (15.12)$$

$$I_i(Z) = e\sqrt{8\pi T_i/m_i}\ R^2 N_i \left(1 - \frac{e\psi}{T_i}\right). \qquad (15.13)$$

Exercise. Carry out the algebra leading to (15.10) and (15.13) which are called the *orbit limited currents.*

Solution: (15.13) is straightly obtained by calculating Maxwellian integrals.

$$I_e = -e4\pi^2 R^2 \left(\frac{m_e}{2\pi T_e}\right)^{3/2} \int_{v_{min}}^{\infty} dvv^3 \left(1 - \frac{2e^2 Z}{R m_e v^2}\right) exp\left(-\frac{m_e v^2}{2T_e}\right)$$

$$\int_{v_{min}}^{\infty} dvv^3\ exp\left(-\frac{m_e v^2}{2T_e}\right) = \frac{1}{2}\int_{v_{min}^2}^{\infty} dx\ x\ exp\left(-\frac{mx}{2T_e}\right) = etc.$$

$$= \frac{T_e}{m_e}\left[\frac{2e^2 Z}{R m_e} + \frac{2T_e}{m_e}\right] exp\left(\frac{e\psi}{RT_e}\right)$$

$$\int_{v_{min}}^{\infty} dvv\ exp\left(-\frac{m_e v^2}{2T_e}\right) = \frac{T_e}{m_e}\ exp\left(-\frac{m_e v_{min}^2}{2T_e}\right) = \frac{T_e}{m_e}\ exp\left(\frac{e\psi}{RT_e}\right).$$

Sum of these results yields (15.10).

15.2. Moment equations of a dusty plasma

A dusty plasma can be considered as a three-species plasma consisting of ions, electrons, and dust grains. They are dynamically interrelated through continuity equation and the Poisson equation. First, we consider the charging collisions between grains and electrons; grains and ions. We use the Boltzmann equation for electrons and ions with a collision term of BGK type [Reference: P. L. Bhatnagar, E. P. Gross and M. Krook, *Phys. Rev.* **94**, 511 (1954)]:

$$\frac{\partial}{\partial t} f_\alpha(\mathbf{r}, \mathbf{v}, t) + \mathbf{v} \cdot \frac{\partial}{\partial \mathbf{r}} f_\alpha + \frac{e_\alpha}{m_\alpha} \mathbf{E} \cdot \frac{\partial}{\partial \mathbf{v}} f_\alpha = \left(\frac{\partial f_\alpha}{\partial t}\right)_c . \quad (\alpha = e,\ i) \tag{15.14}$$

The RHS is the rate of change of f_α due to capture of electrons and ions by grains. We can write the capture collision integral in the form

$$\left(\frac{\partial f_\alpha}{\partial t}\right)_c = -v\sigma_\alpha(v, q) n_d(\mathbf{r}, t) f_\alpha(\mathbf{r}, \mathbf{v}, t) + v\sigma_{\alpha 0}(v, q_0) N_d f_{\alpha 0}(v) \tag{15.15}$$

where $n_d(\mathbf{r}, t)$ is the grain number density, $q_0 = -eZ_0$ (grain charge at the equilibrium), $q = -eZ$, N_d denotes the equilibrium grain number density. Now we integrate (15.14) over the whole velocity space to obtain

$$\frac{\partial}{\partial t} n_\alpha(\mathbf{r}, t) + \nabla \cdot (n_\alpha \mathbf{u}_\alpha) = -n_d \frac{I_\alpha}{e_\alpha} + N_d \frac{I_{\alpha 0}}{e_\alpha} \tag{15.16}$$

with

$$n_\alpha(\mathbf{r}, t) = \int f_\alpha(\mathbf{r}, \mathbf{v}, t) d^3 v, \tag{15.17}$$

$$\mathbf{u}_\alpha(\mathbf{r}, t) = \frac{\int \mathbf{v} f_\alpha d^3 v}{\int f_\alpha d^3 v} = \frac{\int \mathbf{v} f_\alpha d^3 v}{n_\alpha(\mathbf{r}, t)}, \tag{15.18}$$

$$I_\alpha(q, \mathbf{r}, t) = e_\alpha \int v\sigma_\alpha f_\alpha(\mathbf{r}, \mathbf{v}, t) d^3 v, \tag{15.19}$$

$$I_{\alpha 0} = e_\alpha \int v\sigma_{\alpha 0} f_{\alpha 0}(v) d^3 v. \tag{15.20}$$

Next, we multiply (15.14) by $\mathbf{v}$ and integrate it over the whole velocity space to obtain

$$\frac{\partial}{\partial t}(n_\alpha \mathbf{u}_\alpha) + \nabla \cdot (n_\alpha \mathbf{u}_\alpha \mathbf{u}_\alpha) + \frac{1}{m_\alpha} \nabla \cdot \overline{P} - \frac{e_\alpha}{m_\alpha} n_\alpha \mathbf{E} = -\int d^3 v\ \mathbf{v}\ v\ \sigma_\alpha n_d f_\alpha \tag{15.21}$$

where $\overline{P}$ is the pressure tensor. It is to be noted that $\int d^3v\mathbf{v}\mathbf{v}v\sigma_{\alpha 0}(v)f_{\alpha 0}(v) = 0$ for an isotropic $f_{\alpha 0}(v)$. In the integral in (15.21), the quantity $n_d v\sigma_\alpha \equiv \nu_\alpha$ is the collision frequency between grain and α-species. To evaluate the integral, we replace it by its average:

$$\nu_\alpha = n_d v\sigma_\alpha \longrightarrow \langle n_d v\sigma_\alpha\rangle = \frac{\int v\sigma_\alpha n_d f_\alpha d^3v}{\int f_\alpha d^3v} = \frac{n_d I_\alpha}{n_\alpha e_\alpha}.$$

Then (15.21) becomes

$$\frac{\partial}{\partial t}(n_\alpha \mathbf{u}_\alpha) + \nabla\cdot(n_\alpha \mathbf{u}_\alpha \mathbf{u}_\alpha) + \frac{1}{m_\alpha}\nabla\cdot\overline{P} - \frac{e_\alpha}{m_\alpha}n_\alpha \mathbf{E} = -n_d\,\frac{I_\alpha}{e_\alpha}\mathbf{u}_\alpha. \tag{15.22}$$

Using (15.16) and the identity $\nabla\cdot(n\mathbf{u}\mathbf{u}) = \mathbf{u}\nabla\cdot(n\mathbf{u}) + n\mathbf{u}\cdot\nabla\mathbf{u}$ in the above equation yields the equation of motion

$$n_\alpha\left(\frac{\partial \mathbf{u}_\alpha}{\partial t} + \mathbf{u}_\alpha\cdot\nabla\mathbf{u}_\alpha\right) = -\frac{T_\alpha}{m_\alpha}\nabla n_\alpha + \frac{e_\alpha}{m_\alpha}n_\alpha\mathbf{E} - N_d\,\frac{I_{\alpha 0}}{e_\alpha}\mathbf{u}_\alpha. \tag{15.23}$$

(15.16) and (15.23) are the continuity and momentum equations whose sink and frictional terms are derived from the charging collision integral.

For the grains, we can assume the sink-free continuity equation,

$$\frac{\partial n_d}{\partial t} + \nabla\cdot(n_d\mathbf{u}_d) = 0 \tag{15.24}$$

and the drag-free momentum equation

$$\frac{\partial \mathbf{u}_d}{\partial t} + \mathbf{u}_d\cdot\nabla\mathbf{u}_d = -\frac{eZ}{m_d}\,\mathbf{E}. \tag{15.25}$$

Next, we derive the charging equation from the charge conservation equation:

$$e\,\frac{\partial}{\partial t}(n_i - n_e - Zn_d) + \nabla\cdot\mathbf{J} = 0, \quad \mathbf{J} = e(n_i\mathbf{u}_i - n_e\mathbf{u}_e - Zn_d\mathbf{u}_d). \tag{15.26}$$

Using (15.16) and (15.24) in the above equation yields

$$-e\left(\frac{\partial}{\partial t} + \mathbf{u}_d\cdot\nabla\right)Z = I_i + I_e \tag{15.27}$$

where we used the current balance equation

$$I_{i0} + I_{e0} = 0. \tag{15.28}$$

Finally, we have the Poisson equation

$$\nabla\cdot\mathbf{E} = 4\pi e(n_i - n_e - Zn_d). \tag{15.29}$$

• Expansion of currents, Linearization of Charging equation

Putting $Z = Z_0 + Z_1$ (subscript 1 denotes the perturbation), we write

$$I_\alpha = e_\alpha \int v \left(\sigma_{\alpha 0} + \frac{\partial \sigma_{\alpha 0}}{\partial Z_0} Z_1 \right) (f_{\alpha 0} + f_{\alpha 1}) d^3 v. \tag{15.30}$$

The zero order current is

$$I_{\alpha 0} = e_\alpha \int v \, \sigma_{\alpha 0} f_{\alpha 0} d^3 v. \tag{15.31}$$

The first order current is

$$I_{\alpha 1} = e_\alpha \int v \left(\sigma_{\alpha 0} f_{\alpha 1} + \frac{\partial \sigma_{\alpha 0}}{\partial Z_0} Z_1 f_{\alpha 0} \right) d^3 v. \tag{15.32}$$

In the first integral, $v\sigma_{\alpha 0}$ is replaced by its average over $f_{\alpha 0}$, $I_{\alpha 0}/(N_\alpha e_\alpha)$, thus we can write

$$I_{\alpha 1} = \frac{n_{\alpha 1}}{N_\alpha} I_{\alpha 0} + \frac{\partial I_{\alpha 0}}{\partial Z_0} Z_1. \tag{15.33}$$

Therefore we have

$$I_{e1} = -I_0 \left(\frac{n_{e1}}{N_e} - \frac{e^2 Z_1}{R T_e} \right), \tag{15.34}$$

$$I_{i1} = I_0 \left(\frac{n_{i1}}{N_i} + \frac{e^2 Z_1}{R(T_i + e^2 Z_0/R)} \right). \tag{15.35}$$

Let us define the zero order and the first order collision frequencies,

$$\nu_{i0} = \frac{I_0}{e} \frac{N_d}{N_i}, \quad \nu_{e0} = \frac{I_0}{e} \frac{N_d}{N_e}, \; \nu_{\alpha 1} = Z_1 \frac{\partial \nu_{\alpha 0}}{\partial Z_0}. \tag{15.36}$$

Then (15.33) can be written as

$$I_{i1} = e \frac{n_{i1}}{N_d} \nu_{i0} + e \frac{N_i}{N_d} \nu_{i1}, \quad I_{e1} = -e \frac{n_{e1}}{N_d} \nu_{e0} - e \frac{N_e}{N_d} \nu_{e1}. \tag{15.37}$$

Linearizing (15.27) gives

$$ie \, \omega Z_1 = I_{i1} + I_{e1}. \tag{15.38}$$

Thus, using the above equations yield

$$Z_1 = \frac{\nu_{i0} n_{i1} - \nu_{e0} n_{e1}}{N_d (i\omega - \eta)} = \frac{1}{i\omega - \eta} \frac{I_0}{e} \left(\frac{n_{i1}}{N_i} - \frac{n_{e1}}{N_e} \right) \tag{15.39}$$

$$\textit{where} \quad \eta = \frac{N_i}{N_d} \frac{\partial \nu_{i0}}{\partial Z_0} - \frac{N_e}{N_d} \frac{\partial \nu_{e0}}{\partial Z_0} = \frac{e I_0}{R} \left(\frac{1}{T_e} + \frac{1}{T_i + \frac{e^2 Z_0}{R}} \right). \tag{15.40}$$

Exercise. Verify the above equality. The quantity η is called the *grain charge relaxation frequency*.

15.3. Waves in a dusty plasma

15.3.1. *High frequency electrostatic wave*

In this Section, we derive the dispersion relation of the high frequency Langmuir wave in which the grain charge fluctuation effects are accounted for. The latter is represented by the sink terms in the continuity equation and the drag term in the momentum equation. The ion and the grain density perturbations can be neglected in the high frequency motion. Linearizing (15.16), (15.23), and (15.29) yields for the waves of the phasor $exp(i\mathbf{k}\cdot\mathbf{r}-i\omega t)$

$$(\nu_{e0} - i\omega)n_{e1} + iN_e\mathbf{k}\cdot\mathbf{u}_e = -N_e\nu_{e1}, \tag{15.41}$$

$$(\nu_{e0} - i\omega)\mathbf{u}_e = \frac{ie\phi}{m_e}\mathbf{k} - i\mathbf{k}\,\frac{T_e}{m_e}\frac{n_{e1}}{N_e} \quad (\mathbf{E} = -\nabla\phi), \tag{15.42}$$

$$k^2\phi = -4\pi e(n_{e1} + N_d Z_1). \tag{15.43}$$

Using the foregoing equations and (15.39) the dispersion relation is obtained as

$$(i\omega - \nu_{e0})^2 + \omega_{pe}^2 + \frac{k^2T_e}{m_e} + \frac{\nu_{e0}}{\eta - i\omega}\left[\omega_{pe}^2 - (i\omega - \nu_{e0})\frac{N_e}{N_d}\frac{\partial\nu_{e0}}{\partial Z_0}\right] = 0. \tag{15.44}$$

This is the dispersion relation of the Langmuir wave in a dusty plasma with the grain charging effect included.

15.3.2. *Dust acoustic wave*

The dust acoustic wave is a very low frequency wave in which the heavy grain inertia plays a role. In this low frequency region, the electrons and the ions can be assumed Boltzmann-distributed:

$$n_e = \frac{eN_e}{T_e}\,\phi, \quad n_i = -\frac{eN_i}{T_i}\,\phi \tag{15.45}$$

where n denotes the perturbation number density. We also have Poisson's equation written in the form

$$\nabla^2\phi = 4\pi e(n_e - n_i + Z_0 n_d + Z_1 N_d). \tag{15.46}$$

For the cold dust fluid, we have

$$\frac{\partial\mathbf{u}_d}{\partial t} = \frac{eZ_0}{m_d}\nabla\phi, \tag{15.47}$$

$$\frac{\partial n_d}{\partial t} + N_d\nabla\cdot\mathbf{u}_d = 0. \tag{15.48}$$

The above equations yield the dispersion relation of dust acoustic wave in the form

$$\frac{\omega^2}{k^2} = \frac{\omega_{pd}^2 \lambda_D^2}{1 + k^2\lambda_D^2 - 4\pi e N_d \lambda_D^2 I_0 (T_e^{-1} + T_i^{-1})/(i\omega - \eta)} \tag{15.49}$$

where

$$\lambda_D^2 = (\lambda_e^{-2} + \lambda_i^{-2})^{-1}, \quad \lambda_\alpha^2 = \frac{T_\alpha}{4\pi N_\alpha e^2} \quad (\alpha = e, i), \quad \omega_{pd}^2 = \frac{4\pi N_d e^2 Z_0^2}{m_d}.$$

15.4. Kinetic theory

For simplicity, we shall consider one-dimensional linear waves, varying only in z and t with phasor $e^{ik_z z - i\omega t}$. Also we confine to the longitudinal wave $\mathbf{E} \| \mathbf{k}$. Then we have

$$\frac{\partial f_\alpha}{\partial t} + v_z \frac{\partial f_\alpha}{\partial z} + \frac{e_\alpha}{m_\alpha} E(z,t) \frac{\partial f_{\alpha 0}(v)}{\partial v_z} = 0 \tag{15.50}$$

where $\alpha = e, i$ for electrons and ions.

$$\frac{\partial f_d}{\partial t} + v_z \frac{\partial f_d}{\partial z} + \frac{eZ(z,t)}{m_d} E(z,t) \frac{\partial f_{d0}(v)}{\partial v_z} = 0. \tag{15.51}$$

The last equation is for the grains; to emphasize the nature of the charge number Z being a dynamical variable it is written separately. The charging equation (15.27) must be supplemented:

$$(-e)\left(\frac{\partial Z}{\partial t} + V_d \frac{\partial Z}{\partial z}\right) = I_i + I_e \quad (cf.\ (15.25)) \tag{15.52}$$

where V_d is the grain fluid velocity,

$$I_i = e \int_0^\infty d^3v\, v\, \sigma_i f_i, \tag{15.53}$$

$$I_e = -e \int_{v_g(Z)}^\infty d^3v\, v\, \sigma_e f_e, \tag{15.54}$$

where $v_g(Z) = \sqrt{\frac{2e^2 Z}{R m_e}}$ is the electron grazing velocity, and $\sigma_{i,e}$ are given by (15.4) and (15.7). Also we have the Poisson equation

$$\frac{\partial E}{\partial z} = 4\pi e(n_i - n_e - Z n_d). \tag{15.55}$$

In linearizing the above equations, we designate the perturbations by subscript 1 and the equilibrium values by 0. We write

$$Z = Z_0 + Z_1, \quad \sigma = \sigma_0 + \sigma_1, \quad I = I_0 + I_1,$$

$$\sigma_{i0} = \pi R^2 \left(1 + \frac{2e^2 Z_0}{R m_i v^2}\right), \tag{15.56}$$

$$\sigma_{e0} = \pi R^2 \left(1 - \frac{2e^2 Z_0}{R m_e v^2}\right), \tag{15.57}$$

$$\sigma_{i1} = \pi R \, \frac{2e^2 Z_1}{m_i v^2}, \tag{15.58}$$

$$\sigma_{e1} = -\pi R \, \frac{2e^2 Z_1}{m_e v^2}, \tag{15.59}$$

$$I_{i0} = e \int_0^\infty d^3v \; v \; \sigma_{i0} f_{i0}, \tag{15.60}$$

$$I_{i1} = e \int_0^\infty d^3v \; v \; (\sigma_{i1} f_{i0} + \sigma_{i0} f_i), \tag{15.61}$$

$$I_{e0} = -e \int_{v_g(Z_0)}^\infty d^3v \; v \; \sigma_{e0} f_{e0}, \tag{15.62}$$

$$I_{e1} = -e \int_{v_g(Z_0)}^\infty d^3v \; v \; (\sigma_{e1} f_{e0} + \sigma_{e0} f_e). \tag{15.63}$$

The equilibrium charge number Z_0 is determined by the charge balance equation

$$I_{i0} + I_{e0} = 0, \tag{15.64}$$

$$I_{e0} = -e \; \sqrt{8\pi T_e/m_e} \; R^2 N_e \; exp \left(-\frac{e^2 Z_0}{R T_e}\right), \quad (cf.\ (15.10)) \tag{15.65}$$

$$I_{i0} = e \; \sqrt{8\pi T_i/m_i} \; R^2 N_i \left(1 + \frac{e^2 Z_0}{R T_i}\right). \quad (cf.\ (15.13)) \tag{15.66}$$

After Fourier transforming, the linear solutions of (15.50) and (15.51) are

$$f_\alpha(k,\omega) = -i \, \frac{e_\alpha}{m_\alpha} E(k,\omega) \, \frac{\frac{\partial f_{\alpha 0}}{\partial v_z}}{\omega - k_z v_z}, \tag{15.67}$$

$$f_d(k,\omega) = -i \, \frac{e Z_0}{m_d} E(k,\omega) \, \frac{\frac{\partial f_{d0}}{\partial v_z}}{\omega - k_z v_z}. \tag{15.68}$$

The charging equation, (15.52), gives

$$-iew\ Z_1 = I_{i1} + I_{e1} \tag{15.69}$$

where we neglected the term $V_d\ \partial Z/\partial z$ since the grains have a zero mobility ($m_d \to \infty$) as compared to electrons and ions. (15.69) gives

$$(i\omega - \eta)Z_1$$
$$= \left[-\frac{ie}{m_i}\int_0^\infty d^3v v\sigma_{i0}\frac{\frac{\partial f_{i0}}{\partial v_z}}{\omega - k_z v_z} - \frac{ie}{m_e}\int_{v_0}^\infty d^3v v\sigma_{e0}\frac{\frac{\partial f_{e0}}{\partial v_z}}{\omega - k_z v_z}\right] E(k_z, \omega) \tag{15.70}$$

where $v_0 \equiv v_{g(Z_0)}$, and η is defined by

$$\eta = 2\pi e^2 R\left[\int_0^\infty \frac{f_{i0}}{m_i v} d^3v - \int_{v_0}^\infty \frac{f_{e0}}{m_e v} d^3v\right]$$
$$= \frac{e}{R} I_{i0}\left(\frac{1}{T_e} + \frac{1}{T_i + e^2 Z_0/R}\right). \tag{15.71}$$

Exercise. Prove (15.71) to show that it agrees with (15.40). Hint: $d^3v = 4\pi v^2 dv$.

Linearizing and Fourier transforming Poisson equation (15.55) gives

$$ik_z E(k_z, \omega) = 4\pi e \int f_i d^3v - 4\pi e \int f_e d^3v - -4\pi e N_d Z_1 - 4\pi e Z_0 \int f_d d^3v. \tag{15.72}$$

In the following we simply put $k_z = k$. Furthermore one can prove that

$$\int d^3v G(v) \frac{\frac{\partial f_0}{\partial v_z}}{\omega - k_z v_z} = \int d^3v G(v) \frac{\frac{\partial f_0}{\partial v}}{\omega - kv}$$

where $G(v)$ is a spherically symmetric function of v. Thus, the subscripts z can be omitted in the above equations (see Exercise below).

Using (15.67)–(15.70) in (15.72) yields

$$1 + \frac{\omega_{pi}^2}{k}\int_{-\infty}^\infty \frac{\frac{\partial f_{i0}}{\partial v}}{\omega - kv} d^3v + \frac{\omega_{pe}^2}{k}\int_{-\infty}^\infty \frac{\frac{\partial f_{e0}}{\partial v}}{\omega - kv} d^3v + \frac{\omega_{pd}^2}{k}\int_{-\infty}^\infty \frac{\frac{\partial f_{d0}}{\partial v}}{\omega - kv} d^3v$$
$$+ \frac{i\ N_d}{\omega + i\eta}\left[\frac{\omega_{pi}^2}{k}\int_0^\infty d^3v v\sigma_{i0}\frac{\frac{\partial f_{i0}}{\partial v}}{\omega - kv} + \frac{\omega_{pe}^2}{k}\int_{v_0}^\infty d^3v v\sigma_{e0}\frac{\frac{\partial f_{e0}}{\partial v}}{\omega - kv}\right] = 0 \tag{15.73}$$

where $\omega_{pd}^2 = 4\pi e^2 N_d Z_0/m_d$, the squared grain plasma frequency. (15.73) is the general dispersion relation of plasma waves in a dusty plasma. We shall evaluate the integrals in the frequency range such as grain thermal velocity $\ll \omega/k \ll$ thermal velocities of ion or electron, or

$$v_{thd} \ll \frac{\omega}{k} \ll v_{the,\, i}.$$

Then the first three integrals give, respectively,

$$\frac{1}{k^2\lambda_e^2}, \quad \frac{1}{k^2\lambda_i^2}, \quad -\frac{\omega_{pd}^2}{\omega^2}$$

with $\lambda_{e,i}$ being the Debye lengths. The integrals in the large bracket can be done straightly (neglect $\omega \ll kv$):

$$\int_0^\infty d^3v v\sigma_{i0}\frac{\frac{\partial f_{i0}}{\partial v}}{\omega - kv} = \frac{m_i}{T_i}\int_0^\infty d^3v v\sigma_{i0}\frac{f_{i0}}{k} = etc.$$

$$= \sqrt{8\pi}\ N_i R^2\sqrt{\frac{m_i}{T_i}}\left(1 + \frac{e^2 Z_0}{RT_i}\right) = \frac{I_{i0}}{e}\frac{m_i}{T_i}$$

Likewise, we obtain

$$\int_{v_0}^\infty d^3v v\sigma_{e0}\frac{\frac{\partial f_{e0}}{\partial v}}{\omega - kv} = \sqrt{8\pi}\ N_e R^2\sqrt{\frac{m_e}{T_e}} exp\left[-\frac{e^2 Z_0}{RT_e}\right] = \frac{|\, I_{e0}\,|}{e}\frac{m_e}{T_e}.$$

Using these results in (15.73) yields

$$\frac{\omega^2}{k^2} = \frac{\lambda_D^2\omega_{pd}^2}{1 + k^2\lambda_D^2 - \frac{4\pi e\ N_d\lambda_D^2 I_0}{i\omega - \eta}(T_e^{-1} + T_i^{-1})} \tag{15.74}$$

where $\lambda_D^{-2} = \lambda_e^{-2} + \lambda_i^{-2}$, $I_0 = I_{i0}$. This equation agrees with (15.49).

Exercise. Prove the identity

$$\int d^3v\ G(v)\ \frac{\mathbf{k}\cdot\frac{\partial f_0(v)}{\partial \mathbf{v}}}{\omega - \mathbf{k}\cdot\mathbf{v}} = \int d^3u\ G(u)\ \frac{k\ \frac{df_0}{du}}{\omega - ku}$$

where $G(v)$ and $f_0(v)$ are spherically symmetric functions of v and the integration variable u is defined by

$$\mathbf{v} = \frac{\mathbf{k}}{k}\ u\ \equiv \mathbf{u}.$$

Proof: First note $v^2 = u^2$, and thus $G(v) = G(u)$ and $d^3v = d^3u$ (Jacobian is 1). Also $\mathbf{k}\cdot\mathbf{v} = ku$.

$$\frac{\partial f_0(v)}{\partial \mathbf{v}} = \frac{\mathbf{v}}{v}\frac{df_0}{dv}, \quad v = u, \quad and \quad f_0(v) = f_0(u).$$

Thus the identity follows.

Exercise. When $\omega/k \gg v_{the}$, v_{thi}, carry out the asymptotic evaluation of the Maxwellian integrals

$$\int_0^\infty d^3vv\, \sigma_{i0}\, \frac{\frac{\partial f_{i0}}{\partial v}}{\omega - kv} = -\frac{4}{3}\sqrt{8\pi}N_i\sqrt{\frac{T_i}{m_i}}\, R^2\, \frac{k}{\omega^2}\left(1+\frac{e^2Z_0}{2RT_i}\right),$$

$$\int_{v_0}^\infty d^3vv\, \sigma_{e0}\, \frac{\frac{\partial f_{e0}}{\partial v}}{\omega - kv} = -\frac{4}{3}\sqrt{8\pi}N_e\sqrt{\frac{T_e}{m_e}}\, R^2\, \frac{k}{\omega^2}\left(1+\frac{e^2Z_0}{2RT_e}\right) exp\left[-\frac{e^2Z_0}{RT_e}\right]$$

$$= -\frac{4}{3}\, \frac{k}{\omega^2}\left(1+\frac{e^2Z_0}{2RT_e}\right)\, \frac{I_0}{e}.$$

Reference: [S. A. Trigger and P. P. J. M. Schram (1999) *Kinetic theory in local charge approximation* J. Phys. D: Appl. Phys. **32** pp 234] where the capture collision integral (15.15) is calculated (without taking the moment) to evaluate the perturbation of the charge number Z_1.

• **Continuity equation** The RHS of (15.16) is the capture integral of electron and ion by the grains, thus depleting electron and ion number density. It serves as a sink term. Linearizing gives

$$\frac{\partial n_{i1}}{\partial t} + N_i\nabla\cdot\mathbf{u}_i = -N_d\frac{I_{i1}}{e} - n_d\frac{I_0}{e}, \tag{15.75}$$

$$\frac{\partial n_{e1}}{\partial t} + N_e\nabla\cdot\mathbf{u}_e = -N_d\frac{|I_{e1}|}{e} - n_d\frac{I_0}{e}. \tag{15.76}$$

The grain continuity equation is sink-free:

$$\frac{\partial n_d}{\partial t} + N_d\nabla\cdot\mathbf{u}_d = 0. \tag{15.77}$$

We use (15.34) and (15.35) in (15.75) and (15.76). Also Z_1 can be eliminated by using (15.39). Then we obtain

$$I_{e1} = -I_0\left[\frac{n_{e1}}{N_e}\left(1-\frac{\Lambda}{1-i\omega/\eta}\right)+\frac{n_{i1}}{N_i}\,\frac{\Lambda}{1-i\omega/\eta}\right], \tag{15.78}$$

$$I_{i1} = I_0\left[\frac{n_{e1}}{N_e}\frac{1-\Lambda}{1-i\omega/\eta}+\frac{n_{i1}}{N_i}\left(1-\frac{1-\Lambda}{1-i\omega/\eta}\right)\right], \tag{15.79}$$

where

$$\Lambda = \frac{T_i + e^2Z_0/R}{T_e + T_i + e^2Z_0/R}. \tag{15.80}$$

In a dusty plasma, n_{i1} and n_{e1} are coupled even in the continuity equations.

• **Dielectric permeability of a dusty plasma.** The relation $\mathbf{D} = \mathbf{E} + 4\pi i\mathbf{J}/\omega = \varepsilon\mathbf{E}$ may be used to obtain ε. $\mathbf{J}$ involves the fluid velocity, and the algebra leading to the expression for ε is very laborious since even the continuity equations (15.75) and (15.76) are complex enough. In the low frequency region where the electrons and ions form the thermalized background with Boltzmann distribution, the dielectric function can be obtained rather easily. In this case, it is convenient to use the relation

$$\mathbf{D} = \mathbf{E} + 4\pi\mathbf{P} = \varepsilon_L\mathbf{E} \tag{15.81}$$

where $\mathbf{P}$ is the polarization. In particular, for longitudinal wave (electrostatic), we don't have to use the equations of motion when the electrons and ions are Boltzmann-distributed. (15.81) gives $-ik\phi + 4\pi P = -ik\varepsilon_L\phi$, or

$$\varepsilon_L = 1 + \frac{4\pi i}{k\phi}P. \tag{15.82}$$

We also have $\nabla \cdot \mathbf{E} = -4\pi\nabla \cdot \mathbf{P}$, or

$$-4\pi ikP = -4\pi e(n_e - n_i + Z_0 n_d + Z_1 N_d). \tag{15.83}$$

(15.82) and (15.83) yield

$$\varepsilon_L = 1 + \frac{4\pi e}{k^2\phi}(n_e - n_i + Z_0 n_d + Z_1 N_d) \tag{15.84}$$

where we use (15.39) for Z_1 and

$$n_e = \frac{eN_e}{T_e}\phi, \quad n_i = -\frac{eN_i}{T_i}\phi \tag{15.85}$$

and the grain equations (15.24) and (15.25). It is left as an exercise to derive

$$\varepsilon_L(k,\omega) = 1 - \frac{\omega_{pd}^2}{\omega^2} + \frac{1}{k^2}\left[\frac{1}{\lambda_D^2} + \frac{4\pi ieN_dI_0}{\omega + i\eta}(T_e^{-1} + T_i^{-1})\right] \tag{15.86}$$

which should be compared with (15.49), which is obtained by setting $\varepsilon_L = 0$.

References

General references for dusty plasma physics are:

Shukla, P. K. in *The Physics of Dusty Plasma*, edited by Shukla, P. K., Mendis, D. A., and Chow, V. W. (World Scientific, Singapore, 1996).

Whipple, E. C., Northrop, T. G. and Mendis, D. A. *J. Geophys. Res.* [Oceans] **90** (1985) p. 7405.

For dust acoustic wave:

Rao, N. N., Shukla, P. K. and Yu, M. Y. *Planet. Space Sci.* **38** (1990) p. 543.

For nonlinear interaction including dusty particulates:

Stenflo L. and Shukla, P. K. Stimulated scattering involving dust particulates, *Phys. Plasmas* **7** (2000) p. 3472.
Stenflo, L. Theory for stimulated scattering of electromagnetic waves, *J. Atmospheric and Terrestrial Phys.* **52** (1990) p. 495–499: this reference presents parametric interactions in a compact form.
Stenflo, L. Stimulated scattering of large amplitude waves in the ionosphere, *Physica Scripta* **T30** (1990) p. 166: compact dispersion relation in a magnetized plasma.
Stenflo, L. Theory of stimulated scattering of large amplitude waves, *J. Plasma Phys.* **53** (1995) Pt. 2, p. 213.

The presentation in this Chapter is drawn from:

Lee, H. J. Comment on "Dust electro-acoustic and Coulomb modes in dense dusty plasmas", *Phys. Plasmas* **9** (2002) pp. 4416–4417.
Lee H. J. and Cho S. H. Acoustic wave in a dusty plasma with frequent grain charging collisions, *Physica Scripta* **68** (2003) pp. 54–57.

Appendix A

Fourier transform, delta function, and useful formulas

A.1. Fourier transform, delta function

We define the Fourier transform pair as follows:

$$\tilde{f}(k) = \int_{-\infty}^{\infty} f(x) e^{-ikx} dx, \tag{A.1}$$

$$f(x) = \frac{1}{2\pi} \int_{-\infty}^{\infty} \tilde{f}(k) e^{ikx} dk. \tag{A.2}$$

We often drop the tilde on Fourier transforms when there is no danger of confusion. Instead, the argument is written explicitly. In case of a space-time function $f(x, t)$, its Fourier transform is defined as

$$f(k, \omega) = \int_{-\infty}^{\infty} \int_{-\infty}^{\infty} f(x, t) e^{-ikx + i\omega t} dx dt.$$

Its inversion is

$$f(x, t) = \left(\frac{1}{2\pi}\right)^2 \int_{-\infty}^{\infty} \int_{-\infty}^{\infty} f(k, \omega) e^{ikx - i\omega t} dk d\omega.$$

Our convention of Fourier transform is consistent with the form of wave expressed by the phasor $e^{ikx - i\omega t}$. With this phasor, it is often tacitly assumed that the imaginary part of the frequency ω_i ($\omega = \omega_r + i\omega_i$) is infinitesimally positive. This signifies the *causality* that the perturbation was non-existent in the infinitely remote past ($t = -\infty$).

The operational property of the delta function is

$$\int_{-\infty}^{\infty} f(x)\delta(x - a) dx = \int_{a-\epsilon}^{a+\epsilon} f(x)\delta(x - a) dx = f(a). \tag{A.3}$$

We take $f(x) = \delta(x)$ in (A.1). Then (A.1) gives $\tilde{f}(k) = 1$ due to (A.3). Then (A.2) gives

$$\delta(x) = \frac{1}{2\pi}\int_{-\infty}^{\infty} e^{ikx}dk = \frac{1}{2\pi}\lim_{n\to\infty}\int_{-n}^{n} e^{ikx}dk = \lim_{n\to\infty}\frac{sin(nx)}{\pi x} \tag{A.4}$$

which is one of the well known expressions for the *Dirac delta function.* Clearly, the RHS of (A.4) is real. So we can write

$$\delta(x) = \frac{1}{2\pi}Re\int_{-\infty}^{\infty} e^{ikx}dk = \frac{1}{2\pi}\int_{-\infty}^{\infty} cos\ kx\ dk. \tag{A.5}$$

This is true because $\int_{-\infty}^{\infty} sinkx\ dk = 0$. (A.5) can be written as

$$\delta(x) = \frac{1}{\pi}\int_{0}^{\infty} coskx\ dk = \frac{1}{\pi}\ Re\ \int_{0}^{\infty} e^{ikx}dk. \tag{A.6}$$

Let us consider the integral

$$\begin{aligned}
&\lim_{t\to\infty}\int_0^t e^{ix(t-t')}dt' + \text{c.c.} \\
&= \lim_{t\to\infty}\int_0^t e^{ix\xi}d\xi + \text{c.c.} \quad per \ \ (t - t' = \xi) \\
&= \int_0^\infty e^{ix\xi}d\xi + \text{c.c.} = 2Re\int_0^\infty e^{ix\xi}d\xi.
\end{aligned}$$

Thus we have the identity

$$\delta(x) = \frac{1}{2\pi}\left[\lim_{t\to\infty}\int_0^t e^{ix(t-t')}dt' + \text{c.c.}\right]. \tag{A.7}$$

One can verify that direct integration of (A.7) yields (A.4). Direct integration of (A.6) also recovers (A.4). It is left as an exercise to prove the asymptotic relation ($t > 0$, $cf.$ (A.39), below)

$$\lim_{t\to\infty}\frac{1 - e^{-ixt}}{x} = P\ \frac{1}{x} + i\pi\delta(x) \tag{A.8}$$

where P denotes the principal value. It is useful to know the following properties of δ-function.

$$\frac{\partial}{\partial x}\delta(x-a) = -\frac{\partial}{\partial a}\delta(x-a), \tag{A.9}$$

$$\frac{d}{dt}\delta[X(t)] = \frac{d}{dX}\delta(X)\,\frac{dX(t)}{dt}. \tag{A.10}$$

We can also prove from the above equation by integration by parts

$$\int_{-\infty}^{\infty} f(x)\delta'(x-a)dx = -f'(a), \tag{A.11}$$

$$\int_{-\infty}^{\infty} f(x)\delta''(x-a)dx = f''(a). \tag{A.12}$$

An important property of the delta function is

$$\delta[g(x)] = \sum_{a_i} \frac{\delta(x-a_i)}{\mid g'(a_i)\mid} \tag{A.13}$$

where a_i $(i = 1, 2, 3, \ldots)$ is a root of $g(x) = 0$. In particular, we have

$$\delta(cx) = \frac{1}{\mid c\mid}\delta(x). \tag{A.14}$$

(A.14) will be proven directly later with the aid of the step function. As an application of (A.13), a formula often used in this book (Chapter 5) is

$$\int_{-\infty}^{\infty} \frac{\partial g}{\partial v}\delta(\omega - kv)dv = \frac{1}{k}\left[\frac{\partial g}{\partial v}\right]_{v=\frac{\omega}{k}}. \tag{A.15}$$

As another application of (A.13), we can write

$$\delta[(x-a)(x-b)] = \frac{\delta(x-a)+\delta(x-b)}{\mid a-b\mid}. \tag{A.16}$$

The delta function is visualized as

$$\delta(x) = \begin{cases} 0 & when\ x \neq 0 \\ \infty & when\ x = 0 \end{cases} \tag{A.17}$$

so that it has unit area when integrated:

$$\int_{-\infty}^{\infty} \delta(x)dx = 1. \tag{A.18}$$

There are many functions (*distribution*) which satisfy the properties (A.17) and (A.18):

$$\delta(x) = \lim_{\beta\to\infty}\sqrt{\frac{\beta}{\pi}}e^{-\beta x^2} = \lim_{\epsilon\to 0}\sqrt{\frac{1}{\pi\epsilon}}e^{-\frac{x^2}{\epsilon}}. \tag{A.19}$$

A distribution of x is a delta function $\delta(x)$ when it is described by

$$\delta(x) = \begin{cases} \frac{1}{2\epsilon} & when \mid x \mid < \epsilon \\ 0 & when\ x \mid > \epsilon \end{cases} \tag{A.20}$$

where $\epsilon \to 0$ is implied. Also, we have

$$\delta(x) = \lim_{\epsilon\to 0}\frac{\epsilon}{\pi(\epsilon^2 + x^2)}. \tag{A.21}$$

(A.21) is useful for deriving *Plemelj's formula* (A.22) below:

$$\frac{1}{\omega \pm i\epsilon} = \frac{\omega \mp i\epsilon}{\omega^2+\epsilon^2} = \frac{\omega}{\omega^2+\epsilon^2} \mp i\pi\delta(\omega) \qquad per\ \ (A.21).$$

The operational product with the first term is

$$\begin{aligned}
&\lim_{\epsilon\to 0}\int_{-\infty}^{\infty}\frac{f(\omega)\omega}{\omega^2+\epsilon^2}d\omega \\
&= \lim_{\epsilon\to 0}\int_{-\infty}^{-\epsilon}\frac{f(\omega)\omega}{\omega^2+\epsilon^2}d\omega + \lim_{\epsilon\to 0}\int_{\epsilon}^{\infty}\frac{f(\omega)\omega}{\omega^2+\epsilon^2}d\omega + \lim_{\epsilon\to 0}\int_{-\epsilon}^{\epsilon}\frac{f(\omega)\omega}{\omega^2+\epsilon^2}d\omega \\
&= P\int_{-\infty}^{\infty} f(\omega)\frac{d\omega}{\omega} + f(0)\lim_{\epsilon\to 0}\int_{-\epsilon}^{\epsilon}\frac{\omega}{\omega^2+\epsilon^2}d\omega.
\end{aligned}$$

The last term vanishes because it is an odd function integral. So we obtain the Plemelj formula

$$\lim_{\epsilon\to 0}\frac{1}{\omega \pm i\epsilon} = P\frac{1}{\omega} \mp i\pi\delta(\omega). \tag{A.22}$$

An alternative derivation of (A.22) is given later in Appendix A.3. In the following, we write useful relations.

$$\delta(-x) = \delta(x), \tag{A.23}$$

$$x\delta(x) = 0, \tag{A.24}$$

$$\delta(x) = -x\delta^{'}(x), \tag{A.25}$$

$$\int_{-\infty}^{\infty} \delta(x-a)\delta(x-b)dx = \delta(a-b), \tag{A.26}$$

$$\sum_{n=-\infty}^{\infty} e^{inx} = 2\pi\delta(x). \tag{A.27}$$

The last formula is the summation version of (A.4): $\sum_n \to \int dn$.

The three dimensional delta functions are

$$\delta(\mathbf{r}-\mathbf{r}') = \delta(x-x')\delta(y-y')\delta(z-z') \quad (\textit{cartesian coordinates}), \tag{A.28}$$

$$\delta(\mathbf{r}-\mathbf{r}') = \frac{1}{\rho}\delta(\rho-\rho')\delta(\varphi-\varphi')\delta(z-z') \quad (\textit{cylindrical coordinates}), \tag{A.29}$$

$$\delta(\mathbf{r}-\mathbf{r}') = \frac{1}{r^2}\delta(r-r')\delta(cos\theta-cos\theta')\delta(\varphi-\varphi') = \frac{1}{r^2}\delta(r-r')\delta(\Omega-\Omega') \tag{A.30}$$

in *spherical coordinates*, where Ω is the solid angle.

From the Gauss theorem we can prove

$$\nabla_r^2 \frac{1}{|\mathbf{r}-\mathbf{r}'|} = -4\pi\delta(\mathbf{r}-\mathbf{r}') = -\nabla_{r'}^2 \frac{1}{|\mathbf{r}-\mathbf{r}'|}. \tag{A.31}$$

A distribution of point charges q_i located at the points $\mathbf{r}_i$ can be represented with a charge density

$$\rho(\mathbf{r}) = \sum_{i=1}^{n} q_i \delta(\mathbf{r}-\mathbf{r}_i). \tag{A.32}$$

In this book, we distinguish two kinds of step function:

$$S(x) = \begin{cases} 1 & if\ x \geq 0 \\ 0 & if\ x < 0 \end{cases}, \tag{A.33}$$

$$H(x) = \begin{cases} 1 & if\ x \geq 0 \\ -1 & if\ x < 0 \end{cases}. \tag{A.34}$$

$H(x)$ has a parity: $H(-x) = -H(x)$; odd function of x. Clearly,

$$S(x) = \frac{1}{2}\left[H(x)+1\right]. \tag{A.35}$$

$S(x)$ has an infinite slope at $x = 0$, and elsewhere the slope is zero. Thus we have

$$\frac{dS}{dx} = \delta(x). \tag{A.36}$$

Integrating the above, we have

$$S(x) = \int_{-\infty}^{x} \delta(t)dt. \tag{A.37}$$

Clearly, this integral is zero when $x < 0$, and 1 when $x > 0$. We also have

$$\frac{dH}{dx} = 2\,\frac{dS}{dx} = 2\,\delta(x). \tag{A.38}$$

We can derive the important identity

$$P\int_{-\infty}^{\infty} \frac{e^{i\omega x}}{\omega} d\omega = \begin{cases} i\pi & if\ x > 0 \\ -i\pi & if\ x < 0 \end{cases}. \tag{A.39}$$

(A.39) can be proven by contour integration around the closed contour consisting of the principal value section plus an infinitesimal semicircle around $\omega = 0$ plus the arc of infinite radius surrounding the upper (lower) ω plane when $x > 0$ ($x < 0$). The *infinitesimal semicircle* can be either convexed-up (clockwise), or concave-down (counter-clockwise). It is immaterial. (A.39) can be extended to the form

$$P\int_{-\infty}^{\infty} \frac{e^{i\omega x}}{\omega - \omega_0} d\omega = \begin{cases} i\pi e^{ix\omega_0} & if\ x > 0 \\ -i\pi e^{ix\omega_0} & if\ x < 0 \end{cases}. \tag{A.40}$$

We can deduce from the above

$$P\int_{-\infty}^{\infty} \frac{e^{-i\omega x}}{\omega - \omega_0} d\omega = \begin{cases} -i\pi e^{-ix\omega_0} & if\ x > 0 \\ i\pi e^{-ix\omega_0} & if\ x < 0 \end{cases}. \tag{A.41}$$

Using (A.39), the step functions can be written in the forms

$$S(x) = \frac{1}{2} + \frac{1}{2\pi i} P \int_{-\infty}^{\infty} \frac{e^{i\omega x}}{\omega} d\omega, \tag{A.42}$$

$$H(x) = \frac{1}{i\pi} P \int_{-\infty}^{\infty} \frac{e^{ikx}}{k} dk, \tag{A.43}$$

$$H(x) = \frac{1}{\pi} P \int_{-\infty}^{\infty} \frac{sin\ kx}{k} dk. \tag{A.44}$$

Differentiating (A.42) gives $dS(x)/dx = \delta(x)$ as expected. Here $k = 0$ is not a singularity, thus we don't need the principal value symbol 'P'. Thus we simply write

$$H(x) = \frac{1}{i\pi}\int_{-\infty}^{\infty} \frac{e^{ikx}}{k} dk \tag{A.45}$$

which in turn gives

$$\frac{dH}{dx} = \frac{1}{\pi}\int_{-\infty}^{\infty} e^{ikx} dk = 2\ \delta(x). \tag{A.46}$$

Example. We have the following two equations for step function $S(t)$:

$$S(at) = S(t) \quad if\ a > 0, \tag{A.47}$$

$$S(at) = 1 - S(t) \quad if\ a < 0. \tag{A.48}$$

Differentiating (A.47) gives

$$a\ \frac{dS(at)}{d(at)} = \frac{dS}{dt} \quad or \quad a\delta(at) = \delta(t) \quad if\ a > 0. \tag{A.49}$$

Differentiating (A.48) gives

$$a\ \frac{dS(at)}{d(at)} = -\frac{dS}{dt} \quad or \quad a\delta(at) = -\delta(t) \quad if\ a < 0. \tag{A.50}$$

Therefore we conclude $\delta(at) = \delta(t)/\mid a \mid$.

Example. Using (A.46) for $H(x)$, we obtain

$$\frac{d^2}{dx^2}|x| = \frac{d^2}{dx^2}[xH(x)] = \frac{d}{dx}[H(x) + 2x\delta(x)] = \frac{dH}{dx} = 2\delta(x). \tag{A.51}$$

A.2. Convolution integral

Convolution of two functions is

$$(f * g)_{\mathbf{r}} = \int_{-\infty}^{\infty} g(\mathbf{r}')f(\mathbf{r} - \mathbf{r}')d^3r' = \int_{-\infty}^{\infty} f(\mathbf{r}')g(\mathbf{r} - \mathbf{r}')d^3r'. \tag{A.52}$$

It can be easily proved that

$$I = \int e^{-i\mathbf{k}\cdot\mathbf{r}}(f * g)_{\mathbf{r}}\ d^3r = \tilde{f}(\mathbf{k})\ \tilde{g}(\mathbf{k}). \tag{A.53}$$

In words, the Fourier transform of a convolution integral is the product of each Fourier transform. The inverse of $\tilde{f}(\mathbf{k})\ \tilde{g}(\mathbf{k}) =$

$$Inverse\ of\ \tilde{f}(\mathbf{k})\ \tilde{g}(\mathbf{k}) = \left(\frac{1}{2\pi}\right)^3 \int e^{i\mathbf{k}\cdot\mathbf{r}} \tilde{f}(\mathbf{k})\ \tilde{g}(\mathbf{k}) d^3k = (f * g)_{\mathbf{r}}.$$

In words, the Fourier transform (or the inverse transform) of the product of two functions is the convolution. We prove the identity (A.53) by means of the technique used repeatedly in the text. We have in (A.53)

$$f(\mathbf{k}) = \int f(\mathbf{r}) e^{-i\mathbf{k}\cdot\mathbf{r}} d^3r,$$

$$g(\mathbf{k}) = \int g(\mathbf{r}) e^{-i\mathbf{k}\cdot\mathbf{r}} d^3r,$$

where the tilde is omitted. Substitute the inverse relations

$$f(\mathbf{r}') = \left(\frac{1}{2\pi}\right)^3 \int f(\mathbf{k}') e^{i\mathbf{k}'\cdot\mathbf{r}'} d^3k',$$

$$g(\mathbf{r} - \mathbf{r}') = \left(\frac{1}{2\pi}\right)^3 \int g(\mathbf{k}'') e^{i\mathbf{k}''\cdot(\mathbf{r}-\mathbf{r}')} d^3k''.$$

Therefore,

$$I = \int d^3r e^{-i\mathbf{k}\cdot\mathbf{r}} \int d^3r' \left[\left(\frac{1}{2\pi}\right)^3 \int f(\mathbf{k}') e^{i\mathbf{k}'\cdot\mathbf{r}'} d^3k'\right] \times \left[\left(\frac{1}{2\pi}\right)^3 \int g(\mathbf{k}'') e^{i\mathbf{k}''\cdot(\mathbf{r}-\mathbf{r}')} d^3k''\right].$$

In the above, use

$$\left(\frac{1}{2\pi}\right)^3 \int d^3r' e^{i(\mathbf{k}'-\mathbf{k}'')\cdot\mathbf{r}'} = \delta(\mathbf{k}' - \mathbf{k}''),$$

$$\left(\frac{1}{2\pi}\right)^3 \int d^3r e^{-i\mathbf{k}\cdot\mathbf{r}+i\mathbf{k}''\cdot\mathbf{r}} = \delta(\mathbf{k} - \mathbf{k}'').$$

Then we have

$$I = \int f(\mathbf{k}') d^3k' g(\mathbf{k}'') d^3k'' \delta(\mathbf{k}' - \mathbf{k}'') \delta(\mathbf{k} - \mathbf{k}'') = f(\mathbf{k}) g(\mathbf{k}) \qquad Q.E.D.$$

- *Fourier Transform of* $[A(x)B(x)]$

$$
\begin{aligned}
&= \int e^{-ikx} A(x)B(x)dx = \frac{1}{2\pi}\int dx e^{-ikx} A(x) \int e^{ik'x} B(k')dk' \\
&= \frac{1}{2\pi}\int dk' B(k') \int e^{-i(k-k')x} A(x)dx = \frac{1}{2\pi}\int dk' B(k')A(k-k') \\
&= \frac{1}{2\pi}\int dk' B(k') \int dk'' A(k'')\delta(k''-k+k'). \qquad \text{(A.54)}
\end{aligned}
$$

A.3. Derivation of Plemelj formula

Consider the integral

$$
I = \lim_{\epsilon\to 0}\int_a^b \frac{f(x)dx}{x-x_0+i\epsilon} \qquad (\epsilon > 0 \quad and \quad a < x_0 < b)
$$

which can be written as

$$
\begin{aligned}
I &= \lim_{\epsilon\to 0}\int_a^b \frac{f(x)-f(x_0)}{x-x_0+i\epsilon}\,dx \; + f(x_0)\,\lim_{\epsilon\to 0}\int_a^b \frac{dx}{x-x_0+i\epsilon} \\
&= \int_a^b \frac{f(x)-f(x_0)}{x-x_0}dx + f(x_0)[ln(b-x_0+i\epsilon) - ln(a-x_0+i\epsilon)].
\end{aligned}
$$

Note that the complex numbers $b-x_0+i\epsilon = (b-x_0)e^{i0^+}$ and $a-x_0+i\epsilon = (x_0-a)e^{i\pi}$.

$$
Therefore, \quad I = \int_a^b \frac{f(x)-f(x_0)}{x-x_0}dx + f(x_0)\left(ln\,\frac{b-x_0}{x_0-a} - i\pi\right). \qquad \text{(A.55)}
$$

On the other hand,

$$
\begin{aligned}
&P\int_a^b \frac{f(x)dx}{x-x_0} \\
&\quad = \int_a^b \frac{f(x)-f(x_0)}{x-x_0}dx + f(x_0)P\int_a^b \frac{dx}{x-x_0} \\
&\quad = \int_a^b \frac{f(x)-f(x_0)}{x-x_0}dx + f(x_0)\lim_{\epsilon\to 0}\left[\int_a^{x_0-\epsilon}\frac{dx}{x-x_0} + \int_{x_0+\epsilon}^b \frac{dx}{x-x_0}\right].
\end{aligned}
$$

The second term $[\cdots]$ gives upon integration, $ln(-\epsilon) - ln(a - x_0) + ln(b - x_0) - ln(\epsilon)$. Note that $ln(-\epsilon) = ln\epsilon + i\pi$ and $ln(a - x_0) = ln(x_0 - a) + i\pi$. Thus the second term equals to $f(x_0)ln\frac{b-x_0}{x_0-a}$. Therefore, we have

$$P\int_a^b \frac{f(x)dx}{x - x_0} = \int_a^b \frac{f(x) - f(x_0)}{x - x_0}dx + f(x_0)ln\,\frac{b - x_0}{x_0 - a}. \tag{A.56}$$

Equations (A.55) and (A.56) give

$$I = \lim_{\epsilon\to 0}\int_a^b \frac{f(x)dx}{x - x_0 + i\epsilon} = P\int_a^b \frac{f(x)dx}{x - x_0} - i\pi f(x_0).$$

Likewise we can prove that for $\epsilon < 0$ the last term changes sign, i.e. $+i\pi f(x_0)$.

A.4. First order linear differential equation

$$\frac{dy}{dx} + a(x)y + b(x) = 0. \tag{A.57}$$

$$Transform: f(x) = y(x)e^{\int^x a(x')dx'}. \tag{A.58}$$

$$\frac{df}{dx} = \left(\frac{dy}{dx} + a(x)y\right)e^{\int^x a(x')dx'}. \tag{A.59}$$

$$\frac{df}{dx} + b(x)e^{\int^x a(x')dx'} = 0. \tag{A.60}$$

Indefinite integral form:

$$y(x) = e^{-\int^x a(x')dx'}\left[C - \int^x dx'b(x')e^{\int^{x'} a(x'')dx''}\right]. \quad (C:\ const.) \tag{A.61}$$

Definite integral form:

$$\int_{f(x_0)}^{f(x)} df(x') + \int_{x_0}^x dx'b(x')e^{\int^{x'} a(x'')dx''} = 0. \tag{A.62}$$

$$f(x) - f(x_0) = -\int_{x_0}^x dx'b(x')e^{\int^{x'} a(x'')dx''}. \tag{A.63}$$

$$y(x)e^{\int^x a(x')dx'} - y(x_0)e^{\int^{x_0} a(x')dx'} = -\int_{x_0}^x dx'b(x')e^{\int^{x'} a(x'')dx''}. \tag{A.64}$$

$$y(x) = e^{-\int^x a(x')dx'}\left[y(x_0)e^{\int^{x_0} a(x')dx'} - \int_{x_0}^x dx'b(x')e^{\int^{x'} a(x'')dx''}\right]. \tag{A.65}$$

A.5. Harmonic series

$$S = \sum_{n=-N}^{N} e^{inx} = e^{-iNx} \sum_{n=0}^{2N} e^{inx} = e^{-iNx} \frac{e^{(2N+1)ix} - 1}{e^{ix} - 1}$$

$$= \frac{e^{(N+1)ix} - e^{-iNx}}{e^{ix} - 1} = \frac{sin\ (N + \frac{1}{2})x}{sin\ \frac{x}{2}}. \tag{A.66}$$

Also we have the equality from the above:

$$\sum_{n=0}^{2N} e^{inx} = \frac{sin\ (N + \frac{1}{2})x}{sin\ \frac{x}{2}} e^{iNx}. \tag{A.67}$$

Let us change $N \to \frac{N'-1}{2}$ in (A.67):

$$\sum_{n=0}^{N'-1} e^{inx} = \frac{sin\ \frac{N'x}{2}}{sin\ \frac{x}{2}}\ e^{ix\frac{(N'-1)}{2}}. \tag{A.68}$$

Separating the real and the imaginary parts gives

$$\sum_{n=0}^{N-1} cosnx = \frac{sin\ \frac{Nx}{2}}{sin\ \frac{x}{2}}\ cos(N-1)\frac{x}{2}, \tag{A.69}$$

$$\sum_{n=0}^{N-1} sinx = \frac{sin\ \frac{Nx}{2}}{sin\ \frac{x}{2}}\ sin(N-1)\frac{x}{2}. \tag{A.70}$$

Integrating both sides of (A.66) gives

$$\int_0^{\pi} \frac{sin\ (N + \frac{1}{2})x}{sin\ \frac{x}{2}} dx = \pi \qquad (regardless\ of\ N) \tag{A.71}$$

where π corresponds to the $n = 0$ term in the series. (A.71) can be used to prove

$$\int_0^{\infty} \frac{sin\ x}{x}\ dx = \frac{\pi}{2}. \tag{A.72}$$

(A.72) is a special case of (A.44) (when $x = 1$ therein), which was proved by contour integration. Here we introduce a simple ingenuous method due to Reference: [N. Wiener *The Fourier integral and certain of its applications* 1958 Dover]. We can write

$$\lim_{N\to\infty} \int_0^{\pi} sin\ \left(N + \frac{1}{2}\right) x\ f(x) = 0 \tag{A.73}$$

for any *integrable* (not blowing up) function $f(x)$ over $(0, \pi)$, in view of (75) in § Introduction. We choose as

$$f(x) = \frac{2}{x} - \frac{1}{sin\ \frac{x}{2}} \tag{A.74}$$

which is certainly bounded over $(0, \pi)$. Then (A.71), (A.73), and (A.74) yield the result (A.72).

- (A.66) gives Lagrange's trigonometric identity [see also Reference: R. V. Churchill, J. W. Brown, and R. F. Verhey, *Complex variables and applications*, International student edition (1960) p. 17]

$$1 + cos\ x + cos\ 2x + \cdots + cos\ Nx = \frac{1}{2} + \frac{sin[(N+\frac{1}{2})x]}{2sin\ \frac{x}{2}}. \tag{A.75}$$

(A.69) can be put into the form

$$1 + cos\ x + cos\ 2x + \cdots + cos\ Nx = \frac{sin[(\frac{N+1}{2})x]}{sin\ \frac{x}{2}} cos \frac{Nx}{2}. \tag{A.76}$$

Equivalence of (A.75) and (A.76) can be proven by mathematical induction.

A.6. Differentiation of an integral: Leibnitz's rule

$$I(\alpha) = \int_{x_1(\alpha)}^{x_2(\alpha)} f(x, \alpha)\ dx, \tag{A.77}$$

$$\frac{dI}{d\alpha} = f(x_2, \alpha)\ \frac{dx_2}{d\alpha} - f(x_1, \alpha)\ \frac{dx_1}{d\alpha} + \int_{x_1(\alpha)}^{x_2(\alpha)} \frac{\partial f}{\partial \alpha}\ dx. \tag{A.78}$$

Logarithm and Exponential

A natural logarithmic function with base e is defined by

$$ln\ x = \int_1^x \frac{1}{t}\ dt \quad x > 0 \tag{A.79}$$

which gives the relation

$$\frac{d}{dx} ln\ x = \frac{1}{x} \quad x > 0. \tag{A.80}$$

By definition, the relation $ln\ x = y$ is equivalent to $x = e^y$. Thus we have from (A.79)

$$x = exp\left(\int_1^x \frac{1}{t}\right) = e^{ln\ x}. \tag{A.81}$$

Let us work with the relation

$$ln\ e^x = x \tag{A.82}$$

which is due to the definition. [Here we have $e^{ln(\cdots)} = ln\ e^{(\cdots)} = (\cdots)$.]

Differentiating (A.82) by chain rule gives

$$\frac{1}{e^x}\frac{d}{dx}\,e^x = 1. \tag{A.83}$$

Therefore we have the important relation

$$\frac{d}{dx}\,e^x = e^x. \tag{A.84}$$

Using (A.84), we can Taylor expand the exponential function itself in the form

$$e^x = \sum_{n=0}^{\infty}\frac{x^n}{n!}. \tag{A.85}$$

Putting $x = 1$ in the above gives

$$e = \sum_{n=0}^{\infty}\frac{1}{n!} = 2.7182818... \tag{A.86}$$

On the other hand, we can prove

$$e = \sum_{n=0}^{\infty}\frac{1}{n!} = \lim_{n\to\infty}\left(1+\frac{1}{n}\right)^n. \tag{A.87}$$

The numerical value of e can be derived also from the second equality. As another way to find the numerical value of e, (A.79) can be made use of by putting $x = e$

$$1 = \int_1^e \frac{1}{t}\,dt. \tag{A.88}$$

e can be determined from the above integral by trial and error. For a proof of (A.87), see W. Rudin *Principles of Mathematical Analysis* McGraw-Hill (1964) p. 55.

Appendix B

Multiple scale perturbation analysis

In usual perturbation analysis, an equation to be solved is expanded straightforwardly in a power series of ϵ, the small parameter involved in the equation under consideration. In this case the higher order solution usually contains secular terms which diverge as the independent variable t becomes large. This means that the naive straight expansion is unable to predict correct approximate solution, and this naive perturbation scheme should be remedied. We consider the time t consists of times of different scales:

$$t_0 = t, \; t_1 = \epsilon t, \; t_2 = \epsilon^2 t, \; t_3 = \epsilon^3 t, \ldots \quad \textit{or}$$

$$\frac{d}{dt} = \frac{\partial}{\partial t_0} + \epsilon \frac{\partial}{\partial t_1} + \epsilon^2 \frac{\partial}{\partial t_2} + \cdots. \tag{B.1}$$

Representing a dynamical change in terms of different time scales can be compared with reading time on a watch (quoted from Nayfeh: A. H. Nayfeh (1981) *Introduction to perturbation techniques*, Wiley). "$\epsilon = \frac{1}{60}$, variations on the scale t_0 can be observed on the second arm of a watch, variations on the scale t_1 can be observed on the minute arm of a watch, and variations on the scale t_2 can be observed on the hour arm of a watch. Thus, t_0 represents a fast scale, t_1 represents a slower scale, t_2 represents an even slower scale, and so on". By this scheme, uniform solutions can be obtained by removing the secularity-causing terms order by order. In this Appendix, this perturbation method is illustrated to present the essential features of the analysis through engaging with two examples. First we consider the *Van der Pol equation* (in dimensionless form)

$$\frac{d^2x}{dt^2} + x - \epsilon(1 - x^2)\frac{dx}{dt} = 0. \quad (\epsilon \ll 1) \tag{B.2}$$

While we expand t according to the scheme (B.1), we also expand the function x as

$$x = x_0(t_0, t_1, t_2, \ldots) + \epsilon x_1(t_0, t_1, t_2, \ldots) + \epsilon^2 x_2(t_0, t_1, t_2, \ldots) + \cdots. \quad \text{(B.3)}$$

One easily obtains

$$\frac{d^2}{dt^2} = \frac{\partial^2}{\partial t_0^2} + \epsilon\, 2\,\frac{\partial}{\partial t_0}\frac{\partial}{\partial t_1} + \epsilon^2\left(2\frac{\partial}{\partial t_0}\frac{\partial}{\partial t_2} + \frac{\partial^2}{\partial t_1^2}\right) + \cdots. \quad \text{(B.4)}$$

Putting (B.3) and (B.4) into (B.2) yields

$$\epsilon^0: \quad \frac{\partial^2 x_0}{\partial t_0^2} + x_0 = 0, \quad \text{(B.5)}$$

$$\epsilon^1: \quad \frac{\partial^2 x_1}{\partial t_0^2} + x_1 = -2\frac{\partial^2 x_0}{\partial t_0 \partial t_1} + (1 - x_0^2)\frac{\partial x_0}{\partial t_0}. \quad \text{(B.6)}$$

(B.5) is solved by

$$x_0 = A(t_1, t_2, \ldots)e^{it_0} + A^*(t_1, t_2, \ldots)e^{-it_0} \quad \text{(B.7)}$$

where $*$ denotes the complex conjugate. If (B.7) is substituted into the RHS of (B.6), various harmonics of $e^{\pm it_0}$ are generated: terms of $e^{\pm it_0}$, $e^{\pm 2it_0}$, and $e^{\pm 3it_0}$. Among these, those terms which are oscillating with the phase $e^{\pm it_0}$ are the secularity causing terms which should be removed by putting the sum of them equal to zero. [Recall that the solution of $\frac{d^2y}{dt^2} + \omega^2 y = cos\omega t$ is $y \sim tsin(\omega t)$. This is the secular solution which diverges as t becomes large.] Thus removal of the secular terms in the RHS of (B.6) requires

$$2\frac{\partial A}{\partial t_1} - A + A^2 A^* = 0, \quad \text{(B.8)}$$

$$2\frac{\partial A^*}{\partial t_1} - A^* + A^{*2} A = 0. \quad \text{(B.9)}$$

Since (B.8) and (B.9) are complex conjugate to each other, it is sufficient to consider only one of them. In fact (B.8) is the condition to determine the undetermined coefficient A in the zero order solution (B.7). In order to solve (B.8), we write A in the form

$$A = \frac{1}{2}\alpha(t_1)e^{i\beta(t_1)} \quad \text{(B.10)}$$

with α and β real functions of t_1. Then (B.7) becomes

$$x_0 = \alpha(t_1)cos(t_0 + \beta(t_1)). \tag{B.11}$$

Substituting (B.10) into (B.8) gives

$$\frac{d\alpha}{dt_1} + i\alpha\frac{d\beta}{dt_1} - \frac{\alpha}{2} + \frac{\alpha^3}{8} = 0.$$

Separating real and imaginary parts, we obtain

$$\frac{d\alpha}{dt_1} = \frac{\alpha}{2} - \frac{\alpha^3}{8} \tag{B.12}$$

$$\frac{d\beta}{dt_1} = 0 \quad or \quad \beta = \beta_0 = const. \tag{B.13}$$

We solve (B.12) by writing as

$$dt_1 = \frac{8d\alpha}{4\alpha - \alpha^3} = \frac{8d\alpha}{\alpha(2-\alpha)(2+\alpha)} = \frac{2d\alpha}{\alpha} + \frac{d\alpha}{2-\alpha} - \frac{d\alpha}{2+\alpha}.$$

Integrating the above equation gives

$$t_1 + c = 2ln\alpha - ln \mid 2 - \alpha \mid - ln(2+\alpha) \quad (c : const.)$$

$$or \quad t_1 + c = ln\frac{\alpha^2}{\mid 4 - \alpha^2 \mid}$$

$$or \quad \frac{\alpha^2}{4-\alpha^2} = e^{t_1 + c} = e^{\epsilon t + c}. \tag{B.14}$$

Solving (B.14) for α^2 gives

$$\alpha^2 = \frac{4e^{\epsilon t + c}}{1 + e^{\epsilon t + c}} = \frac{4}{1 + e^{-\epsilon t - c}}$$

$$or \quad \alpha(t) = 2(1 + e^{-\epsilon t - c})^{-\frac{1}{2}}.$$

The constant c can be determined in terms of the 'initial' amplitude $\alpha_0 = \alpha(t_1 = 0)$, giving the final expression for $x(t)$:

$$x(t) = \frac{2}{1 + (\frac{4}{\alpha_0^2} - 1)e^{-\epsilon t}} cos(t + \beta_0). \tag{B.15}$$

Equation (B.15) shows the secularity-free behavior that $x \to 2cos(t + \beta_0)$ as $t \to \infty$, regardless of the value of α_0.

As another example we shall consider the *Duffing equation*:

$$\frac{d^2u}{dt^2} + f(u) = 0$$

where $f(u)$ is a polynomial of u. We consider a special case

$$\frac{d^2u}{dt^2} + u + \epsilon u^3 = 0. \qquad (\epsilon \ll 1) \tag{B.16}$$

We point out that (B.16) allows for exact solutions by the technique of 'energy integral' (see Chap. 7). Here we use multiple time scale perturbation method. Expand u as

$$u(t) = u_0(t_0, t_1, \ldots) + \epsilon u_1(t_0, t_1, \ldots) + \cdots. \tag{B.17}$$

Substituting (B.10) and (B.17) into (B.16) and breaking down the resultant equation order by order, we obtain

$$\epsilon^0 : \quad \frac{\partial^2 u_0}{\partial t_0^2} + u_0 = 0, \tag{B.18}$$

$$\epsilon^1 : \quad \frac{\partial^2 u_1}{\partial t_0^2} + u_1 = -2\frac{\partial^2 u_0}{\partial t_0 \partial t_1} - u_0^3. \tag{B.19}$$

The general solution of (B.18) can be written as

$$u_0 = \alpha(t_1, t_2, \ldots) cos[t_0 + \beta(t_1, T_2, \ldots)]. \tag{B.20}$$

[We can write the solution of (B.18) in terms of the exponential functions as in (B.7).] Substituting (B.20) into (B.19) yields

$$\frac{\partial^2 u_1}{\partial t_0^2} + u_1 = 2\frac{\partial \alpha}{\partial t_1} sin(t_0 + \beta) + \left(2\alpha\frac{\partial \beta}{\partial t_1} - \frac{3}{4}\alpha^3\right) cos(t_0 + \beta)$$
$$- \frac{1}{4}\alpha^3 cos(3t_0 + 3\beta). \tag{B.21}$$

To eliminate the secular terms in the RHS of the above equation, we put

$$\frac{\partial \alpha}{\partial t_1} = 0,$$

$$2\alpha \frac{\partial \beta}{\partial t_1} - \frac{3}{4}\alpha^3 = 0.$$

Without the secularity-causing terms, (B.21) gives the particular solution

$$u_1 = \frac{1}{32}\alpha^3 cos(3t_0 + 3\beta). \tag{B.22}$$

Appendix C

Strained coordinate perturbation method

This is another perturbation method which is appropriate to extract the Korteweg–de Vries equation from the nonlinear ion acoustic equations (7.64)–(7.66) in Chapter 7. Here we write them again

$$\frac{\partial^2 \phi}{\partial x^2} = 4\pi e(n_0 e^{\frac{e\phi}{T_e}} - n_i), \tag{C.1}$$

$$\frac{\partial n_i}{\partial t} + \frac{\partial}{\partial x}(n_i v) = 0, \tag{C.2}$$

$$\frac{\partial v}{\partial t} + v\frac{\partial v}{\partial x} = -\frac{e}{m_i}\frac{\partial \phi}{\partial x}. \tag{C.3}$$

It is more convenient to write the above equations in dimensionless form by introducing appropriate scaling factors:

$$n_i = n_0 n', \quad \phi = \frac{T_e}{e}\phi', \quad x = \lambda_e x', \quad t = \omega_{pi}^{-1} t' \tag{C.4}$$

where n_0 is the equilibrium electron (or ion) number density, $\lambda_e = \sqrt{\frac{T_e}{4\pi n_0 e^2}}$ is the electron Debye length, $\omega_{pi} = \sqrt{\frac{4\pi n_0 e^2}{m_i}}$ is ion plasma frequency, and the primed variables are dimensionless. Naturally the dimensionless velocity is $v' = \frac{v}{\lambda_e \omega_{pi}}$ where $\lambda_e \omega_{pi} = \sqrt{\frac{T_e}{m_i}}$ is the ion acoustic wave speed. Then (C.1)–(C.3) become

$$\frac{\partial^2 \phi'}{\partial x'^2} = e^{\phi'} - n', \tag{C.5}$$

$$\frac{\partial n'}{\partial t'} + \frac{\partial}{\partial x'}(n'v') = 0, \tag{C.6}$$

$$\frac{\partial v'}{\partial t'} + v'\frac{\partial v'}{\partial x'} = -\frac{\partial \phi'}{\partial x'}. \tag{C.7}$$

The boundary conditions are now $n' = 1,\ v' = 0,\ \phi' = 0$ as $x' \to \pm\infty$. In the following, we omit the primes with the understanding that all the quantities are dimensionless. We expand

$$n = 1 + \epsilon n^{(1)} + \epsilon^2 n^{(2)} + \cdots,$$

$$\phi = \epsilon\phi^{(1)} + \epsilon^2\phi^{(2)} + \cdots,$$

$$v = \epsilon v^{(1)} + \epsilon^2 v^{(2)} + \cdots.$$

We break down (C.5)–(C.7) order by order:

$$\frac{\partial n^{(1)}}{\partial t} + \frac{\partial v^{(1)}}{\partial x} = 0, \tag{C.8}$$

$$\frac{\partial v^{(1)}}{\partial t} + \frac{\partial \phi^{(1)}}{\partial x} = 0, \tag{C.9}$$

$$\frac{\partial^2 \phi^{(1)}}{\partial x^2} - \phi^{(1)} + n^{(1)} = 0, \tag{C.10}$$

$$\frac{\partial n^{(2)}}{\partial t} + \frac{\partial v^{(2)}}{\partial x} = -\frac{\partial}{\partial x}(n^{(1)}v^{(1)}), \tag{C.11}$$

$$\frac{\partial v^{(2)}}{\partial t} + \frac{\partial \phi^{(2)}}{\partial x} = -v^{(1)}\frac{v^{(1)}}{\partial x}, \tag{C.12}$$

$$\frac{\partial^2 \phi^{(2)}}{\partial x^2} - \phi^{(2)} + n^{(2)} = \frac{1}{2}[\phi^{(1)}]^2. \tag{C.13}$$

The linear equations (C.8)–(C.10) are combined to yield the linear Boussinesq equation

$$\left(\frac{\partial^4}{\partial x^2 \partial t^2} + \frac{\partial^2}{\partial x^2} - \frac{\partial^2}{\partial t^2}\right)\phi^{(1)} = 0. \tag{C.14}$$

The nonlinear (ϵ^2 order) equations (C.11)–(C.13) yield nonlinear Boussinesq equation

$$\left(\frac{\partial^4}{\partial x^2 \partial t^2} + \frac{\partial^2}{\partial x^2} - \frac{\partial^2}{\partial t^2}\right)\phi^{(2)} = \frac{1}{2}\frac{\partial^2}{\partial t^2}[\phi^{(1)}]^2,$$

$$-\frac{1}{2}\frac{\partial^2}{\partial x^2}[v^{(1)}]^2 + \frac{\partial^2}{\partial x \partial t}[n^{(1)}v^{(1)}]. \tag{C.15}$$

The foregoing algebra shows that a straight expansion in the series of ϵ gives the Boussinesq equation. In order to derive the KdV equation, we should stretch the x and t coordinates *appropriately.* In order to guess the appropriate stretching, we consider the dispersion relation that is derived from the linear equation (C.14).

$$\omega^2 = \frac{k^2}{1+k^2} \quad (dimensionless), \quad or \ \omega^2 = \frac{\lambda_e^2\omega_{pi}^2 k^2}{1+\lambda_e^2 k^2} \quad (dimensional). \tag{C.16}$$

Examination of the linear dispersion relation (C.16) can give a clue as to how we stretch x and t coordinates in (C.5)–(C.7) to isolate the KdV equation in the limit of long wavelength. The x and t dependence of the linear wave is through the phasor $\theta(x,t) = kx - \omega t$. Expanding ω in (C.16) in the powers of k (small in the long wave region), we obtain $\theta = k(x \pm t) \pm \frac{1}{2}k^3 + \cdots$. This suggests that the x and t coordinates are stretched as

$$\xi = \epsilon^{\frac{1}{2}}(x - \lambda t), \quad \tau = \epsilon^{\frac{3}{2}} t. \tag{C.17}$$

As might be expected, the acceptable wave speed λ turns out to be ± 1. The fractional power on ϵ is introduced to terminate perturbation equation at the lowest order (guided by algebraic scrutiny). In accordance with (C.17), we transform as

$$\frac{\partial}{\partial x} = \epsilon^{\frac{1}{2}}\frac{\partial}{\partial \xi}, \quad \frac{\partial}{\partial t} = \epsilon^{\frac{3}{2}}\frac{\partial}{\partial \tau} - \epsilon^{\frac{1}{2}}\lambda\frac{\partial}{\partial \xi}. \tag{C.18}$$

Then (C.5) gives in the lowest two orders of ϵ

$$\phi^{(1)} = n^{(1)}, \tag{C.19}$$

$$\frac{\partial^2}{\partial \xi^2}\phi^{(1)} = -n^{(2)} + \phi^{(2)} + \frac{1}{2}[\phi^{(1)}]^2. \tag{C.20}$$

From (C.6) we obtain

$$\epsilon^{\frac{3}{2}}: \quad \frac{\partial v^{(1)}}{\partial \xi} - \lambda\frac{\partial n^{(1)}}{\partial \xi} = 0, \tag{C.21}$$

$$\epsilon^{\frac{5}{2}}: \quad \frac{\partial n^{(1)}}{\partial \tau} - \lambda\frac{\partial n^{(2)}}{\partial \xi} + \frac{\partial}{\partial \xi}[v^{(2)} + v^{(1)}n^{(1)}] = 0. \tag{C.22}$$

Next, (C.7) yields

$$\epsilon^{\frac{3}{2}}: \quad -\lambda\frac{\partial v^{(1)}}{\partial \xi} + \frac{\partial \phi^{(1)}}{\partial \xi} = 0, \tag{C.23}$$

$$\epsilon^{\frac{5}{2}}: \quad \frac{\partial v^{(1)}}{\partial \tau} - \lambda\frac{\partial v^{(2)}}{\partial \xi} + \frac{\partial \phi^{(2)}}{\partial \xi} + v^{(1)}\frac{\partial v^{(1)}}{\partial \xi} = 0. \tag{C.24}$$

Equations (C.19), (C.21), and (C.23) give $\lambda^2 n^{(1)} = n^{(1)}$ or $\lambda \pm 1$. We first consider the case $\lambda = 1$. Then we have

$$\phi^{(1)} = n^{(1)} = v^{(1)}. \tag{C.25}$$

Eliminating $n^{(2)}$ and $v^{(2)}$ from (C.20), (C.22), and (C.24), and using (C.25), we obtain the evolution equation for $n^{(1)}$:

$$\frac{\partial n^{(1)}}{\partial \tau} + n^{(1)} \frac{\partial n^{(1)}}{\partial \xi} + \frac{1}{2} \frac{\partial^3 n^{(1)}}{\partial \xi^3} = 0 \tag{C.26}$$

which is the KdV equation. It is to be noted that the equations (C.19)–(C.24) are closed: no equations and no terms are left over.

Appendix D

Equation of motion in a rotating frame of reference

In geophysical fluid dynamics, the rotation of the earth has an important effect and the equation of the fluid motion needs to be written in the rotating frame. Instability theory of a rotating star is another example which requires the rotating frame description. This is a well-known elementary problem of coordinate transformation, but it is worth becoming familiar with the basic concept without confusion. Here we present the theory on the basis of a simple concept of coordinate transform but with some labor of algebra. The author suspects that the rotating frame description might be of some usefulness in the analysis of magnetized plasma dynamics. We designate the fixed frame (inertial frame) as (x, y, z) and the rotating frames as (x', y', z'), which rotate with angular speed Ω about the vertical z–axis: $\overrightarrow{\Omega} = \hat{\mathbf{z}}\Omega$. We can easily establish the following relations between the unit vectors along the fixed frame and the unit vectors along the rotating frame:

$$\begin{aligned} \hat{\mathbf{x}} &= \hat{\mathbf{x}}' cos\ \Omega t - \hat{\mathbf{y}}' sin\ \Omega t, \\ \hat{\mathbf{y}} &= \hat{\mathbf{x}}' sin\ \Omega t + \hat{\mathbf{y}}' cos\ \Omega t, \\ \hat{\mathbf{z}} &= \hat{\mathbf{z}}'. \end{aligned} \tag{D.1}$$

The inverse relation is

$$\begin{aligned} \hat{\mathbf{x}}' &= \hat{\mathbf{x}}\ cos\ \Omega t + \hat{\mathbf{y}}\ sin\ \Omega t, \\ \hat{\mathbf{y}}' &= -\hat{\mathbf{x}}\ sin\ \Omega t + \hat{\mathbf{y}}\ cos\ \Omega t, \\ \hat{\mathbf{z}}' &= \hat{\mathbf{z}}. \end{aligned} \tag{D.2}$$

A point P in the space can be represented by two ways, using the unprimed coordinates and the primed coordinates, and we have

$$\hat{\mathbf{x}}' x' + \hat{\mathbf{y}}' y + \hat{\mathbf{z}}' z = \hat{\mathbf{x}} x + \hat{\mathbf{y}} y + \hat{\mathbf{z}} z. \tag{D.3}$$

Using (D.2) in (D.3) yields the transformation law of the position vector:

$$x = x' cos\ \Omega t - y' sin\ \Omega t,$$
$$y = x' sin\ \Omega t + y' cos\ \Omega t,$$
$$z = z'. \tag{D.4}$$

The inverse relations are:

$$x' = x cos\ \Omega t + y sin\ \Omega t,$$
$$y' = -x sin\ \Omega t + y cos\ \Omega t,$$
$$z' = z. \tag{D.5}$$

The rotation matrix in (D.5) is

$$\mathbf{R} = \begin{pmatrix} cos\ \Omega t & sin\ \Omega t & 0 \\ -sin\ \Omega t & cos\ \Omega t & 0 \\ 0 & 0 & 1 \end{pmatrix}$$

with $\mathbf{r}'(t) = \mathbf{R}(t) \cdot \mathbf{r}(t)$. The inverse is

$$\mathbf{R}^{-1} = \begin{pmatrix} cos\ \Omega t & -sin\ \Omega t & 0 \\ sin\ \Omega t & cos\ \Omega t & 0 \\ 0 & 0 & 1 \end{pmatrix}$$

with $\mathbf{r}(t) = \mathbf{R}^{-1}(t) \cdot \mathbf{r}'(t)$. The rotation matrix $\mathbf{R}$ is orthogonal which preserves the length of a vector and the product of $\mathbf{R}$ and its *transpose* yields the unit matrix.

Let us write the velocity in the rotating frame as

$$\mathbf{v}_r = \hat{\mathbf{x}}' \frac{dx'}{dt} + \hat{\mathbf{y}}' \frac{dy'}{dt} + \hat{\mathbf{z}}' \frac{dz'}{dt}. \tag{D.6}$$

The RHS of the above equation can be written in terms of the unprimed quantities (the fixed frame quantities) by using (D.2) and (D.5). After some algebra, we obtain

$$\mathbf{v}_r = \frac{d\mathbf{r}}{dt} + \mathbf{r} \times \vec{\Omega} = \mathbf{v} + \mathbf{r} \times \vec{\Omega} \tag{D.7}$$

with $\vec{\Omega} = (0, 0, \Omega)$. The components of acceleration in the rotating frame is defined by the relation

$$\mathbf{a}_r = \hat{\mathbf{x}}' \frac{d^2x'}{dt^2} + \hat{\mathbf{y}}' \frac{d^2y'}{dt^2} + \hat{\mathbf{z}}' \frac{d^2z'}{dt^2}. \tag{D.8}$$

This definition derives from (D.6) by $\mathbf{a}_r = d\mathbf{v}_r/dt$ with regarding $\hat{\mathbf{x}}'$, $\hat{\mathbf{y}}'$, $\hat{\mathbf{z}}'$ as constant unit vectors. Again, using (D.2) and (D.5) in (D.8) gives (after a lengthy algebra)

$$\mathbf{a}_r = \frac{d^2\mathbf{r}}{dt^2} - 2\,\vec{\Omega} \times \frac{d\mathbf{r}}{dt} + \vec{\Omega} \times (\vec{\Omega} \times \mathbf{r}). \tag{D.9}$$

Substituting $\mathbf{v} = d\mathbf{r}/dt = \mathbf{v}_r + \vec{\Omega} \times \mathbf{r}$, which is (D.7), in the above yields

$$\mathbf{a}_r = \mathbf{a} - 2\vec{\Omega} \times \mathbf{v}_r - \vec{\Omega} \times (\vec{\Omega} \times \mathbf{r}) \tag{D.10}$$

where $\mathbf{a} = \frac{d^2\mathbf{r}}{dt^2}$ is the acceleration in the inertial frame. If the acceleration and velocity are measured in the rotating frame, the acceleration in the inertial frame equation of motion should be replaced by

$$\mathbf{a} = \mathbf{a}_r + 2\vec{\Omega} \times \mathbf{v}_r + \vec{\Omega} \times (\vec{\Omega} \times \mathbf{r}). \tag{D.11}$$

If we write the usual fluid equation of motion in a rotating frame, it takes the form

$$\frac{d\mathbf{v}_r}{dt} = -\frac{1}{\rho}\nabla p + \nu\nabla^2\mathbf{v}_r + \mathbf{f} - 2\vec{\Omega} \times \mathbf{v}_r - \vec{\Omega} \times (\vec{\Omega} \times \mathbf{r}) \tag{D.12}$$

where the last two terms are, respectively, the Coriolis force and the centrifugal force.

Appendix E

Wiener–Hopf method

The Wiener–Hopf method is useful for solving a certain integral equation [with a displaced kernel, $K(x-x')$] and boundary value problems in a semi-infinite domain with half-range boundary condition; a boundary condition given over $x = (0,\infty)$ or $(-\infty,0)$. This method is useful for problems involving the half-range Fourier integrals such as in Van Kampen's integral equation or possibly the surface wave equation in a semi-infinite plasma. The method is based upon the principle of analytic continuation, and begins with Fourier transform. We consider the Fourier transform of $f(x)$:

$$f(k) = \int_{-\infty}^{\infty} f(x)\, e^{-ikx} dx \tag{E.1}$$

for complex $k = k_r + ik_i$. Then, on the complex k-plane

$$f(k) = \int_{-\infty}^{\infty} f(x)\, e^{k_i x}\, e^{-ik_r x} dx. \tag{E.2}$$

The analytic region of $f(k)$ in the complex k-plane is restricted by the asymptotic behavior of $f(x)$ as $x \to \pm\infty$. In order for the integral not to diverge at $x = \pm\infty$, we need

$$If \quad f(x) \sim e^{-\alpha x} \; as \; x \to \infty, \quad k_i < \alpha, \tag{E.3}$$

$$If \quad f(x) \sim e^{-\beta x} \; as \; x \to -\infty, \quad k_i > \beta. \tag{E.4}$$

Therefore the analytic region in the k-plane in (E.2) is confined to the strip $\beta < k_i < \alpha$ if $\alpha > \beta$.

If $f(x) = 0$ for all $x < 0$, the Fourier transform in (E.2) becomes

$$f(k) = \int_{0}^{\infty} f(x)\, e^{k_i x}\, e^{-ik_r x} dx \tag{E.5}$$

whose analytic region is the plane below $k_i = \alpha$, or $-\infty < k_i < \alpha$.

If $f(x) = 0$ for all $x > 0$, the Fourier transform in (E.2) becomes

$$f(k) = \int_{-\infty}^{0} f(x)\, e^{k_i x}\, e^{-ik_r x} dx \tag{E.6}$$

whose analytic region is the plane above $k_i = \beta$, or $\beta < k_i < \infty$.

Let us write (E.1) as the sum of two parts: $f(k) = f_+(k) + f_-(k)$ with

$$f_+(k) = \int_{0}^{\infty} f(x)\, e^{-ikx} dx, \tag{E.7}$$

$$f_-(k) = \int_{-\infty}^{0} f(x)\, e^{-ikx} dx. \tag{E.8}$$

In the above equations we substitute the Fourier expression $f(x) = \int_{-\infty}^{\infty} f(k')\, e^{ik'x} dk'/2\pi$ to write

$$f_+(k) = -\frac{1}{2\pi i}\int_{-\infty}^{\infty} \frac{f(k')}{k'-k} dk', \qquad (Im\ k' < \alpha) \tag{E.9}$$

$$f_-(k) = \frac{1}{2\pi i}\int_{-\infty}^{\infty} \frac{f(k')}{k'-k} dk'. \qquad (Im\ k' > \beta) \tag{E.10}$$

Since the integration path can be changed inside the analytic region (here the strip, $\beta < Im\ k' < \alpha$) (E.9) and (E.10) can be written as

$$f_+(k) = -\frac{1}{2\pi i}\int_{-\infty+i\alpha}^{\infty+i\alpha} \frac{f(k')}{k'-k} dk', \tag{E.11}$$

$$f_-(k) = \frac{1}{2\pi i}\int_{-\infty+i\beta}^{\infty+i\beta} \frac{f(k')}{k'-k} dk'. \tag{E.12}$$

Adding (E.11) and (E.12), we can write

$$f(k) = f_+(k) + f_-(k) = \frac{1}{2\pi i}\oint \frac{f(k')}{k'-k} dk' \tag{E.13}$$

where the closed integral path is formed by connecting the two parallel paths at the ends of the strip ($Re\ k = \pm\infty$) because the line integrals over the connected parts vanish. (E.13) is the Cauchy integral formula where k is any point inside the strip.

Summary: A Fourier transformed function $F(k)$ can be decomposed into two subscripted functions F_+ and F_-. Suppose $F(k)$ has two singular points whose imaginary parts are α and β. For definiteness we assume $\alpha > \beta$. Then

F_+ is analytic in the region $-\infty < k_i < \alpha$. F_- is analytic in the region $\beta < k_i < +\infty$. $F(k)$ is analytic in the region of the strip $\beta < k_i < \alpha$. The singularities of $F(k)$ are picked up by the exponential function e^{ikx} in the Fourier inversion integral, and the asymptotic behavior of $F(x)$ is determined as in (E.3) and (E.4).

Now the Wiener–Hopf technique will be explained by solving specific problems.

Example. Solve the integral equation

$$\psi(x) = \lambda \int_0^\infty e^{-|x-x'|}\, \psi(x')\, dx'. \tag{E.14}$$

Let us calculate the Fourier transform

$$\begin{aligned}\psi(k) &= \int_{-\infty}^\infty e^{-ikx}\, \psi(x)\, dx \\ &= \int_{-\infty}^\infty e^{-ikx} dx\, \lambda \int_0^\infty e^{-|x-x'|}\, \psi(x')\, dx' \\ &= \lambda \int_0^\infty dx' \psi(x') \int_{-\infty}^\infty e^{-ikx}\, e^{-|x-x'|} dx \\ &= \lambda \int_0^\infty dx' \psi(x') e^{-ikx'} \int_{-\infty}^\infty e^{-|y|} e^{-iky} dy.\end{aligned}$$

Thus, the given integral equation takes the form in the k-plane

$$\psi(k) = \frac{2\lambda}{k^2+1} \int_0^\infty dx \psi(x) e^{-ikx}. \tag{E.15}$$

Let us introduce the $(\pm)$ functions in accordance with (E.7), (E.8), and (E.13) to write

$$\psi_+(k)\left[1 - \frac{2\lambda}{k^2+1}\right] + \psi_-(k) = 0. \tag{E.16}$$

In (E.16), we have two unknowns ψ_+ and ψ_- in one equation. To deduce a meaningful conclusion from the equation, the function in the bracket should be *factorized* into (+) and (–) type functions. By inspection we can write

$$1 - \frac{2\lambda}{k^2+1} = \Gamma_+(k)\Gamma_-(k) \tag{E.17}$$

with

$$\Gamma_+(k) = \frac{k^2+1-2\lambda}{k-i}, \quad \Gamma_-(k) = \frac{1}{k+i} \tag{E.18}$$

where Γ_+ is analytic in the region $k_i < 1$, hence we associate it with the (+) type, and Γ_- is analytic in the region $k_i > -1$, hence the (–) type. [We can identify the singularities of $\psi(k)$, $k = \pm i$, immediately from (E.15).] Thus let us write (E.16) in the form

$$\psi_+(k)\Gamma_+(k) = -(k+i)\psi_-(k). \tag{E.19}$$

The LHS of (E.19) is analytic in the region $-\infty < Im\ k < 1$ while the RHS is analytic in the region $-1 < Im\ k < \infty$. Since they have a common analytic region $-1 < Im\ k < +1$ where they are equal, we conclude that $-(k+i)\psi_-(k)$ is the analytic continuation of $\psi_+(k)\Gamma_+(k)$ in the upper k-plane. Hence $\psi_+(k)\Gamma_+(k)$ is analytic throughout the entire plane, i.e., an entire function over the whole plane. According to the Liouville theorem, it must be a constant (=C). Thus we have

$$\psi_+(k) = C\,\frac{k-i}{k^2+1-2\lambda}, \quad \psi_-(k) = -\frac{C}{k+i}. \tag{E.20}$$

Now it remains to invert (E.20) to obtain $\psi_\pm(x)$. Let us calculate

$$\psi_-(x) = \int_{-\infty}^{\infty} \psi_-(k)\, e^{ikx}\frac{dk}{2\pi} = -C\int_{-\infty}^{\infty} \frac{e^{ikx}}{k+i}\,\frac{dk}{2\pi}.$$

This integral can be done easily by contour integration. The integration path can be any straight horizontal line between $-1 < Im\ k < 1$. When $x < 0$ ($x > 0$), the contour is completed by encircling the lower(upper) plane. Thus we obtain

$$\psi_-(x) = \begin{cases} iCe^{x} & (x<0) \\ 0 & (x>0) \end{cases}. \tag{E.21}$$

We can ascertain this result as follows:

$$\begin{aligned}\psi_-(x) &= \int_{-\infty}^{\infty} \psi_-(k)\, e^{ikx}\frac{dk}{2\pi} \\ &= \int_{-\infty}^{\infty} e^{ikx}\frac{dk}{2\pi}\int_{-\infty}^{0} \psi(x')\, e^{-ikx'}dx' \quad per\ (E.8) \\ &= \int_{-\infty}^{0} \psi(x')\,\delta(x-x')dx'.\end{aligned}$$

Therefore,

$$\psi_-(x) = \begin{cases} \psi(x) & (x<0) \\ 0 & (x>0) \end{cases}. \tag{E.22}$$

Likewise,

$$\psi_+(x) = \begin{cases} 0 & (x < 0) \\ \psi(x) & (x > 0) \end{cases}. \tag{E.23}$$

Next, let us invert $\psi_+(k)$.

$$\psi_+(x) = C \int_{-\infty}^{\infty} \frac{k - i}{k^2 - \gamma^2}\, e^{ikx}\, \frac{dk}{2\pi} \qquad (\gamma = \sqrt{2\lambda - 1}) \tag{E.24}$$

where we assume $2\lambda - 1 > 0$ and the integration path is any horizontal line between $-1 < Im\ k < 1$, i.e., inside the strip. But a meaningful result is obtained by setting the horizontal path below the real axis because the poles are located on the real k axis. The contour is completed by encircling the upper plane for $x > 0$. Picking up the residues at $k = \gamma$ and $k = -\gamma$, one obtains

$$\psi_+(x) = iC\left[cos\ \gamma x + \frac{sin\ \gamma x}{\gamma}\right] \qquad (x > 0). \tag{E.25}$$

$\psi(x) = \psi_+(x) + \psi_-(x)$ solves the given integral equation (E.14).

Exercise. Obtain from (E.7) and (E.8) by direct calculation

$$\psi_-(k) = \frac{\lambda}{1 - ik} \int_0^{\infty} \psi(x)\, e^{-x} dx,$$

$$\psi_+(k) = -\frac{\lambda}{1 - ik} \int_0^{\infty} \psi(x)\, e^{-x} dx + \frac{2\lambda}{k^2 + 1} \psi_+(k).$$

Use these equations to cross check the solutions.

Example. Solve the boundary-value problem by Wiener–Hopf method:

$$\left(\frac{\partial^2}{\partial x^2} + \frac{\partial^2}{\partial y^2}\right) \phi(x, y) = 0 \tag{E.26}$$

in the upper plane $y > 0$, subject to boundary conditions

$$\phi \to 0 \quad as \quad y \to \infty, \tag{E.27}$$

$$\phi(x, 0) \equiv f(x) = \begin{cases} e^{-ax} & (x > 0) \\ f_-(x) = ? & (x < 0) \end{cases}, \tag{E.28}$$

$$\frac{\partial \phi}{\partial y}(x, 0) \equiv g(x) = \begin{cases} g_+(x) = ? & (x > 0) \\ c\ e^{bx} & (x < 0) \end{cases}, \tag{E.29}$$

where a, b, c are positive constants. The functions marked by (?) are unknown or not given. This is a boundary-value problem of mixed type. Specifying (?) functions will be over-specification which confuses unique determination of the solution. Our task is to find $\phi(x, y)$ for $y > 0$ with simultaneous determination of (?) functions in consistence with the specified conditions. Let us Fourier transform (E.26) with respect to the variable x. Putting in (E.26)

$$\phi(x, y) = \int_{-\infty}^{\infty} \frac{dk}{2\pi}\, e^{ikx} \phi(k, y)$$

yields

$$\frac{\partial^2}{\partial y^2}\phi(k, y) - k^2\phi(k, y) = 0 \tag{E.30}$$

whose solution may be assumed, in view of (E.27), to take the form

$$\phi(k, y) = \Phi(k)\, e^{-|k|y} \tag{E.31}$$

from which we can write

$$\phi(x, y) = \int_{-\infty}^{\infty} \frac{dk}{2\pi}\, e^{ikx}\Phi(k)\, e^{-|k|y}. \tag{E.32}$$

Putting $y = 0$ in (E.32) gives

$$f(x) = \int_{-\infty}^{\infty} \frac{dk}{2\pi}\, e^{ikx}\Phi(k). \tag{E.33}$$

Fourier transforming both sides of (E.33) yields

$$\Phi(k) = \int_{-\infty}^{0} f_-(x)\, e^{-ikx} dx + \frac{1}{a + ik}. \tag{E.34}$$

Differentiating (E.32) and using (E.29) give

$$g(x) = -\int_{-\infty}^{\infty} \frac{dk}{2\pi}\, e^{ikx}|k|\Phi(k). \tag{E.35}$$

Fourier transforming both sides of (E.35) yields

$$|k|\Phi(k) = \frac{c}{ik - b} - \int_{0}^{\infty} g_+(x)\, e^{-ikx} dx. \tag{E.36}$$

Eliminating $\Phi(k)$ in (E.34) and (E.36) gives

$$|k|f_-(k) + g_+(k) = i\,\frac{|k|}{k - ia} - i\frac{c}{k + ib} \tag{E.37}$$

where

$$f_-(k) = \int_{-\infty}^{0} f_-(x)\, e^{-ikx} dx, \tag{E.38}$$

$$g_+(k) = \int_{0}^{\infty} g_+(x)\, e^{-ikx} dx. \tag{E.39}$$

In order to solve (E.37) in which two unknown functions are involved, we apply the Wiener–Hopf logic. Before doing this, the k-plane that contains the double-valued function $|k|$ should be rectified. We devise a scheme:

$$|k| = \lim_{\lambda \to 0} \sqrt{k^2 + \lambda^2} = \sqrt{k + i\lambda}\sqrt{k - i\lambda}.$$

We represent the function $|k|$ as $\sqrt{k^2 + \lambda^2}$ and put $\lambda = 0$ at the end. Then the k-plane should be cut by two cut lines emanating from the branch points $k = \pm i\lambda$ and extending to $\pm\infty$ respectively. Any contour integration, if necessary, must be carried out in this cut plane. (E.37) is rewritten as

$$\sqrt{k + i\lambda}\; f_-(k) + \frac{g_+(k)}{\sqrt{k - i\lambda}} = i\,\frac{\sqrt{k + i\lambda}}{k - ia} - \frac{i\,c}{(k + ib)\sqrt{k - i\lambda}}. \tag{E.40}$$

In (E.40), the branch point $k = -i\lambda$ doesn't belong to the region where $f_-(k)$ is analytic ((−) region). The branch point $k = +i\lambda$ doesn't belong to the region where $g_+(k)$ is analytic ((+) region). Thus the LHS of (E.40) has been separated in (+) and (−) type functions. The Wiener–Hopf method requires similar separation of the RHS. As a preliminary of this separation job, we consider the following simple problem.

Problem. Decompose

$$f(k) = \frac{1}{(k + i)\sqrt{k - i}} = f_+(k) + f_-(k). \tag{E.41}$$

The (+) function is analytic as $k_i \to -\infty$. The corresponding analytic region of (+) function is $k_i < 1$, below the branch point $k = i$, extending down to $k_i = -\infty$. This region is the (+) region. The (−) function is analytic as $k_i \to +\infty$. The corresponding analytic region of (−) function is $k_i > -1$, above the simple pole at $k = -i$, extending up to $k_i = +\infty$. This region is the $(-)$ region. We form a closed loop between $-1 < k_i < 1$, extending to $Re\ k \to \pm\infty$. Applying the Cauchy integral formula to this loop, we can write

$$f(k) = \frac{1}{2\pi i}\int_{-\infty+i\beta}^{\infty+i\beta}\frac{f(k')}{k'-k}dk' + \frac{-1}{2\pi i}\int_{-\infty+i\alpha}^{\infty+i\alpha}\frac{f(k')}{k'-k}dk' \tag{E.42}$$

where $\beta > \alpha$, and the first term is $f_-(k)$ and the second term is $f_+(k)$. If we need to work on the $(-)$ region, a branch cut emanating from $k = i$ and extending to $+\infty$ must be introduced. The line integral $f_-(k)$ can be evaluated by contour integration by encircling the lower plane (thus the cut line is irrelevant). We have

$$f_-(k) = -2\pi i\,\frac{1}{2\pi i}\ Res\ of\ \frac{1}{(k'-k)(k'+i)\sqrt{k'-i}}\ at\ (k' = -i)$$

$$f_-(k) = \frac{1}{k+i}\,\frac{1}{\sqrt{2}\,\sqrt{-i}} = \frac{1+i}{2}\,\frac{1}{k+i}. \tag{E.43}$$

It follows that

$$f_+(k) = f(k) - f_-(k) = \frac{1}{k+i}\left(\frac{1}{\sqrt{k-i}} - \frac{1+i}{2}\right). \tag{E.44}$$

Note that $f_+(k)$ in (E.44) is regular at $k = -i$. This is a hard way of separating $\pm$ parts of $f(k)$.

• **Easier way.** This result can be obtained by inspection in a simple case like this one. To remove the singularity at $k = -i$ in (E.41), a term is subtracted and added:

$$f(k) = \frac{1}{(k+i)\sqrt{k-i}} - \frac{1}{k+i}\,\frac{1}{\sqrt{-2i}} + \frac{1}{k+i}\,\frac{1}{\sqrt{-2i}}.$$

The sum of the first and the second terms on the RHS is regular at $k = -i$, which is $f_-(k)$ in (E.44).

Problem. Obtain $f_+(k)$ by evaluating the second line integral in (E.42). This integral requires contour integration around the loop along the branch cut and the branch point $k = i$. Carry out the integration. This is the hard way to obtain $f_+(k)$.

Returning to (E.40), the RHS is to be decomposed into (+) and (–) type terms. We invoke the inspection method, the easy way. The decomposition can be attained after reshuffling the singular terms by subtraction-addition technique. We write

$$\frac{RHS}{i} = \frac{1}{k-ia}\left[\sqrt{k+i\lambda} - \sqrt{ia+i\lambda}\right] + \frac{\sqrt{ia+i\lambda}}{k-ia}$$
$$- \frac{c}{k+ib}\left[\frac{1}{\sqrt{k-i\lambda}} - \frac{1}{\sqrt{-ib-i\lambda}}\right] - \frac{c}{k+ib}\,\frac{1}{\sqrt{-ib-i\lambda}}. \quad \text{(E.45)}$$

Note that: (–) region is $-b < k_i < \infty$, (+) region is $-\infty < k_i < a$. Thus the RHS of (E.40) is decomposed as

$$RHS_- = \frac{i}{k-ia}\left[\sqrt{k+i\lambda} - \sqrt{ia+i\lambda}\right] - \frac{ic}{k+ib}\,\frac{1}{\sqrt{-ib-i\lambda}}, \quad \text{(E.46)}$$

$$RHS_+ = i\,\frac{\sqrt{ia+i\lambda}}{k-ia} - \frac{ic}{k+ib}\left[\frac{1}{\sqrt{k-i\lambda}} - \frac{1}{\sqrt{-ib-i\lambda}}\right]. \quad \text{(E.47)}$$

In RHS_-, $k = ia$ is no longer a singularity, and $k = -ib$ doesn't belong to the (–) region. In RHS_+, $k = -ib$ is no longer a singularity, and $k = ia$ doesn't belong to the (+) region.

In (E.40), the (–) type functions on the two sides are equated to give

$$f_-(k) = \frac{RHS_-}{\sqrt{k+i\lambda}}. \quad \text{(E.48)}$$

(E.34) gives

$$\Phi(k) = f_-(k) + \frac{1}{a+ik}. \quad \text{(E.49)}$$

Thus we obtain

$$\Phi(k) = \frac{1-i}{\sqrt{2}}\,\frac{1}{\sqrt{k+i\lambda}}\left[\frac{\sqrt{a+\lambda}}{k-ia} + \frac{c}{\sqrt{b+\lambda}}\,\frac{1}{k+ib}\right]. \quad \text{(E.50)}$$

It is worth remembering $\sqrt{\pm i} = (1 \pm i)/\sqrt{2}$. Now we let $\lambda \to 0$. Note that

$$\sqrt{k+i\lambda} \to \begin{cases} \sqrt{k} & (k>0) \\ i\sqrt{-k} & (k<0) \end{cases}. \quad \text{(E.51)}$$

Thus it follows that

$$\Phi(k) = \begin{cases} \dfrac{1-i}{\sqrt{2}} \dfrac{1}{\sqrt{k}} \left(\dfrac{\sqrt{a}}{k-ia} + \dfrac{c}{\sqrt{b}} \dfrac{1}{k+ib} \right) & (k>0) \\ \dfrac{-1-i}{\sqrt{2}} \dfrac{1}{\sqrt{-k}} \left(\dfrac{\sqrt{a}}{k-ia} + \dfrac{c}{\sqrt{b}} \dfrac{1}{k+ib} \right) & (k<0) \end{cases}. \tag{E.52}$$

Using this in (E.32) gives

$$\phi(x,y) = \int_{-\infty}^{0} \frac{-1-i}{2\pi\sqrt{2}} \frac{1}{\sqrt{-k}} \left(\frac{\sqrt{a}}{k-ia} + \frac{c}{\sqrt{b}} \frac{1}{k+ib} \right) e^{ikx}\, e^{ky} dk$$
$$+ \int_{0}^{\infty} \frac{1-i}{2\pi\sqrt{2}} \frac{1}{\sqrt{k}} \left(\frac{\sqrt{a}}{k-ia} + \frac{c}{\sqrt{b}} \frac{1}{k+ib} \right) e^{ikx}\, e^{-ky} dk. \tag{E.53}$$

The integrals are the complex conjugate of each other. Thus we can write

$$\phi(x,y) = 2Re \int_{0}^{\infty} \frac{1-i}{2\pi\sqrt{2}} \frac{1}{\sqrt{k}} \left(\frac{\sqrt{a}}{k-ia} + \frac{c}{\sqrt{b}} \frac{1}{k+ib} \right) e^{ikx}\, e^{-ky} dk. \tag{E.54}$$

To evaluate (E.54) let us consider the integral in the form

$$I(u,a) = \int_0^\infty \frac{dx}{\sqrt{x}} \frac{e^{-ux}}{x+a}. \tag{E.55}$$

This integral can be evaluated by employing an analogous method to that used for the plasma dispersion function. Differentiating (E.55) gives

$$\frac{\partial I}{\partial u} = -\int_0^\infty \frac{dx}{\sqrt{x}} \frac{x}{x+a} e^{-ux} = -\int_0^\infty \frac{dx}{\sqrt{x}} e^{-ux} \left(1 - \frac{a}{x+a}\right) \tag{E.56}$$

$$\text{or} \quad \frac{\partial I}{\partial u} - aI(u,a) = -\int_0^\infty \frac{dx}{\sqrt{x}} e^{-ux} \tag{E.57}$$

$$\text{or} \quad \frac{\partial J}{\partial u} = -e^{-au} \int_0^\infty \frac{dx}{\sqrt{x}}\, e^{-ux} \tag{E.58}$$

$$\text{where} \quad J(a,u) = I\, e^{-au} = \int_0^\infty \frac{dx}{\sqrt{x}} \frac{e^{-u(x+a)}}{x+a}. \tag{E.59}$$

RHS of (E.58) can be evaluated as follows:

$$\frac{\partial J}{\partial u} = -\int_a^\infty \frac{dy}{\sqrt{y-a}}\, e^{-uy} = -\sqrt{\pi}\, \frac{e^{-au}}{\sqrt{u}} \tag{E.60}$$

where we changed $x = y - a$ and $y = z^2 + a$. Both sides of (E.60) can be integrated, giving

$$J(a,u)\Big|_0^u = -\sqrt{\pi}\int_0^u \frac{e^{-au'}}{\sqrt{u'}}du' = -2\sqrt{\frac{\pi}{a}}\int_0^{\sqrt{au}} e^{-\xi^2}d\xi. \tag{E.61}$$

On the other hand, (E.59) gives

$$J(a,u=0) = \int_0^\infty \frac{dx}{(x+a)\sqrt{x}} = \int_a^\infty \frac{dy}{y\sqrt{y-a}} = 2\int_0^\infty \frac{dz}{z^2+a} = \frac{\pi}{\sqrt{a}}. \tag{E.62}$$

Therefore,

$$J(a,u) = \frac{\pi}{\sqrt{a}}\left(1 - erf\sqrt{au}\right), \tag{E.63}$$

$$I(a,u) = e^{au}\frac{\pi}{\sqrt{a}}\left(1 - erf\sqrt{au}\right), \tag{E.64}$$

where $erf\ w = 2\int_0^w dx\ e^{-x^2}$, the error function. Using the above result in (E.54) yields

$$\phi(x,y) = Re\ \left[e^{-ax}\left(1 - erf\sqrt{-az} - i\ \frac{c}{b}\ e^{bx}(1 - erf\sqrt{bz}\right)\right] \tag{E.65}$$

where $z = x + iy$.

References for Wiener–Hopf method:

Mathews J. and Walker, R. L. *Mathematical Methods of Physics*, Benjamin New York, 1970 Chapter 8.

Morse, P. M. and Feshbach, H. *Methods of Theoretical Physics*, McGraw-Hill New York, 1953 Vol. 1 Chapter 8.

Bibliography

The following list of references includes the papers and the books that are referred in the text as well as materials for general background and further reading. The relevant Chapters are not indicated but the *titles* of the material are furnished.

Akhiezer, A., Akhiezer, I., Polovin, R., Sitenko, A., and Stepanov, K. *Collective Oscillations in a Plasma*, Pergamon, Oxford, 1967.

Al'tshul L. M. and Karpman V. I. The kinetics of waves in a weakly turbulent plasma, *Sov Phys. JETP* **20** (1965) pp. 1043–1056.

Al'tshul', L. M. and Karpman, V. I. Theory of nonlinear oscillations in a collisionless plasma, *Sov. Phys. JETP* **22** (1966) pp. 361–369.

Alexandrov, A. F., Bogdankevich, L. S. and Rukhadze, A. A., *Principles of Plasma Electrodynamics*, Springer-Verlag, New York, 1984.

Arfken, G. B. *Mathematical Methods for Physicists*, Academic Press (1985).

Arfken, G. B. and Weber, H. J. *Mathematical Methods for Physicists* 5th ed., Academic Press (Harcourt), 2001.

Alfven, H and Fälthammar, C.-G. *Cosmical Electrodynamics*: Clarendon, Oxford, 1963.

Barr, H. C. and Boyd, T. J. M. Surface waves in hot plasmas, *J. Phys. A: Gen. Phys.* **5** (1972), pp. 1108–1118.

Bekefi, G. *Radiation Processes in Plasmas*, John Wiley, New York, 1966.

Bellan, P. M. *Fundamentals of Plasma Physics*, Cambridge University Press Cambridge, 2006.

Bernstein, I. B., Greene, J. M. and Kruskal, M. D. Exact nonlinear plasma oscillations, *Phys. Rev.* **108** (1957) p. 546.

Bernstein, I. B. Waves in a plasma in a magnetic field, *Phys. Rev.*, **109** (1958) p. 10.

Bhatnagar, P. L., Gross, E. P. and Krook, M. *Phys. Rev.* **94** (1954) p. 511.

Bhatnagar, P. L. *Nonlinear Waves in One-Dimensional Dispersive Media*, Oxford University Press Oxford, 1979.

Bodner, S. E. and Frieman, E. A. *The Quasi-Linear Theory of Plasma Oscillations and Instability*, in the eighth Lockheed symposium of magnetohydrodynamics, Stanford University Press, Stanford, California, 1964.

Bohm, D. and Gross, E. P. Theory of plasma oscillations. A. Origin of medium-like behavior, *Phys. Rev.* **75** (1949) p. 1851.

Bohm, D. and Gross, E. P. Theory of plasma oscillations. B. Excitation and damping of oscillations, *Phys. Rev.* **75** (1949) p. 1864.

Bohm, D. and Gross, E. P. Effects of plasma boundaries in plasma oscillations, *Phys. Rev.* **79** (1950) p. 992.

Braginskii, S. I. and Kazantsev, A. P. (1957). *MHD eaves in a rarefied plasma* — the Journal is unknown to the author. (Kinetic derivation of Alfven wave).

Braginskii, S. I. Transport phenomena in a completely ionized Rwo-temperature plasma, *Soviet Phys. JETP* **6** (1958) p. 358.

Braginskii, S. I. *Transport Process in a Plasma*, Reviews of plasma physics Vol. 1: Consultant's Bureau, New York, 1965.

Brambilla, M. *Kinetic Theory of Plasma Waves*, Clarendon Oxford, 1998, p. 43.

Briggs, R. J. *Electron-Stream Interaction with Plasmas*, MIT Press, Cambridge, Massachusetts, 1964.

Buti, B. *Nonlinear Surface Phenomena in Plasmas*, Advances in space plasma physics, World Scientific Publishing Co. Pte. Ltd: Singapore, 1985, p. 167.

Camac, M., Kantrowitz, A. R., Litvak, M. M., Patrick, R. M., and Petschek, H. E. Shock waves in collision-free plasmas, *Nuc. Fusion Supplement* **Pt. 2** (1962) pp. 423–444.

Case, K. M. Plasma oscillations, *Annals Phys.* **7** (1959) p. 349.

Chandrasekhar, S. *Adiabatic Invariants in the Motions of Charged Particles: The Plasma in a Magnetic Field*: A symposium on magnetohydrodynamics, Ed. Landshoff, R. K. M., Stanford University Press, 1958.

Chandrasekhar, S., Kaufman, A. N., and Watson, K. M. Properties of an ionized gas at low density in a magnetic field, *Annals Phys.* **5** (1958) p. 1.

Chandrasekhar, S. *Plasma Physics*; Notes compiled by S. K. Trehan, Phoenix Books, University of Chicago Press, 1960.

Chandrasekhar, S. *Hydrodynamic and Hydromagnetic Stability*, Dover New York, 1961 Chapter X, XI.

Chapman, S. and Cowling, T. G. *The Mathematical Theory of Non-uniform Gases* 2nd ed., Cambridge University Press, 1964.

Chew, G. F., Goldberger, M. L., and Low, F. E. The Boltzmann equation and the one-fluid hydromagnetic equations in the absence of particle collisions, *Proc. Royal Soc. London* **236. A** (1956) p. 112.

Clemmow, P. C. and Daugherty, J. P. *Electrodynamics of Particles and Plasmas*, Addison-Wesley London, 1969, p. 405.

Cohen, B. I. and Max, C. E. Stimulated scattering of light by ion modes in a homogeneous plasma: space-time evolution, *Phys. Fluids* **22** (1979) p. 1115.

Cohen, B. I. Compact dispersion relations for parametric instabilities of electromagnetic waves in magnetized plasmas, *Phys. Fluids* **30** (1987) p. 2676.

Courant, R. and Friedrichs, K. O. *Supersonic Flow and Shock Waves*, Springer New York, 1948.

Davidson, R. C. *Methods in Nonlinear Plasma Theory*, Academic Press, New York, 1972.

Dawson, J. and Oberman, C. High frequency conductivity and the emission and absorption coefficients of a fully ionized plasmas *Phys. Fluids* **5** p. 517.

Debnath, L. *Nonlinear Partial Differential Equations for Scientists and Engineers*, Birkhauser Boston, 1997.

Dodd, R. K., Eilbeck, J. C., Gibbon, J. D., and Morris, H. C. *Solitons and Nonlinear Wave Equations*, Academic Press New York, 1982.

Drake, J. F., Kaw, P. K., Lee, Y. C., Schmidt, G., Liu, C. S., and Rosenbluth, M. N. Parametric instabilities of electromagnetic waves in plasmas, *Phys. Fluids* **17** (1974) p. 778.

Drummond, W. E. and Pines, D. Nonlinear stability of plasma oscillations, *Nucl. Fusion: 1962 Supplement*, Pt 3 (1962) p. 1049.

Engel, R. D. Nonlinear stability of the extraordinary wave in a plasma, *Phys. Fluids* **8** (1965) p. 939.

Fejer, J. A. and Leer, E. Purely growing parametric instability in an inhomogeneous plasma, *J. Geophys. Res.* **77** (1972) p. 700.

Fejer, J. A. and Leer, E. Excitation of parametric instabilities by radio waves in the ionosphere, *Radio Science* **7** (1972) p. 481.

Fejer, J. A. and Kuo, Y.-Y. Structure in the nonlinear saturation spectrum of parametric instabilities, *Phys. Fluids* **16** No. 9 (1973) pp. 1490–1496.

Frieman, E. A. On a new method in the theory of irreversible processes, *J. Math. Phys.* **4** (1963) p. 410.

Galeev, A. A., Karpman, V. I., and Sagdeev R. Z. *Nucl. Fusion* **5** (1965) p. 20.

Gardner, C. S., Greene, J. M., Kruskal, M. D., and Miura, R. M. Method for solving the Korteweg-deVries equation, *Phys. Rev. Lett.* **19** (1967) p. 1095.

Gould, R. W., O'Neil, T. M., and Malmberg, J. H. Plasma wave echo, *Phys. Rev. Lett.* **19** (1967) pp. 219–222.

Gradov, O. M. and Stenflo, L. Linear theory of a cold bounded plasma, *Phys. Reports* **94** (1983) pp. 111–137.

Guernsey, R. I. Surface waves in hot plasmas, *Phys. Fluids* **12** No. 9 (1969) pp. 1852–1857.

Hall, L. S. and Heckrotte, W. Instabilities: convective versus absolute, *Phys. Rev.* **166** (1968) p. 120.

Hildebrand, F. B., *Advanced Calculus for Applications*, Prentice-Hall, Englewood Cliffs, N. J., 1976.

Hildebrand, F. B., *Methods of Applied Mathematics*, Prentice-Hall, Englewood Cliffs, N. J., 1965.

Infeld, E. Nonlinear evolution of the Kruskal-Schwarzschild instability of a plasma above a vacuum, *Phys. Rev. A* **39** (1989) pp. 723–727.

Jackson, J. D. Longitudinal plasma oscillations, *J. Nucl. Energy Part C* **1** (1960) p. 171.

Jeffrey, A. and Kawahara, T. *Asymptotic Methods in Nonlinear Wave Theory*, Pitman Advanced Publishing Program Boston, 1982.

Jung, Y.-D. and Lee, H. J. *Causality and collisionless damping in a plasma* (in Korean), New Physics: Sae Mulli **63** (2013) pp. 438–443.

Kadomtsev, B. B. *Plasma Turbulence*, Academic Press, 1965, p. 18.

Kadomtsev, B. B. *Cooperative Effects in Plasmas*, Reviews of Plasma Physics Vol 22 Ed. V. D. Shafranov, Consultant Bureau New York, 2001.

Karpman, V. I. *Nonlinear Waves in Dispersive Media*, Pergamon, 1975.

Kaw, P. K. and McBride, J. B. Surface waves on a plasma, *Phys. Fluids* **13** (1970) pp. 1784–1793.

Kaw, P. K. *Parametric Excitation of Electromagnetic Waves in Magnetized Plasmas*, Advances in Plasma Physics Vol. 6, Ed. by A Simon and W. B. Thompson, Wiley New York, 1976.

Klimontovich, Y. L., *The Statistical Theory of Non-equilibrium Processes in a Plasma*, MIT Press Cambridge Mssachusetts, 1967.

Korteweg, D. J. and de Vries, G. On the change of form of long waves advancing in a rectangular canal, and on a new type of long stationary waves, *Phil. Mag.* **39** (1895) pp. 422–443.

Krall, N. A. and Trivelpiece, A. W. *Principles of Plasma Physics*, San Francisco Press, 1986.

Kruer, W. L. *The Physics of Laser Plasma Interactions*, Addison-Wesley, 1988.

Kruskal, M. and Schwarzschild, M. Some instabilities of a completely ionized plasma, *Proc. Royal Soc. London* **2** (1954) p. 348.

Kubo, R. Statistical-mechanical theory of irreversible processes I. General theory and simple applications to magnetic and conduction problems, *J. Phys. Soc. Japan* **12** (1957) pp. 570–586.

Kulsrud, R. M. Adiabatic Invariant of the Harmonic Oscillator, *Phys. Rev.* **106** (1957) pp. 205–210.

Kurlsrud, R. M. *Adiabatic Invariant of the Harmonic Oscillator*, Proceedings of the International School of Physics Enrico Fermi Course XXV Advanced Plasma Theory Academic Press, 1964.

Kuo, Y.-Y. and Fejer, J. A. Spectral-line structures of saturated parametric instabilities, *Phys. Rev. Lett.* **29** No. 25 (1972) pp. 1667–1670.

Landau, L. D. On the vibrations of the electronic plasma, *J. Physics (USSR)* **10** (1946) p. 25.

Landau, L. D. and Lifshitz, E. M. *Electrodynamics of Continuous Media*, Pergamon, 1960 p. 288

Landau, L. D. *Mechanics*, Pergamon Oxford, 1960.

Landau, L. D. *The Transport Equation in the Case of Coulomb Interactions* in Collected papers of L. D. Landau, Ed. D. Ter Haar Pergamon Oxford, 1965 p. 163.

Laval and Pellat (1972). *Quasilinear Theory*, Plasma physics, Universite de Grenoble, Summer school of theoretical phys. Les Houches Ed. C. Dewitt and J. Peyraud.

Lee, H. J. Kinetic theory of coherent resonant interactions of electrostatic plasma waves, *J. Korean Phys. Soc.* **7** (1984) pp. 256–265.

Lee, H. J. Surface waves in magnetized thermal plasma, *J. Korean Phys. Soc.*, **28** (1995) pp. 51–59.

Lee, H. J. Electrostatic surface waves in a magnetized two-fluid plasma, *Plasma Phys. Control. Fusion* **37** (1995) pp. 755–762.

Lee, H. J. and Cho, S. H. Boundary conditions for surface waves propagating along the interface of plasma flow and free space *J. Plasma Phys.* **58** pt. 3 (1997) pp. 409–419.

Lee, H. J. and Lim, Y. K. Modulation of a TM mode surface wave on a free boundary between plasma and vacuum, *Plasma Phys. and Control. Fusion*, **40** (1998) pp. 1619–1633.

Lee, H. J. and Cho, S. H. Acoustic wave in a dusty plasma with frequent grain charging collisions *Physica Scripta* **68** (2003) pp. 54–57.

Lee, H. J. and Lim, Y. K. Kinetic theory of surface waves in a plasma slab, *J. Korean Phys. Soc.* **50** (2007) pp. 1056–1061.

Lee, H. J. and Kim, C.-G. Kinetic theory of electrostatic surface waves in magnetized plasmas, *J. Korean Phys. Soc.* **54** (2009) pp. 85–93.

Lee, M. J. and Lee, H. J. Kinetic theory of electrostatic surface waves in a magnetized plasma slab, *The Open Plasma Phys. J.* **3**, (2010) pp. 131–137.

Lee, M. J. and Lee, H. J. Landau damping of surface plasma waves, *The Open Plasma Phys. J.* **4** (2011) pp. 55–62.

Lee, H. J. (2013). Addendum to 'Causality, Kramers-Kronig Relations, and Landau Damping' *The Open Plasma Phys. J.* **5**, pp. 36–40, *The Open Plasma Phys. J.* **6**, pp. 13–14.

Lee, H. J. and Song, M. Y. Causality and collisionless damping in plasma, *J. Modern Phys.* **4** (2013) pp. 555–558.

Lee, H. J. and Lee, M. J. Echo in a semi-bounded plasma, *The Open Plasma Phys. J.* **8** (2015) pp. 1–7.

Lee, H. J. Fluctuation of the electric field in a plasma *J. Korean Phys. Soc.* **66** (2015) pp. 1167–1185.

Lee H. J. and Lee, M. J. Surface wave echo Echo in a semi-bounded plasma, *J. of Modern Phys.* **7** (2016) pp. 1400–1412.

Lee, H. J. and Lim, Y. K. Surface wave echo Echo in a plasma slab, Science domain International www.sciencedomain.org, *Physical science international J.* **12** No. 2 (2016) pp. 1–16. Article number PSIJ 28647.

Lee, H. J. Nonlinear analysis of the cold fluid-Poisson plasma by using the characteristic method, *J. of the Korean Phys. Soc.* **69** No. 7 (2016) pp. 1191–1211.

Lee, H. J. Causality in plasma electrodynamics, *J. of the Korean Phys. Soc.* **73** No. 1 (2018) pp. 65–85.

Leibovich, S. and Seebass, A. R. Examples of dissipative and dispersive systems leading to the Burgers and the Korteweg-deVries equations, *Nonlinear Waves*, Cornell University Press Ithaca (1974).

Lifshitz, E. M. and Pitaevskii, L. P. *Physical Kinetics*, Vol. 10 Landau and Lifshitz Course of Theoretical Physics, Butterworth-Heinenann, Oxford, 1981.

Lighthill, E. J. *Waves in Fluids*, Cambridge University Press, 1978.

Lim, Y. K. and Lee, H. J. Causality, Kramers-Kronig relations, and Landau damping, *The Open Plasma Phys. J.* **5** (2012) pp. 36–40.

Ma, J. X. and Shukla, P. K. Compact dispersion relation for parametric instabilities of electromagnetic waves in dusty plasmas, *Phys. Plasmas* **2** (1995) p. 1506.

Mathews and Walker, R. L. *Mathematical Methods of Physics*, Benjamin New York, 1970.

Matsumoto, H. and Kimura, I. Linear and nonlinear cyclotron instability and VLF emissions in the magnetosphere, *Planet. Space Sci.* **19** (1971) p. 567.

McBride, J. B. and Kaw, P. K. Low frequency surface waves on a warm plasma *Phys. Lett.* **33A** (1970) pp. 72–73.

Montgomeery, D. C. and Tidman, D. A. *Plasma Kinetic Theory*, McGraw-Hill, New York, 1964.

Montgomery, D. C. *Theory of the Unmagnetized Plasma*, Gordon and Beach, New York, 1971.

Morse, P. M. and Feshbach, H. *Methods of Theoretical Physics*, McGraw-Hill, New York, 1953, Vol. 1, 2.

Nayfeh, A. H. *Introduction to Perturbation Techniques*, Wiley New York, 1981.

Nicholson, D. R. *Introduction to Plasma Theory*, John Wiley, New York, 1983.

Nishikawa, K. Parametric excitation of coupled waves I. General formulation, *J. Phys. Soc. Japan* **24** (1968) pp. 916–922.

Nishikawa, K. Parametric excitation of coupled waves II. Parametric plasmon-photon interaction, *J. Phys. Soc. Japan* **24** (1968) pp. 1152–1158.

Nishikawa, K. and Wakatani, M. *Plasma Physics*, Springer-Verlag, Heidelberg, 1990.

O'Neil, T. Collisionless damping of nonlinear plasma oscillations, *Phys. Fluids* **8** (1965) p. 2255.

Porkolab, M. and Chang, R. P. H. Instabilities and induced scattering due to nonlinear Landau damping of longitudinal plasma waves in a magnetic field, *Phys. Fluids* **15** (1972) p. 283.

Porkolab, M. and Chang, R. P. H. Nonlinear wave effects in laboratory plasmas: A comparison between theory and experiment, *Rev. Modern Phys.* **50** No. 4 (1978) pp. 746–794.

Rao, N. N., Shukla, P. K. and Yu, M. Y. Dust-acoustic waves in dusty plasmas, Planet. *Space Sci.* **38** (1990) pp. 543–546.

Reif, F. *Fundamentals of Statistical and Thermal Physics*, McGraw-Hill International Ed., 1985.
Rosenbluth, M. N. Parametric instabilities in inhomogeneous media, *Phys. Rev. Lett.* **29** (1972) p. 565.
Rostoker, N. *Nucl. Fusion* **1** (1960) p. 101.
Rostoker, N. and Rosenbluth, M. N. *Phys. Fluids* **3** (1960) p. 1.
Sagdeev, R. Z. and Galeev, A. A. *Nonlinear Plasma Theory*, Benjamin New York, 1969.
Sagdeev, R. Z. The 1976 Oppenheimer lectures: Critical problems in plasma astrophysics I. Turbulence and nonlinear waves, *Rev. Mod. Phys.* **51** No. 1 (1979) pp. 1–9.
Sanmartin, J. R. Electrostatic plasma instabilities excited by a high frequency electric field, *Phys. Fluids* **13** (1970) pp. 1533–1542.
Schram, P. P. J. M., *Kinetic Theory of Gases and Plasmas*, Kluer Academic, Dordrecht, 1991.
Scott, A. C., Chu, F. Y. F. and McLaughlin, D. W. *The soliton: a new concept in applied science* IEEE **61** (1973) pp. 1443–1483 –Gel'fand-Levitan equation derived.
Sedov, L. I. *A Course in Continuum Mechanics* Vol. 2 Wolters-Noordhoff Publishing Groningen, 1972, Chapter 7.
Shivarova, A. and Zhelyazkov, I. Surface waves in a homogeneous plasma sharply bounded by a dielectric, *Plasma Physics* **20** (1978) pp. 1049–1073.
Shkarofsky, I. P., Johnston, T. W., and Bachynski, M. P. *The Particle Kinetics of Plasmas*, Addison-Wesley, London, 1966.
Shukla, P. K. in *The Physics of Dusty Plasma* Ed. P. K. Shukla, D. A. Mendis, and V. W. Chow World Scientific, Singapore, 1996.
Silin, V. P. Parametric resonance in a plasma *Sov. Phys. JETP* **21** (1965) p. 1127; *JETP* **24** (1967) p. 1242. .
Sitenko, A. G., Nguyen-Van-Chong, and Pavlenko, V. N. Contribution to the theory of echo phenomena in a plasma, *Nucl. Fusion* **10** (1970) pp. 259–267.
Sitenko, A. G., Pavlenko, V. N. and Zasenko, V. I. Echo in a half-space plasma, *Phys. Letters* **53A** (1975) pp. 259–260.
Sitenko, A. G., Pavlenko, V. N. and Zasenko, V. I. Echo surface waves in plasmas, *Soviet J. Plasma Phys.* **2** (1976) pp. 448–450.
Sitenko, A. G., *Fluctuations and Non-linear Wave Interactions in Plasmas*, Pergamon, Oxford, 1982.
Sitenko, A. and Malnev, V. *Plasma Physics Theory*, Chapman and Hall, London, 1995.
Stenflo, L. Nonlinear interaction of electrostatic waves in a plasma, *Ann. Physik.* **7**. Folge, Bd. 27, Heft 3, (1971) 289–295.
Stenflo, L. Stimulated scattering of electromagnetic waves by magnetosonic modes in a plasma, *J. Plasma Phys.* **34**, Pt. 1 (1985) p. 95.
Stenflo, L. Theory of nonlinear plasma surface waves, *Physica Scripta* **T63** (1996) pp. 59–62.
Stenflo, L. Resonant three-wave interactions in plasmas, *Physica Scripta* **T50** (1994) pp. 15–19.
Stenflo L. and Shukla, P. K. *Theory of stimulated scattering of large-amplitude waves*, *J. Plasma Phys.* **64** part 4 (2000) pp. 353–357.
Stenflo, L. Stimulated scattering of large amplitude waves in the ionosphere, *Physica Scripta* **T30** (1990) pp. 166–169.
Stix, T. H. *The Theory of Plasma Waves*, McGraw-Hill: New York (1962) p. 47.
Stix, T. H. *Waves in Plasmas*, AIP New York, 1992.
Sturrock, P. A. *Plasma Physics*, Cambridge University Press, 1994.

Swanson, D. G. *Plasma Waves*, Academic Press New York, 1989.

Titchmarsh, E. C. *Introduction to the Theory of Fourier Integrals*, Clarendon Press Oxford, 1937.

Trigger, S. A. and Schram, P. P. J. M. Kinetic theory of the charging process in dusty plasmas *J. Phys. D: Appl. Phys.* **32** (1999) pp. 234–239.

Tzoar, N. Parametric excitations in plasma in a magnetic field, *Phys. Rev.* **178** (1969) pp. 356–362.

Van Kampen, N. G. and Felderhof, B. U. *Theoretical Methods in Plasma Physics*, John Wiley, 1967, Chapter XV.

Vedenov, A. A. Quasi-linear plasma theory (theory of a weakly turbulent plasma), *Plasma Phys. (J. Nucl. Energy Part C)* **5** (1963) p. 169.

Vedenov, A. A. *Theory of Turbulent Plasma*, ILIFFE, Ltd., 1968.

Weinstock, R. *Calculus of Variations*, McGraw-Hill, 1952.

Whitham, G. B. *Linear and Nonlinear Waves*, Wiley New York, 1974, Chapter 2, Fig. 2.2.

Whipple, E. C. Potentials of surfaces in space, *Rep. Prog. Phys.* **44** (1981) pp. 1197–1250.

Whipple, E. C., Northrop, T. G. and Mendis, D. A. The electrostatics of a dusty plasma *J. Geophys. Res. [Oceans]* **90** (1985) pp. 7405–7413.

Wu, T.-Y. *Kinetic Equations of Gases and Plasmas*: Addison-Wesley (1966), Chapter 2.

Yourgrau, W and Mandelstam, S. *Variational Principles in Dynamics and Quantum Theory* Dover, 1968.

Index